160 960942 7

D0636149

Reservoir Characterization— Recent Advances

Edited by

Richard A. Schatzinger

and

John F. Jordan

AAPG Memoir 71

Published by

The American Association of Petroleum Geologists

Tulsa, Oklahoma, U.S.A. 74101

Published 1999
Printed and bound in the United States of America

ISBN: 0-89181-351-9

Reservoir Characterization—Recent Advances/edited by Richard Schatzinger and John Jordan.
p. cm. -- (AAPG memoir: 71)
Includes bibliographical references and index.
ISBN 0-89181-351-9
1. Hydrocarbon reservoirs. 2. Oil reservoir engineering.
I. Schatzinger, Richard A. II. Jordan, John F. III. Series.
TN870.57.R43 1999
622'.3382—dc21 99-25177
CIP

Association Editor: Neil F. Hurley
Science Director: Jack Gallagher
Publications Manager: Kenneth M. Wolgemuth
Managing Editor, Publications: Anne H. Thomas

This book and other AAPG publications are available from:

The AAPG Bookstore
P.O. Box 979
Tulsa, OK 74101-0979
Telephone: (918) 584-2555;
or (800) 364-AAPG (U.S.A.—book orders only)
Fax: (918) 560-2652
or (800) 898-2274 (U.S.A.—book orders only)

Geological Society Publishing House
Unit 7, Brassmill
Enterprise Centre
Brassmill Lane
Bath BA1 3JN
United Kingdom
Tel 1225-445046
Fax 1225-442836

Australian Mineral Foundation
AMF Bookshop
63 Conyngham Street
Glenside, South Australia 5065
Australia
Tel (08) 379-0444
Fax (08) 379-4634

Canadian Society of Petroleum Geologists
#160, 540 5th Avenue S.W.
Calgary, Alberta T2P 0M2
Canada
Tel (403) 264-5610
Fax (403) 264-5898

Affiliated East-West Press Private Ltd.
G-1/16 Ansari Road Darya Ganj
New Delhi 110 002
India
Telephone: 91 11 3279113
Fax: 91 11 3260583

Preface

Reservoir characterization is the process of creating an interdisciplinary high-resolution geoscience model that incorporates, integrates, and reconciles various types of geological and engineering information from pore to basin scale. The reservoir data are then conceptually and quantitatively modeled and compared to the historical production data and fluid flow distribution patterns within and beyond the limits of the reservoir to match well production histories and predict their future behavior. Ultimately, the team effort must provide improved reservoir management strategies, which will maximize recovery and profits.

The goals of reservoir characterization are to simultaneously (1) maintain high displacement efficiency, (2) optimize high sweep efficiency, (3) provide reliable reservoir performance predictions, and (4) reduce risk and maximize profits. Notice that in addition to the technical concepts that we normally associate with "characterization," maximizing profits is an essential element of this process. This is especially true in these times of falling crude oil prices. Reservoir characterization is neither pedantic nor "ivory tower." It must be pragmatic or it will be of little use. This requirement, however, provides a basic strength to the process and, based on the growing interest in reservoir characterization by industry and academia alike, will continue to spark new tools and methods for improved reservoir management.

Basic building blocks for the technical side of reservoir characterization can be descibed in a simplified scheme as construction of four models. The geological model (structural, depositional-stratigraphic, and diagenetic-geochemical components) has historically been rather qualitative. Quantification of geological models for easier assimilation into the subsequent phases of reservoir characterization remains an area of continued interest and needed improvement. The geological model and petrophysical data are combined to create the permeability layer model. Combining of the individual permeability layer models in the development of the flow unit (hydrodynamic) model unifies and quantifies geological components and reservoir fluid flow properties. Finally, the simulation model incorporates historical production data with the hydrodynamic model to create realizations that can be used to test various improved recovery processes prior to their start-up in the reservoir. Simulation can also be of great use in "postmortem" studies where the representation of the previous models and the success of production strategies can be evaluated.

Each of the four building blocks mentioned helps to successively provide a more clear description of the reservoir only if there is iterative feedback. It is only through the process of iterative feedback, interdisciplinary integration, and model improvement that reservoir characterization really works. Without feedback, the process is incomplete; each building block merely describes (some in much greater detail than others) and prediction is based on incomplete information. It is the predictive capability of a well-done, interdisciplinary reservoir characterization effort that can make reservoir management more of a science than a black art.

This book is based on papers presented at the Fourth International Reservoir Characterization Technical Conference held in Houston, Texas, in March 1997. This was one of a series of conferences sponsored by the Department of Energy (and the affiliated tutorials, DOE Class Workshop, and poster sessions); these conferences have been focal points for the discussion of new ideas in reservoir characterization since the first conference was held in 1985.

This book covers the broad spectrum of topics within reservoir characterization. Following the overview paper by Fowler and others, the book has been divided into six sections: (1) Reservoir Description, (2) Enhanced/Improved Oil Recovery (EOR/IOR) Characterization, (3) Methods and Techniques, (4) Fracture Analysis, (5) Upscaling and Simulation, and (6) Modeling.

Acknowledgments

This book is based on the Proceedings of the Fourth International Reservoir Characterization Technical Conference held March 2-4, 1997, in Houston, Texas. The conference was one of a series of Department of Energy-sponsored conferences (and the affiliated tutorials, DOE Reservoir Class Workshop, and poster sessions) that have been focal points for the discussion of new ideas in reservoir characterization since the first conference was held in 1985. This most recent conference and its predecessors could not have happened without the support, financial and otherwise, of the U.S. Department of Energy (DOE)-Fossil Energy, Oil Program.

We would like to express our sincere thanks to the National Petroleum Technology Office (NPTO) of the DOE and particularly to Robert Lemmon (Conference Project Manager) for their unflagging support. Co-sponsors of the Conference included the Department of Energy Bartlesville Project Office (now NPTO), BDM-Oklahoma, Inc., and the American Association of Petroleum Geologists. The theme for the Conference was Advances in Reservoir Characterization for Effective Reservoir Management. Thomas C. Wesson (DOE) and Thomas E. Burchfield (BDM-Oklahoma, Inc.) were co-chairmen. Rick Schatzinger was the Conference Coordinator. Keynote speakers included Olivier Guillon (Elf Aquitaine Production), Mark McElroy (Phillips Petroleum Company), Leif Hinderaker (Norwegian Petroleum Directorate), Larry Lake (University of Texas at Austin), Ganesh Thakur (Chevron), and Betty Felber (DOE/NPTO). A workshop featuring summaries of several DOE Reservoir Class projects was co-chaired by Susan Jackson and Michael Fowler.

A great deal of thanks is due to the Steering Committee for all of their work in organizing the conference. The Steering Committee members included:

Steve Begg (BP Exploration)
Tom Burchfield (BDM-Oklahoma, Inc.)
Jack Caldwell (Geco-Prakla)
Lifu Chu (University of Tulsa)
Robert Finley (Bureau of Economic Geology, University of Texas)
Michael Fowler (BDM-Oklahoma, Inc.)
Dominique Guerillot (IFP)
Neil Humphreys (Mobil Oil Corporation)
Susan Jackson (BDM-Oklahoma, Inc.)
Jerry Jensen (Heriot-Watt University)
Mohan Kelkar (University of Tulsa)
Robert Lemmon (DOE/NPTO)
Charles Mankin (Oklahoma Geological Survey)
Yi Nan Qiu (RIPED)
Rick Schatzinger (BDM Petroleum Technologies)
Eve Sprunt (Mobil Oil Corporation)
Ray Sulak (Phillips Petroleum Company)
Ganesh Thakur (Chevron Technology Company)
Min Tham (BDM-Oklahoma, Inc.)
Lynn Watney (Kansas Geological Survey)
Paul Worthington (Gaffney, Cline & Associates)
David Zornes (Phillips Petroleum Company)

We also gratefully acknowledge the diligent efforts of the large number of unnamed peer reviewers.

Finally, we would also like to thank Lyle Baie, Ken Wolgemuth, and Anne Thomas for their encouragement, as well as the entire AAPG publications staff for making this book a reality.

Richard A. Schatzinger
John F. Jordan

AAPG
Wishes to thank the following
for their generous contributions
to

Reservoir Characterization—
Recent Advances

◆

Conoco (U.K.) Ltd.

◆

Kansas Geological Survey

◆

Statoil Research Centre

◆

University of Texas at Austin,
Center for Petroleum and Geosystems Engineering

◆

◆

◆

Contributions are applied against the production costs of publication, thus directly reducing the book's purchase price and making the volume available to a greater audience.

About the Editors

Richard A. Schatzinger

Richard A. Schatzinger has recently co-founded Fowler, Schatzinger, and Associates. Prior to this, he was a principal geologist in the Reservoir management and Characterization Group of BDM Petroleum Technologies, Bartlesville, Oklahoma. He holds B.S. (1971) and M.S. (1975) degrees in geology from San Diego State University and a Ph.D. (1987) in geology from the University of Texas at Austin. He was a carbonate research scientist for five years at Phillips Petroleum prior to joining BDM. His professional experience includes the sedimentology and diagenesis of reservoirs. His current research includes quantification of outcrop geological models and their application to improved recovery techniques.

John F. Jordan

John F. Jordan received B.S. (1992) and M.S. (1997) degrees in geology from the University of Georgia. His background is in sequence stratigraphy and petrology of the southern Appalachians. He worked for two years on basin analysis and basin modeling projects as an associate geoscientist in the Reservoir Management and Characterization Group of BDM Petroleum Technologies, Bartlesville, Oklahoma. He recently joined Statoil Energy, where he is actively involved with hydrocarbon exploration and development of the southern Appalachian Basin.

Table of Contents

Section I
Overview

Fowler, M.L, et al, The role of reservoir characterization in the reservoir management process (as reflected in the department of energy's reservoir management demonstration program), 1999, *in* R. Schatzinger and J. Jordan, eds., Reservoir Characterization-Recent Advances, AAPG Memoir 71, p. 3-18.

Chapter 1

The Role of Reservoir Characterization in the Reservoir Management Process (as Reflected in the Department of Energy's Reservoir Management Demonstration Program)

Michael L. Fowler
BDM Petroleum Technologies
Bartlesville, Oklahoma, U.S.A.

Mark A. Young
Michael P. Madden
BDM-Oklahoma
Bartlesville, Oklahoma, U.S.A

E. Lance Cole
Petroleum Technology Transfer Council
Sand Springs, Oklahoma, U.S.A.

ABSTRACT

Optimum reservoir recovery and profitability result from guidance by an effective reservoir management plan. Success in developing the most appropriate reservoir management plan requires knowledge and consideration of (1) the reservoir system, including rocks, fluids, and rock-fluid interactions, as well as wellbores and associated equipment and surface facilities; (2) the technologies available to describe, analyze, and exploit the reservoir; and (3) the business environment under which the plan will be developed and implemented. Reservoir management plans *de-optimize* with time as technology and the business environment change or as new reservoir information becomes available. Reservoir characterization is essential for planning appropriately scaled reservoir management plans.

BDM-Oklahoma and the U.S. Department of Energy (DOE) encourage operators with limited resources and experience to implement sound reservoir management techniques through cooperative research and development projects. In the three projects awarded, careful attention to reservoir context promotes a reservoir characterization effort that is sufficient for, but not in excess of what is necessary for, creating an effective reservoir management plan.

INTRODUCTION

At the first meeting of the organizing committee for the first International Reservoir Characterization Technical Conference in Dallas, Texas, in 1985, the attendees chose the following definition of reservoir characterization: "Reservoir characterization is a process for quantitatively assigning reservoir properties, recognizing geologic information and uncertainties in spatial variability" (Lake and Carroll, 1986). Traditionally, its goal has been to transfer quantitative information on reservoir property distribution with enough detail and accuracy to a numerical simulator so that fluid-flow simulation predictions would match reservoir performance. Through simulation, an appropriately detailed and accurate model of reservoir property distribution enables the operator to avoid the deleterious effects of heterogeneities and exploit them to best economic advantage. To the extent that information supplied to the simulator is incomplete, inaccurate, or at an inappropriate scale, the ability to predict reservoir performance and maximize economic returns is lessened.

Reservoir characterization, when appropriately done, is a valuable tool for avoiding costly errors in reservoir decision making. Powerful tools and techniques have evolved since 1985, and new tools and techniques continue to be developed, tested, and refined, but efficient and cost-effective reservoir characterization is still more commonly viewed as art rather than science. Operators often find it difficult to prescribe and construct an appropriate model at an acceptable cost.

Characteristics of the model grow out of its purpose, which in turn depends strongly on the context in which it is applied. A proper understanding of reservoir management therefore is fundamental to performing appropriate reservoir characterization. The goal of this paper is to identify the role of reservoir characterization in reservoir management and to clarify what factors determine the type of reservoir characterization model necessary.

The relationship between reservoir characterization and reservoir management has been explored by reviewing the extensive literature and by participating in reservoir management projects with a variety of small independent operators under DOE's Reservoir Management Demonstration Program. Two projects serve as examples to show appropriate scaling of efforts. Much also was learned about the relationship between reservoir characterization and reservoir management from observing and studying field demonstration projects being carried out under the DOE's Reservoir Class Program and from the authors' experiences in industry projects with major and independent oil companies.

RESERVOIR MANAGEMENT: PERCEPTIONS VS. REALITIES

Misconceptions about reservoir management abound. Some operators think of it as something done only on large reservoirs where large improved recovery targets can justify the large investments in time and dollars they see as necessary. Others think of it as the day-to-day reservoir problem-solving activities of reservoir engineers. Still others, usually those with limited familiarity with new technologies, view it as a strictly "high-tech" venture involving the application of state-of-the-art new techniques and technologies. Numerous operators of marginally profitable mature reservoirs think of reservoir management as an unaffordable expenditure. Not all reservoir operators fall under these categories, but an atmosphere of conservatism and reluctance prevails; consequently, the full potential of effective reservoir management has not been realized.

Definition of Reservoir Management

Just as there are many publications on the subject, there are many definitions of reservoir management in them. Thakur (1991) defined reservoir management as the "judicious use of available resources to maximize economic recovery"; Cole et al. (1993) specified that "resources" in the Thakur definition include people, equipment, technology, and money. Other definitions, such as that offered by Wiggins and Startzman (1990), which defines reservoir management as "application of state-of-the-art technology to a known reservoir system within a given management environment," take a slightly different view. Most definitions, however, identify the components of reservoir management much as if an automobile were to be defined as consisting of engine, wheels, steering mechanism, etc.

Nearly all discussions of reservoir management agree on the following as general characteristics.

- It requires and makes use of resources.
- It is continuous and long term, over the life of a reservoir.
- It concentrates on optimizing economics.

From this listing, we can surmise that the main activity of reservoir management is a sequence of resource-deployment decisions made to optimize the economic recovery of petroleum.

The Plan as a Central Concept

Definitions serve to enlighten us on the critical considerations, but taking a slightly different perspective may convey the core concept more completely. Every reservoir being operated today, similar to every business being operated today, is being managed. Some are managed well; some are, without question, poorly managed. We can think of well and poorly managed reservoirs and businesses as those that are and are not realizing their maximum potential to their operators. Every reservoir operator is taking some kind of approach; that is, has some kind of philosophy, guidelines, or strategy that is used to guide interactions with the reservoir. Formulating these guidelines or plans and following them are the essence of reservoir management (Cole et al., 1994).

The spectrum of possible approaches, strategies, or

plans is extremely wide. Some plans may be very simply conceived or literally just assumed. Such a simple, straightforward approach could amount to a stark "produce the reservoir until the total cost of production becomes greater than the revenue obtained, then quit" approach where "quit" implies either selling the property to an organization having lower overhead costs that can continue to operate the reservoir at a profit or simply abandoning the reservoir. The opposite extreme might be a case in which all the latest improved oil recovery technologies are periodically screened, and selected technologies are carefully applied in the context of a complete and detailed 3-D description of the physical and chemical aspects of the subsurface reservoir. Intermediate between the extremes are plans that consist of informal guidelines that may or may not be regularly reviewed for appropriateness. Reservoir management can be thought of as the decision-making process that matches the approach or plan to the reservoir and its operator in such a way as to maximize the profitability of the reservoir to the operator.

ELEMENTS OF CONTEXT INFLUENCING THE RESERVOIR MANAGEMENT PLAN

How can we select from the vast number of possible approaches to managing reservoirs? The answer is "it is a matter of how well we know (1) the business environment, both internal and external to our company, under which the plan will be constructed and implemented, (2) the availability and use of proven and developing technologies, and (3) the reservoir and its facilities" (Fowler et al., 1996). Knowing the context under which reservoir management is to be performed is as important as knowing about the reservoir itself. The importance of this contextual information will carry over to reservoir characterization planning as well.

Reservoir Management Business Environment

These factors fall into one of two categories: those that are external to the operator's organization (i.e., those that affect all operators equally) and those that are internal and perhaps unique to the operator's organization. No plan can be optimized without paying close attention to both types of factors. Some projects fail because of oversights in this area. Any of the factors may change at any time, either while the plan is formulated or after it has been implemented. Keeping aware of the consequences of such changes enables plans to be revised to maintain optimum performance.

Examples of external factors include market economics, taxes, operational laws and regulations, safety and environmental laws and regulations, and less tangible items, such as public opinion. Internal factors include the organization's ability to raise or commit capital and a number of items related to the organization's "corporate culture." The cultural factors include items such as how the company measures value (e.g., is the objective to increase reserves? to obtain a certain rate of return on investment? or to achieve some other measure of success?), the organization's attitude toward risk, its organizational structure, and its ability to commit to long-term plans (Wiggins and Startzman, 1990; Cole et al., 1993). Some internal factors may be capable of being changed to accommodate the best interests of a reservoir management plan, but some factors may be difficult or impossible to alter. The plan must acknowledge those factors that cannot be altered.

The larger structure of the organization may have a minimal direct impact on reservoir management activities, but there is a consensus that success requires a team approach. Ideally the team will include all persons who have anything to do with the reservoir (Satter et al., 1994). An organizational structure that encourages the formation of multidisciplinary teams will be much more conducive to creation of optimal plans than one that dictates that plans be created by geologists, engineers, and others working sequentially and independently. Satter and Thakur (1994) present an excellent discussion on the structure and function of reservoir management teams.

At project inception, all members should share in developing common project goals and objectives, and aid in developing and assigning project responsibilities for each team member. A team leader with the multidisciplinary insight and management skills to encourage cooperative participation in these and subsequent project activities is necessary.

The dynamic interaction of the group comprising the team contributes strongly to the success of the effort. The team leader must be aware that team members may have varying degrees of technical skill and experience in their own disciplines, as well as varying experience in working closely with people from other disciplines. The leader must monitor and nurture the daily interaction of team members. To do so, the team leader must be aware of individual personality traits and differences in rank, must be aware that certain team members may have commitments to other projects that may compete for their time and dedication at inconsistent times and often at inconvenient intervals (although management should do everything possible to minimize conflicts in priorities!), and must realize that occasional disruptions, such as loss or addition of team members, inevitably occur.

Available Technologies

A second consideration necessary for creating optimal strategies is a familiarity with existing and newly developing technologies that are available to improve hydrocarbon recovery, increase operational efficiencies, and characterize reservoirs. Maintaining an awareness of appropriate technologies in such diverse areas as recovery, wellbore and facilities, and reservoir characterization is a difficult task, especially for smaller organizations. The difficulty of the task is compounded by the rapid evolution of technology in almost every area. Smerdon (1996) estimates the half-life of an engineer's technical skills—the time it takes for half of everything an engineer knows about his field to become obsolete—at less time than it takes to earn an undergraduate degree. Similar trends probably apply in all professions. Reservoir management requires lifelong learning.

It is not realistic to assume that any organization will have (or should have) the necessary knowledge and experience in all areas that may be required. A realistic goal would be to obtain enough of a general or screening-level knowledge of available technologies to know when an expert should be consulted for in-depth evaluation. Professional societies and organizations, such as the Petroleum Technology Transfer Council, are key resources for maintaining technical awareness and accessing the data, information, and contacts required for operators to develop confidence in their technology-deployment decisions.

Improved Recovery Technologies

It is important to be aware of routine techniques, as well as of new techniques and technologies, associated with improved recovery. Secondary recovery techniques include injection of water or gas (immiscible) for pressure maintenance or displacement of hydrocarbons. Advanced secondary recovery techniques include those aimed at improving contact with mobile oil, such as conformance and recompletion considerations; infill drilling using vertical, deviated, and horizontal wells; and employing polymers for profile modification and mobility control. Enhanced oil recovery techniques include application of processes to recover immobile oil, processes such as microbial, alkaline and alkaline-surfactant-polymer, surfactant, steam, in-situ combustion, and miscible and immiscible gas injection. A moderately comprehensive review of technologies associated with advanced secondary and enhanced recovery is presented by Cole et al. (1994).

Wellbore and Facilities Technologies

Familiarity with existing and evolving techniques and technologies associated with wellbore equipment and surface facilities can also influence formulating an appropriate plan. The ability of the mechanical equipment both within wells and on the surface to handle changes in fluid types, volumes, and relative volumes under different pressures and temperatures because of possible changes brought about by implementing a project can be critical to its success. New or different technologies and techniques may be required to accommodate such changes. The focus of some plans may be to incorporate new technologies or approaches just to improve the performance of facilities and equipment.

Reservoir Characterization Technologies

Successful reservoir management also depends on familiarity with existing and newly developing technologies that are available to characterize reservoirs. Technical knowledge on how to build reservoir models from both analog and deterministic sources remains appropriate, but it is also becoming increasingly important to be aware of technologies for collecting, handling, integrating, and analyzing large volumes of data. This does not imply that a high-tech approach is always appropriate. It is important to be aware of the wide range of both traditional and newly developing technologies available and to have some familiarity with the economics involved in assessing and implementing those technologies.

Reservoir System

A third consideration for plan building is knowing the reservoir system. The reservoir system is composed of subsurface reservoir rock, the reservoir's contained fluids, all its wellbores and downhole equipment, and its surface equipment and facilities. Man's activities may have affected components of the reservoir system and should be considered also.

The Physical Reservoir

The result of reservoir characterization is a complete conceptual picture or model of the reservoir, a representation or estimate of reservoir reality. It represents not only the 3-D extent or bounds of the reservoir, but also the qualitative (presence or absence) and quantitative (magnitude) values of rock, fluid, and other reservoir parameters affecting fluid flow at every location in the volume of the reservoir. The degree of uncertainty associated with placement and magnitude of fluid-flow properties is an important characteristic of this model.

An important objective of model construction is to accurately represent and minimize, as far as economically feasible, the uncertainty in our knowledge of reservoir parameters. In the past, the aim generally was to create a single "most probable" representation of the reservoir to be used as input to subsequent decision making, but the need for a small number of more extreme yet reasonably probable representations should be recognized as a useful, if not critical, addition. This approach allows bracketing the range of reasonably expected recovery and economic outcomes. We can think of the goal of reservoir characterization as the construction of a model or a small number of models that will aid in predicting by simulation or other means the outcome or probable range of outcomes of potential reservoir management plans (i.e., projects, processes, or operating plans and procedures) in order to evaluate their relative economic merits.

Deterministic data (i.e., data derived by actual measurement of reservoir properties rather than derived by analogy from similar reservoirs or deposits) can come from a wide variety of technologies and cover a wide range of scales. Although data are taken directly from the reservoir, uncertainties in many types of information gathered are inherent because of the resolving power of the tool used and the inability of many tools to measure desired properties directly; however, the degree of uncertainty associated with the deterministic, quantitative measurement of reservoir properties is often less than that associated with properties assigned by a model of conceptual or analog origin. Figure 1 schematically depicts the scale of measurement and associated resolving power associ-

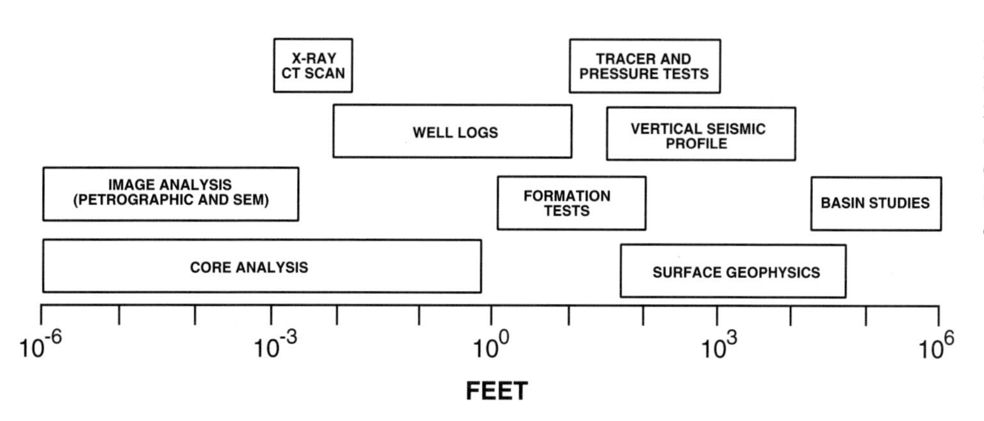

Figure 1—Scales of measurement and relative resolution of some common deterministic reservoir characterization tools (modified from Jackson and Tomutsa, 1991).

ated with several common deterministic tools.

Several geological and engineering tools are commonly available to gather deterministic information from which to formulate a reservoir model. Samples of reservoir rocks and fluids collected early in a reservoir's development history may be the best source of information available to predict fluid-flow patterns that may be critical when the reservoir reaches maturity. Reservoir production and injection data are often-overlooked and inexpensive sources of information on larger scale reservoir architecture and heterogeneities. A wide variety of wireline logging tools, in addition to being useful in establishing structural and stratigraphic frameworks, are available to make direct measurements of reservoir properties, such as rock composition, porosity, and fluid content. Single- or multiple-well pressure transient tests may be run to qualitatively or quantitatively discern variations in the properties of the reservoir's pore system. Tracer testing (i.e., the addition of small quantities of an easily detectable material to an injected fluid and the subsequent monitoring of its temporal and volumetric appearance in the same or adjacent wells) can yield measurements of critical well-to-well flow characteristics or estimates of residual oil saturation.

Approaches based on seismic methods perhaps have the greatest potential to provide reliable information on reservoir property variations in the interwell region. Recent advances in several areas of seismic technology are making this possible. The accuracy, resolution, and usefulness of 3-D seismic are being realized thanks to continuing improvements in data acquisition instrumentation, field procedures, data processing software, and 3-D data visualization techniques. Development of procedures for extracting useful information from attributes of the seismic signal (e.g., amplitude, phase, and frequency) facilitates the delineation of stratigraphic traps and other subtle geological features and makes it possible to determine the distribution of porosity and fluid content in reservoir rocks. Developments in downhole seismic surveys (e.g., cross-well tomography and vertical seismic profiling) are increasing resolution in the interwell area.

Other deterministic tools and techniques include electromagnetic mapping, remote sensing, surface geochemical sampling surveys, and numerous other approaches that may add to our knowledge of the distribution and movement or potential movement of fluids in the reservoir.

Because a single reservoir characterization model or a small number of such representations is desired, and because the necessary data have both engineering and geological origins, the need for close cooperation among geoscientists, engineers, and other professionals (i.e., the members of the reservoir management team) in formulating such models is paramount. Data from various individual technological sources often suggest a number of nonunique interpretations of reservoir reality. The reservoir management team has the duty to understand and use the various technological data types in complementary and supplementary ways to arrive at the most probable range of possible reservoir realities upon which to base future reservoir performance predictions. Model construction is not a trivial task, and its successful completion requires the continual cooperation and interchange of information and ideas among team members. The task cannot be efficiently accomplished (indeed, it might not be accomplished at all!) if geologists and engineers work on the task sequentially and independently.

Reservoir Infrastructure and History

An additional important aspect of reservoir knowledge is familiarity with the production/injection infrastructure. Natural processes in the subsurface can interact with wellbore equipment, resulting in problems such as corrosion, scaling, paraffin deposition, etc. Surface processes, such as erosion or flooding, can affect wells and facilities. Knowledge of the history of drilling, completion, recompletion, and workover practices employed in field development, as well as familiarity with current surface and wellbore facilities, is also necessary. Equally important is a knowledge of past production and injection practices.

Human cultural development may certainly affect surface facilities and the use of wellbores. It is also important, however, to know of alterations in the natural properties of the reservoir that have resulted from past human activities. Development and depletion procedures can have a profound influence on basic reservoir characteristics and performance. In some cases, introduced changes are equivalent to introducing whole new and often extreme episodes of diagenesis, tectonics, and fluid exchange. The nature of these changes is unexpected in many instances and can result in decreased reservoir performance and permanent reservoir damage. Examples include situations where stimulation practices have led to communication between reservoir units behind pipe, or where long periods of water injection above formation parting pressure have led to channeling between injection and production wells.

Dynamic Nature of the Reservoir Management Process

The world of reservoir management is dynamic. Technologies are evolving rapidly on numerous fronts, and the petroleum business environment is ever changing. Changing business and technological contexts de-optimize previously existing plans. The formulation, implementation, and revision of plans therefore can be considered to be a fundamental reservoir management process. It is by necessity an iterative process (Figure 2), requiring regular, if not continuous, attention or monitoring for every reservoir being managed. Once a plan has been implemented, monitoring both reservoir performance and the current status of knowledge of the reservoir, technologies, and business environment should begin immediately.

The plan itself may specify a condition or set of conditions that, when met, indicates that the plan should be reevaluated. These criteria may include items such as cumulative volume, relative volume, rate of production or injection of a specified fluid, passage of a specific period of time, or attaining a particular stage of reservoir development. At any time, however, reservoir performance anomalies of any kind (e.g., production or injection volumes, facilities usage, or regulatory compliance) with respect to plan expectations or predictions may indicate an immediate need for plan revision. Ideally, the plan should specify guidelines for tolerance in variation from plan prediction in all critical performance areas.

New information of various forms also may be just cause for plan revision at any time. It may be new reservoir information, perhaps extracted from data collected under specifications of the current plan. It may be in the form of new technologies, ideas, or procedures not available or known at the time the plan was formulated. The critical new information may even be in the form of performance anomalies arising in analogous reservoirs. Unexpected or unpredicted changes in circumstances or opportunities related to the general or operator-specific business environment also may be cause to reevaluate the plan. Examples of factors that may be significant include market economics, new laws and regulations, changes in key personnel, and decisions to buy, sell, or trade reservoirs.

A periodic requirement for revising reservoir management plans means that revised reservoir characterization models may need to be built. Reservoir characterization, therefore, becomes a consideration that must be addressed numerous times throughout the life of a reservoir.

CASE-STUDY EXAMPLES FROM DOE'S RESERVOIR MANAGEMENT DEMONSTRATION PROGRAM

Because a generally conservative attitude has prevailed toward reservoir management in the past, there is potential for increasing profitability in a large number of petroleum reservoirs. In most cases, this increased profitability could be accompanied by increased recovery as well. Recognizing these facts, and continuing its mission to increase profitability and recovery from domestic reservoirs to prevent or fore-

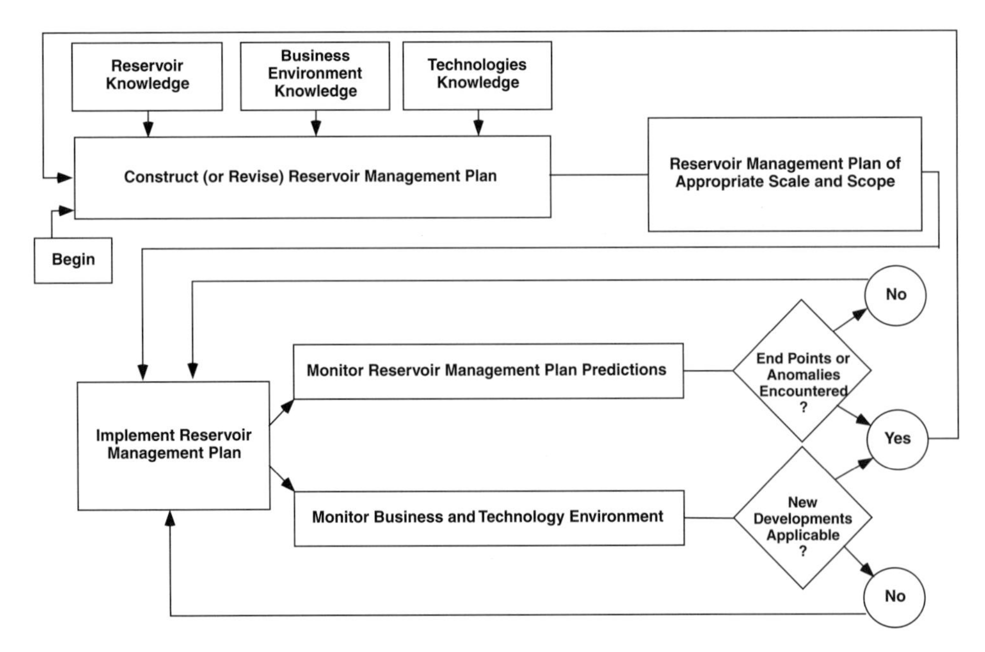

Figure 2 —The reservoir management process is an interactive procedure involving plan construction (or revision), plan implementation, and monitoring of reservoir performance, technological advancements, and the reservoir management business environment (from Fowler et al., 1996).

stall their abandonment, the DOE has been funding, since 1995, a program of reservoir management demonstration projects whose primary objective is to improve understanding through demonstration and technology transfer. Program projects encourage operators, especially small independent operators, to learn, apply, and disseminate sound techniques. Projects are now underway to develop plans for a variety of reservoirs managed by small operators. Technology transfer plays a vital role in each of these projects.

Projects supported by DOE under the Reservoir Management Demonstration Program are limited to oil reservoirs operated by small business organizations. Multiple operators must be involved in proposed projects, although participation by research organizations, state government agencies, service companies, consultants, etc., is encouraged. Projects must address resources significant to the region in which they occur and also address a major technological need. Projects are performed under a Cooperative Research and Development Agreement (CRADA) at a total level of funding up to about $500,000, at least 50% of which must be cost-shared by industry partners. DOE's contribution is mainly in the form of professional labor supplied through BDM-Oklahoma, but data acquisition and analysis, additional consulting expertise, etc., also may be included in DOE's contribution. Projects are intended to be short (12–18 months maximum duration), and technology transfer must be a major focus of each. All projects have a common goal: to develop a comprehensive strategy to improve the operational economics and optimize the oil recovery from the target field.

As initially conceived, the Reservoir Management Demonstration Program would include 15 projects, one in each of the ten Petroleum Technology Transfer Council (PTTC) Regions (Figure 3), three projects involving Native American reservoirs, and two offshore projects. At the present time, three projects have been initiated under the program. One project, begun in early 1995 in East Randolph field in Ohio, is complete. This project in the Appalachian PTTC Region deals with a small, recently discovered oil reservoir in a newly developing play in an area of mostly mature production. A second project, begun in late 1995 in Citronelle field in Alabama, is more than half completed. This project in the Eastern Gulf PTTC Region involves developing a strategy for a mature domestic giant oil reservoir (160 million bbl cumulative production). A third project was launched in late 1996 in Bainville North field in Montana. This project in the Rocky Mountain PTTC Region will deal with the unique problems and opportunities related to multiple producing zones within a field.

Summary of the Plan-Building Process for the East Randolph Field

Since 1992, PEP Drilling Company and Belden & Blake Corporation have developed this unique, but significant, oil reservoir in the Cambrian Rose Run formation in Portage County, Ohio (Figure 4). One of only a few fields to produce oil from the Rose Run, East Randolph field covers about 1500 ac (607.5 ha), lies at a depth of about 7200 ft (2196 m), and contains an average of about 15 ft (4.5 m) of pay in the upper three of five marginal marine sandstone zones typically present. The field contains just over 30 wells and had produced about 450,000 bbl of 42° API oil and 1.2 Gcf of gas as of June 1996. Two factors have been important for the development of a reservoir management plan for the reservoir. First, East Randolph field has been and continues to be developed by small independent operators. Second, the field has been entirely developed in the 1990s. In fact, rapid development is still going on as efforts continue to define the limits of the field.

The proposal submitted by the operators listed potential targets or opportunities to pursue as goals. The list included defining development and infill well locations, selecting and designing implementation of a

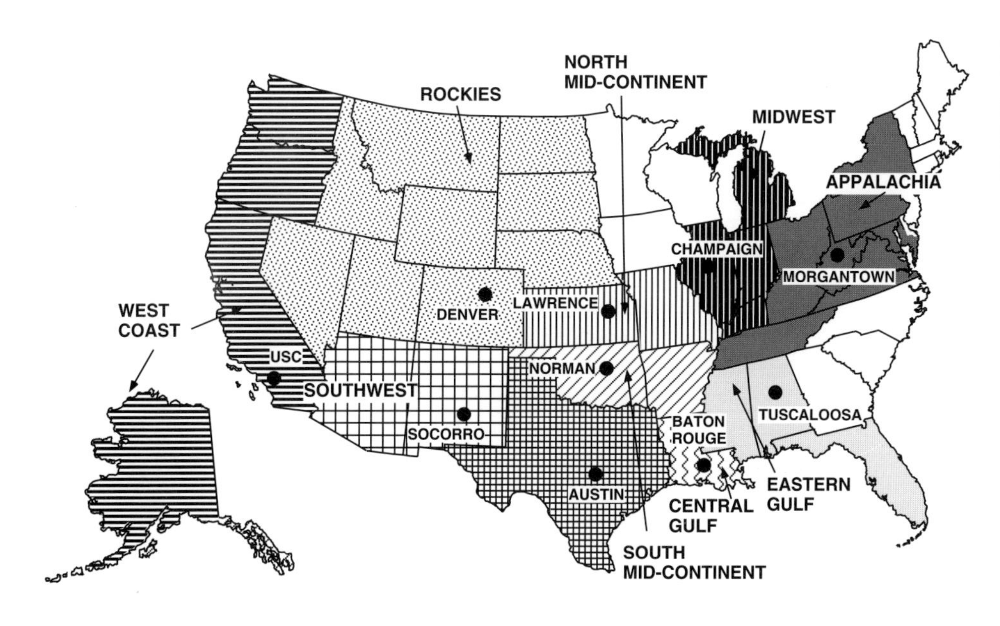

Figure 3—The ten Petroleum Technology Transfer Council regions and the locations of their Regional Lead Organizations.

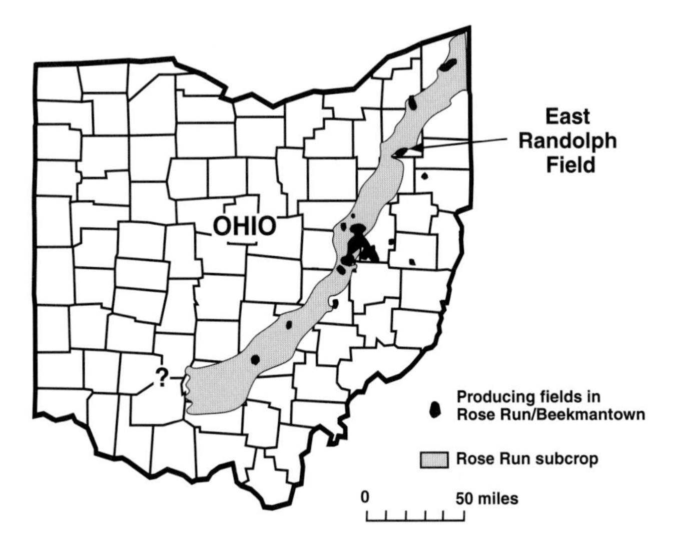

Figure 4—East Randolph field is located in eastern Ohio in a northeast-southwest trend of reservoirs producing mostly gas from the Rose Run (from Riley and Baranoski, 1992; Fowler et al., 1996).

secondary recovery method, selecting optimum hydraulic fracturing techniques, and addressing a paraffin buildup problem in producing wells. The proposal also suggested project tasks and teaming arrangements that might be used to execute those tasks. A kickoff meeting of all project participants was held at the project outset to further prioritize targets and to assign specific plan development tasks to teams and subteams.

Role of Reservoir Characterization

Reservoir characterization played a major role in arriving at a reservoir management plan for this project and in pursuing the targets selected as highest priority; i.e., defining development and infill well locations and selecting an optimum secondary recovery method. A series of incremental and sometimes iterative steps was performed in arriving at the final model (Salamy et al., 1996). The steps involved analyzing existing data, identifying data insufficiencies, obtaining and incorporating new information into the emerging model, and testing the predictive limits of the model.

At the project outset, field limits were not yet accurately defined; preproject estimates of original oil in place (OOIP) were in the neighborhood of 4.4 million bbl of oil. Although the three productive sandstone intervals in the Rose Run were recognized, the high gas/oil ratio (GOR) observed for most wells (1500–2000) was attributed to conditions in all three sandstone zones.

Initial geologic work with neutron and density logs digitized in the project suggested that the uppermost sandstone zone had a much higher gas saturation than the other zones. Analysis of production data showed a correlation between high initial GORs and occurrence of a well-developed upper sandstone zone, further suggesting a possible gas cap. Fieldwide work based on digital logs and previously existing sidewall-core analysis data determined structural heterogeneities (faults), vertical layering of rock properties, and horizontal variations in rock properties (Thomas and Safley, 1996). Zone mapping and volumetric analysis based on this geologic model yielded an OOIP figure of approximately 11 million bbl of oil. This discrepancy in OOIP estimates is important to resolve because of its potential impact on continued development and future recovery.

In parallel with the initial geological work and as an initial and potentially cost-effective check on reservoir parameters, a single-well reservoir model was developed on one of the highest GOR wells in the field. Pressure-volume-temperature (PVT) parameters input to this model were derived from initial reservoir parameters using published correlation techniques. Relative permeability and capillary pressure data input to the model were taken from analogous fields nearby. Model results were unstable in predicting production and indicated the need for more representative values for PVT and relative permeability parameters, additional field pressure data, and field volumetric information. As a result, a pressure buildup was run, and surface-recombined fluid samples were obtained from an existing field well.

The new PVT data (which indicated a GOR of only 485), new pressure data, and production data were then used in a material balance calculation. A sensitivity analysis done on gas/oil volume ratios indicated that gas/oil volume ratios in the range of 0.16 to 0.2 would yield OOIP in the observed range of 11 to 13 million bbl of oil. Using a gas/oil volume ratio of 0.17 and OOIP of 12 million bbl of oil yielded a reasonable match with observed field pressure history. This analysis confirmed that the field's high GOR was not a result of gas coming out of solution.

A second single-well simulation was run using a three-layer (one gas layer, two oil layers) model, the new PVT data, and, again, relative permeability data from analogous fields. This modeling confirmed the upper zone as predominantly a gas zone and accurately predicted reservoir pressure encountered by a subsequently drilled well at the edge of the modeling area. Predictions, however, were sensitive to relative permeability data, so it was recommended that these new pieces of information be obtained.

An infill well was drilled and cored, and relative permeability and capillary pressure data were obtained on samples from the core. A Combinable Magnetic Resonance (CMR) log was run to better define water-saturation distribution, and a Formation Micro-Imager (FMI) log was run to investigate distribution and orientation of natural fractures. Cleaning the samples for special core analysis also gave insight into the nature of the paraffin deposition problem and anomalously low measured permeabilities from routine core analysis samples.

As a final step, a full-field simulation was undertaken

using all the newly collected information. The simulation study was completed in two stages. First, history matching of field production and pressure data was done, holding constant all known field and experimental data. Results showed a good match with oil and gas production and field pressure data, thus validating the basic model. Second, the model was used to predict waterflood and gas injection results as potential secondary recovery methods for the field. Simulation results indicated that a high producing water/oil ratio coupled with low oil recovery makes waterflooding a less favorable option than gas injection for secondary recovery. The base-case simulation indicates that ultimate primary recovery is 8% of OOIP. Waterflooding adds only an additional 0.5% recovery, whereas gas injection results in an additional 7.8% recovery of OOIP.

Discussion

The fact that the field is operated by small independents governed not only the nature of the implementations recommended by the plan, but also the expenditure of effort and capital in collecting and analyzing data to arrive at the plan. Collection of new information had to be adequately justified. Although mutually supportive evidence from different reliable and cost-effective sources was sought, highly redundant confirmations were avoided as unnecessary and unjustifiable. Continued rapid development of the field while the plan was being formulated meant that new information had to be considered and incorporated continually, and that rapid development of a plan was necessary to optimize field development and definition activities.

The incremental approach to reservoir description for plan development employed in this project resulted in efficient data collection. Existing data were analyzed at each step with the objective of determining whether the uncertainty associated with the predictive power of the models based on those data was acceptable. If not, the type and quantity of new data needed to constrain the modeling efforts were identified and obtained after first considering the potential cost-effectiveness of the new information. This approach avoids the collection of unnecessary data and fits very well with the typical independent operator's economic constraints.

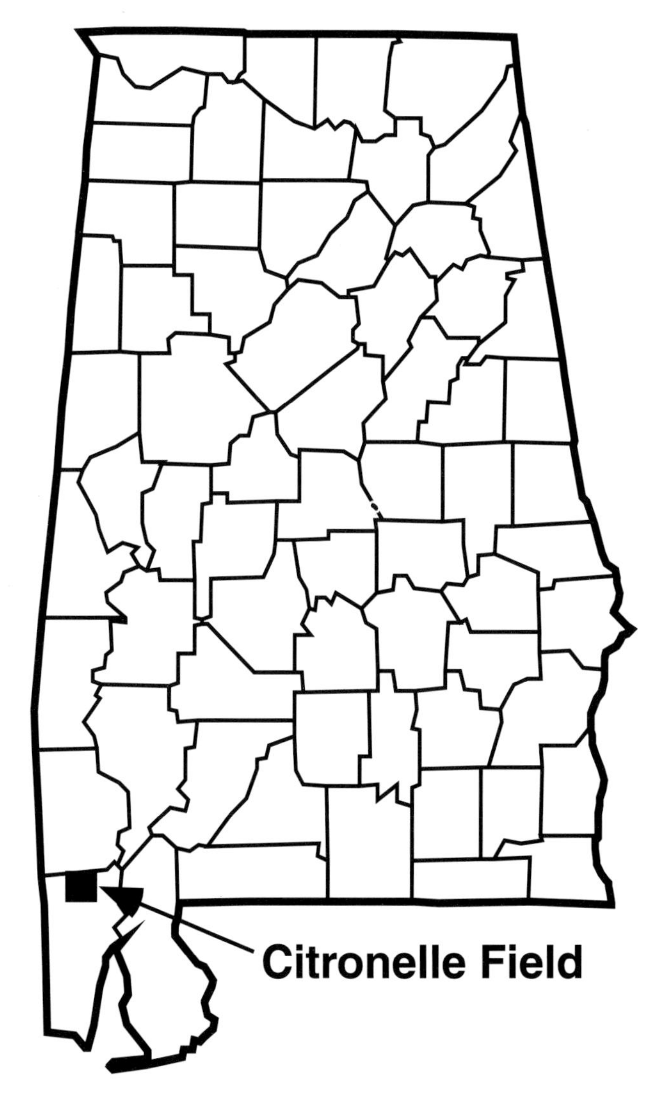

Figure 5—Citronelle field is located on the eastern edge of the Mississippi Interior Salt Basin in Mobile County, Alabama (from Fowler et al., 1996).

Summary of the Plan-Building Process for the Citronelle Field

Citronelle field, in Mobile County, Alabama (Figure 5), has been producing since its discovery in 1955 from fluvial sandstones of the Cretaceous Rodessa Formation at depths greater than 10,000 ft (3050 m). The field, located over a deep-seated salt intrusion, was developed and essentially remains today on 40-ac (16.2-ha) spacing, covering a surface area of 16,400 ac (6642 ha) with 468 wells. An 800-ft-thick (244-m) gross pay interval contains at least 42 productive sandstone zones that form more than 300 separate reservoirs, each with the highly variable permeability characteristic of fluvial deposition. Field pressure declined relatively rapidly, leading to the inception of waterflooding in 1961. By early 1995, approximately 15,000 bbl of water was being injected daily into 50 injection wells to produce about 3600 bbl of oil from about 175 producers. Cumulative recovery is about 160 million bbl oil, 120 million bbl water, and 1.2 Gcf gas. Both major and independent companies have conducted studies and collected voluminous data (several versions of some parameters) from the field over the course of its 40-yr history. The field is currently being managed in its mature waterflood stage for the most part by small operators.

In the proposal submitted, the operators identified the following possible means for improving the economic performance of the field.

- Create a computerized field database for future analyses.
- Identify untapped, incompletely drained, and new pool targets.
- Optimize the waterflood strategy.
- Investigate alternate recovery methods.

- Evaluate drilling and completion techniques.
- Investigate casing leak prevention and repair.
- Investigate production problems related to paraffin, chlorite, scaling, etc.
- Investigate improvement of downhole hydraulic pump life.

The operators also suggested a list of tasks to address the targets listed and made tentative assignment of team members to the tasks. In addition to appropriate personnel from BDM-Oklahoma, the team for this project included geoscience, engineering, management, and other professionals representing operators of the 341 Tract, East, Southeast, and Northwest units of Citronelle field; operators of geologically analogous reservoirs in the area; the Alabama Geological Survey; the State Oil and Gas Board of Alabama; and the University of Alabama. The team also included a private engineering consultant with a long history of association with Citronelle field.

Several major reservoir management decisions were made by the team very early in the project, even before a detailed blueprint existed for constructing a strategy. All team members recognized that, under current operations, the economic limit for the field was approaching within a few years. There also was a general agreement that, based on what was already known about the field and its performance, a substantial oil target remains in Citronelle field, justifying an effort for its recovery. Discussions focused on the necessity of achieving a cost-effective approach through careful matching of the limited resources available for investment by the operators to the probability of improving production and profitability. It was further agreed that rather than considering the entire field, the most economically reasonable approach would be to concentrate (initially, at least) on geographic areas where certain significant problems or opportunities existed. Solutions developed in these areas should have the most significant impact on profitability per dollar expended.

Although the operators had done considerable work in identifying potential opportunities for improving profitability, a kickoff workshop with the full reservoir management team in attendance was held at project inception to further identify and delineate problems and opportunities to be addressed. The following actions were recommended at the workshop.

(1) Identify and prioritize geographic areas in the field where opportunities are best developed or most prominent.
(2) Review the data available to address opportunities in the areas identified.
(3) Identify additional data requirements.
(4) Develop a detailed procedure and schedule to govern plan-building activities based on the results of the previous steps.
(5) Assign team personnel responsible for executing the tasks delineated in the previous step.
(6) Identify opportunities for technology transfer.

The highest priority opportunity identified at the kickoff workshop was that of waterflood optimization. Boundary areas (Figure 6) between the Citronelle 341 Tract unit and the East and Southeast units were identified as areas where the current waterflood has been least efficient in recovering oil reserves because no unified effort has been made in those areas in the past to optimize injection or production strategies.

Role of Reservoir Characterization

The long history of Citronelle field has included numerous fieldwide geological and engineering studies. These studies, which were based on a strong foundation of core and wireline log data, include most of the current wells in the field. Computerized databases were originally associated with some of these studies, whereas other studies were based on work done strictly by hand. Data from earlier studies were available only in hard-copy form at project inception, however. The quality of much of the work done in the past was judged to be sufficient to justify using it as a basis for the needs of the current study.

A decision was made early in the project to pursue two parallel approaches. The first approach used interpreted sandstone geometries and associated volumetric parameters derived from past studies along with past pay determinations and preliminary estimates of permeabilities, porosities, and water saturations from past engineering studies to perform preliminary characterization of the reservoir in the areas of interest along the unit boundaries. A second, simultaneous approach involved entering gross and net pay data along with gradually refined estimates of permeabilities, porosities, and water saturations into a fieldwide database that could be used with modern mapping tools to create minimally biased volumetric and other estimates. Production data were obtained from the state of Alabama, and cumulative values at various points in time also were entered into this database.

As a first step in determining the best areas and approaches for obtaining additional waterflood recovery, cross sections in the areas of interest defined along the unit boundaries were used to define potential flow units as isolated targets for improved recovery. Flow units were defined by combining into one package the sandstones that are likely to be in vertical and horizontal communication with each other across the area of interest, but at the same time separated from adjoining sandstones or packages by substantial shale barriers. Because of past hydraulic fracturing practices in the field, a 30-ft (9.2-m) minimum thickness was used to define effective shale barriers separating sandstone packages. As a next step toward identifying potential recovery targets, floodable OOIP volumetrics were calculated for each of the approximately 20 sandstone packages or flow units identified.

Flow units do not occur in isolation, however. Several flow units are commonly present in a single well, as well as numerous sandstones not identified as belonging to discrete flow units. The next step

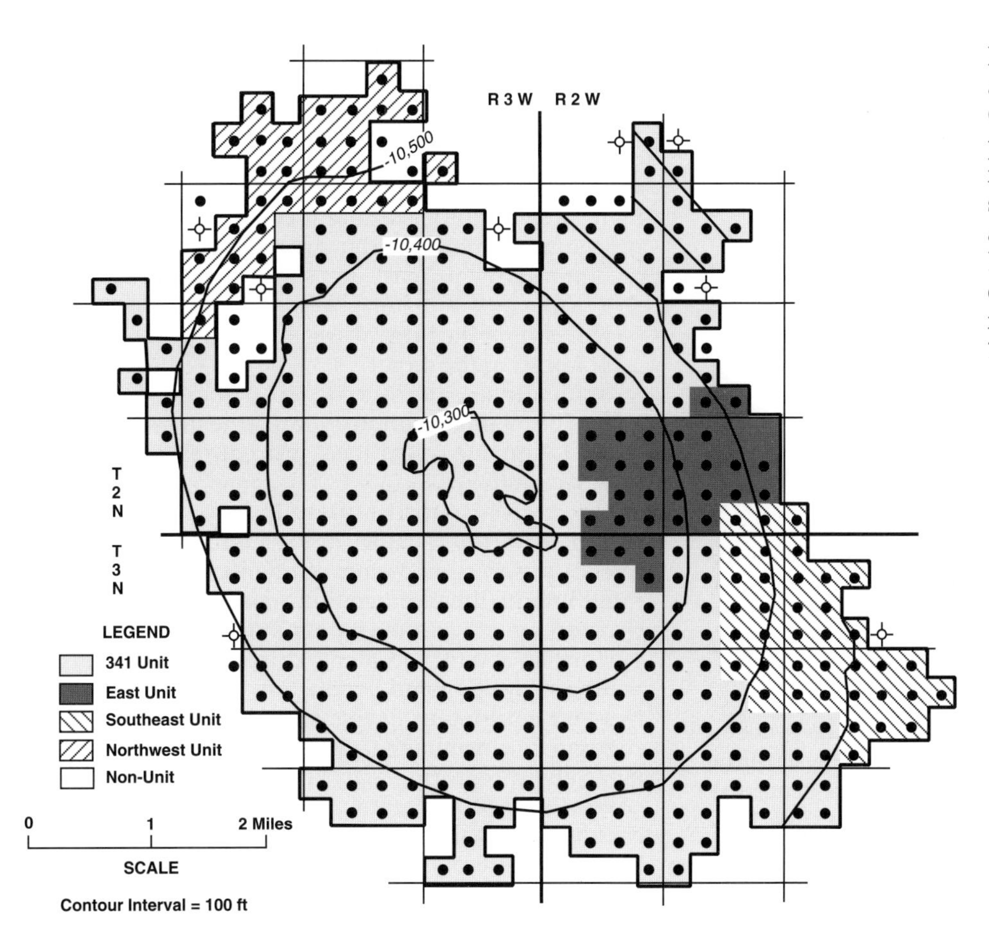

Figure 6—The boundary areas between the units in Citronelle field are the primary focus of the current reservoir management study. Structural countours are on the base of the Ferry Lake Anhydrite immediately overlying the Citronelle reservoir (from Fowler et al., 1996).

involved looking at groups of wells characterized by the presence of multiple flow units in common. These wells form logical groups to consider for optimizing production through altering injection/production strategies. Identifying these groups of wells and prioritizing them on the basis of their OOIP volumetrics is now nearing completion.

The next anticipated step is to review the completion, production, and injection histories of the associated wells for the top-priority groups to determine the size of the remaining potential recovery target. The final step will be to evaluate and strategize recompletions and injection/production geometries to maximize recovery for all sandstone packages and flow units in the groups.

Discussion

The series of steps involving progressive prioritization first on the basis of OOIP and then on the basis of remaining potential is meant to ensure that areas with the best recovery economics are addressed first. In its refined form, this general methodology can be applied throughout the field to maximize economic recovery from other sandstone packages. Such an approach will allow the untapped profitability potential of the field to be developed in small incremental steps that are more financially feasible for small operators than examining the entire field in detail at once.

Summary of the Plan-Building Process for Bainville North Field

Bainville North field was discovered in 1979 in the Williston Basin in Roosevelt County, Montana (Figure 7), with establishment of production from the Devonian Winnipegosis Formation at a depth of 11,500 ft (3507.5 m). Production has since been added from four additional zones between the Mississippian Ratcliffe Formation at 8500 ft (2592.5 m) and the Ordovician Red River Formation at 12,500 ft (3812.5 m). Fifteen wells currently produce in the field, whose cumulative production is 2.8 million bbl

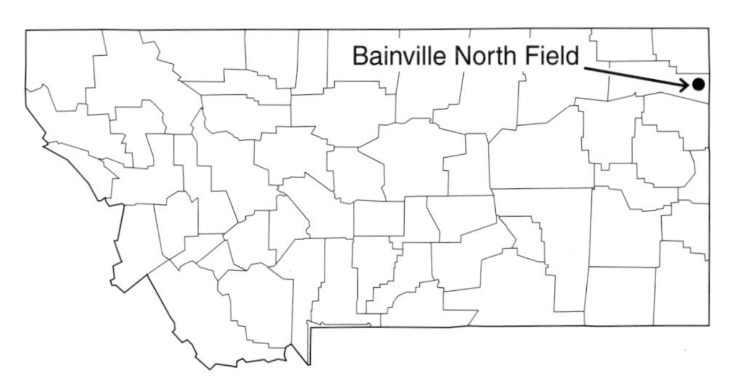

Figure 7—Bainville North field is located in Roosevelt County, Montana, in the Williston Basin.

oil, 2.8 Gcf gas, and 1.4 million bbl water through early 1996.

In the 1990s, Nance Petroleum Corporation substantially increased production from the Red River Formation through use of 3-D seismic and other advanced rock, fluid, and high-resolution wireline log data to locate infill wells. Although only seven wells are currently producing from the Red River interval, cumulative production through early 1996 from the interval was 1.0 million bbl oil and 762 Mcf gas.

In the operators' proposal, reservoir characterization of the highly heterogeneous dolomitic Red River reservoir to improve recovery was presented as a top priority. The strategy developed in the project will address the design and implementation of improved recovery processes within the context of the field's multiple productive zones. This will include considerations of well spacing and utilization, artificial lift optimization, and facilities requirements. Additional problems to be addressed may include paraffin deposition, scaling, corrosion, and casing collapse.

Role of Reservoir Characterization

Reservoir characterization is expected to proceed by integrating existing core data, fluid data, wireline log data, pressure data, 3-D seismic data, and any additional data deemed necessary into reservoir models for use in simulation. Historical performance of the field will be simulated, and predictions will be made for implementing improved recovery techniques. From these results, a reservoir management strategy will be formulated.

Discussion

This project is just beginning, but it has potential for significant impact on Williston Basin petroleum development. Because of the geologic similarities between the various producing zones in the basin and because very few advanced recovery projects have been implemented in the basin, the strategies developed to address the opportunities at Bainville North will be applicable to most reservoirs in the region.

STEPWISE APPROACH TO PERFORMING RESERVOIR MANAGEMENT AND RESERVOIR CHARACTERIZATION

One of the key objectives in the Reservoir Management Demonstration Program being sponsored by DOE was to resolve the sequence of considerations that goes into developing an effective reservoir management plan. Five major steps in plan development have been recognized in the studies performed. Achieving each of these steps requires an appropriate degree of accompanying reservoir characterization. Five separate reservoir characterization studies do not have to be performed, however, so the scale of the reservoir characterization effort can be logically matched with each step in plan building.

Reservoir Management Plan Construction Methodology

At this time, only the broadest categories in plan construction have been identified, but it is hoped that subsequent work on a variety of reservoir management projects under different contexts will enable the procedures to be defined in greater detail. As currently recognized, the primary steps in plan construction are

(1) Defining the target size
(2) Locating the target
(3) Identifying appropriate technologies
(4) Optimizing technology implementation
(5) Optimizing operational procedures and technologies

A plan ideally specifies its own limitations based on the conditions and assumptions that were incorporated into its development. It may predict by simulation or by other means all aspects of reservoir performance over the plan's duration (e.g., reservoir wellbore injection and production performance, facilities and equipment usage, environmental and other regulatory compliance, etc.). It may specify surveillance and monitoring activities, including data types, collection protocol, database construction, data processing and analysis, and performance variance to be tolerated. The plan also may specify or recommend future plan revisions based on specific criteria, such as timing or volume performance of reservoir fluid production or injection. The plan should be flexible enough to accommodate potential modifications.

These steps are very general and should be applicable whether or not improved recovery is being considered. In each step, careful attention must be paid to the complete context of reservoir system, available technologies, and the business environment. Specific objectives of any plan also will include consideration of the current stage of reservoir development and the type and scale of the decisions required (e.g., evaluation of a potential new process implementation, local production and injection optimization, new facilities or equipment technologies, etc.). A comprehensive plan initiated at the time of reservoir discovery will ensure early collection of native-state reservoir data vital to implementing advanced recovery processes many years in the reservoir's future. On the other hand, reservoirs in which data collection has been neglected and reservoirs acquired without adequate accompanying data require a plan designed to correct or alleviate the effects of information deficiencies. Ideally, a reservoir management plan or series of plans will provide guidelines over the life of the reservoir.

Target Definition

Defining the target size, whether that target may be the recovery of additional petroleum resources or merely saving dollars lost to chronic production problems or inefficient operating procedures, will help to determine the scale and scope of the plan being developed, as well as help scale the effort expended in constructing the plan. Multiple targets of the same or different types may be addressed by the same plan. In fact, this approach should lead to a plan that will optimize the profitability of the reservoir on several fronts. Often, the target or targets can be defined adequately with existing data, but additional information may have to be collected to reduce uncertainty about the target size to an acceptable level. If at least one of the targets is additional recovery, reservoir characterization, at least at some general level, will be needed to estimate the quantity of petroleum potentially present, unless reliable information is already available from previous work.

Target Location

In some cases the focus of the plan will be the entire field, but more often certain reservoir zones or areas of the field will present the best opportunities. Additional data may have to be gathered on a field-wide scale to locate the target or targets accurately. If additional recovery is the target, questions such as whether the oil is mobile or immobile also may have to be addressed. Information required to answer the questions asked at this stage in plan development will probably be more detailed than that required to estimate the size of the recovery target in the previous step.

Technology Selection

Identifying appropriate technologies to achieve the target may involve gathering yet more information not only in order to evaluate the technical appropriateness of potential technologies, but also to arrive at an economic prioritization of potentially acceptable technologies. For example, when the target is improved recovery, this step will include a first-pass screening evaluation of a wide variety of technologies, followed by an in-depth evaluation of the top-ranked recovery technologies.

Technology Optimization

Optimizing the implementation of selected technologies can require major data collection and analysis efforts, especially if the focus is recovery technologies. Reservoir characterization may need to be done in great detail to allow development of models to predict recovery and economic results with a sufficiently low degree of uncertainty. Well placement and completion configurations will strongly depend on the results of this modeling.

Operational Optimization

Implementing new technologies in a reservoir will likely mean that operating procedures and related technologies will need to be adjusted to produce the best reservoir performance.

Scaling the Reservoir Characterization Effort

If additional recovery is an objective, reservoir characterization plays a role in each of the five major steps in plan development; however, five separate reservoir characterizations may not have to be performed, as project participants in the DOE's Reservoir Management Demonstration Program have learned. If estimates of the size of the recovery target are highly uncertain (i.e., are uncertain to the point of being unable to determine whether an economic target is present), basic data should be collected to reduce that uncertainty before proceeding. This approach avoids expending large amounts of effort and money to build a highly detailed model for a reservoir that turns out to have an unjustifiably small target.

Initial location of a recovery target and post-screening selection of an appropriate recovery method or process may require more accurate and more detailed information about the reservoir than that needed for a rough estimate of target size. Optimizing the implementation of a selected methodology, however, will require a degree of detail and uncertainty dictated by the method selected. When recovery processes involving large investments in chemicals are involved, greater detail and much lower uncertainty than that for initial process selection will be required. This step might be thought of as a refinement of the initial target-location step. Even optimizing operational procedures requires a knowledge of reservoir characteristics and capabilities.

The general methodology involved in constructing a reservoir characterization model to fit any stage of reservoir management plan preparation may be summarized in a small number of sequential steps. In practical applications, the steps may need to be followed through several iterations to achieve an acceptable product. A flowchart summarizing typical activities in these steps is shown in Figure 8.

Specify Appropriate Resolution and Uncertainty Characteristics

This step consists of making an initial estimate of the detail and degree of uncertainty necessary to perform the evaluation. The activity will be a function of the stage of plan development under which the reservoir characterization is being performed. The required level of model detail may vary from place to place within the reservoir, or the model may be entirely focused on a segment of the reservoir.

Incorporate Information

Information that goes into building a model originates in (1) data collected from different sources (e.g., rock and fluid samples, wireline logs, seismic, well tests, production data, etc.), (2) data collected at different resolutions, and (3) data collected with inherently different degrees of uncertainty (including both deterministic and conceptual or analog information). The model must provide information throughout the 3-D volume of that part of the reservoir that is of interest, but this does not mean that each deterministic data type needs to be collected throughout the entire volume of interest. Instead, information derived from various data sources is incorporated by interpolation, extrapolation, or correlation with other data of deterministic or conceptual/analog origin. This diverse

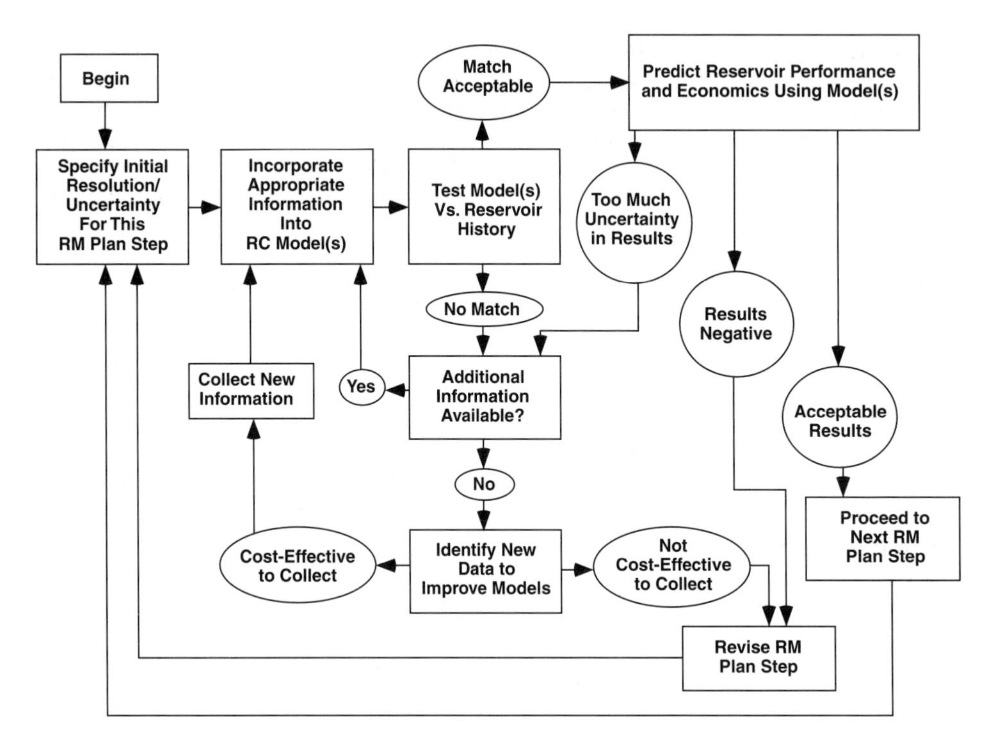

Figure 8—Construction of appropriate reservoir characterization (RC) models is an interactive procedure strongly controlled by reservoir management (RM) context and the cost-effective availability of information.

information is incorporated in mutual support of a single model, or a small number of models, that expresses the probable range of variation. Information derived from geoscience, reservoir engineering, and other data is continually compared and contrasted to test various details of the emerging model(s) until indications from all existing sources are in agreement. Inconsistent information or conflicting information derived from different data sources or scales may have to be addressed with additional data collection before full confidence is achieved. Al-Hussainy and Humphreys (1996) provide an enlightening discussion of uncertainty in reservoir management.

Test the Model Against Known Reservoir Performance

Models from the previous step are tested either qualitatively or quantitatively using simulation or simpler approaches to measure their ability to serve as a basis for predicting reservoir pressure and production. Failure of the models to support a reasonable match with actual reservoir past performance indicates that the models need correcting by incorporating new information. If the models perform well, they are ready for use in predicting the future performance of the reservoir under the plan being formulated.

Identify, Collect, and Incorporate New Information

Our knowledge of the reservoir will never be perfect, but the collection of additional information can be viewed as an attempt to reduce model uncertainty and, ultimately, risk when implementing the reservoir management plan. Several guidelines govern the judicious collection of information. The main goal is to obtain the most information affordable by project economics to reduce uncertainty in predicting the project outcome. If the cost of reducing uncertainty to an acceptable level is too high, the plan may have to be redesigned. Information from all potential sources should be evaluated in terms of possible contribution to reducing model uncertainty. Potential contributors of low-cost information (such as conceptual models from the published literature) should not be overlooked. After a model has been revised or refined by incorporating new information, it should be retested.

Use the Model to Predict Future Reservoir Performance

Production performance predictions obtained from reservoir characterization models by simulation or other means will form the basis for the economic analysis of various plan scenarios. This step enables plan optimization by selecting those activities, processes, implementation strategies, operational strategies, etc., offering the best economic performance. If economic analysis indicates that results will be unacceptable, or if the range of probable outcomes includes unacceptable economic performance, the plan may need to be redesigned.

CONCLUSIONS

Reservoir management is not optional; anyone who is responsible for a reservoir is managing it. It is not a luxury reserved for large, recently discovered reservoirs or for major oil companies. It does not necessarily imply a high-tech approach.

The heart of reservoir management is a plan or strategy that guides the operator's interaction with the reservoir. Reservoir management is the creation of an efficient plan that maximizes the profitability of a reservoir to its operator. An appropriate plan cannot be bought off the shelf or transported from one reservoir to another reservoir or from one operator to another operator; it must be custom built. To accomplish this requires a knowledge of the business environment; a knowledge of technologies available to describe, analyze, and exploit the reservoir; and a knowledge of the reservoir system, including its rocks, fluids, wellbores, and surface facilities.

Reservoir characterization is an important technique to gain the knowledge of the reservoir needed for plan building. Approaches can be appropriately scaled by considering the reservoir management context, the context both within the plan-building process and within the business environment and existing technology.

Reservoir management is a dynamic process. Existing plans become inappropriate as the technology and the business environment change, or as new reservoir information indicates that the models on which current plans are based is inadequate. Continual monitoring will identify plans that are no longer optimum. De-optimized plans should be revised to achieve maximum profitability and recovery. Revision of plans commonly calls for a revised reservoir characterization model.

The generally risk-averse approach to reservoir management (especially in the management of domestic reservoirs) taken by industry means that many reservoirs are operated under less-than-optimum plans. The potential to increase profitability and recovery from domestic reservoirs is therefore great. The DOE's Reservoir Management Demonstration Program focuses on this concept. The program is seeking to improve domestic reservoir recovery and profitability and to delay abandonment by encouraging operators to develop, apply, and disseminate efficient reservoir management techniques.

Work already accomplished in two program projects demonstrates the scaling of reservoir characterization efforts in developing plans for reservoirs under widely contrasting contexts. Both projects recognized that small operators have limited capital resources to invest. The East Randolph field project took an incremental approach to defining critical reservoir parameters while this small, newly discovered field was still under active development and definition. In the Citronelle project, the focus was on developing an improved recovery approach or methodology for small areas of high potential in this already mature waterflood. The operators can then repeat the methodology developed for the improved recovery targets remaining in this large field.

ACKNOWLEDGMENTS

The authors and other researchers participating in the DOE's program of reservoir management demonstrations appreciate the department's foresight in supporting this methodological research and demonstration. This project would not have been possible without the dedication and support received from all members of the East Randolph, Citronelle, and Bainville North field reservoir management teams. We also are grateful to the BDM-Oklahoma Information Services Department for its support.

REFERENCES CITED

Al-Hussainy, R., and N. Humphreys, 1996, Reservoir management: principles and practices: JPT, v. 48, no. 12, p. 1129–1135.

Cole, E. L., M. L. Fowler, S. P. Salamy, P. S. Sarathi, and M. A. Young, 1994, Research needs for strandplain/barrier island reservoirs in the United States: U.S. Department of Energy Report DE95000118, 186 p.

Cole, E. L., R. S. Sawin, and W. J. Weatherbie, 1993, Reservoir management demonstration project: University of Kansas Energy Research Center Technology Transfer Series 93-5, 236 p.

Fowler, M. L., M. A. Young, E. L. Cole, and M. P. Madden, 1996, Some practical aspects of reservoir management: SPE Paper 37333, Eastern Regional Meeting, Columbus, Ohio, October 23–25.

Jackson, S. R., and L. Tomutsa, 1991, Reservoir characterization—state-of-the-art review, *in* Research needs to maximize economic producibility of the domestic oil resource, part I—Literature review and areas of recommended research: U.S. Department of Energy Report NIPER-527, p. 143–172.

Lake, L. W., and H. B. Carroll, Jr., 1986, Preface, *in* L. W. Lake and H. B. Carroll, Jr., eds., *Reservoir Characterization*: Orlando, Florida, Academic Press, 659 p.

Riley, R. A., and M. T. Baranoski, 1992, Reservoir heterogeneity of the Rose Run Sandstone and adjacent units in Ohio and Pennsylvania: Ohio Oil and Gas Association Winter Meeting, Canton, Ohio, October 20.

Salamy, S. P., M. A. Young, L. E. Safley, J. L. Wing, and J. B. Thomas, 1996, Application of reservoir management techniques to the East Randolph field, Portage County, Ohio; reservoir engineering study: SPE Paper 37334, Eastern Regional Meeting, Columbus, Ohio, October 23–25.

Satter, A., and G. C. Thakur, 1994, Integrated petroleum reservoir management—a team approach: Tulsa, Oklahoma, PennWell Publishing Co. , 335 p.

Satter, A., J. E. Varnon, and M. T. Hoang, 1994, Reservoir management: Journal of Petroleum Technology, v. 46, no. 12, p. 1057–1064.

Smerdon, E. T., 1996, Career-long education: Journal of Petroleum Technology, v. 48, no. 11, p. 1059.

Thakur, G. C., 1991, Waterflood surveillance techniques—a reservoir management approach: Journal of Petroleum Technology, v. 43, no. 10, p. 1180–1188.

Thomas, J. B., and L. E. Safley, 1996, Improved reservoir characterization of the Rose Run Sandstone in East Randolph field, Portage County, Ohio: Fourth Annual Technical Canton Symposium, Canton, Ohio, October 8–9.

Wiggins, M. L., and R. A. Startzman, 1990, An approach to reservoir management: SPE Paper 20747, reservoir management panel discussion, 65th Annual Technical Conference and Exhibition, New Orleans, Louisiana, September 23–26, p. 327–333.

Section II
Reservoir Description

Clark, M.S., et al., 1999, Characterization of the distal margin of a slope-basin (class-III) reservoir, ARCO-DOE slant well project, Yowlumne Field, California, *in* R. Schatzinger and J. Jordan, eds., Reservoir Characterization-Recent Advances, AAPG Memoir 71, p. 21–28.

Chapter 2

Characterization of the Distal Margin of a Slope-Basin (Class-III) Reservoir, ARCO-DOE Slant Well Project, Yowlumne Field, California

Michael S. Clark[1]
John D. Melvin
Rick K. Prather
Anthony W. Marino
ARCO Western Energy
Bakersfield, California, U.S.A.

James R. Boles
Douglas P. Imperato
Geology Department, University of California
Santa Barbara, California, U.S.A.

ABSTRACT

Yowlumne is a giant oil field in the San Joaquin Basin, California, that has produced over 16.7 million m^3 (105 million bbl) of oil from the Stevens Sandstone, a clastic facies of the Miocene Monterey Shale. Most Yowlumne production is from the Yowlumne Sandstone, a layered, fan-shaped, prograding Stevens turbidite complex deposited in a slope-basin setting. Well log, seismic, and pressure data indicate seven depositional lobes with left-stepping and basinward-stepping geometries.

Log-derived petrophysical data, constrained by core analyses, indicate trends in reservoir quality. Concentration of channel and lobe facies along the axis and western (left) margin of the Yowlumne fan results in average net/gross sandstone ratios of 80%, porosity (ϕ) of 16%, and liquid permeability (K_{liquid}) of 10–20 md. By contrast, more abundant levee and distal margin facies along the eastern margin result in shale-bounded reservoir layers with higher clay contents and lower net/gross sandstone ratio (65%), porosity (12%), and permeability (2 md). Although a waterflood will enable recovery of 45% of original oil in place along the fan axis, reservoir simulation indicates 480,000 m^3 (3 million bbl) of oil trapped at the thinning fan margins will be abandoned with the current well distribution. Economic recovery of this bypassed oil will require high-angle wells with multiple hydraulic fracture stimulations to provide connectivity between the reservoir layers.

[1] Present affiliation: Chevron USA Production Co., Bakersfield, CA, U.S.A.

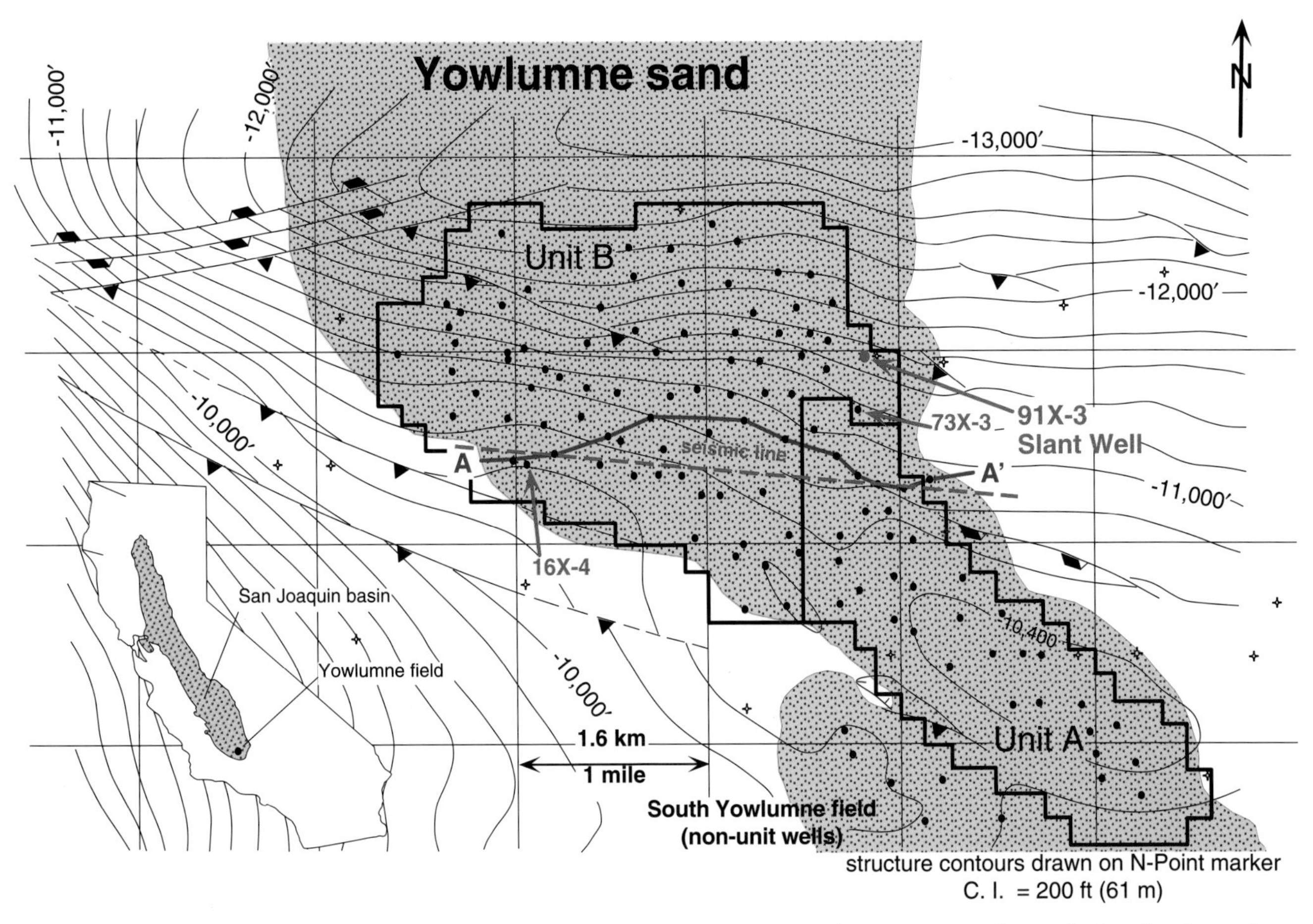

Figure 1—Structure map of Yowlumne field drawn on the N-Point marker, a regional correlation horizon that marks the approximate top of the Stevens sandstone. The map shows the relationship of Units A and B to the small anticlinal closure at Yowlumne.

INTRODUCTION

A significant number of petroleum reservoirs in the United States are turbidite sandstones deposited in slope-basin settings (DOE class-III reservoirs). Although these reservoirs represent a wide spectrum of sedimentary environments, most are characterized as submarine fans or fan-shaped turbidite complexes (Weimer and Link, 1991).

Many class-III reservoirs are located in mature oil fields characterized by declining production. Typically, reservoir quality, measured as increasing porosity and permeability, improves with increasing sandstone thickness. Because fans thicken in the middle, production tends to increase toward the depositional axes. By comparison, production tends to decrease toward the margins where reservoir thickness and reservoir quality decrease. Consequently, it is more difficult to produce oil economically from the margins, and oil reservoired there is more likely to be bypassed and abandoned than oil stored elsewhere in the fan, yet many waterflood patterns sweep oil toward the margins where it is banked against the zero edge of reservoir. Thus, bypassed oil stored along fan margins and oil trapped against them by waterflooding represent significant remaining reserves in many slope-basin fields.

Economic recovery of oil trapped at thinning fan margins requires a detailed understanding of the reservoir to design a cost-effective exploitation strategy. This paper presents an analysis of Stevens sandstones (upper Miocene) at Yowlumne field in the San Joaquin Basin of California (Figure 1). Geologic modeling, reservoir characterization, and flow simulation are used to locate high-angle wells along a thinning fan margin in the Yowlumne Stevens. Also, this analysis enabled the design of multiple hydraulic fracture stimulations ("frac jobs") that maximized well flow rates by improving the connectivity of flow units in a layered, low-permeability turbidite reservoir.

GEOLOGIC SETTING

Slope-basin reservoirs are abundant in the southern San Joaquin Basin (Figure 1), where much of the oil and gas production comes from turbidite sandstones (Weimer and Link, 1991). The Stevens sandstone is an informal unit, known primarily from the subsurface,

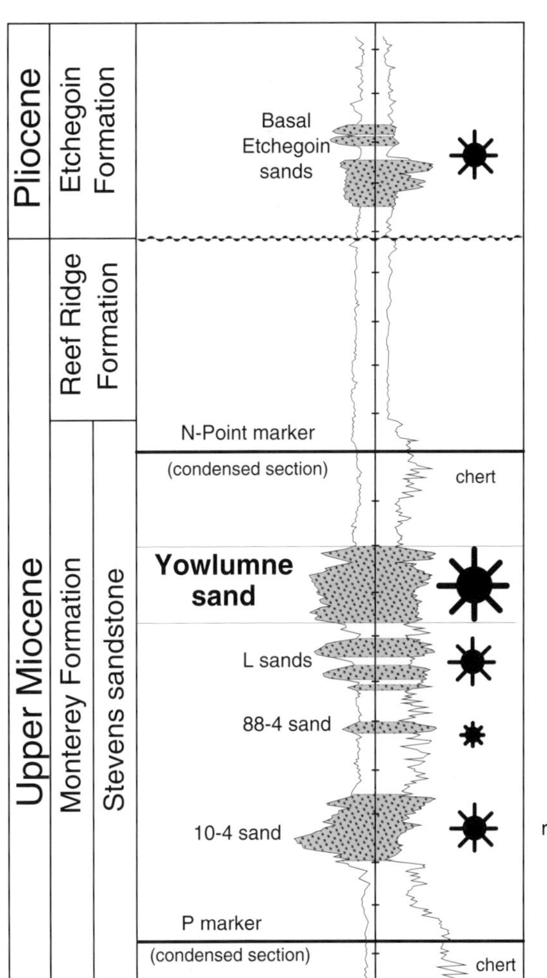

relative magnitude of oil production

Reservoir Statistics

Discovery Date	1974
Reservoir Depth	3,410-4,050 m (11,200-13,300 ft)
Ave.Thickness (gross)	69 m (225 ft)
Max. Thickness (gross)	150 m (493 ft)
Area	1,243 ha (3,070 acres)
Producing Wells	48 active & 40 SI
Water Injection Wells	30 active & 9 SI
Cumulative Production (all zones through 1996)	16.7×10^6 m^3 of oil (105 million bbl) 2.6×10^9 m^3 of gas (92 billion ft^3)

Rock Properties

	fan axis	east margin
Effective Porosity	16%	12%
Air Permeability	50-100 md	5-10 md
Liquid Permeability	10-20 md	2 md
Net/Gross Sandstone	80%	65%
Clay Volume (Vsh)	<6%	>12%

Fluid Properties

Oil Gravity	882-855 kg/m^3 (29-34∞ API)
Fm Water Density	1019 kg/m^3 (22,000 ppm)
Oil Viscosity	.52 cps @ reservoir temperature
Drive Mechanism	fluid expansion
Res.Temperature	113-132∞C (235-270∞F)
Original Res. Pressure	39.02 MPa (5,660 psi)
G.O.R.	107 m^3/m^3 (600 cf/bbl)
Rw @ 25°C	.41 ohm-m

Figure 2—Type log for Yowlumne field showing producing reservoirs, and the rock and fluid properties of the main reservoir (the Yowlumne sandstone).

that represents a deep-water clastic facies of the upper Miocene Monterey Shale (Figure 2). It is also one of the most prolific turbidite reservoirs in the basin and has contributed nearly 15% of the more than 1.9 billion m^3 (12 billion bbl) of oil produced here since 1864. Yowlumne is a giant oil field in the basin that has produced over 16.7 million m^3 (105 million bbl) of oil from Stevens turbidites. Because Stevens oils derive from Monterey Shale source rocks, Yowlumne field is part of a Monterey-Stevens petroleum system (Graham and Williams, 1985).

The oil accumulation at Yowlumne is controlled in part by a small anticlinal closure that formed during Mio–Pliocene deformation of the south margin of the basin (Metz and Whitworth, 1984; Graham and Williams, 1985). The field was discovered in 1974, and Yowlumne Unit A was set up in 1978 to waterflood the area of structural closure. Because subsequent drilling established production on the north-dipping flank of the structure as well (Figure 1), Unit B was created in 1982 to flood the flank accumulation (Metz and Whitworth, 1984).

METHODOLOGY

Field exploitation is enhanced with a thorough knowledge of the reservoirs. Most production at Yowlumne is from the Yowlumne sandstone, one of several discontinuous sandstone bodies collectively referred to as the Stevens sandstone (Figure 2). A better understanding of this reservoir is needed to reduce a 35% annual decline in the field production, and to develop remaining reserves before aging facilities require abandonment. A five-step analysis was performed to achieve these goals:

(1) Description of the reservoir architecture (i.e., stratal geometries, flow units, and facies distributions) using well log correlations, pressure data, and existing 3-D seismic data.
(2) Determination of reservoir properties using core, log, borehole breakout, and micro-seismic data.
(3) Construction of a digital petrophysical database (e.g., porosity, permeability, and water saturations).

Table 1. Software and Data Used to Characterize the Yowlumne Reservoir.

Vendor/Software	Use	Cost (6 mo)
TerraScience (Unix)	Log analysis/correlation	$5,000
Microsoft Excel (Windows)	Spreadsheet analysis	$300
ECL Grid (Unix)	Map-to-grid conversion	$5,000
ARCO Acres (Unix)	Flow simulation	Proprietary
–	Well-digitizing	$50/well
–	Core analyses	Preexisting data
–	3D-seismic survey	Preexisting data
Bolt Technology	Micro-seismic data	$32,600

Dashes indicate that the vendor varies or that a variety of vendors are used.

(4) Contour mapping of petrophysical properties by individual flow units.
(5) Location and quantification of bypassed reserves using a model to simulate fluid flow in the reservoir.

A digital petrophysical database was constructed from well log data using Unix-based log correlation and log analysis software (Table 1). Total porosity (ϕ_{total}) is derived from sonic data, and shale volumes (V_{sh}) are derived from gamma ray values. Effective porosity (ϕ_{eff}) is obtained from $\phi_{eff} = \phi_{total} - (\phi_{sh} \times V_{sh})$, where ϕ_{sh} is the average porosity of the shale fraction ($V_{sh} > 90\%$). Liquid permeability (K_{liquid}) is calculated from ϕ_{eff} using an algorithm derived from core data, and validity of log-derived K_{liquid} and ϕ_{eff} distributions (of net sandstone) is confirmed by comparison to core data corrected for overburden pressure. Net sandstone is defined as sandstone with V_{sh} = 0% to 30% and ϕ_{eff} = 8% to 30%. Water saturation (S_w) is calculated from ϕ_{eff} using the Archie equation with water resistivity values corrected for formation temperatures.

RESULTS

Lithology

The Yowlumne sandstone is a clay-bearing arkosic wacke containing quartz, potassium feldspar, and plagioclase feldspar with subordinate plutonic and metamorphic rock fragments. Most plagioclase is albitized and was altered in the source area before transport. Porosity is primarily intergranular and results from compaction of an original depositional porosity that is further reduced by minor carbonate and quartz cementation, and slightly enhanced by plagioclase dissolution. The clay component is dominantly authigenic kaolinite, which lines pores and is derived from feldspar diagenesis. Some mixed-layer smectite/illite, probably detrital in origin, is also present. The mineralogy indicates derivation from plutonic and metamorphic terranes in the nearby Tehachapi-San Emidio Mountains of the California Transverse Ranges (Whelan, 1984; Tieh et al., 1986).

Reservoir Architecture

Well log and 3D-seismic data indicate that the Yowlumne sandstone is a fan-shaped body up to 150 m (493 ft) thick that is buried to depths of 3410 to 4050 m (11,188 to 13,288 ft) (Figure 1). Bouma sequences observed in cores indicate deposition by turbidity currents. Also, the fan is lens shaped in cross-section and does not significantly incise underlying strata (Figure 3). Because large-scale channeling is absent, deposition was primarily as sheet sands transported by sediment-gravity flows.

Thin shales divide the Yowlumne fan into lobe-shaped reservoir layers. Five of these layers, called sands A, B, C, D, and E, produce oil from Unit B (Figure 3). A basal sixth layer, called the W sand, is pressure isolated from the overlying sandstone and is wet. Layers A–E merge into homogeneous, clean sandstone on the west margin of the fan, yet contain interbedded shale layers on the east. For example, the 16X-4, a horizontal well on the west side of the field (Figure 1), penetrates a thick interval of clean sandstone (Marino and Schultz, 1992). By contrast, the 73X-3, a vertical well on the east side (Figure 1), penetrates shale layers some 2+ m (7 ft) thick that are interbedded with the reservoir sandstones.

Reflection geometries on 3D-seismic indicate downlap within the fan, with basinward progradation to the north and lateral progradation to the west (Figure 3). In other words, lobe-shaped, shale-bounded reservoir layers in Unit B step to the left when facing basinward in the direction of sediment transport (Jessup and Kamerling, 1991; Clark et al., 1996b). Thus, the basal productive layer (sand E) is thickest on the right (east) side of the fan and the top layer (sand A) is thickest on the left (west).

Compartmentalization and Flow Units

Variations in reservoir pressures and injection of radioactive tracers indicate weak compartmentalization

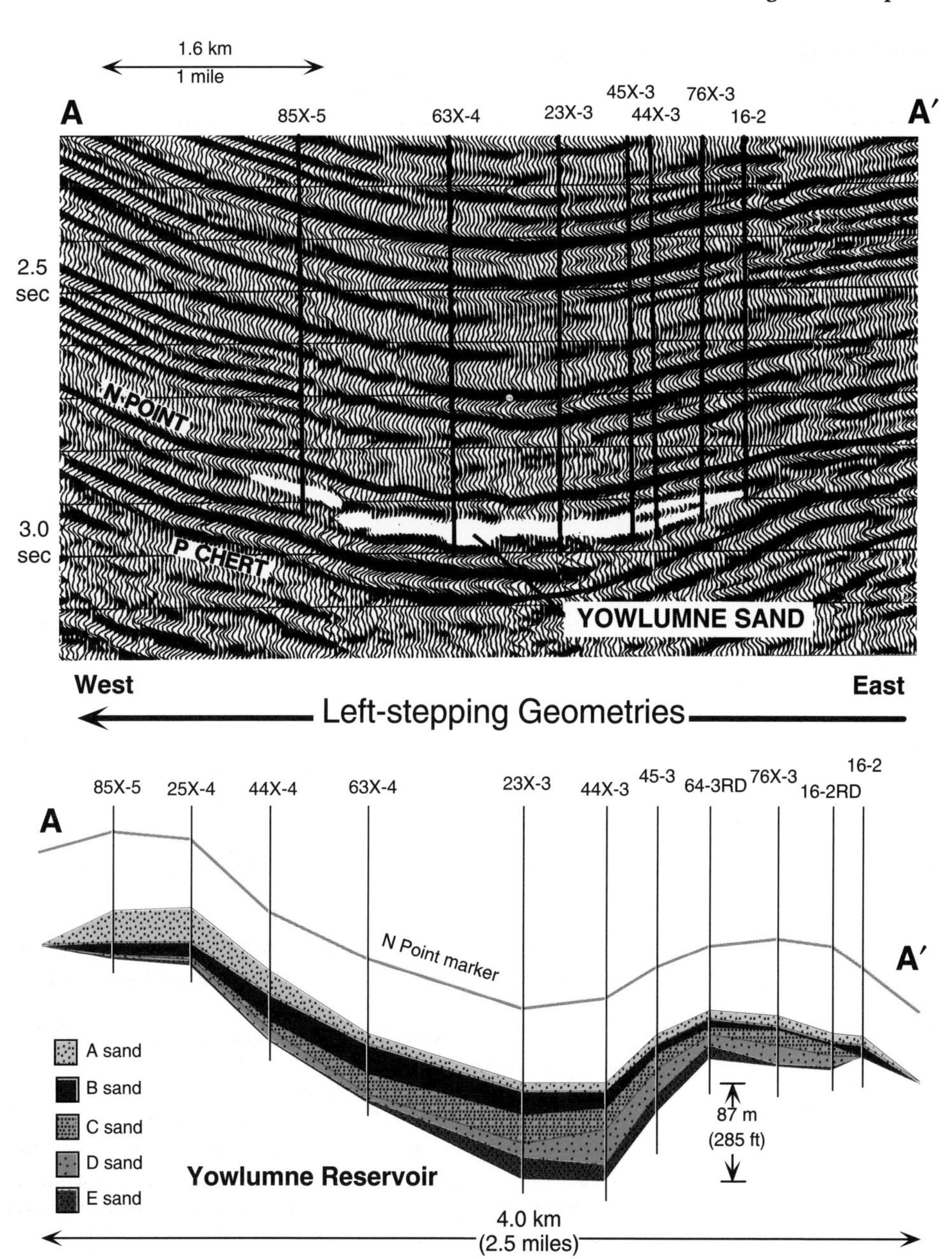

Figure 3—Seismic line and cross-section D-D′ showing the lens shape of the Yowlumne fan. The line and section cut from west to east across the body of the fan, perpendicular to the direction of sediment transport.

of the reservoir, which results in separate permeability pathways along which fluids flow at different rates (Metz and Whitworth, 1984; Berg and Royo, 1990). Most likely, these pathways represent different flow units that, for the most part, are in pressure communication over geologic time. Consequently, these compartments, which correlate to the six shale-bounded, lobe-shaped reservoir layers already discussed, develop the same oil-water contact over thousands of years, yet acquire slightly different pressures as the field is rapidly produced over tens of years.

Compartmentalization exists on an even larger scale. (1) The gross reservoir interval in Yowlumne Unit A has, over time, consistently exhibited reservoir pressures that differ from those measured in the same interval in Unit B. (2) Gross sandstone isopach maps indicate separate northern (Unit B) and southern (Unit A) depocenters (loci of thickening), which appear to represent different depositional accumulations. (3) Because sandstones in Unit B have more quartz, less clay, and higher original porosity than those in Unit A, different depositional histories may be indicated (Whelan, 1984). (4) The oil-water contact in Unit B is 660 m (2165 ft) structurally lower than the contact in Unit A. Although a few studies (e.g., Berg and Royo, 1990) interpret a single contact that is steeply tilted to 5° (480 ft/mi; 90.9 m/km) by hydrodynamic influences, a large density difference of 0.154 g/cc between the oils (32° API) and formation waters (TDS of 22,000 ppm) results in buoyant oils unlikely to support a contact tilted more than 1° (100 ft/mi; 18.9 m/km), even in the presence of a strong water drive. Thus, Units A and B appear to be separate compartments. If so, basinward progradation to the north indicates that

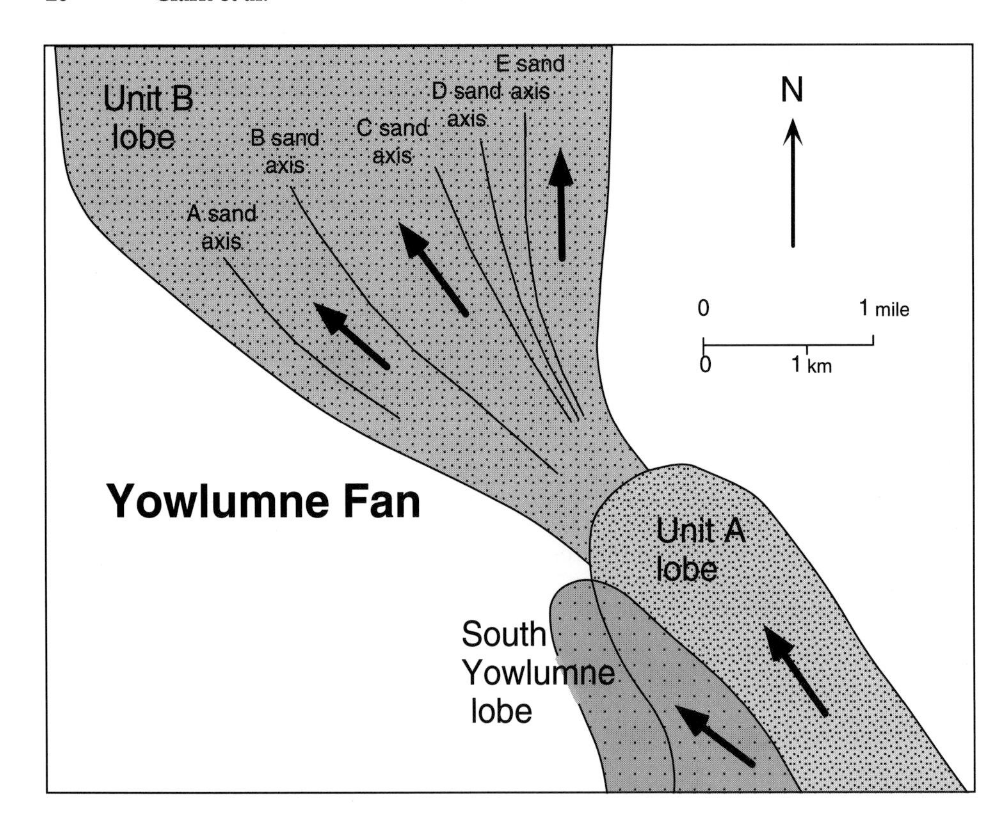

Figure 4—Map showing left-stepping and basinward-stepping geometries exhibited by depositional lobes in the Yowlumne fan.

compartments in Unit B are younger than those in Unit A (Figure 4).

A common oil-water contact, tilted less than 1º (100 ft/mi, 18.9 m/km), appears to exist between Unit A and South Yowlumne field, a lobate sandstone accumulation located west of the main Yowlumne field (Figure 1). If a common contact exists, then there is depositional continuity with Unit A and South Yowlumne field representing adjacent depositional lobes. If the system progrades laterally to the west, as comparison to Unit B indicates, then sandstones in the South Yowlumne lobe are younger than those in Unit A (Figure 4).

Reservoir Rock Properties

Log-derived petrophysical data, constrained by core analyses, indicate decreasing reservoir quality toward the east margin of the fan. For example, values of net/gross sandstone = 80%, $V_{sh} < 6\%$, $\phi_{eff} = 16\%$, and K_{liquid} = 10–20 md along the fan axis decrease to net/gross sandstone = 65%, $V_{sh} > 12\%$, $\phi_{eff} = 12\%$, and K_{liquid} = 2 md along the eastern fan margin (Figure 2). Most likely, this eastward degradation in reservoir quality results from an increase in the frequency and thickness of interbedded shales (net/gross sandstone), increasing clay content (V_{sh}), and decreasing grain size.

Close proximity of Yowlumne field to the San Andreas Fault is probably the main influence on stress orientations in the reservoir. Borehole-breakout data indicate maximum horizontal stresses oriented north-south (Castillo and Zoback, 1994), whereas micro-seismic data collected during a fracture stimulation indicate northwest–southeast fracture propagation in the main reservoir. Thus, induced fracturing is likely to be most effective in deviated well bores oriented northeast-southwest (i.e., perpendicular to the direction of fracture propagation) and least effective in wells oriented northwest-southeast.

DISCUSSION

Depositional Model

The Yowlumne sandstone lobes are part of a trend of Stevens sandstones that includes Landslide field, a northward-prograding complex located about 6.4 km (4 mi) south of Yowlumne field. Seismic data indicate continuity between these fields (Stolle et al., 1988; Quinn, 1990); therefore, depositional lobes should exist between Landslide and Yowlumne, and additional lobes may exist basinward (to the north) of Unit B.

Most likely, basinward-stepping geometries in the Yowlumne-Landslide system resulted, in part, from additional sediment input into the southern San Joaquin Basin during renewed uplift of the basin margins. Also, these geometries probably resulted, in part, from decreasing accommodation (i.e., tectonic subsidence plus eustatic sea-level change). Decreasing subsidence rates along the Bakersfield Arch, a regional high in the central part of the basin, resulted in decreased accommodation, which, in turn, had a profound effect on the aggradational and progradational geometries of Stevens sandstones (Clark et al., 1996a). Because the Miocene Yowlumne fan was located close to a tectonically active transform margin represented by the modern San Andreas Fault, subsidence-driven accommodation was probably just as important at Yowlumne as along the arch.

Left-stepping geometries probably resulted from Coriolis forces, which in northern hemisphere basins cause ocean currents to circulate clockwise around basin margins. Consequently, northern hemisphere fans preferentially deposit levees on the right sides of channels, forcing lobe deposition to the left (Menard, 1964). Abundant interbedded shales, characteristic of levee facies, on the east side of the Yowlumne fan (e.g., the 73X-3 well), and homogeneous sandstones, characteristic of lobe facies, on the west (e.g., the 16X-4 well) are consistent with this interpretation (Clark et al., 1996b). Alternatively, the clean sandstones may represent channel facies instead of lobes (Metz and Whitworth, 1984; Berg and Royo, 1990); however, a channel complex is inconsistent with progradational geometries and a lack of major incision beneath the fan. Also, left-stepping geometries in Unit A and South Yowlumne field indicate lateral shifting of sandstone lobes where the alternative model requires a channel axis; furthermore, decreasing reservoir quality toward the east (right) margin of the fan is more consistent with the lobe model.

Flow Simulation

Remaining field reserves are identified by further subdividing reservoir layers A, B, C, D, E, and W into ten flow units, and exporting digital contour maps (grids) of net/gross sandstone, ϕ_{eff}, K_{liquid}, and S_w for net sandstone in each flow unit to a computer model that simulates fluid flow in the reservoir (Table 1). Although a full-field model indicates an ultimate recovery of 45% of the original oil in place, 480,000 m^3 (3 million bbl) of bypassed oil remains along the thinning fan margins. These reserves represent oil banked against the margins by waterflood and reserves not exploited due to decreasing reservoir quality and thickness.

From detailed (partial-field) modeling of a prospective area indicated by the full-field model, a high-angle well, the 91X-3, was located to exploit 74,000 m^3 (465,460 bbl) of bypassed oil banked up against the east margin of the Yowlumne fan (Figure 1). This well was deviated up to 85º to the west, resulting in the tangential penetration of 335 m (1100 ft) of reservoir with a true stratigraphic thickness of only 55 m (180 ft). Three planned fracture stimulations, spaced 76 m (250 ft) apart, would improve connectivity across shale-bounded reservoir layers by inducing fractures northwest–southeast, oblique to the well path. Thus, one deviated well provides the same productive capacity as several vertical wells, resulting in more cost-effective exploitation of the field margin.

CONCLUSIONS

The interaction of climate with rapid uplift dramatically increased coarse sediment input into the southern San Joaquin Basin during the late Miocene and resulted in northward progradation of a layered, fan-shaped turbidite complex called the Yowlumne sandstone. Deposition of lobe-shaped reservoir compartments with basinward-stepping geometries is attributed to decreasing accommodation and high sediment flux. Deposition in Unit B of seven lobe-shaped compartments with left-stepping geometries is attributed to Coriolis forces.

More abundant shale-bearing levee facies on the east (right) margin of the fan and lobe facies on the west (left) resulted in decreasing reservoir quality toward the east. Cost-effective exploitation of bypassed oil trapped in tight sandstones along the east side is achieved using horizontal to high-angle deviated wells with multiple fracture stimulations. A thorough understanding of the reservoir architecture, and computer programs that model rock properties and fluid flow, helped to effectively locate wells and design completion programs that maximized productivity from the layered, low-permeability turbidite reservoir that characterizes the thinning fan margin in this part of Yowlumne field.

ACKNOWLEDGMENTS

This study was funded in part by a grant from the Class-III Reservoir Field Demonstration Program of the U.S. Department of Energy, Contract Number DE-FC22-95BC14940. We would like to thank Tom Berkman, Bill Fedewa, Mark Kamerling, Mike Laue, Bruce Niemeyer, Mike Simmons, and George Stewart for their contributions and many thoughtful suggestions. Comments by Don Lewis and an anonymous reviewer greatly improved the quality of this paper. Also, we extend our sincere thanks to ARCO Western Energy and the Department of Energy for permission to publish this paper.

REFERENCES CITED

Berg, R. R., and Royo, G. R., 1990, Channel-fill turbidite reservoir, Yowlumne field, California in (Barwin, J. H., MacPherson, J. G., and Studlick, J. R., eds.) Sandstone Petroleum Reservoirs: Casebooks in Earth Science, Springer-Verlag, New York, p. 467-487.

Castillo, D. A., and Zoback, M. D., 1994, Systematic variations in stress state in the southern San Joaquin Valley: inferences based on well-bore data and contemporary seismicity: Bulletin of the American Association of Petroleum Geologists, v. 78, p. 1257-1275.

Clark, M. S., Beckley, L. M., Crebs, T. J., and Singleton, M. T., 1996a, Tectono-eustatic controls on reservoir compartmentalization: An example from the upper Miocene, California: Marine and Petroleum Geology, v. 13, p. 475-491.

Clark, M. S., Melvin, J., and Kamerling, M., 1996b, Growth patterns of a Miocene turbidite complex in an active-margin basin, Yowlumne field, San

Joaquin basin, California (abstract): American Association of Petroleum Geologists, Annual Convention Official Program, San Diego, 1996, v. 5, p. A27.

Graham, S. A., and Williams, L. A., 1985, Tectonic, depositional, and diagenetic history of Monterey Formation (Miocene), central San Joaquin basin, California: American Association of Petroleum Geologists Bulletin, v. 69, p. 385-411.

Jessup, D. D., and Kamerling, M., 1991, Depositional style of the Yowlumne sands, Yowlumne oil field, southern San Joaquin Basin, California (abstract): Bulletin of the American Association of Petroleum Geologists, v. 75, p. 368.

Marino, A. W., and Schultz, S. M., 1992, Case study of Stevens sand horizontal well: Society of Petroleum Engineers, SPE 24910, p. 549-563.

Menard, H. W., 1964, Marine geology of the Pacific: McGraw-Hill, New York, 271 p.

Metz, R. T., and Whitworth, J. L., 1984, Yowlumne oil field in (Kendall, G. W., and Kiser, S. C., eds.) Selected Papers presented to San Joaquin Geological Society, v. 6, p. 3-23.

Quinn, M. J., 1990, Upper Miocene sands in the Maricopa depocenter, southern San Joaquin Valley, California in (Kuespert, J. G., and Reid, S. A., eds.) Structure, stratigraphy and hydrocarbon occurrences of the San Joaquin basin, California: Pacific Section of AAPG, Guidebook 65, p. 97-113.

Stolle, J., Nadolny, K. A., Collins, B. P., Greenfield, D. S., and March, K. A., 1988, Landslide oil field, Kern County, California, a success story—an exploration/development case history in (Randall, J. W., and Countryman, R. C., eds.) Selected Papers presented to the San Joaquin Geological Society, v. 7, p. 1-13.

Tieh, T. T., Berg, R. R., Popp, R. K., Brasher, J. E., and Pike, J. D., 1986, Deposition and diagenesis of upper Miocene arkoses, Yowlumne and Rio Viejo fields, Kern County, California: Bulletin of the American Association of Petroleum Geologists, v. 70, p. 953-969.

Weimer, P., and Link, M. H., 1991, Global occurrences in submarine fans and turbidite systems in (Weimer, P., and Link, M. H., eds.) Seismic facies and sedimentary processes of submarine fans and turbidite systems: Springer-Verlag, New York, p. 9-67.

Whelan, H. T. M., 1984, Geostatistical estimation of the spatial distributions of porosity and percent clay in a Miocene Stevens turbidite reservoir: Yowlumne field, California: unpublished Master's thesis, Stanford University, California, 126 p.

Ye, L., D. Kerr, K. Yang, Facies architecture of the Bluejacket Sandstone in the Eufaula Lake area, Oklahoma: implications for reservoir characterization of the subsurface Bartlesville Sandstone, 1999, *in* R. Schatzinger and J. Jordan, eds., Reservoir Characterization-Recent Advances, AAPG Memoir 71, p. 29-44.

Chapter 3

Facies Architecture of the Bluejacket Sandstone in the Eufaula Lake Area, Oklahoma: Implications for Reservoir Characterization of the Subsurface Bartlesville Sandstone

Liangmiao "Scott" Ye
ARCO EPT
Plano, Texas, U.S.A.

Dennis Kerr
Kexian Yang
Smedvig Technologies, Inc.
Houston, Texas, U.S.A.

ABSTRACT

Outcrop studies of the Bluejacket Sandstone (Middle Pennsylvanian) provide useful insights to reservoir architecture of the subsurface equivalent Bartlesville sandstone in Glenn Pool field. Quarry walls and road cuts in the Lake Eufaula area offer excellent exposures for detailed facies architectural investigations using high-precision surveying and photomosaics. Subsurface studies include conventional logs, borehole image log, and core data.

Reservoir-scale facies architecture is reconstructed in four hierarchical levels: multistory discrete genetic intervals, individual discrete genetic interval, facies within a discrete genetic interval, and subfacies of the meandering channel-fill facies. From the Eufaula Lake to Glenn Pool field areas, the Bluejacket (Bartlesville) Sandstone, taken as a whole, represents an incised valley fill above a type-1 sequence boundary. It comprises two distinctive architectures: a lower braided channel-fill–dominated interval, regarded as representing the lowstand systems tract, and an upper meandering channel-fill–dominated interval, regarded as the transgressive systems tract.

Braided channel-fill facies are typically 30–80 ft (9–24 m) thick and are laterally persistent, filling an incised valley wider than the largest producing fields. The lower contact is irregular, with local relief of 50 ft (15 m). The braided-fluvial deposits consist of 100–400 ft (30–122 m) wide, 5–15 ft (1–4 m) thick channel-fill elements. Each channel-fill interval is limited laterally by an erosional contact or overbank deposits, and is separated vertically by discontinuous mudstones or highly concentrated mudstone interclast lag conglomerates.

Low-angle parallel-stratified or trough cross-stratified medium- to coarse-grained sandstones volumetrically dominate. This section has a blocky well log profile.

Meandering fluvial deposits are typically 100–150 ft (30–45 m) thick and comprise multiple discrete genetic intervals. Meandering channel-fill facies successions include basal trough cross-stratified medium-grained sandstones, medial low-angle–stratified fine-grained sandstones with numerous mudstone drapes, and an upper mudstone. Well log profile is typically a serrated bell shape. Splay facies is up to 20 ft (6 m) thick, and consists of ripple-stratified and lesser trough cross-stratified, medium-grained sandstones separated by laterally persistent thin mudstones. Floodplain mudstones laterally and vertically segment, with the exception of very limited areas, channel-fill and splay sandstones into reservoir compartments.

Porosity and permeability values are mostly influenced by the discrete genetic interval (DGI) level. Lower DGIs, dominated by braided channel-fill facies, have the highest porosity and permeability, and show very weak permeability anisotropy. The upper DGIs, dominated by meandering channel-fill and splay facies, have progressively lower porosity and permeability vertically upward through the DGIs. Meandering channel-fill facies, particularly the middle channel-fill subfacies, have strongly developed permeability anisotropy.

INTRODUCTION

The Middle Pennsylvanian Bluejacket (Bartlesville) Sandstone is known to extend from southeastern Kansas across the northeastern Oklahoma platform and into the Arkoma foreland basin (Howe, 1951; Weirich, 1953; Visher, 1968). The Bluejacket Sandstone is a member of the Boggy Formation of the Krebs (Cherokee) Group of the Desmoinesian Series (see Hemish, 1986, 1989a, for the latest stratigraphic discussion). Although the surface extent of the Bluejacket Sandstone has been established for many decades (e.g., Dane and Henricks, 1936; Miser, 1954), the formalization of a type section (area) has been only recently established by Hemish (1989a) for exposures 10 mi (16 km) north of Vinta, Oklahoma (Figure 1).

The Bartlesville sandstone is the subsurface equivalent to the Bluejacket Sandstone, and is arguably the name most familiar to the oil and gas industry (Weirich, 1968; Visher et al., 1971; Ebanks, 1979). The Bartlesville sandstone has also been given local names within producing fields. For example, in Glenn Pool field, the subsurface location reported in this study, the Bartlesville sandstone is known as the Glenn sand (Kuykendall and Matson, 1992). In this report, the formal name "Bluejacket" is reserved for discussion of outcrop occurrences, and the informal name "Bartlesville" is used for subsurface occurrences.

The Bartlesville sandstone has been a major oil producer in Oklahoma for more than 90 yr, producing more than 2 billion bbl of oil since the discovery of a number of fields back to the early 1900s; however, estimates of recovery of original oil in place are as low as 20% for many Bartlesville-producing fields/reservoirs, despite post-primary recovery attempts (e.g., Welch, 1989; Kerr et al., 1999). The reason for poor recoveries in the Bartlesville sandstone is clearly attributed to complex reservoir architecture, which is very difficult to properly characterize with conventional subsurface information alone (Ye, 1997). Information obtained from analogous outcrop is critical to reconstructing reliable reservoir models and developing effective reservoir management strategies.

This study offers detailed description of Bluejacket Sandstone exposures in the Eufaula Lake area and characterization of selected reservoir units in Glenn Pool field, Oklahoma. The intent is to illustrate the usefulness of outcrop information to improve subsurface reservoir modeling.

Study Areas and Data Availability

Outcrop Study Areas

The Bluejacket Sandstone forms a prominent cliff throughout much of its exposure. Its resistant habit is one characteristic used in mapping this lithostratigraphic unit (e.g., Oakes, 1967, 1977, Hemish, 1989b); however, much of the natural exposures is covered with lichens or weathered surfaces that obscure sedimentary structures and other details. Thus, man-made exposures are preferred for studying the details necessary for facies architecture reconstruction. Although many sites were examined in the study of the Bluejacket Sandstone (Martinez, 1993; Ye, 1997), two sites

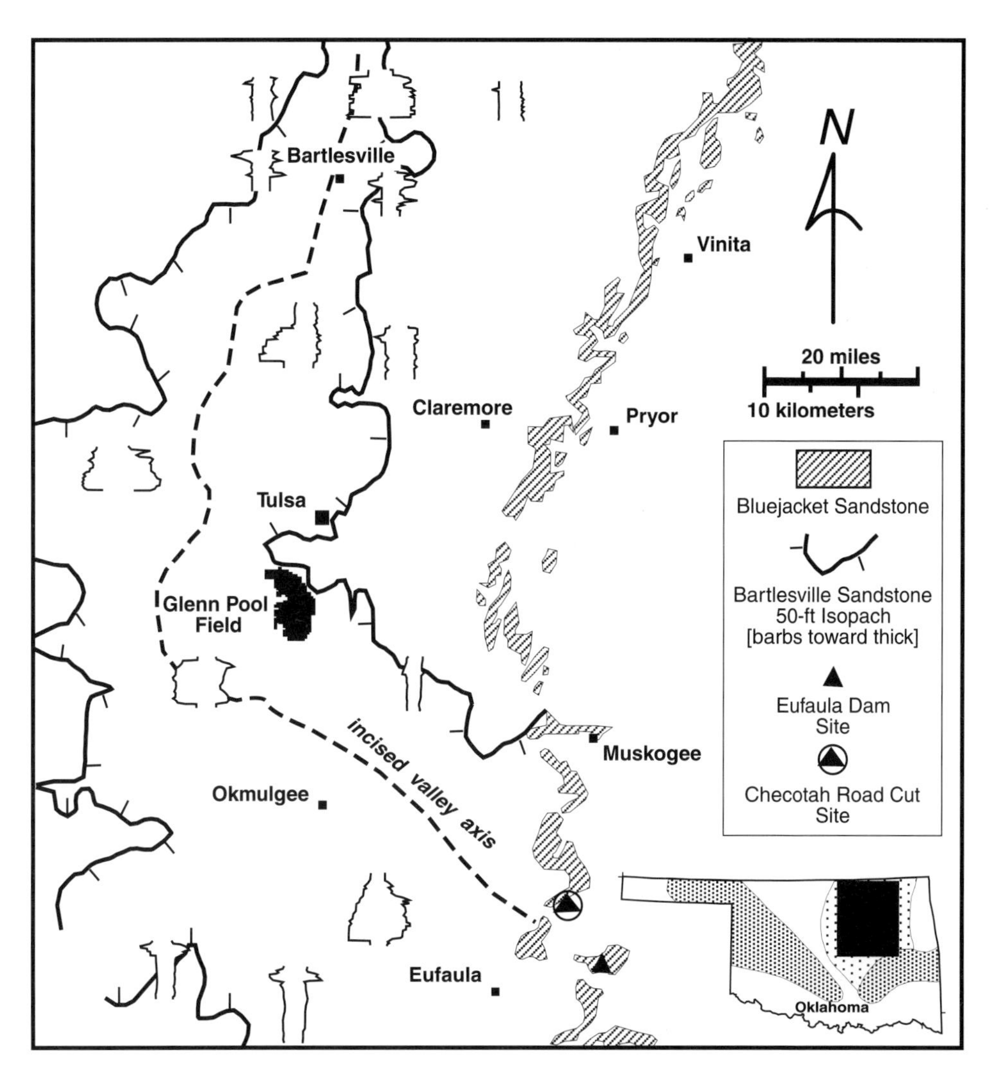

Figure 1—Index map of study area. Locations include Bluejacket Sandstone outcrop study sites and Glenn Pool field (Figure 2). Also shown is the incised valley margin (50-ft isopach) and axis. For other locations studied see Martinez (1993), Mishra (1996), and Ye (1997). For features in Oklahoma inset map see Figure 2.

in the vicinity of Eufaula Lake are reported here: Checotah road cut and Eufaula Lake dam.

The Checotah road cut is located along Oklahoma Highway 266 (parallels Interstate 40), 6.5 mi (10.5 km) east of Checotah, Oklahoma (Figure 1). The road cut extends 680 ft (207 m) in an east-west direction and is about 50 ft (15 m) high. Being normal to the general paleocurrent direction (north to south), the road cut exposes a transverse cross section. The cut has a slope of 55°; however, the numerous ledges across the cut provide ready access for detailed investigation. Detailed facies interpretation was performed based on observations on the lithology, physical structures, and current orientations. Spatial configuration of facies element boundaries were measured using an infrared-beam–based precision survey system and photomosaics (Ye, 1997).

The second outcrop site is located at the north abutment of the Eufaula Lake dam (Figure 1). The lower part of the Bluejacket Sandstone forms a vertical cliff in two segments: a west leg, extending northwest-southeast, 850 ft (259 m) long and 70 ft (21 m) high; and an east leg, extending approximately southwest-northeast, 400 ft (122 m) long and 65 ft (20 m) high. Both legs are oblique to the general north-south paleoflow direction. For this cliff exposure, detailed observations can be done only for the lower part of the section because of limited accessibility; however, architectural element boundaries down to the level of single-channel cut-and-fill successions can be identified and surveyed precisely from a distance.

Subsurface Study Area

Parts of the Glenn Pool field were selected from an opportunity under the U.S. DOE Class I initiative, rather than for any other reason (Kerr et al., 1999). Glenn Pool field is a giant, mature (discovered in 1905) field located in Creek and Tulsa counties, Oklahoma (Figure 2). The areas of investigation were chosen based on the operator's (Uplands Resources) interest and the overall project objectives (Kerr et al., 1999). Kuykendall and Matson (1992) provide an overview of the field history and a conventional view of the geology.

The Glenn Pool field study area encompasses essentially all of Uplands Resources' operating area and selected adjacent areas in the southern part of Glenn Pool field. The area covers about 1700 ac (688 ha) (T17N, R12E). The production tracts include all of Tracts 6, 7, 9, 10, 11, 12, 13, 16, Wm. Berryhill Unit, and

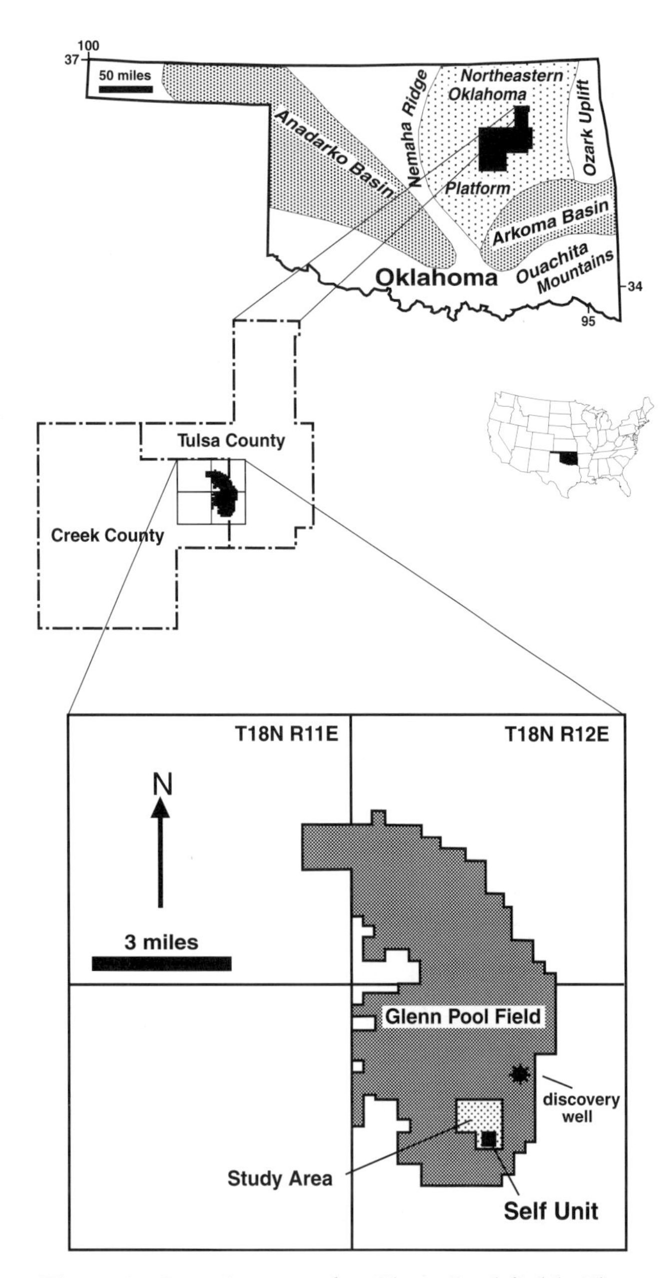

Figure 2—Location map for Glenn Pool field. Glenn Pool field is shown relative to Pennsylvanian paleogeographic elements of Oklahoma. The study area includes much of Uplands Resources' operating acreage. See Kuykendall and Matson (1992) for an overview of the field history.

Self Unit, and parts of 3, 4, 33, and Corbray Unit. The best estimate from historical records indicates that 382 wells were drilled to the Bartlesville in the study area; however, only 161 logs were available for use. The top of the Bartlesville is 1400–1900 ft (427–580 m) below the surface, and gross thickness reaches 180 ft (55 m), with net pay typically being 65–100 ft (20–30 m) thick. Detailed well log correlation and subsurface mapping resulted in the recognition of seven units compared to the conventional three-part division of the Bartlesville in Glenn Pool field.

The Self Unit, a 160 ac (65 ha) tract, received close attention. A project-cosponsored well, the Uplands Resources Self No. 82 (API #35-037-28444; Sec. 21, T21N, R12E) is one of the few wells in the field known to have a conventional well log suite and core through the Bartlesville sandstone. Self No. 82 also includes a microresistivity borehole image log and detailed collection of porosity and permeability measurements.

Outcrop and Subsurface Study Areas as Analogs

Comparative study of outcrop Bluejacket Sandstone and subsurface equivalent Bartlesville sandstone suggests that they are indeed analogous, despite the 60-mi (97-km) distance between the locations (Ye, 1997). The Checotah road cut is similar to the upper two-thirds of the Bartlesville in Glenn Pool field. The Eufaula Lake dam site is similar to the lower one-third of the Bartlesville in Glenn Pool field.

Regional Depositional Framework

Extensive previous geological studies on the Bartlesville sandstone have been conducted since late in the nineteenth century. Most recent systematic study of the depositional origin and framework was completed in the late 1960s (Saitta, 1968; Saitta and Visher, 1968; Phares, 1969; Visher et al., 1971). These studies concluded that the Bluejacket (Bartlesville) Sandstone is part of a large fluvial-deltaic complex deposited essentially in a single regressive interval; however, more recent studies have challenged this conventional assessment (cf. Martinez, 1993; Mishra, 1996; Ye, 1997; Kerr et al., 1999).

Regional-scale to reservoir-scale investigations in both outcrop and subsurface in a 2000 mi^2 (5180 km^2) area in northeastern Oklahoma reveal that the Bluejacket (Bartlesville) Sandstone is an incised valley-fill deposit. Incised valley-fill section consists of two sequence stratigraphic architectural elements: (1) the lower lowstand systems tract, 40–150 ft (12–45 m), composed of braided fluvial deposits; and (2) the upper transgressive systems tract, 70–100 ft (21–30 m), composed of tidal-influenced meandering fluvial, estuarine tidal, and shallow marine deposits (Ye, 1997). This new interpretation is supported by the following general observations.

(1) Outcrop survey illustrates that a type-1 sequence boundary exists at the base of the Bluejacket (Bartlesville) Sandstone, indicated by subaerial erosion of underlying Savanna Shale and a basinward facies shift from marine, dark-gray shale below to braided fluvial deposits above. Similar relationships were found in core descriptions and were interpreted from well log facies.

(2) Well log correlation shows Bartlesville sandstone thickens at the expense of the Savanna Shale. The incised valley extends from north to south over 120 mi (193 km) within Oklahoma. Width is 4–5 mi (6–8 km) near the Oklahoma–Kansas border in Washington County to over 40 mi (64 km) in the south near Eufaula Lake (Figure 1). Thickness varies from 120 to

250 ft (37–76 m) along the erosional axis of the valley. The 50-ft (15-m) gross thickness isopach essentially delimits the valley margin. Outside the valley the thickness is less than 30 ft (9 m).

(3) Well log character is typically a very sharp basal contact overlain by a blocky profile. The upper part of the Bartlesville is noted for its bell-shaped log profiles and also its thick funnel-shaped profiles south of Okmulgee (Mishra, 1996). Core and outcrop facies interpretation indicates sedimentary facies transition upward from braided fluvial to meandering, to tidal-influenced meandering, to tidal estuarine and shallow marine deposits (Ye, 1997). The same succession of facies may also be observed horizontally from north to south.

This pattern of facies association clearly indicates that the Bluejacket (Bartlesville) Sandstone was deposited in a transgressive manner. The regional Inola Limestone marker, capping the Bartlesville sandstone interval, represents maximum flooding (Figure 3) (Ye, 1997).

RESERVOIR ARCHITECTURE RECONSTRUCTION

An important concept in modern reservoir characterization is that many fluvial reservoirs are composed of a hierarchy of architectural elements, each of which may influence pore fluid flow (e.g., Miall, 1985, 1996; Miall and Tyler, 1991; Jordan and Pryor, 1992). Fluvial reservoirs such as the Bartlesville sandstone usually consist of multiple genetic units (stories), one story is composed of several different facies, and a facies may be subdivided into different subfacies and/or smaller scale elements. Proper documentation of heterogeneity at each level of reservoir architectures is critical to better management planning for mature reservoirs such as the Bartlesville (Kerr et al., 1999).

Facies Architecture Hierarchy

The architecture of the Bluejacket (Bartlesville) Sandstone has up to four spatial hierarchy levels, in

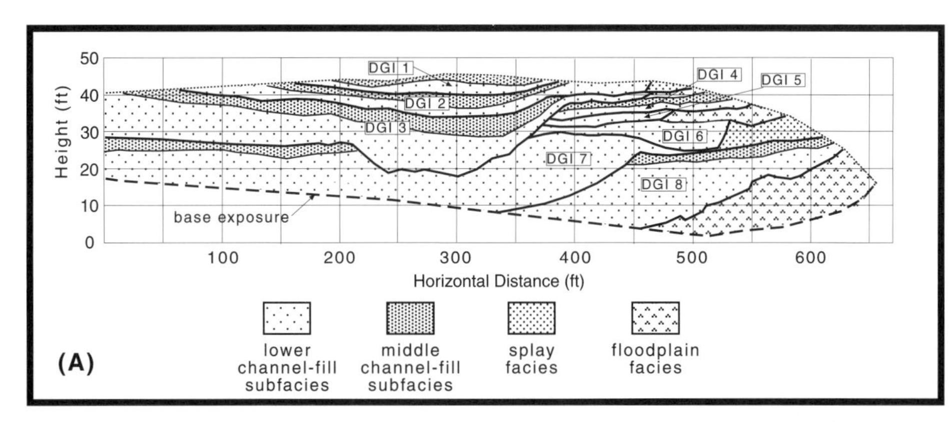

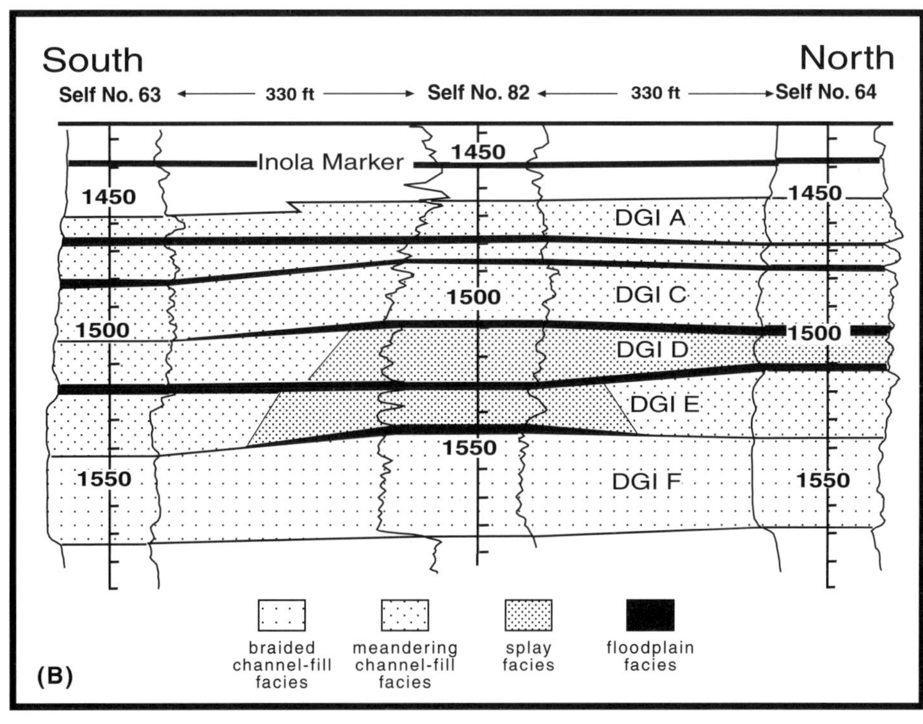

Figure 3—Comparison of discrete genetic intervals (DGI) from the Checotah road cut site (A) and in the vicinity of the Uplands Self No. 82 well (B) in Glenn Pool field. Note the overall similarities in scale. By contrast, the Checotah road cut shows a high degree of downcutting into older channel-fill deposits compared to Glenn Pool field.

descending volumetric order: multistory discrete genetic interval, discrete genetic interval, facies, and subfacies. Although other approaches to defining fluvial architecture levels and elements are available (e.g., Miall, 1996), the four-level approach is more appropriate for the Bluejacket (Bartlesville). This is particularly the case when faced with subsurface correlating and mapping of the Bartlesville given conventional well log suites.

A discrete genetic interval (DGI) is defined as a collection of genetically related, contiguous facies deposited during a discrete increment of time (Kerr and Jirik, 1990). Kerr and Jirik (1990) developed the concept of a DGI for the purposes of subsurface mapping of Tertiary fluvial deposits in the south Texas onshore Gulf Coast region. Thus, a DGI is an operational unit for subsurface mapping. A single DGI is correlated as having a common elevation from the top of gross channel-fill thickness to a stratigraphic datum (marker bed). After correlation through a cross section network has been completed, the facies and their thicknesses are plotted at each well site for each DGI. If the facies genetic relationships are logical, then the task is complete. If, however, the facies genetic relationships do not make sense, then the correlation must be reconsidered, and another iteration is required. Once one is satisfied with the DGI correlation, facies-based isopach mapping is performed.

Level 1: Multistory Discrete Genetic Intervals

The first level, involving the largest rock volume, of hierarchical reservoir architecture deals with the stacking pattern of multiple discrete genetic intervals (see level 2 for description of a DGI). Figure 3A shows the vertical relationship of different DGIs surveyed from the Checotah road cut. Eight DGIs are present here. One DGI basically represents one channel cut-and-fill succession and its time-equivalent deposits. For most of the DGIs, the upper part (i.e., upper channel subfacies and floodplain mudstone) has been eroded by overlying DGIs. Middle channel subfacies are mostly preserved except for DGIs 5 and 6. Horizontally, DGI 8 is truncated by DGI 7, and DGIs 4, 5, and 6 are truncated by DGI 3. Top of DGI 7 is partially eroded by DGI 3. All these indicate that different discrete genetic intervals are not well separated vertically; however, the mudstone rip-up clast pebble conglomerates common at the bottom of channel-fills may function as flow barrier/baffles between different DGIs.

By comparison, vertical separation of different DGIs is much more evident in the subsurface. Figure 3B is a cross section through the middle of the Self Unit to illustrate that for most cases, different DGIs of Bartlesville sandstone are separated by 1–5 ft (0.3–1.5 m) floodplain mudstones; however, one DGI cutting down through the intervening floodplain mudstone also occurs. These erosional windows typically take up about 20 ac (8 ha); however, they can extend for some distance along a channel course, depending on local accommodation conditions (Ye, 1997).

Other important observation in this level of hierarchical architectures is that the thickness of each DGI decreases upward for both outcrop and subsurface, as shown in Figure 4. For Bartlesville sandstone reservoirs in the Self Unit, maximum thickness of one DGI decreases upward from 33 ft (10 m) of DGI E to 22 ft (7 m) of DGI A. For the Checotah road cut, thickness of a individual DGI decreases upward from 23 ft (7 m) (DGI 8) to 7 ft (2 m) (DGI 1). Thinning upward in both outcrop and subsurface implies one aspect of their similarities, and also supports the interpretation that Bartlesville sandstone was deposited in a transgressive manner (Ye, 1997).

Level 2: Individual Discrete Genetic Interval

From a sedimentological point of view, one DGI represents a brief episode of sedimentation that deposited a genetically related 3-D volume of sedimentary rocks. From an oil development point of view, DGI represents the fundamental subsurface mapping unit. The DGI concept is critical for detailed and accurate reservoir characterization.

In the case of the Bartlesville sandstone, a DGI is correlated by common elevation relative to the Inola marker (Figure 3B); however, it was found, through iterations of correlation and mapping, that a 5-ft (1.5-m) tolerance was necessary to arrive at logical genetic relationships over the subsurface study area. Such a tolerance is likely the result of local variations in accommodation (i.e., differential compaction or slumping along the valley margin).

A Bartlesville sandstone DGI in Glenn Pool field is commonly made of three facies: channel-fill sandstone, splay sandstone, and floodplain mudstone. These three

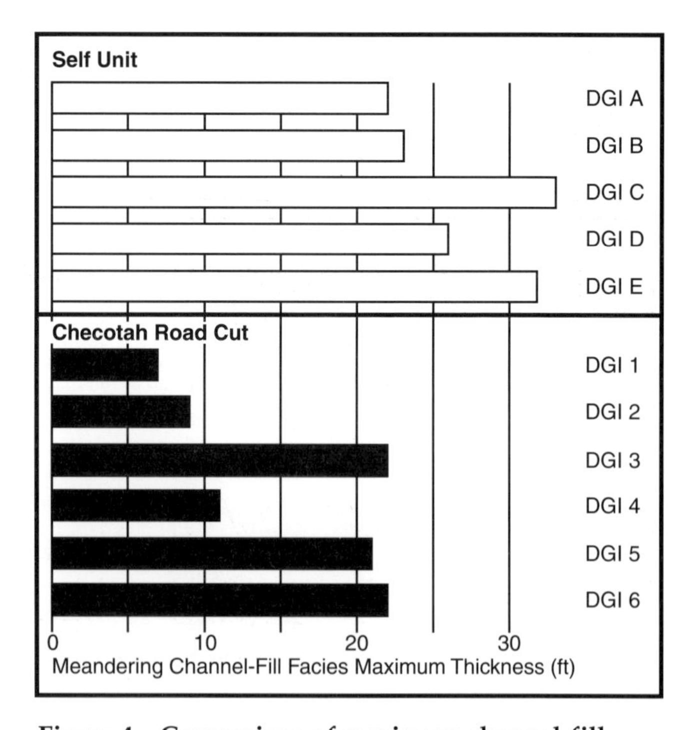

Figure 4—Comparison of maximum channel-fill deposit thickness from the Self Unit (Glenn Pool field) and Checotah road cut exposure. Both show an upward decrease in maximum thickness. Compare with Figure 3.

facies are identified in all meandering DGIs recognized in the subsurface (Figures 5, 6). At the outcrop study site (Figure 3A), splay deposits only appear in DGIs 2 and 6; floodplain mudstone is recognized in DGI 5 only. The lower presence of floodplain mudstone and splay sandstone in outcrop is probably due to the limited expanse and 2-D nature of the road cut, and they are commonly covered with soil.

The relationship among these three facies elements is better shown in the facies map constructed for the Bartlesville sandstone reservoirs. As shown in these maps (Figures 5, 6), channel-fill facies constitutes the main framework of the system. Channels extend generally from north to south, which is consistent with regional paleocurrent direction. Splay, which is deposited as a result of bank breaking during flood discharge, develops along the margin of channels and spreads outward in a fan-like shape. Floodplain mudstone occurs between channels and splays, or on top of channels; therefore, floodplain mudstones laterally and vertically segment channel-fill and splay sandstones into reservoir compartments.

Each upper Bartlesville DGI in southern Glenn Pool field is made up of multiple channel courses (DGI A–E). It is tempting to interpret this character of level 2 elements as either an anastomosing fluvial system or distributary channels of a delta system. As discussed under the section on level 3 elements, an anastomosing system is not reasonable because of the prominence of lateral accretion bar deposits and noted absence of vertical accretion and truncated channel-fill facies successions (cf. Miall, 1996) in the upper Bartlesville sandstone. The more likely candidates for distributary channel-fill deposits in the upper Bartlesville sandstone are those mapped by Mishra (1996) in the vicinity of Okmulgee, south of Glenn Pool field (Figure 1). These channel-fills are narrow (500–1000 ft; 153–305 m) and have width:thickness ratios ranging from 50 to 150, and clearly bifurcate and rarely rejoin in a downvalley (southward) direction.

Level 3: Facies Within a Discrete Genetic Interval

The third level of the facies architecture hierarchy deals with individual channel-fill, splay, and floodplain

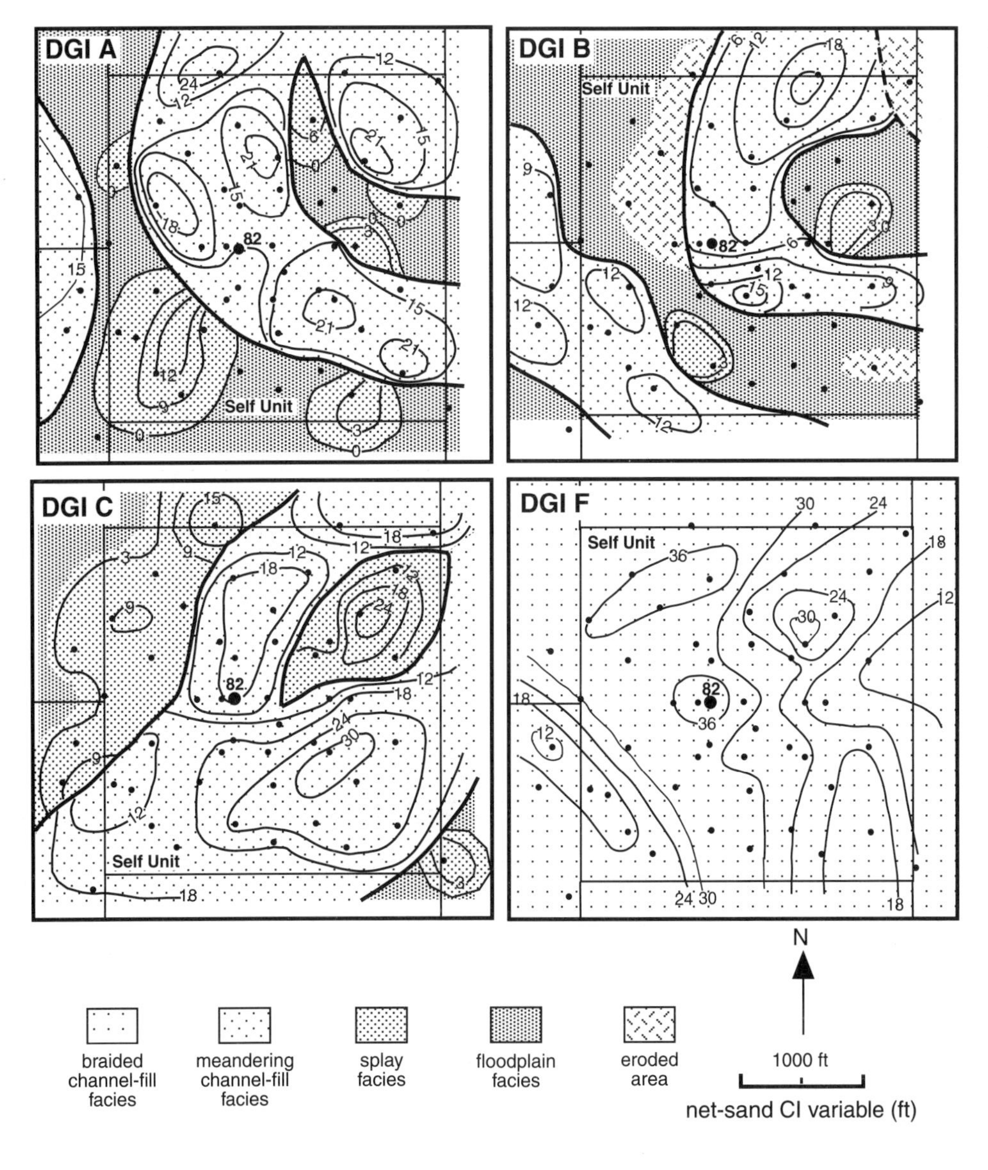

Figure 5—Facies and net-sand isopach maps for selected discrete genetic intervals (DGI) for the Self Unit in southern Glenn Pool field (Figure 2).

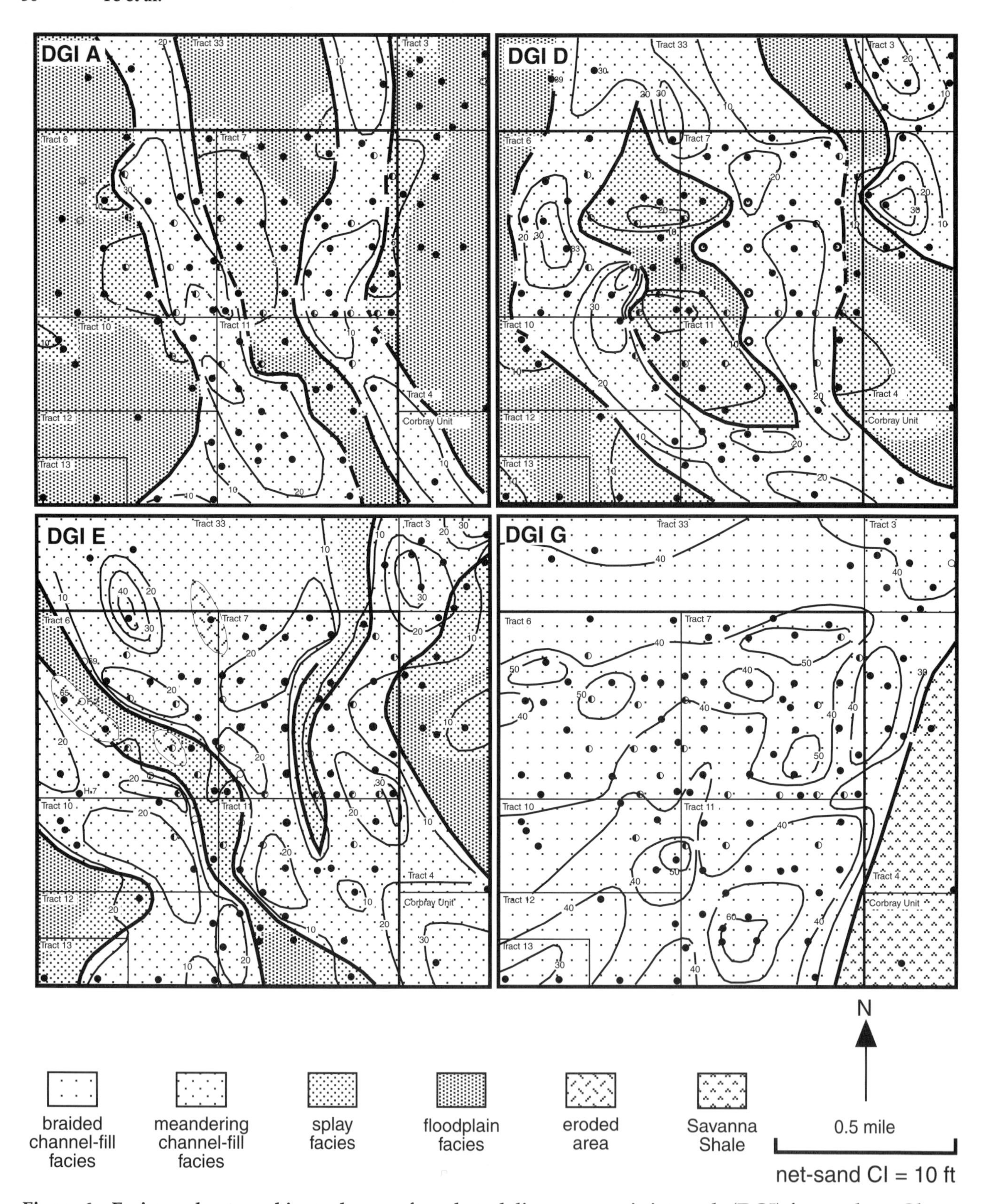

Figure 6—Facies and net-sand isopach maps for selected discrete genetic intervals (DGI) for southern Glenn Pool field, exclusive of the Self Unit (Figure 2).

facies. Core and outcrop examples are provided in Figure 7.

Meandering Channel-Fill Facies—The Checotah road cut and the upper part of Bartlesville sandstone are mainly composed of meandering channel-fill deposits. General characteristics include fining-upward texture profile, upward decrease in scale of physical sedimentary structures, and upward increase in proportion of mudstone interbeds. Well log character is the familiar bell-shaped profile. The Bartlesville profile is also highly serrated in response to the abundant mudstone interbeds, which are thought to represent the influence of tidal fluctuations (Ye, 1997).

Estimates of meandering channel-belt width to maximum net-sand thickness ratios come from subsurface mapping (Figures 5, 6). Although highly variable within a given discrete genetic interval (DGI), width to maximum net-sand thickness ratios decrease upward from 100 to 300 in DGI D and E to 40 to 55 in DGI A, B, and C. This geometry falls within the broad range of meandering channel-belt geometries suggested from the compilation of Fielding and Crane (1987). The vertical decrease in meandering channel-belt width to maximum net-sand thickness ratio is compatible with increasing accommodation as the Bartlesville incised valley was filling during baselevel rise (also see Figure 4).

From core and outcrop descriptions, the meandering channel-fill facies was subdivided into into lower, middle, and upper channel-fill subfacies (see level 4).

Splay Facies—Splay facies is characterized by coarsening-upward texture profile from fine-grained to medium-grained sandstone. Upward increase in stratal thickness is observed in lower levels only; otherwise, irregular vertical stacking of thick to thin beds (3–25 in; 8–64 cm) is observed. Ripple lamination and

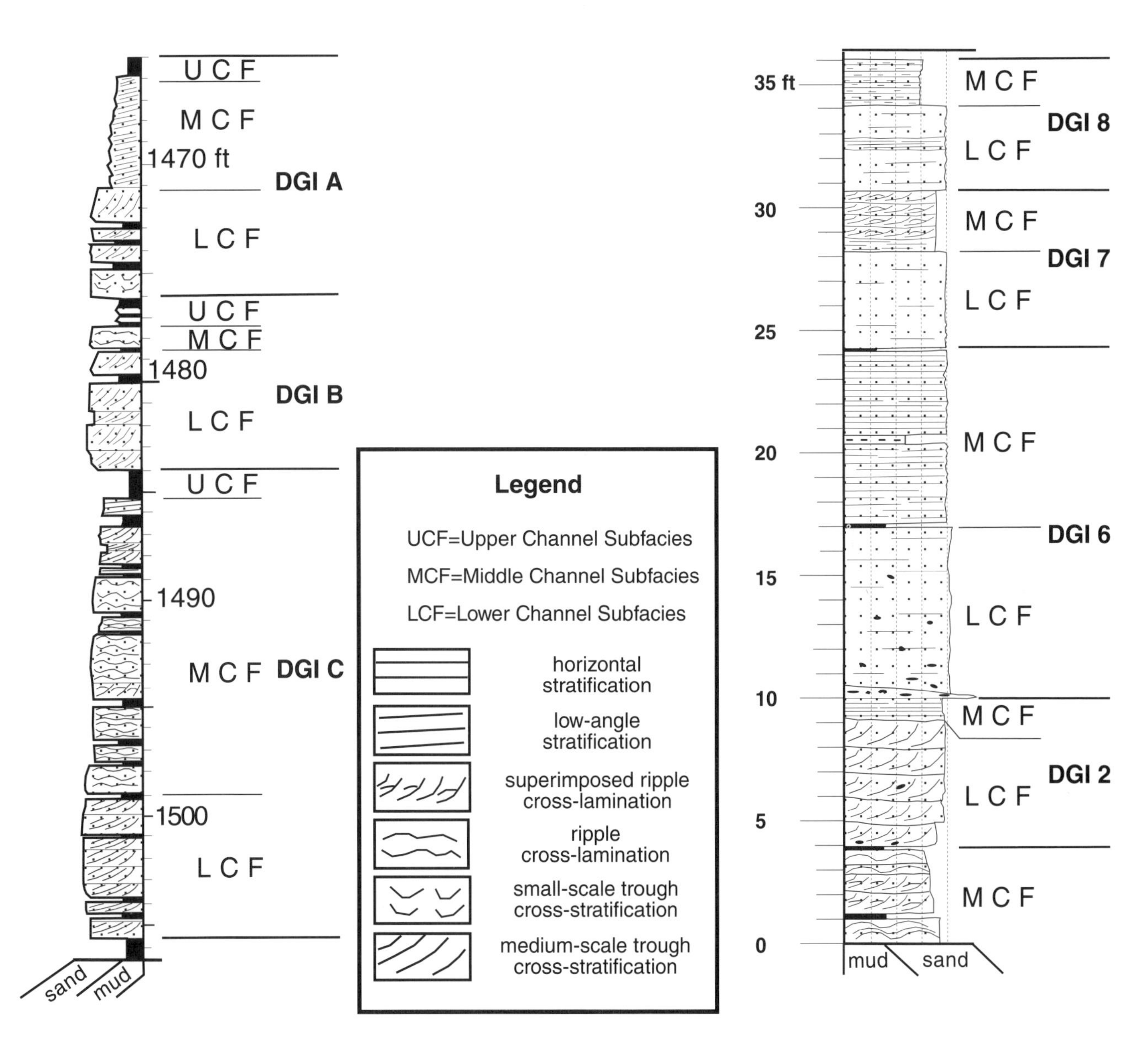

Figure 7—Comparison of core graphic from Uplands Self No. 82 in Glenn Pool field (left) with measured section from Checotah road cut (right).

low-angle parallel bedding dominate. Medium-scale cross-stratification and contorted bedding are less common. Thin-bedded (1–3 in; 3–8 cm) mudstones interstratified with sandstones are observed throughout. This results in a funnel-shaped or stacked symmetrical well log profile (Figure 3B).

Floodplain Mudstone Facies—Deposits interpeted as floodplain facies consist of strata ranging from siltstone with ripple cross-lamination to dark-gray mudstone and carbonaceous shale. South and northwest of Glenn Pool field, the floodplain facies also includes coal.

The floodplain facies lithologies usually function as impermeable barriers to fluid flow between the DGIs, as discussed (level 2).

Braided Channel-Fill Facies—DGI F of Self No. 82 core and DGI G elsewhere in southern Glenn Pool field are interpreted as braided fluvial channel-fill deposits (Ye, 1997). Braided channel-fill facies is characterized by moderately to well sorted, generally structureless (but less commonly parallel-bedded), upper medium-grained to lower coarse-grained sandstones. In Glenn Pool field, thickness of braided fluvial deposits varies from 30 to 80 ft (9 to 24 m) depending on local relief of the basal contact (i.e., base of the incised valley). This section has a blocky well log profile (Figure 3B) and is laterally persistent, filling the relief along the lower elevations of the paleovalley. Figures 5 and 6 show thickness pattern of DGI F and G braided sandstones. Internal architecture could not be resolved through well log analysis because of its vertical and horizontal continuity and lack of continuous mudstones.

Interpretation of this massive DGI F sandstone as braided fluvial deposits was verified by outcrop study. Similar thick-bedded, structureless rocks have been observed in the lower part of Bluejacket Sandstone outcrops in the Eufaula Dam area (Figure 8). These "massive" thick sandstones undergo lateral transition to well cross-stratified and parallel-bedded (with locally developed parting lineation) thick sandstones over a short distance. They are highly channelized, showing unidirectional paleoflow, and vary greatly in thickness according to the position relative to the channel thalwegs. Braided fluvial deposits are actually made of many individual channel cut-and-fill successions, which are stacked together and cross-cut each other horizontally and vertically. The result is a widespread amalgam of individual successions (i.e., a sheet geometry). Each channel-fill is characterized by relatively uniform lithology and sedimentary structures. Most channel-fills are limited laterally by erosional contacts, and rarely by overbank fines. There is little shale occurring between and within channels, even though discontinuous mudstones or concentrated mudstone interclast lag conglomerates at the bottom of channels may act as vertical flow baffles. Carbonized log cast accumulations are also common along the basal fill of larger channels.

A total of 25 individual braided channel cut-and-fill successions are recognized in the Eufaula Dam cliff. Figure 8 indicates that the braided channel-fill is 100–400 ft (30–122 m) wide and 5–15 ft (1–4 m) thick. The widespread, continuous blocky interval of the lower Bartlesville sandstone is actually composed of numerous braided channel cut-and-fill successions that are smaller than the meandering channels that constitute the more isolated upper Bartlesville sandstones; however, the mapped subsurface extent of the braided channel-fill facies in southern Glenn Pool field

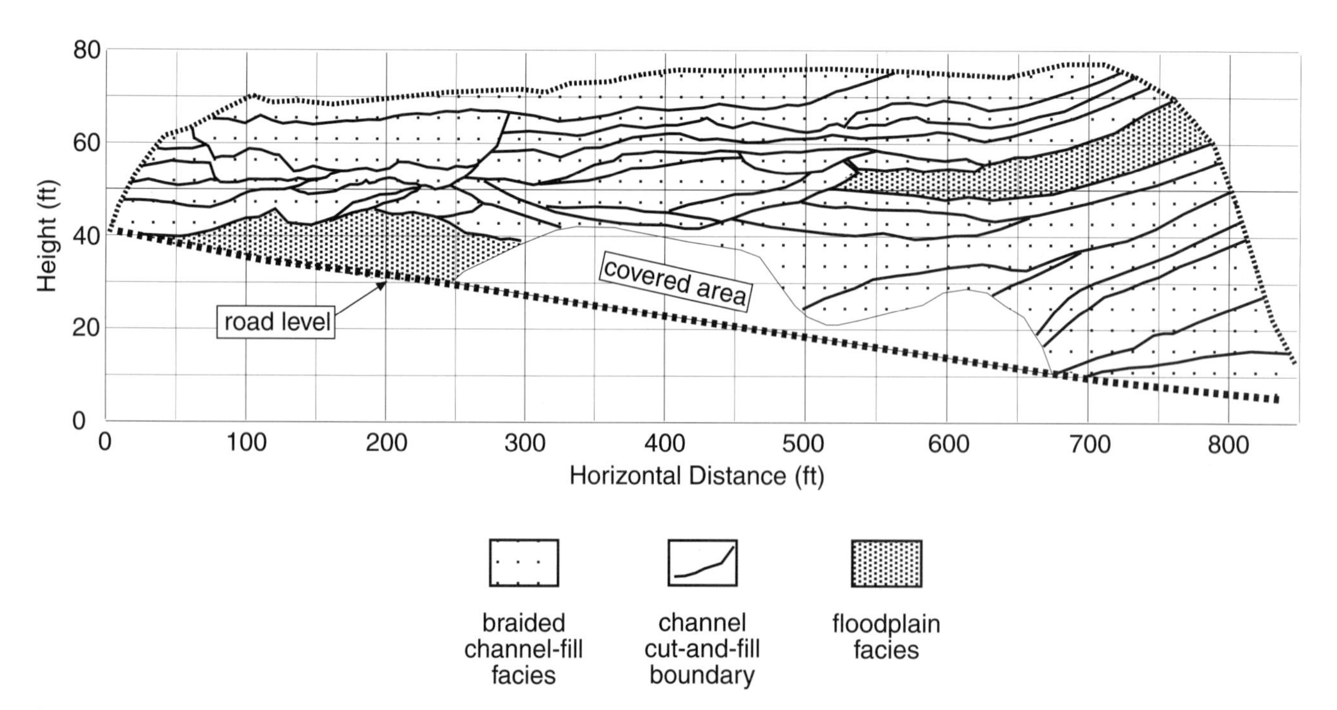

Figure 8—Surveyed cliff exposure west-leg Eufaula dam site (Figure 1). Note numerous channel cut-and-fills in the braided fluvial facies and limited lateral continuity of the floodplain facies.

indicates a much higher width to maximum net-sand thickness ratio (see DGI F and G in Figures 5 and 6).

Level 4: Subfacies of Channel-Fill Facies

As stated, meandering channel-fill facies were subdivided into lower, middle, and upper channel-fill subfacies. Lower channel-fill subfacies is composed of well to moderately sorted, medium-grained sandstone with medium-scale cross-stratification. Mudstone drapes are common on cross strata; pebbly mudstone rip-up clasts commonly occur along the bottom of the subfacies. Middle channel-fill subfacies is made up of moderately sorted, lower medium-grained to poorly sorted, silty fine-grained sandstones with horizontal to low-angle parallel stratification and ripple lamination. Medium- to very thin-bedded mudstones drape lateral accretion surfaces. Upper channel-fill subfacies is composed of mudstone to silty claystone.

The thickness proportion and well log character of the meandering channel-fill subfacies depend on the location within a channel filling and discrete genetic interval. The lower channel-fill subfacies forms a fairly uniform blanket across the bottom 20% of the total channel-fill thickness. Its gamma-ray log response is the lowest of the three subfacies and is usually less than 60% of the gamma-ray mudline. The middle and upper channel-fill thickness proportion varies based on location within the channel-fill. The middle channel-fill subfacies is thickest, about 75% of the total channel-fill, in the central part of a lateral accretion bar deposit. The upper channel-fill, where preserved, is thickest along the abandoned thalweg, making up about 80% of the total fill (the original thickness is difficult to estimate because of the differential compaction between mud and sand). The middle channel-fill subfacies gamma-ray response is between about 60% and 30% of the gamma-ray mudline. The upper channel-fill facies gamma-ray response is greater than 30% of the gamma-ray mudline. If mapping of the individual subfacies is of interest, then division of the gamma-ray log into thirds through the channel-fill facies is a ready guide. The absolute values of the gamma-ray log for each subfacies increase upward through the DGIs, reflecting the vertical increase in mud content typical of the Bartlesville sandstone.

In a meandering fluvial system, lateral accretion bar deposits of middle channel-fill subfacies are the most prominent component of channel-fill deposits (Miall, 1996). The lateral accretion bars develop by intermittent lateral deposition on the inside of thalweg bends. Mud typically drapes the depositional surface, resulting in the potential for highly heterogeneous reservoirs. Thus, the distribution of lateral accretion mudstone drapes is a key factor for the reservoir heterogeneity, potentially leading to retarding or blocking of pore fluid flow.

Core and log facies interpretation indicates that DGI C is made up of lateral accretion bar deposits at the location of the Self No. 82 well (Ye and Kerr, 1995). The dip azimuths of the lateral accretion surfaces as indicated in a microresistivity image acquired for this well show a progressive upward rotation from 200° to 146°. Such rotation suggests that the channel thalweg locally increased in sinuosity with growth of the bar. Further evaluation of the microresistivity log allows for extrapolation of lateral accretion mudstone drapes in DGI C (Ye and Kerr, 1995; Ye, 1997). Average vertical spacing of lateral accretion surfaces and average structure-corrected dip angle can be estimated from the image. Based on a simple geometric argument, the number of lateral accretion mudstone drapes in a given horizontal distance can be predicted. The expected distribution of lateral accretion mud drapes can be reconstructed assuming lateral continuity and reasonable curvature development of the bar (Ye and Kerr, 1995; Ye, 1997).

Porosity and Permeability

Figure 9 is the profile of porosity and permeability measured for the Self No. 82 core. Both core-plug measurements and minipermeameter measurements are shown for comparison. Figure 9 shows that the porosity and permeability are strongly DGI related. Descending from DGI A to F, porosity increases from about 8% to about 23%; core-plug permeability increases from less than 0.1 md to more than 300 md. These data suggest that channel-fill facies in one DGI are of poorer reservoir quality than splay sandstone in other DGIs.

From the nine cored wells in the Self Unit, the relationship between facies/subfacies and petrophysical properties within a given DGI suggests that splay sandstones may have higher porosity and permeability than channel-fill sandstones; however, this difference may not be significant because samples are biased toward high porosity and permeability intervals.

Minipermeameter measurements were taken at a spacing of 0.2–0.5 ft (0.1–0.2 m) (Figure 9), in comparison to 1–3 ft (0.3–0.9 m) for plug measurements. All the minipermeability values reflect only horizontal permeability due to fixed orientation of the core slab surface. Two points can be made from comparison of minipermeability and core-plug permeability. First, the core-plug measurement and minipermeameter measurements are close to each other. Statistical analysis illustrates that when permeability is higher than 1 md, the minipermeability values are systematically 25% lower than core-plug permeabilties measured in the core at the same depth. This difference indicates that the minipermeameter is a reliable tool for measuring permeability for rocks with permeability greater than 1 md. The lower values can be attributed to occlusion by the residual oil in the core. Second, the general trends of permeability variations are exactly matched by both measurements; however, different levels of detailed variations are represented because of the different sampling scales of these two measurements, and therefore minipermeabilities potentially can represent reservoir varitoins better than core-plug pemeabilities. Also,

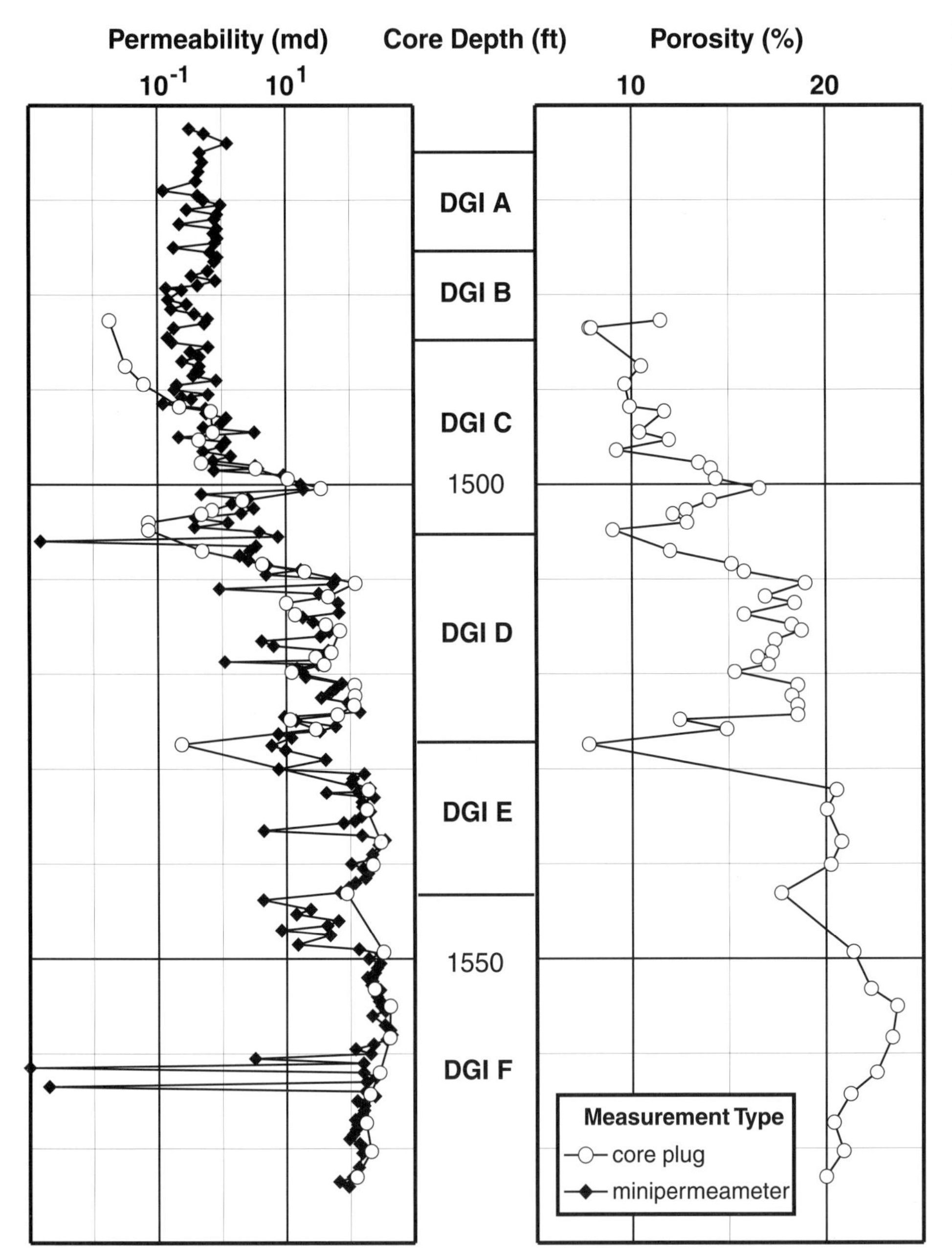

Figure 9—Core porosity and permeability measurements from Uplands Self No. 82 well, southern Glenn Pool field. Note vertical decrease in porosity and permeability.

core-plug permeability measurements are generally biased toward good reservoir qualities.

Analysis of the relationship between core-plug–measured vertical and horizontal permeabilities (Kv and Kh) has shown that Kv/Kh ratio increases downward from about 0.3 to 0.9 (Figure 10), indicating that the lower Bartlesville sandstone is more homogeneous than the upper Bartlesville sandstone. Kv/Kh ratios for both meandering channel-fill sandstones and splay sandstones are similar (0.3–0.5 for both). However, braided fluvial deposits (DGI F) have the highest Kv/Kh ratios (about 0.8–0.9), indicating that braided fluvial deposits are less heterogeneous. These differences in Kv/Kh ratio also imply that fluid flow in a vertical direction in lower Bartlesville sandstone reservoirs is more important than that for upper Bartlesville sandstone reservoirs.

CONCLUSIONS

The Bluejacket (Bartlesville) Sandstone is described in terms of facies architecture hierarchy of elements. Four hierarchical levels of reservoir architectures are recognized in both outcrop and subsurface studies. These four levels are as follows, in descending rock volume: multistory discrete genetic intervals; discrete genetic interval; meandering channel-fill, braided channel-fill, splay, and floodplain facies; and meandering channel-fill subfacies.

Outcrop studies have provided useful insights to reservoir-scale facies architecture; however, spatial resolution differences between outcrop and subsurface data likely influence subsurface correlation and mapping results. For example, the presence of more

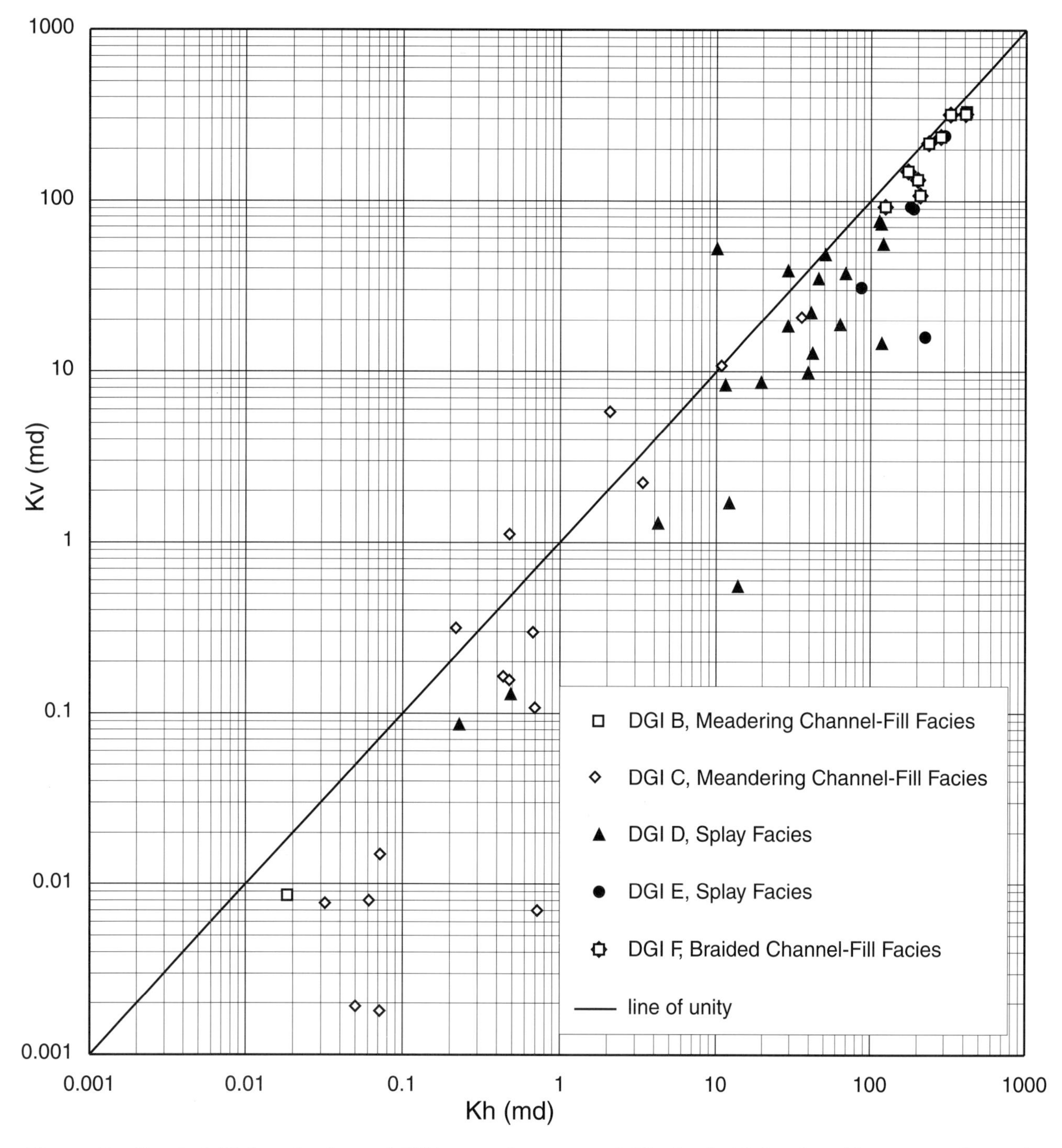

Figure 10—Vertical (Kv) vs. horizontal (Kh) core-plug permeability in the Uplands Self No. 82 well, southern Glenn Pool field. Note the strong permeability anisotropy for meandering channel-fill facies compared to that for braided channel-fill facies.

overbank deposits in outcrop examples of the braided-fluvial–dominated DGIs is not well represented in subsurface architecture representations.

Minipermeameter measurements are quite reliable in comparison with conventional core-plug measurements: reliability, convenience, and low cost allow a detailed characterization of permeability variation. The most important influence on permeability in the Bartlesville sandstone is the DGI element: braided channel-fill facies in DGI F and G (lowstand systems tract) have higher permeability and porosity compared to the splay and meandering channel-fill facies of DGI A–E (transgressive systems tract). Floodplain mudstone facies developed between meandering channel-fill and splay facies of DGI A–E serve to vertically compartmentalize (except for locallized erosional windows) the Bartlesville sandstone in Glenn Pool field.

ACKNOWLEDGMENTS

The results presented here were partially supported by an academic-industry-government partnership funded under U.S. Department of Energy Contract No. DE-FC22-93BC14951, and by the Research Office of The University of Tulsa. We are grateful to Uplands Resources and the U.S. Bureau of Land Management office in Tulsa for providing data, and to Amoco Production Company Technology Center for allowing access to their facilities. We also thank L. Watney and an anonymous reviewer for their reviews of an earlier version of this manuscript.

REFERENCES CITED

Dane, C.H., and Henricks, T.A., 1936, Correlation of Bluejacket Sandstone, Oklahoma: AAPG Bulletin, v. 20, p. 312-314.

Ebanks, W.J., Jr., 1979, Correlation of Cherokee (Desmoinesian) sandstones of the Missouri-Kansas-Oklahoma tristate area, in Hyne, N.J., ed., Pennsylvanian sandstones of the mid-continent: Tulsa Geological Society Special Publication 1, p. 295-312.

Fielding, C.R., and Crane, R.C., 1987, An application of statistical modelling to the prediction of hydrocarbon recovery factors in fluvial reservoir sequences, in Ethridge, F.G., Flores, R.M., and Harvey, M.D., eds., Recent developments in fluvial sedimentology: Society of Economic Paleontologists and Mineralogists Special Publication 39, p. 321-327.

Hemish, L.A., 1986, Coal geology of Craig County and eastern Nowata County, Oklhoma: Oklahoma Geolgical Survey Bulletin 140, 131 p.

Hemish, L.A., 1989a, Bluejacket (Bartlesville) Sandstone Member of the Boggy Formation (Pennsylvanian) in its type area: Oklahoma Geology Notes, v. 49, p. 72-89.

Hemish, L.A., 1989b, Coal geology of Rogers County and western Mayes County, Oklahoma: Oklahoma Geological Survey Bulletin 144, 118 p.

Howe, W.B., 1951, Bluejacket Sandstone of Kansas and Oklahoma: AAPG Bulletin, v. 35, p. 2087-2093.

Jordan, D.W., and Pryor, W.A., 1992, Hierarchical levels of heterogeneity in a Mississippi River meander belt and application to reservoir systems: AAPG Bulletin, v. 76, p. 1601-1624.

Kerr, D.R., and Jirik, L.A., 1990, Fluvial architecture and reservoir compartmentalization in the Oligocene middle Frio Formation, south Texas: Gulf Coast Association of Geological Societies Transactions, v. 40, p. 373-380.

Kerr, D.R., Ye, S.L., Bahar, A., Kelkar, B.M., and Montgomery, S.L., 1999, Glenn Pool field, Oklahoma: A case of improved production from a mature reservoir: AAPG Bulletin, v. 83, p. 1–18.

Kuykendall, M.D., and Matson, T.E., 1992, Glenn Pool field, northeast Oklahoma platform: AAPG Treatise of Petroleum Geology—Atlas of Oil and Gas Fields (Stratigraphic Traps III), p. 155-188.

Martinez, G., 1993, Reservoir heterogeneity in portion of the Bartlesville Sandstone (Desmoinsian), Mayes County, northeastern Oklahoma: M.S. thesis, The University of Tulsa, 147 p.

Miall, A.D., 1985, Architectural-element analysis: a new method of facies analysis applied to fluvial deposits: Earth Science Review, v. 22, p. 261-308.

Miall, A.D., 1996, The geology of fluvial deposits: New York, Springer-Verlag, 582 p.

Miall, A.D., and Tyler, N., eds., 1991, The three-dimensional facies architecture of terrigenous clastic sediments, and its implications for hydrocarbon discovery and recovery: Society of Economic Paleontology and Mineralogy Concepts in Sedimentology and Paleontology Series, v. 3, 309 p.

Miser, H.D., 1954, Geologic map of Oklahoma, scale 1:500,000: U.S. Geological Survey.

Mishra, J.K., 1996, Sequence stratigraphy of the Bartlesville Sandstone in parts of Okmulgee, Creek, and Okfushee counties: M.S. thesis, The University of Tulsa, 68 p.

Oakes, M.C., 1967, Geology and petroleum of McIntosh County, Oklahoma: Oklahoma Geological Survey Bulletin 111, 88 p.

Oakes, M.C., 1977, Geology and mineral resources (exclusive of petroleum) of Muskogee County, Oklahoma: Oklahoma Geological Survey Bulletin 122, 78 p.

Phares, R.S., 1969, Depositional framework of the Bartlesville Sandstone in northeastern Oklahoma: M.S. thesis, The University of Tulsa, 56 p.

Saitta, B.S., 1968, Bluejacket Formation — A subsurface study in northeastern Oklahoma: M.S. thesis, The University of Tulsa, 142 p.

Saitta., B.S., and Visher, G.S., 1968, Subsurface of the southern portion of the Bluejacket delta: a guidebook to the Bluejacket-Bartlesville Sandstone: Oklahoma City Geological Society, 72 p.

Visher, G.S., 1968, Depositional framework of the Bluejacket-Bartlesville Sandstone, Oklahoma: Oklahoma City Geological Society, p. 32-42.

Visher, G.S., Saitta, S.B., and Phares, R.S., 1971, Pennsylvanian delta patterns and petroleum occurrence in eastern Oklahoma: AAPG Bulletin, v. 55, p. 1206-1230.

Weirich, T.E., 1953, Shelf principle of oil origin, migration, and accumulation: AAPG Bulletin, v. 37, p. 2027-2045.

Weirich,T.E., 1968, History of the Bartlesville oil sand, in Visher, G.S., ed., A guidebook to the Bluejacket-Bartlesville Sandstone, Oklahoma: Oklahoma City Geological Society, 72 p.

Welch, R.A., 1989, Berryhill Glenn Sand Unit: ARCO Reservoir Engineering Report.

Ye, L., 1997, Reservoir characterization and sequence stratigraphy of the Pennsylvanian Bartlesville Sandstone, Oklahoma: Ph.D. dissertation, The University of Tulsa, 256 p.

Ye, L., and Kerr, D.R., 1995, Use of microresistivity image logs in the detailed reservoir architecture reconstruction of Glenn Sandstone, Glenn Pool field, northeastern Oklahoma: Transactions of the 1995 AAPG Mid-Continent Section Meeting, Tulsa, OK, Tulsa Geological Society, p. 203-213.

Schatzinger, R. A., L. Tomutsa, Multiscale heterogeneity characterization of tidal channel, tidal delta, and foreshore facies, Almond Formation outcrops, Rock Springs Uplift, Wyoming, 1999, *in* R. Schatzinger and J. Jordan, eds., Reservoir Characterization-Recent Advances, AAPG Memoir 71, p. 45-56.

Chapter 4

Multiscale Heterogeneity Characterization of Tidal Channel, Tidal Delta, and Foreshore Facies, Almond Formation Outcrops, Rock Springs Uplift, Wyoming

Richard A. Schatzinger [1]
BDM Petroleum Technologies
Bartlesville, Oklahoma, U.S.A.

Liviu Tomutsa [2]
BDM Petroleum Technologies
Bartlesville, Oklahoma, U.S.A.

ABSTRACT

In order to accurately predict fluid flow within a reservoir, variability in the rock properties at all scales pertinent to the specific depositional environment needs to be taken into account. The present work describes rock variability at scales from hundreds of meters (facies level) to millimeters (laminae) based on outcrop studies of the Upper Cretaceous Almond Formation. Tidal channel, tidal delta, and foreshore facies were sampled on the eastern flank of the Rock Springs uplift, southeast of Rock Springs, Wyoming. The Almond Formation was deposited as part of a mesotidal Upper Cretaceous transgressive systems tract within the greater Green River Basin.

Bedding style, lithology, lateral extent of beds of bedsets, bed thickness, amount and distribution of depositional clay matrix, bioturbation, and grain sorting provide controls on sandstone properties that may vary more than an order of magnitude within and between depositional facies in outcrops of the Almond Formation. Permeability along these surfaces is often decreased by cementation, smaller pores, tighter grain packing, and compaction of sand-size rock fragments. These features can be mapped on the scale of an outcrop. Application of outcrop heterogeneity models to the subsurface is generally hindered by differences in diagenesis between the outcrop and the reservoir, poorly defined interwell subsurface continuity and facies architecture, and different absolute values of petrophysical properties (which often includes scaling problems) between the outcrop and the reservoir. In this paper we emphasize linkage between lateral cyclicity of petrophysical properties and the scale of

[1] Present address: Fowler, Schatzinger & Associates, Bartlesville, Oklahoma, U.S.A.; e-mail: rschatzi@ionet.net

[2] Present address: Lawrence Berkeley National Laboratory, Berkeley, California, U.S.A.; e-mail: ltomutsa@lbl.gov

primary bedding features. Such relationships can be transferred from outcrops directly into the subsurface because scaling problems are avoided.

The measurements for this study were performed both on drilled outcrop plugs and on blocks. One-inch-diameter plugs were taken at lateral spacing from 15 cm (6 in.) to 16.5 m (50 ft) and vertical spacing from 8 cm (3 in.) to 1.5 m (5 ft) to capture hierarchically stacked patterns of variations on the scale of meters to hundreds of meters. Probe permeameter permeability and x-ray computed tomography (CT) porosity from outcrop blocks captured variations at the scale of a few mm to a few hundred mm. Conventional gas porosity and permeability measurements were performed on the plugs and were integral to mapping the distribution of petrophysical properties at the scale of the facies (tens to hundreds of meters). Microscopic-scale heterogeneities such as grain size, pore distribution, authigenic cement content, and paragenetic stages were recorded using thin-section point-count methods and semi-automated petrographic image analysis.

In this study we found that permeability decreased 50–60% across bedding surfaces, by about 50% across bedset boundaries, and by 1–2 orders of magnitude across sandstone facies contacts. Permeability distribution tends to map parallel the "grain" of bedding within bedsets. Mapping also indicates that bedset boundaries are essentially always inclined to upper and lower facies boundaries. Fluid flow through facies must cross bedset boundaries. Lateral cyclicity of permeability is primarily related to bedding surfaces and the periodicity of individual sandwaves within major bedsets. The frequency of bedset boundaries encountered can then be a significant controlling factor to fluid flow and recovery efficiency.

CT and minipermeameter analysis map petrophysical properties at a scale approximately two orders of magnitude finer than that mapped using plugs. In our study, large-scale plug data and the detailed minipermeameter maps of sandstone blocks indicate similar ranges of permeability for similar facies. Therefore, when the architecture of depositional facies within this system is correctly described, data from small-sized samples are acceptable for modeling the reservoir at a larger scale.

INTRODUCTION

The Upper Cretaceous Almond Formation was deposited along a dominantly transgressive shoreline along the western margin of the North American Cretaceous Western Interior Seaway. It marks the uppermost phase of Mesaverde Group sedimentation (Roehler, 1990) (Figure 1). In the area of the Rock Springs Uplift and eastward in the Washakie Basin, the Almond Formation has typically been informally divided into lower (dominantly estuarine) and upper (dominantly marine) units (e.g., Van Horn, 1979). Almond sedimentation was markedly cyclical, particularly within the lower, estuarine section. Even within the upper marine section, multiple periods of eastward regression of the shoreline can be mapped into the Washakie Basin (Krystinik and Mead, 1996). Within the greater Green River Basin area, especially the Washakie Basin, these sandstones comprise important gas and oil reservoirs (see table 1 of Martinsen and Christensen, 1992). Rapid fluctuations of relative shoreline positions are typical in the upper Almond. Each phase of shoreline buildout was terminated by a rapid flooding event as the western margin of the Lewis Shale transgressed farther west. Fluvial to alluvial plain deposits of the Ericson Sandstone lie immediately beneath the Almond Formation in the study area (Figure 2), where the Ericson's upper portion records the earliest phases of the overall shoreline retreat that occurred during the Late Cretaceous (Campanian).

The study area for this paper is located on the eastern flank of the Rock Springs Uplift in southwestern Wyoming (Figure 2). Excellent outcrops of Almond

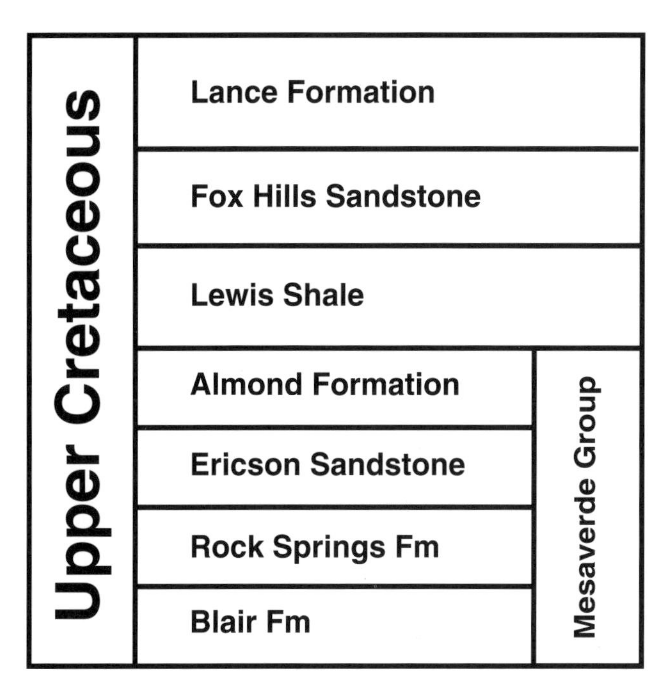

Figure 1. Stratigraphic column showing Upper Cretaceous units in the study area.

Formation that exhibit the lateral relationship between depositional facies are located immediately north of Highway 430 in the SE 1/4 of the SW 1/4 of Sec. 33, T16N, R102W, approximately 32 mi (51 km) southeast of Rock Springs, Wyoming. This area was called Outcrop G in a detailed study of Almond coals and depositional environments by Roehler (1988). The area during Almond deposition was an embayed shoreline created by progradation of deltas north and south of the Rock Springs Embayment. Tidal forces were amplified within the Rock Springs Embayment so that the Almond in this area was deposited under mesotidal conditions (Van Horn, 1979; Roehler, 1988; Schatzinger et al., 1992).

ROCK PROPERTY VARIATIONS AT VARIOUS SCALES

Outcrop Stratigraphic Cross Section

The coarsest scale of heterogeneity within the Almond Formation was evaluated by mapping the facies relationships at Outcrop G. At this location, eight detailed stratigraphic sections were measured over the lateral distance of more than 305 m (1000 ft). Outcrop gamma-ray profiles were collected at and between stratigraphic sections. The profiles were correlated to produce a cross section whose lateral extent approximates the distance between wells with 40 ac (16 ha) spacing. Photomosaics were made to help trace out facies architecture, and critical contacts were "walked out" on the outcrop. The critical southern half of this work is reproduced here as a stratigraphic cross section (Figure 3). The lower portion of this cross section illustrates the northward pinchout of a major tidal inlet containing reservoir-quality sandstones that are nearly 6.1 m (20 ft) thick. The inlet channel overlies and cuts into adjacent foreshore sands (beach and welded swash bar). Just above the Pintail coal bed is a 9 m (30 ft) thick section of nonreservoir marsh/lagoonal shales and thin, silty tidal creek

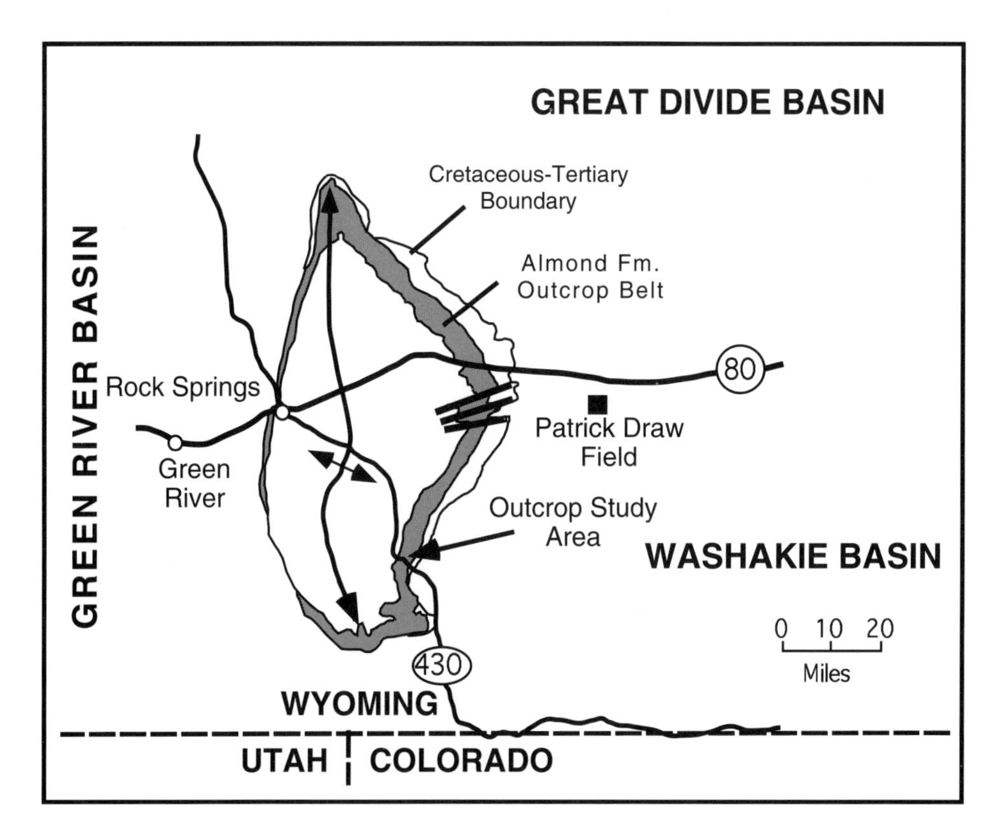

Figure 2. Location map of the study area on the eastern flank of the Rock Springs Uplift, southwestern Wyoming.

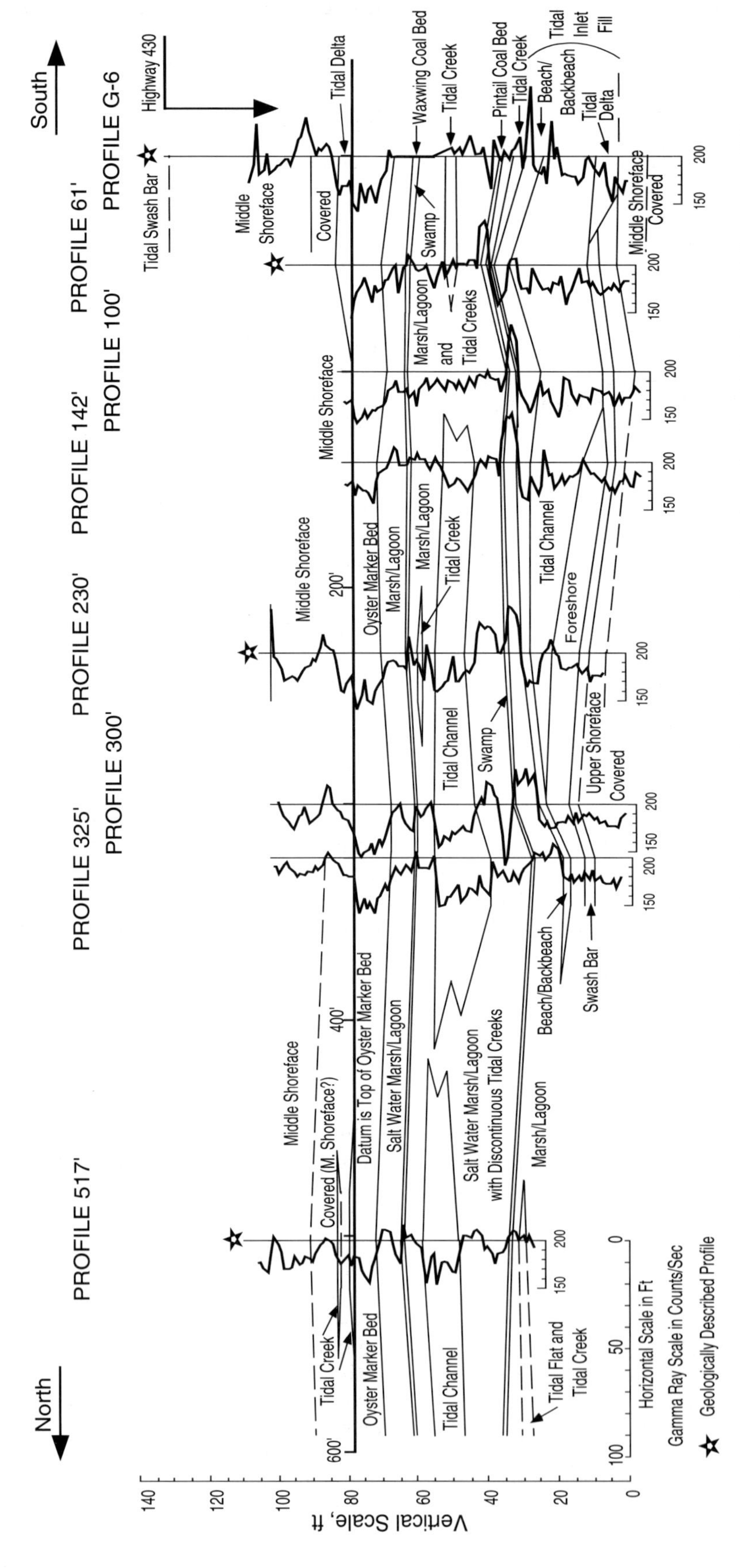

Figure 3. Outcrop stratigraphic cross section is nearly parallel to depositional strike. Note the breaks in tidal channel facies and the complexity of tidal inlet fill that can be found between distances much less than that of 40 ac (16 ha) well spacings.

deposits that contains a horizon with discontinuous, but laterally extensive, reservoir-quality tidal channel sandstones.

In the strike-parallel cross section presented here (Figure 3) tidal channel units have geometries with average lateral extents of only 91.5–106.8 m (300–350 ft). If tidal channels of this scale were present in a reservoir, log correlations for wells with as close as 10 ac (4 ha) spacing would almost certainly erroneously indicate a continuous reservoir unit; however, as can be seen in the stratigraphic cross section (Figure 3), the discontinuity of tidal channels in the plane of this cross section creates a scale of heterogeneity that could lead to significant bypassed oil. One would clearly expect little sweep of this zone if an injection well were placed at the southernmost limit of this cross section.

At the interfacies scale, additional complexity to modeling fluid flow is created by a tightly cemented sandstone with up to 30% oyster shells and shell fragments (Oyster marker bed). Calcite for cement was supplied by the abundant oyster shells and filled virtually all pore space in this bed very early in its diagenetic history, creating a field-to-regional–scale impermeable barrier to vertical flow of fluids (except along fractures). Several reservoir-quality sandstones are present below the Oyster marker bed; however, one would expect little cross flow between permeable sandstones above and below the cemented sandstone.

Petrophysical Properties from Plugs

The measurements for this study were performed both on drilled outcrop plugs from nine depositional facies and on blocks collected from the Almond Formation at Outcrop G. One-inch- (2.54 cm-) diameter plugs were taken at lateral spacing from 15 cm (6 in) to 16.5 m (54 ft) and vertical spacing from 8 cm (3 in) to 1.5 m (5 ft) to capture hierarchically stacked variations on the scale of meters to hundreds of meters. Plugs were subjected to standard core analysis to determine porosity and permeability and to aid in mapping petrophysical property distribution within depositional facies.

When porosity and permeability for the various depositional facies were analyzed, it was recognized that in terms of reservoir-quality, sandstones, there were only two groups: tidal delta sandstones and all other facies combined. A representative chart (Figure 4) comparing the porosity and permeability of tidal delta, tidal channel, and foreshore facies (beach and beach-welded swash bar) illustrates the significantly higher values for the tidal delta compared to the other facies that all have similar petrophysical properties. The plug data indicate average values of just greater than 1 d (darcy) permeability and 31.5% porosity for tidal delta sandstones compared to permeability and porosity values, respectively, of 226 md and 25.6% for tidal channels, 422 md and 27.3% for the beach, and 219 md and 25.3% for the beach-welded swash bar.

The structure of the porosity and permeability distribution (lamination, bedding, bedsets), however, is different in each of the facies. In this case the structure of the petrophysical property rather than the scale of the heterogeneity would cause each of the major reservoir-quality facies to react differently to fluid flow.

Beach Semi-Variograms

Semi-variograms of outcrop plug permeabilities from the foreshore (beach) facies (not including the beach-welded swash bar) indicate horizontal correlation lengths of less than 0.3 m (1 ft) and 2.1–2.4 m (7–8 ft). The data were collected essentially parallel to depositional strike. In a forebeach setting with negligible diagenetic effects, lateral correlation length would be expected to be much greater than 2.4 m (8 ft). It is believed that the Almond data indicate shorter lateral correlation length because the flat-lying semiparallel laminations are not well preserved. Diagenetic compartmentalization is related to variations in clay cement and relative differences in compaction, which are currently being investigated as primary causes of the 2.1–2.4 m- (7–8 ft-) scale heterogeneity. Lateral correlation lengths of less than 0.3 m (1 ft) are believed to be due to grouping of laminae with similar permeability values, but may have to do with the minimal horizontal sample distance.

Beach-Welded Tidal Swash Bar Semi-Variograms

Semi-variograms of outcrop permeability data taken exclusively from the beach-welded tidal swash bars indicate a horizontal correlation length of approximately 2.1 m (7 ft). Planar-tabular foreset cross-bedding with millimeter-thick laminae are well preserved within this subfacies. It is believed that the horizontal correlation length for the beach-welded swash bars correlates to the width of individual sand waves that were washed shoreward across the swash platform just downdrift of a tidal inlet to become welded onto the beachface.

Tidal Channel Semi-Variograms

Semi-variograms of tidal channel permeability indicate a vertical correlation length at approximately 1.2 m (4 ft) and two lateral correlation lengths at approximately 0.6 m (2 ft) and 3.1 m (10 ft). It is interesting, but probably accidental, that the correlation length of 1.2 m (4 ft) is twice the thickness of major bedsets within the measured tidal channels. It is likely that the lateral correlation length of 0.6 m (2 ft) reflects the scale of poorly preserved foreset bedding within major bedsets. Well-preserved large-scale lateral accretion surfaces mark the significant bedset boundaries within the tidal channel facies (illustrated below). Bedding within major tidal channel bedsets is poorly preserved; however, there is evidence from the photomosaics that the correlation length of 3.1 m (10 ft) may represent the distance that bedding (ripple to dune-size sand waves) can be traced laterally within bedsets. Large, compacted clay clasts, clay-rich laminae, thin clay, and carbonate-cemented laminae and beds, as well as imprints of organic material such as plant stems and other vegetation, are abundant at the base of the tidal inlet channel where it cuts into laminated,

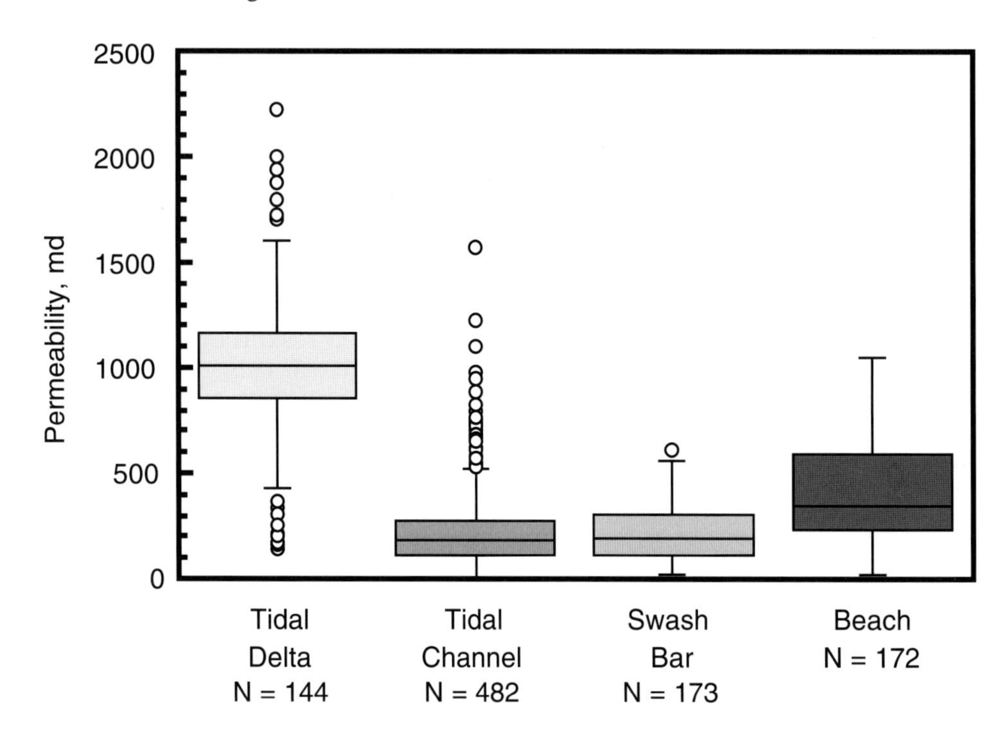

Figure 4. Comparison of porosity and permeability for outcrop facies samples. Box encloses 50% of the data centered about the median value. Lines extend to 95% value. Outliers beyond these ranges are indicated by circles. N = number of samples analyzed.

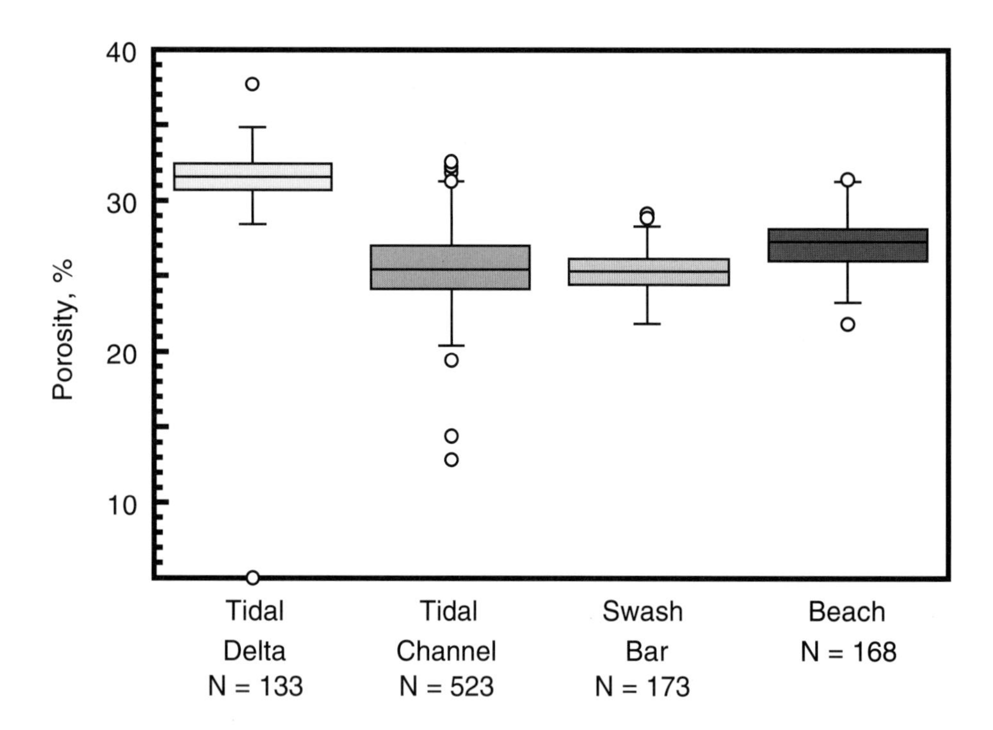

white beach sandstones. Oysters and other molluscan remains are not abundant in the base of the inlet channel that was studied.

Tidal Delta Semi-Variograms

Semi-variograms of permeability from the tidal delta show a wide scatter; however, in both lateral and vertical directions, a correlation length of approximately 1.5 m (5 ft) is found. Additional samples are necessary to better resolve the variogram range. Permeability data with a lateral spacing of 1.5 m (5 ft) were taken along a single bedset boundary 137 m (450 ft) wide. A well-developed 6.1 m (20 ft) cyclicity was observed. This scale of cyclicity is only a fraction of the lateral extent of mapped major bedsets in the tidal delta facies. It is believed that the 6.1 m (20 ft) cyclicity seen on raw permeability data reflects the frequency of beds (i.e., dune-size sand waves) within the major bedsets.

Bidirectional cross-lamination between major beds and bedsets is moderately to well preserved within tidal delta sandstones. Broad, low-angle festoon bedding is well preserved at major bedset surfaces. Iron

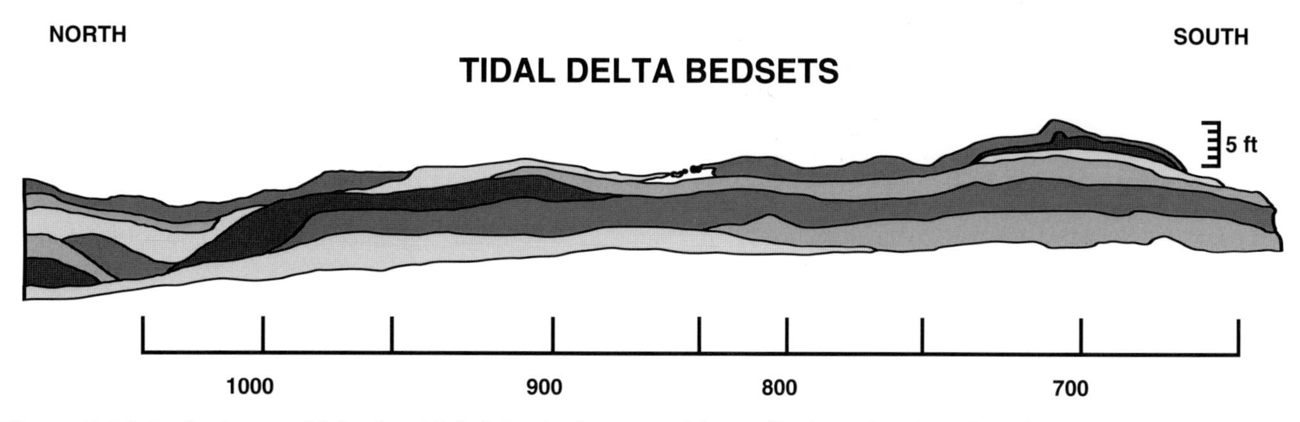

Figure 5. Major bedsets within the tidal delta facies extend laterally for a few hundred feet. Uneven horizontal scale is due to construction of cross section from a photomosaic. Shading is used to assist visual discrimination of bedsets.

oxide cements accentuate bedset boundaries and form concretionary layers that extend laterally for distances up to 15.3 m (50 ft).

Based on the permeability distribution of outcrop plug data and process-driven geological interpretations, the differences in permeability structure between depositional facies are primarily related to the differing bedding styles present and enhancement of these heterogeneities by diagenetic processes. Bedset boundaries with permeability cyclicity on the scale of 6.1 m (20 ft) in the tidal delta, 3.1 m (10 ft) in the tidal channel, and 2.1 m (7 ft) in the swash bar appear to be the primary control on lateral self-similarity within the facies studied. Lateral cyclicity of 2.1–2.4 m (7--8 ft) in the beach and 1.5 m (5 ft) in the tidal delta is also indicated by semi-variogram analysis and may be related to diagenetic compartmentalization. Vertical correlation lengths of 1.2 m (4 ft) in the tidal channel and 1.5 m (5 ft) in the tidal delta, and lateral correlation lengths of less than 0.6 m (2 ft) in all facies, are most likely related

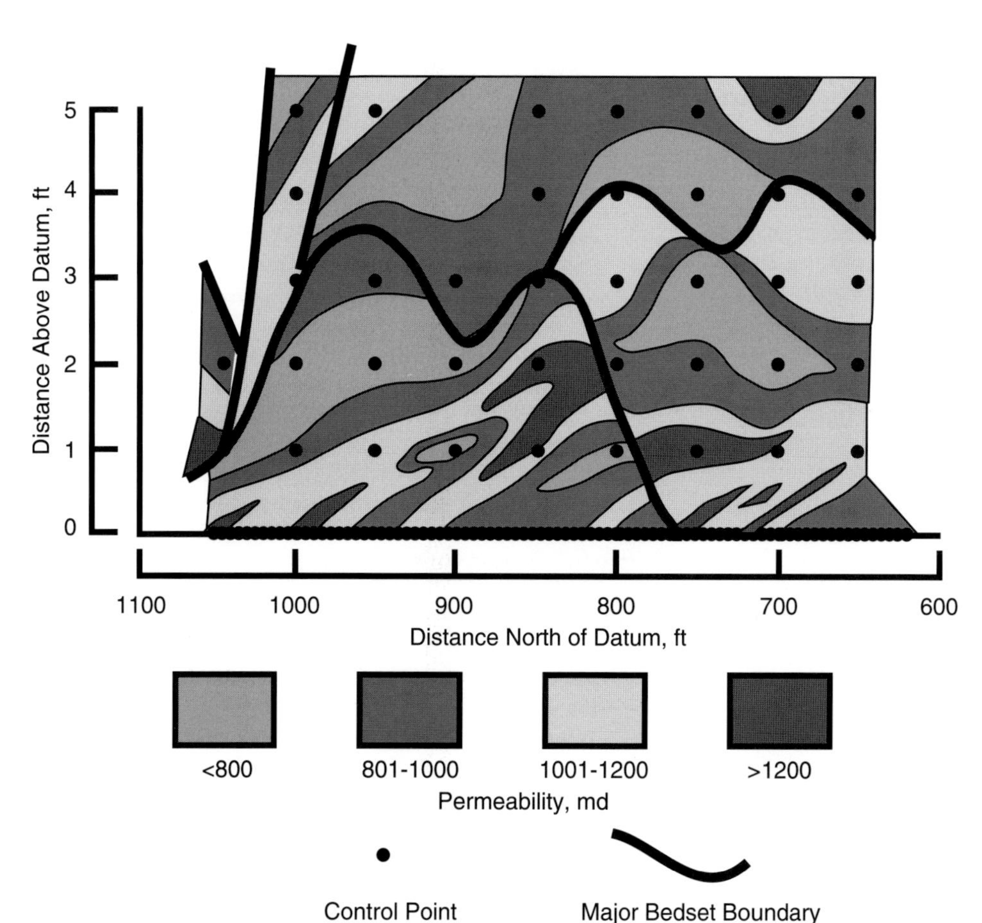

Figure 6. Distribution of permeability within selected bedsets comprising the tidal delta facies. Data were derived from 1-in- (2.54-cm-) diameter sandstone plugs analyzed

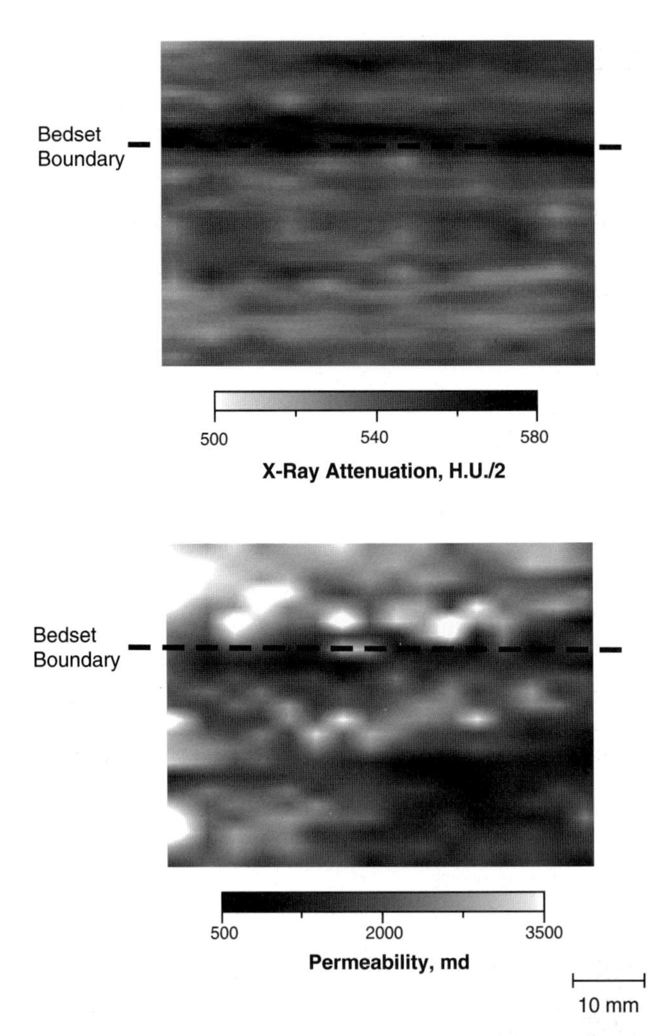

Figure 7. X-ray CT and probe permeameter maps of a tidal delta bedset boundary. The CT image was reconstructed from sections taken at 5 mm (0.2 in) spacing. The probe minipermeameter measurements were performed on a grid 4 mm (0.16 in) along laminations and 2 mm (0.08 in) across laminations using a 3 mm (0.12 in) ID tip.

to the sample distribution rather than any fundamental sedimentological or diagenetic feature.

Mapping Bedsets and Permeability Distribution

Tidal Delta Facies

Mapped bedsets within the tidal delta facies tend to extend laterally for 30.5–91.5 m (100–300 ft) or greater (Figure 5) and are up to 0.9 m (3 ft) thick. Permeability within bedsets follows the "grain" of the primary bedding direction and often exhibits significant differences across bedset surfaces (Figure 6). We found that the permeability trend within major bedsets of this facies and others could not be properly interpreted based on widely spaced plugs (1.5–15.3 m; 5–50 ft spacing) without first understanding the location of the bedset boundaries and the bedding direction within the bedsets.

Thin sections indicate that the bedset boundaries, which are well preserved and very obvious on the outcrop, are enhanced by iron oxide and kaolinite cements, have tighter packing with more long-grain contacts, and have smaller pores than the sandstone beds on either side of the surface. The beds away from the bedset surfaces tend to have larger, clean pores that appear well connected; better grain and pore size sorting; and only local "clumps" of compacted rock fragments or partly clay-filled interparticle porosity.

Figure 7 displays X-ray CT and probe permeameter surface views from a block containing a tidal delta bedset boundary. The CT image was reconstructed from sections taken at 5 mm (0.2in) spacing. The probe permeameter measurements were performed on a 4 mm (0.16 in) grid along laminations and 2 mm (0.08 in) across laminations using a 3 mm (0.12 in) ID (inner diameter) and 6 mm (0.24 in) OD (outer diameter) tip. The X-ray attenuation values are consistent with the relatively high porosity values determined by plug measurements. They indicate that the bedset boundary has slightly lower porosity than the rest of the rock. The probe permeameter measurements indicate that the bedset boundary also has slightly lower permeability.

Tidal Channel Facies

Laterally accreting major bedsets within the tidal channel facies have average lateral dimensions (parallel to depositional strike) of approximately 46 m (150 ft) (Figure 8). Sandstone in thin sections from the basal portion of a tidal inlet channel that cut into an adjacent beach deposit exhibits very poor grain size sorting, widespread centimeter-thick layers containing iron oxide cements, finely crystalline calcite and siderite, and small, dirty-appearing pores. The "dirty" appearance of the pores is caused by the presence of disseminated clays of two origins. Detrital clay-size particles within interparticle pore space originated from partly leached framework grains and compacted rock fragments. Authigenic kaolinite is also present within the pores, further reducing the permeability. Multiple layers of centimeter-thick channel fill were noted at the base of the inlet channel, and all were poorly to very poorly sorted.

Figure 9 displays X-ray CT and probe permeameter surface views from a block containing a tidal channel/beach facies contact. The CT image was reconstructed from 4 mm (0.16 in) thick sections taken at 5 mm (0.2 in) spacing. The probe minipermeameter measurements were performed on a 5 × 5 mm (0.2 × 0.2 in) grid using a 3 mm (0.12 in) ID and 6 mm (0.24 in) OD tip. The X-ray attenuations are higher than in Figure 7, indicating a lower average porosity and greater content of minerals with increased CT densities (such as iron oxide cement). The higher attenuations for the tidal channel correspond to both a lower average porosity and more CT-dense mineralization. The beach sandstone displays a higher porosity and less diagenetic alteration. The probe permeameter measurements indicate a large permeability contrast

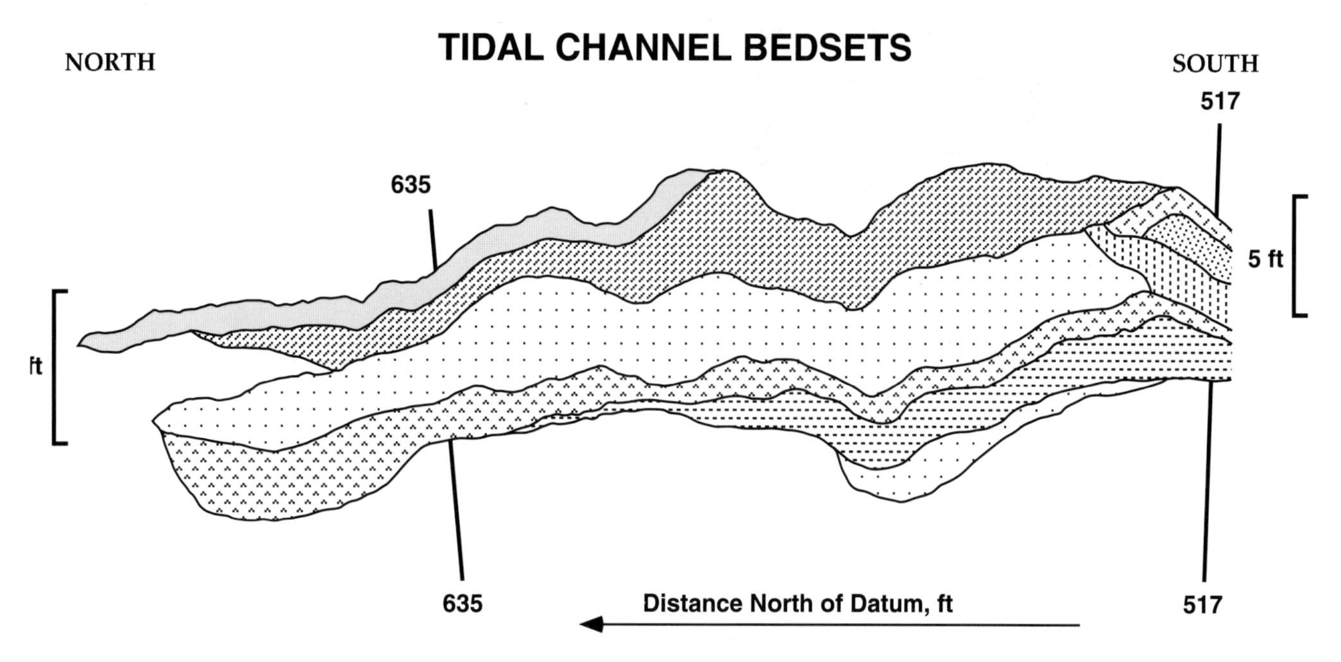

Figure 8. Laterally accreting major bedsets within the tidal channel facies extend laterally no more than about 200 ft (61 m). This image was constructed from a photomosaic. Note that fluids moving laterally through this facies would have to pass through more than one bedset boundary. Patterns are used to assist visual discrimination of individual bedsets.

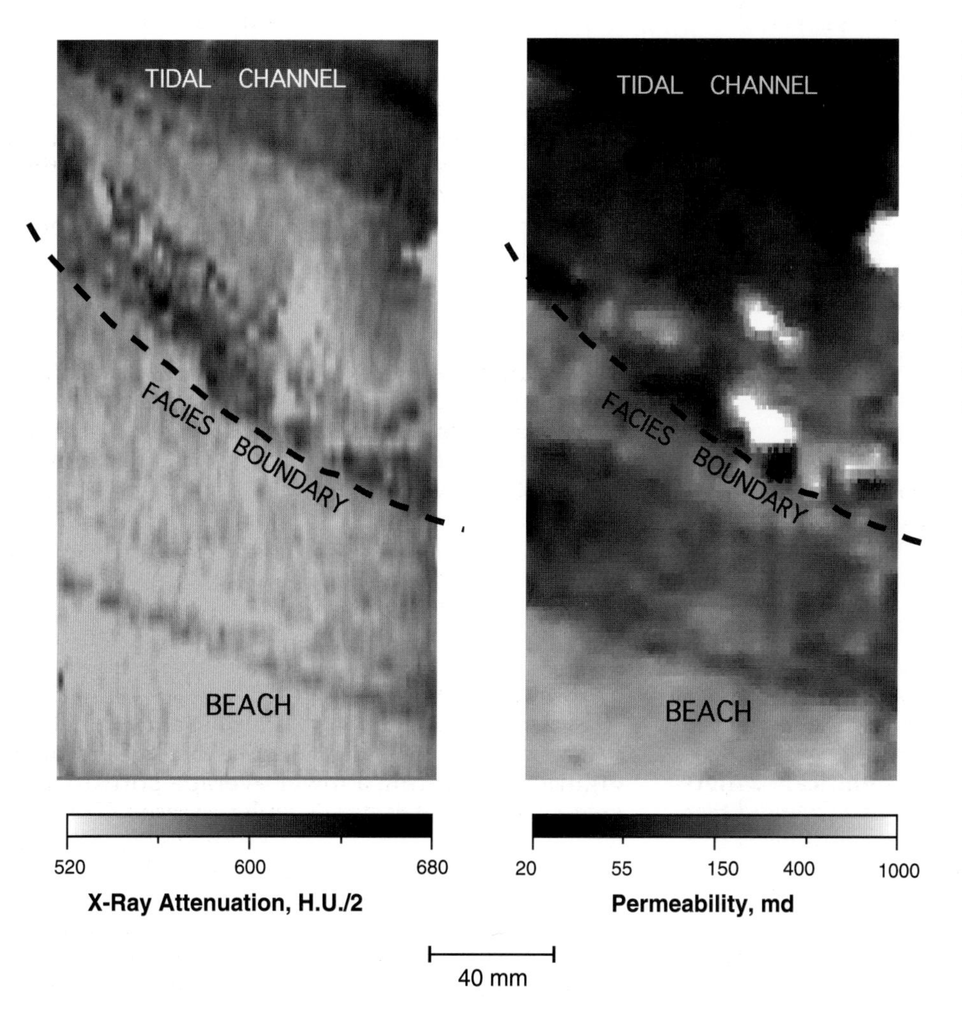

Figure 9. X-ray CT and probe permeameter maps of a tidal channel/beach facies contact. The CT image was reconstructed from sections taken at 5 mm (0.2 in) spacing. The probe minipermeameter measurements were performed on a 5×5 mm (0.2×0.2 in) grid, using a 3 mm (0.12 in) ID (inner diameter) tip.

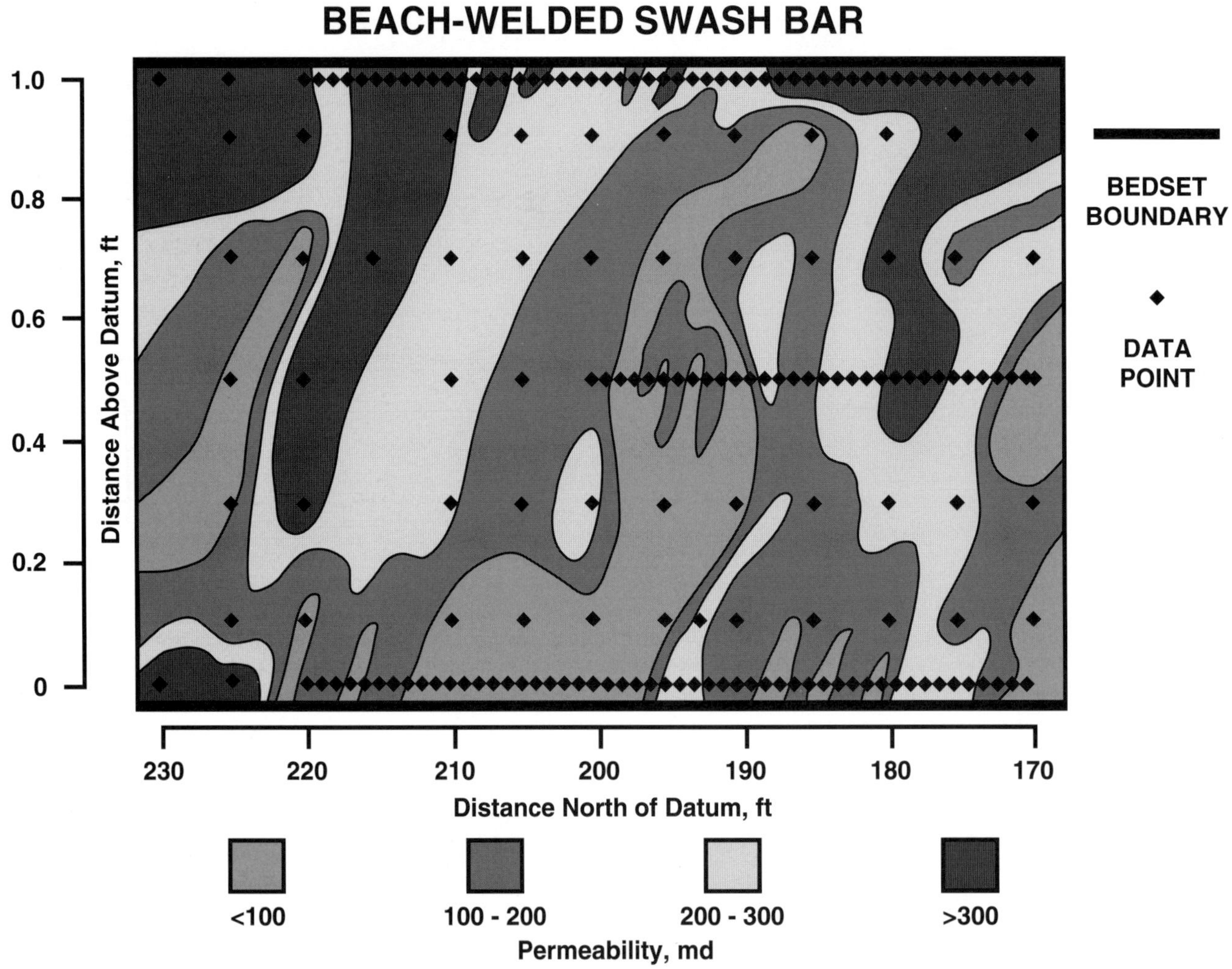

Figure 10. Map of the distribution of permeability within the tidal swash bar based on closely spaced 1- in- (2.54- cm-) diameter outcrop plugs. Note northward dip (to the left) of permeability trend is in the same direction as the planar-tabular cross-bedding.

between the tidal channel and the beach facies, with the beach displaying permeabilities in the 150–500 md range and the tidal channel displaying permeabilities of 20–150 md.

Foreshore Facies

Bedsets were not recognized within the beach facies; however, within the 0.3-m-(1-ft-) thick beach-welded swash bar, closely spaced samples allowed us to map the permeability distribution (Figure 10). Once again a primary knowledge of the bedding style (planar-tabular foreset cross-bedding) was necessary to contour the permeability map correctly. Bedding within this unit comprises millimeter-scale foreset cross-lamination. The map of permeability distribution (Figure 10) indicates that groups of inclined laminations extend laterally on the order of 4.6 m (15 ft) and comprise permeability layers that dip in the same direction as the bedding, but less steeply.

Thin sections from the tidal swash bar facies show large variations in grain size, a relatively clean pore system with only minor amounts of interparticle clay cement, and an abundance of partly compacted soft rock fragments that tend to block pore throats; therefore, compaction is primarily responsible for the lower permeability of the swash bar sandstones compared to the tidal delta sandstones.

CONCLUSIONS

Based on 1-in- (2.54-cm-) diameter plugs of outcrop sandstones collected from a transgressive barrier system, permeability distribution between bedsets, as well as between facies, is primarily related to the differing bedding styles present, bedset bounding surfaces, and enhancement of these heterogeneities by diagenetic processes. Bed and bedset boundaries provide important controls on permeability because permeability along these surfaces is often decreased by cementation, smaller pores, tighter grain packing, and compaction of sand-size rock fragments.

Permeability drops significantly across bedding surfaces. In our study we found that permeability decreased from approximately 1 d ("matrix" values) to

400–500 md across tidal delta bedset boundaries. Matrix permeability for the tidal channel and foreshore (beach and beach-welded swash bar) are both on the order of 200–500 md. The drop in permeability across the facies contact between a tidal inlet channel deposit and the adjacent beach sandstones is one to two orders of magnitude.

Field checking of bedding features and maps based on the plug data indicate that permeability distribution parallels the "grain" of bedding within bedsets. There is a loss of up to approximately one-half of the matrix permeability across some bedding surfaces and most bedset boundaries. Mapping also indicates that bedset boundaries are essentially always inclined to upper and lower facies boundaries. In addition, depositional facies have differing scales of lateral continuity. Outcrop mapping has indicated, for example, that tidal creek deposits may extend laterally only a few tens of feet in the strike direction, tidal channel facies may extend only a few hundred feet, but distinct tidal delta lobes may be traced from 0.5 to 1+ mi (0.8 to 1.6 km) in the strike direction. Fluid flow between wells and flow through a given reservoir interval must cross bedset boundaries. The frequency of bedset boundaries encountered can then be a significant controlling factor to fluid flow and recovery efficiency.

Semi-variogram analysis of our data indicates that lateral cyclicity of permeability is primarily related to bedset boundaries within facies and is predictable. Self-similarity created by bedset boundary distribution on the scale of 6.1 m (20 ft) in the tidal delta, 3.1 m (10 ft) in the tidal channel, and 2.1 m (7 ft) in the swash bar sandstones has been recognized. Diagenetic enhancement of bedding features, as recognized in this study, must be handled separately in order to maximize the significance (predictive capability) of lateral cyclicity of bedset surfaces.

Distribution of petrophysical properties interpreted from reservoir-quality sandstone blocks using CT and minipermeameter analysis can be mapped at a scale approximately two orders of magnitude finer than that mapped using plugs. The CT and minipermeameter data distribution can be explained through petrographic analysis. Both the large-scale plug data and the detailed minipermeameter maps of sandstone blocks show similar ranges of permeability for the equivalent depositional facies based on our study of outcropping Almond Formation barrier island sandstones. This correspondence implies that for this system, if one correctly identifies the depositional facies and has an understanding of the facies architecture, the small-size samples are not a major hindrance to modeling the reservoir at a larger scale. Thus it is possible to provide a good approximation of scaled fluid flow within this system from cuttings, thin sections, or core plugs.

ACKNOWLEDGMENTS

This project was funded by the financial support of the Department of Energy under agreement DE-AC22-96PC91008. The support of Union Pacific Resources Company for access to the outcrop is greatly appreciated. We also wish to thank the staff of BDM-Oklahoma, Inc., and two anonymous reviewers for their assistance with this paper.

REFERENCES CITED

Krystinik, L. F., and R. H. Mead, 1996, Sequence stratigraphic and synsedimentary tectonic controls on reservoir compartmentalization in a transgressive sequence set: Almond Formation, southwest Wyoming: AAPG Annual Convention, Program with Abstracts, p. A79.

Martinsen, R. S., and G. Christensen, 1992, A stratigraphic and environmental study of the Almond Formation, Mesaverde Group, Greater Green River Basin, Wyoming, *in* C. E. Mullen, ed., Wyoming Geological Association Guidebook, Forty-Third Field Conference, p. 171–190.

Roehler, H., 1988, The Pintail coal bed and barrier bar G-A model for coal of barrier bar-lagoonal origin, Upper Cretaceous Almond Formation, Rock Springs Coal Field, Wyoming: USGS Professional Paper 1398, 60 p.

Roehler, H., 1990, Stratigraphy of the Mesaverde Group in the Central and Eastern Greater Green River Basin, Wyoming, Colorado, and Utah: USGS Professional Paper 1508, 52 p.

Schatzinger, R. A., M. J. Szpakiewicz, S. R. Jackson, M. M. Chang, B. Sharma, and M. K. Tham, 1992, Integrated geological-engineering model of Patrick Draw field and examples of similarities and differences among various shoreline barrier systems: DOE Report NIPER-575, 146 p.

Van Horn, M., 1979, Stratigraphy relationships and depositional environments of the Almond and associated formations, east-central flank of the Rock Springs Uplift, *in* Rocky Mountain Section SEPM Field Trip, Cretaceous of the Rock Springs Uplift, Wyoming, Sept. 21–23, p. 51–63.

Knox, P.R., M.D. Barton, Predicting interwell heterogeneity in fluvial-deltaic reservoirs: effects of progressive architecture variation through a depositional cycle from outcrop and subsurface observations, 1999, *in* R. Schatzinger and J. Jordan, eds., Reservoir Characterization-Recent Advances, AAPG Memoir 71, p. 57–72.

Chapter 5

Predicting Interwell Heterogeneity in Fluvial-Deltaic Reservoirs: Effects of Progressive Architecture Variation Through a Depositional Cycle from Outcrop and Subsurface Observations

Paul R. Knox
Bureau of Economic Geology
The University of Texas at Austin
Austin, Texas, U.S.A.

Mark D. Barton
Shell Development Company
Houston, Texas, U.S.A.

ABSTRACT

Early recognition of stratigraphic heterogeneity can lead to improved estimates of hydrocarbon reserves and more efficient development strategies. Although good models based on studies of modern and ancient fluvial-deltaic deposits exist for predicting interwell heterogeneity, a single model applies only to genetically related deposits of a single pulse of progradation and retreat, which describes only one, or a part of one, reservoir. Recent studies document significant changes in depositional style in adjacent stratigraphic intervals, and a model for predicting this variability is needed; therefore, we have carried out studies of outcrop reservoir analogs and of reservoirs that suggest that variability is predictable if assessed within the framework of depositional cycles and that differences between adjacent high-frequency cycles can substantially impact production behavior and reserve-growth potential.

Outcrop studies of the Cretaceous Ferron sandstone of Utah have demonstrated that (1) incised fluvial deposits in progradational parts of low-frequency depositional cycles tend to be narrow, deep, and internally homogeneous, whereas those in retrogradational parts of such cycles tend to be wider, internally heterogeneous, and display lateral channel migration, and (2) river-dominated delta-front styles are more common in progradational parts of intermediate-frequency cycles, whereas wave-dominated delta-front styles are more common in retrogradational parts. Detailed investigations of two fluvial-influenced upper delta-plain reservoirs in the Oligocene Frio Formation of Tijerina-Canales-Blucher (T-C-B) field, south

Texas, demonstrate that (1) a fluvial reservoir deposited early in an intermediate-frequency cycle contains several narrow laterally isolated channel belts, whereas an adjacent reservoir deposited late in that cycle contains a single broad channel belt that is internally heterogeneous, and (2) this difference in styles is paralleled by a strong difference in production behavior, with a gas completion in the former reservoir draining 40 ac (16 ha), whereas the typical oil completion in the latter reservoir is approximately 1.5 ac (0.61 ha), and less than 10% of the original oil in place in the latter reservoir has been recovered.

The application of these findings to other reservoir intervals and other basins should be made with appropriate caution, but they have implications for project economics throughout field life, from the exploration phase through mature-field redevelopment. Benefits range from improved prospect ranking to improved development efficiency and better prioritization of mature fields for acquisition or characterization. Application of these techniques to U.S. fluvial-deltaic reservoirs could improve near-term reserve-growth potential, preventing permanent loss of strategic resources by curtailing premature field abandonments.

INTRODUCTION

Unsuspected interwell-scale stratigraphic heterogeneity has contributed to a situation in which more than one-third of the mobile oil in domestic fluvial-deltaic reservoirs remains in place despite many decades of production. According to data from the U.S. Department of Energy TORIS data base, 14.4 billion barrels (Gbbl) of oil have been produced from these reservoirs, but 10.8 Gbbl of mobile oil remain, with currently proved reserves equaling less than one-tenth of the remaining volumes (U.S. Department of Energy, 1991). A key to recovering the large volume of remaining mobile oil is to identify those reservoirs that have been poorly drained and apply integrated, advanced characterization methods to locate specific untapped or incompletely drained compartments. Recognizing those reservoirs that are poorly drained will be dependent on identifying those most likely to be internally heterogeneous and, therefore, compartmentalized. The problem of stratigraphic heterogeneity is particularly pronounced in fluvial-dominated deltaic reservoirs because fluvial- and distributary-channel, mouth-bar, and splay sandstone bodies may appear laterally continuous, but are commonly separated by impermeable abandoned-channel, floodplain, marsh, bayfill, and marine mudstones. These isolated sandstone bodies, many of which are long and narrow, are ineffectively contacted by conventional patterned drilling grids and, in some cases, are internally compartmentalized by low-permeability layers that prevent efficient reservoir drainage.

Models developed through studies of modern deltas and ancient deltas exposed in outcrops have improved fluvial-deltaic reservoir characterization; however, these models apply only to lateral variability within a genetically related depositional unit, which constitutes the sediments deposited during a single pulse of shoreline progradation and retreat. Deposition of these sediments is governed by a shared set of controlling forces, such as fluvial, wave, and tidal energies, as well as tectonic subsidence and climate. Because the balance of these controlling forces may shift over a time frame that spans multiple progradational pulses, the mix of controlling forces from pulse to pulse may be different, resulting in changes in depositional systems. As a result, size and geometry of sand bodies may change in vertically successive depositional units, leading to a situation in which vertically adjacent sand bodies (reservoirs) may have different patterns of lateral variability. In other words, a depositional unit perhaps 10 m thick and containing the deposits of a fluvial-dominated delta with large lateral variability of reservoir quality might be directly overlain by a depositional unit containing the deposits of a wave-dominated delta with much lower lateral variability of reservoir quality. These vertically adjacent reservoirs would require much different strategies to efficiently recover the hydrocarbon resources within them.

To evaluate the potential cycle-to-cycle changes in lateral reservoir variability in different fluvial-deltaic settings and their impacts on reservoir production characteristics, we present two complementary studies. The first study investigated cycle-to-cycle variability using well-exposed delta-front, lower delta-plain, and incised fluvial deposits of the Upper Cretaceous Ferron sandstone in Utah. Ferron findings were then used to help predict cycle-to-cycle variability in upper delta-plain fluvial reservoirs of the Oligocene Frio

Formation in south Texas, as well as to help explain the impact of this variability on reservoir performance.

The primary objectives of this paper are to (1) summarize observations from Ferron sandstone outcrops that demonstrate changes in facies associations through two scales of depositional cycles and (2) underscore the potential applications to oil and gas reservoir characterization using examples from mature Oligocene-age reservoirs in the Texas Gulf Coast. The following sections will review previous findings on reservoir characteristics within the framework of depositional cycles, clarify terminology used in our discussions, and present the methodology used in the two studies.

Previous Work in Depositional Cyclicity

Previous work indicates that prediction of sandstone geometries and heterogeneity at the interwell scale is dependent on a clear understanding of the depositional setting and the position of the reservoir within a depositional cycle. The concepts of sequence stratigraphy developed during the 1970s and 1980s improved the reliability of facies interpretation by more clearly defining the effects of changing relative sea level and sediment supply on depositional cyclicity, facies geometry, and stacking patterns (Vail et al., 1977; Jervey, 1988; Posamentier et al., 1988; Posamentier and Vail, 1988; Galloway, 1989a, b). The recent documentation of high-frequency depositional cycles and a hierarchy of cycle scales has increased the resolution of depositional models (Goldhammer et al., 1990; Mitchum and Van Wagoner, 1991) and underscored the potential of high-frequency eustatic changes to affect stratal geometries.

The advances provided by the above concepts are being widely applied to predict lithology in exploration studies. Only recently, however, has the predictive framework of sequence and cyclic stratigraphy been thoroughly incorporated with facies models to go beyond the prediction of stratal geometries and begin to document progressive changes in facies associations through a high-frequency depositional cycle. Researchers in outcrops of carbonate sequences (Sonnenfeld, 1991; Kerans and Fitchen, 1995) and clastic successions (Cross et al., 1993; Gardner, 1993; Barton, 1994) recognized that the continually changing balance of depositional controls throughout a depositional cycle also resulted in changes in facies associations and their preservation potential within a depositional system. These insights into facies associations in a high-frequency stratigraphic framework provide sufficiently detailed information from which to begin the process of predicting reservoir architecture and heterogeneity at the interwell scale.

Methodology

To investigate changes in architecture through depositional cycles, the hierarchy of depositional cyclicity must first be identified and the depositional facies assessed at the scale of the smallest identifiable cycle, ideally a single cycle of allocyclic progradation and retreat. Only at this scale can facies be associated genetically and be said to be a consequence of the same balance of allocyclic forces. These relationships were first investigated in Ferron outcrops, which contain sufficient vertical and lateral resolution to evaluate surfaces, stacking patterns, facies, and their interdependence. The thesis of interdependency, supported by the outcrop work, was then applied to subsurface Frio reservoirs to evaluate (1) the potential to predict reservoir geometries at the interwell scale and (2) the effect of differing geometries and internal heterogeneities on reservoir performance.

It is the shared conditions of high sediment supply and high accommodation rate that existed both on the west margin of the Cretaceous Interior Seaway and the north margin of the Gulf of Mexico basin that make the Ferron an acceptable analog for the Frio and other Gulf Coast clastic reservoirs. In conditions of high accommodation, marine flooding surfaces and their equivalents are easiest to identify in well-log data. As a consequence, our generic depositional cycle (independent of scale) begins at maximum flooding and progresses from progradation through aggradation to retrogradation, culminating with a maximum flooding surface. In this sense, the resulting depositional units are akin to genetic depositional sequences as defined by Galloway (1989a).

Both the Ferron and Frio are ideal for demonstrating these relationships because they were deposited under conditions of high accommodation and high sediment supply; consequently, they provide the best record of short-term (high-frequency scale) changes in the balance of depositional controls that result in brief marine flooding events. These flooding events are more clearly recorded because of the abundant influx of sediment and are also then preserved by the rapid burial. The high accommodation rates result in limited incision of these surfaces during the following lowstand.

Differing sets of terminology have been applied to describe the hierarchy of scales of depositional cycles, such as, from smallest scale to largest, parasequences, high-frequency sequences, sequences and composite sequences (Mitchum and Van Wagoner, 1991), versus cycles, high-frequency sequences, and composite sequences (Kerans and Fitchen, 1995). The term "order" has been used to describe the hierarchy, such as 3rd-, 4th-, and 5th-order (Goldhammer et al., 1990; Mitchum and Van Wagoner, 1991), and specific time spans have been assigned to various orders. Unless a cycle is clearly constrained by age dating, we avoid the use of order terminology. Instead, we have applied the generic terms of "low," "intermediate," and "high frequency" to describe observed hierarchy of cyclicity.

Channelized deposits are observed in both the Ferron and Frio studies. We use the term "channelform" to describe the deposits of a single-story scour-and-fill feature of Allen (1983) or the macroform of Miall (1985). We use the term "channel belt" to describe the deposits of a feature containing multiple laterally or vertically stacked channelforms sharing a common fluvial axis. Broadly stratigraphic equivalents resulting from deposition following a major avulsion event would be considered separate channel belts.

The first step of establishing the stratigraphic framework of cyclicity and its hierarchy was accomplished for the Ferron outcrop area on the basis of previous studies (Hale, 1972; Ryer, 1981) and this study. Previously identified coals and widespread shales were assumed to mark significant flooding events, and thinner widespread marine shales were traced in this study in order to define boundaries of higher frequency depositional cycles. Lithofacies within each high-frequency (HF) unit were identified, and depositional facies were interpreted and mapped. The stacking pattern of HF units was then determined on the basis of changes in shoreline position. Depositional facies associations (e.g., wave-dominated delta) were defined and correlated with their position within an intermediate-frequency (IF) cycle. Facies associations, in the context of position within an IF cycle, were then compared between different IF cycles.

In the Frio subsurface study, a 100-km-long, dip-oriented, well-log cross section was constructed to identify regional stratigraphic surfaces, general depositional settings, and stacking patterns. The cross section stretches from updip of the study area in Jim Wells County basinward to the downdip limit of readily available well control at the Texas coast. Well spacing is generally 3–5 km or less. Marine flooding surfaces defining IF units were readily interpretable in the downdip position, general depositional facies were interpreted from log response, and approximate shoreline position in each IF unit was identified. Stacking patterns of Frio IF units indicated three low-frequency (LF) units similar to the operational members of the Frio established by Galloway (1986). This gross framework of LF and IF units was then correlated into the study area and subdivided into HF units within the field on the basis of throughgoing surfaces, identified in a fieldwide grid of stratigraphic cross sections, that were assumed to correspond to minor flooding surfaces. Net sandstone maps were then prepared for each HF unit of interest and combined with maps of log pattern to interpret the lateral distribution of facies and identify reservoir architecture. Structure and net pay maps were created for each unit, and fluid contacts were determined and annotated. Petrophysical parameters, such as porosity and water saturation, were estimated in order to assess original hydrocarbons in place and the amount of area drained by each completion. Past completions were then mapped and tabulated by HF unit to document past reservoir drainage and production behavior. The variability of production behavior is potentially a measure of differences in heterogeneity within each architectural unit. Facies architecture and internal heterogeneity were then compared with the facies and position within an IF depositional unit to evaluate any cycle-to-cycle variability that would parallel outcrop observations.

CYCLE-TO-CYCLE VARIABILITY IN OUTCROP

Ferron Sandstone Setting

The Upper Cretaceous (Turonian) Ferron Sandstone Member of the Mancos Shale crops out along the Molen Reef and Coal Cliffs fringing the San Rafael Swell of eastern Utah (Figure 1). Cliff faces of more than 100 m in height and 100 km in length expose the entire 250-m thickness of the Ferron. This outcrop occurs where present landforms and original depositional trends combine to form good three-dimensional exposures, in that long continuous sections parallel to depositional dip are complemented by shorter strike-oriented exposures in cross-cutting canyons.

Early studies of the Ferron Sandstone Member identified it as a major sandstone body composed of two distinct clastic wedges, an early wedge derived from the northwest, typically referred to as the Clawson and Washboard sandstones, and a later wedge derived from the southwest (Hale, 1972), referred to as the Ferron sandstone. Ryer (1981) recognized the upper (Ferron) wedge as being composed of a series of sandstone tongues partially bounded by marine shales. Initial sandstone tongues step seaward, whereas the overlying tongues step landward. Later studies (Ryer, 1993; Gardner, 1993; Barton, 1994) recognized that each of these tongues is further subdivided by marine shales into bodies similar to the commonly used "parasequence" (Van Wagoner et al., 1988).

Stratal successions (high-frequency units) bounded by minor flooding surfaces are stacked in a systematic fashion to form five IF units bounded by major or more regionally extensive flooding surfaces (Figure 2). Within each IF unit, the stacking pattern of HF units progresses from an initial aggradational-to-progradational set to a downstepping set, then to an aggradational set, followed in some cases by a retrogradational set (IF units 3, 4, and 5 in Figure 2).

Accompanying changes in the stacking pattern of IF and HF units are progressive changes in facies associations. The variability of lithofacies geometry and heterogeneity were documented within delta-front deposits of IF units by Gardner (1993). Subsequent studies by Barton (1995) recognized similar changes in delta-front deposits of HF units, as well as within incised fluvial deposits of IF units. This documentation of variability in vertically adjacent HF units is critical because it supports subsurface observations of HF unit variability in Frio subsurface fluvial-influenced upper delta-plain reservoirs that will be presented later. The following discussion summarizes this observed variability and compares it with the position of the deposits within a depositional cycle.

Summary of Ferron Outcrop Observations

Outcrop observations make possible a comparison of similar depositional settings at different times in a depositional cycle. Summarized below are (1) comparisons between incised fluvial deposits from a

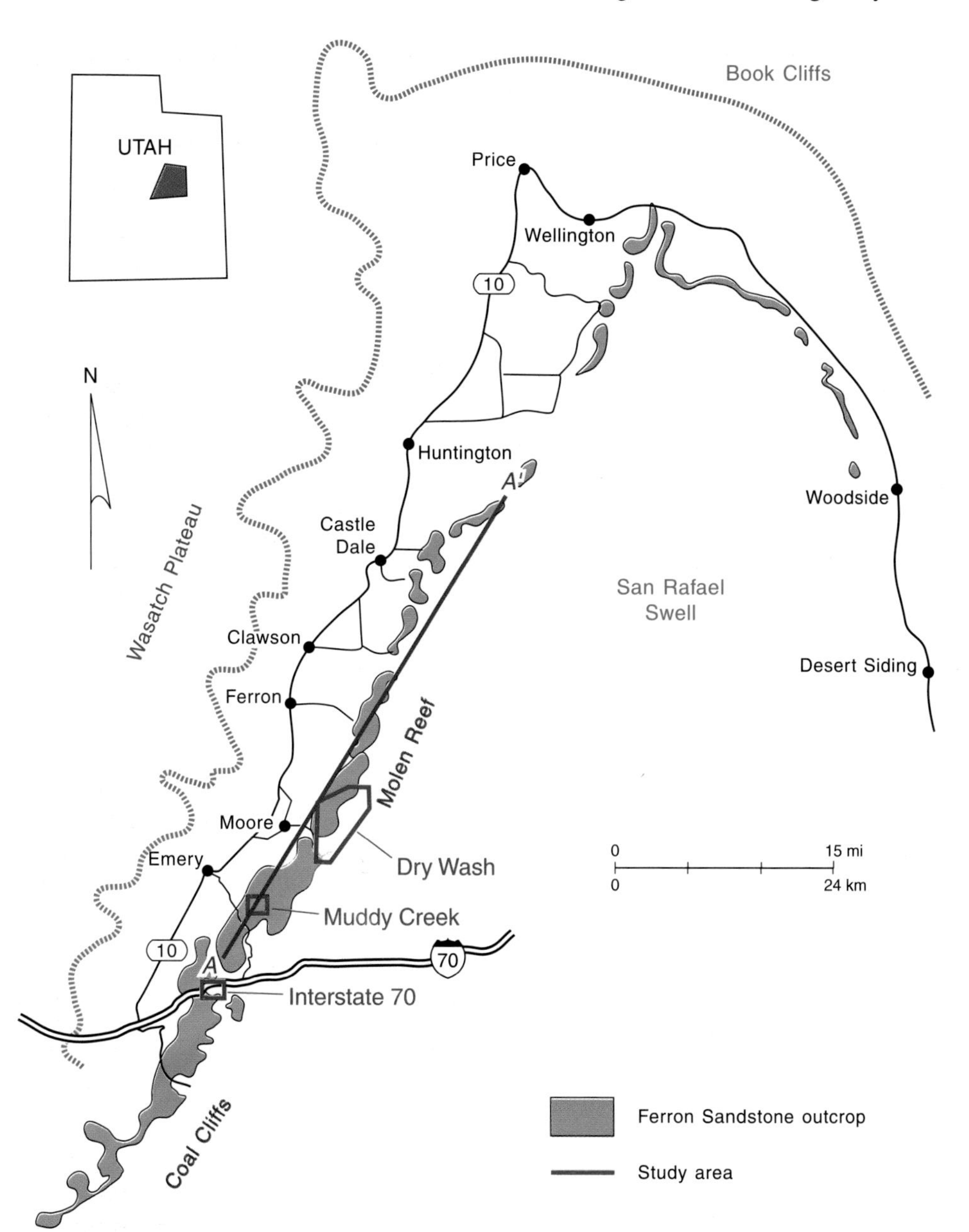

Figure 1. Location of Ferron sandstone outcrops along the northwest flank of the San Rafael Swell in central Utah. Case studies are discussed in the text. Cross section A-A′ is shown in Figure 2. From Barton (1997), reprinted here with permission of GCS SEPM Foundation.

seaward-stepping (low-accommodation) IF unit and those from a landward-stepping (high-accommodation) IF unit, and (2) comparisons between delta-front deposits from two vertically adjacent HF units occupying early and late portions of a lowstand within a seaward-stepping IF unit (Figure 2).

Incised fluvial deposits identified by Barton (1995) in unit 2 exposed along I-70 (Figure 1) are typical of incised channels observed throughout units 2 and 3, which are IF units within the seaward-stepping portion of the Ferron low-frequency depositional unit. Lithofacies, channelform boundaries, and permeability within this exposure are shown in Figure 3, along with the map-view geometry of the incised fluvial system. These incised fluvial deposits tend to be relatively internally homogeneous in both lithology and permeability, with many vertically stacked channel-bar and channel-fill beds composed nearly exclusively of trough-cross-stratified sandstones that lack evidence of marine or tidal influence. Channel-on-channel boundaries exhibit thin intervals (0.1–1 m) of slightly reduced permeabilities consisting of basal channel-lag deposits containing rounded mud clasts and dispersed clay. Incised fluvial systems are narrow (<250 m wide) and deep (>20 m), with typical width:depth ratios of approximately 7:1.

Incised fluvial deposits identified by Barton (1995) in the landward-stepping (high-accommodation) IF unit 5 are typified by outcrops at Muddy Creek (Figure 1). A cross section of this fluvial fill, shown in Figure 3, illustrates that in contrast to incised features in units 2 and 3, these deposits consist of laterally stacked channelforms (each approximately 100–300 m wide, 7–12 m deep) containing a heterolithic, upward-fining

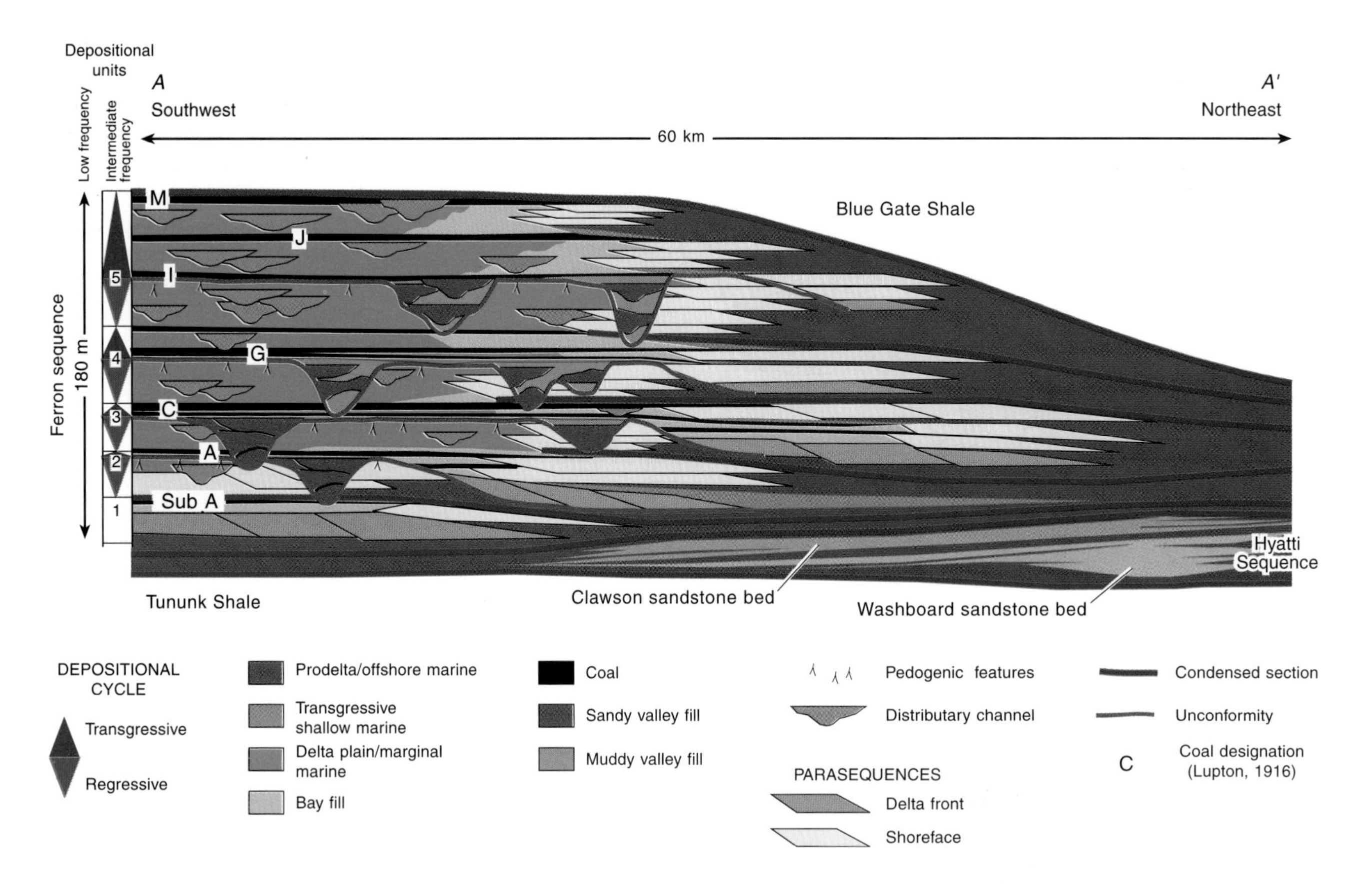

Figure 2. Schematic cross section depicting stratigraphic relationships within the five intermediate-frequency cycles of the Ferron sandstone. Location is shown in Figure 1.

succession of trough cross strata through rippled strata typical of medium- to high-sinuosity rivers. Permeability within this fluvial fill is somewhat more heterogeneous than in unit 2 fluvial fills. Channel-lag deposits drape lateral accretion surfaces and exhibit pronounced reduced permeability arising from a greater volume of mud clasts and dispersed clay. A greater number of stratigraphically equivalent incised fluvial fills are seen in map view (Figure 3), and each fill (from 0.5–1 km wide and 15–20 m deep) tends to have a higher width:depth ratio, approximately 40:1, relative to those in units 2 and 3.

Delta-front deposits in two HF units deposited during early and late lowstand periods in unit 2 are seen at exposures in Dry Wash (Figure 1). Although these two units are vertically adjacent, they exhibit vastly different lithofacies and permeability distributions, typical of opposing deltaic styles. Delta-front deposits of the early lowstand HF unit show characteristics common to fluvial-dominated deltas, such as prominent mouth-bar facies and growth faulting, and permeability is consequently highly variable in the lateral dimension. In contrast, delta-front deposits of the late lowstand HF unit show characteristics common to wave-dominated deltas, such as extensive strike-elongate shoreface facies and lagoon/washover facies, resulting in limited lateral variations in permeability. Previously, Gardner (1993) had documented this style of variability at the scale of successive Ferron IF-scale units.

Barton (1997) also documented the effect of wave- versus fluvial-dominated delta styles on production from reservoirs in the Ferron gas field, located within 5 km of the Dry Wash outcrops (Figure 1). Reservoirs in the aggradational to backstepping shallow marine successions consisted of wave-dominated units that were effectively drained by conventional well spacings, but might contain stratigraphically trapped gas near their landward pinch-outs. In contrast, offlapping and downstepping units contain more fluvial-dominated architectural styles, and low production from past completions indicates that they have contacted only small reservoir volumes. Downstepping fluvial-dominated delta-front architectural styles may therefore contain larger reserve-growth potential than aggradational reservoir intervals.

Depositional models for Ferron IF depositional cycles, supported by detailed observations across the spectrum of HF units, are shown in Figure 4. Deposition progresses from (1) wave-dominated highstand deltas through (2) river-dominated deltas fed by incised fluvial systems during early lowstand, to (3) wave-dominated deltas fed by sinuous fluvial systems in the late lowstand, and culminates with (4) barrier bars, lagoons, and estuaries during transgression. Lowstand deposits are better developed in seaward-stepping IF units

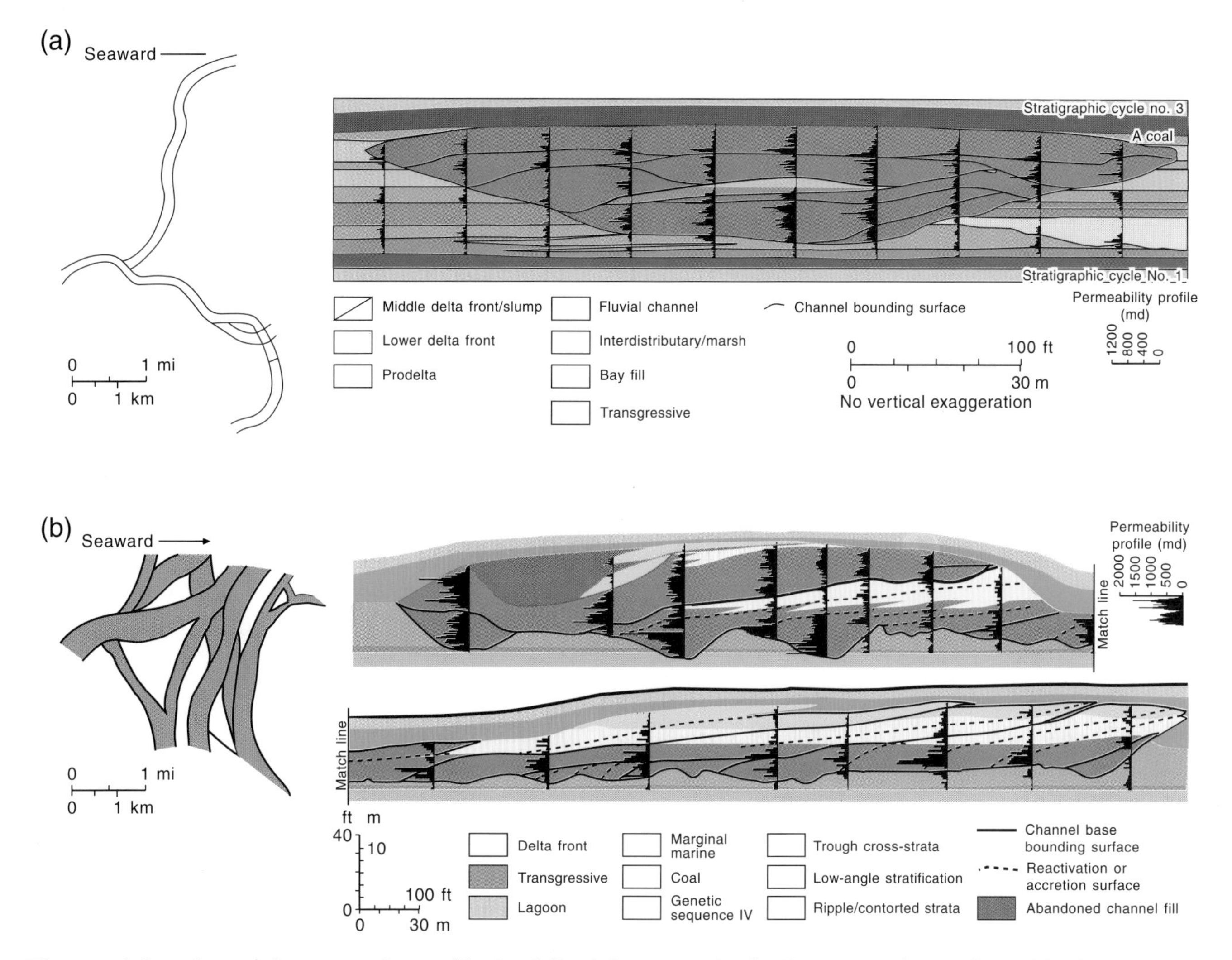

Figure 3. Map view and cross sections of incised fluvial systems in the Ferron sandstone from (a) a low-accommodation (cycle 2) setting at the I-70 outcrop and (b) a high-accommodation (cycle 5) setting at the Muddy Creek outcrop. Note vertical stacking of channelforms in (a), contrasted with lateral stacking in (b). From Knox and McRae (1995), reprinted here with permission of GCAGS.

(units 1, 2, and 3) and transgressive deposits are thin or absent within these units. This progressive change in depositional style results in changes in sandstone volume, architecture, and heterogeneity in both the fluvial and deltaic depositional systems throughout both intermediate- and low-frequency depositional cycles.

CYCLE-TO-CYCLE VARIABILITY IN THE SUBSURFACE

Frio Sandstone Setting

The Oligocene-age Frio reservoirs of Tijerina-Canales-Blucher (T-C-B) field, Jim Wells County, Texas, lie within the Norias deltaic system and Gueydan fluvial system (Figure 5), as identified by Galloway (1982). The Frio has been divided into the upper, middle, and lower informal members (Galloway, 1986). Each member spans 1–2 m.y., corresponds to a 3rd-order depositional unit as defined by Mitchum and Van Wagoner (1991), and is considered a low-frequency (LF) cycle in this study. Further subdivision, down to the approximate 4th-order level, was accomplished by correlating prominent maximum flooding surfaces. On the basis of their occurrence from 6 to 10 times within a 3rd-order unit, the IF units bounded by these surfaces are assumed to span approximately 0.1–0.6 m.y. These flooding surfaces are correlated from the downdip marine interval, where they can be more easily identified on well logs, into the updip, nonmarine area of T-C-B field using the 100-km-long dip-oriented stratigraphic cross section shown in Figure 6. Confidence in these long-distance well-log correlations is moderate to high. Marine flooding surfaces identified downdip correlate to the widespread floodplain shales in the upper delta plain with an estimated accuracy of ±5 m stratigraphically. On the basis of these correlations, the Scott and Whitehill reservoirs were found to lie within a single IF unit, with the approximate base of the IF unit placed below

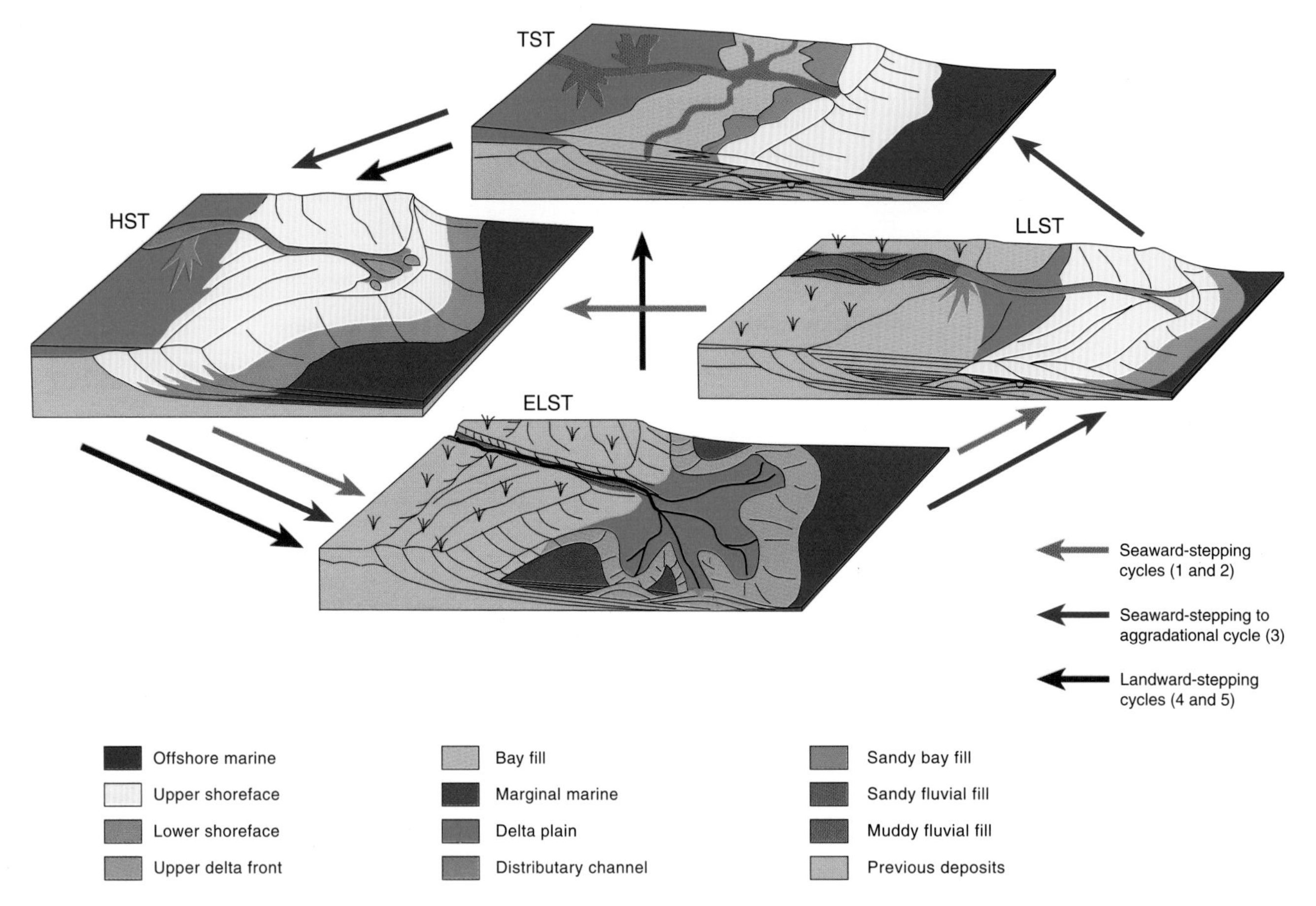

Figure 4. Model showing evolution of Ferron depositional system during an intermediate-frequency relative sea-level cycle in the Ferron sandstone. Wave-dominated shoreline systems develop during rising relative sea level, whereas fluvial incision and river-dominated shoreline systems develop during falling relative sea level. Intermediate-frequency genetic units in seaward-stepping portions of the Ferron lack substantial transgressive deposits. TST = transgressive systems tract, HST = highstand systems tract, ELST = early lowstand systems tract, LLST = late lowstand systems tract. From Barton (1997), reprinted here with permission of GCS SEPM Foundation.

the Whitehill and its approximate top being just above the Scott ("Study interval" in Figure 6); therefore, the Whitehill represents the earliest preserved deposits of sandstone in the T-C-B area during this depositional cycle following maximum flooding, and the Scott represents the last deposits prior to the next maximum flooding. Knox and McRae (1995) interpreted both the Scott and underlying Whitehill reservoirs as having been deposited in a fluvial-influenced upper delta-plain environment on the basis of regional setting, lack of brackish or marine microfauna within interbedded shales, and blocky to upward-fining log signature.

Architecture of Reservoirs within the Scott/Whitehill IF Unit

The general stratigraphy and architecture of the Scott/Whitehill reservoir interval have been deduced from detail-scale well-log correlation across the area of T-C-B field. The interval is subdivided into four HF units by laterally continuous surfaces that may correspond to minor marine flooding or climate-change-induced rejuvenation of the fluvial system (Figure 7). Each HF unit ranges in thickness from 6 to 15 m, with each successive unit generally thickening from the lower Whitehill at the base (6 m thick) through the upper Scott (15 m thick) at the top. Assuming equivalent time spans for each unit, this would suggest persistently increasing rates of accommodation.

Sandstones within the Scott/Whitehill interval display symmetrical, blocky, or upward-fining log patterns, range in thickness from 1 to more than 15 m, and are separated by siltstones and mudstones of similar thickness. Thicker sandstones consist of amalgamated individual channel deposits, each of which reaches a maximum of 6 m in thickness. Depositional facies identified on the basis of log character include sandy channel deposits, silty to muddy abandoned channel fill, rare sandy splay deposits, silty levee deposits, and fine-grained floodplain mudstones (Figure 7).

The lowermost HF unit, referred to as the lower Whitehill unit, is the earliest preserved deposit of the depositional cycle. It is composed entirely of floodplain mudstone throughout the study area. Correlation

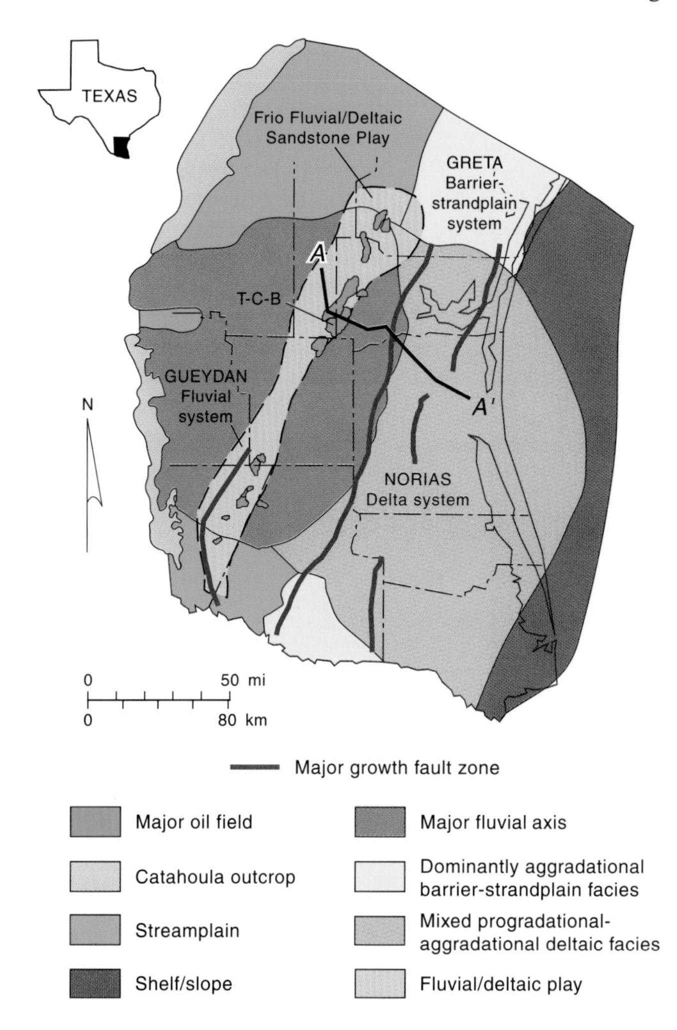

Figure 5. Location of the Tijerina-Canales-Blucher (T-C-B) field within the Frio fluvial-deltaic sandstone (Vicksburg fault zone) play in south Texas. Also shown are the major Frio depositional systems, as defined by Galloway (1982). Cross section A-A′ is shown in Figure 6. From Knox and McRae (1995), reprinted here with permission of GCAGS.

outside of the T-C-B field area shows no widespread sandstones at this stratigraphic level, but the existence of narrow localized channel deposits has not been ruled out because channel bodies may be narrower than the well spacing used for regional correlation (approximately 2 km apart).

The overlying IF unit, the upper Whitehill unit, consists of three relatively narrow (1.5 km wide), but generally thin, fluvial channel-belt deposits (Figure 8) separated by large areas of floodplain mudstone. These channel belts are generally dip oriented and are typically less than 6 m in thickness. Greater thicknesses are the result of vertical stacking of channel belts. One anomalously narrow channel belt (0.5 km wide) is observed at the base of the interval (stratigraphically lowest sandstone in Figure 7). The overlying, comparatively broader, channel belts in the middle and upper portions are interpreted to contain two to three incomplete, vertically amalgamated channel deposits, each ranging from 1.5 to 3 m in thickness. Abandoned channel mudstones are more common in the uppermost channel deposits.

The overlying HF unit, the lower Scott, is similar to the upper Whitehill except that the channel belts tend to be broader (2.5 km wide, compared to the 1.5-km width of the upper Whitehill). Overall, the lower Scott contains a greater percentage of sandstone than the underlying upper Whitehill interval.

The upper Scott HF unit, at the top of the IF unit and, thus, presumably in the most strongly landward-stepping portion of the depositional cycle (high accommodation), differs markedly from the underlying intervals. It is distinctly thicker and sandier than underlying units, with a single broad channel belt (5.5 km wide) that covers the entire study area and consists of vertically amalgamated channel deposits (Figure 8). Dip-oriented bodies of sandstone having thicknesses in excess of 9 m are the result of two or three vertically amalgamated channel deposits. The geometry of individual channelforms is not resolvable with well logs, probably because the width of channels is less than the typical well spacing. The uppermost portion of the upper Scott unit is dominated by siltstones and mudstones of abandoned channel, levee, and floodplain deposits (Figure 7).

In summary, from the lower Whitehill HF unit at the base of the IF depositional cycle (comparatively lower accommodation) to the upper Scott HF unit at the top (highest accommodation?), there is a progressive change in unit thickness, net sandstone percent, and channel architecture. Individual HF units thicken upward through the IF unit, and net sandstone percentage increases. Additionally, channel belts become wider in each stratigraphically higher unit, and volumes of fine-grained channel-fill deposits, such as upper point bar and abandoned channel fill, increase upward. This variability of vertically adjacent HF units occurs at the same scale as the delta-front variability discussed in Ferron outcrops.

Internal Heterogeneity of Scott/Whitehill Reservoirs

Although the gross architecture of the reservoir compartments (channel belts) can be identified with reasonable accuracy from well-log control, intracompartment heterogeneity cannot. This is because both lateral and vertical boundaries between individual channelforms may occur between wells (Figure 7). In outcrop, these surfaces can be clearly traced out at the interwell scale, whereas in a subsurface setting, internal heterogeneity is difficult to recognize and sometimes can only be inferred from production histories. The relative size of areas drained in a series of reservoirs is a measure of the internal complexity of the reservoirs, assuming similar drive mechanisms and fluid viscosities.

Reservoir status maps of the upper Whitehill and upper Scott intervals (Figure 8) document potential compartment boundaries and past completions. Structural closure occurs in areas of subtle highs on a gradually plunging anticlinal nose. The following

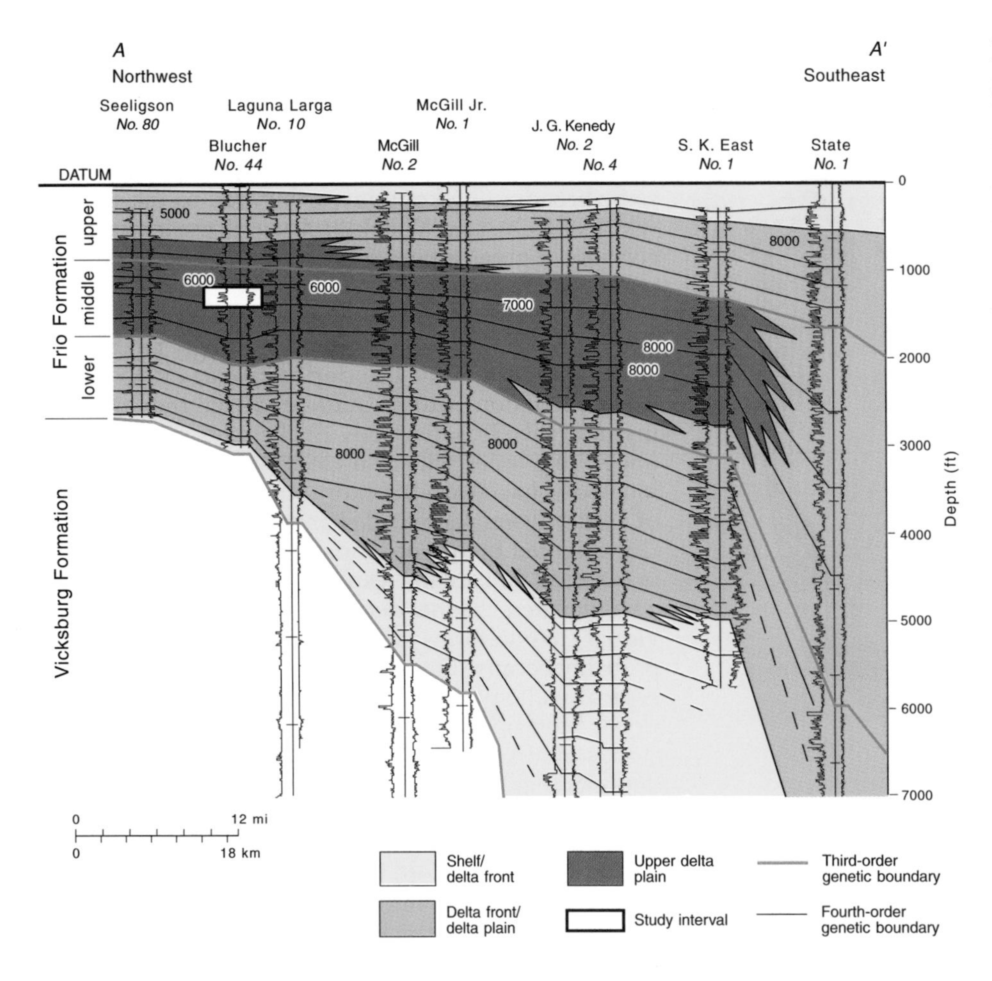

Figure 6. Dip-oriented stratigraphic cross section A-A′ through the Frio Formation from updip of T-C-B field down to the present coastline. The Frio consists of three third-order genetic units, each containing many fourth-order units (not all of which are shown). See Figure 5 for cross-section location. From Knox and McRae (1995), reprinted here with permission of GCAGS.

discussion summarizes the production history of the two reservoirs and estimates drainage areas for completions.

The upper Whitehill has produced gas from two wells, one on the northern structural crest and another on the north flank (Figure 8c). Perforations in the well on the north flank were structurally below the documented gas/water contact on the structural crest, indicating a permeability barrier or compartment boundary between the two areas. Mapping of channel belts and evidence of tightly carbonate-cemented sandstone in one well located at a channel-belt boundary between the crest and north flank (Figure 8c) suggest that the northernmost of the two completions may be stratigraphically isolated from the structural crest. Volumetric calculations and evidence from recent wells postdating production in the crestal completion, which accumulated 6 Gcf of gas, indicate that the crestal completion drained approximately 40 ac (16 ha). The well on the north flank accumulated only 0.2 Gcf, probably because of its proximity to the gas/water contact. Resistivities suggesting untested oil or gas accumulations occur in wells within isolated channel belts at depths below the crestal accumulation, indicating a component of stratigraphic trapping may be present in the upper Whitehill.

Eight wells have produced oil from the upper Scott HF unit on the main structural crest (Figure 8d). Cumulative production has ranged from less than 1,000 bbl to more than 54,000 bbl per well, with all but one well accumulating less than 20,000 bbl. Initial water cuts are highly variable and are independent of structural position and offset production history, indicating a lack of communication between well locations. Volumetric analyses suggest that upper Scott unit completions have drained areas ranging from less than 0.5 ac (0.2 ha) to a maximum of approximately 5 ac (2 ha), averaging about 1 ac (0.4 ha). This area is significantly less than the 40-ac (16-ha) area drained by the completion in the Whitehill. Calculations indicate that despite completion at a 20–40-ac (8–16-ha) spacing, and the fact that all current completions are either abandoned, idle, or nearly watered out, less than 10% of the original oil in place has been recovered from the Scott.

Although some of the difference in recovery from the Whitehill and the Scott can be attributed to the different mobility ratios of oil and gas, a significant part is attributed to smaller compartment sizes in the Scott zone. This indicates that the upper Scott channel belt (upper portion of the depositional cycle, higher accommodation) is much more internally heterogeneous than the upper Whitehill channel belts (lower portion of the depositional cycle, lower accommodation). In other words, boundaries between individual channelforms in upper Scott channel belts act as baffles or barriers to fluid flow and impede reservoir drainage.

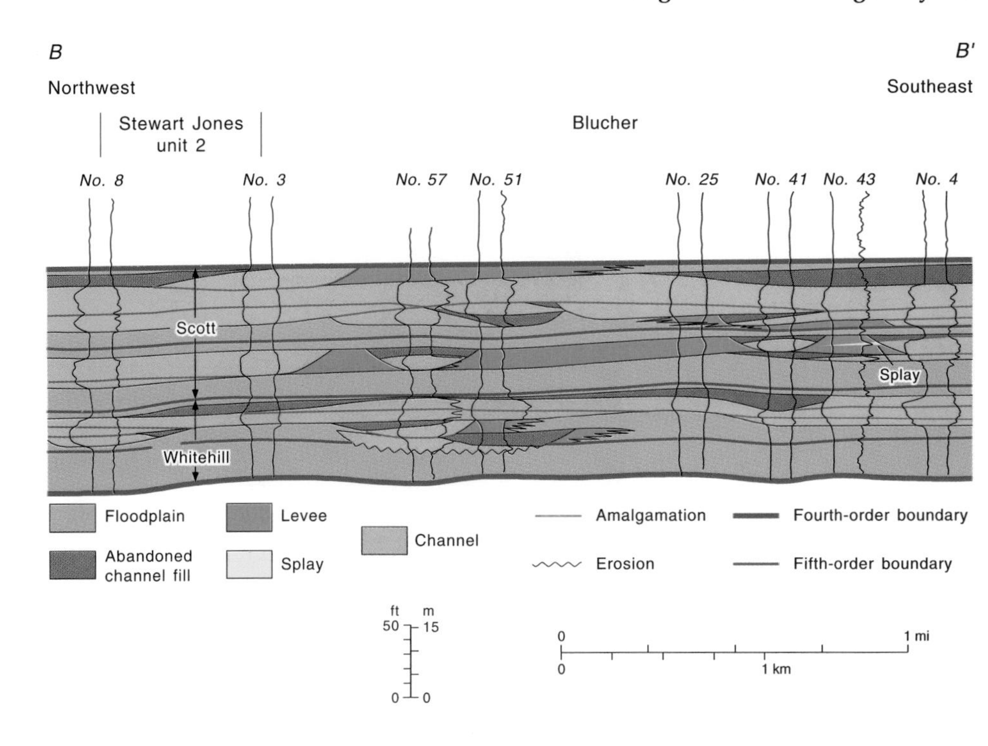

Figure 7. Dip-oriented cross section B-B′ showing the Scott/Whitehill fourth-order genetic unit, which contains at least four fifth-order genetic units. These are, from top down, the upper and lower Scott and the upper and lower Whitehill reservoir intervals. Cross-section location is given in Figure 8. From Knox and McRae (1995), reprinted here with permission of GCAGS.

Scott/Whitehill Summary

As expected from outcrop observations of the Ferron Sandstone, a spectrum of reservoir styles exists within the upper delta-plain deposits of a single IF unit in T-C-B field (Figure 9). Reservoirs within HF units range from moderately narrow, internally homogeneous channel belts deposited in the lowest part of an IF depositional cycle to broad, internally heterogeneous channel belts in the upper part of an IF depositional cycle, just as Ferron delta-front deposits varied from fluvial- to wave-dominated in style within a single IF unit. These reservoir styles contain varying reserve-growth potential and require very different strategies for optimal development. In general, fluvial-influenced upper delta-plain reservoirs in the uppermost HF units contain the greatest reserve-growth potential and may require the tightest well spacings for optimal reservoir drainage, whereas those in the lower HF units may contain narrow stratigraphically isolated accumulations that are internally homogeneous and present stepout opportunities in mature fields.

DISCUSSION

Potential Causes of Ferron and Frio Variability

In combination, observations from Ferron outcrop and Frio subsurface studies document changes in sandstone geometry and internal heterogeneity from high-frequency cycle to high-frequency cycle resulting from changes in depositional style in delta-front/lower delta-plain, upper delta-plain, and incised fluvial systems. To be able to predict these changes in the subsurface, we must attempt to understand the root causes of this variability. Potential causes include (1) changing preservation potential related to raising or lowering of relative sea level and (2) the changing balance of energies in fluvial input and the receiving basin. The changing energy balance may be related to (1) a changing climate that alters fluvial volumes and loads; (2) altered wave or tidal energies caused by varying ocean basin physiography, which is driven by antecedent shelf/slope/basin physiography that governs water depths as relative sea level changes; (3) changing wave-energy patterns, which may be related to changing climate patterns; or (4) some combination of all three factors.

Variability in Ferron incised fluvial systems from seaward-stepping to landward-stepping IF units has been suggested by both Gardner (1993) and Barton (1995) as being a consequence of varying preservation potential that is ultimately tied to relative sea-level fluctuations. In IF cycles dominated by progradation, stacking patterns indicate limited accommodation. Here, successive scour removes the upper, finer grained part of any channel fill deposited, resulting in a more lithologically and petrophysically homogeneous incision fill. Stacking patterns in landward-stepping IF units indicate generally greater accommodation. Increased aggradation preserves more deposited mud, and decreased fluvial gradients may result in greater lateral migration (meandering), increasing the width of the incised fluvial feature. Lateral migration into any preserved overbank mud contributes a high volume of poorly indurated mud clasts into the channel lag, further increasing permeability contrasts at channelform-on-channelform contacts. Alternately, climate changes in the source area could alter the balance of bedload and suspended load and change channel sinuosity; however, lack of extensive cannibalization of channelforms in landward-stepping fluvial fills indicates that accommodation represents a strong control on heterogeneity within incised fluvial systems.

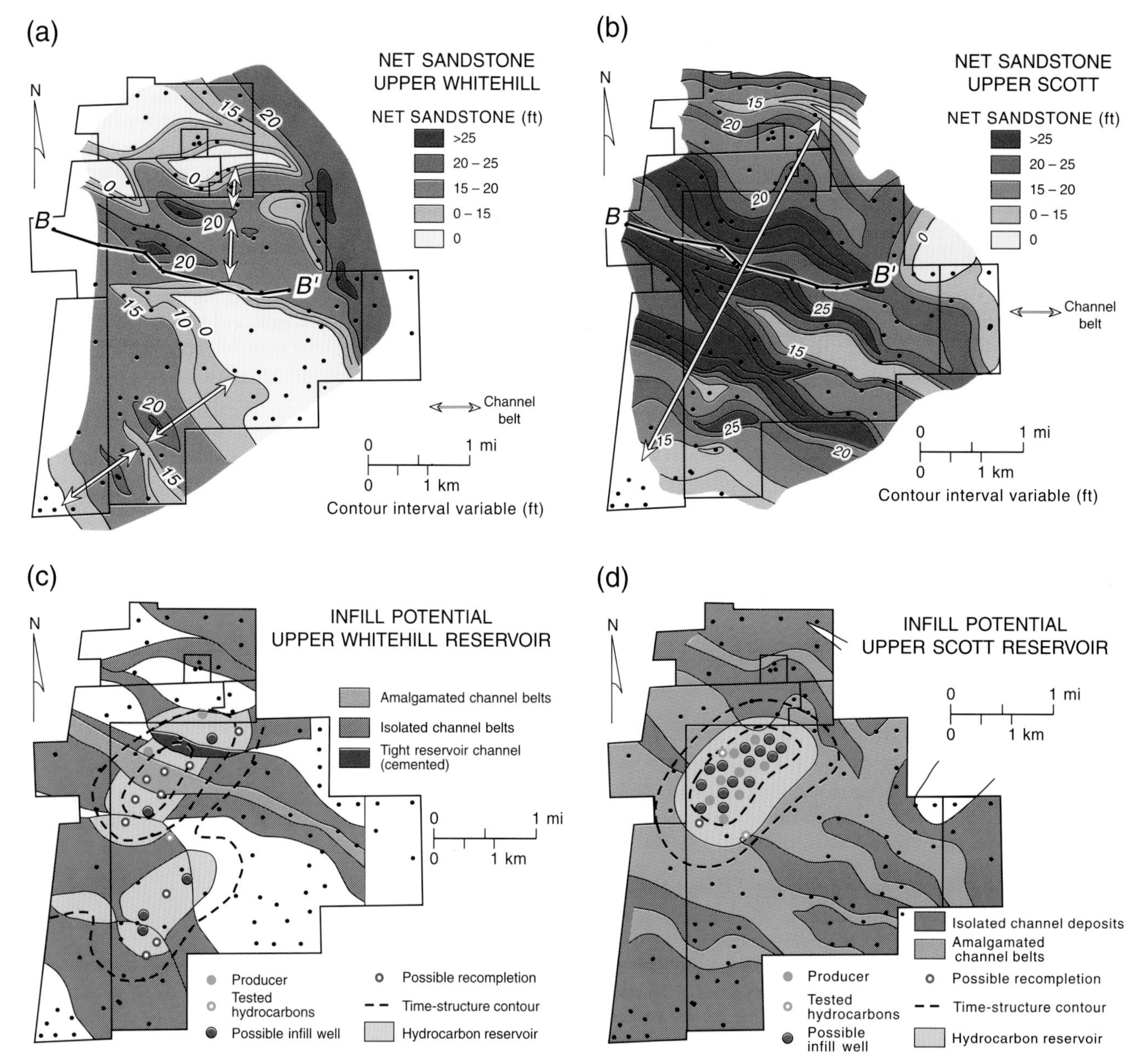

Figure 8. Net sandstone maps for the (a) upper Whitehill and (b) upper Scott reservoirs and compartment maps of the (c) upper Whitehill and (d) upper Scott reservoirs. Narrow, stratigraphically isolated channel belts of the upper Whitehill interval were deposited in the proximal fluvial-influenced upper delta plain during the early part of a depositional cycle, whereas the broad, internally complex channel belt of the upper Scott interval was deposited during the latest portion of a depositional cycle. From Knox and McRae (1995), reprinted here with permission of GCAGS.

In Frio upper delta-plain channel belts, variability between HF units is also probably a function of changing accommodation or, more accurately, floodplain aggradation potential possibly linked to changing base level (relative sea level). The interpretation of widespread floodplain shales correlating to maximum marine flooding in the Frio parallels observations made in Ferron coastal-plain deposits (Gardner, 1993). The lower part of these Frio shales may represent floodplain aggradation of a highly sinuous system active during highstand, with the middle representing limited to negative floodplain aggradation and sand bypass during subsequent falling and lowstand of relative sea level. Mud with minor, stratigraphically equivalent, sandstones in the upper parts of these shales may represent initial sea-level rise that resulted in increased coastal-plain aggradation and limited sand deposition. Successive HF units (e.g., upper Whitehill through upper Scott) would exhibit greater floodplain aggradation, greater preservation potential of overbank shales, and, consequently, more sinuous channelforms. This would contribute to progressively wider channel belts and greater internal heterogeneity, as observed in Ferron incised fluvial systems. Another possible explanation

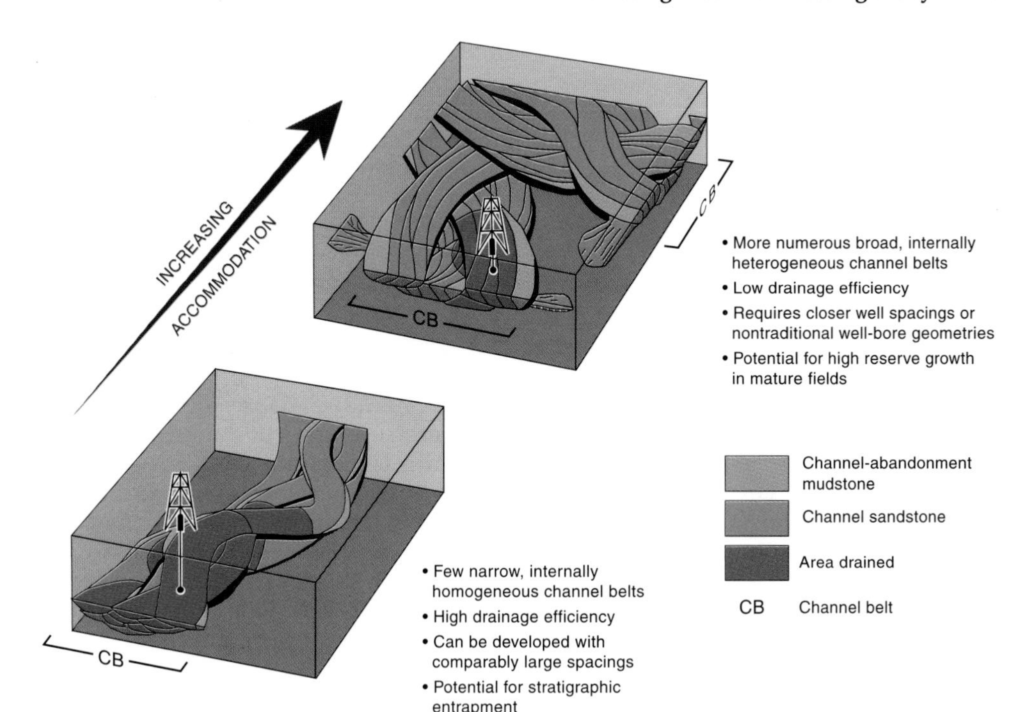

Figure 9. Spectrum of fluvial-influenced upper delta-plain reservoir architecture, internal heterogeneity, production characteristics, and reserve-growth potential in the Scott/Whitehill interval. From Knox and McRae (1995), reprinted here with permission of GCAGS.

for channel-belt width variability may be climatic changes resulting in reorganization of discharge patterns. Blum et al. (1995) attributed greater muddy overbank deposition in Holocene Texas coastal fluvial systems to increases in discharge flashiness resulting from short-term (20 k.y.) climate changes in the source area. A greater frequency of tropical storms during interglacial periods resulted in short periods of high runoff, which led to progressive removal of soil mantles that had been built during less erratic rainfall patterns of glacial periods. Flashy discharge and increased suspended load in rivers may have increased overbank deposition, increasing sinuosity. Similar changes in the Frio source area may have contributed to increased sinuosity, more lateral migration, and wider channel belts. These relationships, however, are difficult to verify in the subsurface study. In any event, successive upward increases in HF unit thickness indicate increasing floodplain aggradation, which is achieved by a rise in base level or, in the case of a coastal plain, relative sea level.

Controls on delta-front style in the Ferron sandstone would seem to be most related to the changing balance of fluvial and wave energy among HF cycles. Although Cross et al. (1993) attributed this style of cycle-to-cycle variability to changing accommodation, other factors, such as decreased preservation potential and climate change, may also contribute to variability. As Cross et al. (1993) argued, decreased preservation potential on the coastal plain increases the sediment volume delivered to the delta front, resulting in a fluvial-dominated style. Alternately, climate changes in the source area might increase discharge or the percentage of suspended load, or lowered relative sea level could create a broad shallow shelf that would effectively dissipate wave energy, either of which would also tip the balance toward fluvial domination. Climate change leading to decreased storm wave energy, and therefore less wave reworking of sediments, may have a similar effect.

Wave-dominated delta fronts deposited during late lowstand might result from (1) progradation into deeper water (and consequently greater wave energy) caused by a more steeply dipping shelf or rising relative sea level; (2) reduced sediment delivery resulting from greater entrapment of sediment on the coastal plain as relative sea level rises, decreasing river gradients; or (3) increased storm wave energy induced by climate change. Gardner (1993) indicated increases in sediment volume within the landward-stepping IF units, suggesting some influence of relative sea level through increased preservation potential and decreased delivery during backstepping phases; however, the strong correlation of wave-dominated delta fronts with increasing shelf water depths suggests, at the least, that increasing water depth is a contributing factor. An evaluation of the influence of changing climate and resulting effects on river discharge and storm wave energy is not possible with the data collected in the Ferron outcrop study.

Application to Other Reservoirs and Stratal Successions

The direct application of Ferron and Frio observations to other fluvial or delta-front stratal successions deposited in other basins, other parts of the same basins, or within the same basins at different times should be made with caution. Although the influence of preservation potential on channel systems (incised fluvial systems and fluvial-influenced upper delta plains) that results from changing base level seems significant, the potential impact of high-frequency climate change can be substantial and should be studied further. Additionally, determining whether an IF unit will have the greatest volume of sand at the base (as in Ferron incised features) or at the top (as in Frio upper delta-plain deposits) depends on identifying the relative dip position and correctly establishing the connection to the marine flooding surfaces. For example, fill in downdip incised fluvial features probably will be dominated by fluvial sands overlain by muddy transgressive deposits, as demonstrated by Shanley and McCabe (1994). In contrast, updip fluvial systems may contain upward-increasing volumes of sand locally, as demonstrated in our Frio example. Another factor important when applying observations to rocks from different time periods is the consequence of varying eustatic amplitude on fluvial style, the most likely impact of which might be on incised feature width:depth ratios and the distance of upstream nickpoint migration.

Caution must also be exercised when applying Ferron observations of varying delta-front style through an IF cycle. As Cross et al. (1993) stated, they commonly observed fluvial-dominated delta-front styles in seaward-stepping units, as was seen in the Ferron; however, they also noted several observations of wave-dominant styles during relative sea-level fall with tide-dominant styles during relative sea-level rise. This underscores the importance of making sound observations from core or other high-resolution techniques across a number of stratigraphic units before beginning to predict subsurface relationships between facies and position within a cycle.

Implications

The successful prediction of interwell reservoir architecture and heterogeneity has implications at all stages of field life, from exploration through initial development on into revitalization of a mature field. In the exploration stage, prediction of reservoir quality is an obvious component of prospect evaluation. In addition to the current practice of predicting sand quality from systems tract information, attention to depositional facies and position in a cycle will provide clues as to the internal heterogeneity of a prospective interval. Thus, if two otherwise equal incised fluvial prospects are being evaluated, the prospect within the seaward-stepping IF unit might be preferred over one in a landward-stepping IF unit because of the greater internal heterogeneity expected, which translates to greater recovery factor.

During initial field development, attention to facies and cycle position can improve the prediction of efficient well spacing. A delta-front reservoir that can be predicted early in the development process to be more wave-dominated in style can be planned with fewer completion locations than adjacent, more heterogeneous, fluvial-dominated reservoirs. Improved development planning, accounting for all reservoirs and their anticipated heterogeneity, may decrease the number of total wells needed in a field. Alternatively, it may lead to earlier recognition of the need for greater completion density in some zones, resulting in the need for more wells and, for example, more slots in the design of a platform. Additionally, the earlier development of more homogeneous reservoirs can increase early cash flow, improving overall field economics.

In mature fields, reservoirs most likely to contain near-term reserve-growth potential are those with significant interwell heterogeneity. Recognition of these reservoirs can help focus characterization studies and future development toward those most likely to contain greater volumes of unrecovered mobile oil. The total U.S. target of such studies stands at more than 10 Gbbl of oil (U.S. Department of Energy, 1991). For companies planning acquisitions of mature fluvial-deltaic reservoirs, the success of such ventures might be significantly improved by taking steps to predict reservoir heterogeneity during the initial evaluation process, enabling operators to accurately identify the reserve-growth potential of the various reservoirs or fields under consideration. In either acquisition or characterization, the recognition of heterogeneous reservoirs likely to contain significant volumes of undrained mobile oil will result in near-term reserve growth and revitalization of mature fluvial-deltaic reservoirs. Such revitalization can prevent the permanent loss of oil resources that occurs through premature abandonment of reservoirs or fields.

CONCLUSIONS

Outcrop observations from the Cretaceous Ferron sandstone of central Utah demonstrate that facies associations within incised-fluvial and delta-front depositional settings vary progressively through both intermediate- and high-frequency depositional units. Systematic facies variability is exemplified by changes from narrow, internally homogeneous incised fluvial fills (characterized by vertical stacking of channelforms in IF units in the seaward-stepping part of a LF unit) to broad, internally heterogeneous incised fluvial fills (that exhibit lateral stacking of channelforms in IF units in the landward-stepping part of a LF unit). HF units vary within single IF units. For example, delta-front deposits in HF units progress through an IF cycle from laterally heterogeneous fluvial-dominated successions in the early lowstand to more homogeneous, strike-oriented, wave-dominated successions in the late lowstand.

Systematic cycle-to-cycle changes in architecture and heterogeneity are also observed in Oligocene Frio Formation reservoirs of the south Texas Gulf Coast. Changes in facies associations in subsurface reservoirs are accompanied by dramatic changes in production behavior, requiring different reservoir management approaches and presenting varying opportunities for significant reserve-growth potential. Upper delta-plain fluvial channel belts deposited early in a depositional cycle are narrow, internally homogeneous, and may present reserve-growth opportunities in off-crest, stratigraphically isolated settings. In contrast, channel belts deposited late in the cycle are broad, internally heterogeneous, and poorly contacted by typical completion spacings, resulting in significant remaining resources following conventional development.

Facies variability in these fluvial-deltaic successions corresponds to the position of a deposit within a depositional cycle. Although accommodation and aggradation potential appear to be strong controlling factors, changes in sediment supply, climate, and physiography also may exert controls on deposition and preservation. Application of Ferron and Frio observations to stratal successions in other basins, other parts of the same basin, or at other geologic times should be made with caution. Relationships between depositional style and position within a cycle may vary across a basin, from basin to basin, or through time because of differing eustatic amplitude, altered balance between fluvial energy and wave/tide energies of the receiving basin, differing climates and source terrain that can change fluvial discharge and load, as well as basin energies, differing basin and shelf topographies, and differing positions along depositional dip. Careful evaluation of applicable outcrop, core, or other high-resolution data within a depositional cycle framework should always be carried out for the specific basin and stratigraphic interval in question before subsurface predictions are made.

Prediction of depositional facies and heterogeneity in a cyclic stratigraphic framework has implications throughout the stages of field life, from exploration to maturity. Prediction of depositional heterogeneity in exploration prospects can focus drilling toward more homogeneous reservoir styles that require lower development costs and provide more rapid recovery of reserves. In initial field development, more accurate prediction of efficient well spacings and identification of more heterogeneous reservoir intervals can improve capital investment projection and field economics. Recognition of heterogeneous reservoirs in mature fields can focus acquisition and characterization efforts to improve near-term reserve growth and prevent permanent loss of resources through premature abandonment. The target for this effort in U.S. fluvial-deltaic reservoirs is estimated at nearly 11 Gbbl of mobile oil (U.S. Department of Energy, 1991).

ACKNOWLEDGMENTS

Studies contributing to this manuscript were partially funded by the U.S. Department of Energy under contract No. DE-FC22-93BC14959 and by the Gas Research Institute under contract No. 5089-260-1902. Mobil Exploration and Producing, U.S., provided subsurface data that made the Frio reservoir characterization study possible. Project supervisors included William Fisher, Noel Tyler, Shirley Dutton, Raymond Levey, and Richard Major of The University of Texas at Austin, Richard Parker, Paul Wescott, and Anthony Garody of GRI, and Edith Allison and Chandra Nautiyal of the U.S. DOE. Assistance was provided by Ted Angle, Douglas Dawson, and Radu Boghici. Technical editing and comments were provided by Steven Seni, Janok Bhattacharya, Brian Willis, Tucker Hentz, and Richard Major. A review by Romeo Flores significantly improved the manuscript. Discussions with Lee McRae, Michael Gardner, and Tom Ryer also benefited the manuscript. Drafting was done by Randy Hitt, Kerza Prewitt, Joel Lardon, and Michelle Bailey under the direction of Richard Dillon and Joel Lardon. Selected figures reprinted with permission of Gulf Coast Association of Geological Societies and Gulf Coast Section of Economic Paleontologists and Mineralogists Foundation (GCS SEPM Foundation). Editing was done by Nina Redmond under the supervision of Susann Doenges, and word processing was done by Susan Lloyd. Publication authorized by the Director, Bureau of Economic Geology, The University of Texas at Austin.

REFERENCES CITED

Allen, J.R.L., 1983, Studies in fluviatile sedimentation: Bars, bar-complexes and sandstone sheets (low-sinuosity braided streams) in the Brownstones (L. Devonian), Welsh Borders: Sedimentary Geology, v. 33, p. 237-293.

Barton, M.D., 1994, Outcrop characterization of architecture and permeability structure in fluvial-deltaic sandstones, Cretaceous Ferron Sandstone, Utah: University of Texas at Austin, Ph.D. dissertation, 259 p.

Barton, M.D., 1995, Sequence stratigraphy, facies architecture, and permeability structure of fluvial-deltaic reservoir analogs: Cretaceous Ferron Sandstone, Central Utah (Ferron GRI Fieldtrip Guidebook), Bureau of Economic Geology, The University of Texas at Austin, Austin, Texas, 139 p.

Barton, M.D., 1997, Application of Cretaceous Interior Seaway outcrop investigations to fluvial-deltaic reservoir characterization: Part I, predicting reservoir heterogeneity in delta front sandstones, Ferron gas field, central Utah, *in* K.W. Shanley and B.F. Perkins, eds., Shallow marine and nonmarine reservoirs: sequence stratigraphy, reservoir architecture and production characteristics: Gulf Coast Section Society of Economic Paleontologists and Mineralogists Foundation Eighteenth Annual Research Conference, p. 33-40.

Blum, M.D., R.A. Morton, and J.M. Durbin, 1995, "Deweyville" terraces and deposits of the Texas Coastal Plain: Gulf Coast Association of Geological Societies Transactions, v. 45, p. 53–60.

Cross, T.A., et al., 1993, Applications of high-resolution sequence stratigraphy to reservoir analysis, *in* R. Eschard and B. Doligez, eds., Subsurface reservoir characterization from outcrop observations: Editions Technip, Paris, p. 11-33.

Galloway, W.E., 1982, Depositional architecture of Cenozoic Gulf Coastal Plain fluvial systems, *in* F.G. Ethridge and R.M. Flores, eds., Recent and ancient nonmarine depositional environments: models for exploration: Society of Economic Paleontologists and Mineralogists, Special Publication 31, p. 127-155.

_________ 1986, Depositional and structural framework of the distal Frio Formation, Texas coastal zone and shelf: The University of Texas at Austin, Bureau of Economic Geology Geological Circular 86-8, 16 p.

_________ 1989a, Genetic stratigraphic sequences in basin analysis I: architecture and genesis of flooding-surface bounded depositional units: American Association of Petroleum Geologists Bulletin, v. 73, p. 125-142.

_________ 1989b, Genetic stratigraphic sequences in basin analysis II: application to northwest Gulf of Mexico Cenozoic Basin: American Association of Petroleum Geologists Bulletin, v. 73, no. 2, p. 143-154.

Gardner, M.N., 1993, Sequence stratigraphy and facies architecture of the Upper Cretaceous Ferron Sandstone Member of the Mancos Shale, East-Central Utah: Colorado School of Mines, Ph.D. dissertation, 528 p.

Goldhammer, R.K., P.A. Dunn, and L.A. Hardie, 1990, Depositional cycles, composite sea-level changes, cycle stacking patterns, and the hierarchy of stratigraphic forcing: examples from Alpine Triassic platform carbonates: Geological Society of America Bulletin, v. 102, p. 535-562.

Hale, L.A., 1972, Depositional history of the Ferron Formation, central Utah, in plateau basin and range transition zone: Utah Geological Association, p. 115-138.

Jervey, M.T., 1988, Quantitative geological modeling of siliciclastic rock sequences and their seismic expression, *in* C.K. Wilgus, B.S. Hastings, C.G.St.C. Kendall, H.W. Posamentier, C.A. Ross, and J.C. Van Wagoner, eds., Sea-level changes: an integrated approach: Society of Petroleum Engineers Special Publication 42, p. 47-69.

Kerans, C., and W.M. Fitchen, 1995, Sequence hierarchy and facies architecture of a carbonate-ramp system: San Andres Formation of Algerita Escarpment and Western Guadalupe Mountains, West Texas and New Mexico: The University of Texas at Austin, Bureau of Economic Geology Report of Investigations No. 235, 86 p.

Knox, P.R., and L.E. McRae, 1995, Application of sequence stratigraphy to the prioritization of incremental growth opportunities in mature fields: an example from Frio fluvial-deltaic sandstones, TCB field, South Texas: Gulf Coast Association of Geological Societies Transactions, v. 45, p. 341-359.

Lupton, C.T., 1916, Geology and coal resources of Castle Valley in Carbon, Emery, and Sevier counties, Utah: U.S. Geological Survey Bulletin 628, 88 p.

Miall, A.D., 1985, Architectural-element analysis: a new method of facies analysis applied to fluvial deposits: Earth-Science Reviews, v. 22, p. 261-308.

Mitchum, R.M., and J.C. Van Wagoner, 1991, High-frequency sequences and their stacking patterns: sequence-stratigraphic evidence of high-frequency eustatic cycles: Sedimentary Geology, v. 70, p. 131-160.

Posamentier, H.W., M.T. Jervey, and P.R. Vail, 1988, Eustatic controls on clastic deposition conceptual framework, *in* C.K. Wilgus, B.S. Hastings, C.G.St.C. Kendall, H.W. Posamentier, C.A. Ross, and J.C. Van Wagoner, eds., Sea-level changes: an integrated approach: Society of Economic-Paleontologists and Mineralogists Special Publication No. 42, p. 109-124.

Posamentier, H.W., and P.R. Vail, 1988, Eustatic controls on clastic deposition II—sequence and systems tract models, *in* C.K. Wilgus, B.S. Hastings, C.G.St.C. Kendall, H.W. Posamentier, C.A. Ross, and J.C. Van Wagoner, eds., Sea-level changes: an integrated approach: Society of Economic Paleontologists and Mineralogists Special Publication No. 42, p. 125-154.

Ryer, T.A., 1981, Deltaic coals of the Ferron Sandstone Member of the Mancos Shale: Predictive model for Cretaceous coal-bearing strata of the Western Interior: American Association of Petroleum Geologists Bulletin, v. 65, no. 11, p. 2323-2340.

Ryer, T. A., 1993, The autochthonous component of cyclicity in shoreline deposits of the Upper Cretaceous Ferron Sandstone, central Utah, American Association of Petroleum Geologists Bulletin (Abs.), v. 77, p. 175.

Shanley, K.W., and P.J. McCabe, 1994, Perspectives on the sequence stratigraphy of continental strata: American Association of Petroleum Geologists Bulletin, v. 78, p. 544-568.

Sonnenfeld, M.D., 1991, High-frequency cyclicity within shelf-margin and slope strata of the upper San Andres sequence, Last Chance Canyon, *in* S. Meader-Roberts, M.P. Candelaria, and G.E. Moore, eds., Sequence stratigraphy, facies, and reservoir geometries of the San Andres, Grayburg, and Queen formations, Guadalupe Mountains, New Mexico and Texas: Permian Basin Section, Society of Economic Paleontologists and Mineralogists Publication 91-32, p. 11-51.

U.S. Department of Energy, 1991, Opportunities to improve oil productivity in unstructured deltaic reservoirs: Technical summary and proceedings of the technical symposium held at Dallas, Texas, January 1991: U.S. Department of Energy, Office of Fossil Energy, Washington D.C., 163 p.

Vail, P. R., R.M. Mitchum, R.G. Todd, J.M. Widmier, S. Thompson, III, J.B. Sangree, J.N. Bubb, and W.G. Hatlelid, 1977, Seismic stratigraphy and global changes of sea-level, *in* C.W. Payton, ed., Seismic stratigraphic applications to hydrocarbon exploration: American Association of Petroleum Geologists Memoir 26, p. 49-212.

Van Wagoner, J.C., H.W. Posamentier, H.W. Mitchum, P.R. Vail, J.F. Sarg, T.S. Loutit, and J. Hardenbol, 1988, An overview of the fundamentals of sequence stratigraphy and key definitions, *in* C.K. Wilgus, B.J. Hastings, H. Posamentier, J.C. Van Wagoner, C.A. Ross, and C.G.St.C. Kendall, eds., Sea-level change: an integrated approach: Society of Economic Paleontologists and Mineralogists Special Publication No. 42, p. 39-46.

Watney, W.L., et al., Petrofacies analysis—a petrophysical tool for geologic/engineering reservoir characterization, 1999, *in* R. Schatzinger and J. Jordan, eds., Reservoir Characterization-Recent Advances, AAPG Memoir 71, p. 73–90.

Chapter 6

Petrofacies Analysis—A Petrophysical Tool for Geologic/Engineering Reservoir Characterization

W. L. Watney
W. J. Guy
J. H. Doveton
S. Bhattacharya
P. M. Gerlach
G. C. Bohling
T. R. Carr
Kansas Geological Survey
Lawrence, Kansas, U.S.A.

ABSTRACT

Petrofacies analysis is defined as the characterization and classification of pore types and fluid saturations as revealed by petrophysical measurements of a reservoir. The word "petrofacies" makes an explicit link between petroleum engineers' concerns with pore characteristics as arbiters of production performance and the facies paradigm of geologists as a methodology for genetic understanding and prediction. In petrofacies analysis, the porosity and resistivity axes of the classical Pickett plot are used to map water saturation, bulk volume water, and estimated permeability, as well as capillary pressure information where it is available. When data points are connected in order of depth within a reservoir, the characteristic patterns reflect reservoir rock character and its interplay with the hydrocarbon column. A third variable can be presented at each point on the crossplot by assigning a color scale that is based on other well logs, often gamma ray or photoelectric effect, or other derived variables. Contrasts between reservoir pore types and fluid saturations are reflected in changing patterns on the crossplot and can help discriminate and characterize reservoir heterogeneity.

Many hundreds of analyses of well logs facilitated by spreadsheet and object-oriented programming have provided the means to distinguish patterns typical of certain complex pore types (size and connectedness) for sandstones and carbonate reservoirs, occurrences of irreducible water saturation, and presence of transition zones. The result has been an improved means to evaluate potential production, such as bypassed pay behind pipe and in old exploration wells, or to assess zonation and continuity of the reservoir.

Petrofacies analysis in this study was applied to distinguishing flow units and including discriminating pore type as an assessment of reservoir conformance and continuity. The analysis is facilitated through the use of color-image cross sections depicting depositional sequences, natural gamma ray, porosity, and permeability. Also, cluster analysis was applied to discriminate petrophysically similar reservoir rock.

INTRODUCTION

Reservoir characterization and modeling are ongoing procedures used as the reservoir is developed. Well log data and occasional cores provide the fundamental stratigraphic information critical to delineating flow units, a primary objective of reservoir characterization. Flow units are correlatable and mappable regions in the reservoir that control fluid flow. Their distinction is usually centered on comparing permeability and porosity information. Flow unit classification is refined as fluid recovery, pressure data, or chemical fingerprinting are obtained. Often, particularly in older fields, production is commingled and cannot be used to substantiate flow units. Moreover, the costs of extensive fluid and pressure testing are not economical. The question examined in this paper is whether the traditional definition of flow units can be modified to include additional information obtained from basic suites of well logs. An approach referred to as "petrofacies analysis" is described that extends the use of well logs to maximize information that relates to pore type and fluid flow. In particular, the utility of distinguishing vertical and lateral trends and patterns of irreducible bulk volume water, water saturation, and porosity is evaluated as a tool to improve the definition of flow units using well logs. Petrofacies are defined as portions of the reservoir that exhibit distinctive geological facies and petrophysical attributes.

Selecting flow units from core and log data is subjective, due to judging whether reservoir conformance (interconnection) and lateral continuity exist without actual fluid flow information (Willhite, 1986). Consistent, explicitly defined methodological steps must be developed to ensure that each well is treated similarly to make the approach robust and to help ensure that the procedure can be repeated and improved as more information becomes available.

The initial task is correlating the reservoir interval and establishing stratigraphic subdivisions and lithofacies. Next, the correlated stratigraphic intervals are mapped with the subsurface control to test coherency of the data. At this stage, porosity and permeability data are integrated with stratigraphic units and lithofacies to define porous and permeable flow units. This information is then compared with the production and well-test data to check for consistency and correlations.

An intermediate step proposed is to extract further information about pore types and fluid saturations using petrofacies analysis. The analysis is based on the Pickett plot and delineation of depth-based trends and patterns in porosity, resistivity, water saturation, and bulk volume water (BVW). Thousands of analyses of this type have demonstrated a well-known fact, that porosity varies considerably due to varying pore type and capillarity; furthermore, the use of rules-of-thumb values for effective porosity and saturation cutoffs has been deemed inadequate to address today's needs for precise descriptions of reservoirs for use in improved oil recovery operations.

While porosity may vary little, saturations and productivity can be considerably different when pore types change. Alternatively, changes in water saturation and BVW may vary closely with elevation of the reservoir, suggesting fluid continuity and reservoir conformance, as well as serving as an additional tool in evaluating lateral reservoir continuity. Added information on pore types, vertical reservoir conformance, and fluid/reservoir continuity provided by petrofacies analysis is important in assessing flow units and ensuring robust reservoir modeling.

Petrofacies analysis is used in this example to extend an initial stratigraphic analysis of a sandstone reservoir in an attempt to define flow units. The ultimate objective of this reservoir characterization is to conduct a reservoir simulation of the field to help evaluate future production options.

STUDY AREA

Petrofacies analysis was applied to a lower Morrowan (Lower Pennsylvanian) sandstone reservoir in Arroyo field, Stanton County, Kansas (T29S, R41W) (Figure 1). Arroyo field, operated by J.M. Huber Corporation, was discovered in 1992 by subsurface methods. Cumulative production exceeds 651,000 bbl of oil and 21 Gcf of natural gas. Arroyo field is a combination structural stratigraphic trap, currently containing 6240 proven productive acres with 24 oil and gas wells and 3 dry holes. The field contains two reservoirs, the lower Morrowan sandstone and the St. Louis Limestone (oolite). The lower Morrowan sandstone is located at approximately 1715 m (5626 ft) below the surface. The sandstone ranges from 0 to 19 m (0 to 62 ft) thick and is lenticular throughout the field (Figure 2). The porosity of the sandstone ranges

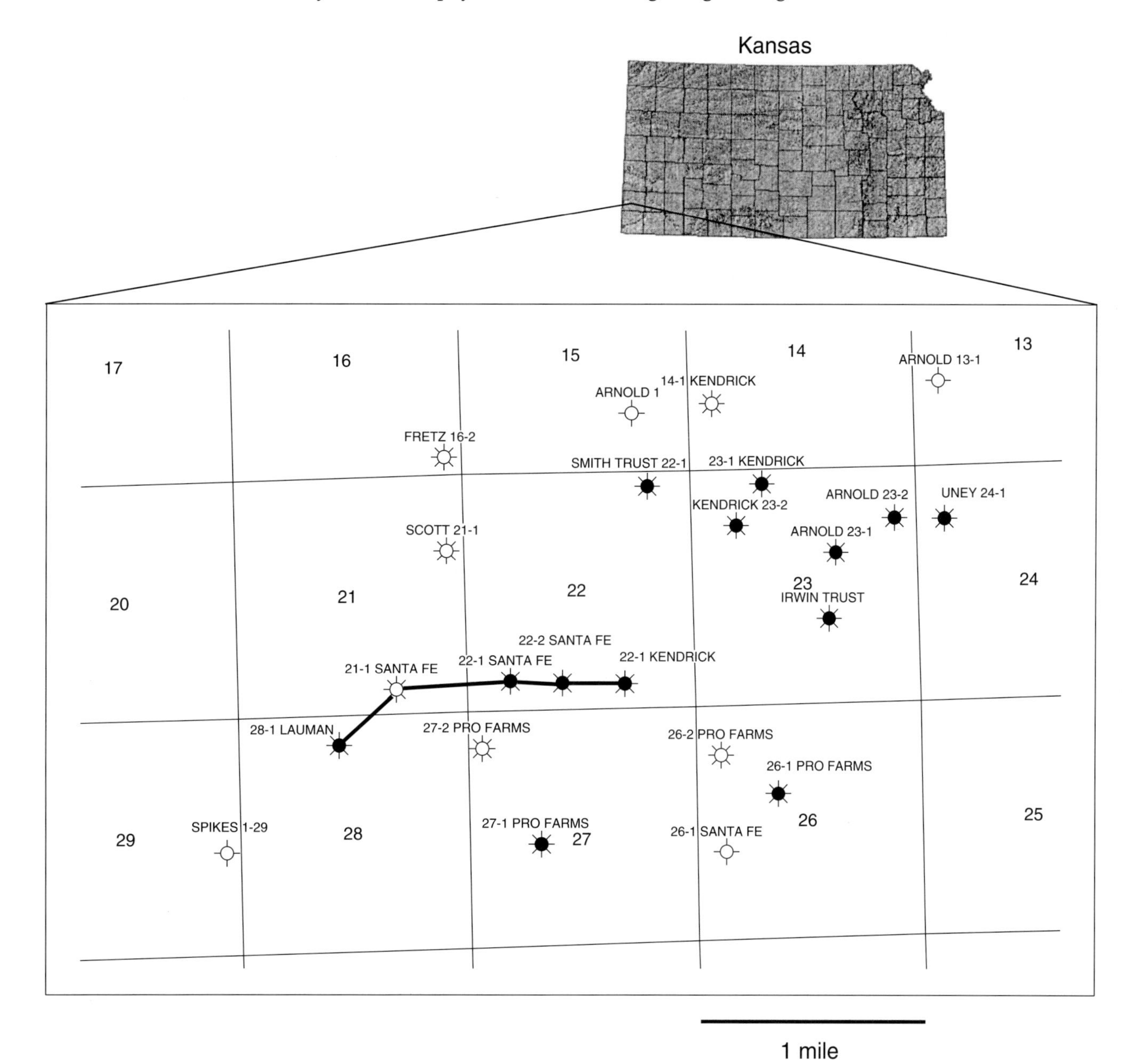

Figure 1. (a) Index map of Arroyo field identifying cross section (shown in Figure 2) and with well names and distribution.

up to 20% and averages 14%. All positions of the sandstone have been perforated in the field, with some wells reported as only gross intervals. Initial reservoir pressure was 1434 psi.

The upper portion of the sandstone has produced only natural gas, and the lowest portion has produced significant amounts of both oil and gas. No water has been produced in any of the wells. Also, no oil-water contact has been recognized. The reservoir drive appears to be gas expansion.

A considerable amount of supporting data on Arroyo field, including digital well logs, completion reports, and interpretive maps, cross sections, and synthetic seismograms, are included in a digital publication of this field on the Kansas Geological Survey's Digital Petroleum Atlas. This atlas is located on the Survey's Internet Home Page (http://crude2.kgs.ukans.edu/DPA/Arroyo/arroyoMain.html).

The lenticular lower Morrowan sandstone is comprised of a series of upward-coarsening, marginal marine shoreface deposits that are mostly confined to within an 0.8 km (0.5 mi) wide meandering valley up to 48 m (157 ft) deep (Figure 1). The sandstone was previously correlated and subdivided into five separate sandstone-dominated genetic units (1, 3, 5, 9, 11) using gamma ray, porosity, and resistivity logs and one spectral gamma ray log. Each genetic unit is delineated by bounding surfaces usually characterized by abrupt

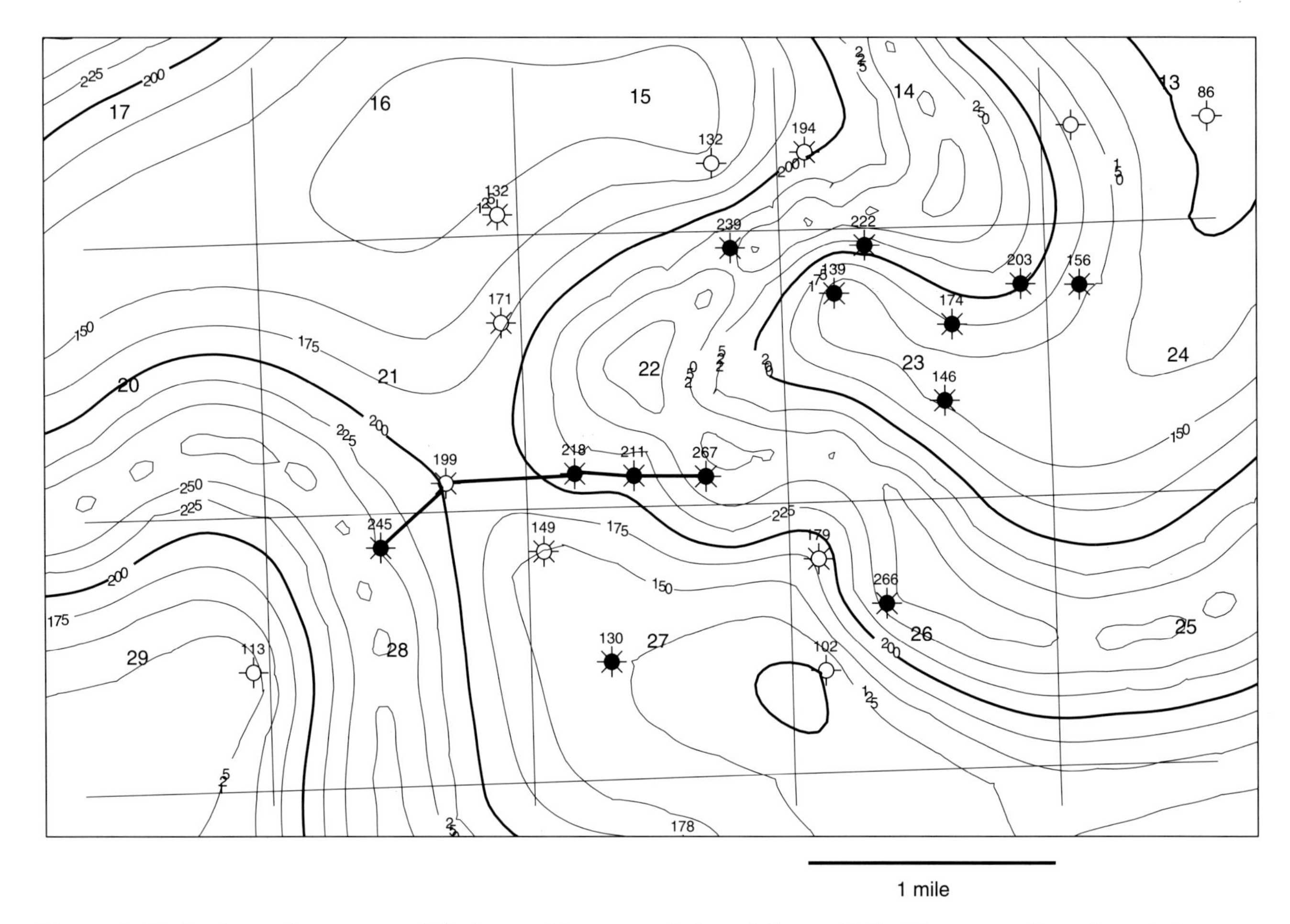

Figure 1. (b) An isopachous map of the lower Morrowan interval (top middle Morrowan limestone to top Mississippian). Contour interval is 25 ft.

changes in lithofacies. The surfaces either represent subaerial exposure or flooding surfaces, or both. Each genetic unit is believed to represent temporally distinct episodic deposition. Only several of the genetic units are developed at any particular location in the field (Figure 2).

The sandstones were deposited in a meandering valley system during overall rise in sea level. Maps of each genetic unit record episodes of infilling of this valley, each unit with varying geometries and sand abundance and quality. The lowest sandstones are more limited in distribution, filling only the lowest (deepest) portions of the valley, while the higher sandstones locally extend beyond the confines of the valley. For these reasons, the stratigraphic distribution was believed to be a controlling factor on flow unit definition.

METHODOLOGY

Volumetric properties of pore space and fluid saturation can be calculated from porosity logs (density, neutron, or sonic) and resistivity logs using the standard Archie equations. When plotted on a double-logarithmic plot of porosity versus resistivity (a Pickett plot), additional information on pore size and fluid producibility may be deduced by the use of pattern recognition informed by basic reservoir engineering principles. A template Pickett plot is shown in Figure 3 for the upper Morrowan in the Arroyo field. A water line (R_o) expresses the theoretical resistivity-porosity coordinates of all zones that are completely saturated with water. The water line is established by the first Archie equation that links the formation factor, F, to the ratio of the resistivity of the completely water-saturated rock, R_o, to the resistivity of the formation water, R_w, to the porosity of the rock:

$$F = R_o / R_w = a / \phi^m$$

using an Arroyo field formation water resistivity, R_w, of 0.04 ohm-m and Archie parameter values of $a = 1$ and $m = 1.8$, which express pore geometry in the Morrowan sandstone. Contours for different values of water saturation parallel the water line, with spacing determined by the saturation exponent, n (generally with a value of about 2 in water-wet rocks) in the second Archie equation:

$$I = R_t / R_o = 1 / S_w{}^n$$

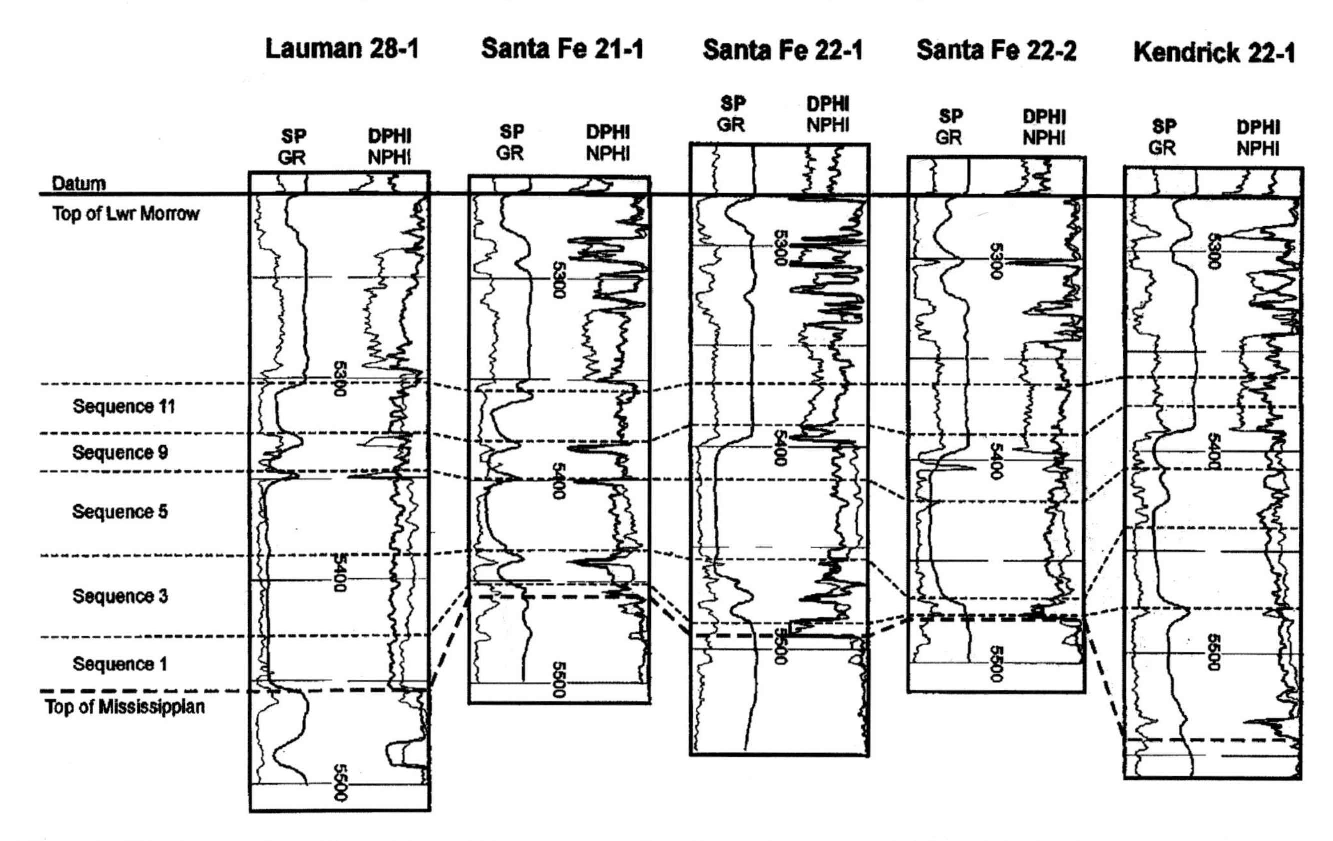

Figure 2. West-to-east stratigraphic well log cross section through Arroyo field containing Lauman 28-1, Santa Fe 21-1, Santa Fe 22-1, Santa Fe 22-2, and Kendrick 22-1 wells. Datum of section is middle Morrowan limestone. Correlated stratigraphic intervals are correlated through the lower Morrowan sandstone interval. Line of cross section is shown on Figure 1a.

where I is the resistivity index and R_t is the actual resistivity of the rock. Lines can also be drawn on the plot that are contours of bulk volume water (BVW), where water content is expressed as a proportion of the total rock, rather than in terms of the pore space as is the case with saturation.

The disposition of crossplotted zones with respect to the log axes of resistivity-porosity and the computed reference axes of water saturation and bulk volume water (BVW) gives useful clues on both pore type and producibility. These properties can be seen when relating Pickett plots to production histories (Figure 4) from some example wells in the Arroyo field. Notice how overall well performance is determined to a large extent by higher porosities and lower water saturations; however, the location of the data-cloud with respect to the BVW contours reflects the pore size and likely water-cut. Lower values of BVW are matched with coarser pores; higher values of BVW are linked with either finer pores or zones with coarse pores and producible water. In terms of data-cloud shape, a classic reservoir profile would show zones high in the reservoir at irreducible water saturation and relatively low BVWs with a progressive increase in bulk volume water with increasing depth and descent into the transition zone. Some aspects of this ideal character are shown in the plots in Figure 4, where the four wells have been arranged from most productive at the top to least productive at the bottom. Notice that the bottom example was not completed for production, but abandoned because a DST (drill-stem test) yielded nothing but saltwater. The associated Pickett plot shows a rather ragged scatter of mostly low-porosity zones with high water saturations that probably reflect residual hydrocarbon saturations. This pattern is common at the margins of fields, as is this well, and contrasts

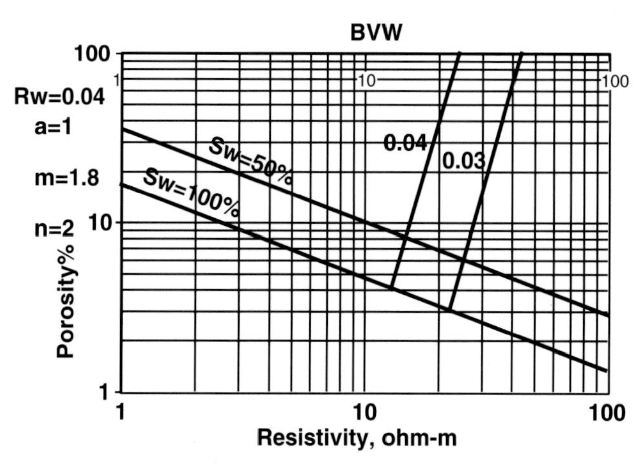

Figure 3. A template Pickett plot for the upper Morrowan sandstone in Arroyo field.

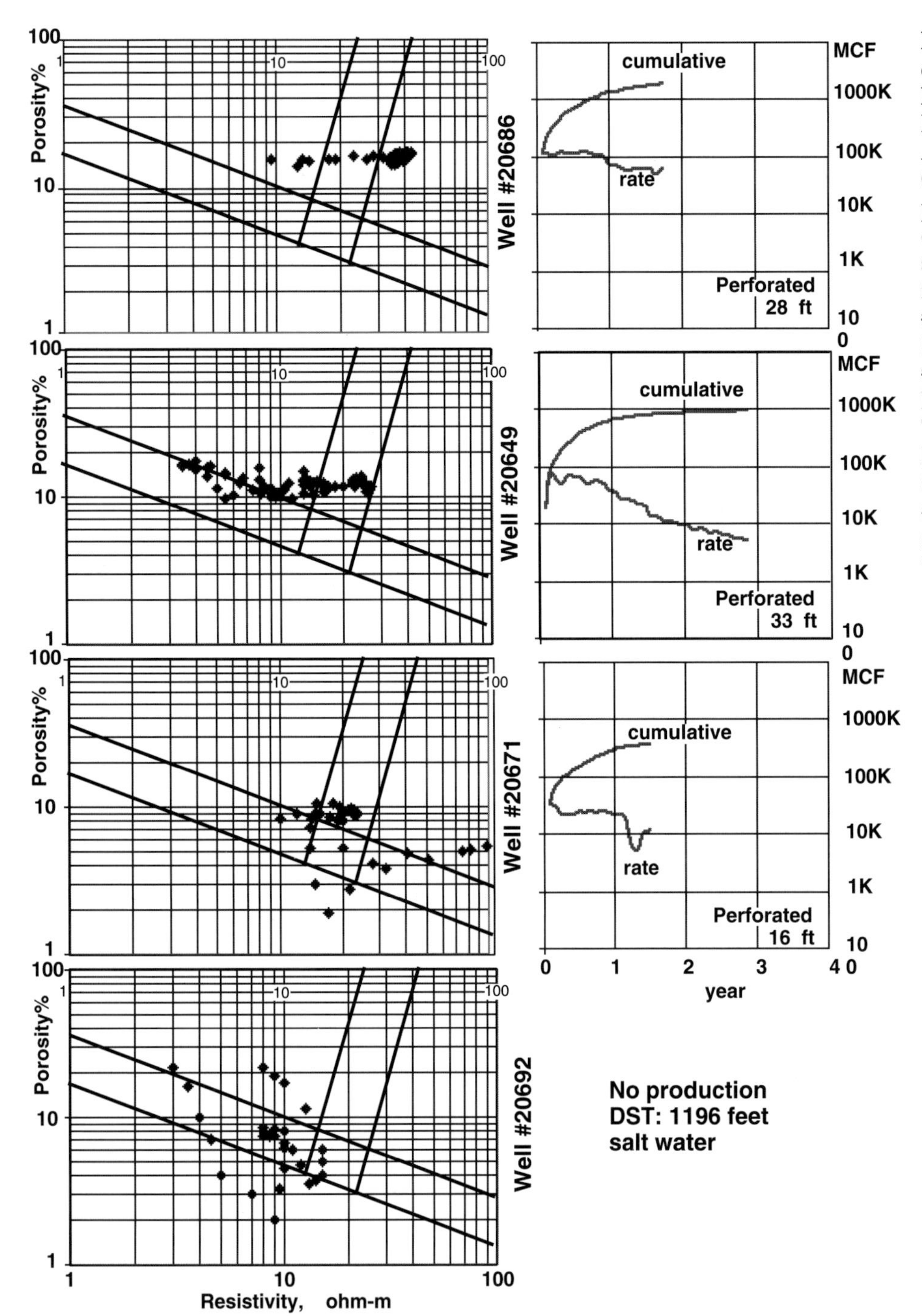

Figure 4. Pickett plots and corresponding production histories for the upper Morrowan sandstone in Arroyo field, Stanton County, Kansas. Each Pickett plot is identified with a five-digit well number. The corresponding well names are as follows: Well #20686: Huber #10-1 Cockreham, SE NE SE Sec. 10-T29S-R41W; Well #20649: Huber #26-2 Pro Farms, SW NW NW Sec. 26-T29S-R41W; Well #20671: Huber #23-2 Kendrick, C NW Sec. 23-T29S-R41W; Well #20692: Petroleum Inc. #1-29 Spikes, NENESE Sec. 29-T29S-R41W.

with an idealized situation far away from fields where zones are completely water-saturated and form a trend on a Pickett plot that conforms closely to the water line (R_o).

The Pickett (porosity-resistivity) crossplots are the fundamental components in the petrofacies analysis. The connection of data points by depth and the ability to annotate the data points with a third variable help establish relationships between the petrophysical response and the geology—lithologies, stratigraphic units, and structure; i.e., the petrofacies. Template lines identify minimum BVW and associated water saturations and porosities on the Pickett plot. This, in turn, helps to correlate the geology to fluid-related parameters and to delineate specific changes in fluids and variations in the pores between the different wells.

The definition of flow units might be refined to include regions of similar or related BVW and pore type using the petrofacies analysis. Often, permeability data are either lacking or are limited to averaging from core-log porosity and permeability correlations. In these cases, assessment of pore type using petrofacies analysis may help to provide novel constraints to flow units lacking other substantial data. Of course, production and transient test data and geochemical

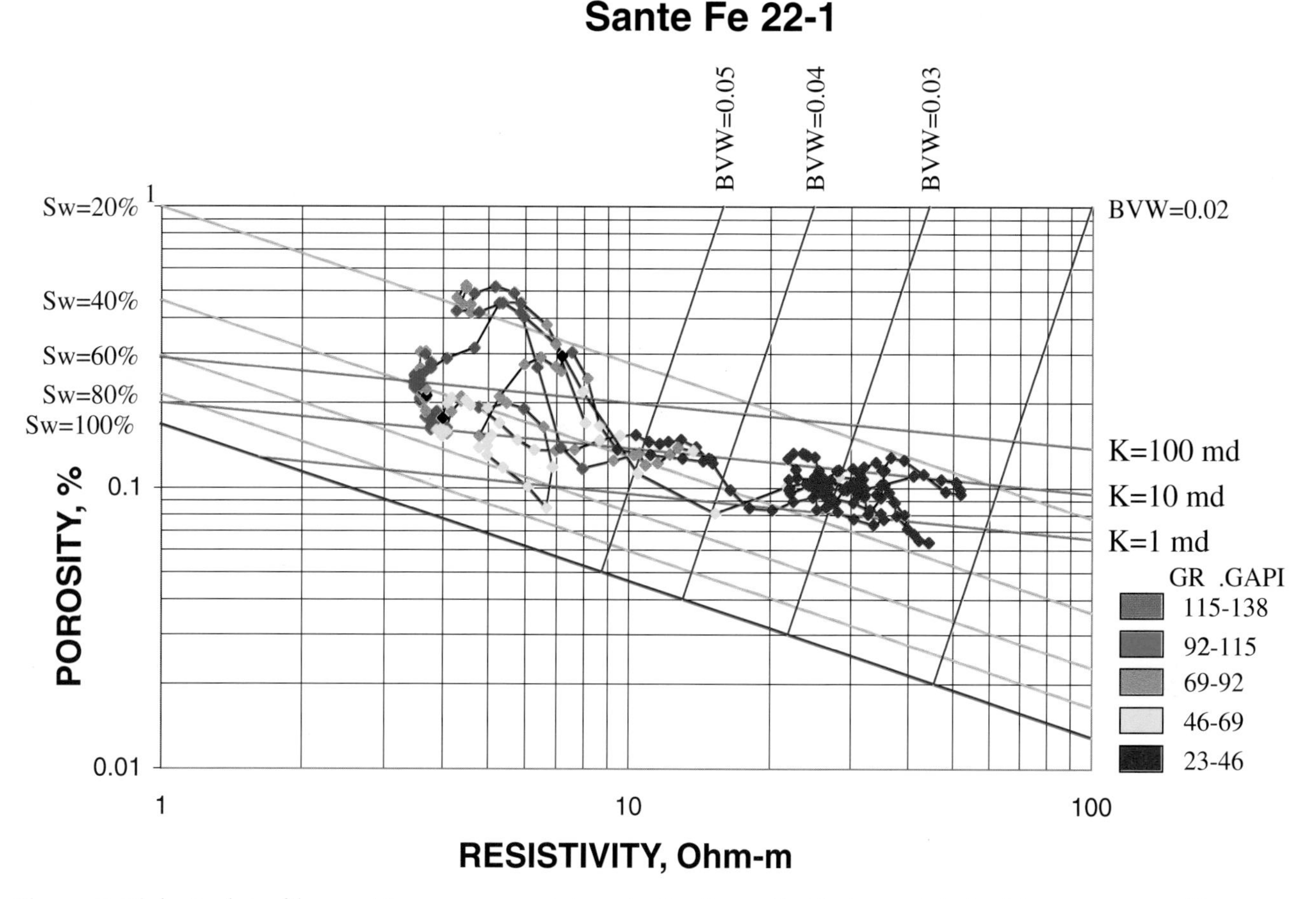

Figure 5. Pickett plot of lower Morrowan sandstone in the Santa Fe 22-1 annotated with gamma ray. Note that this Pickett plot has contours of bulk volume water and permeability, the latter estimated from Timur equation.

tests are necessary to more definitively constrain the definition of flow units.

A west-to-east cross section was chosen to further characterize the sandstone reservoir using the petrofacies analysis approach (Figure 2). The cross section crosses the valley in two places, separated by an intervening high area.

The questions addressed in this analysis include the following: (1) Is additional evidence available to confirm or reject the continuity of sandstones across the intervening high region residing between the valleys? (2) What is the evidence of vertical conformance and lateral continuity? (3) How do properties of the sandstones compare on either side of the valley? (4) Can the definition of flow units be improved? (5) How do the flow units compare with the detailed stratigraphic subdivisions?

In addition to the stratigraphic analysis, the procedure included four operations: (1) construct Pickett plots for each of the wells on the cross section, (2) perform a cluster analysis of basic petrophysical data to independently define similar reservoir properties, (3) prepare a series of color cross sections of selected petrophysical variables with datums on sea level elevation and a stratigraphic marker (middle Morrowan limestone located above the sandstone), and (4) integrate this information to define flow units by comparison with stratigraphic zonation, Pickett plots, color log cross sections, cluster analysis, and well productivities.

RESULTS

The digital data from five well logs comprising a west-to-east cross section in Arroyo field were examined using petrofacies analysis. The objective was to compare well data on the cross section to test for evidence of lateral continuity. Correlations shown in Figure 2 suggest that most of the units are continuous. Units that are not shown as continuous may also be connected from outside of the plane of the cross section.

Pickett Plots

LAS (log ASCII standard) digital well log files were read into an Excel-Visual Basic® program called PfEFFER™ to generate the Pickett crossplots. The initial Pickett crossplots provide a visual differentiation of the variation in the porosity, resistivity, water saturation, and bulk volume water. Permeability lines are annotated on the crossplots, estimated using the empirical relationship between water saturation and

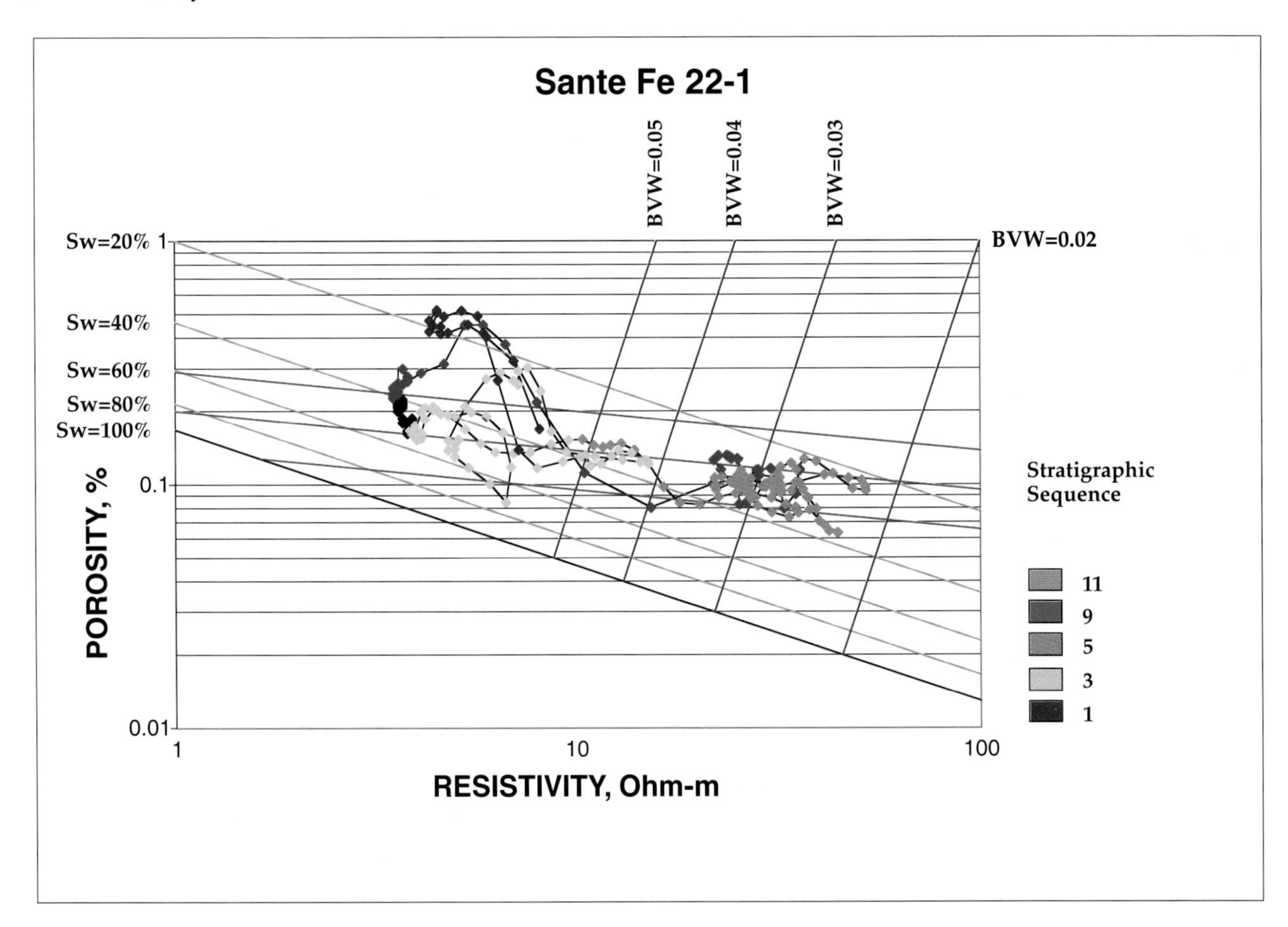

Figure 6. Pickett plot of lower Morrowan sandstone in the Santa Fe 22-1 annotated with stratigraphic units shown in type log in Figure 2.

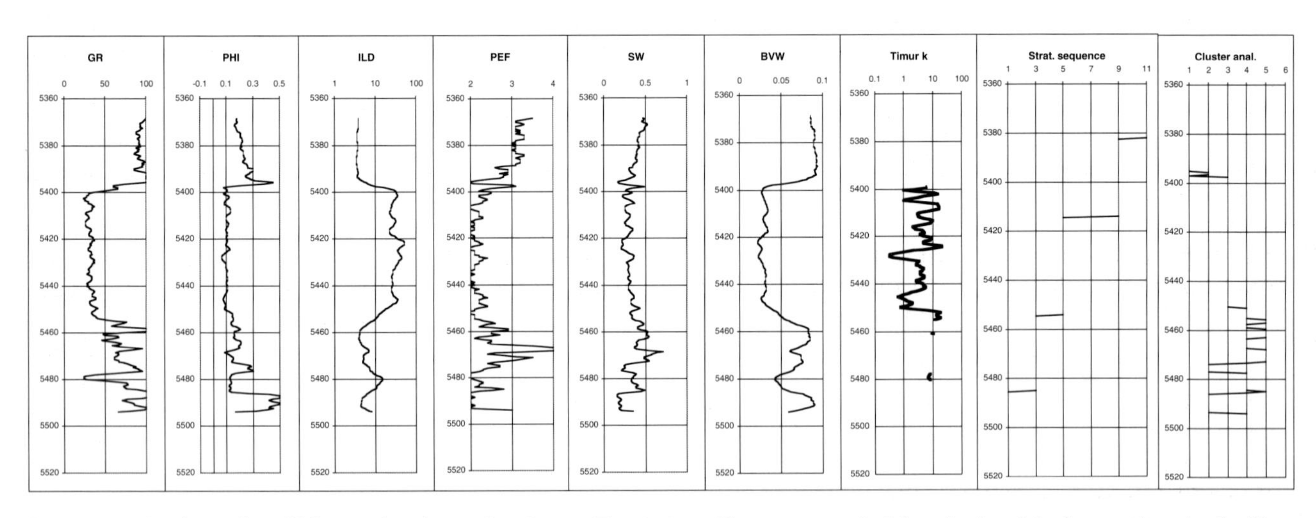

Figure 7. Display of well log suite from the Santa Fe 22-1 well accompanied by derived information including water saturation (S_w), bulk volume water (BVW), permeability derived from Timur equation (Timur k), stratigraphic units, and amalgamation groups from cluster analysis.

porosity developed by Timur. The relationship holds for clean sandstones when water saturations are at irreducible values. We believe that to be the case here.

The crossplot (Figure 5) with points annotated with gamma ray values indicates that points with higher gamma ray values are located on the left side of the crossplot at BVWs in excess of 0.06 (Figure 5). This location presumably represents more shaly and finer pores. The permeability lines are not applicable to these points.

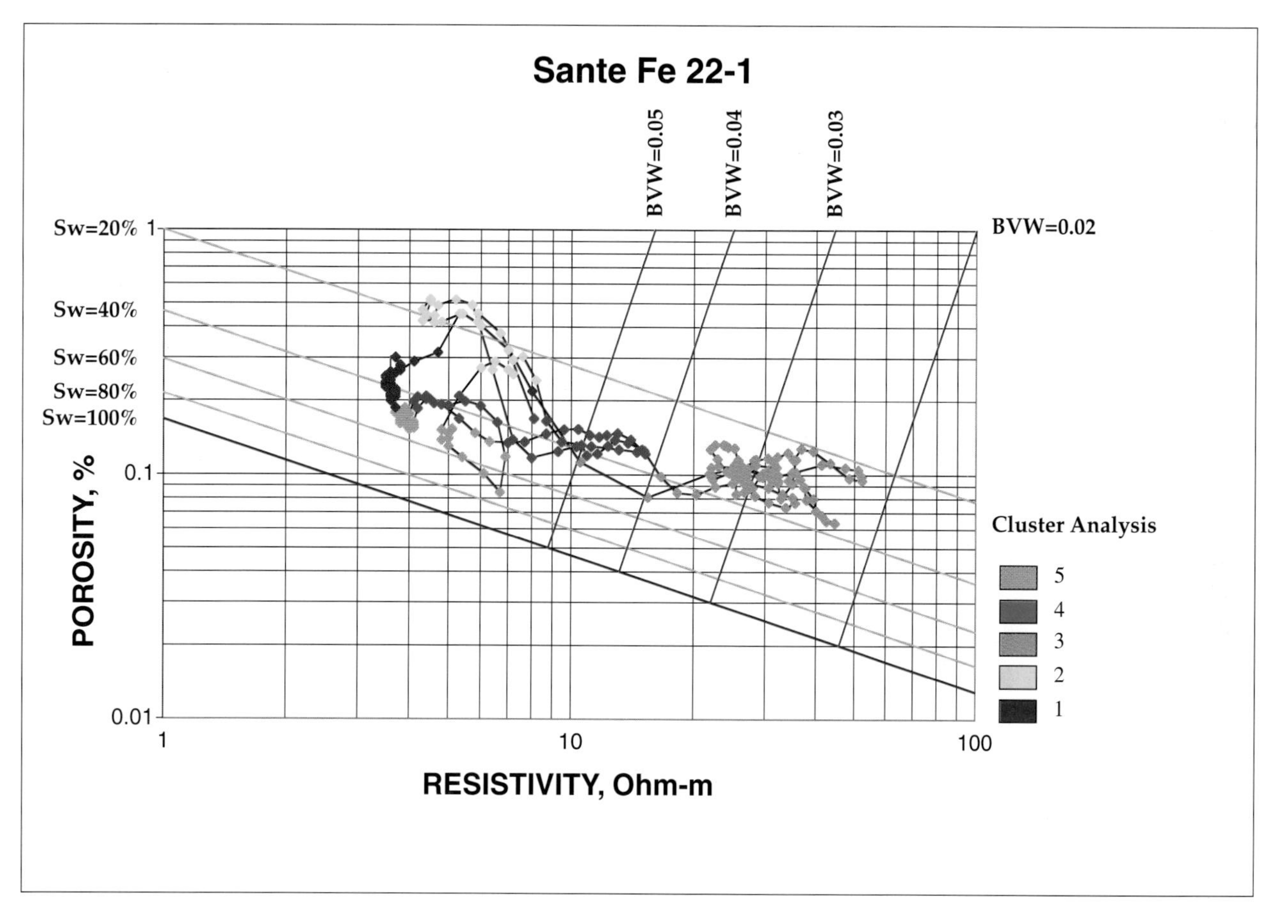

Figure 8. Pickett plot of lower Morrowan sandstone in the Santa Fe 22-1 annotated with amalgamation groups from cluster analysis.

The reservoir has no water leg, and no water has been produced. The points are annotated by stratigraphic interval and form rather tight clusters or bands for each stratigraphic unit (Figure 6). The clustering of points in distinct stratigraphic intervals at lower water saturations suggests that these zones are near their respective minimum BVWs and closely correspond to particular stratigraphic zones. Also, the bands parallel water saturation lines. This pattern is ascribed to changing minimum BVW and pore size within a zone, which has implications to fluid flow. If any portions of the reservoir were in a gas or oil/water transition zone, the bands of points may have more likely paralleled porosity lines, if the pore type were not changing; however, this is not seen and no wells have experienced any water-cut oil or gas production. The variations suggest possible changes in pore type and evidence for reservoir continuity or lack thereof.

Cluster Analysis

Some of the boundaries between stratigraphic units involve sandstone on sandstone and may not present barriers to flow, but do cause changes in transmissibility. Also, the internal variability in sandstone units may create additional heterogeneity that can retard fluid flow. Cluster analysis was used to examine the similarity among petrophysical data. The method provides a consistent automated treatment of the data to aid in comparing considerable amounts of data among the zones and wells. Ward's Method was selected as the clustering technique. The method consists of a series of clustering steps that begins with "t" clusters, each containing one object. The clustering ends with one cluster containing all objects. At each step, a merger of two clusters is made that results in the smallest increase in the variance (Romesburg, 1984).

The petrophysical variables included in the cluster analysis are gamma ray, deep induction resistivity, Pe (photoelectric index), S_w (water saturation), BVW (bulk volume water), and apparent permeability using the Timur equation

$$k_a = 1 \times 10^4 * \phi^{4.5} / S_w^{\ 2}$$

This apparent permeability, k_a, is a minimum estimate when S_w is greater than irreducible S_w. Porosity (ϕ) and S_w are fractional values in this equation. Shaly intervals were assigned to zero permeability (removing depth intervals where gamma ray exceeded 60 API units and neutron minus density porosity was greater than one). Depth was also included within the cluster analysis as an adjacency constraint to enhance spatial continuity.

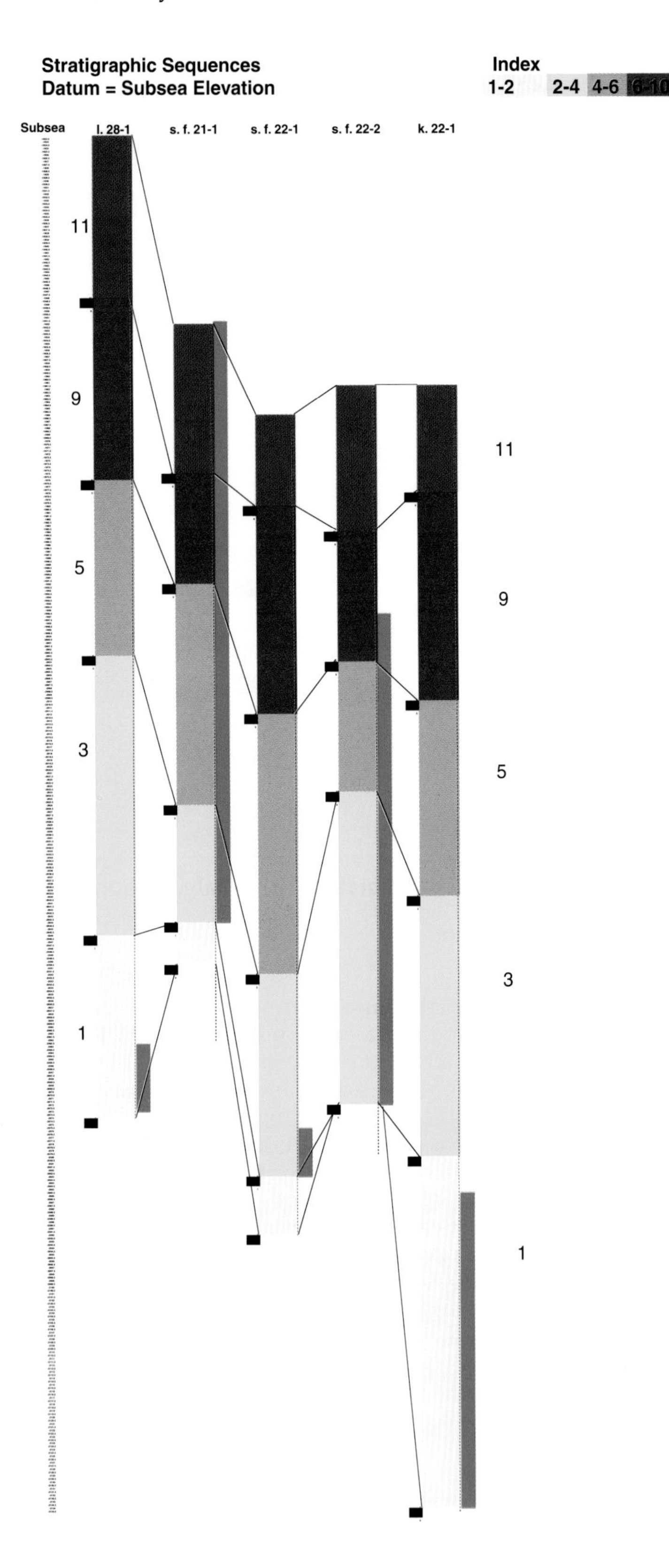

Figure 9. East-to-west color cross section depicting stratigraphic sequences (1, 3, 5, 9, and 11) corresponding with same sequences as defined in Figure 2. Sections are annotated with perforations as bars along the right margin of each well. Index map in Figure 1. (a) [left] Structural presentation with sea level datum.

Six separate groups of points were selected from the cluster analysis of each well. Several criteria were used to determine this number. First, the number is not large enough to produce too many groups, which could complicate reservoir modeling. Second, the cluster dendrogram for each well showed good separation of groups at this level. Third, the number is comparable to the stratigraphic divisions and might show useful groupings and comparison.

The assigned groupings derived from cluster analysis were first compared by depth with the petrophysical data and stratigraphic zonation. The

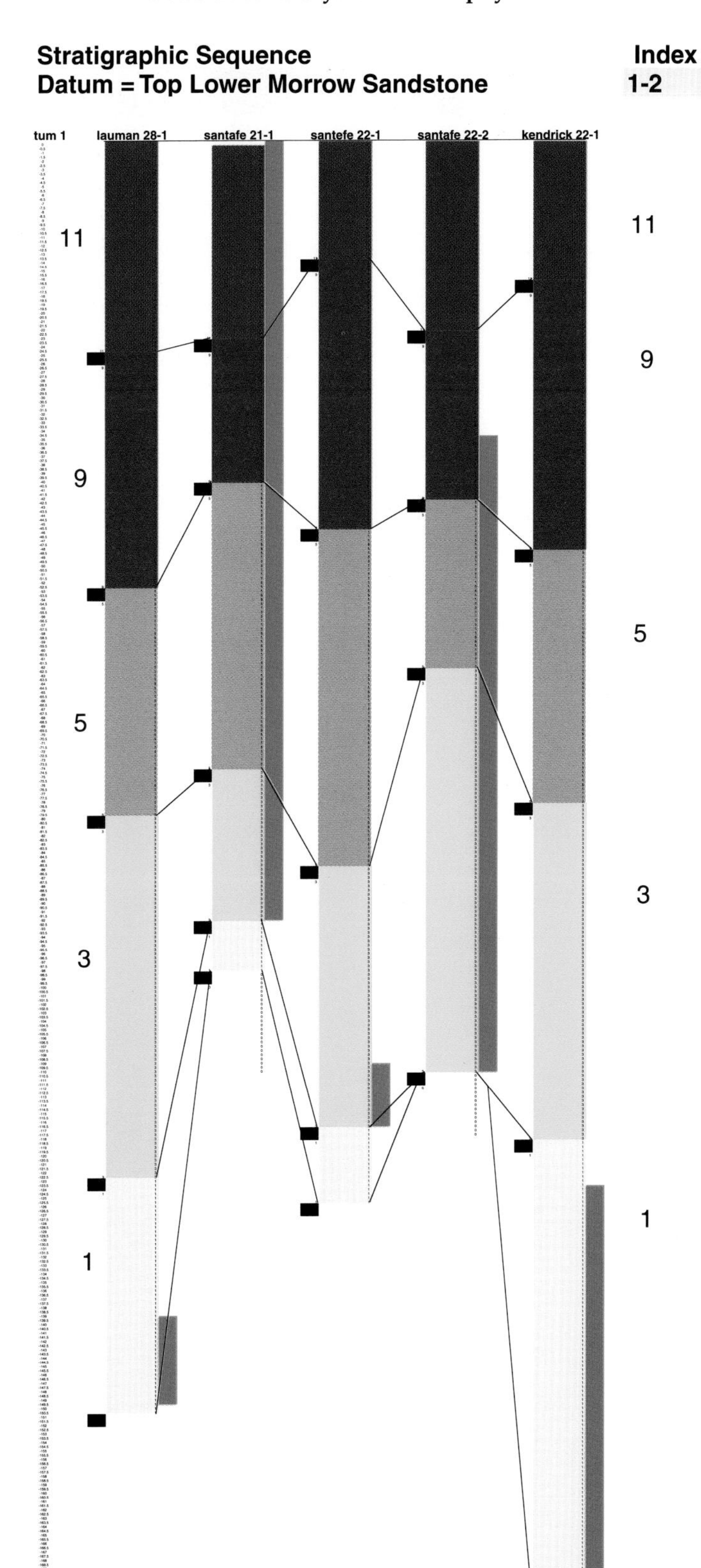

Figure 9. (b) Stratigraphic cross section with datum at top lower Morrowan sandstone.

boundaries between the stratigraphic intervals and the cluster assigned groupings generally coincide. The clustering identified a moderate amount of smaller scale heterogeneity within each stratigraphic interval (Figure 7). This internal variation includes having the same cluster group in different stratigraphic units, e.g., groups 4 and 5 in stratigraphic sequences 5 and 9. It would be anticipated that similar sandstone properties would transcend sandstone intervals. However, the general finding is that each stratigraphic interval is dominated by only one or two assigned cluster groupings.

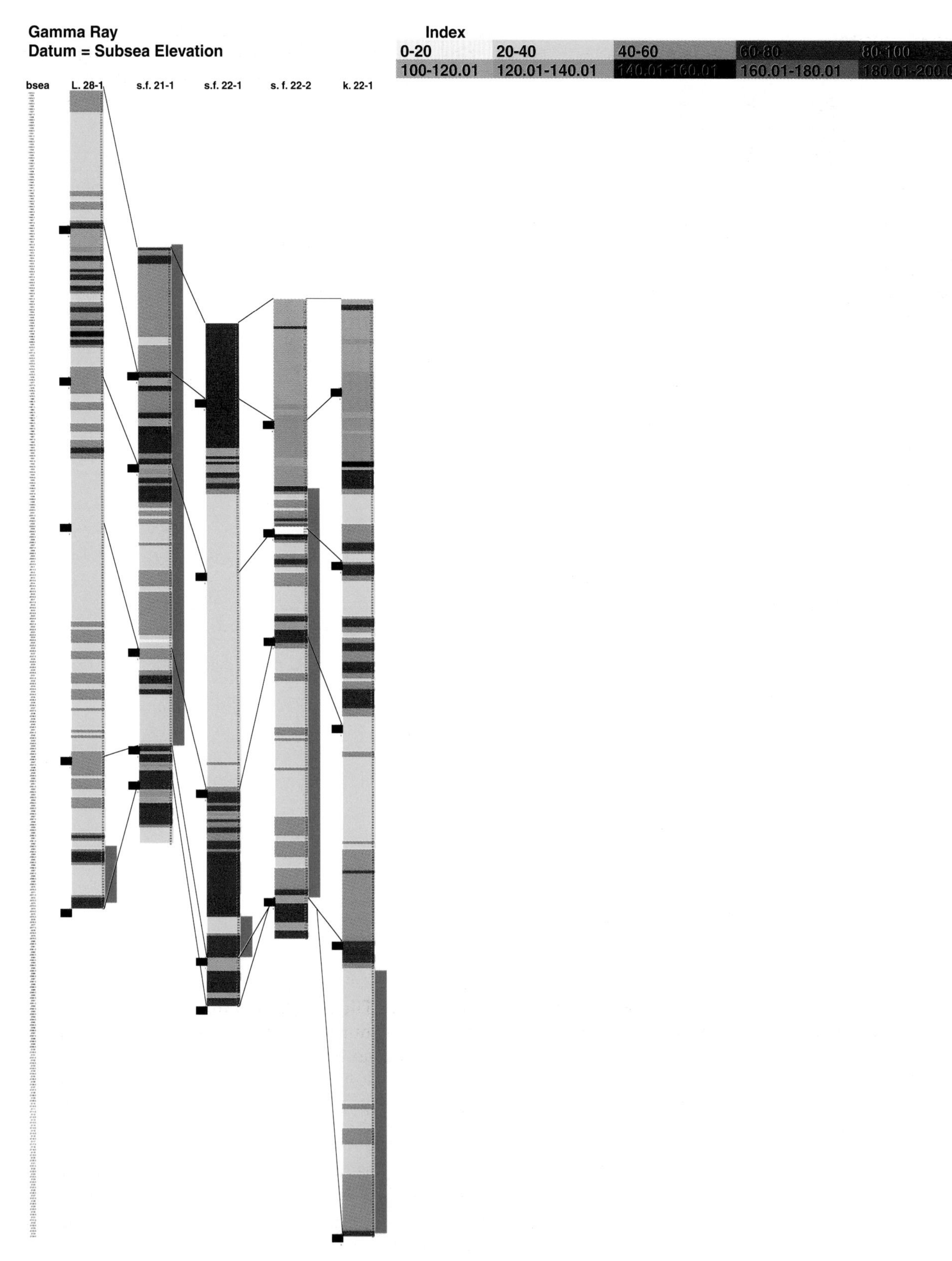

Figure 10. East-to-west color-image cross section depicting gamma ray variation across Arroyo field. The section includes same wells as in Figure 2. Wells are annotated with perforations as bar along right side of each well. Section also shows correlations of stratigraphic sequences. Cross section is part of an Excel spreadsheet and is at the resolution of the digitized data (0.5 ft in this example). Vertical scale bar shown on this and ensuing structural sections. No horizontal scale (wells are equally spaced). (a) [above] Structural version of cross section with a sea level datum.

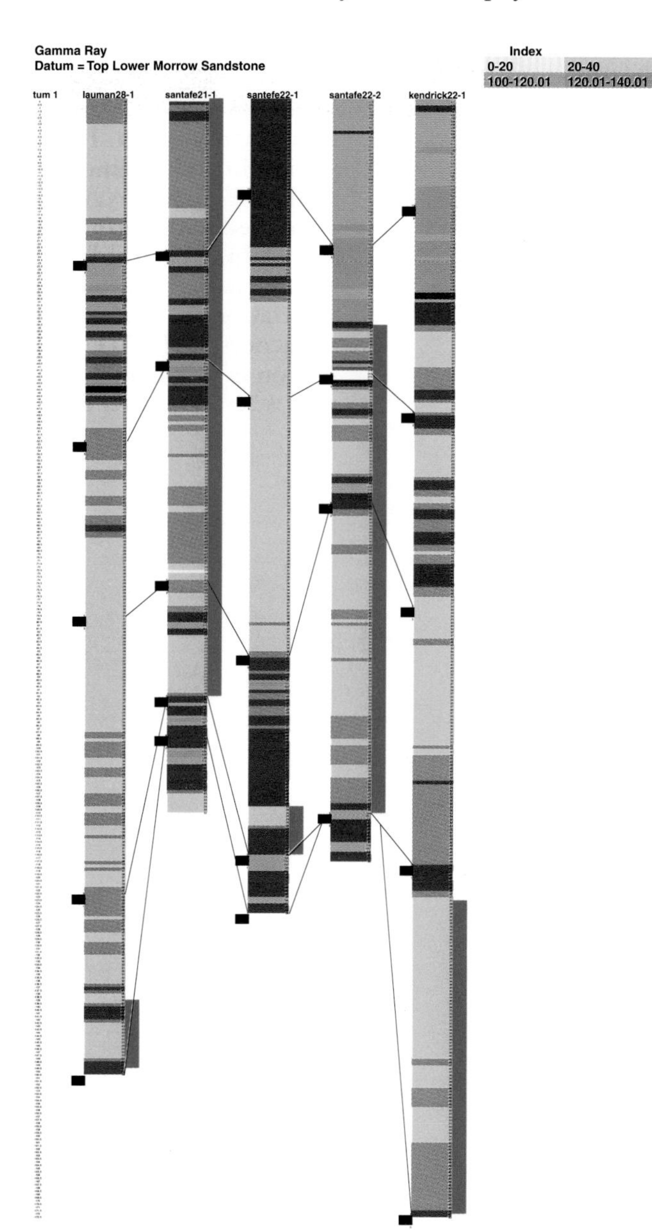

Figure 10. (b) Cross section with a stratigraphic datum on the top of the lower Morrowan sandstone.

The posting of assigned cluster groups as an attribute on the Pickett plot further indicates a close correspondence between stratigraphic units and assigned cluster grouping (as in Figure 8). The cluster analysis can be adapted in the spreadsheet environment to help facilitate consistent, rapid assignment of cluster groups and further aid in flow unit assessment.

Color Cross Sections

In general, flow units are assigned to zones in the reservoir with similar permeability and porosity, and that also exhibit lateral continuity. Flow units are inferred to control fluid flow, and confirmation was sought to substantiate these units. Petrophysical variation within individual well profiles has been described up to this point, focused on vertical conformance. The question remains as to the extent of lateral continuity. Flow units are not fieldwide in extent, but are anticipated to be correlatable to some degree. This continuity is ultimately established using petrophysical data, fluid recovery, pressure, and fluid chemistry. The suites of petrophysical variables including BVW can be used to evaluate lateral continuity. Continuous trends or constancy of properties of the sandstone and correlations with structural elevation suggest possible fluid continuity in the reservoir.

Computer-assisted generation of color cross sections based on original digital well log sampling of 0.16 m (0.5 ft) provides the means to observe and evaluate detailed subtle changes in reservoir character and substantially assists in assessing continuity and assigning flow units. The cross sections are generated with an elevation (subsea) or stratigraphic datum.

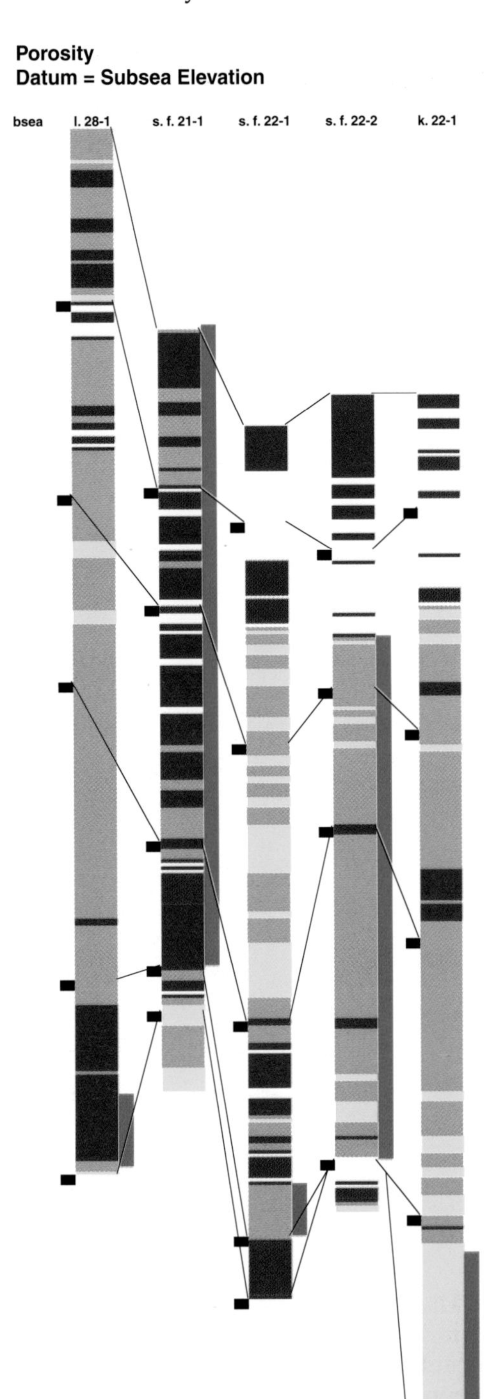

Figure 11. East-to-west color image cross section depicting porosity variation across Arroyo field. The section includes same wells as Figure 2. Wells are annotated with perforations as bar along right side of each well. Section also show correlations of stratigraphic sequences. (a) [left] Cross section is a structural version with a sea level datum.

Cross sections for key petrophysical parameters were generated, including gamma ray, permeability calculated with the Timur equation (apparent permeability filtered on gamma ray and neutron-density shale indicators), Pe (photoelectric effect) from the lithodensity log, porosity, water saturation, BVW, and deep induction resistivity (Figures 9–11). Each petrophysical variable is presented as a structural and stratigraphic cross section placed side by side, the latter with a datum at the top of the middle Morrowan limestone. Stratigraphic units are identified and correlated. Perforated intervals are shown alongside each well profile.

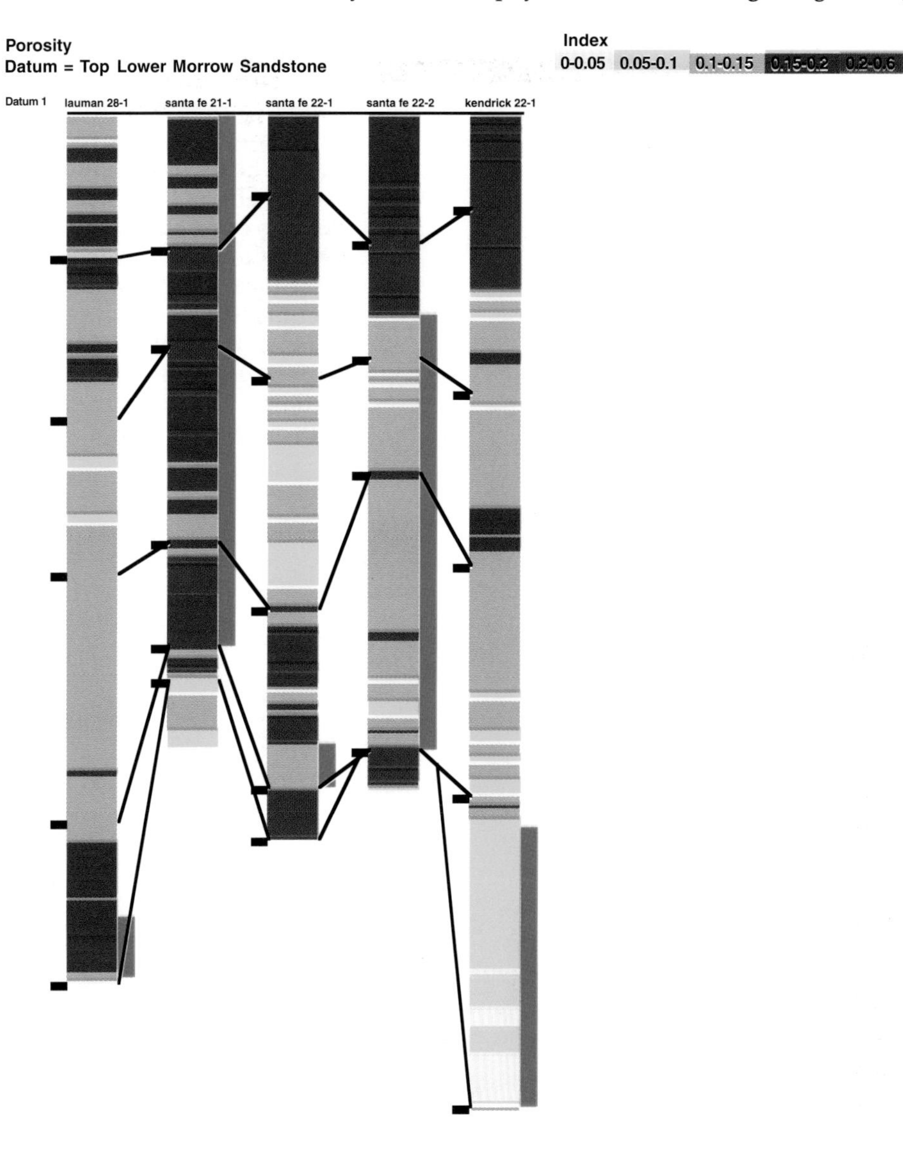

Figure 11. (b) Cross section with a stratigraphic datum on the top of the lower Morrowan sandstone.

The five wells in the cross section are perforated in two distinct intervals, a lower interval restricted to stratigraphic unit #1 in the Lauman 28-1 and Kendrick 22-1. Also, the lowest part of unit #3 in the Santa Fe 2-21 is suggested to be part of the lower interval and may possibly be recorrelated with unit #1 (Figure 9).

The lower interval produces significant amounts of oil and gas. The upper perforated interval includes stratigraphic units #3, #5, #9, and #11. The upper interval produces natural gas and minor amounts of oil from the Santa Fe 2-11 and Santa Fe 2-22. This difference suggests that the reservoirs are separate. The lower and upper sandstones are isolated by a prominent shaly interval, according to the gamma ray and photoelectric logs.

On closer inspection of the gamma ray cross section, the stratigraphic units can be distinguished with the help of the correlation lines; however, there is considerable variation in the internal properties of the stratigraphic units (Figure 10). This variation persists in the other parameters. The stratigraphic units generally appear to delineate most of the petrophysical variation except for several possible re-correlations. These re-correlations are based on further analysis.

Porosity varies from 15 to 20% in the Lauman well to 0 to 8% in the Kendrick (Figure 11). Apparent permeability calculated from the Timur equation and filtered on gamma ray and neutron shale indicators shows considerable changes on the cross section (Figure 12). The permeability and porosity are both higher on the west. Permeability ranges between 10 and 100 millidarcies (md) in the Lauman 28-1 on the west side to between 0.1 to 1 md in the Kendrick 22-1 on the east side.

High permeability and porosity correlate well with a trend of increased natural gas production and decreased oil production in the Lauman 28-1, with 2.5 Gcf of gas and 75,000 bbl of oil. In comparison, the Kendrick 22-1 well on the east and lowest side of the cross section recovered less gas, 1 Gcf, but more oil, over 120,000 bbl. Permeability varies considerably in thin streaks near the base of the Kendrick 22-1 well. Unit #1 is separated from the overlying sandstones by a thicker shaly interval.

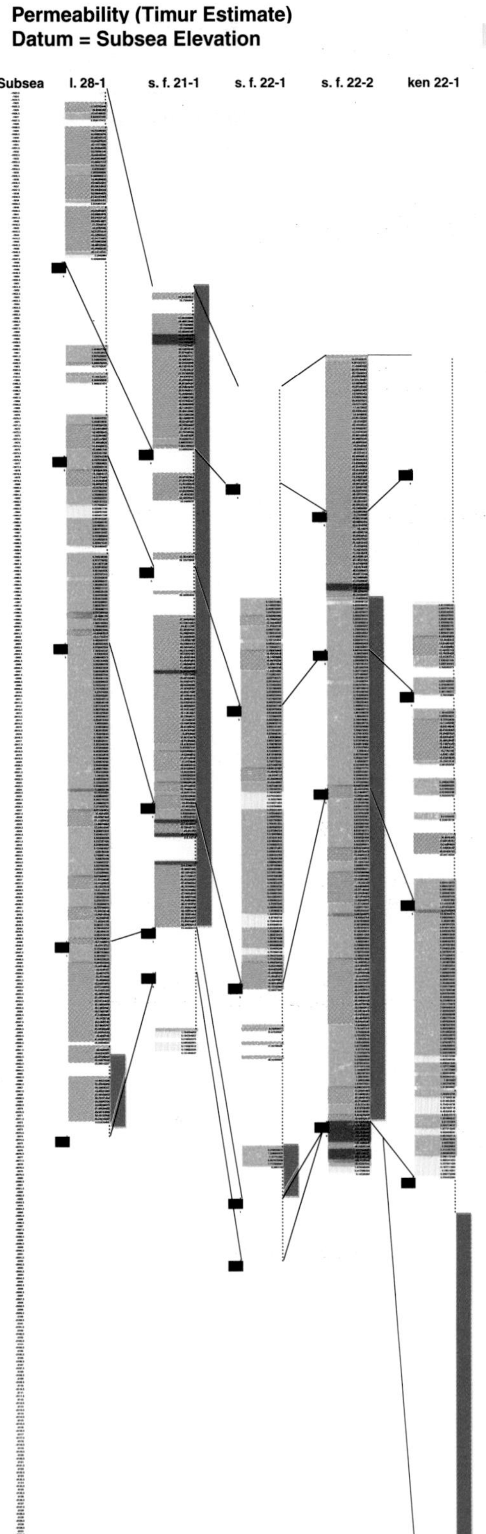

Figure 12. East-to-west color image cross section depicting permeability variation across Arroyo field. The section includes same wells as Figure 2. Wells are annotated with perforations as bar along right side of each well. Section also show correlations of stratigraphic sequences. (a) [left] Cross section is a structural version with a sea level datum.

Unit #3 is also thick in the paleovalleys on either side of a central high. Santa Fe 2-21 is perforated in the basal part of a thin sandstone that is in close proximity to the lower interval, unit #1. This zone in Santa Fe 2-21 has produced 450 Mcf of gas and nearly 50,000 bbl of oil. Production values are less than Lauman and Kendrick, but the mixed production is similar to the recoveries noted in unit #1, suggesting that they are a common reservoir in the deeper portions of the paleovalleys.

Production from perforations in the upper interval in Santa Fe 21-1 and Santa Fe 22-2 is notably different. Santa Fe 21-1 has realized 570 Mcf of gas from the

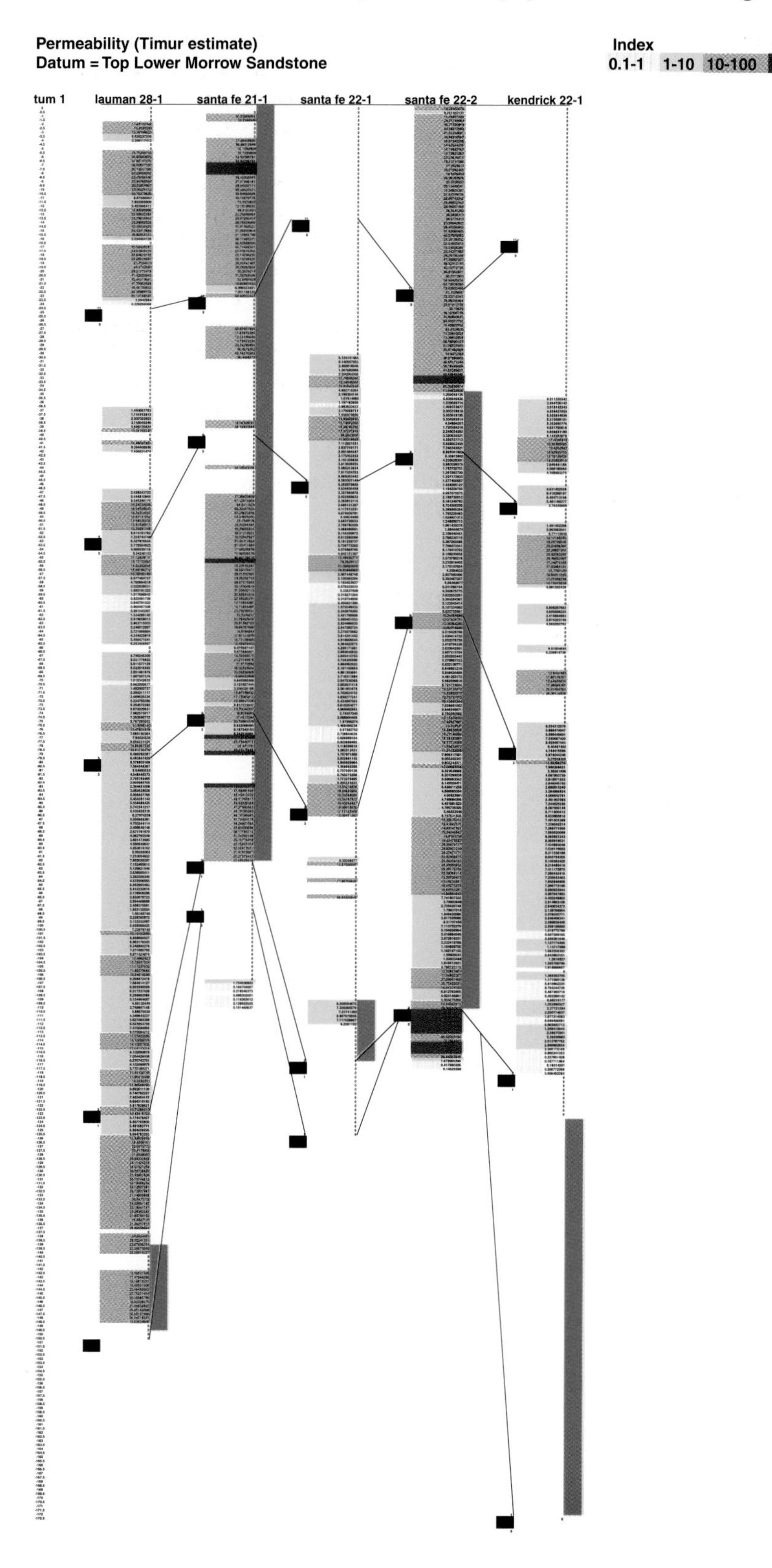

Figure 12. (b) Cross section with a stratigraphic datum on the top of the lower Morrowan sandstone.

upper zone and has been declining relatively rapidly to 1 Mcf per month. In contrast, the Santa Fe 22-2 well has produced nearly three times more gas at 1.8 Gcf, and its production has declined to only 18 Mcf per month. Both wells had produced for nearly 3 yr at the time the production figures were reported.

CONCLUSIONS

The stratigraphic units serve as adequate means to classify flow units in this reservoir, with added refinements using petrofacies analysis. Petrofacies analysis uses Pickett plots to decipher reservoir properties of each stratigraphic unit. Cluster analysis provides a consistent means to further delineate reservoir properties. The boundaries of the clustered groups are commonly those of the stratigraphic units. The clustered groups provide further subdivisions of the reservoir rock that could be used to classify finer scale flow units. Color cross sections further substantiate the use of the stratigraphic divisions as basic templates for distinguishing flow units. The color cross sections are representative of the original digitized well log data and provide the means to precisely subdivide the stratigraphic units. Petrofacies analysis should prove useful for evaluating improved petroleum recovery options.

ACKNOWLEDGMENTS

Several grants have supported the PfEFFER log analysis project at the Kansas Geological Survey and used its developments over the past few years, including the Petrofacies Analysis Project with the Kansas Technology Enterprise Corporation and the industrial consortium, "Development and Demonstration of An Enhanced, Integrated Spreadsheet-based Well Log Analysis Software," Subcontract No. G4S60821 with BDM-Oklahoma and industry consortium. PfEFFER application and testing has been conducted with support from "Shaben Field—Class II Field Demonstration Project," Contract No. DE-FC22-94PC14987, and "Digital Petroleum Atlas," Contract No. DE-FG22-95BC14817, both supported by the Department of Energy.

REFERENCES CITED

Romesburg, H.C., 1984, Cluster Analysis for Researchers: Belmont, California, Lifetime Learning Publications, 334 p.

Willhite, G.P., 1986, Waterflooding: SPE Textbook Series Volume 3: Richardson, Texas, Society of Petroleum Engineers, Richardson, 326 p.

Section III

Enhanced/Improved Oil Recovery (EOR/IOR) Characterization

Martin, F.D., et al, Advanced reservoir characterization for improved oil recovery in a New Mexico Delaware basin project, 1999, *in* R. Schatzinger and J. Jordan, eds., Reservoir Characterization-Recent Advances, AAPG Memoir 71, p. 93–108.

Chapter 7

Advanced Reservoir Characterization for Improved Oil Recovery in a New Mexico Delaware Basin Project

F. David Martin
Richard P. Kendall
Earl M. Whitney
Dave Martin and Associates, Inc.
Albuquerque, New Mexico, U.S.A.

Bob A. Hardage
The University of Texas at Austin
Bureau of Economic Geology
Austin, Texas, U.S.A.

Bruce A. Stubbs
Pecos Petroleum Engineering, Inc.
Roswell, New Mexico, U.S.A.

Bruce Uszynski
Territorial Resources, Inc.
Roswell, New Mexico, U.S.A.

William W. Weiss
New Mexico Petroleum Recovery Research Center
Socorro, New Mexico, U.S.A.

ABSTRACT

The Nash Draw pool in Eddy County, New Mexico, is a field demonstration site in the Department of Energy Class III (Slope Basin Clastic Reservoirs) program. Production is from the basal Brushy Canyon zones of the Permian (Guadalupian) Delaware Mountain Group. The basic problem at the Nash Draw pool is the low recovery typically observed in similar Delaware fields. By comparing production performance for a control area using standard infill drilling techniques to a pilot area developed using advanced reservoir characterization methods, the goal of the project is to demonstrate that advanced technology can significantly improve oil recovery.

During the first two years of the project, six new producing wells were drilled for data acquisition wells. Vertical seismic profiles and a 3-D seismic survey were acquired to assist in interwell correlations and facies prediction.

Restricted surface access at the Nash Draw pool, caused by the proximity of underground potash mining and surface playa lakes, limits field development with conventional drilling. Combinations of vertical and horizontal wells combined with selective zone completions are being evaluated to improve production performance.

Based on the production response of similar Delaware fields, pressure maintenance is a likely requirement at the Nash Draw pool. A detailed reservoir model of the pilot area was developed, and enhanced recovery options, including waterflooding, lean gas, and carbon dioxide injection, were considered.

INTRODUCTION

The Nash Draw Brushy Canyon pool, operated by Strata Production Company (Strata), is located in Sections 12, 13, and 14 T23S, R29E, and Section 18 T23S, R30E, in Eddy County, New Mexico. General characteristics of this slope-basin clastic reservoir are listed in Table 1. Production at the Nash Draw pool (NDP) is from the basal Brushy Canyon zones of the Delaware Mountain Group of Permian, Guadalupian age.

The primary concerns at the NDP are (1) the primary oil recovery is in the order of 10% of the OOIP (original oil in place), (2) a steep initial oil production decline rate, and (3) rapidly increasing gas-oil ratios. This low recovery is caused by low reservoir energy, and low permeabilities and porosities. Initial reservoir pressure is just above the bubblepoint pressure and declines to below the bubblepoint around the wellbore after a few months of production. With the solution gas drive reservoir, oil production declines 50% in the first year, and gas/oil ratios increase dramatically. These concerns point out the importance of considering various reservoir management strategies to maximize the economic recovery of oil at the NDP. Production characteristics of similar Delaware fields indicate that pressure maintenance is a likely requirement at the NDP.

Table 1. General Characteristics of the Nash Draw Delaware Field, Eddy County, New Mexico.*

Discovery Date	1992
Trapping Mechanism	Stratigraphic Trap
Current Number of Wells	17
Current Production	490 BOPD + 2.4 MCFGPD + 500 BWPD**
Reservoir Depth	6600–7000 ft
Pay Thickness, K & L Ss.	20–50 ft
Reservoir Porosity	12–20%
Reservoir Permeability	0.2–6 md
Initial Reservoir Pressure	2963 psi
Bubble Point Pressure	2677 psi
Drive Mechanism	Solution gas drive
Oil Gravity	42.4° API
Primary Recovery Factor	10–15% oil in place
Estimated Oil in Place	25–50 Mbbl
Reserves, Primary Recovery	2.5–5 Mbbl

*From Martin et al. (1997).
**BOPD = bbl of water per day, MCFGPD = million ft^3 of gas per day, BWPD = bbl of water per day, Mbbl= million barrels.

Early in the NDP development, Strata identified three basic constraints: (1) limited areal and interwell geologic knowledge, (2) lack of an engineering method to evaluate the various producing strategies, and (3) restricted surface access that will prohibit development with conventional drilling. The limited surface access at the NDP is caused by the proximity of underground potash mining and surface playa lakes (Figure 1). The objectives of the project are (1) to demonstrate that a development drilling program and pressure maintenance program, based on advanced reservoir management methods, can significantly improve oil recovery compared with existing technology applications and (2) to transfer the advanced methodologies to oil and gas producers, especially in the Permian Basin.

Typical of small independent producers, Strata lacked the in-house expertise to address all of the needs of the project and, therefore, assembled a diverse team of experts to manage and analyze the NDP. One challenge to this type of organization is providing communication and coordination between the team members located in a diverse geographic area. Five subcontractors use e-mail, the Internet, and high capacity data transfer to successfully exchange data and conclusions between each group.

GEOLOGICAL BACKGROUND

The structural trend at the NDP is north-south to northeast-southwest, and there were at least three depositional events. The sandstone reservoirs of the basal Brushy Canyon sequence of the Delaware Mountain Group in this study lie above the Permian Bone Spring formation (Figure 2). The top of the Bone Spring formation is marked by a regionally persistent limestone, ranging from 15.2 to 30.5 m (50 to 100 ft) in thickness that provides an excellent regional mapping horizon. Regional dip is to the east-southeast in the area of the NDP. The structural dip resulted from an overprint of post-depositional tilting that is reflected in reservoir rocks of the Delaware formation and impacts the trapping mechanism in the sandstones.

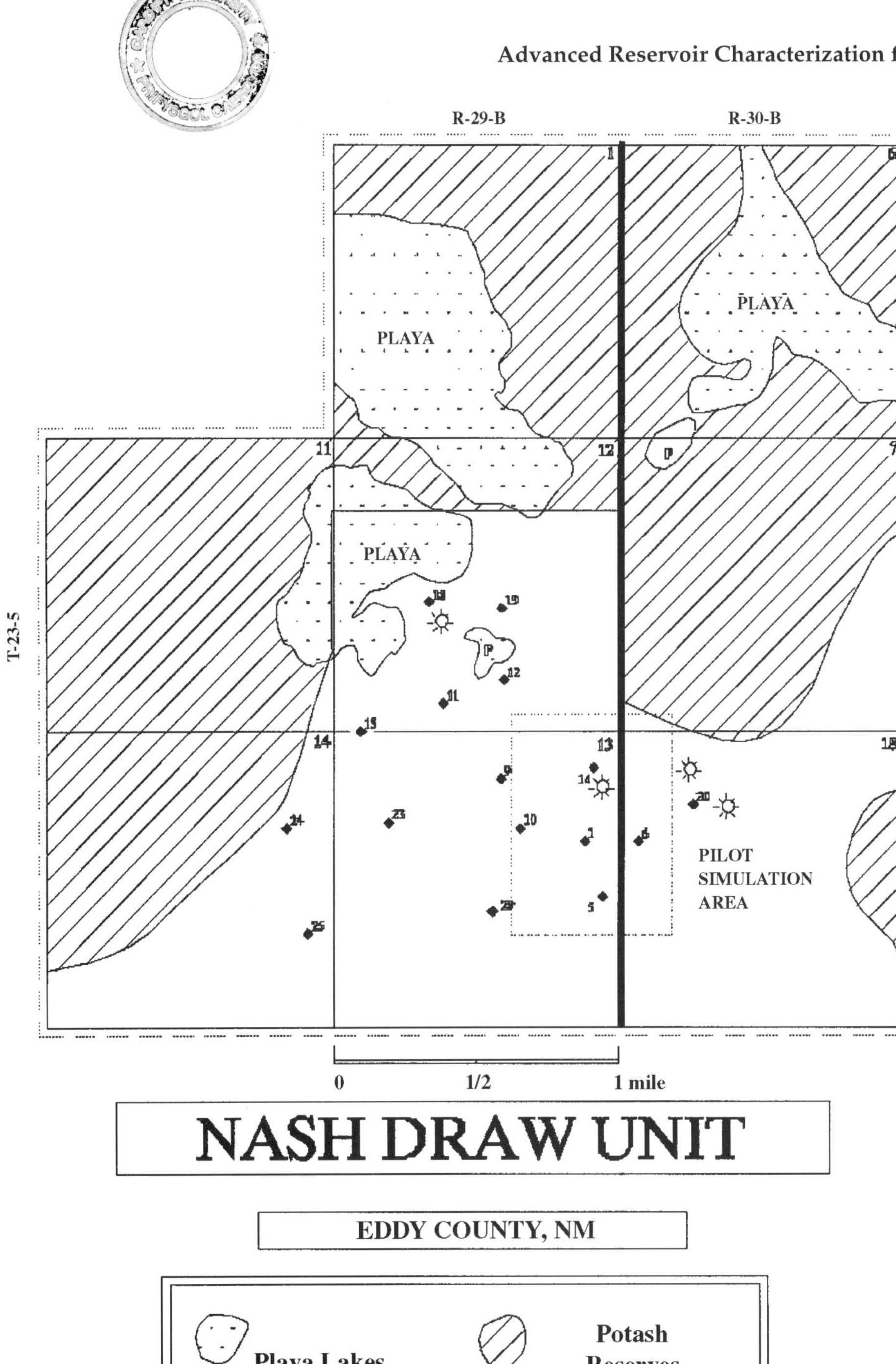

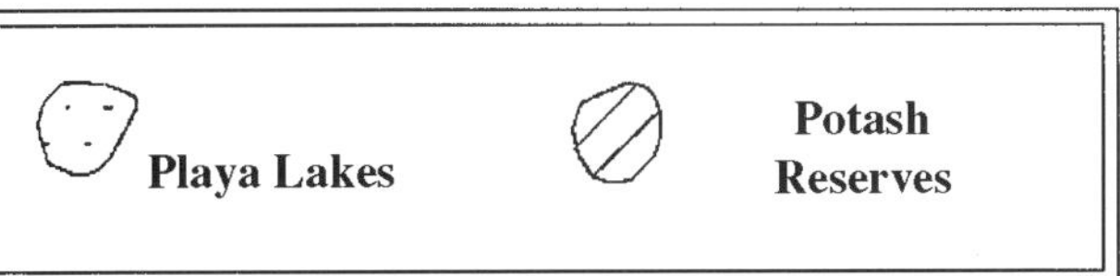

Figure 1. Map of the Nash Draw Unit area (from Martin et al., 1997).

The sandstone units of the Brushy Canyon sequence represent the initial phase of detrital basin fill in the Delaware Basin during Guadalupian time. The Delaware sandstones are deep-water marine turbidite deposits. Depositional models (Kerans et al., 1992; Kerans and Fisher, 1995) suggest that the sands were eolian-derived and were transported across an exposed carbonate platform to the basin margin. Interpretations of the associated transport mechanisms (Gardner, 1992; Fisher and Sarrthein, 1998) suggest that the clastic materials were deposited episodically, and were transported into the basin through shelf bypass systems along an emergent shelf-edge margin.

The Brushy Canyon reservoir consists of thin stacked sandstones; vertical permeability is extremely low, and horizontal permeability is poor to fair. Locally, the three intervals of interest are referred to as the "K", "K-2", and "L" sandstones, which can be correlated from well to well over large distances. The "K" and "L" sandstones, the main producing intervals of the Brushy Canyon formation, have multiple lobes, and both sandstones can be divided into four subunits. Each of the sands is a composite of a series of stacked micro-reservoirs (Figure 2) that range from 0.3 to 1.8 m (1 to 6 ft) in thickness. Lateral extent of reservoir quality facies at the NDP may only be 0.4 km (0.25 mi) or less in some areas (Murphy et al., 1996).

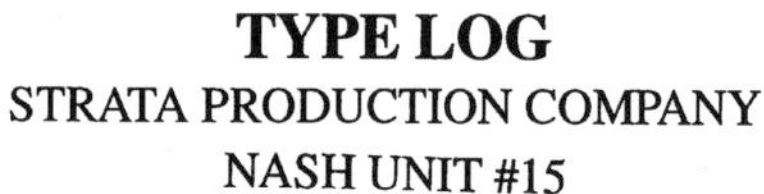

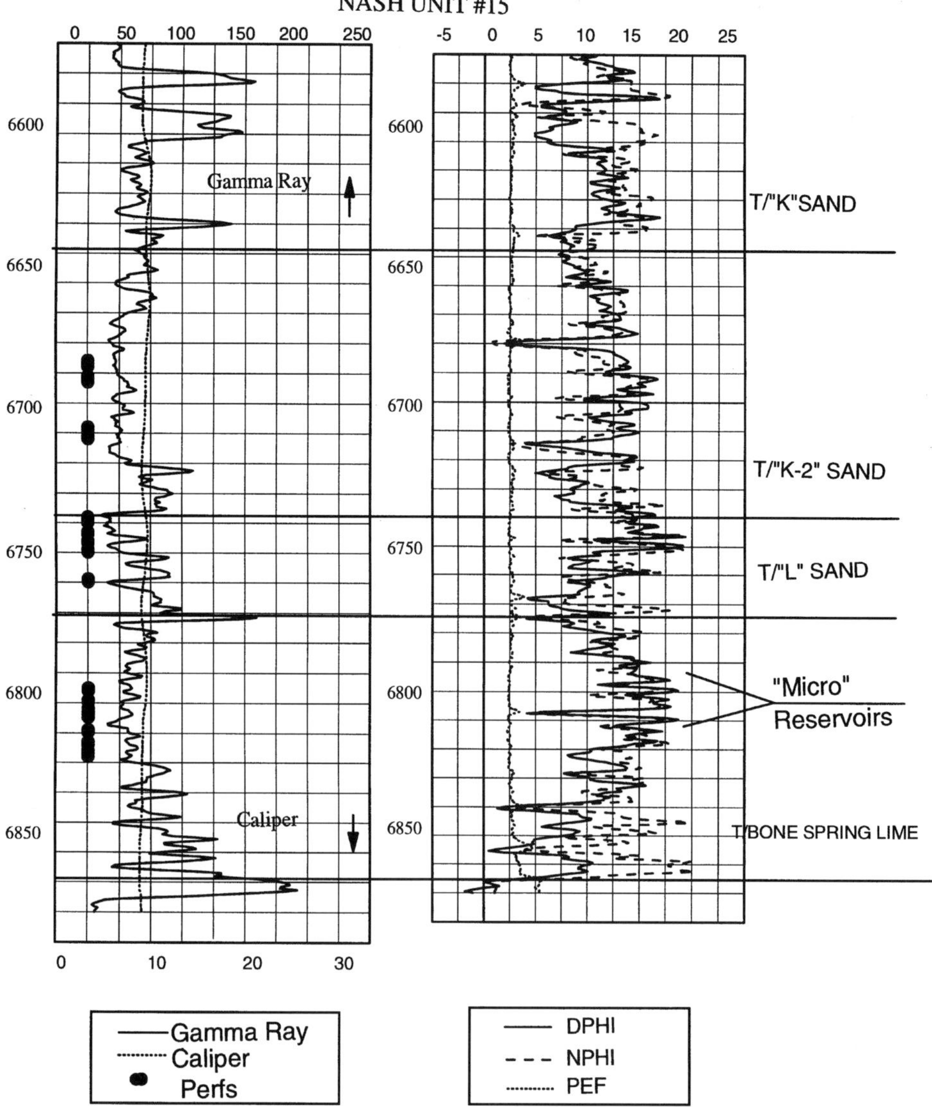

Figure 2. Type log from Nash Draw Well No. 15 showing Brushy Canyon interval as stacks of thin, multiple "micro" reservoirs separated by vertical permeability barriers.

METHODOLOGY

Data from Nearby Delaware Fields

Log and core data from wells in the E. Loving Delaware pool, from Texaco wells southeast of the NDP, and from Maralo wells offsetting the NDP were obtained and analyzed. Structure maps and cumulative oil, gas, and water production for the E. Loving pool, and logs and available core data from all three fields, were analyzed and compared to data from the NDP.

Sixteen wells in the E. Loving pool in Section 14, T23S, R28E were selected as an analogy to the NDP. These wells represent varying structural positions and corresponding production characteristics. Logs were obtained from each well, structure maps were constructed, and available core data were obtained for the wells in the study area. Structure and isopach maps were developed in the analog area using the same criteria that were used in the NDP area.

Data Acquisition at the NDP

The data acquisition portion of the project included compiling new and existing reservoir and engineering data. As part of the project, six new wells have been drilled for data acquisition, and the NDP now consists of 17 producing wells and one salt water disposal well (see Figure 1). Multiple sidewall cores were obtained for analysis from each new well, and 61.9 m (203 ft) of full core was cut for laboratory analysis from Well No. 23. The whole core obtained from Well No. 23 was cut from the "J" zone through the "L" zone. Basic core data, including porosity, permeability, oil and water saturations, grain density, show description, and lithology description, were measured for each foot of core. Conventional suites of logs (neutron porosity, formation density, gamma ray, caliper, dual lateral log, microresistivity log) were obtained in all of the wells, and a magnetic resonance tool was run in Well No. 23 for comparison to the core analysis.

The core data were used to calibrate the logs and determine pay distribution in each zone. A detailed core-calibrated log analysis of S_{or}, S_w, and porosity was applied to the digitized logs to determine the productive and the water zones in each interval. The application of porosity/permeability transforms and relative permeability data to each zone yielded flow capacity data for each interval.

By applying the core-calibrated log analysis to the entire basal Delaware section, oil-productive zones could be identified, reserves could be estimated, production rates could be predicted, and water-productive zones could be avoided. This procedure has proven to be an accurate predictive tool for new wells in the NDP.

Production, transmissibility, and capillary pressure data were combined with geological interpretations to develop reservoir maps. A detailed correlation of the basal Brushy Canyon sandstones was performed in order to better understand the lateral and vertical distribution of the reservoirs. Detailed correlations also provide a more accurate geological model for use in the reservoir simulation phase of the study. Wireline log and core data were compiled for each of the wells within and directly adjacent to the NDP for the purposes of constructing the maps for the initial structural and stratigraphic model.

Seismic Data

A vertical seismic wavetest and two vertical seismic profiles (VSPs) were recorded in Well No. 25, and a 3-D seismic survey was acquired to aid the characterization of the NDP reservoir. The VSP data, including a zero-offset and a far-offset image, were obtained to assess seismic noise caused by nearby subsurface mining and to determine the optimum Vibroseis™ parameters for the 3-D survey. As a result of the VSP work, the 3-D seismic grid at the NDP was redesigned to produce acquisition bins measuring 16.8 × 33.5 m (55 × 110 ft). During data processing, a trace interpolation was done in the source direction to create interpretation bins measuring 16.8 × 16.8 m (55 × 55 ft). For the 3-D survey at the NDP, a total of 917 source points were recorded to create a 3-D coverage across an area of 20.4 km^2 (7.875 mi^2).

Reservoir Modeling

The structure and isopach maps were loaded into Landmark's Stratamodel7® program, and a preliminary 3-D geological layer model was developed. Digitized maps of the interpreted horizons ("J," "K," "K-2," "L," and the top of the Bone Spring formation) were imported into SGM™ to create a stratigraphic framework model of the NDP. Initially, both the "K" and "L" sandstones were divided into four subunits. The sandstones were correlated laterally from well to well in the NDP. Gross isopach, net porosity isopach, and log-derived net pay maps were constructed for each of the subunits of the "K" and "L" sands, as well as the "K-2" and "J" sands. The maps were contoured to conform to the overall gross interval isopach maps for the respective pay zones that were used to construct the geological model. Because the producing zones and subzones are relatively thin, great care had to be exercised to prevent intersections of the horizons. It is also critical that the surfaces tie to the well picks of the lithological markers in the well traces. In general, the most successful approach to this problem was based on the use of gross isopach thickness interpretations building from the structural top of the Bone Springs Formation up to the structural top of the "J" sandstone.

The next major step was the development of a well attribute model. This activity was supported by the engineering database. For each of the 17 NDP wells, the following attributes were imported into the well model: neutron porosity and gamma ray, interpreted porosity and permeability, perforated interval and fractured interval, net pay, and water saturation. In some instances, these attributes were available on a foot-by-foot basis for one or more of the producing zones. Not all of the attributes were available for each well. For reservoir simulation, the most important reservoir attributes are fluid conductivity and rock matrix storage capacity. The distribution of these properties throughout the NDP has been based on the well attribute model. Within SGM, these distributions are interpolated deterministically, that is, weighted by the reciprocal of the square of the distance between the location of interest and nearby wells in the reservoir model.

A detailed reservoir model of the basal Brushy Canyon sandstones in the proposed pilot injection area (which contains the oil lobe supporting the pilot) was developed for reservoir simulation studies. Reservoir attributes including porosity, relative permeability, and oil and water saturations were distributed vertically and laterally throughout the layers in the simulation model.

Reservoir Simulation Forecasts

A reservoir simulation model for the pilot area envisioned that a single well in the pilot area would be converted to injector status to test the efficacy of injecting water, lean gas (immiscible), or CO_2 (immiscible or miscible) to improve oil recovery at the NDP. Eclipse 100® was used for the immiscible hydrocarbon gas cases and VIP-COMP® for the CO_2 cases; however, the reservoir description was the same for all forecasts, and was based on the history match obtained with Eclipse 100. The following tasks were required to complete the pilot simulation phase: possible scale-up of lithological units, interpolation of geological attributes on the simulation grid, validation of pilot simulation model, and design and execution of prediction cases.

RESULTS AND DISCUSSION

A number of planning sessions of the "virtual project team" were held during the early stages of the NDP project. As a result of these meetings, a project plan was developed for the first phase of the project.

This paper will highlight early results of the NDP project. Detailed results are contained in the annual project reports (Murphy et al., 1996, 1997, 1998) and in a recent technical paper (Martin et al., 1997).

Comparison of NDP Data to Other Nearby Delaware Fields

Core data from all three nearby Delaware fields, the E. Loving pool, the Texaco, and the Maralo fields offsetting the NDP, correlate very well in the "L" zone, but there is less agreement in the data from the "K" and "K-2" zones (see Figure 3). Rock characteristics in the analog area are similar enough to those in the NDP to allow accurate comparisons of the production data and characteristics of the two areas.

Petrophysical Data from the NDP

Sidewall core data from each well in the NDP were compiled, and porosity/permeability (ϕ/k) relationships were determined. These relationships were compared to the whole-core data and found to be in good correlation (see Figure 4).

The core data were used to prepare a transform to correct the log cross-plot porosity to yield a true porosity based on the whole-core porosity. The relationship between cross-plot log porosity (logs run on a limestone matrix) and core porosity was determined to be

$$\phi_{corr} = \left(\phi_{x-plot} - 3.7685\right)/0.848294 \qquad (1)$$

Permeability was plotted against porosity, and a regression analysis was performed to generate equations to fit the data. These relationships were used to predict the permeability of each zone based on corrected log porosities. Permeability/porosity distributions were prepared for each zone as presented in Figure 5.

Analysis of whole-core and drilled sidewall core data have shown that individual Brushy Canyon micro-reservoirs may be oil bearing, water-bearing, or transitional in nature (Murphy et al., 1996). In addition, the sandstones have been found to have little or no vertical permeability from one micro-reservoir to the next.

Examination of the whole core showed that the reservoir rock is fine to very fine grained, massive to very thinly laminated. There is some evidence of high energy turbulence as exhibited by sets of low to medium angle cross bedding within some of the sandstone units. Evidence of bioturbation occurs in some of the shaly and silty zones. There is also carbonate clastic debris present in some intervals within the core. Examination of the core under ultraviolet light shows the discontinuous character of the hydrocarbon distribution throughout the reservoir. This correlates with the erratic vertical distribution of calculated oil and water saturations seen in the log analysis.

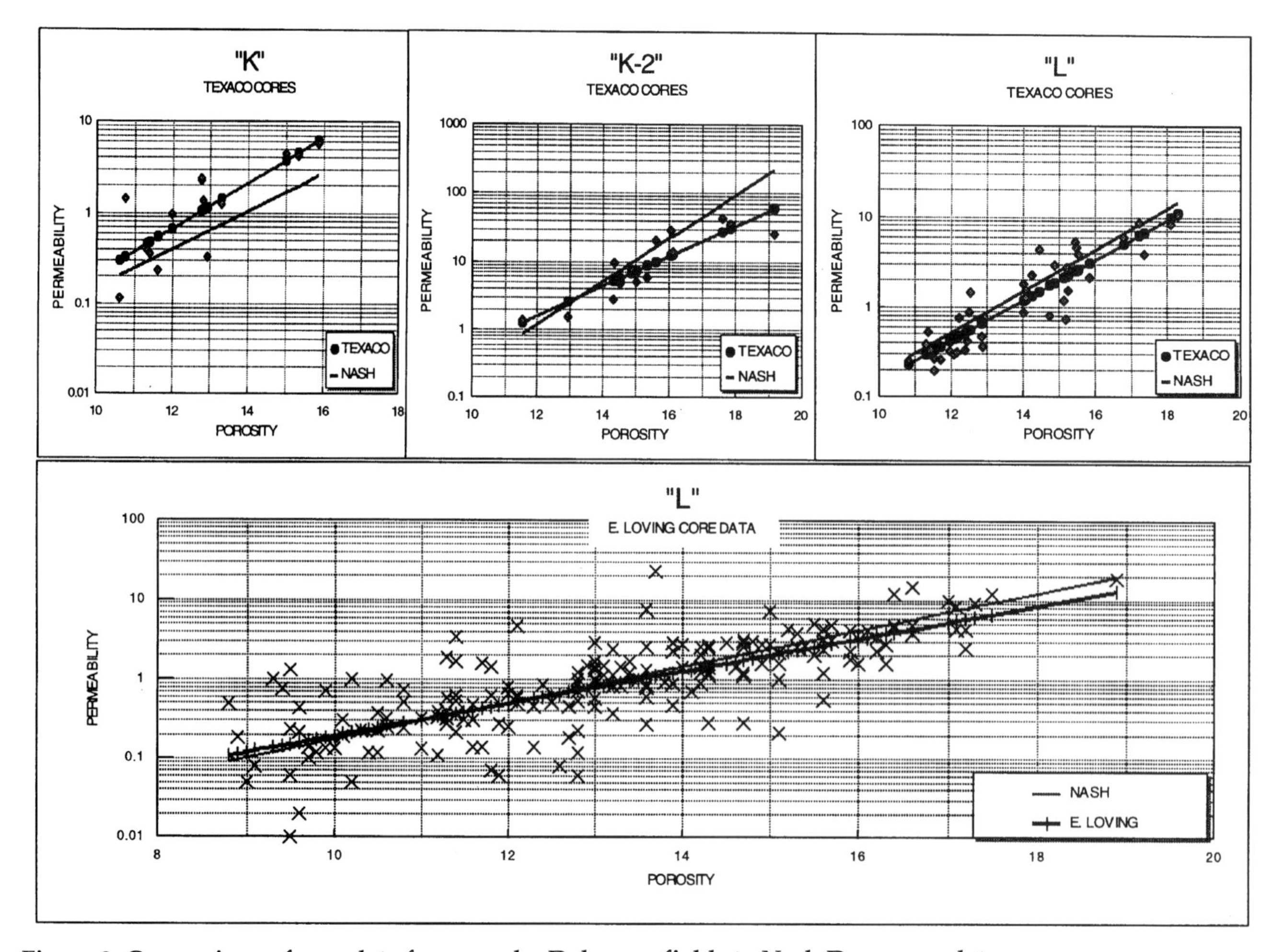

Figure 3. Comparison of core data from nearby Delaware fields to Nash Draw core data.

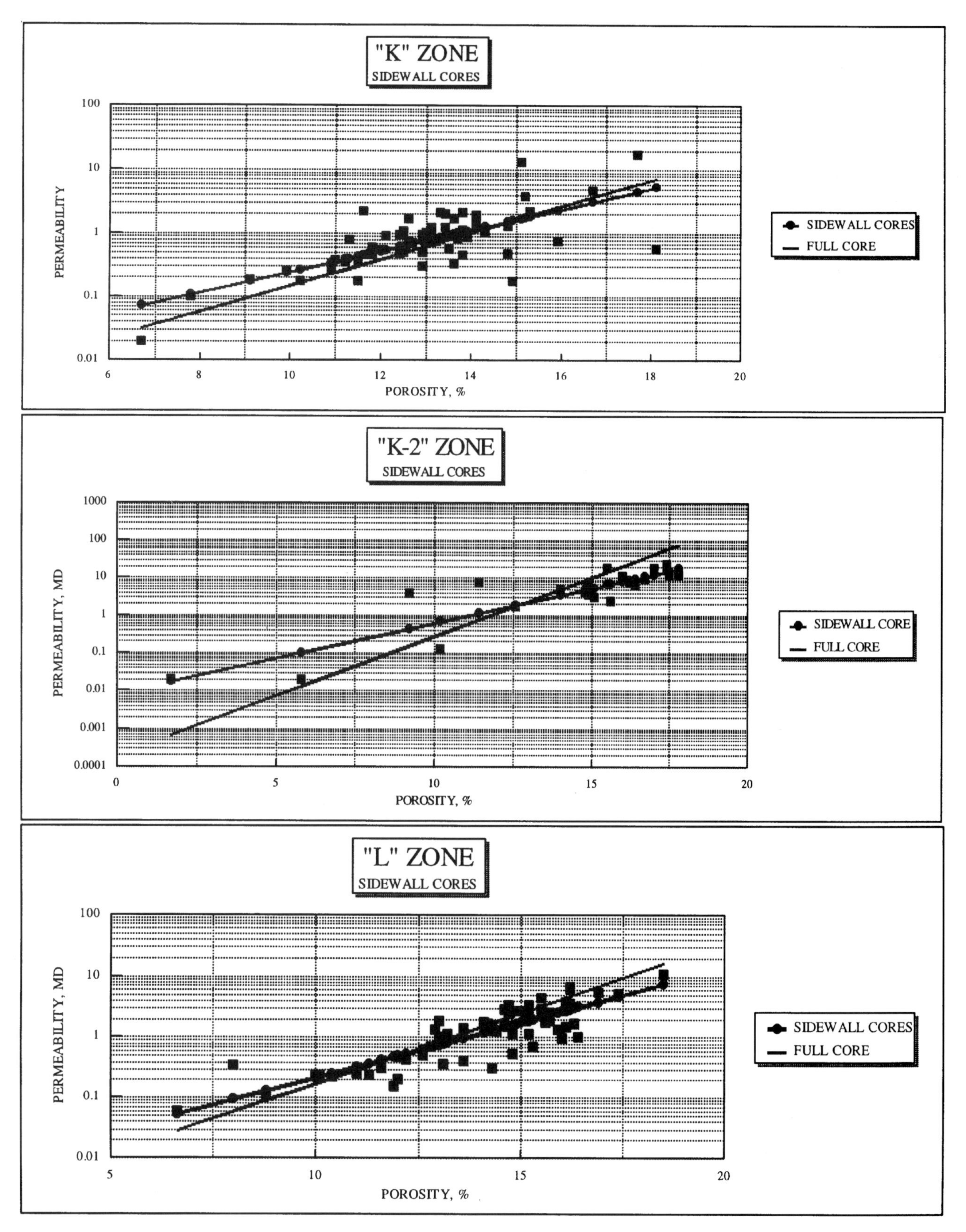

Figure 4. Comparison of Nash Draw sidewall core data to full core data acquired in Nash Draw Well No. 23.

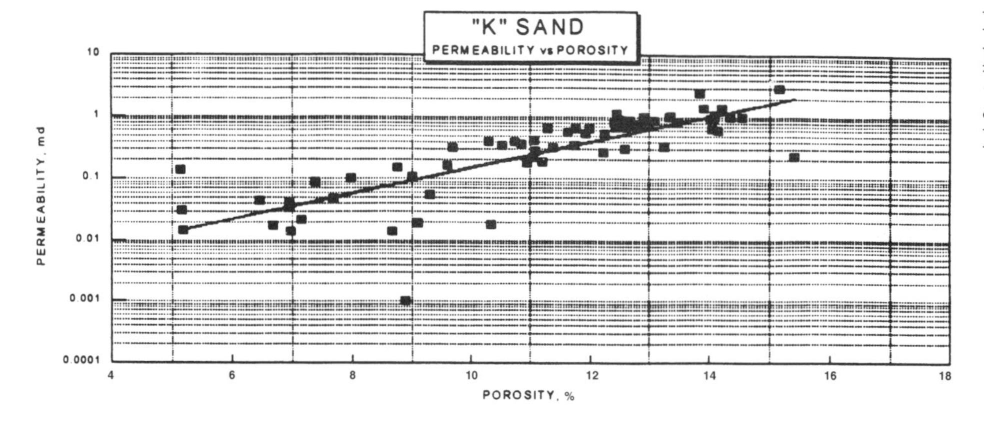

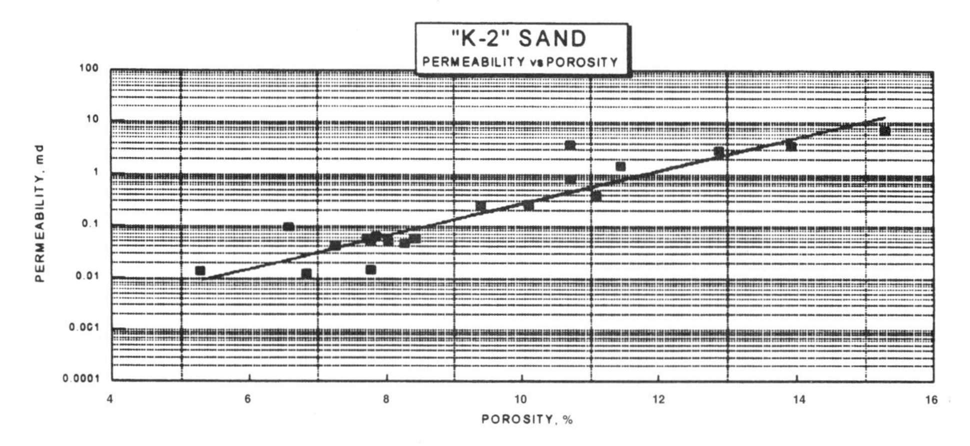

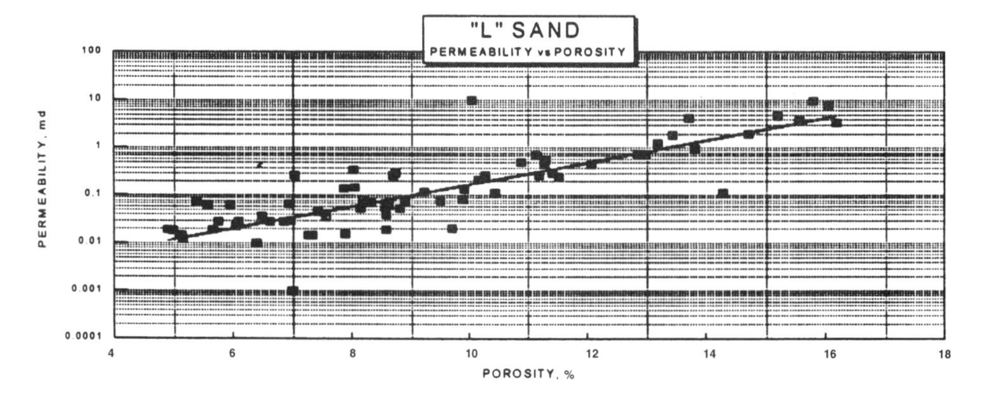

Figure 5. Correlation of permeability to porosity for Brushy Canyon sandstones at the Nash Draw pool (from Martin et al., 1997).

Mineralogy of the "K" and "L" sandstones are similar. Both zones contain some clays—illite and chlorite. Compared to the "L" sandstone, the "K" sandstone has up to 2% more chlorite that occludes permeabilities and may have influenced higher initial water saturations.

The whole-core data were used to calibrate the logs and determine pay distribution in each zone. By performing a detailed core calibrated log analysis of S_{xo}, S_w, and porosity, a detailed analysis was applied to the digitized logs to determine the productive and water zones in each interval. The application of porosity/permeability transforms and relative permeability data to each zone yielded flow capacity data for each interval. These data were summed for each layer and input into the reservoir simulator.

Permeability (k_a)/Porosity Relationships for Each Interval

Porosity/permeability relationships were developed from the sidewall cores and full core analyses. The flow unit variables *a* and *b* are given in Table 2 for the power function

$$k = 10^{a\phi - b} \tag{2}$$

A data file was prepared for each well that included digitized log files, perforations, cement programs, tracer logs, completion information, and frac treatments. These data were used to allocate production, estimate drainage areas, determine productivity, estimate saturations for each interval, and prepare data files for reservoir simulation.

Table 2. Permeability/Porosity Correlations for Variables a and b in Equation 2.

Flow Unit	Sidewall Core a	Sidewall Core b	Full Core a	Full Core b
K	0.164915	2.25338	0.207675	2.8858
K-2	0.186535	2.06872	0.315038	3.69966
L	0.179787	2.45666	0.231250	3.06330

Using core and log data, each well was calibrated to match production, net pay, and transmissibility. By calculating a kh/μ value for each interval, production rates and cumulative production were allocated to each interval. The transmissibility for each layer was used as input into the reservoir simulation model along with saturation data to determine the producing characteristics of each layer. Figure 6 shows the transmissibility values used to calculate production from the various zones for all of the 16 wells in the NDP. These data show that the bulk of the oil production at the NDP comes from the "L" sandstone, but much of the water is produced from the "K" and "K-2" sandstone, if the latter zone is present.

Reservoir Model of the NDP

The structural relationship between the five major producing horizons in the NDP is illustrated in Figure 7. The step-bench sequence is a typical depositional characteristic of the basal Delaware zones in this area. Typical benches are 0.8–1.6 km (0.5–1.0 mi) wide with dip rates of 0.8% to 1.9%. Typical steps are 0.4–0.8 km (0.25–0.5 mi) wide with dip rates of 3.3% to 8.0%. Better producing wells are located on the benches, and poorer producers are located on the steps in the NDP as well as in nearby Delaware fields (Murphy et al., 1997).

Initially, it was believed that the NDP was composed of thinly bedded channel sandstones more or less continuously distributed between wells. The initial geological interpretation suggested that the Brushy Canyon sandstones in the NDP appeared to be blanket type sands; however, data and analyses obtained in the project suggest the sandstones at the NDP are laterally discontinuous and complex in nature. Over the course of the first year of the project, three "generations" of geological models were developed based on evolving interpretations of the structure of the NDP. In the first generation, a full NDP model was developed from the initially available geological interpretation based on logs and cores. The second generation model was based on these data plus newly interpreted pressure transient data. The latest version reflects a geological interpretation, based on the 3-D seismic data, that indicated there is considerable compartmentalization within the NDP. The integration of the log, core, and pressure transient data led to an interpretation of the NDP with three noncommunicating lobes of oil. The proposed pilot injection area is confined to one of these

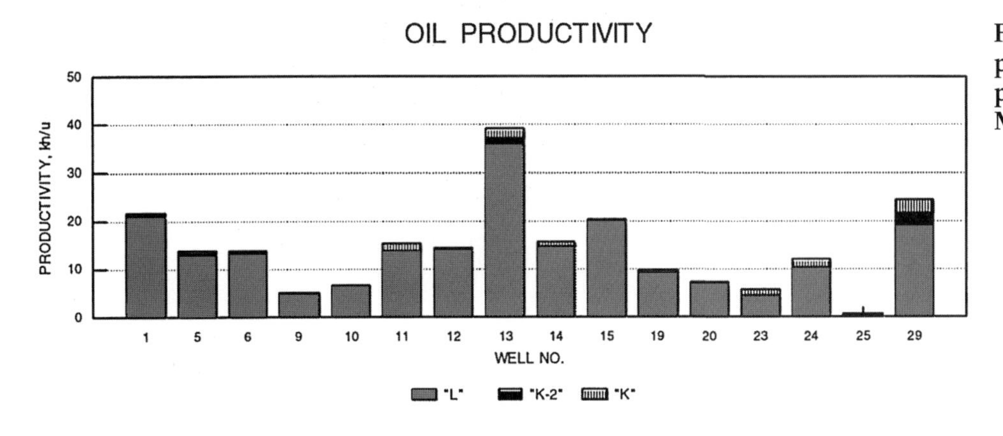

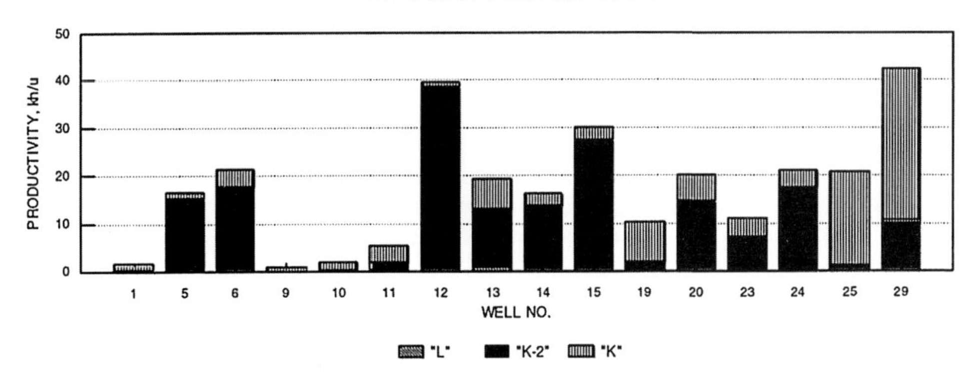

Figure 6. Oil and water productivity values used for production allocations (from Martin et al., 1997).

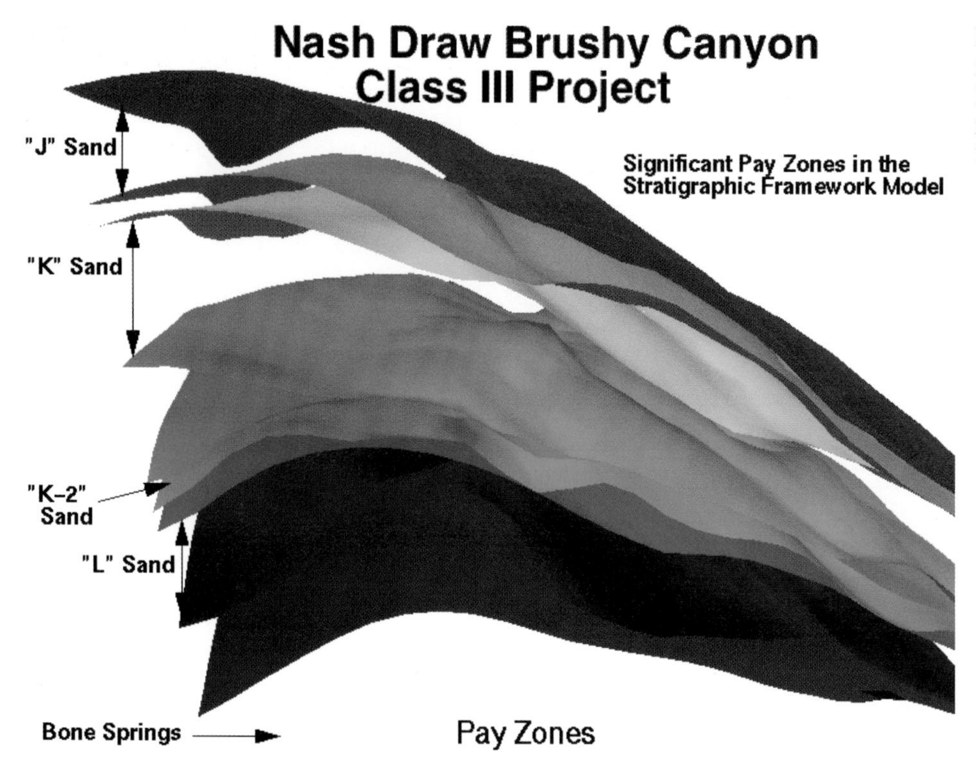

Figure 7. Stratigraphic framework model of the Nash Draw Brushy Canyon reservoir (from Martin et al., 1997).

lobes. Since approximately 90% of the oil produced from the five wells in this lobe can be traced back to the "L" zone (see Figure 6), only this zone is included in the simulation model studies.

Due to the highly lenticular distribution of oil within the four subzones identified in the "L" zone, a 20 layer simulation model (Figure 8) was chosen for the pilot area; that is, five proportional layers for each of the "La," "Lb," "Lc," and "Ld" subzones. This resolution was the minimum required to capture the nature of the thin beds of the sandstones in the "L" zone. The distributions of porosity and water saturation are shown in Figures 9 and 10, respectively. These figures illustrate the highly lenticular nature of the Brushy Canyon sandstones in the NDP pilot area.

VSP and 3-D Seismic Results

There were multiple reasons for shooting the 3-D seismic survey at the NDP. One reason was to develop a more refined geological model that gave better resolution of the structural aspects of the trap. A second reason was to try to determine whether the reservoirs in the basal Brushy Canyon sequence could be imaged using thin-bed seismic techniques.

The VSP calibration data acquired in Well No. 25 established the top of the Bone Spring as a robust reflection peak, the "L" sequence was associated with the first reflection trough immediately above the Bone Spring, and the "K" sequence began just above the first reflection peak above the Bone Spring (Figure 11). The reflection character of both "K" and "L" changes significantly north of the well, which implies variation in the reservoir system and is a direct indication of stratigraphic changes or facies changes, or both.

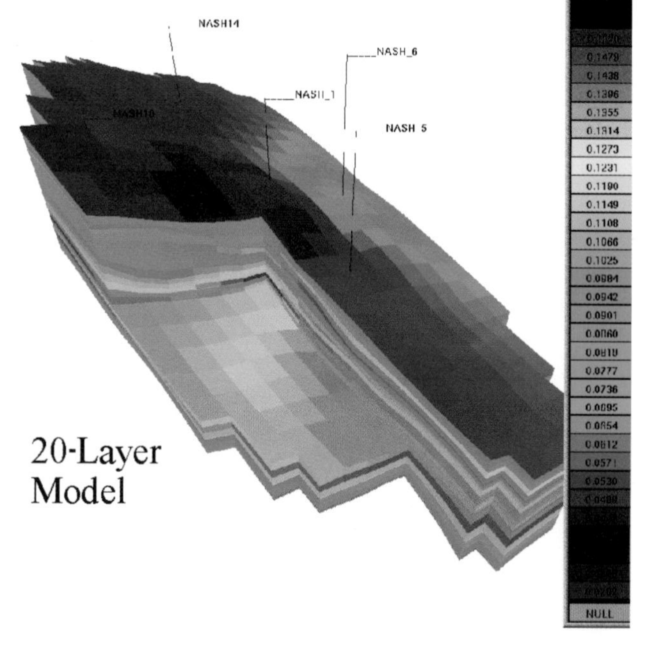

Figure 8. Twenty layer model used in the simulation studies of the Nash Draw pilot area.

Figure 12 shows the source-receiver line geometry used for the 3-D survey. Results from the 3-D seismic data were used to generate amplitude maps (Figures 13, 14) that show high amplitude areas and the producing trends. Well productivity appears to be directly correlatable to the amplitude of the dominant "K" reflection peak and "L" reflection trough. Future wells will be drilled to confirm this analogy and evaluate targeted drilling of the seismic anomalies. Details

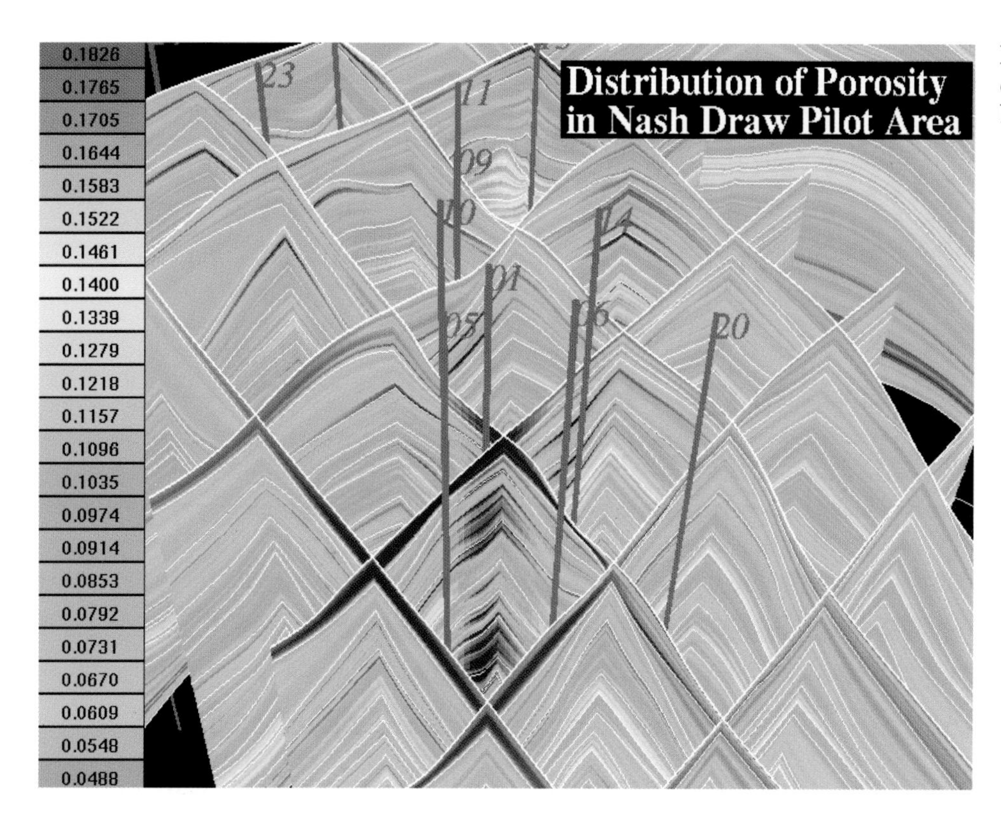

Figure 9. Porosity distribution in the Nash Draw pilot area.

of the acquisition and interpretation of the seismic results are presented elsewhere (Murphy et al., 1996; Hardage et al., 1998a, 1998b).

When the 3-D seismic dataset was interpreted, it became apparent that the original conception of the NDP as a collection of thin channel sands continuously distributed between wells probably was not correct. In particular, on the basis of the interpreted seismic amplitude data, the area around the proposed pilot centered at NDP Well No. 1 was reduced to a "lobe" of approximately 121 hectares (300 acres) containing NDP Wells Nos. 1, 5, 6, 10, and 14 (Murphy et al., 1997; Hardage, et al., 1998b). Moreover, the interpreted seismic data indicate that the NDP may be highly compartmentalized, and that some of the compartments, for some sand sequences in the "L" zone, may be much smaller than 121 ha (300 ac).

Reservoir Simulation

Reservoir simulation results indicate that the permeabilities of the Brushy Canyon sandstones are too low for waterflooding to be effective. Forecasts were made for two possible enhanced recovery scenarios: immiscible gas (both hydrocarbon and CO_2) injection and miscible CO_2 injection. Details of the reservoir simulation studies are presented elsewhere (Murphy et al., 1997, 1998).

Because the pilot area is the most developed area of the NDP, this area has a greater well density and the wells in this area have been producing for a longer period of time; consequently, there is very little natural energy left in the pilot node at the inception of injection, and any fractures or zones with free gas would provide a ready conduit for early breakthrough

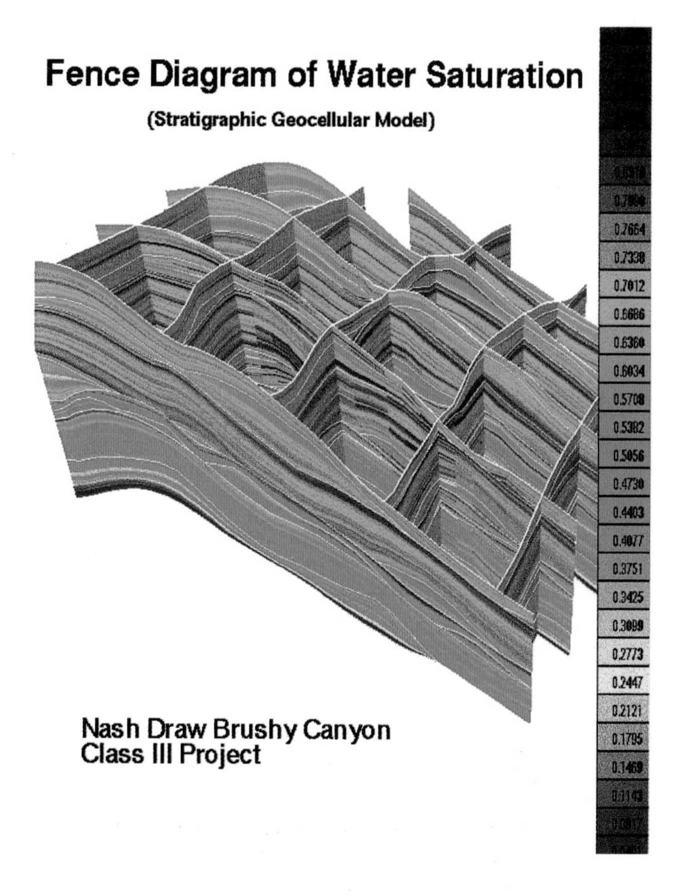

Figure 10. Water saturation distribution in the Nash Draw pilot area.

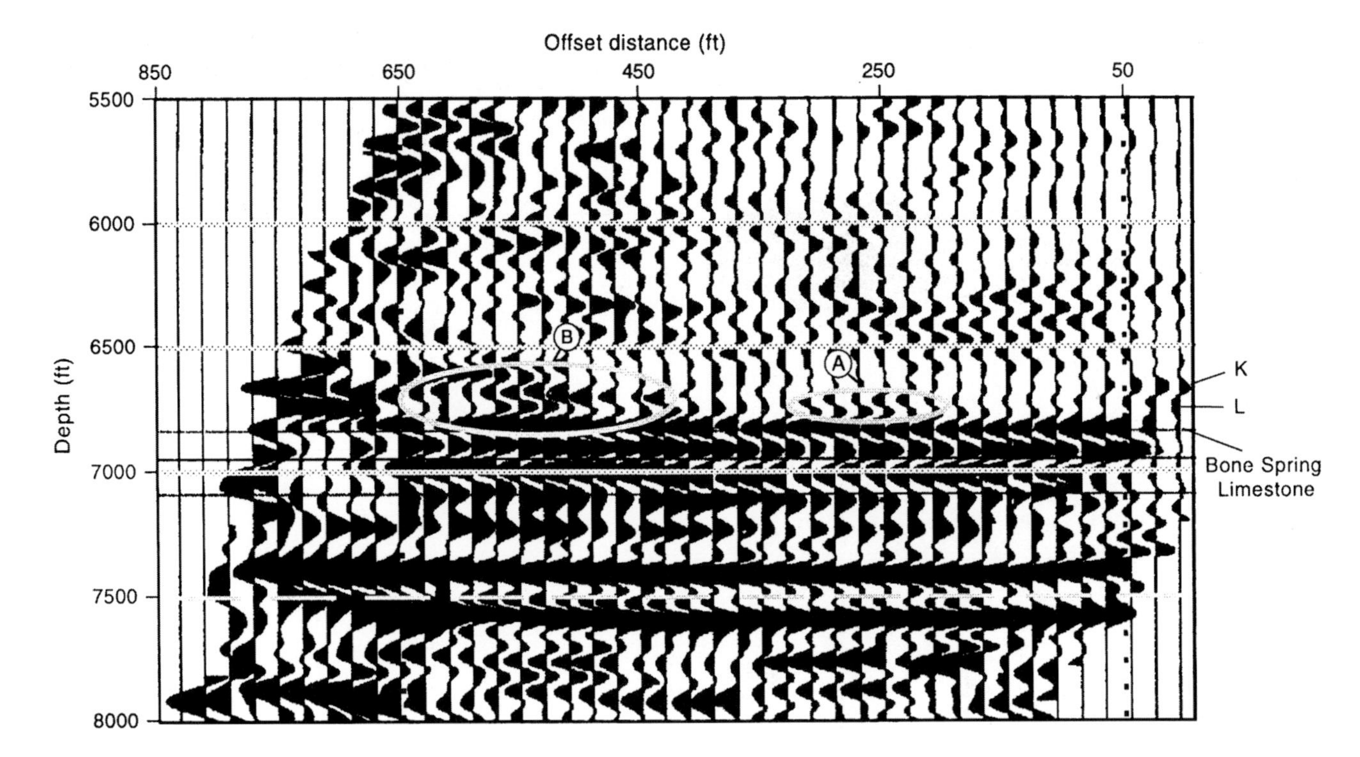

Figure 11. VSP (vertical seismic profile) image showing wavelet peak in the "K" sandstone and wavelet trough in the "L" sandstone.

of injected gas. The simulation results indicate that implementation of a gas injection pressure maintenance scheme after the drainage area pressures have declined below 3,447 kPa (500 psi) will not be successful in improving oil recovery. Even for higher initial drainage area pressures around 10,342 kPa (1500 psi), the simulation studies indicate that immiscible gas injection will be of marginal value. On the other hand, the implementation of pressure maintenance, specifically gas injection, early in the development of the pattern could have doubled the recovery of oil in the five spot pattern during the early years of production.

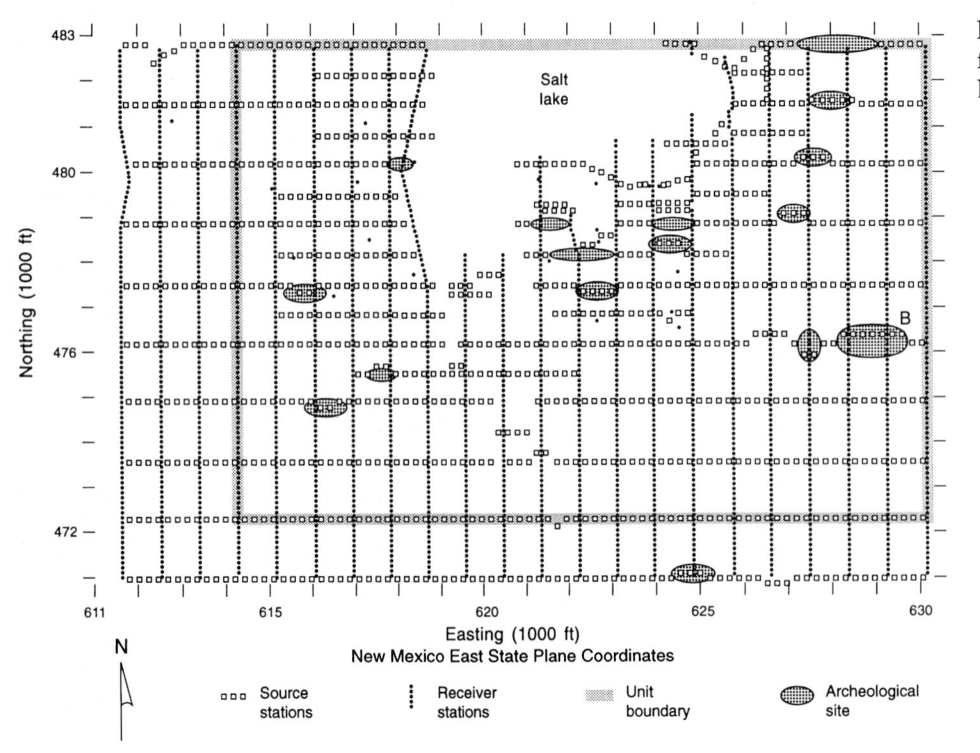

Figure 12. Geometry used for acquisition of the Nash Draw 3-D seismic survey.

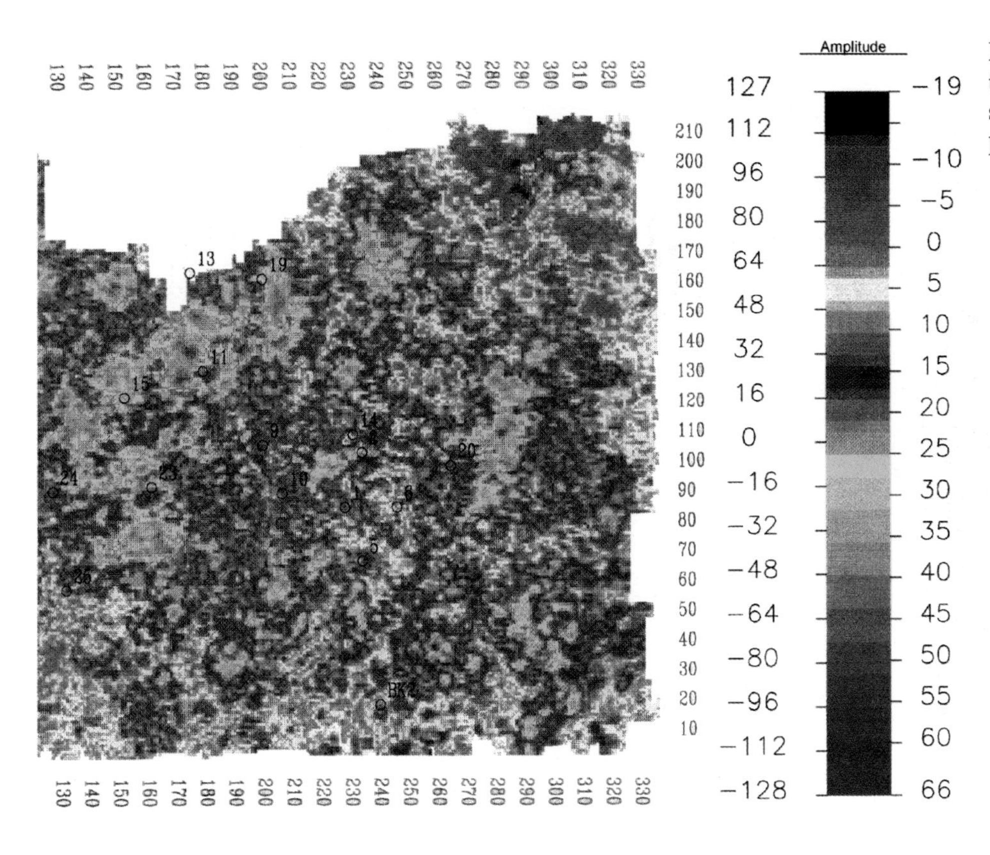

Figure 13. Seismic amplitude map for the "K" sandstone at the Nash Draw pool.

Immiscible gas injection at pressures above 17,237 kPa (2500 psi) might lead to the recovery of enough additional oil to merit at look at economics.

Simulation results for miscible injection in the NDP pilot area indicate that CO_2 injection may be a viable alternative for improved oil recovery for this field. Compared to a continued operations case, increased oil recovery in the range of 2% to 5% of OOIP was observed for several different CO_2 miscible scenarios during the 10 yr of the forecast. CO_2 injection, if implemented before the pressure has declined to below about 10,342 kPa (1500 psi), might be successful, but economics of the process would need to be evaluated.

Thus, a summary of the reservoir simulation results is as follows:

(1) The low permeabilities at the NDP will preclude waterflooding, but immiscible gas injection may be viable if initiated early, and if undeveloped regions of the field can be found that have not been pressure depleted.

(2) Areas of the field already under production may be candidates for CO_2 injection if pressures have not declined too much.

STATUS

The reservoir characterization, geological modeling, seismic interpretation, and simulation studies provided a detailed model of the Brushy Canyon zones. This model is being used to predict the success of different reservoir management scenarios and to aid in determining the most favorable combination of targeted drilling, pressure maintenance, well stimulation (Stubbs et al., 1998), and well spacing to improve recovery from the Nash Draw pool.

The proposed pressure maintenance injection was not conducted because the pilot area was pressure depleted and the seismic results suggest the pilot area is compartmentalized. Because reservoir discontinuities would reduce the effectiveness of any injection scheme, the pilot area will be reconsidered in a more continuous part of the reservoir if areas can be located that have sufficient reservoir pressure.

CONCLUSIONS

Some conclusions can be drawn from this result of this study:

(1) The Brushy Canyon reservoir at the NDP is much more complex than initially indicated by conventional geological analysis, and, while the original concept pictured the NDP as a collection of thin channel sands continuously distributed between wells, the results from this work show the subzones within the sandstones are lenticular and are not always continuous from well to well.

(2) Although the original evaluation was that both the "K" and "L" sandstones were the major oil producing intervals, the results of this study show the primary oil productive zone at the NDP is the "L" sandstone.

(3) Portions of the NDP reservoir, including the initial pilot injection area, appear to be compartmentalized, which could adversely affect production response resulting from pressure maintenance or enhanced recovery projects.

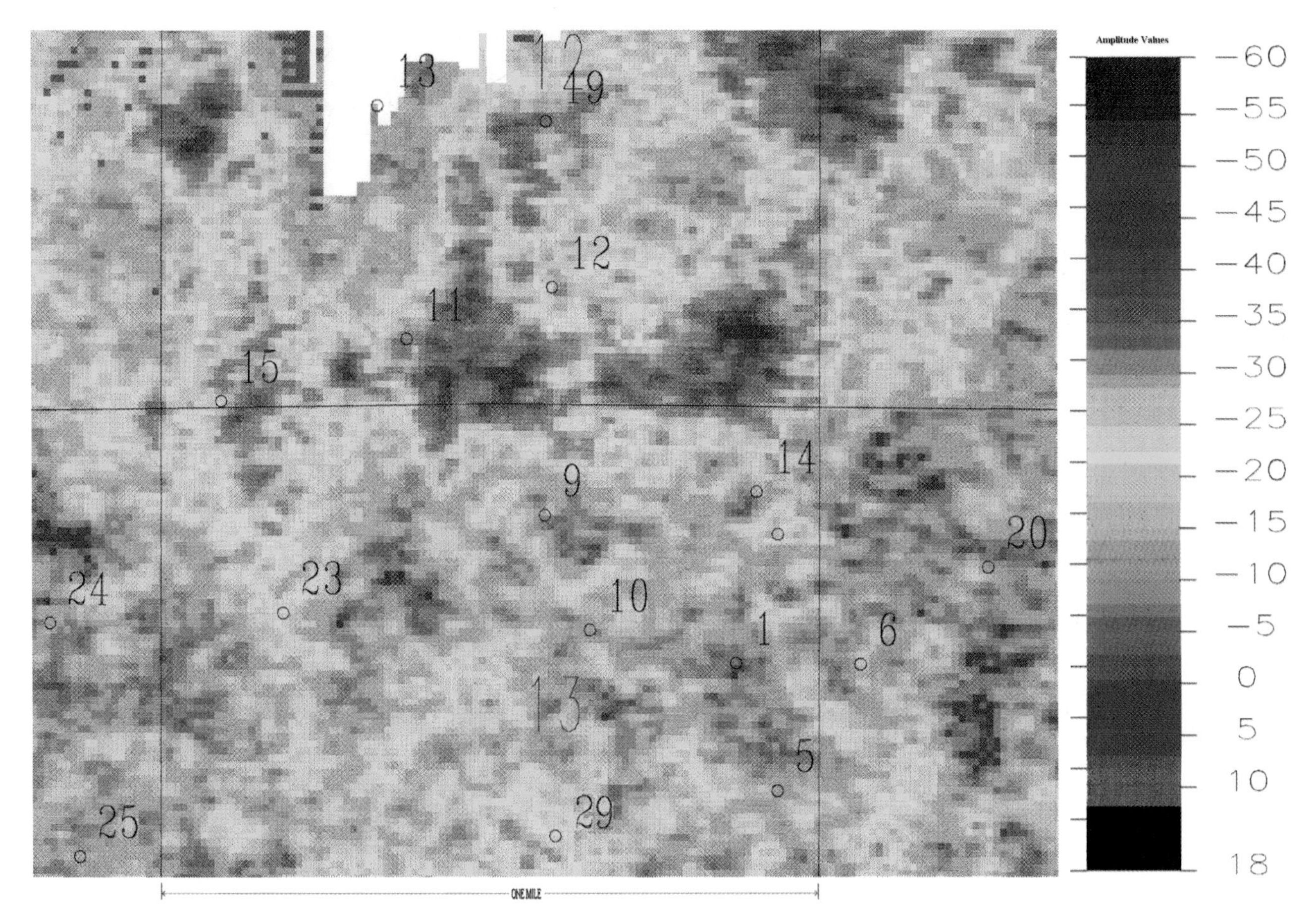

Figure 14. Seismic amplitude map for the "L" sandstone at the Nash Draw pool (from Martin et al., 1997).

(4) By conducting pre-survey VSP wave testing and by careful processing of 3-D seismic data, the thin-bed turbidite reservoirs at the NDP could be imaged, and the individual Brushy Canyon sandstones could be resolved.

ACKNOWLEDGMENTS

The authors acknowledge the support and leadership of Mark B. Murphy, president of Strata Production Company. Because the advanced reservoir management project at the Nash Draw pool would not have been initiated without the partial project funding from the U.S. Department of Energy, that financial support is gratefully acknowledged.

REFERENCES CITED

Fisher, A.G., and Sarrthein, M., 1998, Airborne silts and dune-derived sands in the Permian of the Delaware Basin: Journal of Sedimentary Petrology 58, p. 637-643.

Gardner, M.H., 1992, Sequence stratigraphy of eolian-derived turbidites: deep water sedimentation patterns along an arid carbonate platform and their impact on hydrocarbon recovery in Delaware Mountain Group reservoirs, Permian Exploration and Production Strategies: Applications of Sequence Stratigraphic and Reservoir Characterization Concepts, D.H. Mruk and B.C. Curran (eds.) West Texas Geological Society, Inc., Publication No. 92-91, p. 7–11.

Hardage, B.A., et al., 1998a, 3-D seismic imaging and interpretation of Brushy Canyon slope and basin thin-bed reservoirs, Northwest Delaware Basin: GEOPHYSICS, Vol. 63, No. 5, Sept.–Oct. 1998, p. 1507–1519.

Hardage, B.A., et al., 1998b, 3-D instantaneous frequency used as a coherency/continuity parameter to interpret reservoir compartment boundaries across an area of complex turbidite deposition: GEOPHYSICS, Vol. 63, No. 5, Sept.–Oct. 1998, p. 1520–1531.

Kerans, C., et al., 1992, Styles of sequence development within Uppermost Leonardian through Guadalupian Strata of the Guadalupe Mountains, Texas and New Mexico, Permian Exploration and Production Strategies: Applications of Sequence Stratigraphic and Reservoir Characterization Concepts, D.H. Mruk and B.C. Curran (eds.): West Texas Geological Society, Inc. Publication No. 92-91, p. 1–7.

Kerans, C., and Fisher, W.M., 1995, Sequence hierarchy and facies architecture of a carbonate-ramp system: San Andres Formation of Algerita Escarpment and Western Guadalupe Mountains: University of

Texas Bureau of Economic Geology, Report of Investigation No. 235, 86 p.

Martin, F.D., et al., 1997, Reservoir characterization as a risk reduction tool at the Nash Draw pool: Proceedings, Reservoir Engineering, 1997 SPE Annual Technical Conference & Exhibition, San Antonio, Oct. 5-8, p. 751-766.

Murphy, M.B., et al., 1996, Advanced oil recovery technologies for improved recovery from slope-basin clastic reservoirs, Nash Draw Brushy Canyon Pool, Eddy County, N.M.: first annual report to the U.S. Department of Energy, DOE Cooperative Agreement No. DE-FC-95BC14941, 79 p.

Murphy, M.B., et al., 1997, Advanced oil recovery technologies for improved recovery from slope basin clastic reservoirs, Nash Draw Brushy Canyon pool, Eddy County, N.M.: second annual report to the U.S. Department of Energy, DOE Cooperative Agreement No. DE-FC-95BC14941, 56 p.

Murphy, M.B., et al., 1998, Advanced oil recovery technologies for improved recovery from slope basin clastic reservoirs, Nash Draw Brushy Canyon Pool, Eddy County, N.M.: third annual report to the U.S. Department of Energy, DOE Cooperative Agreement No. DE-FC-95BC14941, 58 p.

Stubbs, B.A., et al., 1998, Using reservoir characterization results at the Nash Draw pool to improve completion design and stimulation treatments: paper SPE 39775 presented at the 1998 Permian Basin Oil and Gas Recovery Conference, Midland, March 23–26, 15 p.

Nevans, J.W., et al., An integrated geologic and engineering reservoir characterization of the North Robertson (Clear Fork) unit, Gaines County, Texas, 1999, *in* R. Schatzinger and J. Jordan, eds., Reservoir Characterization-Recent Advances, AAPG Memoir 71, p. 109–124.

Chapter 8

An Integrated Geologic and Engineering Reservoir Characterization of the North Robertson (Clear Fork) Unit, Gaines County, Texas

Jerry W. Nevans
Fina Oil and Chemical Company
Midland, Texas, U.S.A.

James E. Kamis
BTA Oil Producers
Denver, Colorado, U.S.A.

David K. Davies
Richard K. Vessell
David K. Davies & Associates, Inc.
Kingwood, Texas, U.S.A.

Louis E. Doublet
Thomas A. Blasingame
Texas A&M University
College Station, Texas, U.S.A.

ABSTRACT

An integrated geological/petrophysical and reservoir engineering study has been performed for a large, mature waterflood project (>250 wells, 80% water cut) at the North Robertson (Clear Fork) Unit, Gaines County, Texas. The primary goal of the study was to develop an integrated reservoir description for "targeted" 10-ac (4-ha) infill drilling and future recovery operations in a low-permeability carbonate reservoir. Integration of geological/petrophysical studies and reservoir performance analyses provided a rapid and effective method for developing a comprehensive reservoir description.

This reservoir description can be used for reservoir flow simulation, performance prediction, infill targeting, waterflood management, and optimizing well developments (patterns, completions, and stimulations). The following analyses were performed as part of this study:

- Geological/petrophysical analyses: (core and well log data)
  - –Rock typing based on qualitative and quantitative visualization of pore-scale features.

–Reservoir layering based on rock typing and hydraulic flow units.
–Development of a core-log model to estimate permeability using porosity and other properties derived from well logs. The core-log model is based on "rock types."

- Engineering analyses: (production and injection history, well tests)
  –Material balance decline type curve analyses performed to estimate total reservoir volume, formation flow characteristics (flow capacity, skin factor, and fracture half-length), and indications of well/boundary interference.
  –Estimated ultimate recovery analyses yield movable oil (or injectable water) volumes, as well as indications of well and boundary interference.
  –Well tests provide estimates of flow capacity, indications of formation damage or stimulation, and estimates of drainage (or injection) volume pressures.

Maps of historical production characteristics (contacted oil-in-place, estimated ultimate recovery, and reservoir pressure) have been compared to maps generated from the geologic studies (rock type, permeability/thickness, hydrocarbon pore volume) to identify the areas of the unit to be targeted for infill drilling. Our results indicate that a close relationship exists between the rock type distribution and permeability calculated using porosity and other properties derived from well logs.

The reservoir performance data also suggest that this reservoir depletes and recharges almost exclusively according to the rock type distribution. This integration of rock data and the reservoir performance attributes uses existing data and can eliminate the need for evaluation wells, as well as avoiding the loss of production that occurs when wells are shut-in for testing purposes.

In short, a comprehensive analysis, interpretation, and prediction of well and field performance can be completed quickly, at a minimal cost, and this analysis can be used to directly improve our understanding of reservoir structure and performance behavior in complex formations.

INTRODUCTION

This study was funded in part by the U.S. Department of Energy as part of the Class II Oil Program for improving development and exploitation of shallow-shelf carbonate reservoirs. According to the DOE, shallow-shelf carbonate (SSC) reservoirs in the U.S.A. originally contained >68 billion bbl of oil (10.81×10^9 m^3), or about one-seventh of all the oil discovered in the lower 48 states. Recovery efficiency in such reservoirs is low, as only 20 billion bbl of oil have been produced, and current technology may yield only an additional 4 billion bbl (Pande, 1995).

The typical low recovery efficiency in SSC reservoirs is not restricted to the U.S.A., but is a worldwide phenomenon. SSC reservoirs such as the North Robertson (Clear Fork) Unit (NRU) share a number of common characteristics (Pande, 1995), including:

- A high degree of areal and vertical heterogeneity, relatively low porosity and low permeability,
- Reservoir compartmentalization, resulting in poor vertical and lateral continuity of the reservoir flow units and poor sweep efficiency,
- Poor balancing of injection and production rates, and early water breakthrough in certain areas of the reservoir, indicating poor pressure and fluid communication and limited repressuring in the reservoir, and
- Porosity and saturation (as determined from analysis of wireline logs) do not accurately reflect reservoir quality and performance.

Production at the NRU is from the Lower Permian Glorieta and Clear Fork carbonates. The reservoir interval is thick, with a gross interval of approximately 1400 ft (427 m), and more than 90% of the interval has uniform lithology (dolostone). Unfortunately, the interval is characterized by a complex pore structure that results in extensive vertical layering. The reservoir is characterized by discontinuous pay intervals and high residual oil saturations (35% to 60%, based on steady-state measurements of relative permeability). The most important, immediate problem in the field is that porosity and saturation determined from logs do not accurately reflect reservoir quality and performance.

Given these reservoir characteristics, the ability to accurately target infill well opportunities is critical because blanket infill drilling will be uneconomic in most cases. For this particular case study, the cost-effective and readily available reservoir characterization tools outlined in this work allowed us to optimally target specific well locations.

HISTORICAL BACKGROUND

The NRU is located in Gaines County, west Texas, on the northeastern margin of the Central Basin Platform (Figure 1). The hydrocarbon-bearing interval extends from the base of the Glorieta to the base of the lower Clear Fork, between correlative depths of approximately 6160 and 7200 ft (1878 and 2196 m). The unit includes 5633 surface ac (2281 ha) containing a total of 270 wells (November 1996). This includes 156 active producing wells, 113 active injection wells, and 1 fresh water supply well.

Development and Production History

Production from the North Robertson field area began in the early 1950s with 40-ac (16-ha) primary well development. This 40-ac primary development resulted in 141 producing wells by 1965. The NRU was formed effective March 1987 for the purpose of implementing waterflood and infill drilling operations, and nominal well spacing was reduced from 40 ac (16-ha) to 20 ac (8 ha). Secondary recovery operations were initiated after unitization and in conjunction with infill drilling. Most of the 20-ac (8-ha) infill drilling was completed between unitization and the end of 1991.

The contacted original oil-in-place from material balance decline type curve analysis is estimated to be 215 million bbl (34.19×10^6 m^3), with an estimated ultimate recovery factor of approximately 20% (primary and secondary) based on the current production and workover schedule. Figure 2 presents the production and injection history of the unit from development in 1956 through November 1996. The total oil and water volumes produced and injected since field development are tabulated as follows:

Time Frame	Oil Produced Million Barrels	Water Produced Million Barrels	Water Injected Million Barrels
As of 1987 (primary)	17.5	8.2	0.0
1987–1996 (secondary)	9.3	28.3	64.2

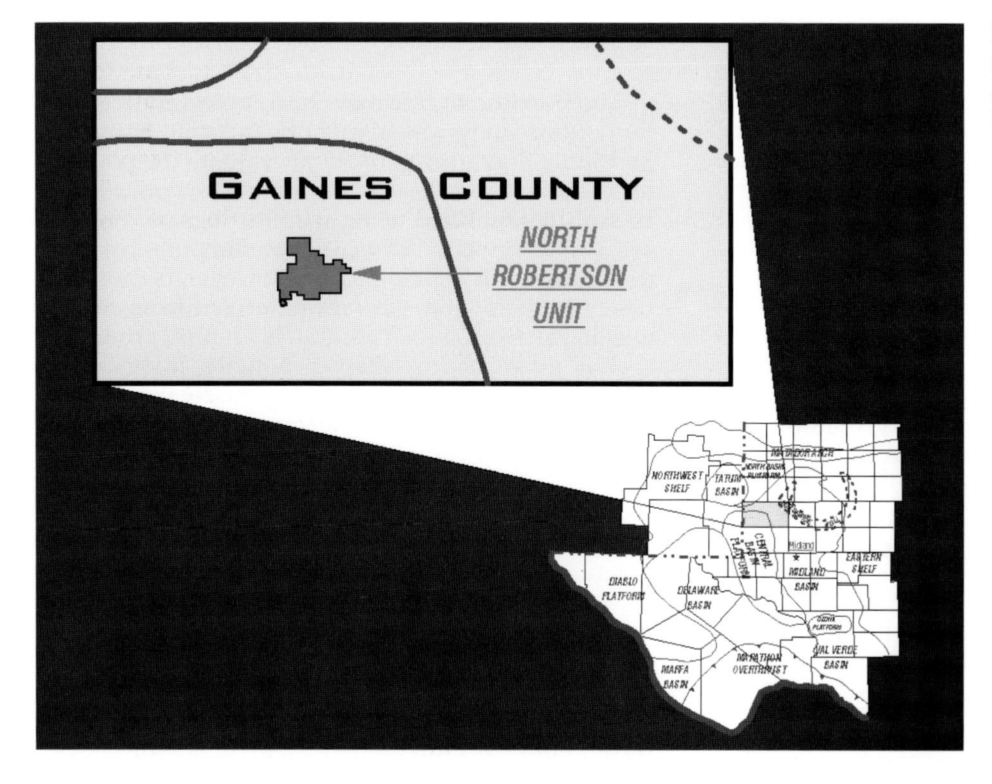

Figure 1. Regional map showing location of the North Robertson (Clear Fork) Unit.

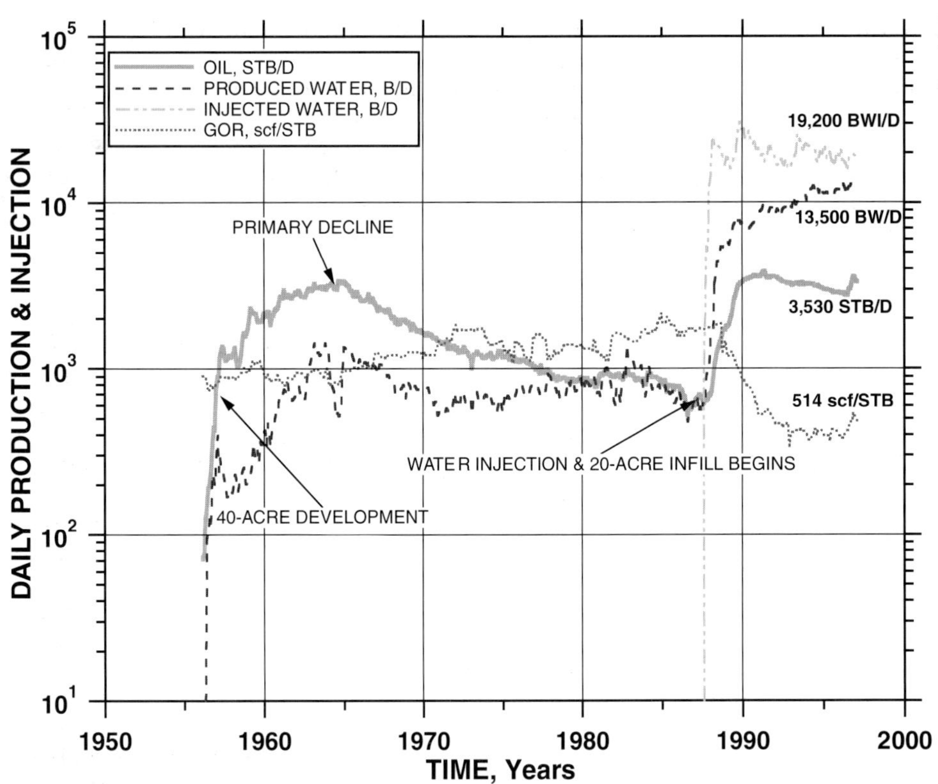

Figure 2. North Robertson Unit production and injection history.

Prior to 10-ac (4-ha) infill drilling, the total unit production rates were approximately 2740 STBOD (stock tank bbl oil/day) (436 m^3/day), 1130 kSCFD (thousand standard cubic feet/day) gas (32,000 m^3/day), and 12,230 BWD (1945 m^3/day). Water injection prior to the infill program was approximately 19,000 BWID (bbl water injected/day) (3021 m^3/day), with injection water comprised of produced water and of fresh water from the Ogallala aquifer obtained from a water supply well within the unit.

The 18 (14 producers, 4 injectors) 10-ac (4-ha) infill wells that were drilled during the second and third quarters of 1996 have increased unit production rates to 3350 STBOD (533 m^3/day), 1700 kSCFD gas (48,000 m^3/day), and 13,500 BWD (2146 m^3/day). Current injection at the unit is approximately 19,600 BWID (3116 m^3/day).

The well configuration at the NRU is an east-west line drive pattern (staggered 5-spot), and was developed for optimum injectivity and pressure support. Sweep efficiency is difficult to quantify due to differences in depositional environments throughout the unit. Fortunately, these differences are easily identified on the basis of reservoir rock type.

GEOLOGIC STUDY

This geological/petrophysical reservoir characterization of the NRU involved the following methodologies (Davies and Vessell, 1996a, b):

- Development of a depositional and diagenetic model of the reservoir,
- Definition of rock types based on pore geometry and development of the rock-log model,
- Rock type extension to noncored wells using the rock-log model,
- Definition of flow units and cross-flow barriers, and
- Mapping of reservoir parameters (thickness [h], porosity thickness [ϕh], permeability thickness [kh], and hydrocarbon pore volume) and rock type for each flow unit.

The purpose of the geological/petrophysical portion of our study was to identify and map individual hydraulic flow units (HFUs) to evaluate the potential for continued development drilling. Flow units had to be readily identified using wireline logs as core data are sparse (8 wells). Thus, the fundamental reservoir description was well log based; however, because values of porosity and saturation derived from routine well log analysis do not accurately identify productive rock in most shallow-shelf carbonates, it was necessary to develop a core- or rock-log model that allowed for the prediction of another producibility parameter—in this case formation permeability.

A geologic model was developed for the reservoir, based fundamentally on the measurement of pore geometrical parameters. Pore-level reservoir modeling gave improved accuracy in the prediction of rock types, permeability, and the identification of flow units. Pore geometry attributes were integrated with well log data to allow for log-based identification of intervals of rock with different capillary characteristics, as well as the prediction of permeability throughout the unit.

No new wells were drilled to aid the reservoir description. The existing database consisted of conventional cores from 8 wells, and relatively complete log suites in 120 wells consisting of

- Gamma ray, GR
- Photoelectric capture cross-section, PE
- Compensated neutron porosity, ϕ_N
- Compensated formation bulk density, ρ_b
- Dual Laterolog® (LLD and LLS, deep and shallow resistivities, respectively)
- Borehole caliper

Pore Geometry Modeling

At the NRU, we found no strong relationship among porosity, permeability, and depositional environment as indicated by the "shot gun" scatter of data points shown in Figure 3. Different environments exhibit similar ranges of porosity and permeability, which is not surprising as the facies have undergone significant diagenetic alteration of pore geometry after deposition. Thus, there is no fundamental relationship between depositional environment and permeability. This is a common problem in many diagenetically altered reservoirs.

Analysis of 3-D pore geometry data allowed our reservoir characterization efforts to be pore system oriented. The resulting reservoir models were based on characteristics of the pore system. Analysis of pore geometry involved the identification of individual pore types and rock types.

In a previous work, Davies and Vessell (1996a) used the quantitative analysis of pore geometry to develop the vertical reservoir layering profile of the reservoir (i.e., to identify vertical compartmentalization) at the NRU. The authors found that integrating pore scale observations with depositional and diagenetic data allows for the determination of the areal compartmentalization and permeability distributions within the reservoir.

Pore Types

The determination of pore types in a reservoir requires the use of rock samples (conventional core, rotary sidewall cores, and cuttings samples in favorable circumstances). In this study, analyses were based on 1 in (2.5 cm) "quick" plugs removed from conventional cores. Individual pore types were classified in terms of the following parameters:

Pore Body Size and Shape: Determined using scanning electron microscope (SEM) image analysis of the pore system (Clelland and Fens, 1991).

Pore Throat Size: Determined through capillary pressure analysis and SEM analysis of pore casts (Wardlaw, 1976).

Aspect Ratio: The ratio of pore body to pore throat size. This is a fundamental control on hydrocarbon displacement (Wardlaw, 1980; Li and Wardlaw, 1986).

Coordination Number: The number of pore throats that intersect each pore (Wardlaw and Cassan, 1978).

Pore Arrangement: The detailed distribution of pores within a sample (Wardlaw and Cassan, 1978).

These parameters were combined to yield a classification of the various pore types in these rocks (Table 1).

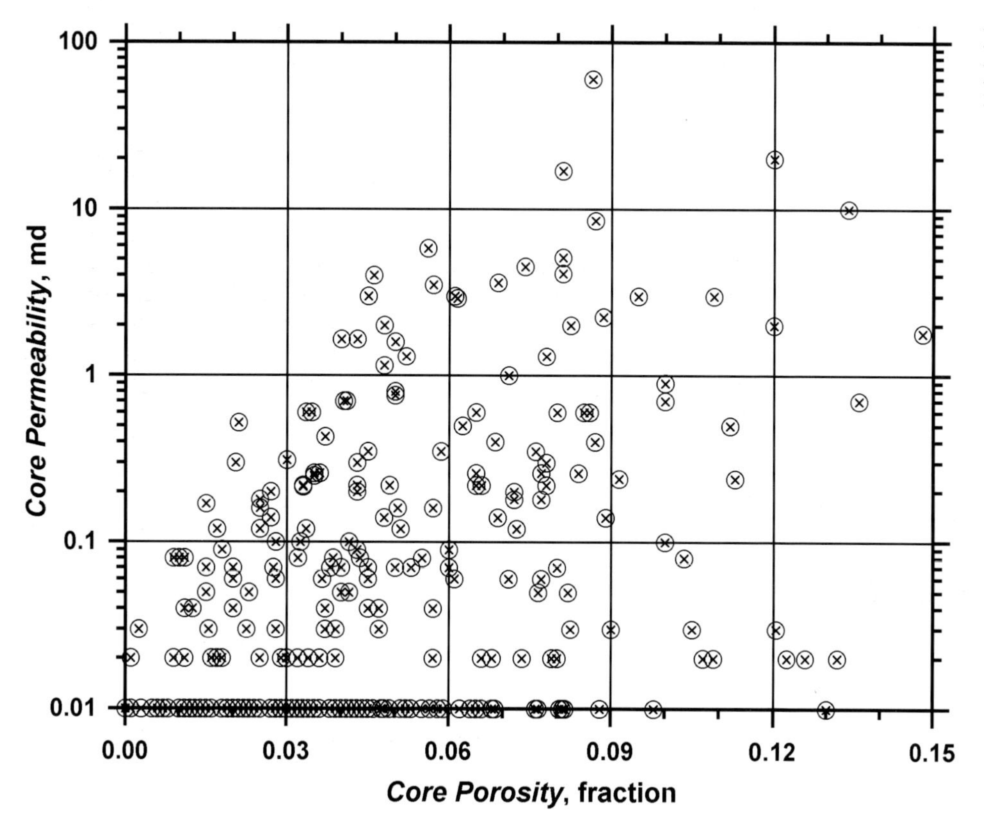

Figure 3. Porosity-permeability relationships for all core data.

Table 1. Pore Type and Classification at the North Robertson Unit.

Pore Type	Size (μm)	Shape	Coordination Number	Aspect Ratio	Pore Arrangement	Geologic Description
A	30–100	Triangular	3–6	50–100:1	Interconnected	Primary interparticle
B	60–120	Irregular	<3	200:1	Isolated	Shell molds and vugs
C	30–60	Irregular	<3	100:1	Isolated	Shell molds and vugs
D	15–30	Polyhedral	~6	<50:1	Interconnected	Intercrystalline
E	5–15	Polyhedral	~6	<30:1	Interconnected	Intercrystalline
F	3–5	Tetrahedral	~6	<20:1	Interconnected	Intercrystalline
G	<3	Sheet/slot	1	1:1	Interconnected	Interboundary sheet & intercrystalline pores

Pore types were identified in each core sample (350 samples in this study). Commonly, each sample (1 in or 2.5 cm diameter core plug) contained several different pore types. It was therefore necessary to group pore types into rock types.

For each sample, the volumetric proportion of each pore type was determined using SEM-based image analysis (Clelland and Fens, 1991). Because the pore throat size is known for each pore type, it was possible to develop a pseudocapillary pressure curve for each sample using the well known relation (Thomeer, 1983):

$$p_c = \frac{214}{d} \quad (1)$$

for which p_c is mercury-air capillary pressure in psia (pounds/in^2 absolute) and d is pore throat diameter in microns (10^{-6} m).

The validity of the geologically determined rock types was evaluated using mercury capillary pressure analysis of selected samples. Our results revealed differences between the rock types in terms of the measured capillary characteristics (Table 2). Such cross-checks allow for independent validation of the pore geometrical classification of rock types. Mercury capillary pressure data are also used to aid in the determination of pore throat sizes (Calhoun, 1960).

Rock Types

A "rock type" is an interval of rock characterized by a unique pore structure (Archer and Wall, 1986), but not necessarily a unique pore type. In this study, 8 rock types were identified, based on the relative volumetric abundance of each pore type (Figure 4). Each rock type is characterized by a particular assemblage (suite) of pore types (Ehrlich and Davies, 1989). For example, Rock Type 1 is dominated by Pore Type A, while Rock Type 2 contains few pores of Type A and is dominated by Pore Types B and C. Identification of rock types is fundamentally important because porosity and permeability are related within a specific pore structure (Calhoun, 1960).

Porosity-Permeability Relationship

At the NRU, the basic relationship between porosity and permeability exhibits a considerable degree of scatter (up to 4 orders of magnitude variation in permeability for a given value of porosity). In contrast, the porosity and permeability are closely related for each rock type (Figure 5). Permeability calculations on the basis of rock type have an error range of less than one-half decade for most samples. Regression equations for permeability as a function of porosity were developed for each rock type to quantitatively define each relationship (using log-log plots to avoid zero-porosity intercepts). These equations were used in the field-wide prediction of permeability using well logs (permeability being a function of porosity and rock type).

Average values of porosity and permeability are given for each rock type in Table 3. We immediately note that the highest porosity rocks at the NRU do not exhibit the highest permeability. The principal pay rock at the NRU is Rock Type 1. Rock Type 1 has significantly lower values of porosity than Rock Type 4 (flow barrier). This characteristic has important implications in terms of selecting completion intervals. Given our observations, the zones with the highest porosity should not always be the principal targets in this Clear Fork reservoir.

Rock-Log Model

Analyses of pore geometries reveal that eight rock types occur in the NRU. Six of the rock types are dolostone, one is limestone (nonpay; structurally low and wet in this field), and one is shale (Table 3). Individual rock types can be recognized using specific cutoff values based on analysis of environmentally corrected and normalized well log responses and using the comparison of core-based determination of rock type.

The well log responses used to isolate the eight different rock types for the NRU study were as follows:

- ρ_{maa} versus U_{maa} with gamma ray (Figure 6)
  - This data plot allows the discrimination of dolostone, limestone, anhydritic dolostone, siltstone, and shale
  - Can be used to identify "pay" vs. "nonpay" reservoir rock
- Shallow and deep laterolog resistivities and porosity (Figure 7)
  - Provides discrimination of "pay" Rock Types 1–3

Table 2. Capillary Characteristics by Rock Type Based on Mercury Injection.

Rock Type	Entry Pore Throat Radius (µm)	Displacement Pressure (psia)	Ineffective Porosity (%)**
1	7.6–53.3	2–10	8.2–29.6
2	2.7–3.6	30–40	23.1–49.5
3	0.4–1.3	80–300	61.6–72.3
4*	1.8	60	88
5	1.1–1.8	60–150	21.7–57.2
6*	0.1	800	100

*Only one measurement.
**Porosity invaded by Hg at p_d > 500 psia.

The rock-log model was first developed using data from only 5 cored wells; subsequently, the model was extended to include the 3 remaining cored wells. Evaluation of cored intervals reveals successful discrimination (>80%) of each of the principal rock types (Rock Types 1–3), despite the fact that wells were logged by different companies at different times. Misidentification of Rock Type 1 results in identification of Rock Type 2, while misidentification of Rock Type 2 results in identification of Rock Type 1. Thus, there is no significant misidentification of the dominant rock types by logs over the cored intervals. The model has been extended to all wells with sufficient log suites in the field (120 wells in the NRU). Specific algorithms allow for rock type identification on a foot-by-foot basis in each well.

As we showed previously (Figure 5), permeability is a function of both rock type and porosity. We have established that rock type and porosity can be determined from well log responses alone; therefore, permeability can, in principle, be predicted using only well log data. This gives us the ability to develop a vertical layering profile based on rock type and permeability in cored and noncored wells.

Improving our understanding of the relationship between porosity and permeability has been one of the key efforts of our geologic work. Our approach has been to integrate all geologic and engineering data in an attempt to refine our existing porosity-permeability algorithms and to redefine flow unit boundaries. Now that we have finished the infill drilling program and loaded all new core and open-hole log data into our database, we can begin to update our porosity-permeability relationships and flow unit definitions.

Early indications are that by using multiple geologic filters it is possible to dramatically reduce the scatter on our porosity versus permeability crossplots, thereby providing us with more robust algorithms. Filters include devices such as depositional environment data, shallowing-upward sequence tops, rock type data, mud log data, and numerous open-hole log responses (PE, spectral gamma ray, invasion profile, etc.).

Hydraulic Flow Units

Individual HFUs were identified based on the integration of the data for the distribution of rock types, petrophysical properties (in particular, permeability and fluid content), and depositional facies. Evaluation of this data for 120 wells revealed that rock types are not randomly distributed, rather the principal reservoir rocks (Rock Types 1, 2, and 3) generally occur in close association, and typically alternate with lower quality rocks (e.g., Rock Types 4, 6, 7, and 8). Correlation of rock types between wells reveals an obvious layering profile in which 12 distinct layers, or HFUs, are distinguishable at the NRU. Correlation of these layers is aided by a knowledge of the distribution of depositional environments as there is a general relationship between depositional environment and rock type. Rock Types 1 and 2 are more common in high energy deposits (shoals, sand flat, forebank). Rock Types 3 and 4 are common in low-energy deposits (supratidal, tidal flat, lagoon).

Maps were prepared for each of the HFUs to illustrate the distribution of important petrophysical parameters.

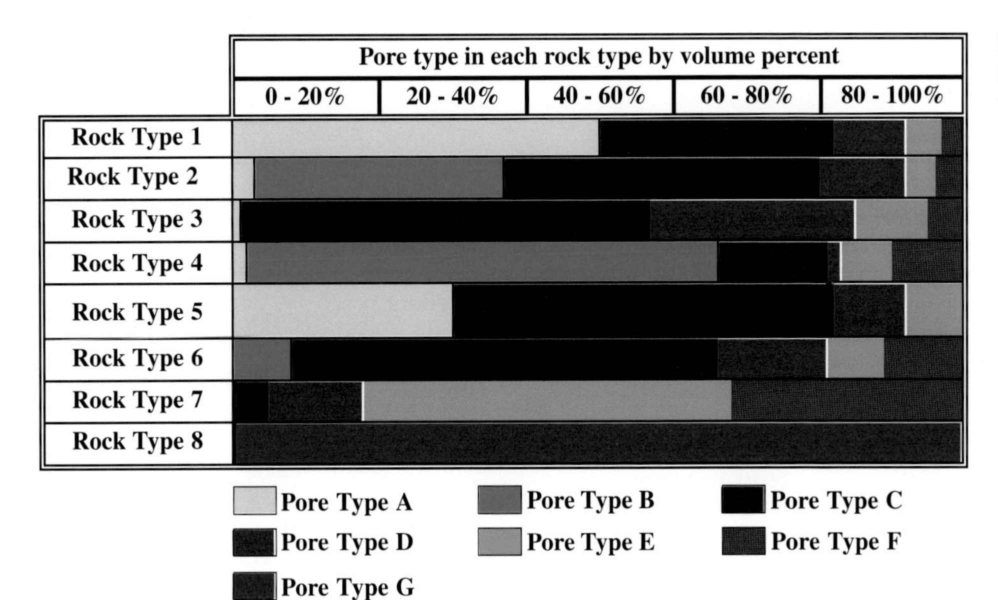

Figure 4. Volumetric proportions of pore types in each rock type.

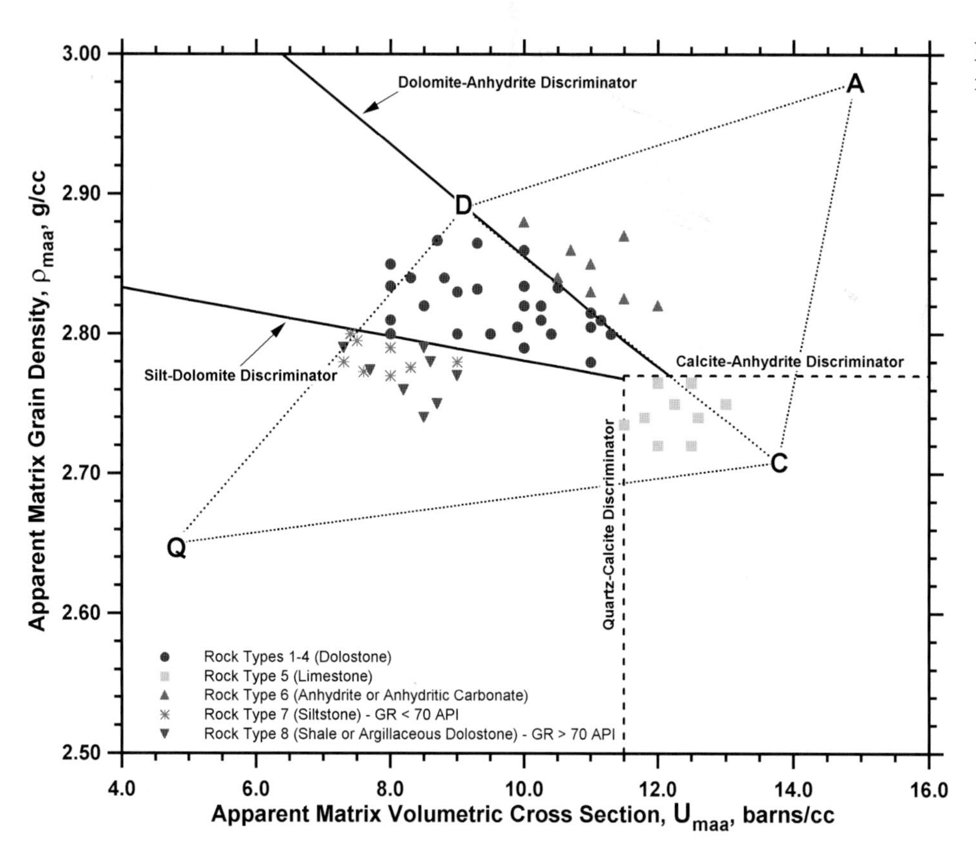

Figure 5. Differentiating pay from nonpay reservoir rocks.

The distribution of the principal rock types for each HFU was also mapped. This allowed for rapid identification of areas of the field dominated by either high or low quality reservoir.

There is a general tendency at the NRU for the higher quality rocks (Rock Types 1 and 2) to occur in discrete trends on the northeast (high energy) edge of the unit, while relatively lower quality rocks (Rock Types 3 and 4) occur in southwest (low energy) portions of the unit. Within these general trends, variations exist in the distributions of permeability. These variations are important, as they result in compartmentalization of the reservoir. There are no faults in the NRU; compartmentalization is entirely stratigraphic and is the result of areal variations in the distributions of individual reservoir rock types.

New Data—Infill Drilling Program

Additional Whole Core

We took 2730 ft (832 m) of core in four new wells as part of an intensive effort to collect needed rock data (Kamis et al., 1997). The data will be used to help quantify the extent of small-scale vertical and lateral heterogeneity, refine the depositional model, improve our understanding of the relationship between porosity and permeability, and help us choose additional 10-ac (4-ha) infill drilling locations within the NRU Clear Fork Formation.

We attempted to cut cores continuously through the entire Clear Fork section. Parts of the section were not cored due to significant mechanical difficulties caused

Table 3. Porosity, Permeability, and Lithology by Rock Type.

Rock Type	Median Porosity (%)	Median (Hg/air) Permeability (md)	Lithology	Reservoir Quality	Cross-Flow Barrier Quality
1	4.0	0.70	Dolostone	Excellent	Poor
2	5.6	0.15	Dolostone	Good	Poor
3	3.5	0.39	Dolostone	Moderate	Moderate
4	7.5	0.01	Dolostone	Poor	Moderate
5*	5.8	0.40	Limestone	Good (waterbearing)	Poor
6	1.0	<0.01	Anhydritic dolostone	None	Good
7	2.3	<0.01	Silty dolostone	None	Good
8	—	—	Shale and argillaceous dolostone	None	Good

*Structurally low and wet at the North Robertson Unit.

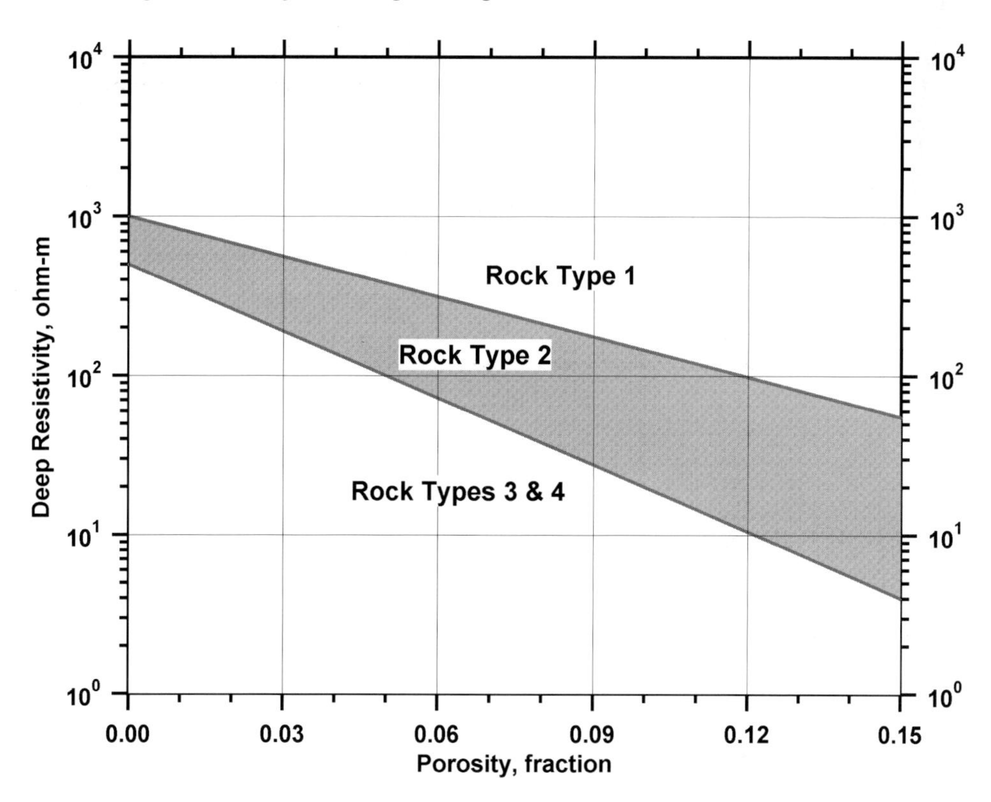

Figure 6. Differentiating between pay reservoir rock types.

by very long core times, often greater than 200 min/ft. This continuous core gives us the ability to make foot by foot comparisons of reservoir quality, rock type, and depositional environment, which ultimately will help us correctly model fluid movement within the reservoir.

Rock Fabrics

Based on our description of the new core, we have described four basic rock fabrics:

(1) Homogeneous: made up of relatively uniformly distributed lateral and vertical porosity and permeability. The best example of this type of rock fabric is found within selected portions of the upper Clear Fork. We are not implying that this zone is perfectly homogeneous, as are some silica clastic sands; however, this layer is much closer to this type of homogeneity than all other zones in the Clear Fork.

(2) Fractured: made up of solution collapse breccias as described above. Fractures are 2–4 in (5–10 cm) in length and very roughly estimated to be 4–6 in (10–15 cm) apart. Not all of these fractures are open, as many have been plugged with anhydrite. Portions of the middle Clear Fork are a good example of this fabric.

(3) Bimodal: made up of two distinct pore sizes. The larger size pores are typically formed from the dissolution of fossil debris, and the smaller pores are typically intercrystalline in origin.

(4) Heterogeneous: made up of anhydrite nodules, and porous dolostone. This fabric is common throughout much of the Clear Fork/Glorieta section. The size and distribution of these anhydrite nodules vary dramatically.

Mud Logs

We have captured continuous reading of all mud gas components while drilling, and loaded the data into a computer database and depth corrected it. In addition, we have started research into the applicability of using mud gas component ratios to estimate fluid content. It is hoped that flushed zones will have uniquely different ratios than previously uncontacted oil zones.

We have noticed also that the mud log is an excellent tool for locating the intervals that contribute most to production. In this particular shallow-shelf carbonate reservoir, it would appear that the pay intervals of interest could probably be cost-effectively defined simply by using the mud log and a base porosity and resistivity log. It is essential that a high-quality and reliable mud logging company be chosen to do the work. Mud loggers with previous experience in the formation being drilled is also an important component in getting reliable data. If an operator requires more detailed information regarding porosity types, rock quality, and fluid distributions, additional information must obviously be gathered.

Open-Hole Logs

The base logging suite for the 10-ac (4-ha) infill wells currently consists of a Dual Laterolog,® Micro Laterolog,® or microspherically focused log (R_{xo} device), compensated neutron log, compensated spectral or litho-density log (with PE), spectral gamma ray log, and a sonic log. We have used several potentially useful modern open-hole logging tools in addition to our normal logging suite in an attempt to more accurately characterize permeability,

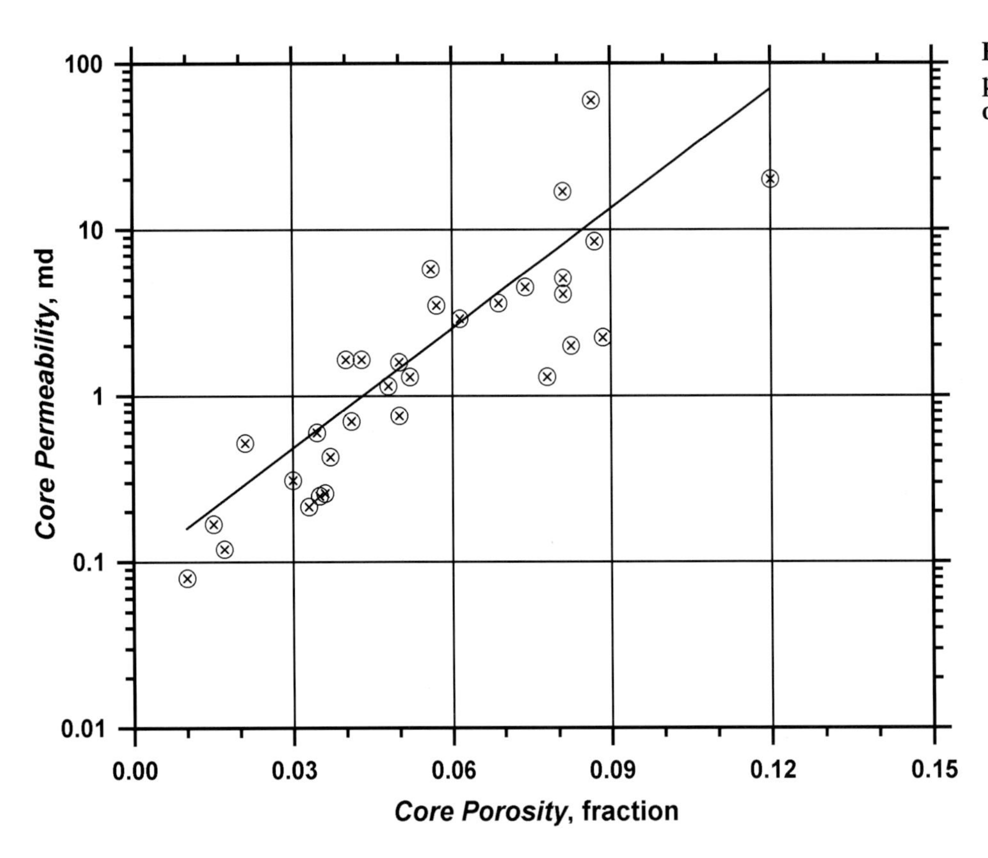

Figure 7. Porosity-permeability relationship for one rock type.

fluid content, and rock fabric. Some of these are summarized below:

High Frequency Dielectric Log: The high frequency (200 MHz) dielectric log yields a salinity-independent measure of fluid distributions in the flushed zone. This is of paramount importance in areas such as the NRU that are currently under active waterfloods. The dielectric tool produces a much more accurate representation of pay intervals than can be obtained from the usual low frequency flushed zone resistivity devices, such as the Micro Laterolog or micro-spherically focused log, which do not always do a good job of differentiating between residual and movable hydrocarbons.

This device works extremely well in formations with mixed lithology, and, if run in combination with a low-frequency flushed zone resistivity device, yields a rock textural parameter that can be used to determine the type of porosity present (intergranular, vuggy, or fracture), which directly affects the producibility of any particular interval.

Nuclear Magnetic Resonance Log: Nuclear magnetic resonance (NMR) logs were recorded over selected sections of two cored wells in an effort to obtain permeability, lithology-independent porosity, pore size distribution, and fluid saturation distribution from a single log. Most of the preliminary NMR work has been performed to differentiate between oil and water in low-resistivity clastics; however, several recent projects performed in the Permian Basin area are adding to the understanding of NMR responses in high-resistivity carbonate reservoirs.

This device gave a good approximation of permeability, pore size, and the location of free hydrocarbons; however, data acquisition and processing are costly, and initially need to be closely calibrated with core. The distribution of free and bound fluids in the rock pores obtained from NMR analysis of selected core samples from the two wells indicated that the reservoir is oil wet. This fluid distribution was then used in the processing of the raw log data to yield a visual representation of the pore and fluid distribution in the reservoir.

ENGINEERING STUDY

Comparing Geological and Petrophysical Properties with Historical Performance

If the pore modeling and rock typing exercises are performed properly, the resulting rock type and permeability distributions associated with the reservoir should mirror the historical production performance. The petrophysical parameter and rock type maps resulting from our geological-petrophysical study were compared to the performance maps derived from the results of decline type curve analyses and pressure transient tests. This provided us with a rapid, cost-effective, and accurate method for targeting infill well locations at the NRU (Davies et al., 1997).

To verify the results of the rock-log modeling, long-term production and injection data were analyzed using material balance decline type curves, and the results were mapped for comparison to the petrophysical parameter and rock type maps generated from the work summarized. In addition, the results of pressure transient tests (average reservoir pressure and flow characteristics) were also incorporated to help correlate the reservoir rock type and historical performance.

Other applications for the decline curve results are

- Reservoir delineation, recovery and rate forecasting,
- Maps of reservoir performance potential,
- Estimating efficiency of reservoir drive mechanisms,
- Analogy with offset properties, and
- Calibrating reservoir simulation models.

Material Balance Decline Type Curve Analysis

An initial study of the 40-ac (16-ha) and 20-ac (8-ha) producing wells, and 20-ac (8-ha) water injection wells was performed using rigorous material balance decline type curve methods (Doublet et al., 1994; Doublet and Blasingame, 1996). The goals of these analyses were

- To analyze long-term production data as well as injection rate and pressure data to evaluate reservoir performance.
- To provide the same flow characteristics associated with the acquisition of field data and without well downtime.

The results of these analyses include the following:

- In-place fluid volumes:
  - Contacted original oil-in-place,
  - Movable oil or injectable water at current conditions, and
  - Reservoir drainage or injection area.
- Reservoir properties (based on performance):
  - Skin factor for near-well damage or stimulation, *s*, and
  - Formation flow capacity based on production performance, *kh*.

An example of a decline type curve match for a hydraulically fractured producing well is shown in Figure 8.

We focused on using data that operators acquire as part of normal field operations (e.g., production rates from sales tickets and pressures obtained from permanent surface or bottomhole gauges). In most cases, these will be the only data available in any significant quantity, especially for older wells and marginally economic wells, where both the quantity and quality of any types of data are limited.

This approach of using production and injection data eliminates the loss of production that occurs when wells are shut in for pressure transient tests, and provides analysis and interpretation of well and field performance at little or no cost to the operator. This technique allows us to evaluate reservoir properties quickly and easily, and provides us with an additional method for locating the most productive areas of the reservoir.

The results of the decline type curve analyses were used to generate reservoir-quality maps of contacted

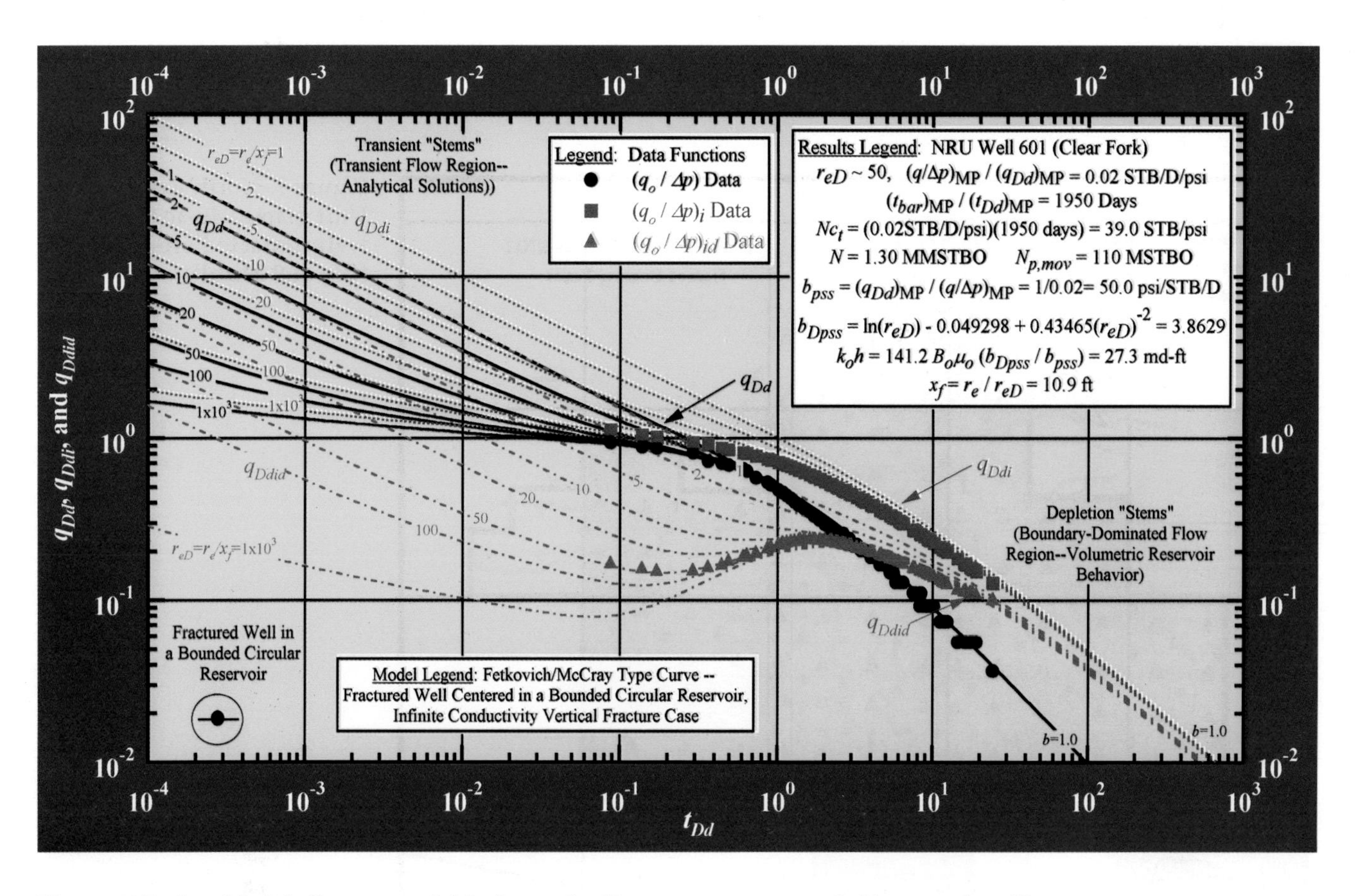

Figure 8. Fetkovich-McCray material balance decline type curve match (fractured well).

original oil-in-place (OOIP), permeability-thickness (*kh*), and estimated ultimate recovery (EUR) for both the original 40-ac (16-ha) (primary), and 20-ac (8-ha) (secondary) producing wells, as shown in Figures 9–12.

Pressure Transient Analysis

A unit-wide pressure transient data acquisition program was initiated prior to 10-ac (4-ha) infill drilling. The purpose of this study was to

- Obtain sufficient data for a representative comparison with tests recorded prior to the initiation of water injection,
- Provide additional data for simulation history matching,
- Estimate completion/stimulation efficiency,
- Identify the best areas of the reservoir with regard to pressure support, and
- Identify any other major problems related to waterflood sweep efficiency.

Our data acquisition program consisted of taking 10 pressure buildup tests (on producing wells) and 13 pressure falloff tests (on injection wells). The locations for these tests were well distributed throughout the unit to obtain a representative, unbiased sampling.

The estimated formation flow characteristics (permeability, skin factor, fracture half-length) compare very well with the results from the prewaterflood transient tests. The major difference we noted was the relative change in average reservoir pressure throughout the unit after 8 yr of continuous water injection. A reservoir pressure map for the 1995 tests (from pressure buildup and falloff tests) is shown in Figure 13.

Optimizing Well Completions and Stimulations at the NRU

Another benefit of the geologic analyses that we performed on the new and existing whole core was the more accurate targeting of discrete producing intervals within the 1400 ft (427 m) Glorieta/Clear Fork section. For the most part these intervals coincide with zones containing a predominance of Rock Type 1 (recall that this is our main pay rock type). These are zones of relatively high permeability and porosity, which are separated by larger intervals of lower permeability and porosity rock that act as barriers or source beds for the higher quality reservoir sections. At North Robertson, the discrete productive intervals we focused on during completion include

- Lower Clear Fork: ±7000–7180 ft (2135–2190 m)
- Middle Clear Fork: ±6350–6500 ft (1936–1982 m); ±6770–6900 ft (2064–2104 m)
- Upper Clear Fork: ±6160–6250 ft (1878–1906 m)

We used three-stage completion designs to keep the treated intervals between 100 and 250 ft (30 and 76 m). We have performed both CO_2 foam fracs and conventional cross-linked borate fracs. All wells' rates have held up extremely well over time for both hydraulic fracture designs. The major factor controlling initial potential appears to be confinement of the vertical completion interval and localized reservoir quality. The advantages of each type of frac design are as follows:

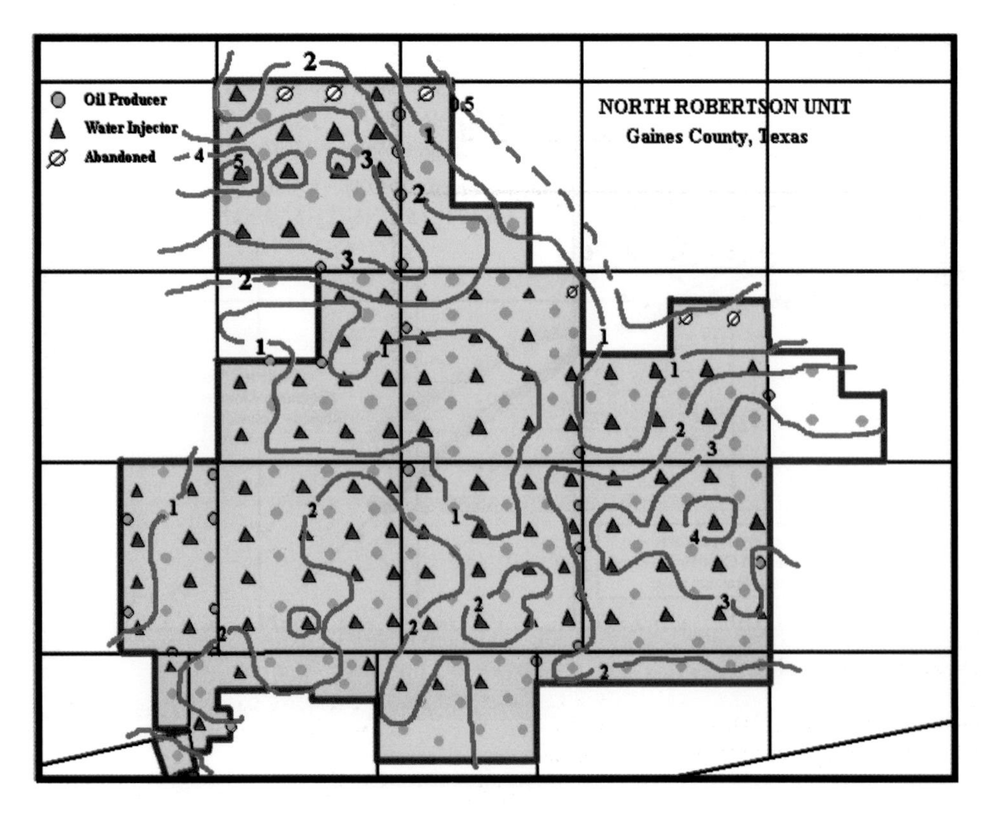

Figure 9. North Robertson Unit map of 40-ac (16-ha) well contacted oil-in-place (CI = 1.0 Mbbl).

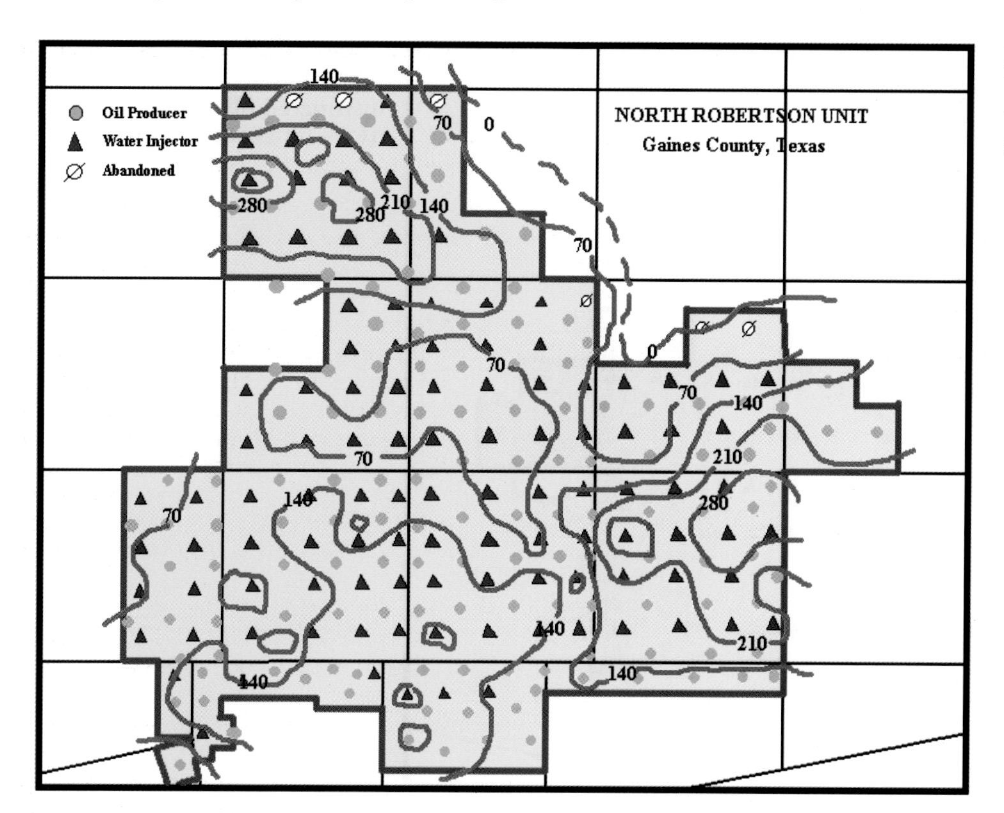

Figure 10. North Robertson Unit map of 40-ac (16-ha) well estimated ultimate recovery (CI = 70 kbbl).

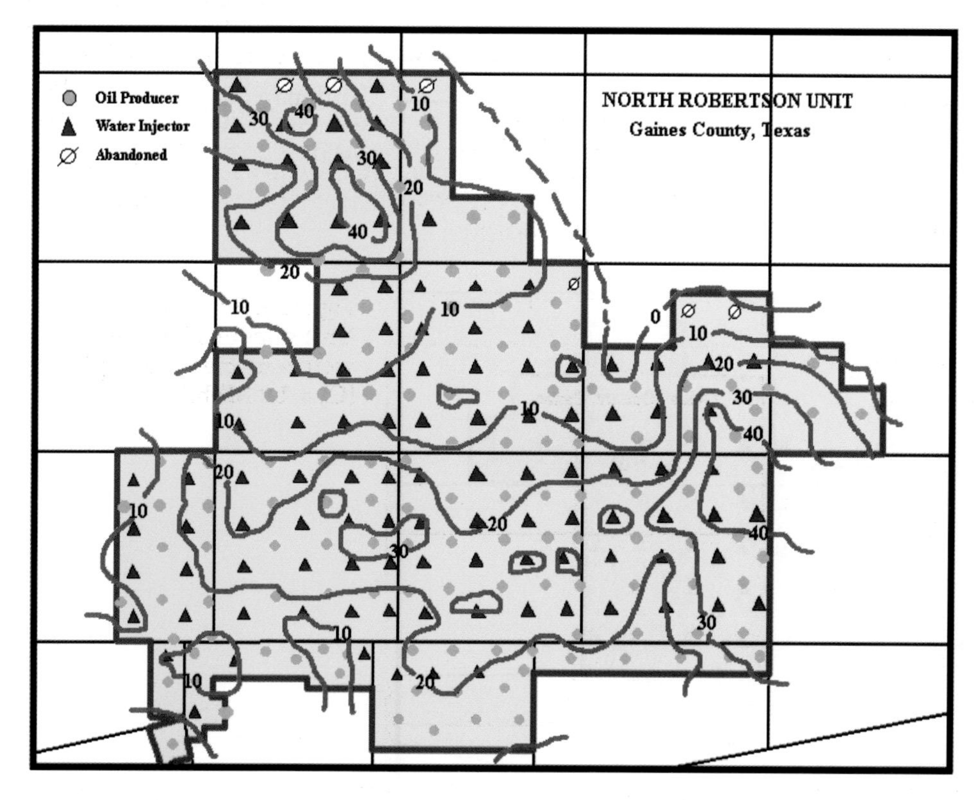

Figure 11. North Robertson Unit map of 40-ac (16-ha) well flow capacity; k_0h (CI = 10 md-ft).

CO_2 Fracs

- Exceptionally clean frac fluid
- Increased relative oil permeability
- Created solution gas drive reduces cleanup requirements
- Formation of carbonic acid for near-well stimulation
- Reduction in interfacial tension helps remove water blocks

Cross-Linked Borate Fracs:

- Exceptionally clean frac fluid
- Low fluid loss without formation-damaging additives
- Excellent proppant-carrying capacity
- Polymer-specific enzyme breaker aids in post-frac cleanup
- 90% of original fracture conductivity retained

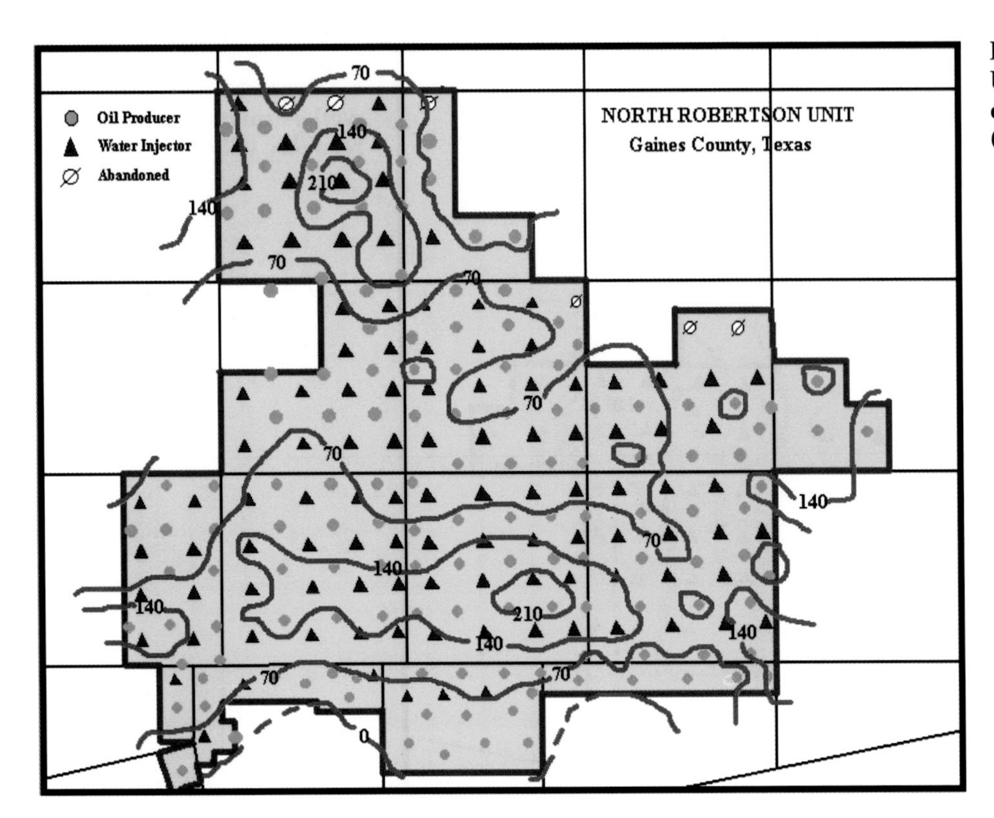

Figure 12. North Robertson Unit map of 20-ac (8-ha) well estimated ultimate recovery (CI = 70 kbbl).

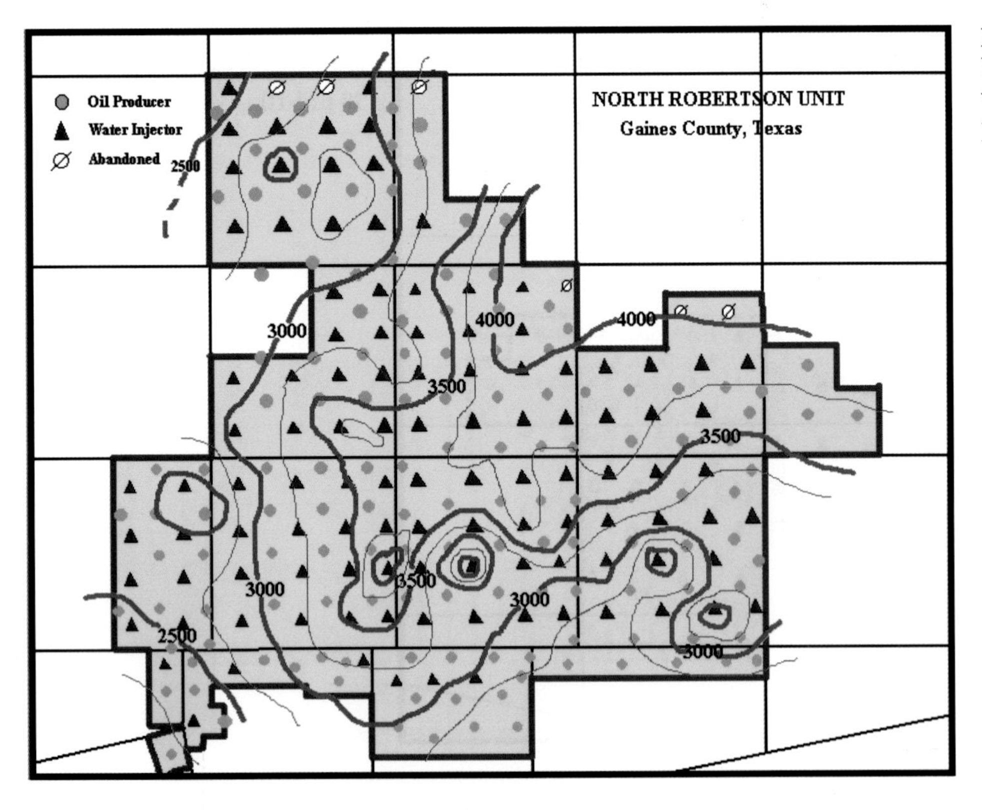

Figure 13. North Robertson Unit map of average reservoir pressure, 1995 (CI = 250 psia).

Pre-frac cleanup acid jobs have been performed to remove near-well damage using between 1000 and 3000 gal (3785 and 11,355 l) of 15% acid. Most intervals have been perforated for limited-entry fracturing (>2 bbl per perforation), with average injection rates between 30 and 40 bbl/min, depending on the size of the interval. The size of the frac jobs has ranged from 35,000 gal (132,475 l) of fluid and 55,000 lbs (24,970 kg) of 16/30 sand to 70,000 gallons (264,950 l) of fluid and 150,000 lbs (68,100 kg) of 20/40 sand. Resin-coated sand has been "tailed-in" for all frac jobs to reduce sand flowback during production. The conventional fracs have been flowed back immediately at 1 bbl/min to induce fracture closure, while the foam fracs have been shut in 2 to 5 days after stimulation to allow the CO_2 to soak into formation.

All hydraulic fracture jobs were designed to yield fracture half-lengths of approximately 200 ft (61 m). Postfrac pressure transient tests performed over specific completion intervals indicated that we are obtaining fracture half-lengths of approximately 120 ft (36 m), with average radial flow skin factors of approximately –5.25.

The result of these improved completion techniques is shown in Figure 14. The average three-month initial potentials (IP) for these new infill wells are 2 to 3 times greater than the IPs from the previous drilling and completion programs at the NRU.

Short-term pressure drawdown tests were used to measure formation flow characteristics in the new producing wells. We recorded drawdown rather than buildup tests to avoid shutting in recently completed wells. These tests were being recorded over individual completion intervals (i.e., lower, middle, or upper Clear Fork) and were used to estimate the completion efficiency and the relative contribution of each zone to total production.

Well Testing

The interval well test results indicated that the middle and upper Clear Fork intervals are much more significant contributors to total production than was previously thought. At this point it appears that each interval's approximate contribution to total oil production is as follows.

	Sec. 327	Sec. 329
Upper Clear Fork	20%	10%
Middle Clear Fork	50%	65%
Lower Clear Fork	30%	25%

Reserves—Incremental vs. Accelerated

Early results indicate that approximately 65% of the production from the new infill wells is incremental, and approximately 35% may be acceleration of existing reserves. The new wells account for approximately 900 STBOD of the total unit production, and the amount of incremental production since the field demonstration was implemented is between 600 and 700 STBOD. On an individual well basis, most of the additional production in Section 329 of the unit appears to be due to acceleration of existing reserves, while most of the additional production in Sections 326 and 327 appears to be incremental. These trends were predicted prior to drilling on the basis of differing reservoir rock types that occur in the two areas. The Section 329 infill area is dominated by grainstone shoal facies with fairly good permeability and porosity characteristics (Rock Type 1). The reservoir within Sections 326 and 327 is dominated by lagoonal facies with good storage capacity (porosity), but relatively lower permeability and connectivity (Rock Type 3). We will continue to monitor and report individual well producing characteristics in an effort to quantify incremental reserves added via infill drilling.

CONCLUSIONS

As a result of this study we hope to identify useful and cost-effective methods for the exploitation of the shallow-shelf carbonate reservoirs of the Permian Basin. The techniques that we outline for the formulation of an integrated reservoir description apply to all oil and gas reservoirs, but are specifically tailored for use in the heterogeneous, low-permeability carbonate reservoirs of west Texas.

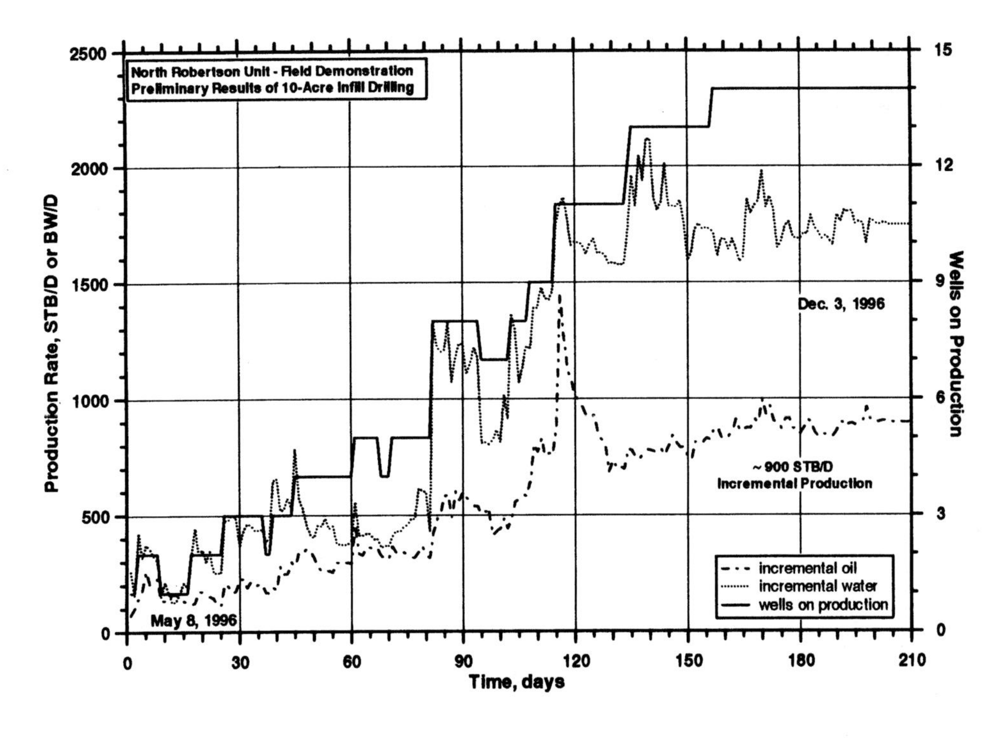

Figure 14. Incremental production increase due to new 10-ac (4-ha) infill wells.

(1) Measurement of pore geometry parameters allows for the improved prediction of permeability and permeability distribution from wireline logs in partially cored intervals and in adjacent uncored wells. This approach improves the prediction of reservoir quality in noncored intervals and results in improved well completions and EOR (enhanced oil recovery) decisions.
(2) Detailed pore geometry attributes allow for better definition of hydraulic flow units. These attributes can be related to well log responses, which allows for the development of a fieldwide, log-based reservoir model.
(3) The material balance decline type curve techniques summarized in this work provide very good estimates of reservoir volumes (total and movable) and reasonable estimates of formation flow characteristics. Using this approach to analyze and interpret long-term production and injection data is relatively straightforward and can provide the same information as conventional pressure transient tests, without the associated costs of data acquisition or loss of production.
(4) From our comparison of geological and petrophysical parameters with historical production performance, we believe that the producing characteristics of individual wells are a direct function of the local rock type distribution. The reservoir depletes and repressures as a function of reservoir quality (rock type) throughout all areas of the unit.
(5) We identified the regions in the unit with additional reserves potential (accelerated or incremental), as well as those areas in which infill drilling is not likely to be economic.
(6) We believe that uniform infill drilling is neither prudent nor warranted, given the stratigraphic compartmentalization and irregular permeability distributions in this reservoir. Infill drilling should be restricted to
   (a) Areas of the field that exhibit Rock Types 1, 2, and 3 as dominant, with good permeability and hydrocarbon pore volume characteristics, high primary and secondary recovery
   (b) Areas of poor reservoir continuity with acceptable porosity and permeability values, with a significant abundance of Rock Types 1, 2, or 3, and good primary, yet poor secondary, recovery characteristics
(7) If the productive intervals within the reservoir can be better defined based on core and well log analyses, then completion efficiency (initial potential and reserves recovery) can be increased and completion costs can be decreased.

REFERENCES CITED

Archer, J.S., and C.G. Wall: Petroleum Engineering Principles and Practice, Graham and Trotman, Ltd. (1986) 362 p.

Calhoun, J.C.: Fundamentals of Reservoir Engineering, Univ. Oklahoma Press, Norman, Oklahoma (1960) 426 p.

Clelland, W.D., and T.W. Fens: Automated rock characterization with SEM/image analysis techniques, SPE Formation Evaluation (1991) v. 6, No. 4, 437-443.

Davies, D.K. and R.K. Vessell: Identification and distribution of hydraulic flow units in a heterogeneous carbonate reservoir: North Robertson Unit, west Texas, SPE Permian Basin Oil and Gas Recovery Conference, Midland, Texas, Paper 35183 (1996a), p. 321–330.

Davies, D.K. and R.K. Vessell: Permeability prediction and flow unit characterization of a heterogeneous carbonate reservoir: North Robertson Unit, west Texas, SPE/DOE Symposium on Improved Oil Recovery, Tulsa, Oklahoma, Paper 35433 (1996b), p. 295–304.

Davies, D.K., R.K. Vessell, L.E. Doublet, and T.A. Blasingame: Improved characterization of reservoir behavior by integrating reservoir performance data and rock type distributions, Fourth International Reservoir Characterization Technical Conference, Houston, TX (1997), p. 645–669.

Doublet, L.E., P.K. Pande, T.J. McCollum, and T.A. Blasingame: Decline curve analysis using type curves—analysis of oil well production data using material balance time: application to field cases, SPE Petroleum Conference and Exhibition of Mexico, Veracruz, Mexico, Paper 28688 (1994).

Doublet, L.E., and T.A. Blasingame: Evaluation of injection well performance using decline type curves, SPE Permian Basin Oil and Gas Recovery Conference, Midland, Texas, Paper 35205 (1996).

Ehrlich, R., and D.K. Davies: Image analysis of pore geometry: relationship to reservoir engineering and modeling, Proc. SPE Gas Technology Symposium, Dallas, Texas, Paper 19054 (1989) 15-30.

Kamis, J., W. Dixon, R. Vessell, and L. Doublet: Environments of deposition for the Clear Fork and Glorieta formations, North Robertson Unit, Gaines County, Texas, Platform Carbonates in the Southern Mid-Continent, Oklahoma Geological Survey Circular, K.S. Johnson, ed., Norman, Oklahoma (in press).

Li, Y. and J.C. Wardlaw: The influence of wettability and critical pore-throat size ratio on snap-off, J. Colloid and Interface Sci. (1986) v. 109, 461-472.

Pande, P.K.: Application of integrated reservoir management and reservoir characterization to optimize infill drilling, DOE continuation application, cooperative agreement No. DE-FCZZ-94BC14989, Fina Oil and Chemical Co., Midland, Texas (1995), p. 8–9.

Thomeer, J.H.: Air permeability as a function of three pore-network parameters, Journal of Petroleum Technology (1983) v. 35, 809-814.

Wardlaw, N.C.: Pore geometry of carbonate rocks as revealed by pore casts and capillary pressure, AAPG Bull. (1976), v. 60, 245-257.

Wardlaw, N.C. and J.P. Cassan: Estimation of recovery efficiency by visual observation of pore systems in reservoir rocks, Bull, Can. Pet. Geol. (1978) v. 26, 572-585.

Wardlaw, N.C.: The effects of pore structure on displacement efficiency in reservoir rocks and in glass micro models, SPE Paper 8843, First Joint SPE/DOE Symposium on Enhanced Oil Recovery, Tulsa, Oklahoma (1980) 345-352.

Section IV
Method and Techniques

Slevinsky, B.A., Well test imaging—a new method for determination of boundaries from well test data, 1999, *in* R. Schatzinger and J. Jordan, eds., Reservoir Characterization-Recent Advances, AAPG Memoir 71, p. 127–148.

Chapter 9

Well Test Imaging—A New Method for Determination of Boundaries from Well Test Data

Bruce A. Slevinsky
Petro-Canada
Calgary, Alberta, Canada

ABSTRACT

A new method has been developed for analysis of well test data that allows the direct calculation of the location of arbitrary reservoir boundaries detected during a well test. The method is based on elements of ray tracing and information theory, and is centered on the calculation of an instantaneous "angle of view" of the reservoir boundaries. In the absence of other information, the relative reservoir shape and boundary distances are retrievable in the form of a diagnostic image. If other reservoir information, such as 3-D seismic, is available, the full shape and orientation of arbitrary (non-straight-line or circular arc) boundaries can be determined in the form of a reservoir image. The well test imaging method can be used to greatly enhance the information available from well tests and other geological data, and provides a method to integrate data from multiple disciplines to improve reservoir characterization. This paper covers the derivation of the analytical technique of well test imaging and shows examples of application of the technique to a number of reservoirs.

INTRODUCTION

Conventional well test analysis methods determine the distance to boundaries using either a simplified technique outlined by Earlougher (1977) or by regression on distance parameters in a mechanistic model of well test response as described by Gringarten (1986). The analytical technique for calculating distance for a single linear sealing fault involves the estimation of an intersection time, t_x, calculated from the intersection point of the two straight line segments of the Horner plot for the pressure buildup, as illustrated in Figure 1. This intersection time is then substituted into equation 1 (from Earlougher, 1977)

$$L = 0.01217\sqrt{kt_x/(\phi * \mu * c_t)} \quad (1)$$

This represents approximately 42 percent of the equivalent radius of investigation at that particular point in time,

$$R_{inv} = 0.029\sqrt{kt/(\phi * \mu * c_t)} \quad (2)$$

Multiple boundaries in the vicinity of a well can cause many different transient-test behaviors depending on the number, type, and orientation of the boundaries. Few general rules are given in the literature to

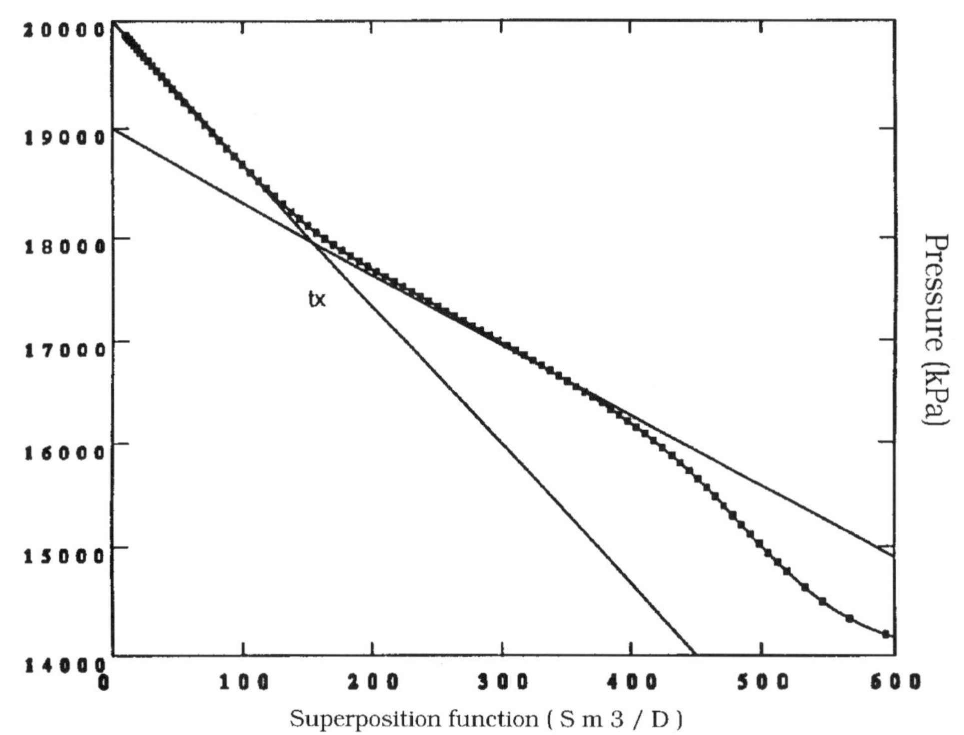

Figure 1. Intersection of Horner curve straight-line segments.

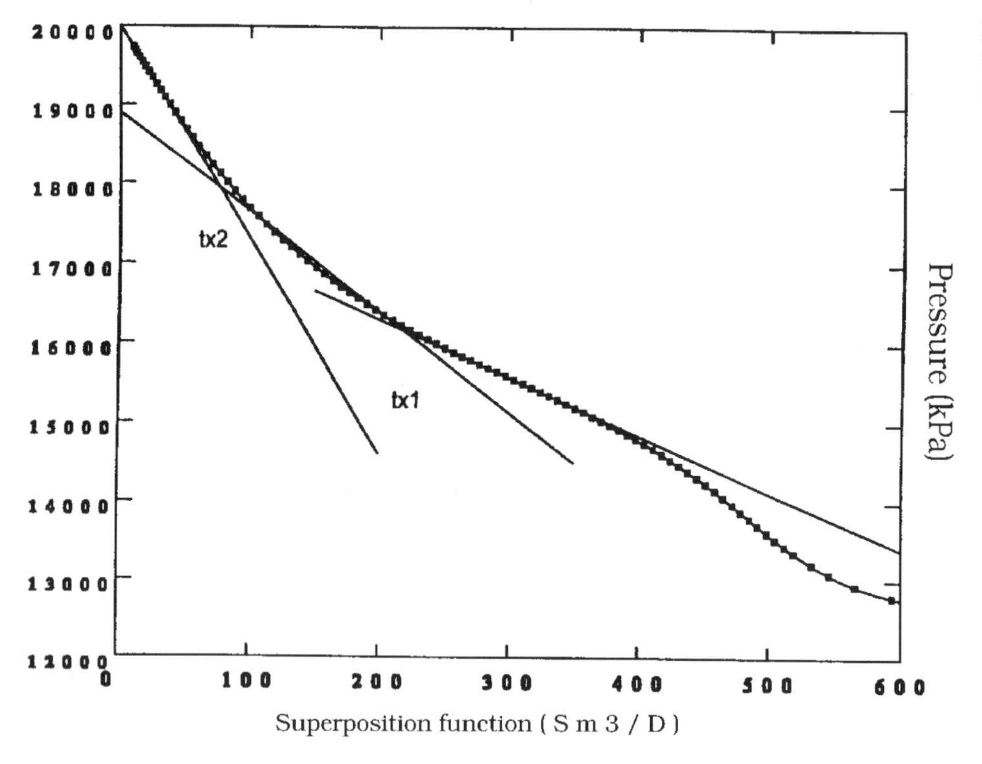

Figure 2. Horner curve for two boundaries at right angles.

resolve these differences. A simple example may show some of the difficulties.

(1) Two linear boundaries intersecting at right angles serve to produce three straight line segments showing slope doubling, and two intersection points on the Horner plot, which can be interpreted to give the distances to the two boundaries (Figure 2), while (2) two linear boundaries arranged parallel to each other produce a continuously increasing Horner plot characteristic of linear flow, which cannot be readily interpreted to give the distances to the two boundaries using the technique described (Figure 3).

In the computer-aided well test analysis technique based on mathematical models and regression analysis of the model parameters; the analyst selects an

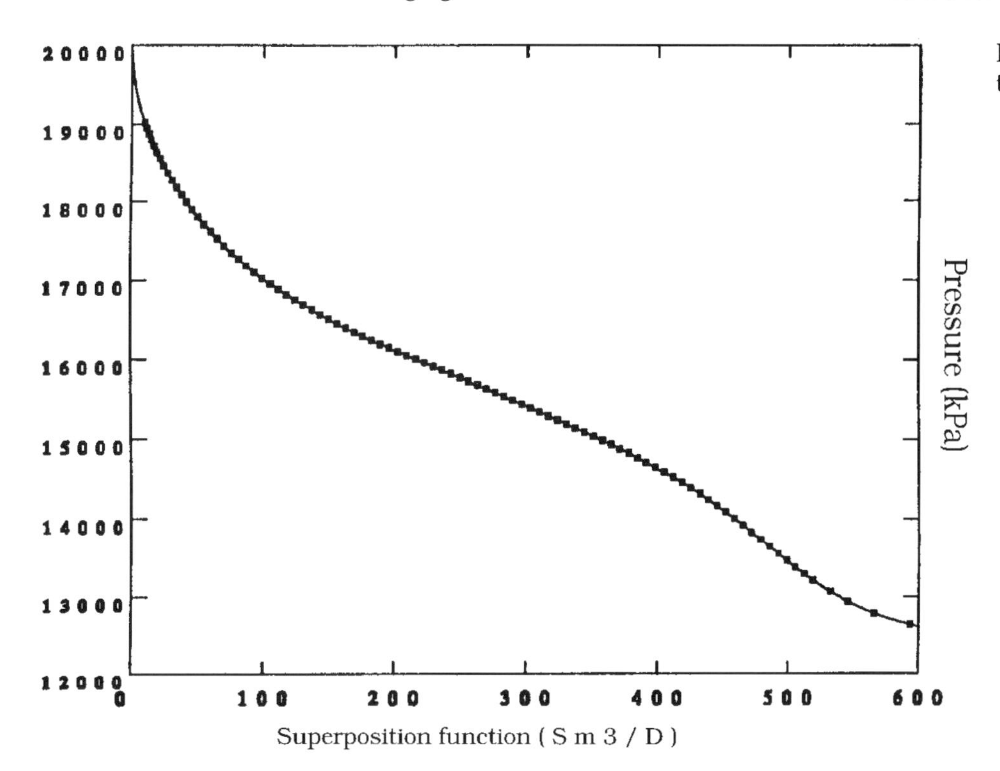

Figure 3. Horner curve for two parallel boundaries.

overall model based on near-wellbore, reservoir flow character, and reservoir boundary parameters, which are then "solved" to fit the observed behavior. The mathematical models usually involve solutions in Laplace space and inversions to real space, using multiple regression passes to converge to values that attempt to characterize the whole system of equations and effects. The mathematical background is complex, the numerical difficulty is significant, and the traditional handling of reservoir boundaries has been simplistic, involving linear bounding elements arranged as wedges or elements of a rectangle. Because the reservoir boundary effects usually occur late in time, the numerical approach above is dominated by near-wellbore and reservoir character parameters, and there is usually significant uncertainty in the choice of the reservoir boundary model and the accuracy of the predicted boundary parameters. The development of well test imaging was targeted at increasing the understanding of boundary effects, removing limitations imposed by arbitrary linear boundary systems and increasing the ease with which boundary data can be interfaced to other forms of data to increase the confidence in boundary distances and orientations.

WELL TEST IMAGING—THEORETICAL DEVELOPMENT

Information Theory and Boundary Type Curve Development

The development of well test imaging began with the use of a conventional well test analysis program, in its test design mode, to produce theoretical well test responses for simple boundary systems to deduce the behavior of the boundary signal. The first case was a single linear boundary system under a variety of conditions of near-wellbore, reservoir flow character, and boundary distance. The models were constructed in pairs with and without the boundary effects to get a visual presentation of the impact of a boundary. A typical simple model with wellbore storage and skin, and a uniform homogeneous reservoir produces the log-log type curve responses of Figures 4 and 5. The boundary effects show up most clearly in the difference in the Bourdet derivative curves at the bottoms of these figures.

Information theory indicates that we should attempt to isolate the effect of any input parameter by filtering out all of the extraneous information in a signal so that we can concentrate solely on the response to the input. In the case of the pairs of well test designs above, a simple way of removing the near-wellbore and reservoir character effects was found by dividing the Bourdet derivative response of the case with boundary effects by the Bourdet derivative of the case without boundaries. This produces a very simple boundary ratio type curve, which has a value of 1.0, up to the point at which boundary interactions take place and a shift upward in response to boundary interactions, as shown on Figure 6, which was derived from Figures 4 and 5. When many well test designs are processed in this way it can be seen by inspection that the curves are all similar (Figure 7) and can be condensed into one curve (Figure 8) by a shift in the time scale. This implies geometric similarity for the single boundary problem with the time, t_0, being representative of the time for the boundary interaction to be perceived at the test well.

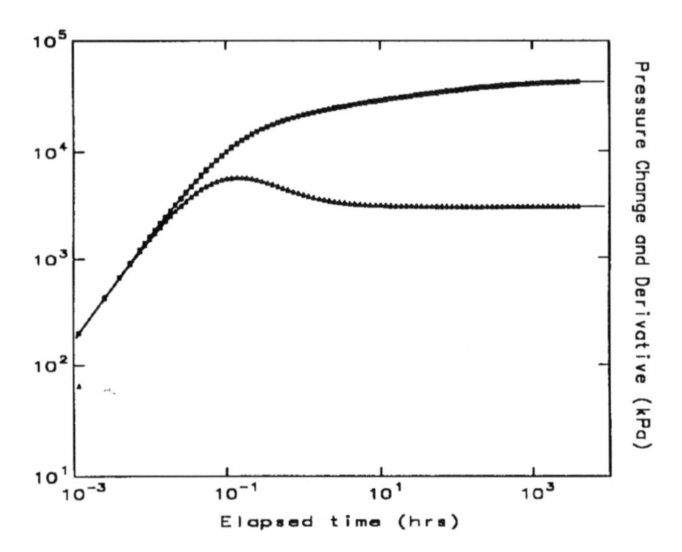

Figure 4. Log-log type curve for homogeneous reservoir.

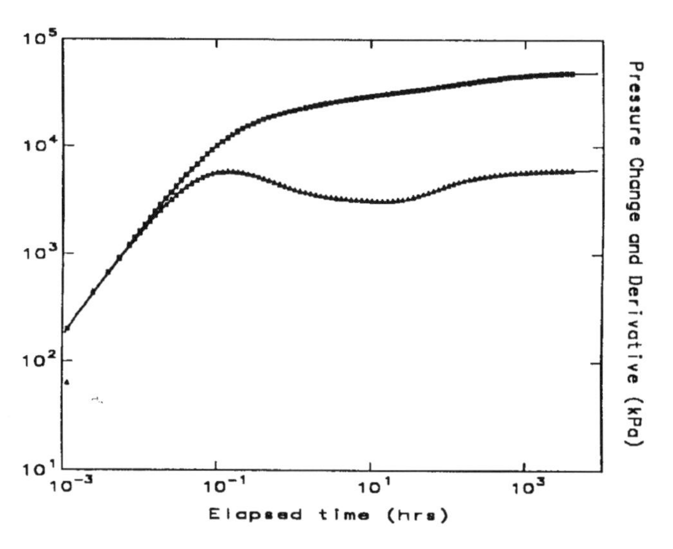

Figure 5. Log-log type curve for homogeneous reservoir with a single linear boundary.

A similar set of well test designs was done for the case of two boundaries in the form of a wedge. The boundary ratio type curves for the case of a 60° wedge with the well on the center-line are shown on Figure 9. The boundary ratio type curve was found to have a geometric similarity relationship similar to that seen in the single boundary case (Figure 10). For all of the wedge angle cases, the boundary ratio type curve approached an asymptotic limit value (wedge boundary ratio limit)

$$\text{Wedge Boundary Ratio Limit} = 360/\text{Wedge included angle} \qquad (3)$$

as shown on Figure 11.

The final set of well test designs studied was the set of two parallel boundary "channel flow" cases, shown on Figure 12. It was found again that only the short time behavior showed differences that could be characterized by the position of the well between the two boundaries, but that all cases approached a linear, half slope on the log-log plot, limit behavior. For a common ratio of distances to the two boundaries, a normalized type curve could be developed for each case, as on Figure 13.

The strong geometric similarity behavior of the sets of cases studied argues strongly that there is an underlying linkage in the physics of how the information about a boundary is sensed by the well test, and that this is linked to the flow and pressure transient behavior of the reservoir. In particular, it can be concluded that the magnitude of the response seen on the boundary ratio type curve was independent of distance to the boundary, and represents a measure of the geometric properties of the boundary. The distance to the boundary appears to be related to the time scaling of the boundary response.

Ray Tracing and Interpretation of Boundary Effects

A set of visualization experiments of a drawdown or buildup well test with a single boundary was conducted using a simple computer program that plotted the position of the radius of investigation as a function of time, under the assumption that the well test could be treated as a non-linear transient wave phenomenon. A composite sequence of these images is shown on Figure 14, illustrating that the effective wave position is dependent on the square root of the test time. The information about the first contact with the boundary ($t = 100$) reaches the well at a time equal to four times the contact time ($t = 400$). All of the information about the boundary, however, does not reach the test well at one time, as seen on the type curves generated in the previous section. To determine just how all of the information reaches the well, the overall wave front was broken down into a number of rays or packets. In order for information on the boundary contact to propagate back to the test well, it was necessary to assume that the boundary behaved in a manner similar to an elastic absorption and re-radiation rather than as a reflection, similar to the concepts used in interpreting seismic data. These subsequently re-radiated rays that pass information back to the source well, illustrated on Figure 15, have wave fronts that combine to form the overall front of Figure 14. Information is carried back to the test well along a straight-line path, from the contact point to the test well, which arrives back at the test well at four times the contact time for each ray. Considering just the array of rays carrying information back to the test well over time, as in Figures 16 and 17, it was observed that the rays passing through the test well formed an angle of view of the boundary which asymptotically approached 180°. Going back to equation 3, this also predicts the

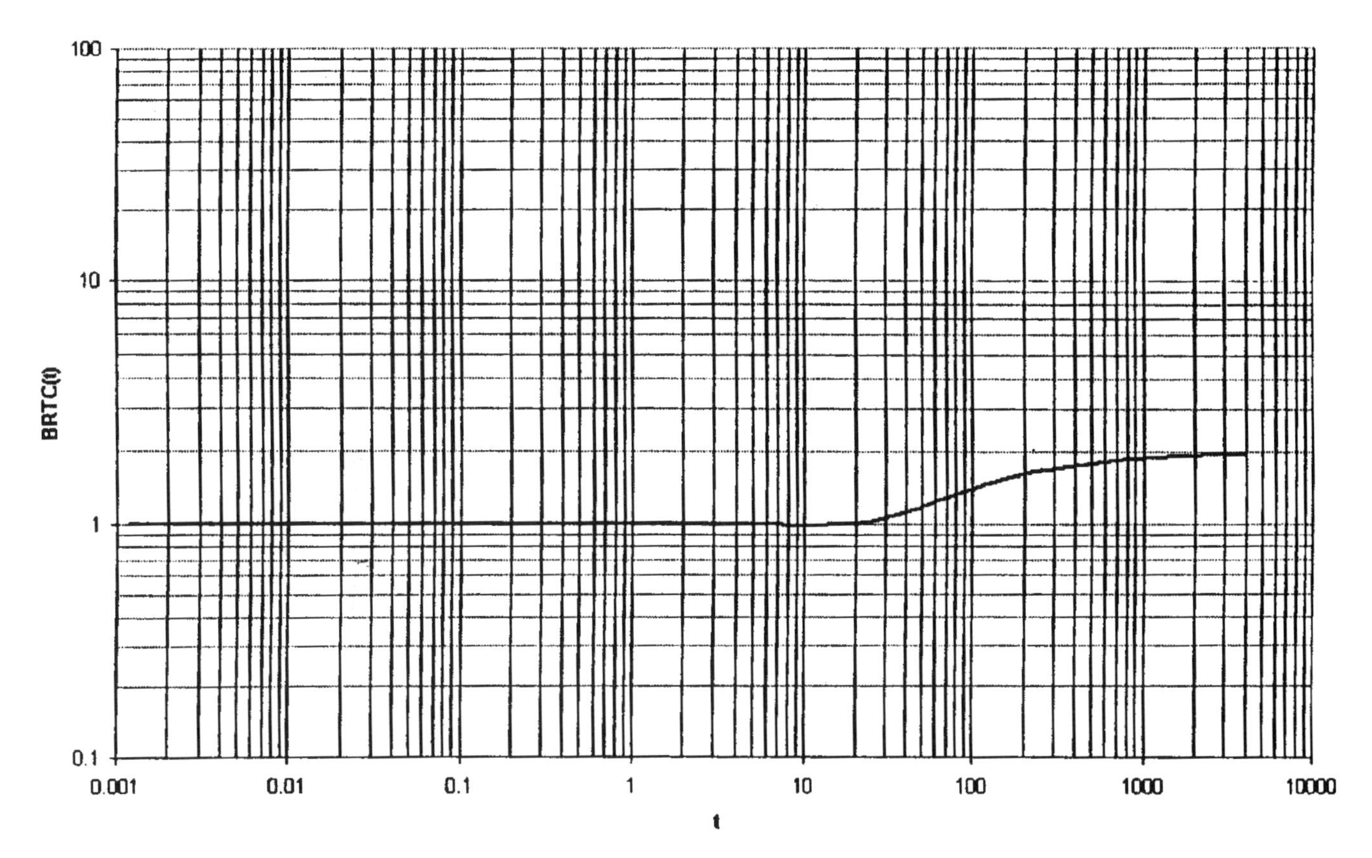

Figure 6. Boundary ratio type curve for a single linear boundary.

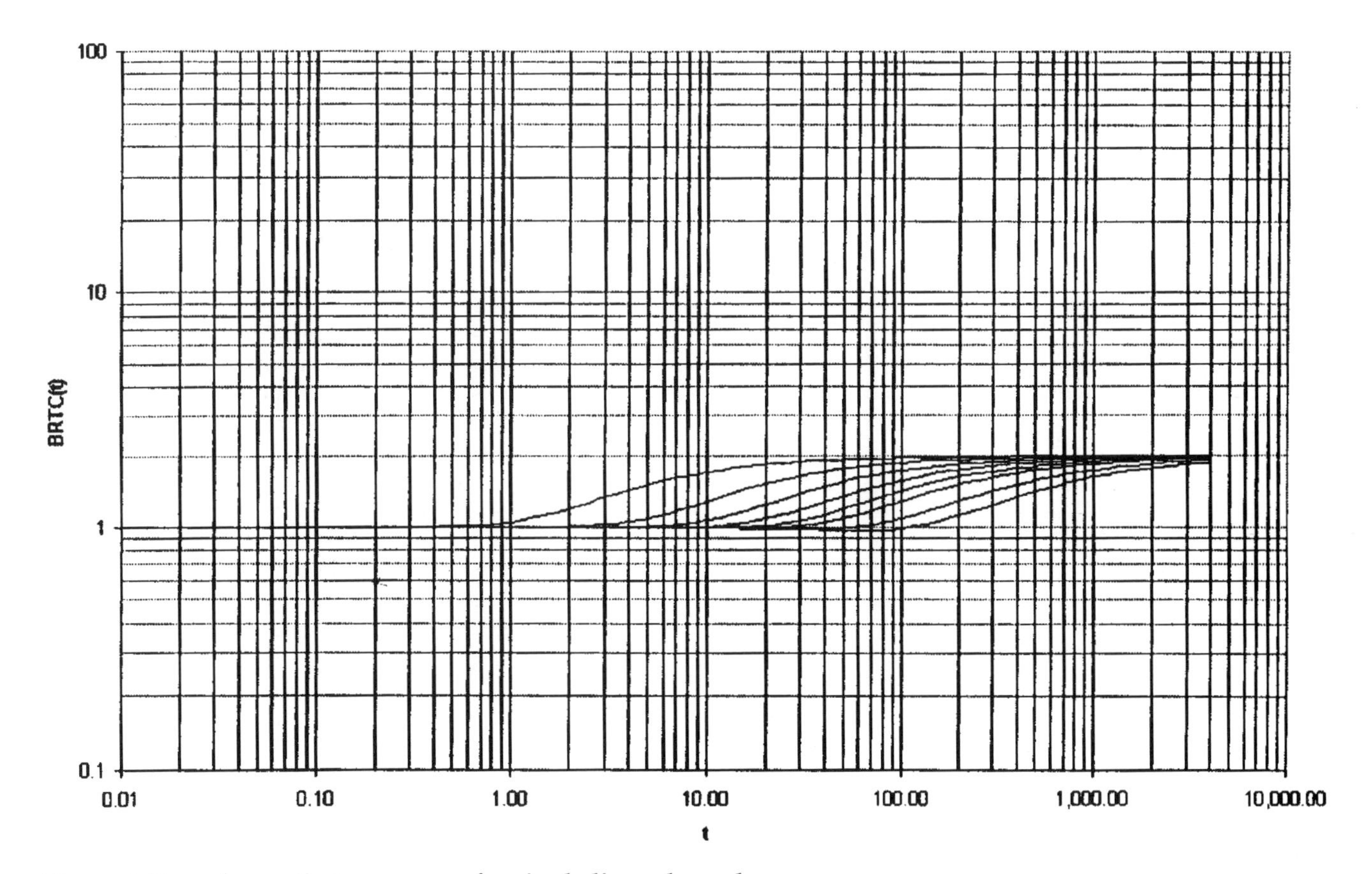

Figure 7. Boundary ratio type curves for single linear boundary cases.

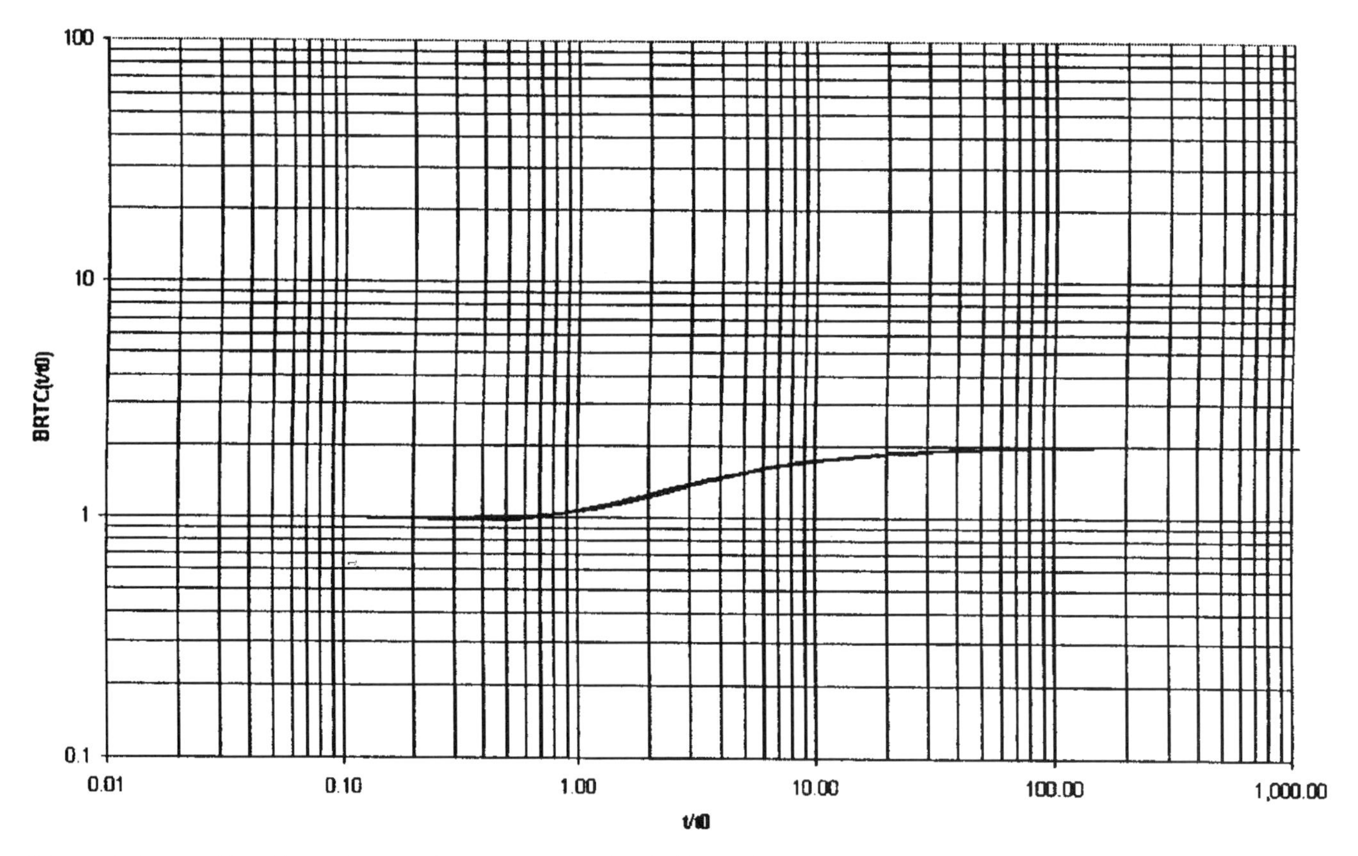

Figure 8. Normalized boundary ratio type curve, single linear boundary.

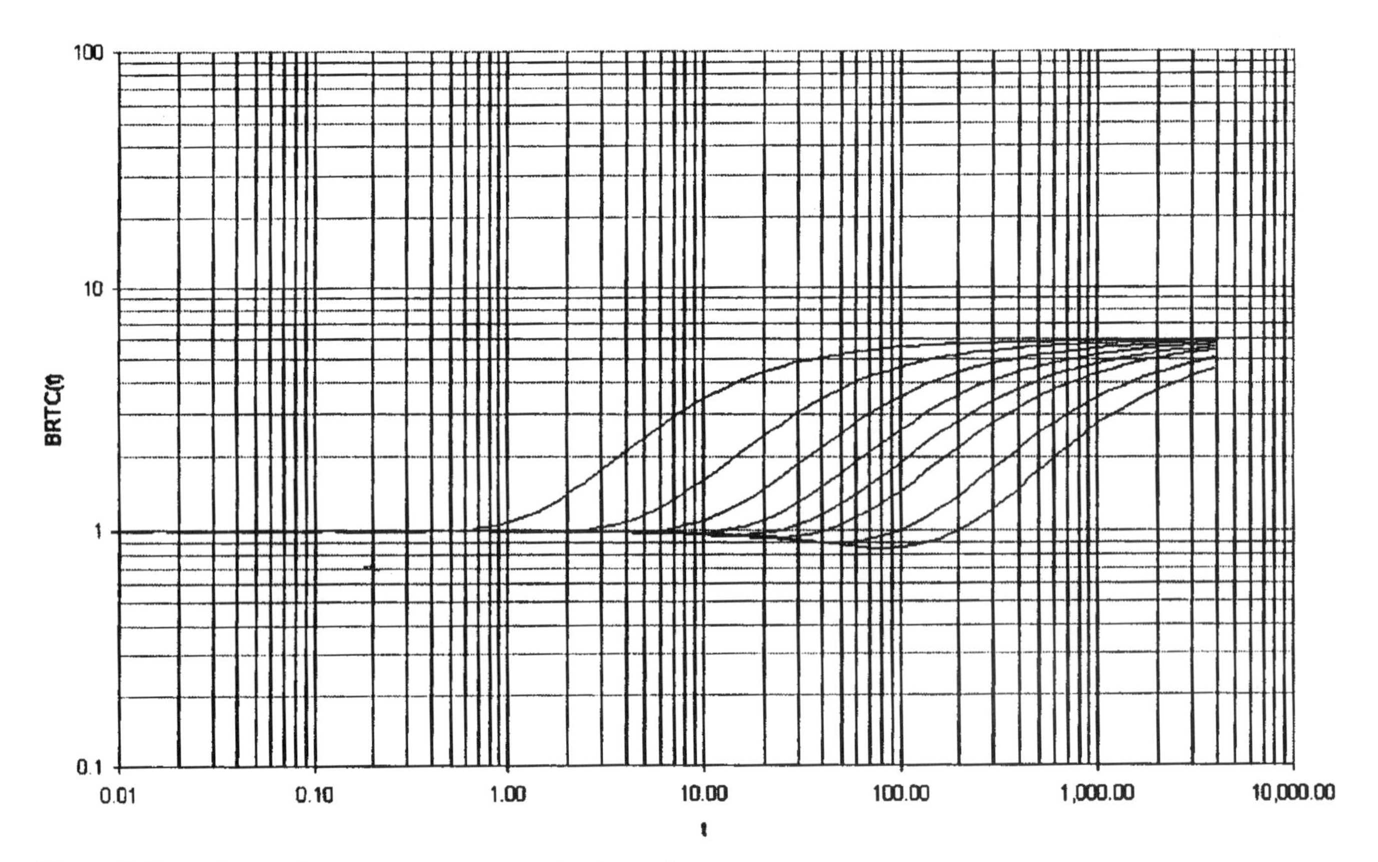

Figure 9. Boundary ratio type curves, 60° wedge boundary.

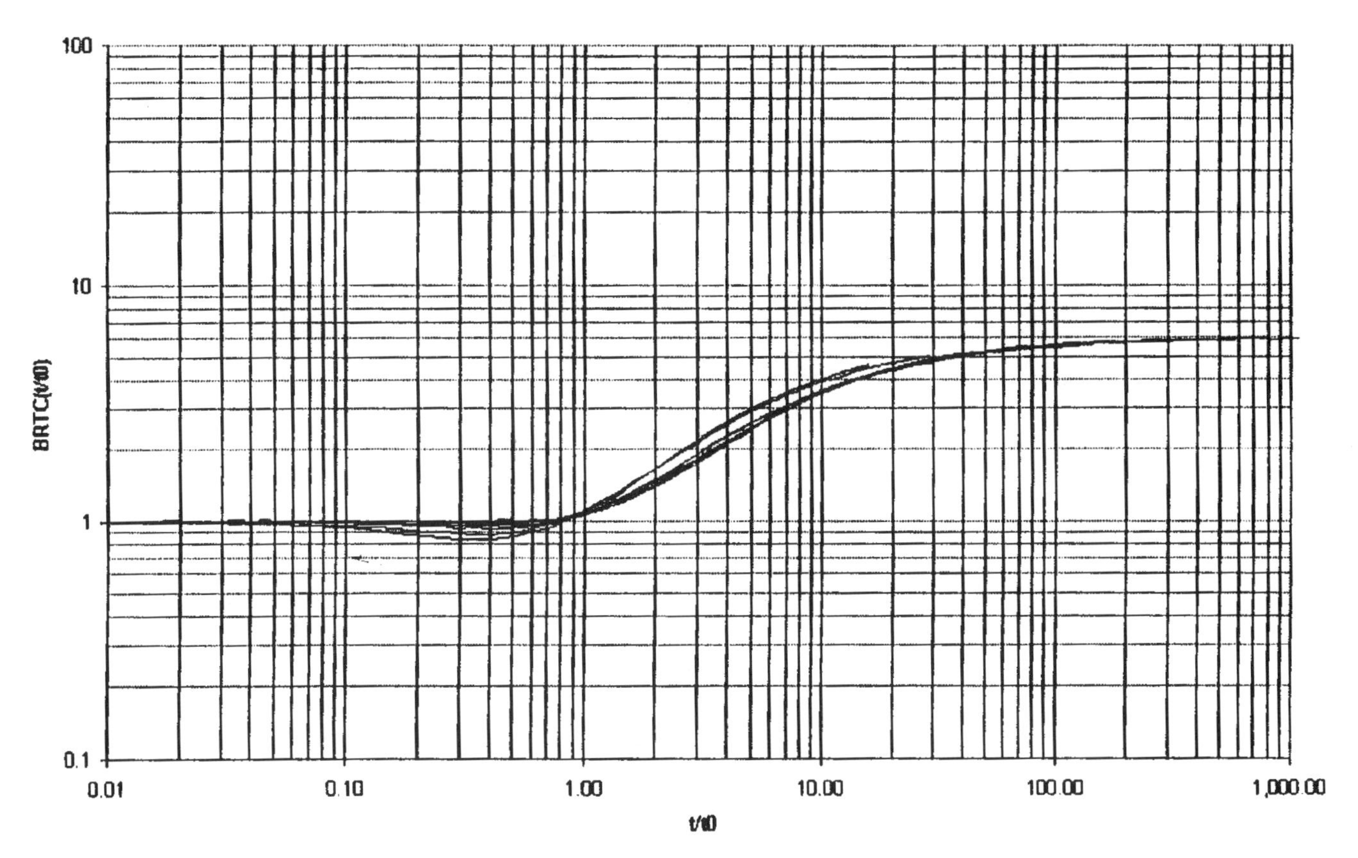

Figure 10. Normalized boundary ratio type curve, 60° wedge boundary.

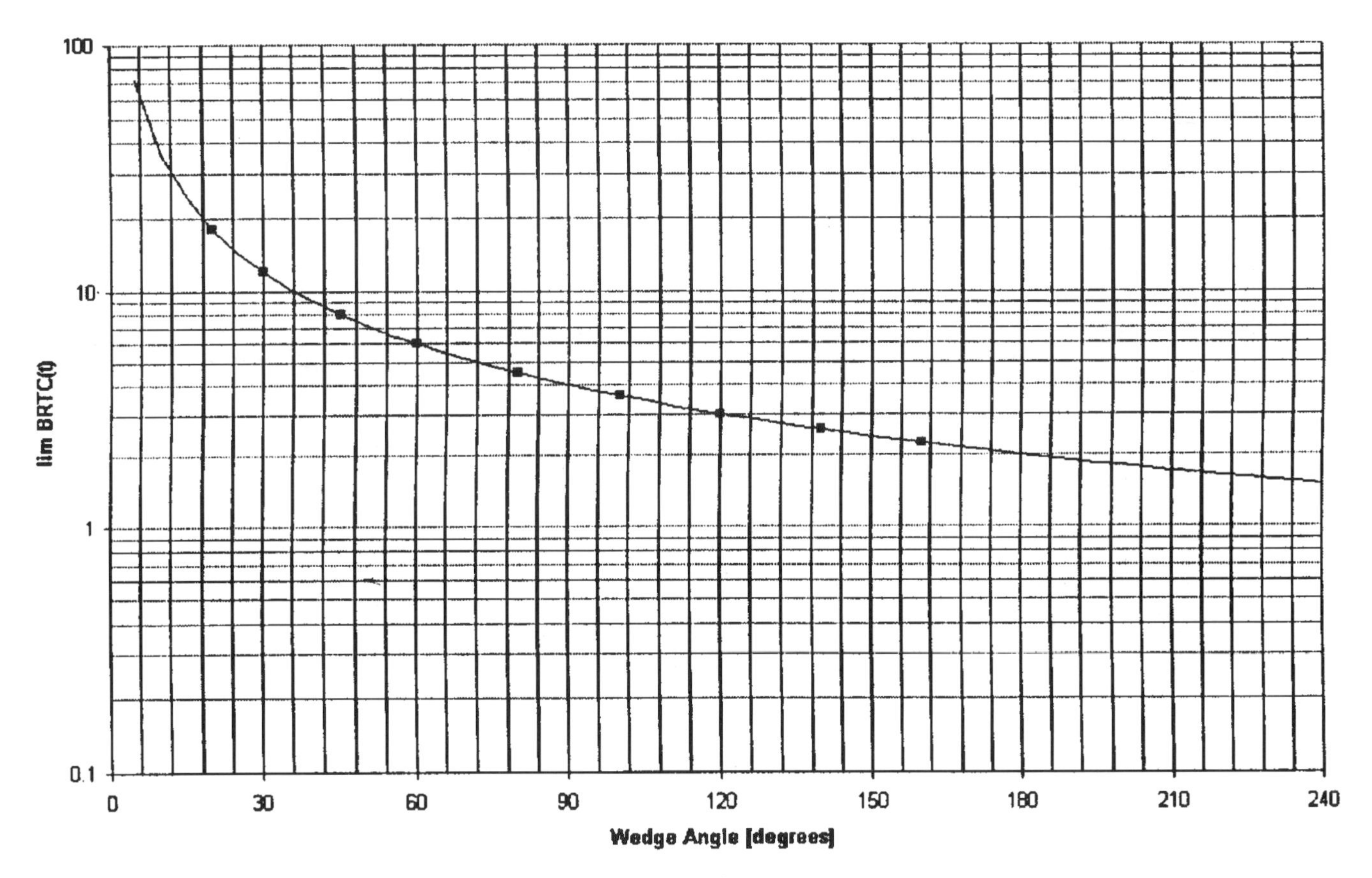

Figure 11. Boundary ratio type curve limit as a function of wedge angle.

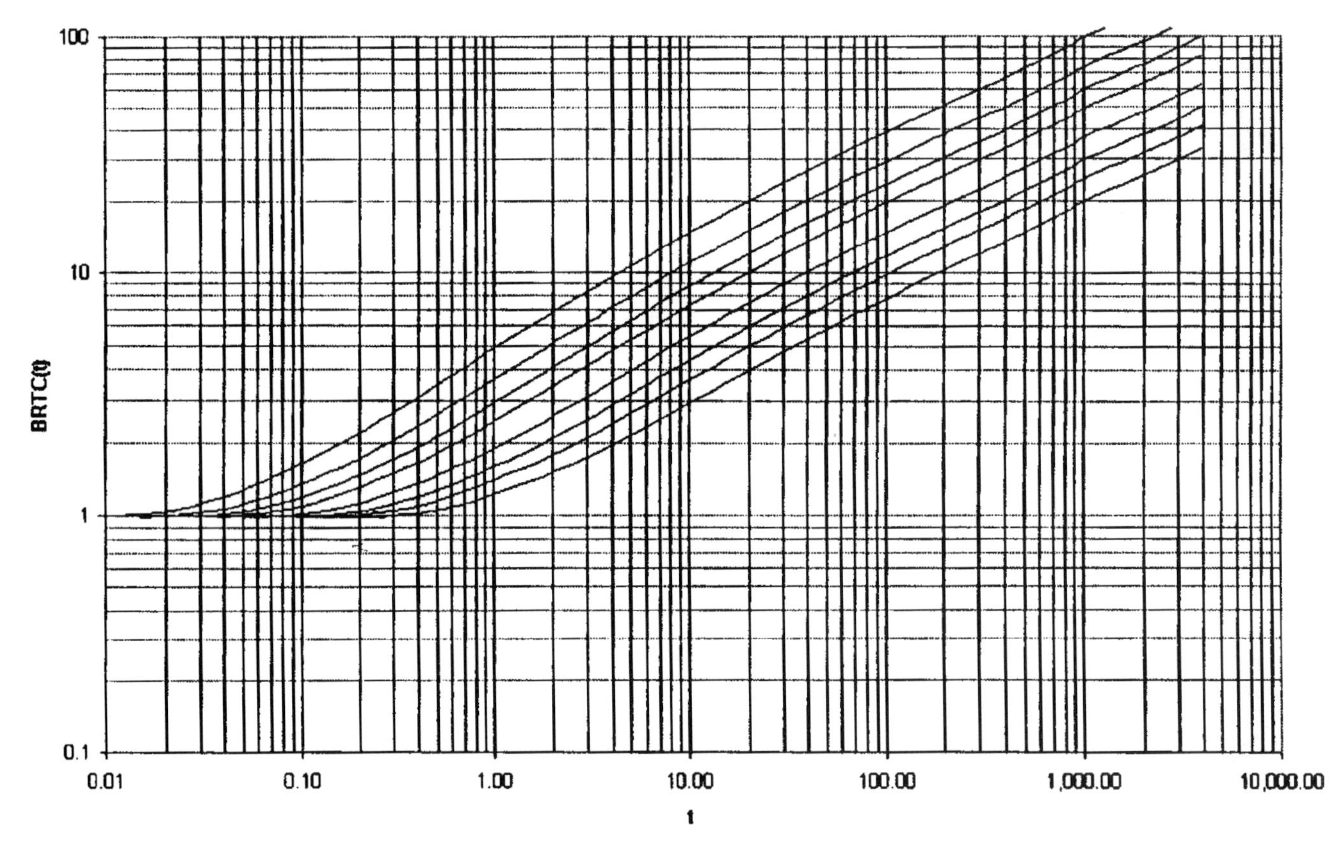

Figure 12. Boundary ratio type curves, 2:1 parallel boundaries.

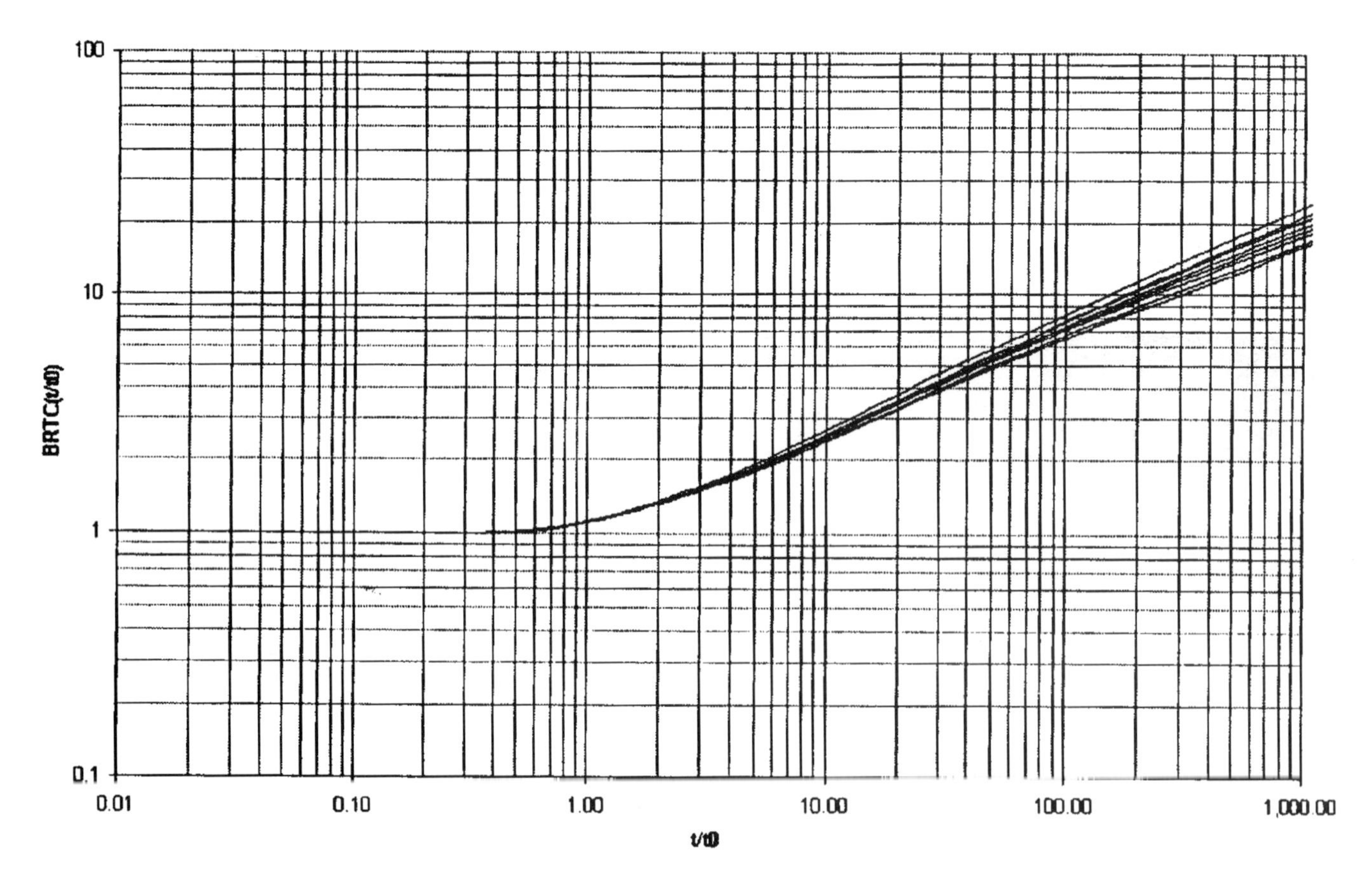

Figure 13. Normalized boundary ratio type curve, 2:1 parallel boundaries.

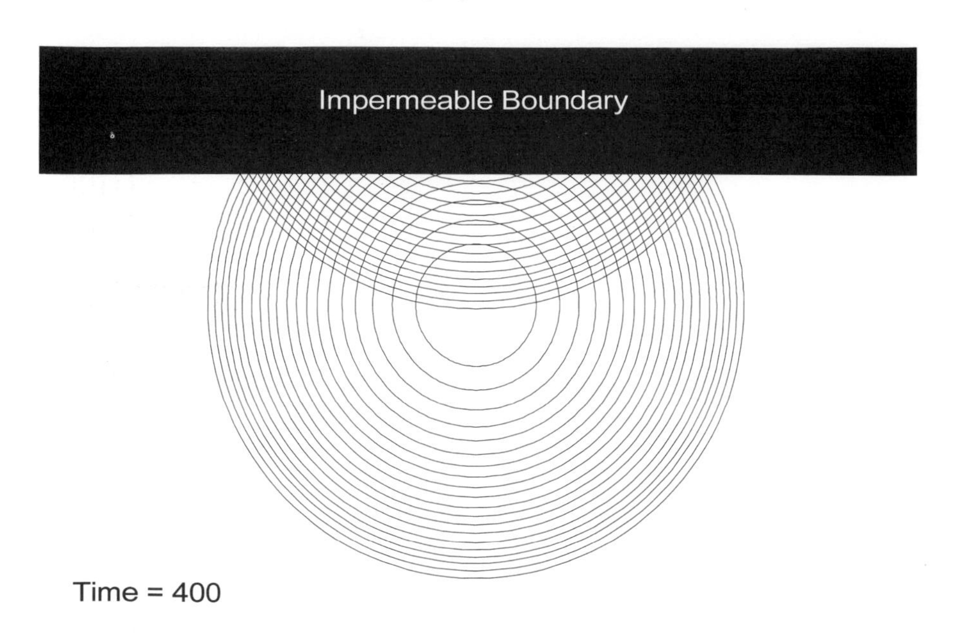

Figure 14. Composite images of R_{inv} as a function of time.

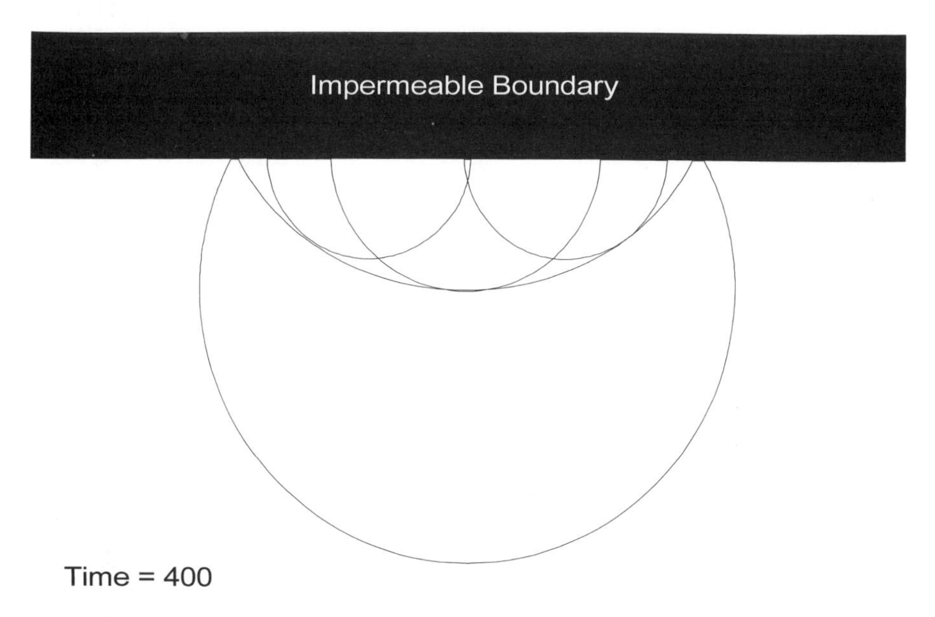

Figure 15. Assumed boundary interaction wave fronts.

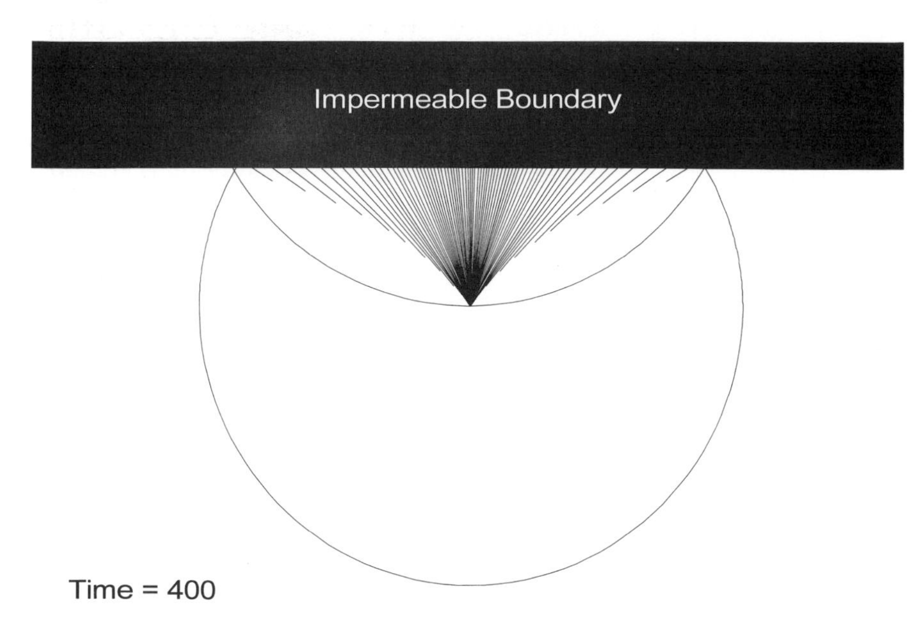

Figure 16. Rays showing boundary interactions at time t_0.

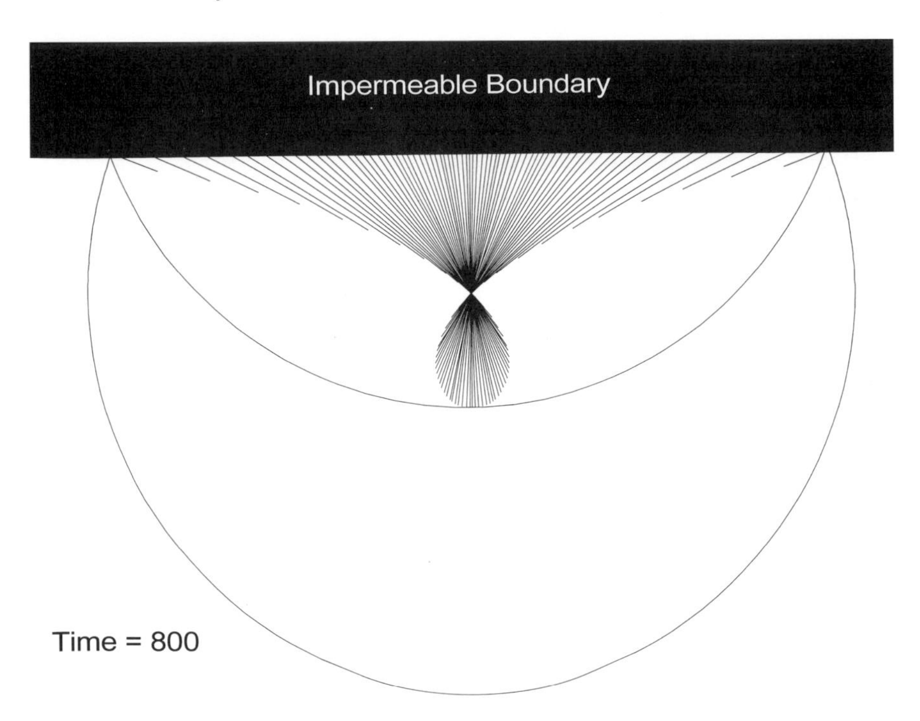

Figure 17. Rays showing boundary interactions at time $t > t_0$.

asymptotic boundary ratio type curve value for the single linear boundary.

If we assume that the instantaneous angle of view of the boundary, AOV(t), is directly related to the boundary ratio type curve value through

$$\mathrm{BRTC}(t) = 360/[360 - \mathrm{AOV}(t)] \quad (4)$$

it is possible to calculate a theoretical type curve for the single linear boundary case. This is accomplished by following the logic of Figure 18, which shows a ray with $R_{\mathrm{inv}}(t)$ just contacting the boundary. The value of the angle of view of the boundary is $2 * \theta$, which can be expressed through relationships with the radius of investigation with time

$$R_{\mathrm{inv}}(t) = d_c \sqrt{t/t_c} \quad (5)$$

as

$$\mathrm{AOV}(t) = 0 \qquad \text{for } t \leq 4 * t_c \quad (6a)$$

where t_c is the time to first contact with the boundary at which time $R_{inv} = d_c$ and therefore,

$$\mathrm{AOV}(t) = 2 * \theta \qquad \text{for } t > 4 * t_c$$
$$= 2 * \arccos[2 * d_c / R_{\mathrm{inv}}(t)] \quad (6b)$$

$$= 2 * \arccos\left(\sqrt{4 * t_c/t}\right) \quad (6c)$$

Therefore, an equation for the BRTC(t) for a single linear boundary can be expressed as

$$\mathrm{BRTC}(t) = 360\Big/\left[360 - 2 * \arccos\left(\sqrt{4 * t_c/t}\right)\right] \qquad \text{for } t > 4 * t_c \quad (7)$$

If we normalize this curve at $t_0 = 4 * t_c$, then the resulting normalized type curve plot is shown on Figure 19, which can be compared with the normalized linear boundary ratio plot of Figure 8, derived from the well test design exercises which began the process. The approximate answer of equation 7 is a very close match to Figure 8 and has several desirable characteristics, in that it:

(1) is easy to evaluate
(2) has a clear physical interpretation
(3) is easily inverted

This same construction process can be used on the 60° wedge example and the parallel channel boundary to reconstruct approximations of those type curves as well, as shown on Figures 20 and 21.

Any arbitrary reservoir boundary system, whether it is composed of straight lines or curves, can be evaluated using the ray tracing methodology to construct a boundary ratio type curve for that boundary system. Combined with any baseline near wellbore and reservoir flow characteristic log-log type curve, this technique can generate an overall type curve for that reservoir and boundary system. Yeung et al. (1993) proposed a method to analyze the transient pressure behavior of an arbitrarily shaped reservoir using a superposition in time approach which is much more time consuming and difficult than the method proposed above.

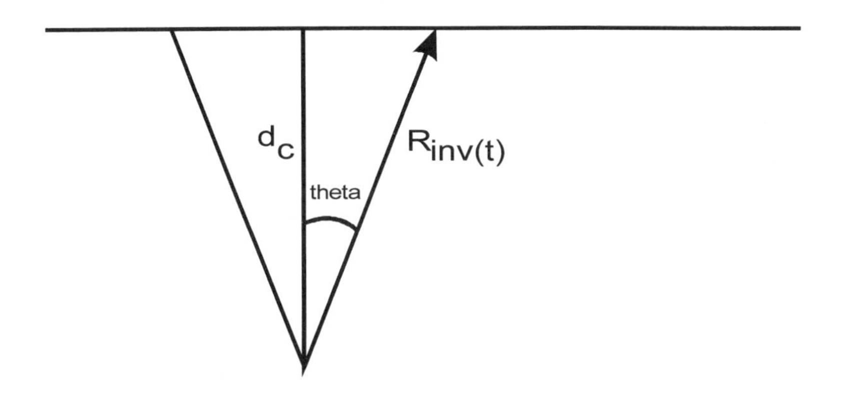

Figure 18. Diagram of ray interaction with boundary at time $t > t_c$.

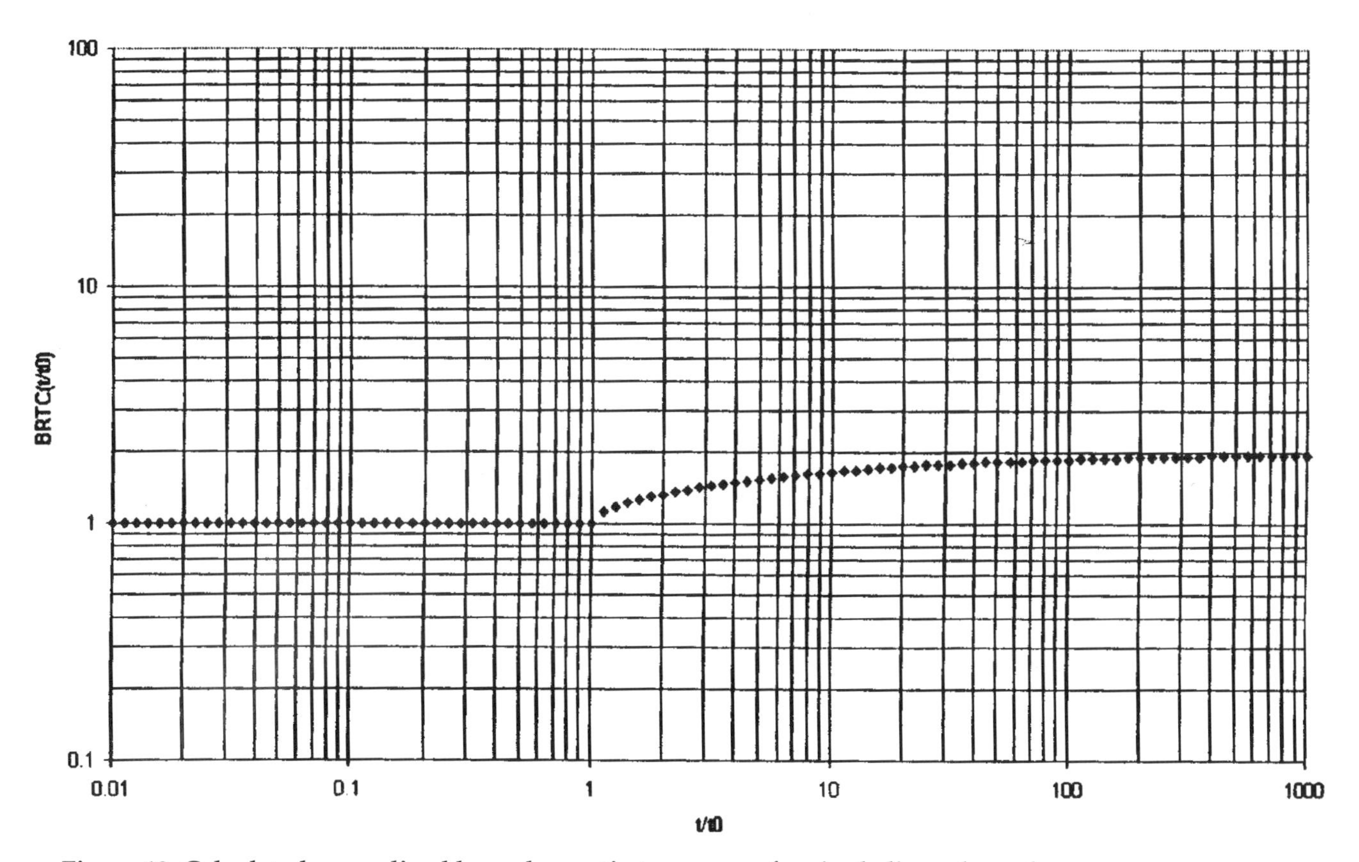

Figure 19. Calculated normalized boundary ratio type curve for single linear boundary.

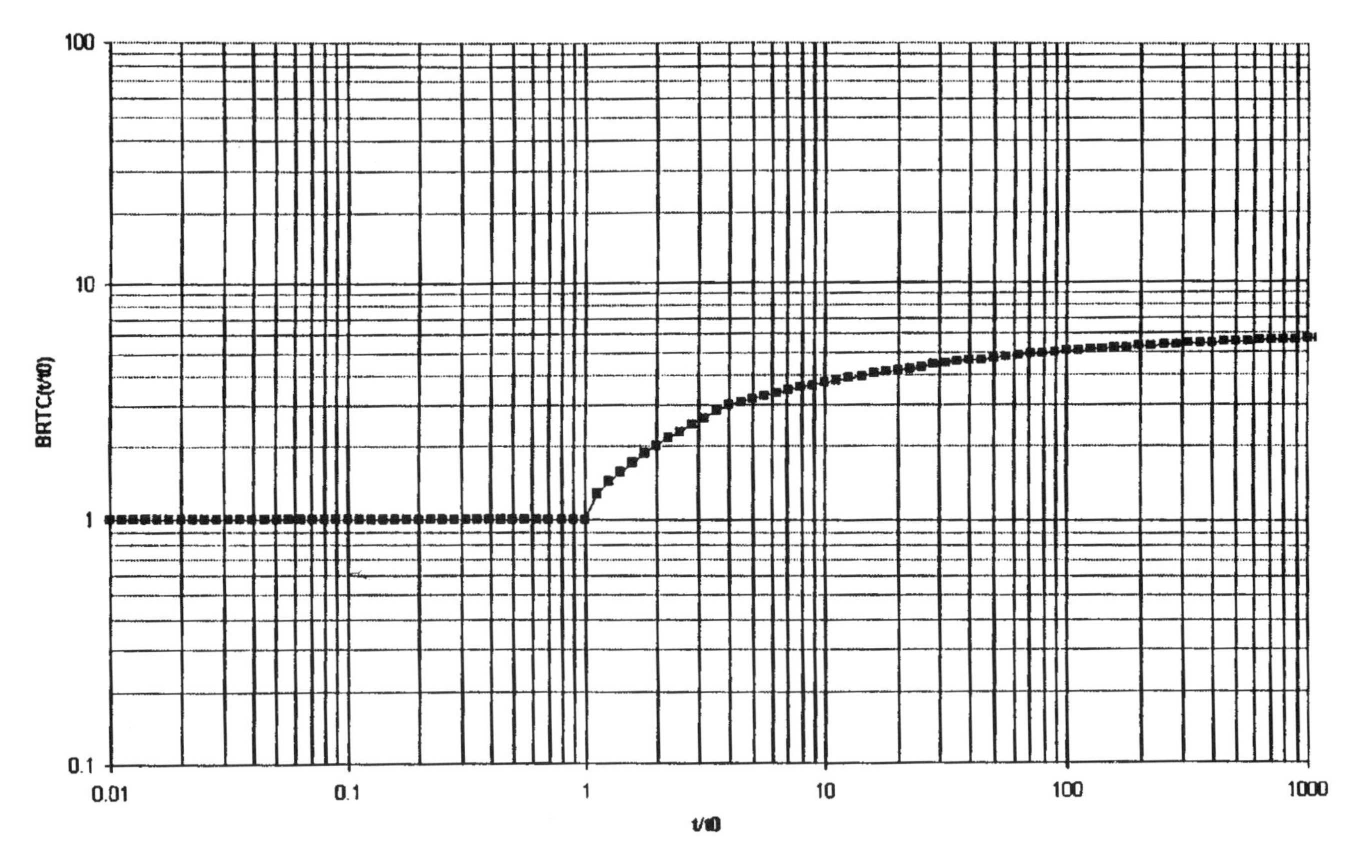

Figure 20. Calculated normalized boundary ratio type curve for 60° wedge boundary.

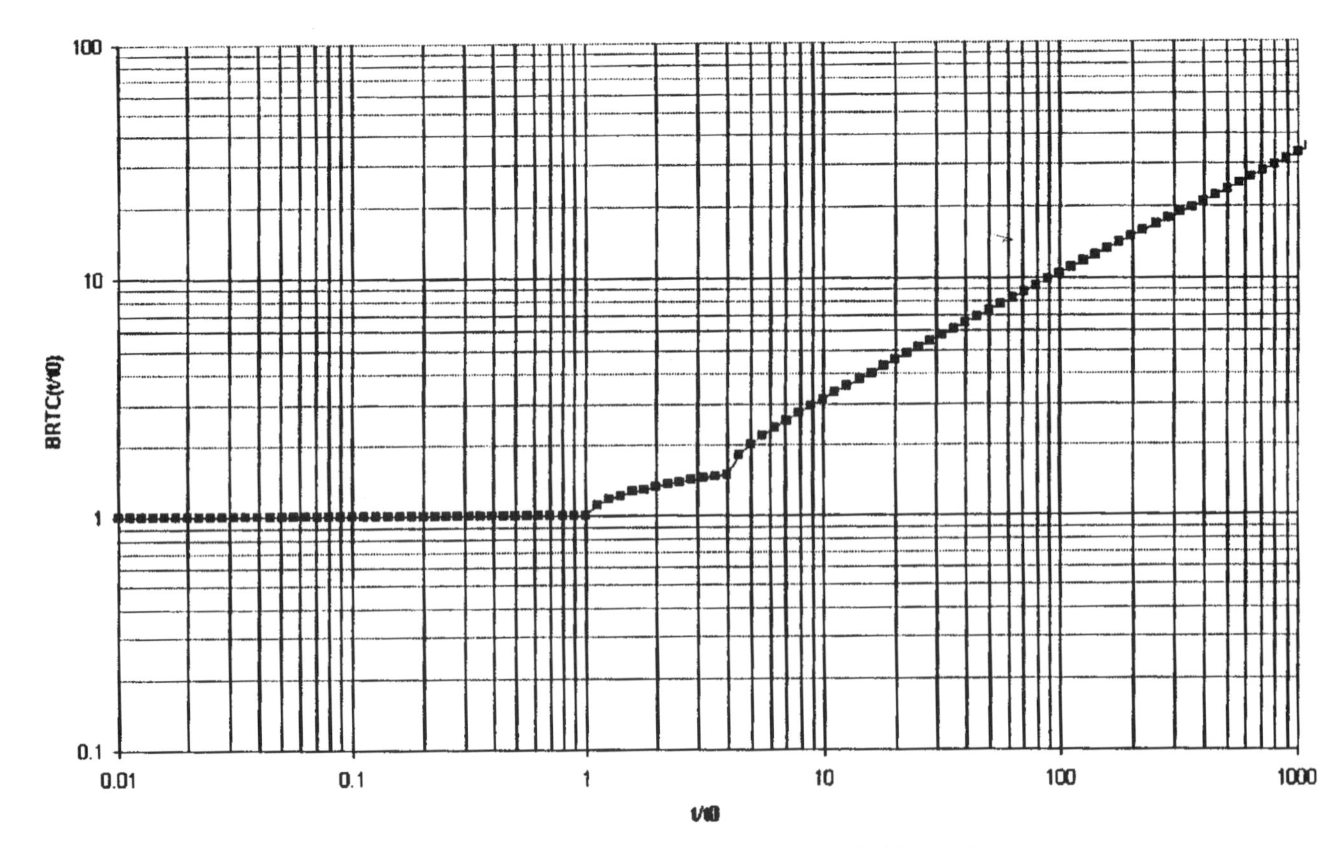

Figure 21. Calculated normalized boundary ratio type curve for 2:1 parallel boundaries.

INVERSION OF A WELL TEST RESPONSE—CALCULATION OF A WELL TEST IMAGE

As useful as being able to construct a type curve for a given geometry may be, the bigger problem in well test analysis is the interpretation of the actual test response to determine the distance to and orientation of the boundaries observed in a well test. The simple form of equation 6 allows simple inversion; that is, for a given well test response, a picture of the boundary system can be created. The process starts by conducting a conventional analysis to produce the best possible match to the early time data to properly calibrate the near-wellbore and reservoir flow characteristics. At this point we either do not match the boundary affected portion of the test or turn off the boundary effects once the best conventional match is achieved. The resulting difference between the model log-log type curve and the actual data represents the information or signal of the boundary. By dividing the actual reservoir response by the nonboundary predicted response we create the boundary ratio type curve as a function of time. In addition, as with most programs, the analysis method without boundaries will predict a radius of investigation for the whole test (i.e., $R_{inv}(t_{max})$), which will be used to calibrate distances for the calculations.

Given the boundary ratio type curve as a function of time, BRTC(t), the angle of view (AOV) of the boundary as a function of time and radius of investigation can be determined by inverting equation 4

$$AOV(t) = 360 - 360/BRTC(t) \tag{8}$$

Given the angle of view of the reservoir boundary system as a function of time, the distance to the reservoir boundary is determined as a function of the radius of investigation as a function of time. The distance to a reservoir boundary that will affect a given well test at a given time is exactly one-half of the radius of investigation at that point (since the information must travel back to the source well, covering twice the distance)

$$R_{inf}(t) = R_{inv}(t)/2 \tag{9}$$

This distance is called the radius of information in the terminology of this paper. Given $R_{inv}(t_{max})$, the $R_{inf}(t)$ is calculated at each point from

$$R_{inf}(t) = 0.5 * R_{inv}(t_{max}) * \sqrt{t/t_{max}} \tag{10}$$

With $R_{inf}(t)$ and the AOV(t), a picture of the reservoir surrounding a well can be drawn in the following manner. Beginning at the last point where the BRTC(t) is less than or equal to 1.0 ($t = t_0$), draw a circular arc at a radius equal to

$$R_{inf}(t_0) = 0.5 * R_{inv}(t_{max}) * \sqrt{t_0/t_{max}} \tag{11}$$

This represents the maximum radius of reservoir behavior without boundary influences. At each point $t > t_0$, draw an arc of radius $R_{inf}(t)$ from equation 10, through an angle of 360 – AOV(t) showing the nonboundary-containing reservoir as a function of time. Continue to draw these arcs until $t = t_{max}$ is reached. The resulting diagram, similar to a sonar display, will be an image or realization of the reservoir and, by difference, the boundary system. A well test image drawn at this stage is called a diagnostic image, in that it can give you a general reservoir shape and distance to a boundary, but the picture may not be realistic or recognizable in terms of geological features you might be able to recognize in other data. To make this picture more understandable in terms of our usual visualization processes and to help distinguish more complex systems when more than one boundary may be present, three standardized interpretation models have been devised.

(1) Angular Model. Draw a tangent line to the circle at $R_{inf}(t_0)$, and draw all remaining arcs with one end placed on the tangent line.
(2) Balanced Model. Draw a radial line from the center of the circle of radius $R_{inf}(t_0)$ and center all of the remaining arcs on this line.
(3) Channel Model. Draw a tangent line to the circle at $R_{inf}(t_0)$ and a radial line perpendicular to it through the test well. Begin to draw arcs centered on the perpendicular line until the remaining arc ends fail to cross or contact the tangent line. At this point, split the arc in half and place the two original ends on the initial tangent line on opposite sides of the well and predict the position of the other boundary by tracing out the position of the newly cut arc ends.

These model descriptions can be expressed mathematically, calculated, and displayed directly from the well test analysis. In many cases, all three models may produce results that are valid and need to be tested against other geological information. This merely illustrates that the well test data (and the information returned by a well test alone) may be insufficient to provide a unique answer to the question of boundary configuration.

WELL TEST IMAGING—EXAMPLES

There are few examples in the literature with enough data to allow comparison of conventional and well test imaging approaches to the prediction of reservoir boundaries. One exception is the paper by Clark et al. (1985) on the Oseberg reservoir, which shows the application of conventional techniques to predict two boundaries at right angles around well 30/6-13 (Figure 22). Application of well test imaging to the raw data from the paper produces the log-log

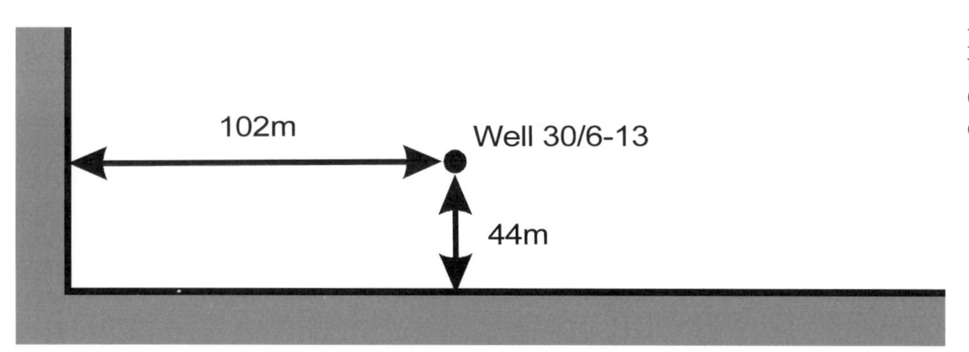

Figure 22. Conventional boundary prediction, Oseberg 30/6-13; after Clark et al. (1985).

type curve of Figure 23, the BRTC(t) of Figure 24, and the well test image (diagnostic image) of Figure 25. The key difference is that the boundaries are found to be neither at right angles nor precisely linear, and the distances to the boundaries are slightly different than those predicted conventionally. These differences could be significant in terms of interpretation of the alignment of the boundaries with known seismic or geological anomalies, and illustrate how our dependence on fixed models of reservoir boundaries could alter our perception of the boundary system.

Another example of the application of the approach is taken from a Petro-Canada–operated well in the Fireweed pool in northeastern British Colombia. This pool is a fractured and dolomitized carbonate reservoir containing gas, where reservoir quality and continuity are an issue. The well test log-log type curve for an initial well in a new area of the pool is shown on Figure 26. The boundary ratio type curve of Figure 27 shows some significant variations, which were difficult to fit with conventional techniques (predicting two boundaries at right angles, but leaving a high residual error in the match). The diagnostic image of the test (Figure 28) shows a system with several "steps" in the boundary system. We believe that these variations depict the underlying fracturing fabric of the reservoir and a conjugate set of lineaments that both create the reservoir and simultaneously

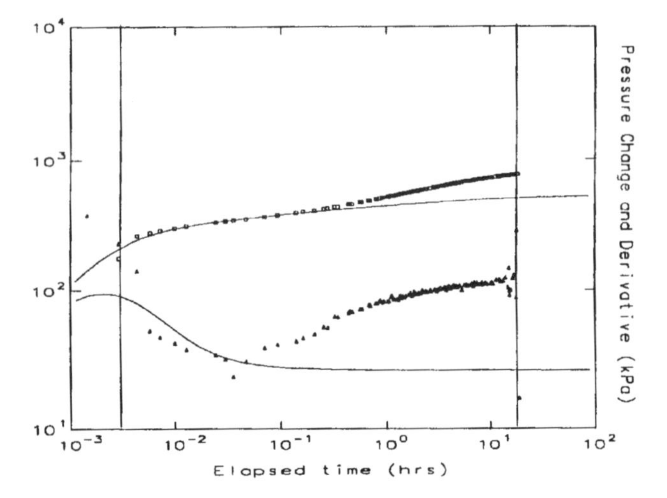

Figure 23. Oseberg 30/6-13, log-log type curve of test B; after Clark et al. (1985).

set up internal reservoir boundaries. These considerations have influenced our application of horizontal well technology to develop the field, using our knowledge of local tectonic history to orient the well test images in the preferred fracturing directions and to orient the horizontal wells along axes which allow the maximum number of lineaments to be contacted.

WELL TEST IMAGE ENHANCEMENT—INTEGRATION OF GEOLOGICAL AND GEOPHYSICAL DATA

The problem of well test image solution uniqueness can be addressed in many ways, including

(1) fitting together of well test images of neighboring wells that may image parts of the same boundary system, much like solving a jigsaw puzzle.
(2) observing a well test at offsetting wells and interpreting the interference data to determine boundary orientations; the ray tracing methodology proposed above can be used to interpret the offset well response.
(3) incorporating geological or geophysical information from other sources to fit against the well test images to provide information on orientation of lineaments or depositional directions.

The experience of the last few years with well test imaging indicates that the integration with geological and geophysical data offers the most potential for significant improvement in the quality of the images. In addition, it can lead to an improvement in the understanding of the characteristics of the boundaries in terms of their expression in the other data types.

Geological maps and 3-D seismic data represent excellent sources of information that can be integrated with well test imaging. As an example, an attribute of the 3-D seismic data (amplitude of a reflector, depth of a surface, slope of a surface), is contoured in the form of a 2-D map. The well location is marked on the map, and the well test image is superimposed at the same scale. As a first approximation, the whole image can be rotated and positioned so that the circular boundary of the radial flow reservoir region just contacts a likely

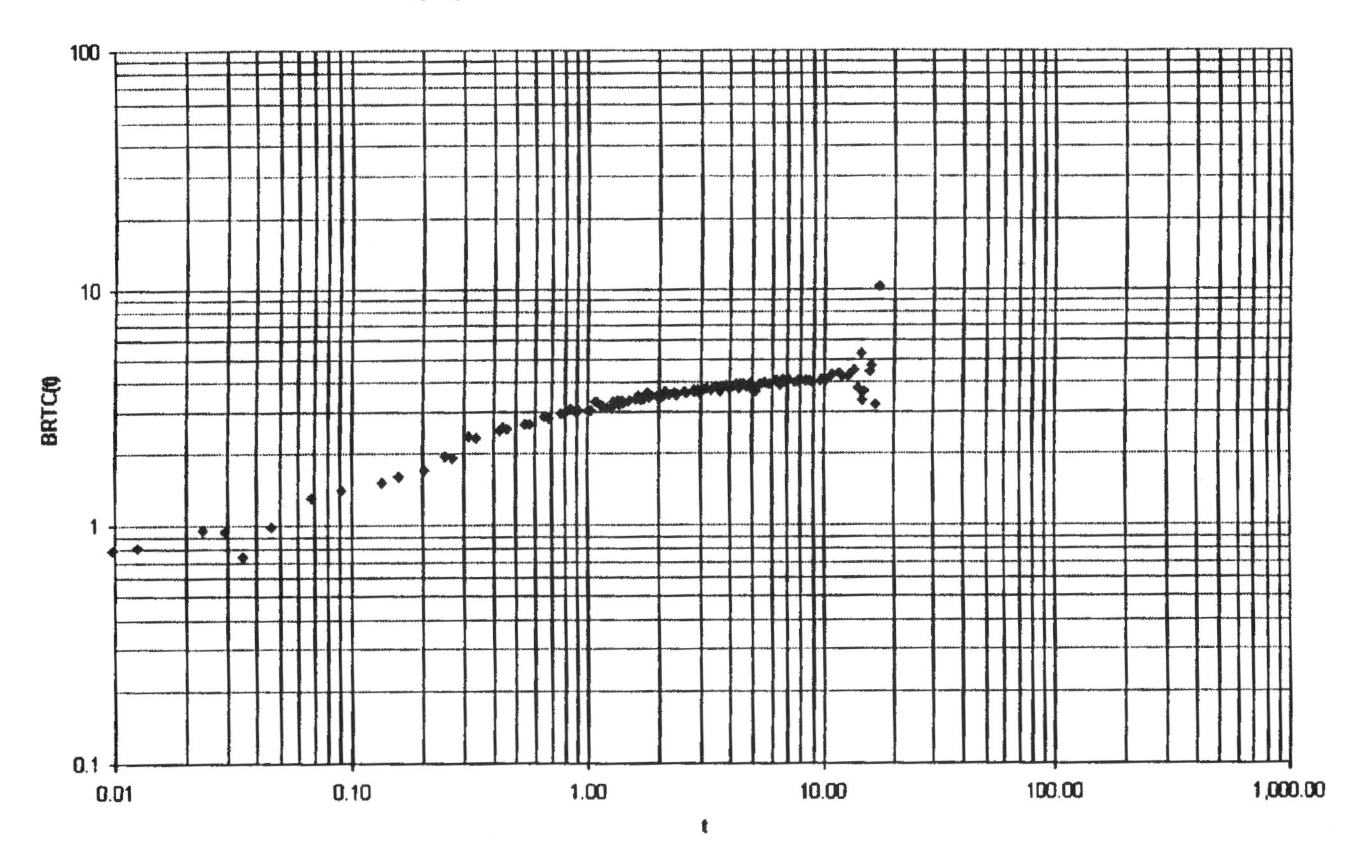

Figure 24. Oseberg 30/6-13, boundary ratio type curve.

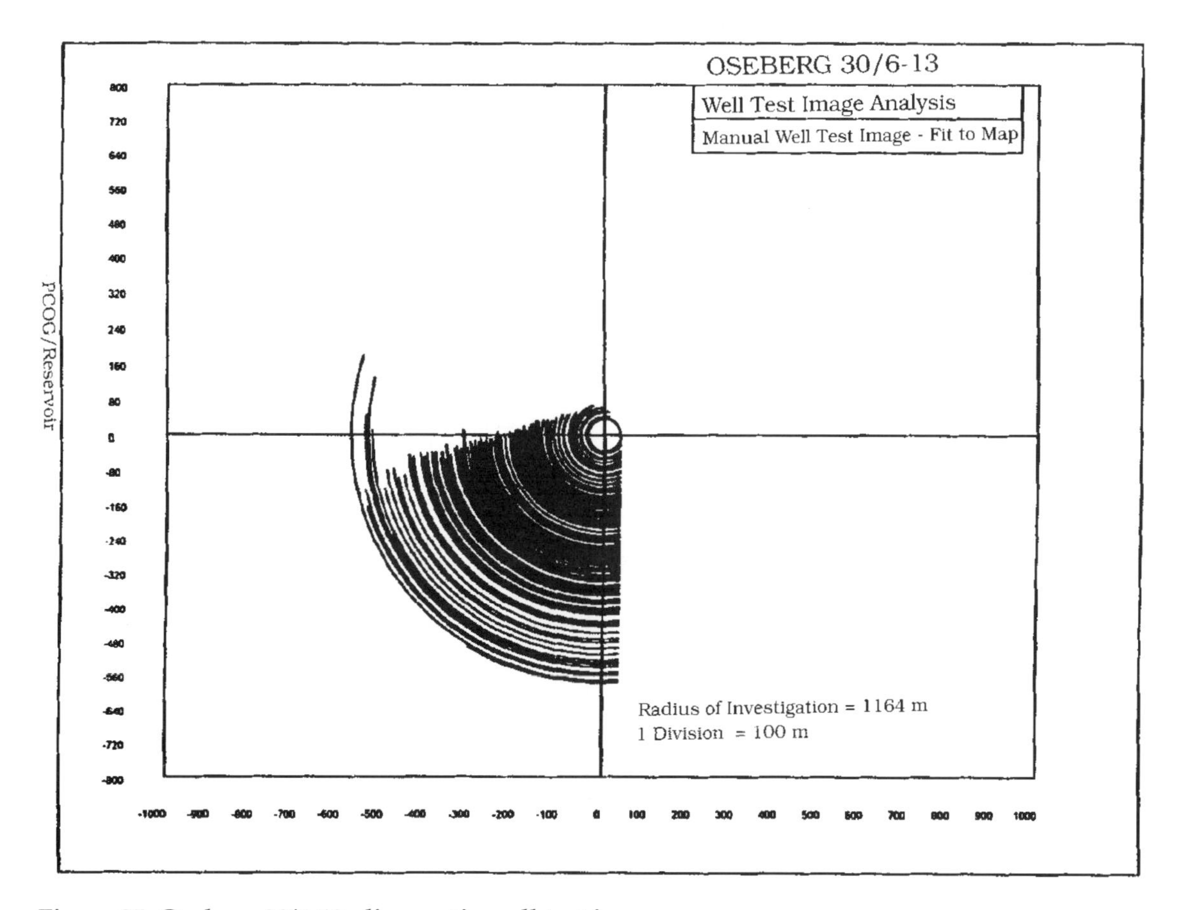

Figure 25. Oseberg 30/6-13, diagnostic well test image.

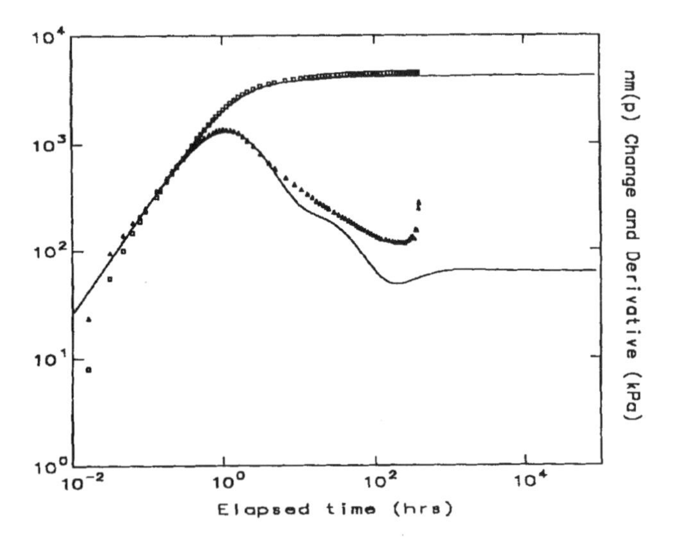

Figure 26. Fireweed b-6-A, log-log type curve.

lineament. This represents identification of the most likely boundary forming element in the region surrounding the well. The image is then rotated until the boundary image tangent to the radial region parallels the mapped lineament. Then the individual arc segments can be rotated to exactly contact the mapped boundary, because in most instances it is not perfectly linear. In many instances several mapped lineaments or potential boundary elements can be made to fit simultaneously, adding to the potential that a more unique match has been determined.

As an example of the application of the process, consider the well test shown on Figure 29, which has been modeled as well as possible to match the early time data. The difference between the match curve and the actual well test was used to create the diagnostic image shown on Figure 30. The angular model indicates a generally open-ended rectangular model with three no-flow boundaries. The slope of a surface near the level of this test was derived from 3-D seismic and is shown on Figure 31. Combining the diagnostic image and the 2-D map of the surface slope produced the reservoir image shown on Figure 32. This image very clearly identified the boundary forming critical slope of 0.12–0.20 m/m, which is characteristic of fault boundaries in this field, and fit two prominent lineaments simultaneously. Although the perception of the general character of the well test boundaries did not change, the additional work did allow the precise orientation of the open end of the system to be deduced and aided in the placement of a potential waterflood support well. This same image can then be superimposed on other data representations; in the case of Figure 33 on the structure top of the zone derived from seismic, and presented in a 3-D mesh perspective to allow visualization of the effective reservoir around the well.

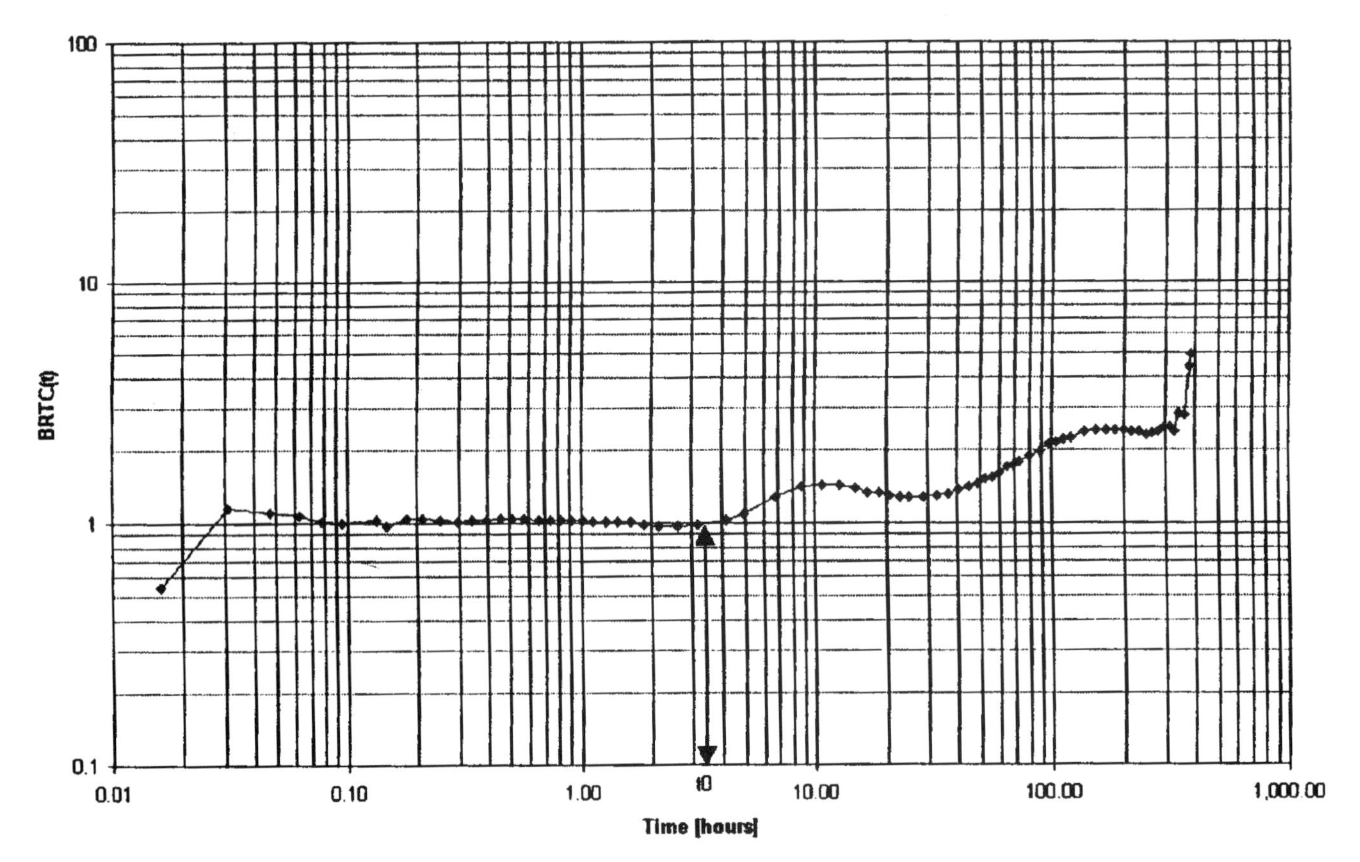

Figure 27. Fireweed b-6-A, boundary ratio type curve.

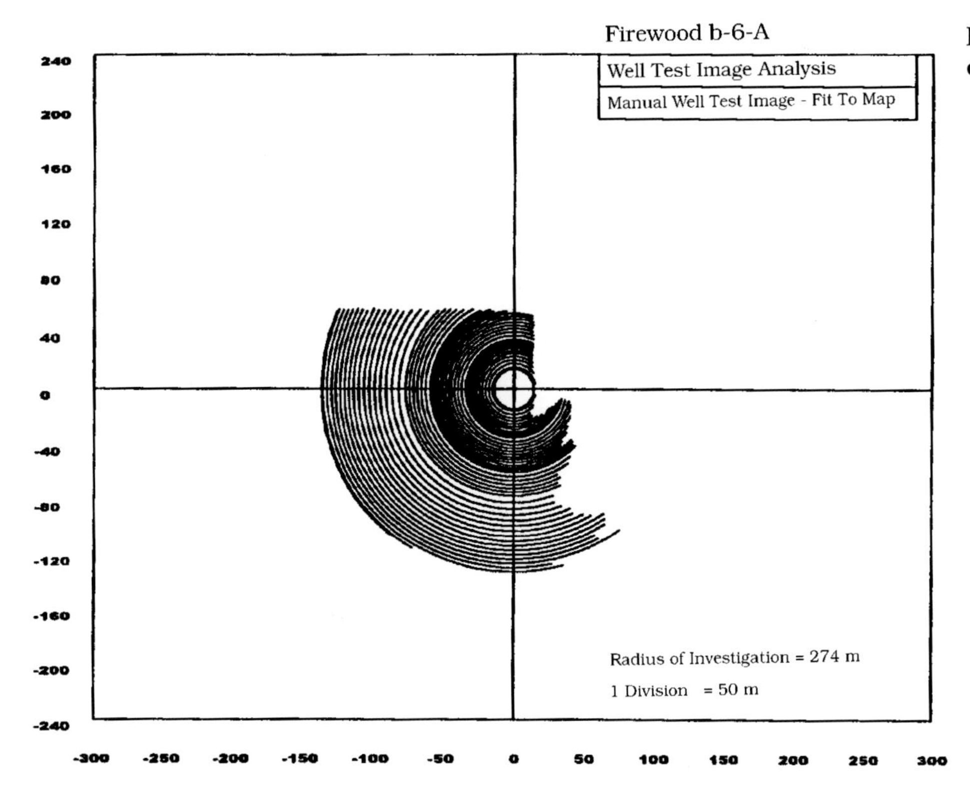

Figure 28. Fireweed b-6-A, diagnostic well test image.

OTHER USES OF WELL TEST IMAGES

The digital nature of the well test imaging process allows a number of calculations to be carried out using the $R_{inf}(t)$ and AOV(t) data generated by the analysis that would otherwise not be possible.

(1) It is possible to integrate the area and volume swept by the test to show the area over which information was generated. This leads to a better descriptor of the volume affected by a well test than the radius of investigation. This can be used to develop an estimate of "proved" reserves due to the test.

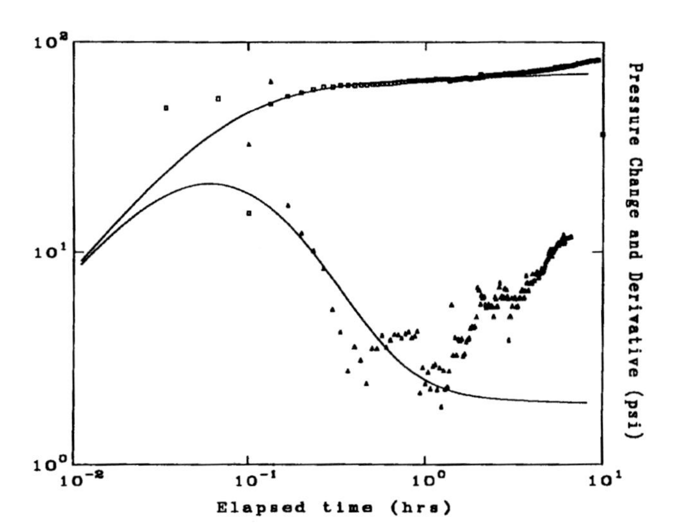

Figure 29. Example well, log-log type curve.

(2) By constructing a line joining nearest neighbor ends of the concentric arcs of the well test, a picture of the boundary alone can be generated.

(3) An automated procedure for generation of simulation model grids with realistic boundaries can be set up because the region of reservoir around the well is known.

(4) When multiple tests of several zones are conducted in a single well which shows fault boundaries, the images of each test can be stacked to provide a 3-D representation of the fault surfaces, as shown on Figure 34.

All of these applications are new to the field of well test analysis, and allow significant input to be provided to 3-D seismic and geological interpretations. Well test imaging provides a natural integrating mechanism for well test data that spans the gamut of reservoir characterization from 3-D seismic to reservoir simulation.

CONCLUSIONS

(1) A new type curve, the boundary ratio type curve, is proposed as a way to characterize boundary influences and to measure the boundary signal in a well test.

(2) Based on observation of the boundary ratio type curve characteristics, the magnitude of the response is concluded to be due to the geometric shape of the boundary, which is characterized by the angle of view (AOV(t)) of the boundary, while the distances to the boundary contacts are represented by the time scaling of the response through the radius of information ($R_{inf}(t)$).

(3) The accuracy of the boundary calculations is limited only by the accuracy with which the underlying

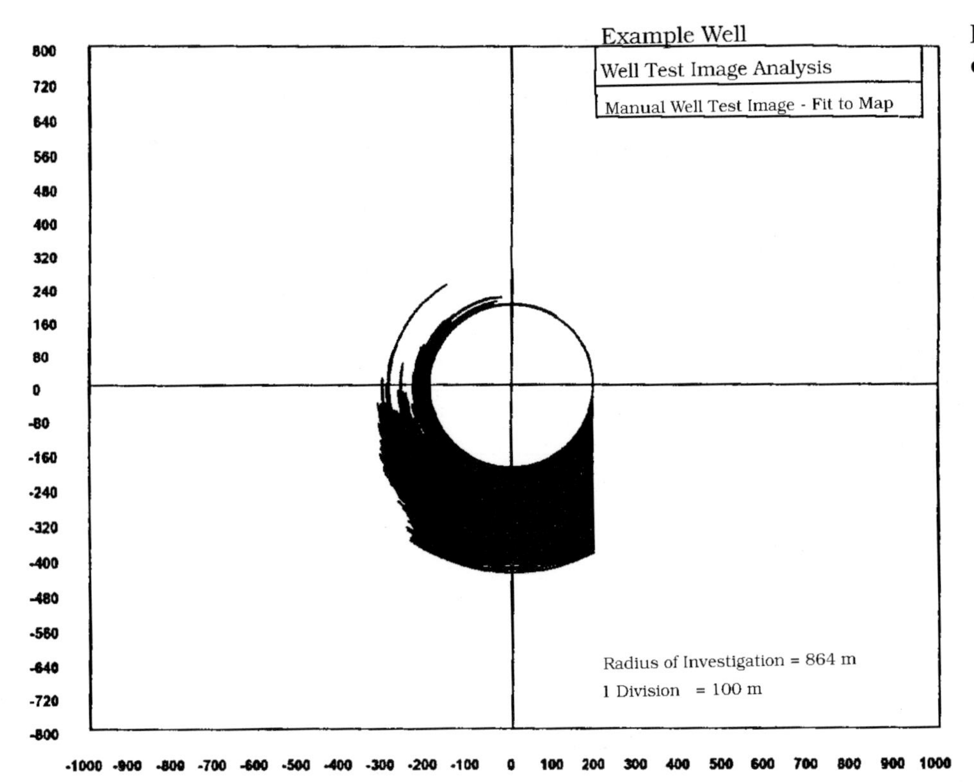

Figure 30. Example well, diagnostic well test image.

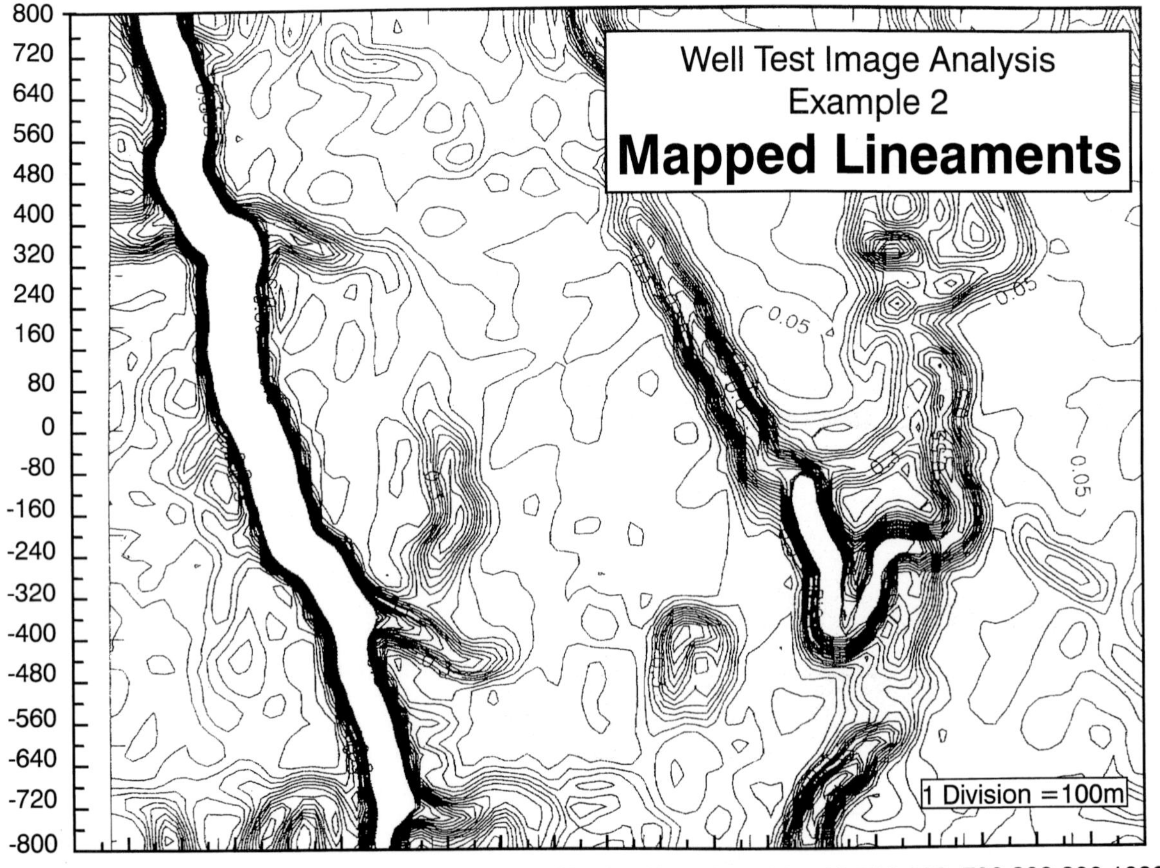

Figure 31. Example well, first derivative of 3-D seismic structure.

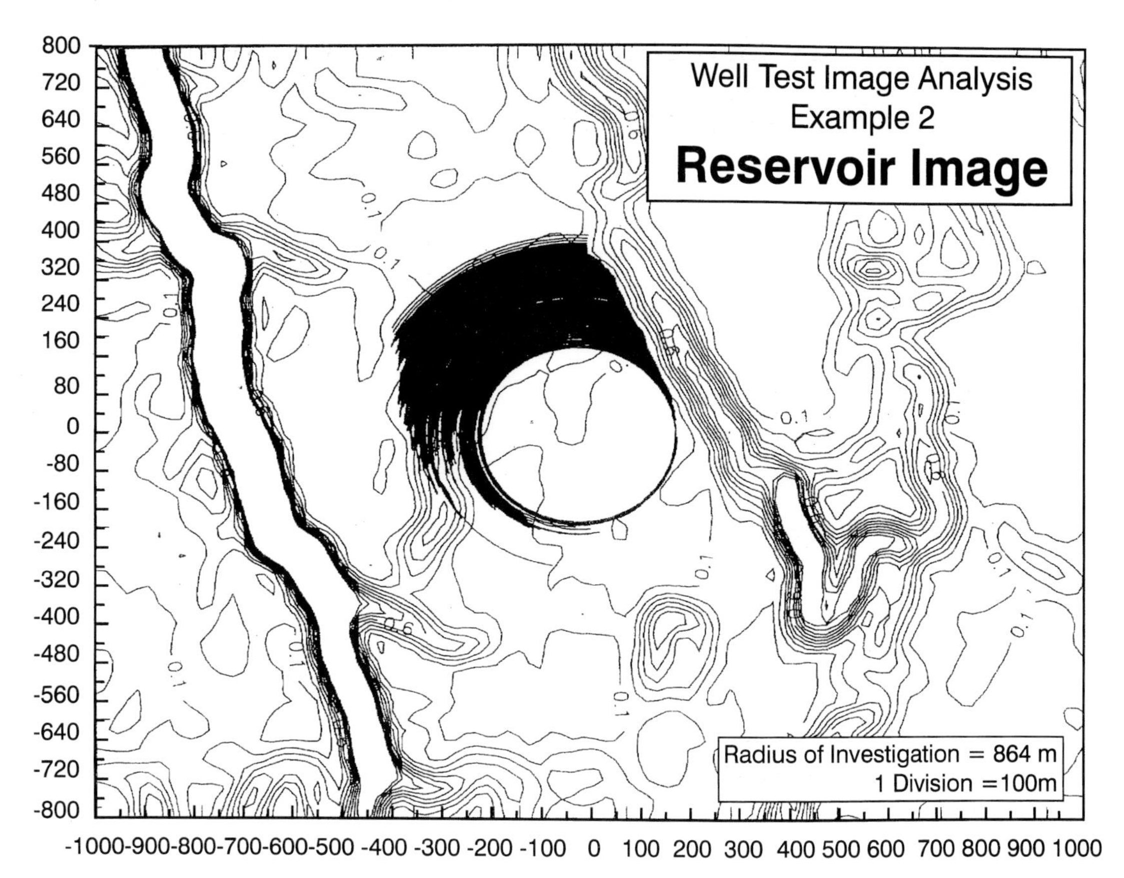

Figure 32. Example well, reservoir image.

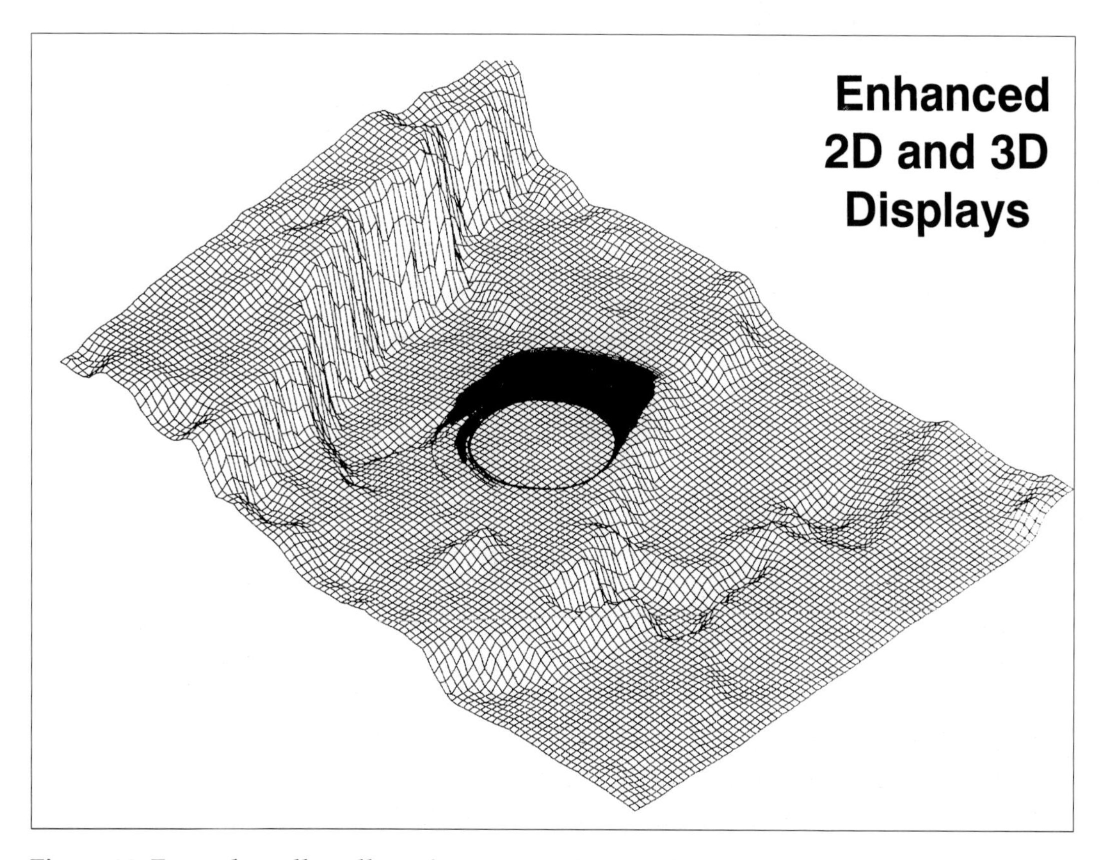

Figure 33. Example well, well test image superimposed on 3-D seismic structure.

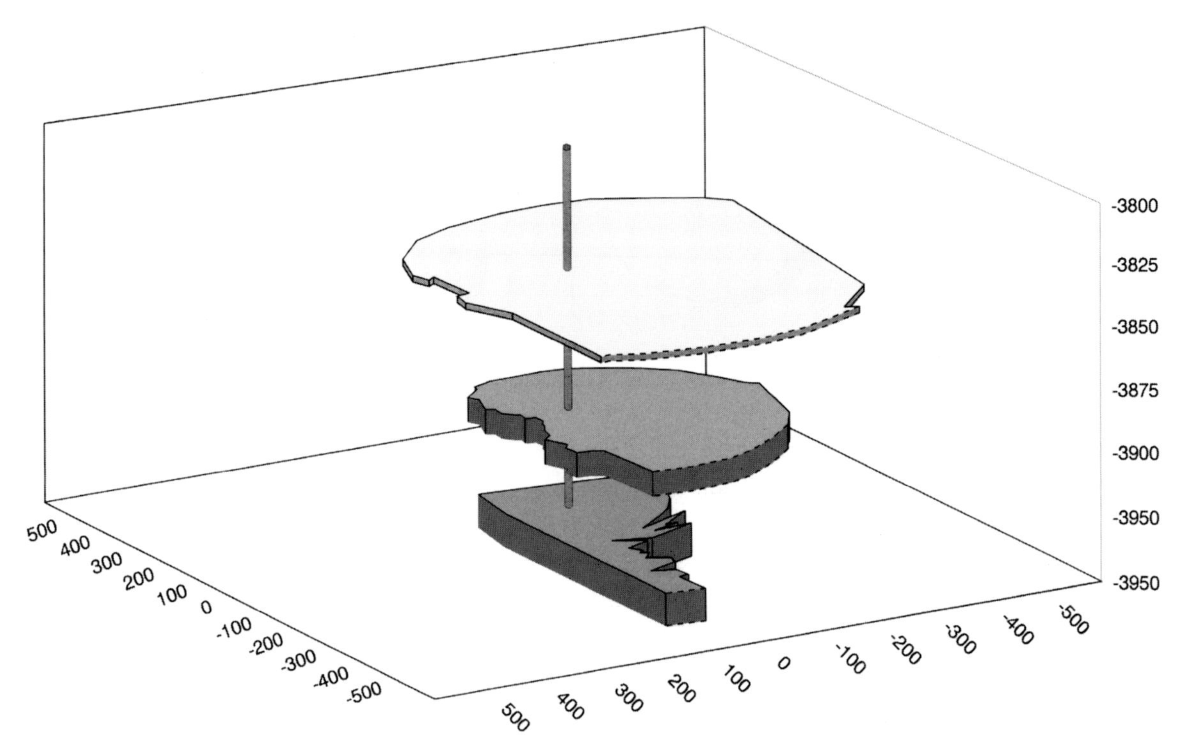

Figure 34. Example well, 3-D representation of multiple tests.

time and pressure measurements are made, and with modern gauges this can be down to the order of meters.

(4) The well test images derived directly from AOV(t) and $R_{inf}(t)$, with perhaps an analytical model as a guide, diagnostic images, are non-unique.

(5) Integration of the well test image with geological or geophysical information can lead to enhanced images (reservoir images), which can resolve the uniqueness problem and increase the quality of the interpretation overall.

(6) Well test imaging can help to extract more information from well test data, enabling better quality interpretations and enhanced reservoir characterization that are unconstrained by simplistic straight-line boundary models.

(7) Well test imaging provides a key element in the integration of well test data with geological and geophysical analyses, and in reservoir engineering simulation in the reservoir characterization process.

ACKNOWLEDGMENTS

I would like to thank Petro-Canada for permission to publish this paper, and particularly Mike Danyluk and Duncan Robertson for their support in the development of the technique as managers of the Technology Improvement Group. As well, I wish to thank all of my colleagues at Petro-Canada for their patience and willingness to share their well tests, geological and geophysical data, and scarce time to make the Well Test Imaging process so much better than it would be otherwise.

NOMENCLATURE

AOV	= angle of view
arccos	= inverse cosine ($\cos^{-1}$)
BRTC	= boundary ratio type curve value
c_t	= total system compressibility [1/psi or 1/kPa]
k	= permeability [md]
L	= distance to a boundary
μ	= oil viscosity [cp or mPa]
ϕ	= porosity [frac]
R_{inf}	= radius of information
R_{inv}	= radius of investigation
t	= time
t_c	= time for radius of investigation to first contact with a boundary
t_0	= time for first boundary interaction to be sensed or measured at the test well

t_x = intersection time, calculated from intersection of straight-line segments on Horner plot

t_{max} = maximum test time (shut-in or drawdown)

θ = angle between the shortest distance between a well and a surface and the current ray contact with the surface

REFERENCES CITED

Clark, D. G., Van Golf-Racht, Pressure-Derivative Approach to Transient Test Analysis: A High-Permeability North Sea Reservoir Example: SPE 12959, 1985, p. 2023–2039.

Earlougher, R. C., Jr., Advances in Well Test Analysis: SPE Monograph Vol. 5, p. 124–125, 1977.

Gringarten, A. C., Computer-Aided Well Test Analysis: SPE 14099, 1986, p. 631–633.

Yeung, Kacheong and Chakrabarty, Chayan; An Algorithm for Transient Pressure Analysis in Arbitrarily Shaped Reservoirs: Computers and Geosciences, Vol. 19, No. 3, p. 391–397, 1993.

Reynolds, A.C., H. Nanqun, D.S. Oliver, Reducing uncertainty in geostatistical description with well-testing pressure data, 1999, *in* R. Schatzinger and J. Jordan, eds., Reservoir Characterization-Recent Advances, AAPG Memoir 71, p. 149–162.

Chapter 10

Reducing Uncertainty in Geostatistical Description with Well-Testing Pressure Data

Albert C. Reynolds
Department of Petroleum Engineering, University of Tulsa
Tulsa, Oklahoma, U.S.A.

Nanqun He
Chevron Petroleum Technology Company
La Habra, California, U.S.A.

Dean S. Oliver
Department of Petroleum Engineering, University of Tulsa
Tulsa, Oklahoma, U.S.A.

ABSTRACT

Geostatistics has proven to be an effective tool for generating realizations of reservoir properties conditioned to static data (for example, core and log data, and geologic knowledge); however, in situations when data are not closely spaced in the lateral directions, there will be significant variability in reservoir descriptions generated by geostatistical simulation (that is, significant uncertainty in the reservoir descriptions). Procedures based on inverse problem theory were previously presented for generating reservoir descriptions (rock property fields) conditioned to pressure data and to geostatistical information represented by prior means and variograms for log-permeability and porosity. Although it has been shown that incorporation of pressure data reduces the uncertainty below the level contained in the geostatistical model based only on static information (the prior model), these previous results did not explicitly account for uncertainties in the prior means and the parameters defining the variogram.

This paper specifically investigates how pressure data can help detect errors in the prior means. If errors in the prior means are large and are not taken into account, realizations conditioned to pressure data represent an incorrect sampling of the a posteriori probability density function for the rock properties; however, if the uncertainty in the prior mean is properly incorporated, realistic realizations of the rock property fields are obtained.

INTRODUCTION

The objective is to generate three-dimensional realizations of rock property fields (that is, simulator gridblock values of log-permeability and porosity) conditioned to a prior model and well-test pressure data. The prior model is based on a multivariate Gaussian distribution with known covariance matrix and uncertain prior means. As in previous related work (Oliver, 1994; Chu et al., 1995b), the prior covariance matrix can be derived from the specified variograms and associated cross-variograms for log-permeability and porosity.

Unlike the past work mentioned, uncertainty in the prior means is specifically incorporated by introducing a partially doubly stochastic prior model using basic ideas described in Tjelmeland et al. (1994). The basic procedure for generating realizations involves generating the joint probability density function (pdf) for the rock property fields and the errors in the prior means conditioned to pressure data, and then sampling this pdf. Although it is not possible to publish the data here, this work was motivated by an actual field study in which the prior mean for permeability was found to be significantly overestimated in the geostatistical model, resulting in unrealistic descriptions of the permeability field when realizations were conditioned to pressure data.

An efficient procedure for sampling the pdf is obtained by adapting ideas and techniques presented in Oliver (1996) and Oliver et al. (1996). A procedure for generating the maximum a posteriori estimates of rock properties and their prior means is also presented. Specific realizations and maximum a posteriori estimates are generated by minimizing an appropriate objective function using the Gauss-Newton method. Sensitivity coefficients are computed using the procedure presented by He et al. (1997).

It is important to note that the goal is to obtain a set of realizations that represent a correct sampling of the probability density function for the rock property fields in question. By making a performance prediction with each realization, the uncertainty in the predicted parameters (for example, breakthrough time or cumulative oil production), can be evaluated; however, if a set of realizations that honors all the data is simply generated, but does not correctly sample the probability density function, the uncertainty in predicted performance cannot be ascertained by simulating reservoir performance with the set of realizations.

RESERVOIR MODEL

The reservoir is assumed to be a rectangular parallelepiped of uniform thickness h. Reservoir boundaries are assumed to prevent flow. Three-dimensional single-phase flow in a Cartesian coordinate system is considered. The reservoir can contain any number of complete-penetration or restricted-entry vertical wells. Each well is produced at a specified rate where the rate may vary with time. Pressure buildup at a well is simulated by setting the rate to zero subsequent to a producing period. Fluid properties are assumed to be known. A slightly compressible fluid of constant compressibility and viscosity is assumed. In all results, the values of the following parameters are fixed: ct = 10–5 psi–1, μ = 0.8 cp, rw = 0.3 ft at all wells.

Permeability and porosity are assumed to be heterogeneous. Permeability may be either isotropic or anisotropic, but its principal axes are assumed to coincide with the directions of the *x-y-z* coordinate system (that is, the only permeability attributes involved are k_x, k_y, and k_z). Permeability attributes (k_x, k_y, and k_z) are assumed to have log-normal distributions with variances given by $\sigma^2_{k_x}$, $\sigma^2_{k_y}$, and $\sigma^2_{k_z}$. Porosity is assumed to be normal with variance given by σ^2_ϕ. The variogram for each rock property attribute may be anisotropic. Each rock property is modeled as a stationary Gaussian random function so that covariance functions are directly related to the variograms (Journel and Huj bregts, 1978). The correlation coefficients between the various properties are assumed to be known, but may be zero. As discussed in more detail later, prior estimates of the means of the properties are assumed to be uncertain.

As indicated in the preceding paragraph, the model is quite general; however, the specific examples presented here pertain to a case in which $k_x = k_y = k$ and $k_z = 0.1k$ at all gridblocks. Thus, in generating realizations of the rock property fields, realizations of porosity and $\ln(k)$ are first produced; and then vertical permeabilities are explicitly determined via the relation $k_z = 0.1k$, which is applied at each gridblock. As in Chu et al. (1995b), the cross-variogram between porosity and log-permeability is generated using the screening hypothesis of Xu et al. (1992). This eliminates the necessity of specific modeling of the cross-variogram.

For given rock property fields, pressure responses are obtained by a standard, purely implicit, seven-point finite-difference simulator, where wellbore pressure is related to the well's gridblock pressure by Peaceman's method (1983) [see Chu et al. (1995a) and He et al. (1997)]. To test the procedures used to obtain conditional realizations, a simulator is used to generate synthetic multiwell pressure data, which is then assumed to represent wellbore pressure measurements.

Prior and A Posteriori Probability Density Functions

Throughout, N denotes the number of simulator gridblocks. For the specific problems considered here, the model, or vector of model parameters, is given by

$$m = \begin{bmatrix} m_\phi \\ m_k \end{bmatrix} \tag{1}$$

where m_ϕ is the N-dimensional column vector of gridblock porosities, and m_k is the N-dimensional column vector of gridblock values of $\ln(k)$. Note the dimension of m is $N_p = 2N$. To use standard notation from stochastic inverse theory, we let M denote the random vector of rock property model parameters, with specific realizations denoted by m in equation 1. For a fully anisotropic representation of permeability, the vector m must be modified so that it includes components for $\ln(k_x)$, $\ln(k_y)$, $\ln(k_z)$, and porosity; the covariance matrix given by equation 2 below must also be modified [see He et al. (1997)].

As in previous work, the prior model is assumed to have a multivariate Gaussian probability density function with prior covariance matrix, C_M. For the specific examples given here, C_M is defined by

$$C_M = \begin{bmatrix} C_\phi & C_{\phi k} \\ C_{\phi k} & C_k \end{bmatrix} \tag{2}$$

where C_ϕ is the covariance matrix for gridblock porosities derived from the porosity variogram, C_k is the covariance matrix for gridblock $\ln(k)$s derived from the variogram for $\ln(k)$, and $C_{\phi k}$ is the cross covariance matrix between porosity and $\ln(k)$.

Throughout, m_{prior} is the vector containing the prior means, defined by

$$m_{prior} = \begin{bmatrix} m_{0,\phi} \\ m_{0,k} \end{bmatrix} \tag{3}$$

If $\ln(k)$ and porosity are modeled as stationary random functions, then $m_{0,\phi}$ and $m_{0,k}$ are treated as constant vectors with each entry of $m_{0,\phi}$ given by $m_{prior,\phi}$ (prior mean for porosity) and each entry of $m_{0,k}$ given by $m_{prior,k}$, which denotes the prior mean for log-permeability; however, the general formulation presented allows each entry of m_{prior} to be different. The random vector Θ represents the error in, or correction to, m_{prior}, with θ denoting specific realizations of Θ. Introduction of Θ allows for the incorporation of uncertainty in the vector of prior means, whereas previously no error in m_{prior} was assumed. The pdf for Θ is assumed to be Gaussian and is given by

$$p_\Theta(\theta) = a\exp\left(-\frac{1}{2}(\theta-\theta_0)^T C_\Theta^{-1}(\theta-\theta_0)\right) \tag{4}$$

where θ_0 is the mean of the random vector Θ, and C_Θ is the associated covariance matrix. Throughout, the term a which appears in the pdf simply represents the normalizing constant. In this work, we assume that errors in the prior means are independent, so C_Θ is a diagonal matrix. Although it is appropriate to choose $\theta_0 = 0$, the derivation presented applies for any value of θ_0. The conditional distribution (pdf) of M given $\Theta = \theta$ is given by

$$p_{M|\Theta}(m|\theta) = a\exp\left(-\frac{1}{2}(m-m_{prior}-\theta)^T C_M^{-1}(m-m_{prior}-\theta)\right) \tag{5}$$

so the joint pdf for M and Θ is given by

$$p_{\hat{M}}(\hat{m}) = p_{\hat{M}}(m,\theta) = p_{M|\Theta}(m|\theta)p_\Theta(\theta) = a\exp\left(\begin{matrix} -\frac{1}{2}(m-m_{prior}-\theta)^T C_M^{-1}(m-m_{prior}-\theta) - \\ \frac{1}{2}(\theta-\theta_0)^T C_\Theta^{-1}(\theta-\theta_0)\end{matrix}\right) \tag{6}$$

where

$$\hat{M} = \begin{bmatrix} M \\ \Theta \end{bmatrix} \tag{7}$$

For simplicity, a realization $\hat{m}$ of $\hat{M}$ is sometimes written as (m,θ) instead of $(m^T,\theta^T)^T$. Throughout, the superscript T is used to denote the transpose of a matrix or vector. For convenience, m_{prior} is referred to as the prior mean; however, one should note that equation 5 indicates that the conditional expectation of M is given by $E[M \mid \Theta = \theta] = m_{prior} + \theta$.

All measured well-test pressure data that will be used as conditioning data are incorporated in the N_d-dimensional column vector d_{obs}. Note N_d is the total number of observed or measured pressure data used as conditioning data. As is standard, d represents the corresponding vector of pressures that will be calculated for a given realization m, and the relationship between the data and m is represented by

$$d = g(m) \tag{8}$$

Given a specific m, equation 8 represents the operation of calculating wellbore pressures by running the reservoir simulator.

The vector of measurement errors is denoted by ε. As in previous work, it is assumed that measurement errors are independent, identically distributed random variables with zero mean and variance σ_d^2, so that the data covariance matrix C_D is a diagonal matrix with all entries equal to σ_d^2. Given the true m, the observed pressure data may be regarded as a realization of the random vector $D = g(m) + \varepsilon$. Thus, the a posteriori pdf for $\hat{M}$ conditional to the observed pressure data, d_{obs}, can be derived as in Tjelmeland et al. (1994) by standard applications of Bayes theorem, and is given by

$$\pi(m,\theta) = p_{\hat{M}|D}(\hat{m}|d_{obs}) = a\exp\left(-\frac{1}{2}(g(m)-d_{obs})^T C_D^{-1}(g(m)-d_{obs}) - \frac{1}{2}(m-m_{prior}-\theta)^T C_M^{-1}(m-m_{prior}-\theta) - \frac{1}{2}(\theta-\theta_0)^T C_\Theta^{-1}(\theta-\theta_0)\right) \tag{9}$$

where the first equality of equation 9 simply defines notation. Equation 9 gives the pdf that is to be sampled to generate realizations (m,θ) of $\hat{M}$. To generate the most probable model (maximum a posteriori estimate) for $\hat{M}$, it is necessary to minimize the objective function $O(\hat{m})$ given by

$$O(\hat{m}) = \frac{1}{2}\left(g(m) - d_{obs}\right)^T C_D^{-1}\left(g(m) - d_{obs}\right) + \frac{1}{2}\left(m - m_{prior} - \theta\right)^T C_M^{-1}\left(m - m_{prior} - \theta\right) + \frac{1}{2}\left(\theta - \theta_0\right)^T C_\Theta^{-1}\left(\theta - \theta_0\right) \tag{10}$$

At this point, the dimension of θ is the same as the dimension of m (specifically, N_p).

NEWTON ITERATION

It is convenient to partition the gradient of $O(\hat{m})$ as

$$\nabla O(\hat{m}) = \begin{bmatrix} \nabla_m O(\hat{m}) \\ \nabla_\theta O(\hat{m}) \end{bmatrix} \tag{11}$$

where ∇_m represents the gradient operator with respect to m, and ∇_θ represents the gradient operator with respect to θ. Using basic vector calculus, it follows that

$$\nabla_\theta O(\hat{m}) = -C_M^{-1}\left(m - m_{prior} - \theta\right) + C_\Theta^{-1}\left(\theta - \theta_0\right) \tag{12}$$

Similarly,

$$\nabla_m O(\hat{m}) = G^T C_D^{-1}\left(g(m) - d_{obs}\right) + C_M^{-1}\left(m - m_{prior} - \theta\right) \tag{13}$$

where G^T is the transpose of the $N_d \times N_p$ sensitivity coefficient matrix G, which is defined as

$$G = \left(\nabla_m\left[g(m)\right]^T\right)^T \tag{14}$$

Using equations 12 and 13 in equation 11 gives the total gradient of the objective function. Again, using basic vector calculus, the Hessian matrix for the Gauss-Newton iteration is given by

$$H = \begin{bmatrix} G^T C_D^{-1} G + C_M^{-1} & -C_M^{-1} \\ -C_M^{-1} & C_M^{-1} + C_\Theta^{-1} \end{bmatrix} \tag{15}$$

Since C_D, C_M, and C_Θ are all positive definite matrices, it is straightforward to show that H is also positive definite. If the objective function of equation 10 has a unique minimum (the maximum a posteriori estimate), the Gauss-Newton method with H replaced by any other positive definite matrix will also converge to the maximum a posteriori estimate [see Fletcher (1987)]. Thus, to obtain a simpler computational scheme, the Hessian H is replaced by the positive definite matrix $\hat{H}$ where

$$\hat{H} = \begin{bmatrix} G^T C_D^{-1} G + C_M^{-1} & O \\ O & C_M^{-1} + C_\Theta^{-1} \end{bmatrix} \tag{16}$$

When $\hat{H}$ is used as the modified Hessian in the Gauss-Newton iteration procedure, the overall iteration can be decomposed as follows:

$$\left(G_l^T C_D^{-1} G_l + C_M^{-1}\right)\delta m^{l+1} = -G_l^T C_D^{-1}\left(g\left(m^l\right) - d_{obs}\right) - C_M^{-1}\left(m^l - m_{prior} - \theta^l\right) \tag{17}$$

$$\left(C_M^{-1} + C_\Theta^{-1}\right)\delta\theta^{l+1} = C_M^{-1}\left(m^l - m_{prior} - \theta^l\right) - C_\Theta^{-1}\left(\theta^l - \theta_0\right) \tag{18}$$

$$m^{l+1} = m^l + \mu_l \delta m^{l+1} \tag{19}$$

$$\theta^{l+1} = \theta^l + \mu_l \delta \theta^{l+1} \tag{20}$$

where l refers to the iteration index, and μ_l is the step size determined by the restricted step method (Fletcher, 1987). Note that, in the spirit of the restricted step, it is important to use the same value of μ_l in both equations 19 and 20; otherwise the search direction is changed. By replacing H by $\hat{H}$, inversion of H is avoided [that is, the iteration on the model (m) is "decoupled" from the iteration on the correction (θ) to the prior mean].

If porosity and log-permeability are modeled as stationary random functions, then the prior mean (given by equation 3) can be written as

$$m_{prior} = \begin{bmatrix} m_{prior,\phi}e \\ m_{prior,k}e \end{bmatrix} \tag{21}$$

where e is the N-dimensional column vector with all entries equal to unity. More generally, e represents a column vector of dimension N_e with all components equal to unity, as given by

$$e = [1,1,\cdots,1]^T \tag{22}$$

and m_{prior} is assumed to have the form

$$m_{prior} = \begin{bmatrix} m_{prior,1}e \\ m_{prior,2}e \\ \vdots \\ m_{prior,N_a}e \end{bmatrix} \tag{23}$$

where each $m_{prior,j}$ is a scalar. In this case, it is reasonable to require that the correction to the prior mean have the same structure as m_{prior}; that is,

$$\theta = \begin{bmatrix} \alpha_1 e \\ \alpha_2 e \\ \vdots \\ \alpha_{N_a} e \end{bmatrix} \tag{24}$$

for some constants, α_j, $j = 1, 2, \ldots, N_a$. Since m_{prior} and θ are both N_p-dimensional column vectors, N_e is required to satisfy $N_a N_e = N_p$. For the case where all attributes are modeled as stationary random functions, N_a is equal to the number of attributes (for example, $N_a = 2$ if equation 3 applies); however, if the mean of each attribute is different in each gridblock, then $N_a = N_p$ (the dimension of the model m). In this case, e is one dimensional and equation 24 does not place any restrictions on the components of θ. When equation 24 applies, C_Θ is defined as a block diagonal matrix with the jth diagonal block given by $\sigma^2_{\theta,j} I$ for $j = 1, 2, \ldots, N_a$, where I is the $N_e \times N_e$ identity matrix.

The $(N_a N_e) \times N_a = N_p \times N_a$ matrix E is defined by

$$E = \begin{bmatrix} e & O & \cdots & O \\ O & e & \cdots & O \\ \vdots & \vdots & \ddots & \vdots \\ O & O & \cdots & e \end{bmatrix} \tag{25}$$

so the transpose of E is given by

$$E^T = \begin{bmatrix} e^T & O & \cdots & O \\ O & e^T & \cdots & O \\ \vdots & \vdots & \ddots & \vdots \\ O & O & \cdots & e^T \end{bmatrix} \tag{26}$$

If $N_a = N_p$ ($N_e = 1$), then E is the $N_p \times N_p$ identity matrix. Defining the N_a-dimensional column vector α by

$$\alpha = \left[\alpha_1, \ \alpha_2, \ \cdots, \ \alpha_{N_a}\right]^T \tag{27}$$

equation 24 can be written as $\theta = E\alpha$.

Partial Subspace Procedure

Reynolds et al. (1996) have implemented subspace methods (Kennett and Williamson, 1988; Oldenberg et al., 1993; Oldenberg and Li, 1994) to significantly enhance the computational efficiency of the Gauss-Newton method. Here, only a partial subspace procedure is considered, where $\delta\theta^{l+1}$ in equation 18 is expanded as

$$\delta\theta^{l+1} = E\delta\alpha^{l+1} \tag{28}$$

at all Newton iterations. Using equation 28 in equation 18 and multiplying the resulting equation by $E^T C_M$ gives

$$E^T\left(I + C_M C_\Theta^{-1}\right)E\delta\alpha^{l+1} = E^T\left(m^l - m_{prior} - \theta^l\right) - E^T C_M C_\Theta^{-1}\left(\theta^l - \theta_0\right) \tag{29}$$

Equation 28 indicates that $\delta\theta^{l+1}$ is a linear combination of the columns of E (that is, the columns of E represent the associated subspace vectors). If the initial guess for θ, $\theta^0 = \theta_0$ is also a linear combination of these subspace vectors, then by mathematical induction it follows that for all l, θ^l is a linear combination of these subspace vectors. This result is apparent because, if θ^l is a linear combination of these subspace vectors (that is, $\theta^l = E\alpha^l$), it follows from equations 20 and 28 that

$$\theta^{l+1} = E\left(\alpha^l + \mu_l \delta\alpha^{l+1}\right) \tag{30}$$

It now follows that when equation 18 is replaced by equation 29, equation 20 can be replaced by

$$\alpha^{l+1} = \alpha^l + \mu_l \delta\alpha^{l+1} \tag{31}$$

and

$$\theta^{l+1} = E\alpha^{l+1} \tag{32}$$

With this modification, the overall computational scheme for estimating the maximum a posteriori estimate (equations 17–20) can now be written as

$$\delta m^{l+1} = m_{prior} + \theta^l - m^l - C_M G_l^T\left(C_D + G_l C_M G_l^T\right)^{-1} \times \left[g\left(m^l\right) - d_{obs} - G_l\left(m^l - m_{prior} - \theta^l\right)\right] \tag{33}$$

$$\left(E^T\left(I + C_M C_\Theta^{-1}\right)E\right)\delta\alpha^{l+1} = E^T\left(m^l - m_{prior} - \theta^l\right) - E^T C_M C_\Theta^{-1}\left(\theta^l - \theta_0\right) \tag{34}$$

$$m^{l+1} = m^l + \mu_l \delta m^{l+1} \tag{35}$$

and equations 31 and 32. Equation 33 was obtained from equation 17 by using basic matrix inversion lemmas (Tarantola, 1987; Chu et al., 1995a). The preceding subspace implementation of the Gauss-Newton iteration is designed to converge to the so-called maximum a posteriori estimate, which is commonly referred to as the most probable model; however, as noted previously, our objective is not to simply generate the most probable estimate of $\hat{m}$ but to generate a suite of realizations that represents a correct sampling of the pdf of equation 9.

The sampling procedure used is presented in the following section.

SAMPLING THE A POSTERIORI DISTRIBUTION

Markov chain Monte Carlo (MCMC) methods produce a correct sampling of a given pdf if a sufficiently large number of states is generated; however, current implementations (Oliver et al., 1997; Cunha et al., 1998) are too computationally intensive for practical applications when the goal is to generate realizations conditioned to production data. The generation of each state in the Markov chain requires an individual run of a reservoir simulator. Procedures based on approximating the a posteriori pdf by a Gaussian distribution centered at the maximum a posteriori estimate require computing either the Cholesky decomposition or the square root of the a posteriori covariance matrix, and do not always result in a correct sampling of the pdf. Thus, a computationally efficient alternative is pursued. For the case where uncertainty in the prior mean is ignored, the basic procedure has been discussed by Oliver et al. (1996) and Kitanidis (1995). This procedure is technically correct only for the case where the data are linearly related to the model; however, Oliver et al. (1996) presented arguments that suggest that the procedure should give an approximately correct sampling in the nonlinear case.

Linear Case

Here, the work of Oliver (1996) and Oliver et al. (1996) is extended to the case where uncertainty is incorporated in the prior mean. Consider the case where the data are linearly related to the model, so equation 8 can be written as

$$d = Gm \tag{36}$$

where G is an $N_d \times N_p$ matrix. For this case, the maximum a posteriori estimate can be obtained by solving the following two equations: $\nabla_\theta O(\hat{m}) = 0$ and $\nabla_m O(\hat{m}) = 0$ (see equations 12 and 13) to obtain m_∞ and θ_∞. It is easy to show that this solution satisfies

$$\begin{bmatrix} G^T C_D^{-1} G + C_M^{-1} & -C_M^{-1} \\ -C_M^{-1} & C_M^{-1} + C_\Theta^{-1} \end{bmatrix} \begin{bmatrix} m_\infty \\ \theta_\infty \end{bmatrix} = \begin{bmatrix} C_M^{-1} m_{prior} + G^T C_D^{-1} d_{obs} \\ -C_M^{-1} m_{prior} + C_\Theta^{-1} \theta_0 \end{bmatrix} \tag{37}$$

Note that the coefficient matrix on the left side of equation 37 is the Hessian matrix defined in equation 15; moreover, when equation 36 applies, it is easy to show that the a posteriori pdf for $\hat{M}$ (equation 9) is Gaussian with covariance matrix given by H^{-1} and expectation given by $(m_\infty, \theta_\infty)$ (Tarantola, 1987).

Next, a procedure for sampling $\pi(m,\theta)$ is presented which does not require the generation, and Cholesky or square root decomposition, of H^{-1}. To construct a realization, an unconditional simulation of m is generated, which is denoted by m_{uc}, and is given by

$$m_{uc} = m_{prior} + C_M^{1/2} Z \tag{38}$$

The components of the N_p-dimensional column vector Z are independent standard random normal deviates. Similarly, unconditional simulations of the data and the correction to the prior mean, respectively, are generated by

$$d_{uc} = d_{obs} + C_D^{1/2} Z_D \tag{39}$$

and

$$\theta_{uc} = \theta_0 + C_\theta^{1/2} Z_\theta \tag{40}$$

where again the components of Z_D and Z_θ are independent standard random normal deviates. The 1/2 superscript on the matrices in the preceding three equations represents the square root of the matrix, but the square roots could also be replaced by the lower triangular matrix arising from the LL^T decomposition of the matrix; however, C_D and C_θ are diagonal matrices, and it is thus trivial to compute their square roots. Because it is desirable to avoid explicit factorization of C_M in the computer implementation, a sequential Gaussian cosimulation (Gomez-Hernandez and Journel, 1993) is used in place of equation 38 to generate m_{uc}. By replacing m_{prior} by m_{uc}, d_{obs} by d_{uc}, and θ_0 by θ_{uc} in equations 12 and 13, and then setting the resulting two equations equal to zero and solving to obtain the solution denoted by (m_s, θ_s), it follows that

$$\begin{bmatrix} G^T C_D^{-1} G + C_M^{-1} & -C_M^{-1} \\ -C_M^{-1} & C_M^{-1} + C_\Theta^{-1} \end{bmatrix} \begin{bmatrix} m_s \\ \theta_s \end{bmatrix} = \begin{bmatrix} C_M^{-1}\left(m_{prior} + C_M^{1/2} Z\right) + G^T C_D^{-1}\left(d_{obs} + C_D^{1/2} Z_D\right) \\ -C_M^{-1}\left(m_{prior} + C_M^{1/2} Z\right) + C_\Theta^{-1}\left(\theta_0 + C_\Theta^{1/2} Z_\theta\right) \end{bmatrix} \tag{41}$$

Subtracting equation 37 from equation 41 gives

$$\begin{bmatrix} G^T C_D^{-1} G + C_M^{-1} & -C_M^{-1} \\ -C_M^{-1} & C_M^{-1} + C_\Theta^{-1} \end{bmatrix} \begin{bmatrix} m_s - m_\infty \\ \theta_s - \theta_\infty \end{bmatrix} = \begin{bmatrix} C_M^{-1} C_M^{1/2} Z + G^T C_D^{-1} C_D^{1/2} Z_D \\ -C_M^{-1} C_M^{1/2} Z + C_\Theta^{-1} C_\Theta^{1/2} Z_\theta \end{bmatrix} = B \tag{42}$$

where the last equality of equation 42 serves to define B. The random vector $\hat{M}_s$ is defined by

$$\hat{M}_s = \left[m_s^T, \theta_s^T\right]^T \tag{43}$$

Since the expectations of Z, Z_D, and Z_θ are all zero, it is clear that the expectation of $\hat{M}_s$ is given by

$$E\left[\hat{M}_s\right] = \begin{bmatrix} m_\infty \\ \theta_\infty \end{bmatrix} \quad (44)$$

that is, $E[m_s] = m_\infty$ and $E[\theta_s] = \theta_\infty$. The covariance of the random vector $\hat{M}_s$ is given by

$$E\left(\left(\hat{M}_s - E\left[\hat{M}_s\right]\right)\left(\hat{M}_s - E\left[\hat{M}_s\right]\right)^T\right) = H^{-1}E\left[BB^T\right]H^{-1} \quad (45)$$

Using the fact that Z, Z_D, and Z_θ are independent vectors, with components of each vector representing independent standard random normal deviates, it is straightforward to show that $E[BB^T] = H$, and thus equation 45 reduces to

$$E\left(\left(\hat{M}_s - E\left[\hat{M}_s\right]\right)\left(\hat{M}_s - E\left[\hat{M}_s\right]\right)^T\right) = H^{-1} \quad (46)$$

Thus, it has been shown that the covariance and expectations of $\hat{M}$ and $\hat{M}_s$ are the same. Since both random vectors satisfy Gaussian distributions when equation 36 applies, it is possible to generate a sampling of $\hat{M}$ by sampling the distribution for $\hat{M}_s$. Samples of $\hat{M}_s$ can be generating by solving equation 41 for m_s and θ_s for a set of independent unconditional simulations, m_{uc}, d_{uc}, and θ_{uc}.

Basic Sampling Procedure, Nonlinear Case

For the nonlinear case of interest, the same type of procedure is applied except that θ is restricted by introducing a subspace method (that is, samples are generated by the computational algorithm of equations 33, 34, 35, 31, and 32, with m_{prior} replaced by m_{uc}, d_{obs} replaced by d_{uc}, and θ_0 replaced by θ_{uc}.

In this process, θ_{uc} must be generated so it lies in the appropriate subspace. To do this, recall that C_Θ is a block diagonal matrix where the jth diagonal block is given by $\sigma^2_{\theta,j}I$ and introduce the associated covariance matrix C_α, which is related to C_Θ by

$$C_\alpha^{-1} = E_T C_\Theta^{-1} E \quad (47)$$

C_α is an $N_a \times N_a$ diagonal matrix with jth diagonal entry denoted by $\sigma^2_{\alpha,j}$. The vector α_{uc} is computed using

$$\alpha_{uc} = \alpha_0 + C_\alpha^{1/2} Z_\alpha \quad (48)$$

where the components of the N_a-dimensional column vector Z_α are independent standard random normal deviates. After applying equation 48, the unconditional simulation of the error in the prior mean is given by

$$\theta_{uc} = E\alpha_{uc} \quad (49)$$

COMPUTATIONAL EXAMPLE

The example considered pertains to a reservoir containing nine completely penetrating wells. A simulation grid with 25 gridblocks in the x and y directions and 10 gridblocks in the z direction is used (that is, 6250 gridblocks are used.) Since it is desired to generate realizations of log-permeability and porosity, there are 12,500 model parameters. The areal grid is 400 × 400 ft (122 × 122 m), and all gridblocks in the z direction are 10 ft (3 m) thick. Figure 1 shows the areal grid and well locations.

The reservoir is areally isotropic, and k_z is equal to $0.1k$. Thus, determination of a distribution for k automatically determines the vertical permeability at each gridblock. An anisotropic spherical variogram for $\ln(k)$ is used with the range in the x-direction equal to 3200 ft (976 m), the range in the y-direction equal to 1600 ft (488 m), and the range in the z-direction equal to 30 ft (9 m). The variance of $\ln(k)$ (sill of the variogram) is specified as $\sigma_k^2 = 0.5$. The anisotropic variogram for porosity is identical to the one for $\ln(k)$ except that the variance for porosity is specified as $\sigma_\phi^2 = 0.002$. The correlation coefficient between log-permeability and porosity is specified as $\rho_{k,\phi} = 0.7$.

The true log-permeability field is shown in Figure 2. This truth case was obtained by unconditional simulation using $m_{prior,k} = 4.0$ and $m_{prior,\phi} = 0.20$. The unconditional simulation also yields the true porosity field. For convenience, $m_{prior,k} = 4.0$ and $m_{prior,\phi} = 0.20$ are referred to as the true prior means. Synthetic well-test pressure data were generated by running the simulator using the true permeability and porosity fields. Well-test pressure data were collected at wells 2, 4, 5, 6, and 8 (see Figure 1) during a period when the other four wells were produced at a specified rate. At the center well (well 5), a two-day drawdown followed by a one-day buildup test was undertaken. At the other four tested wells (wells 2, 4, 6, and 8), pressure data were measured during three-day drawdown tests. This synthetic pressure data are referred to as measured pressure data from this point onward.

In the following discussion, the procedures developed here for sampling the a posteriori pdf (equation 9) are applied. A case is considered where $m_{prior,k} = 5.0$ and $m_{prior,\phi} = 0.25$ (referred to as the incorrect prior means), with and without allowing for uncertainty (errors) in the prior means.

Figure 3 shows an unconditional simulation of log-permeability generated from Gaussian cosimulation using the true prior means. Figure 4 shows an unconditional simulation of the log-permeability obtained from Gaussian cosimulation using the incorrect prior means. As expected, the gridblock values of log-permeability tend to be much higher when the incorrect mean is used (compare Figures 3 and 4). Similar results were obtained for porosity since the incorrect mean for porosity is higher than its true mean.

Figure 5 shows a conditional simulation of the log-permeability field obtained by applying the method of Oliver et al. (1996), using true prior means for $\ln(k)$ and porosity. This is equivalent to the basic procedure

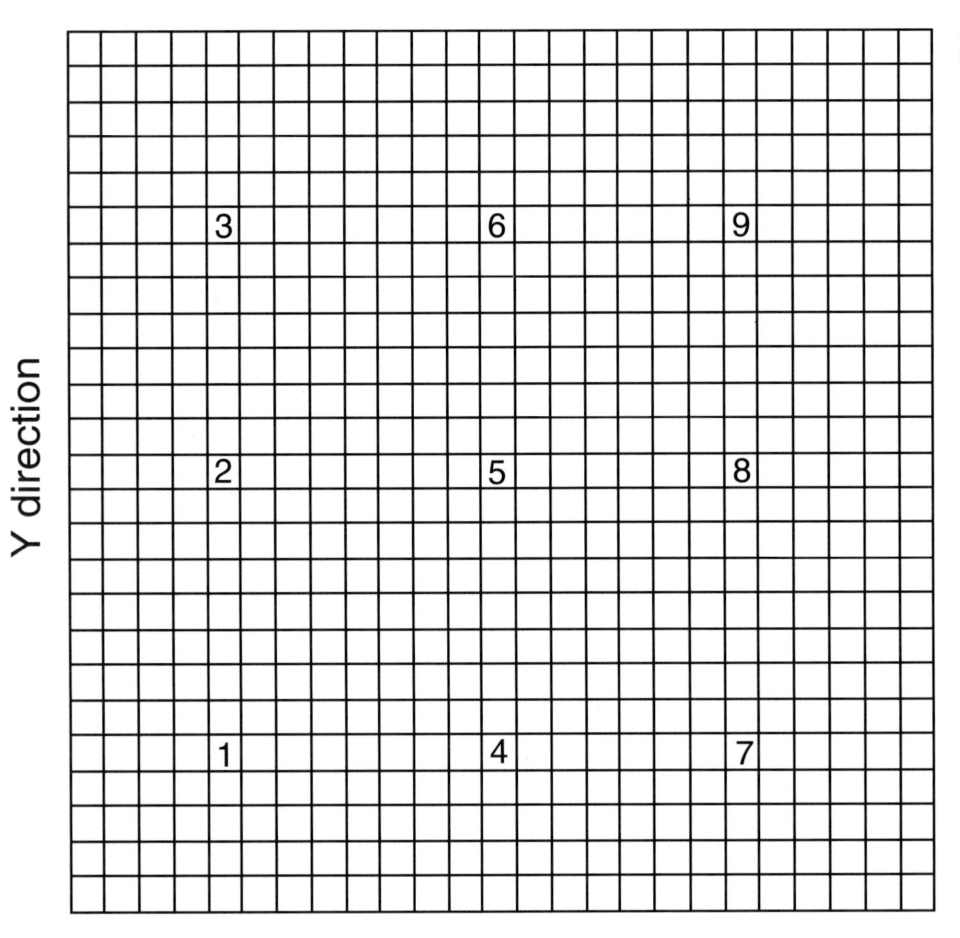

Figure 1. Areal grid, well locations, and well numbers.

Figure 2. True log-permeability field.

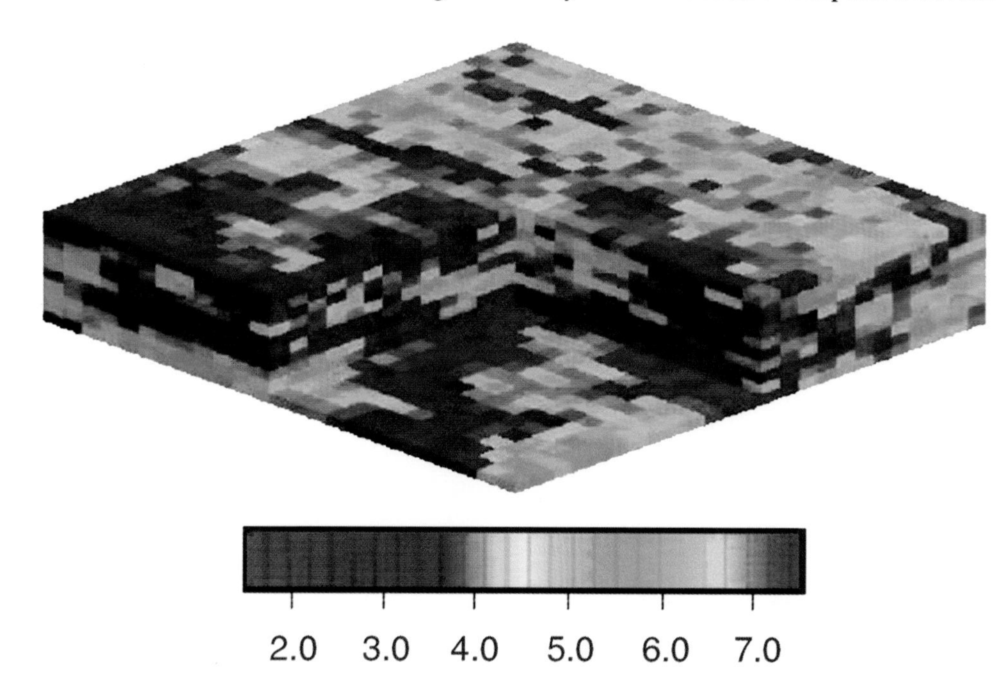

Figure 3. One unconditional realization of log-permeability field with true prior means.

established here with θ equal to zero at all iterations (that is, uncertainty is not incorporated in the prior mean). Figure 6 shows a conditional simulation obtained by the same procedure except, in this case, the incorrect prior means are used. Note that the log-permeability values obtained in Figure 6 tend to be much higher than those obtained in Figure 5. Conditioning to pressure data has reduced the permeability only in regions around the wells. This is the expected result because the incorrect prior means are much higher than the true values.

Figure 7 shows a conditional realization obtained via the basic procedure developed here. In this case, the incorrect prior means are used, where the 2×2 diagonal covariance matrix C_α (see equation 47) has as its two entries $\sigma^2_{\alpha 1} = 0.001$ and $\sigma^2_{\alpha 2} = 0.2$. Note that the realization in Figure 7 is almost identical to the one of Figure 5, which was generated using the true prior means and assuming no errors in the prior means. Although they are not presented here, similar results were obtained for porosity.

The results shown in Figures 5–7, and the corresponding results for porosity (not shown), illustrate that this procedure for accounting for uncertainty in the prior means is viable, and that it yields reasonable realizations of rock properties. The values of θ_s obtained by the basic procedure, which gave the results of Figure 7, indicate that the correction to the prior mean for $\ln(k)$ was –1.041, and the correction to the prior mean for porosity was –0.047. Note that these values are very close to the true error in the incorrect prior means.

The permeability values and associated porosity values corresponding to the results of Figure 7 were input to the simulator to predict pressure data at the five wells tested. Figure 8 shows that the pressure data predicted at well 5 from this realization are in good agreement with the measured pressure data. Equally good agreement was obtained at the other tested wells. The dashed curve in Figure 8 represents the pressure data predicted using the corresponding unconditional simulation as input in the reservoir simulator. Because the unconditional simulation is used as the initial guess in the Gauss-Newton method when constructing the corresponding conditional simulation, the results of Figure 8 give a qualitative measure of how the incorporation of pressure data changes the unconditional simulation.

Fifty conditional simulations of the rock properties were generated using the basic simulation procedure. As discussed previously, this suite of realizations of rock properties represents an approximate sampling of the a posteriori pdf of equation 9. For each realization, reservoir performance is simulated for 1000 days, where each of the nine wells was produced at a specified bottom-hole pressure. Reservoir performance was also predicted from the set of m_{uc} values generated by Gaussian cosimulation. The set of curves in Figure 9 represents the field cumulative oil production predicted from the 50 history-matched realizations (that is, Figure 9 illustrates the uncertainty in predicted reservoir performance). The curve through the solid dots represents the field cumulative oil production generated using true permeability and porosity as simulator input. The prediction from the true reservoir falls very close to the center of the set of predictions generated from the conditional realizations. The set of curves in Figure 10 represents the predictions of cumulative oil production obtained from the set of 50 m_{uc} models, which were used as initial guesses in the Gauss-Newton procedure. Note that these realizations tend to predict erroneously high values of cumulative oil production because the incorrect prior means are much higher than the true prior means. Also note that conditioning realizations to pressure data significantly reduces the uncertainty in predicted performance.

A histogram of the cumulative oil production at 1000 days and an associated cumulative distribution function constructed from the results of Figure 9 are

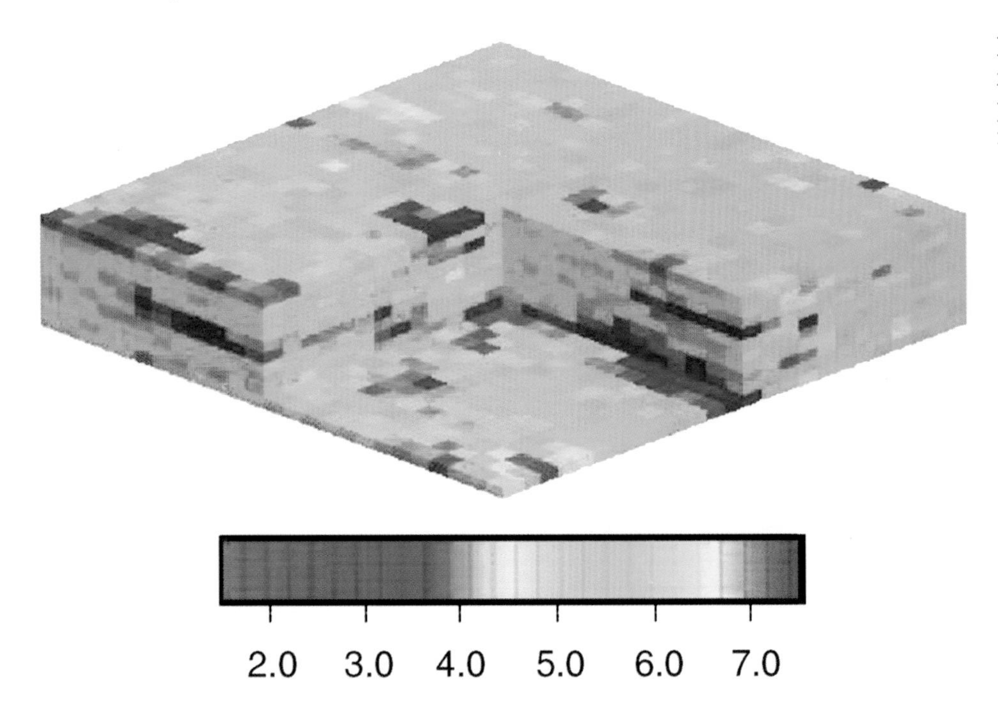

Figure 4. One unconditional realization of log-permeability field with incorrect prior means.

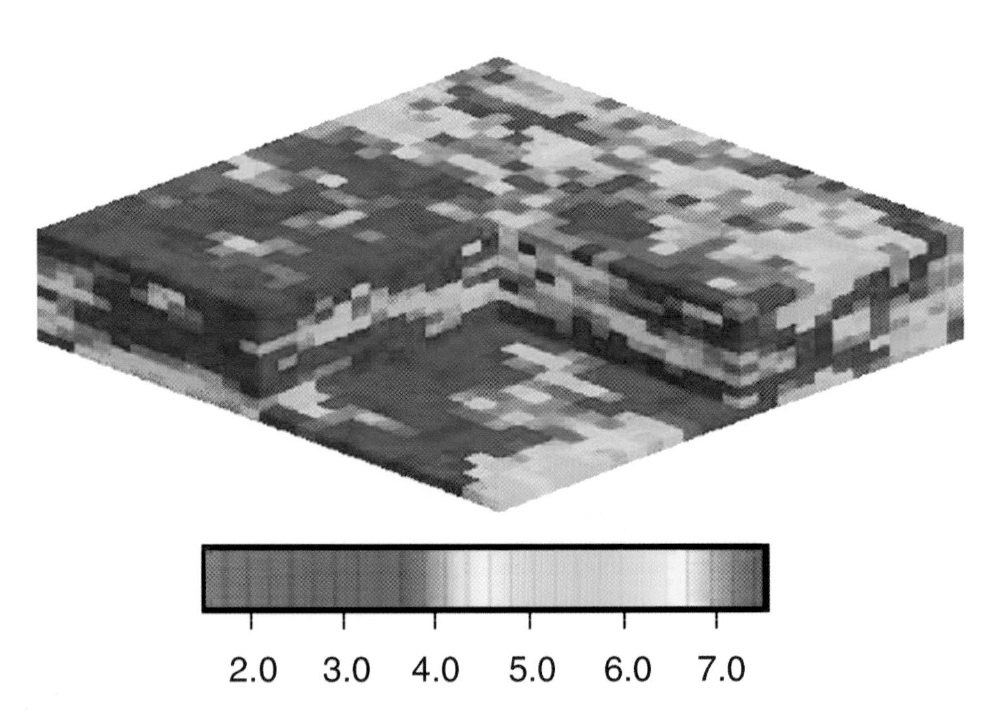

Figure 5. One realization of log-permeability field conditioned to pressure data using true prior means.

shown in Figure 11. The expected value (mean) is 5.70×10^6 STB (stock tank bbl), the median is 5.74×10^6 STB, and the standard deviation is 1.68×10^5 STB. Note that the bar in the histogram over 5.80×10^6 STB represents the number of outcomes (15) between 5.70×10^6 STB and 5.80×10^6 STB. The predicted cumulative oil production at 1000 days using the true rock properties was 5.69×10^6 STB.

CONCLUSIONS

Errors in the prior means can be properly taken into account by using the partially doubly stochastic model developed here. An automatic history procedure has been presented that can be applied to generate a set of realizations conditioned to pressure data and a prior geostatistical model. This set of realizations

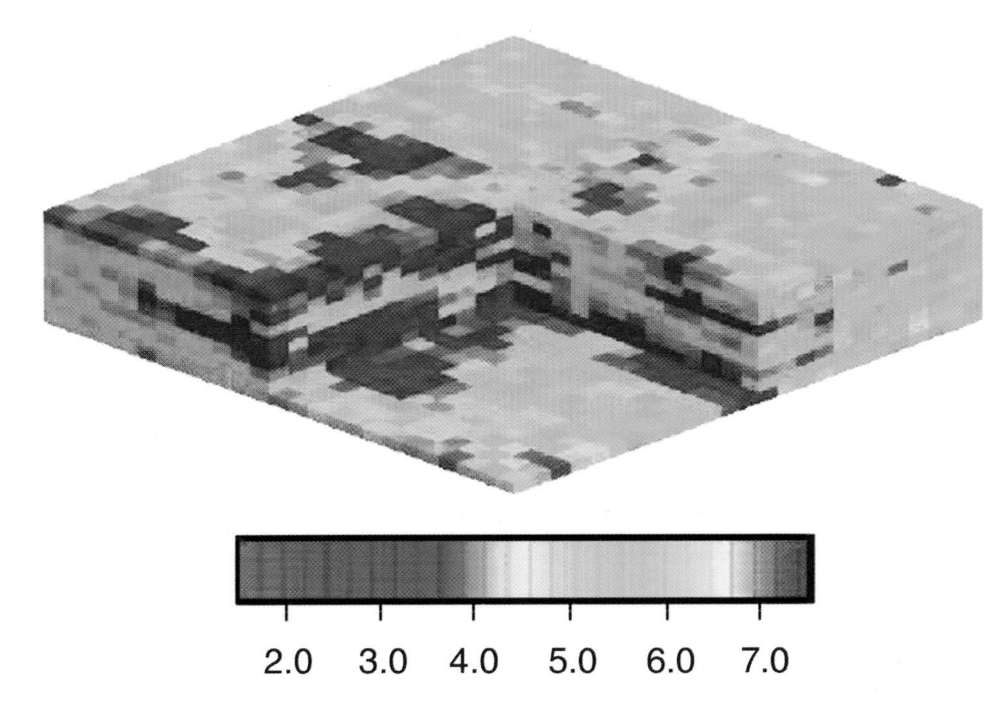

Figure 6. One realization of log-permeability field conditioned to pressure data without adjustment to incorrect prior means.

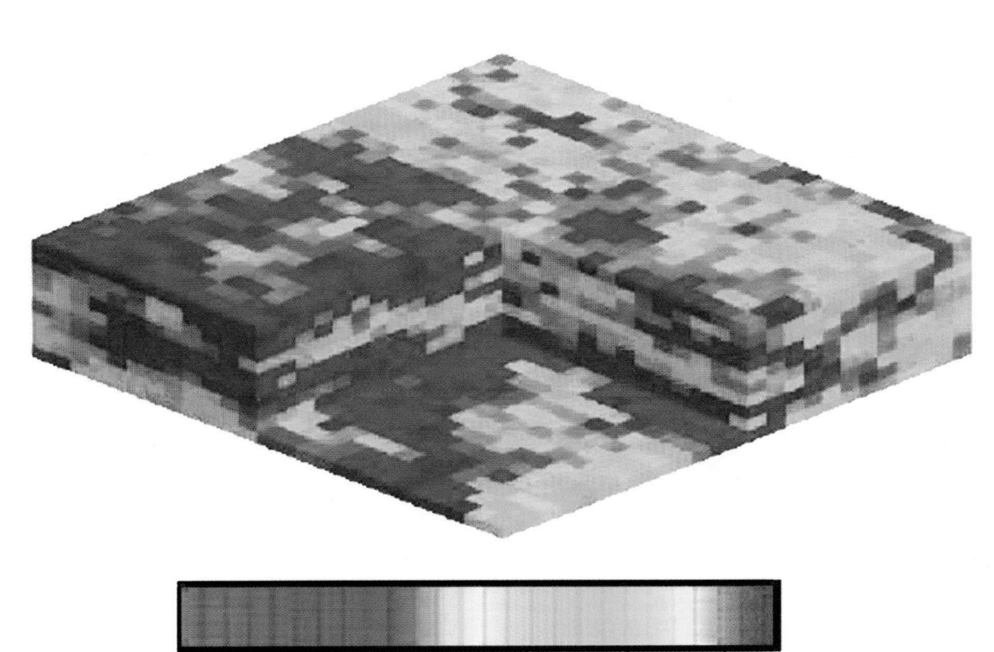

Figure 7. One realization of log-permeability field conditioned to pressure data with adjustment to incorrect prior means.

represents an approximate sampling of the a posteriori probability density function for the rock property fields. It has been rigorously proven that the basic procedure for sampling the pdf is correct if measured data are linearly related to the model. It has been shown that if estimates of prior means are inaccurate and the errors in the prior means are not accounted for, one cannot obtain a correct sampling of the pdf.

Using the basic pdf sampling procedure developed here, one can predict reservoir performance for each realization and evaluate the uncertainty in predicted reservoir performance from the set of outcomes.

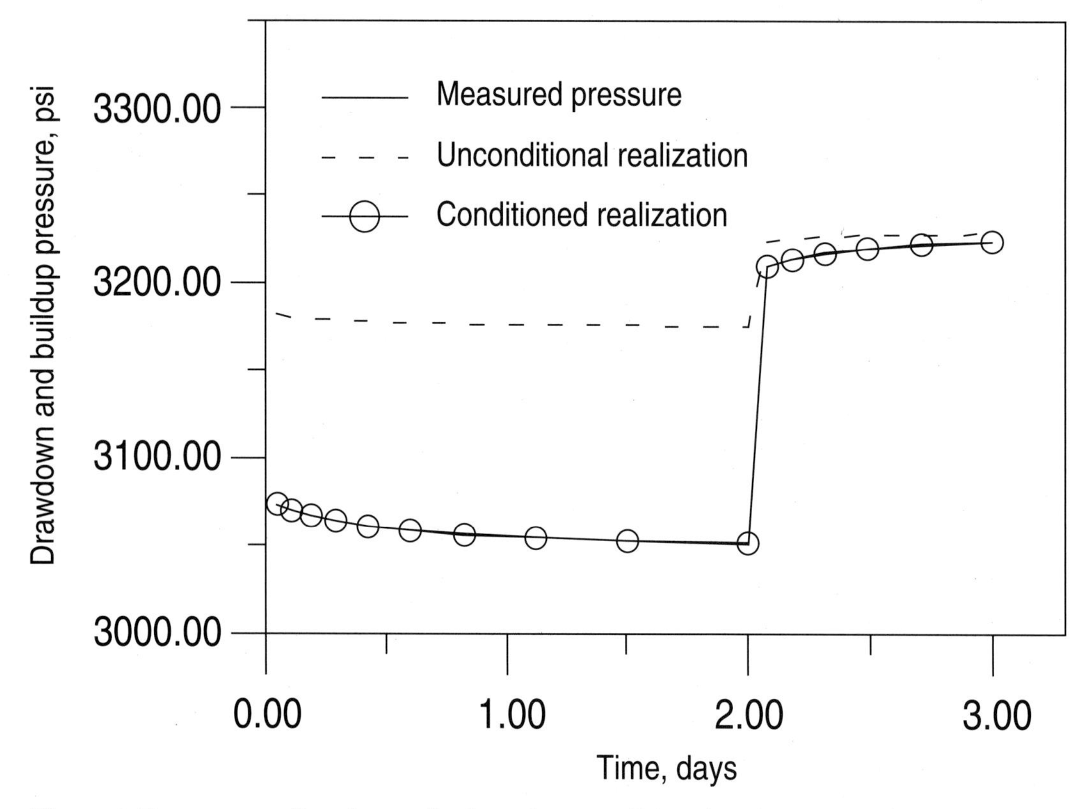

Figure 8. Pressure predicted at well 5 based on conditional and unconditional simulations of rock property fields compared to measured pressure data.

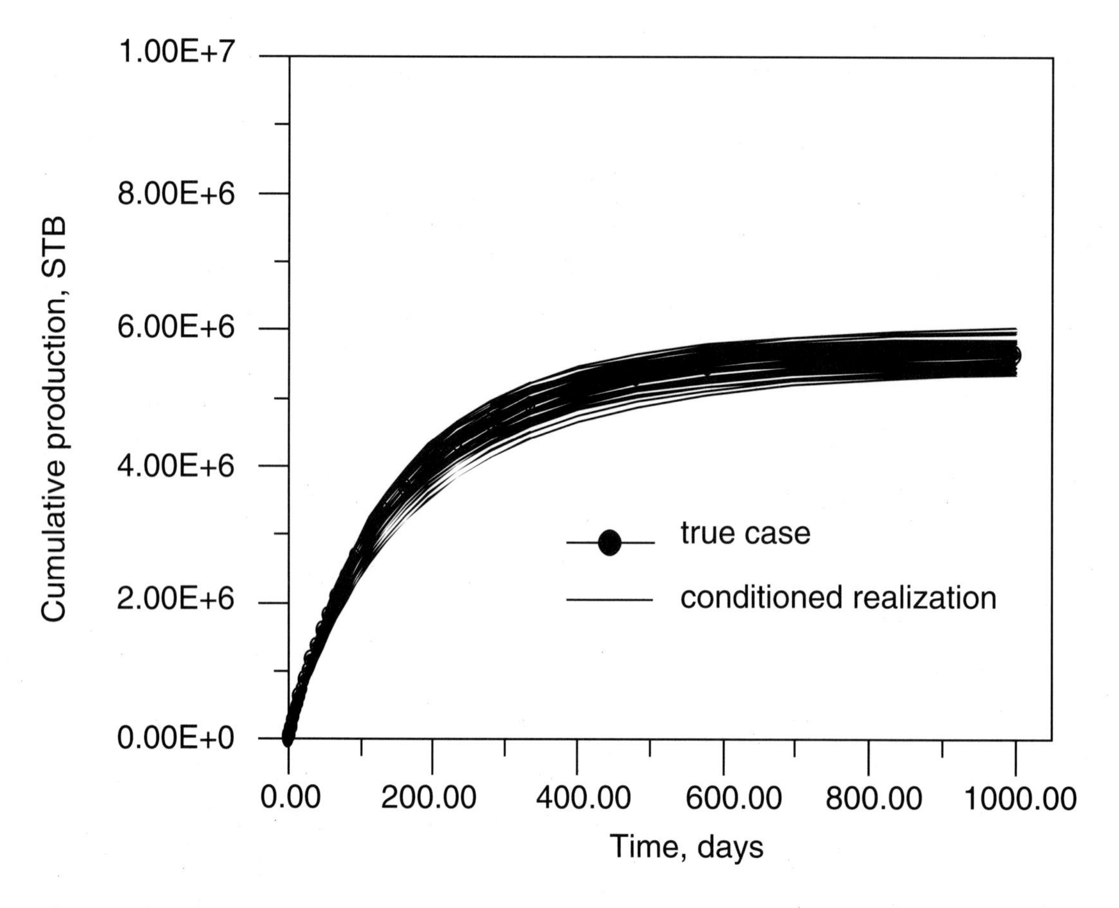

Figure 9. Predicted cumulative oil production from a suite of conditional simulations.

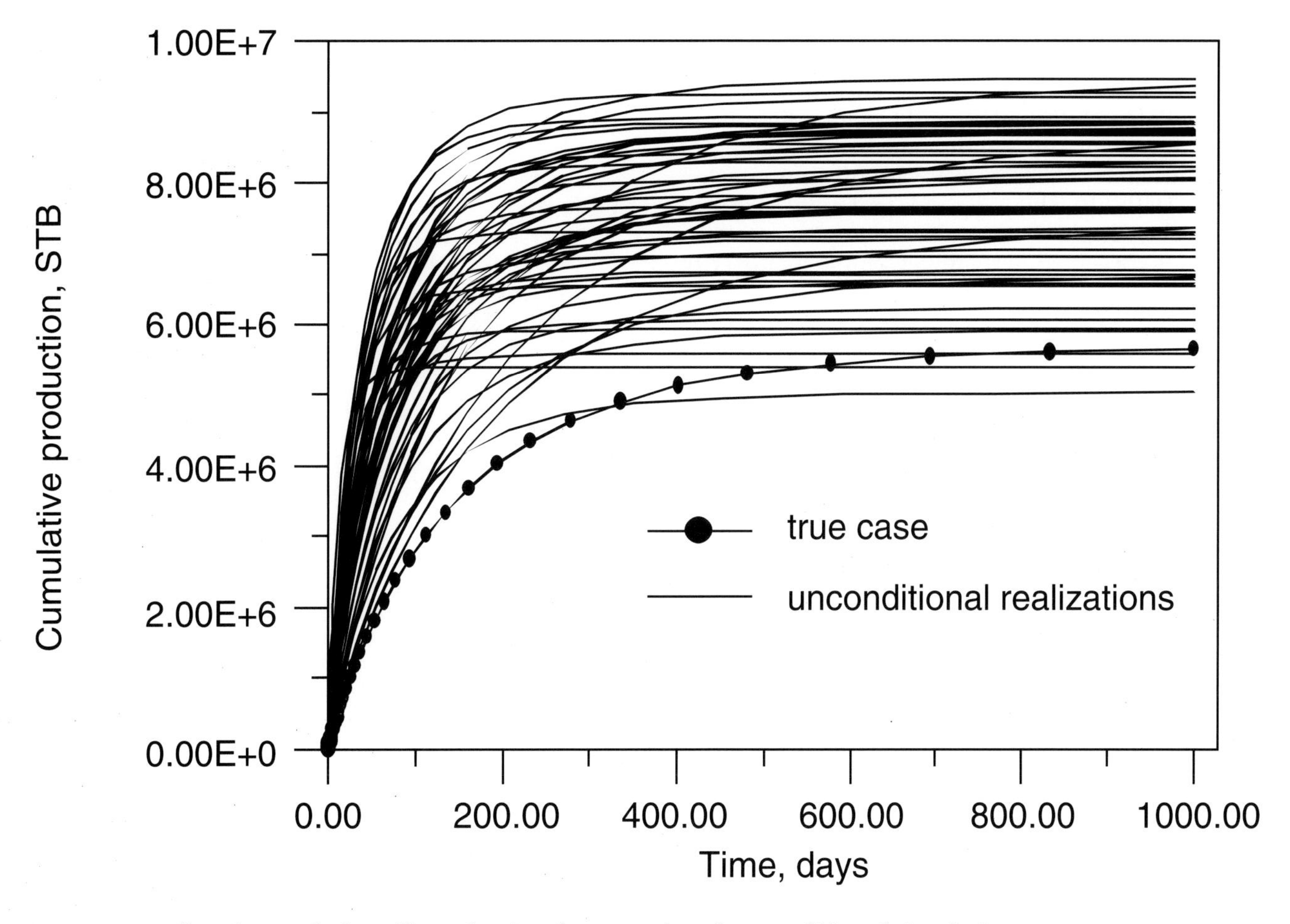

Figure 10. Predicted cumulative oil production from a suite of unconditional simulations.

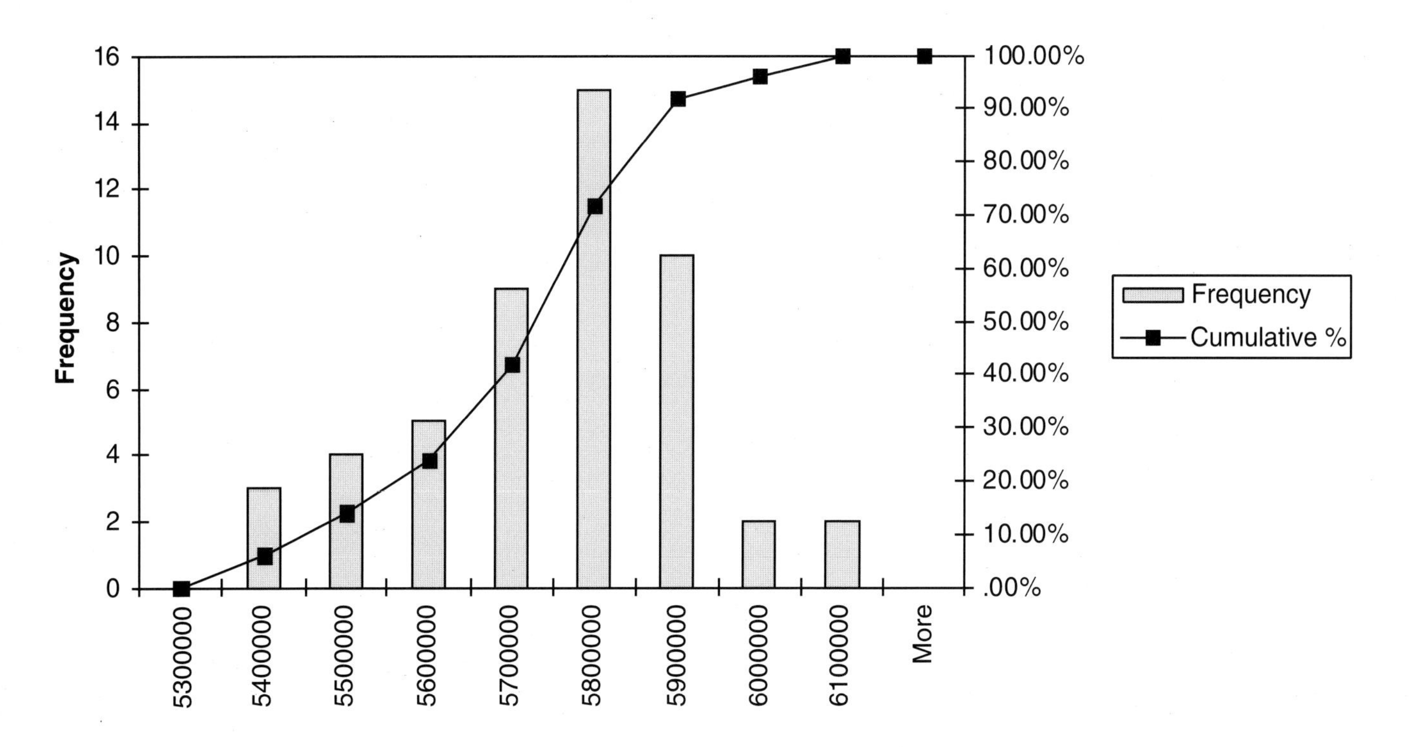

Fig, 11. Histogram and cumulative distribution of cumulative oil production at 1000 days.

ACKNOWLEDGMENTS

This work was supported by a grant from the Strategic Research Department of Chevron Petroleum Technology Company and the member companies of Tulsa University Petroleum Reservoir Exploitation Projects (TUPREP). This support is gratefully acknowledged. Yafes Abacioglu, who also has implemented the techniques presented here, reviewed the manuscript and offered constructive comments.

REFERENCES CITED

Chu, L., A. C. Reynolds, and D. S. Oliver, 1995a, Computation of sensitivity coefficients for conditioning the permeability field to well-test pressure data, In Situ, v. 19, p. 179–223.

Chu, L., A. C. Reynolds, and D. S. Oliver, 1995b, Reservoir description from static and well-test pressure data using efficient gradient methods, paper SPE 29999, presented at the 1995 SPE International Meeting on Petroleum Engineering, Beijing, Nov. 14–17.

Cunha, L. B., D. S. Oliver, R. A. Redner, and A. C. Reynolds, 1996, A hybrid Markov chain Monte Carlo method for generating permeability fields conditioned to multiwell pressure data and prior information, Soc. Pet. Eng. J., v. 3, p. 261–271.

Fletcher, R., 1987, Practical Methods of Optimization, John Wiley & Sons, Inc., New York, 436 + xiv p.

Gomez-Hernandez, J. J., and A. G. Journel, 1993, Joint sequential simulation of multi-Gaussian random variables, in Soares, A., editor, Geostatistic Troia 92, Kluwer, p. 133–144.

He, N., A. C. Reynolds, and D. S. Oliver, 1997, Three-dimensional reservoir description from multiwell pressure data, Soc. Pet Eng. J., v. 2, p. 312-327.

Journel, A. G., and Ch. J. Huijbregts, 1978, Mining Geostatistics, Academic Press Limited, London, 600 + xp.

Kennett, B. L. N., and P. R. Williamson, 1988, Subspace methods for large-scale nonlinear inversion, in Mathematical Geophysics: A Survey of Recent Developments In Seismology and Geodynamics, eds. Vlarr, N. J., et al., D. Reidel Publishing Company, Dordrecht, p. 139–154.

Kitanidis, P. K., 1995, Quasi-linear geostatistical theory for inversing, Water Resources Res., v. 31, p. 2411–2419.

Oldenberg, D. W., P. R. McGillivray, and R. G. Ellis, 1993, Generalized subspace methods for large-scale inverse problems, Geophysics J. Int., v. 10, p. 12–20.

Oldenberg, D. W., and Y. Li, 1994, Subspace linear inverse method, Inverse Problems, v. 10, p. 915–935.

Oliver, D. S., 1994, Incorporation of transient pressure data into reservoir characterization, In Situ, v. 18, p. 243–275.

Oliver, D. S., 1996, On conditional simulation to inaccurate data, Math. Geology, v. 28, p. 811–817.

Oliver, D. S., N. He, and A. C. Reynolds, 1996, Conditioning permeability fields to pressure data, Proceedings of the 5th European Conf. on the Mathematics of Oil Recovery, p. 259–269.

Oliver, D. S., L. B. Cunha, and A. C. Reynolds, 1997, Markov chain Monte Carlo methods for conditioning a permeability field to pressure data, in Math. Geology, v. 29, p. 61–91.

Peaceman, D. W., 1983, Interpretation of well-block pressures in numerical reservoir simulation with non-square grid blocks and anisotropic permeability, Soc. Pet. Eng. J., v. 23, p. 531–543.

Reynolds, A. C., N. He, L. Chu, and D. S. Oliver, 1996, Reparameterization techniques for generating reservoir descriptions conditioned to variograms and well-test pressure data, Soc. Pet. Eng. J., v. 4, p. 413–426.

Tarantola, A., 1987, Inverse Problem Theory, Methods for Data Fitting and Model Parameter Estimation, Elsevier Science Publishers, Amsterdam, 613 + xvip.

Tjelmeland, H., H. Omre, and B. J. Hegstad, 1994, Sampling from Bayesian models in reservoir characterization, Technical Report Statistics No. 2, University of Trondheim, 15 p.

Xu, W., T. T. Tran, R. M. Srivastava, and A. G. Journel, 1992, Integrating seismic data in reservoir modeling: the collocated cokriging approach, paper SPE 24742, presented at the 1992 SPE Annual Technical Conference and Exhibition, Washington D.C., Oct. 4–7.

Blanc, G., et al., Transient productivity index for numerical well test simulations, 1999, *in* R. Schatzinger and J. Jordan, eds., Reservoir Characterization-Recent Advances, AAPG Memoir 71, p. 163–174.

Chapter 11

Transient Productivity Index for Numerical Well Test Simulations

G. Blanc
D. Y. Ding
A. Ene
T. Estebenet
D. Rahon
Institute Français du Pétrole
Pau, France

ABSTRACT

The most difficult aspect of numerical simulation of well tests is the treatment of the bottom-hole flowing (BHF) pressure. In full-field simulations, this pressure is derived from the well-block pressure (WBP) using a numerical productivity index that accounts for the grid size and permeability, and for the well completion. This productivity index is calculated assuming a pseudo-steady-state flow regime in the vicinity of the well and is therefore constant during the well production period.

Such a pseudo-steady-state assumption is no longer valid for the early time of a well test simulation as long as the pressure perturbation has not reached several grid-blocks around the well. This paper offers two different solutions to this problem:

- The first solution is based on the derivation of a transient numerical productivity index (TNPI) to be applied to Cartesian grids.
- The second solution is based on the use of a corrected transmissibility and accumulation term (CTAT) in the flow equation.

The representation of the pressure behavior given by both solutions is far more accurate than the conventional one, as shown by several validation examples presented in the following pages.

INTRODUCTION

Need for Better Numerical Well Test Simulation

Analytical well test solutions have been extensively used for a long time by reservoir engineers to interpret their well test results. In the field of hydrogeology, C.V. Theis (1935) offered one of the first solutions to the problem of the lowering of the piezometric surface under the effect of the rate and duration of a discharge well. Theis' solution was based on the use of the Exponential Integral, or Ei, function and is still being used by petroleum engineers. Hundreds of papers have been published since that date to account for more complex reservoir models, well completion, and test designs. An SPE monograph has summarized the practical aspects of pressure transients (Volume 5, by R. C. Earlougher Jr.). Two SPE Reprint Series have been also published (Volume 9, Pressure Analysis Methods; Volume 14, Pressure Transient Testing Methods). Published more recently, R. Horne's book gives the status of the knowledge in the 90s.

Despite this abundance of solutions, there are many cases where the analytical solutions no longer apply. Heterogeneous reservoirs with complex in boundaries and complex well geometry are examples where numerical model simulation is the only way to compute the pressure transient. Geostatistical modeling, which is now widely used by geoscience engineers to fill in reservoir simulation grids, also requires numerical reservoir simulation (Blanc, 1995, 1996).

In the meantime, the well testing technology has been improved, allowing bottom-hole shutdown and early time pressure recording. This is where the need for more core and more precise numerical BHF (bottom-hole flowing) pressure simulation becomes evident.

Numerical Simulation of Well Tests

The pressure behavior of a single-phase flow in a 3-D (three-dimensional) reservoir is governed by the following equation:

$$\frac{\partial}{\partial x_i}\left(\frac{k_i}{\mu}\frac{\partial(P+\rho g z)}{\partial x_i}\right) = \phi c_t \frac{\partial P}{\partial t} + q \qquad (1)$$

where P is the pressure, i is the axis index, x_i coordinates (x_1, x_2, x_3, = for x, y, z, respectively), k_i is the directional permeability, μ is the fluid viscosity, ρ is the fluid density, g is the gravity, ϕ is the porosity, c_t is the total rock and fluid compressibility, and q is a sink term.

SIMTESTW is the IFP numerical well test simulation program that uses the seven-point finite-difference scheme of equation 1. The grid-block system is Cartesian irregular, and dead cells are used to simulate complex shapes.

Discretization of equation *1* in space and time, for a grid-block i and at time step n, is

$$\sum_j \frac{T_{ij}}{\mu}\left[\left(P_j^n - P_i^n\right) + \rho g\left(z_j - z_i\right)\right] = \phi c_t Vi \frac{P_i^n - P_i^{n-1}}{t^n - t^{n-1}} + \delta_{ip} Q_p \qquad (2)$$

where j is the index of all neighbors of grid-block i, Δx_i, Δy_i, and Δz_i are the size of the rectangular grid-block i, T_{ij} is the transmissibility between grid-blocks i and j (harmonic average of $\Delta z_i \times k_i$ products), V_i is the volume of grid-block $i = \Delta x_i \times \Delta y_i \times \Delta z_i$, Q_p is the rate of the well p, and δ_{ip} is equal to 1 when grid-block i contains the well p and 0 elsewhere.

P^n is then the solution vector of the following matrix equation:

$$A^n \times P^n = B^n \qquad (3)$$

where A^n is the seven-band matrix of the system of discretized equations at time n, P^n is the vector of the pressure map at time n, and B^n is the vector column of the right side of the system of discretized equations at time n.

This program can handle vertical, horizontal, or complex wells operated at an imposed rate or at an imposed BHF pressure. The management of the well completion is flexible enough to allow the user to open or close any grid-block to the flow during the simulation time.

Previous Works: Peaceman Numerical Productivity Index

Steady-State Flow

The problem of getting the BHF pressure from the well grid-block pressure has been addressed by D. W. Peaceman (1978).

Peaceman has shown that the steady-state grid-block pressure P_b at the well location, computed using a square grid-block system ($\Delta xi = \Delta yi = \Delta L$) in a homogeneous, isotropic, and infinite reservoir (permeability = k, porosity = ϕ, compressibility = c_t) produced at constant flow rate by a vertical well, is the same as would be observed at a dummy well of radius r_0 given by

$$r_0 = 0.198 \times \Delta L \qquad (4)$$

When these conditions apply, the BHF pressure can be obtained using a numerical productivity index (NPI), which relates the grid-block pressure P_0 to the BHF pressure P_{wf} according to

$$P_{wf} = P_0 - \frac{Q_b}{NPI} \qquad (5)$$

NPI can be computed using the steady-state radial flow equation between r_0 and r_w

$$NPI = \frac{2\pi \times \Delta z_i k}{\mu \ln\left(\frac{r_w \times e^{-s}}{r_0}\right)} \qquad (6)$$

where r_w is the well radius, and S is the well skin.

Each well can then be assigned a well-bore storage C, which relates the well rate Q to the sand face rate Q_{sf} applied to the reservoir as follows:

$$Q_{sf} = Q + C \times \frac{\Delta P_{wf}}{\Delta t} \qquad (7)$$

where ΔP_{wf} is the pressure drop within the well during the time step Δt.

Unsteady-State Flow

Peaceman (1978) has also shown that, for unsteady-state flow and when a pseudo-steady-state flow regime is reached near the well, r_0 can be derived from the following equation:

$$r_0 = \Delta L\left[4t_D \exp\left(-\gamma - 4\pi p_{Db}\right)\right] \tag{8}$$

where γ is the Euler's constant;

$$t_D = \frac{k}{\phi \mu c_t \Delta L^2} \times t \tag{9}$$

is the dimensionless time related to ΔL; and

$$p_D = \frac{kh}{Q\mu} \times p \tag{10}$$

is the dimensionless pressure, p_{Db} reply to rate Q_b.

Peaceman (1978) concluded that when $t_D > 1$, r_0 computed from equation 4 is close enough to the value given by relation 8, and that equation can therefore be applied in most full-field simulation cases.

Previous Works: IFP Corrected Transmissibility

Peaceman's work has been extended to many more complex conditions by the author himself (Peaceman, 1983, 1990, 1995) and by several others. All these works are based on the derivation of a better NPI or on the evaluation of numerical pseudo-skin to be used with relations 5 and 6.

Recently a new method has been proposed by Ding et al. (1994, 1995) for improving the well modeling in nonstandard configurations such as nonuniform Cartesian grids, flexible grids, and off-center wells. The principle of this method is to improve the flow representation around the well by modifying the transmissibility in the near-well region. This approach, referred to as the corrected transmissibility approach, is particularly suited to 3-D well modeling as shown by Ding (1995).

Within this approach the pressure p_0 corresponding to a radius r_0 is obtained using equivalent transmissibility T_{eq}, which links p_0 to the pressure of adjacent grid-blocks through equivalent distances (Figure 1). For a homogeneous and isotropic reservoir, the following transmissibilities must be used instead of the conventional ones:

$$T_{eq,i} = k\Delta z_i \frac{\Delta y_0}{L_{eq,i}}$$

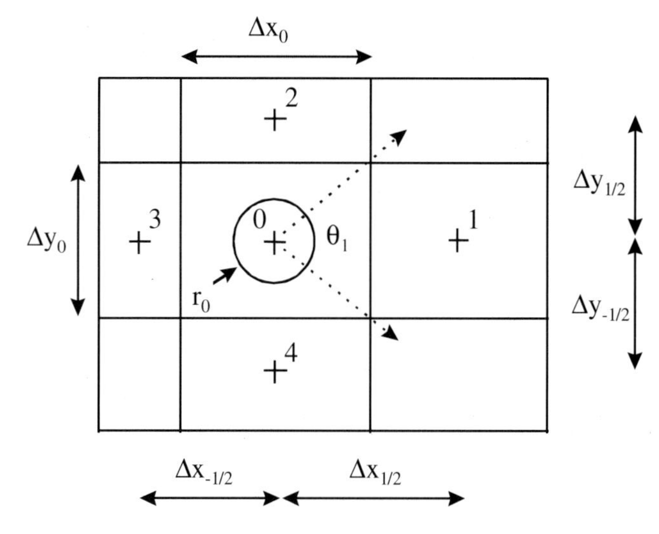

Figure 1. Well-block pressure calculation (corrected transmissibility method).

and with

$$L_{eq,i} = \Delta y_0 \frac{\ln\left(\dfrac{\Delta x_{1/2}}{r_0}\right)}{\theta_i} \tag{11}$$

for i = 1 and 3

$$T_{eq,i} = k\Delta z_i \frac{\Delta x_0}{L_{eq,i}}$$

and with

$$L_{eq,i} = \Delta x_0 \frac{\ln\left(\dfrac{\Delta y_{1/2}}{r_0}\right)}{\theta_i} \tag{12}$$

for i = 2 and 4, where Δx_i, Δy_i, Δz_i, and θ_i are described in Figure 1.

The radius r_0 can be chosen by the user, and the BHF pressure is then obtained using relation 6.

This method can be easily implemented in conventional numerical simulators, provided that they allow transmissibility multipliers.

TRANSIENT NUMERICAL PRODUCTIVITY INDEX (TNPI)

Generalities

In order to investigate how the Peaceman approach behaves early in the test, we simulated a pumping test on a vertical well producing a homogeneous "infinite" reservoir using a regular square (10 m × 10 m) grid system. Table 1 shows all the properties of the test, referred to as Reference Case 1.

Table 1. Reference Case 1, Characteristics.*

Reservoir (one layer, infinite)				
	k (md)	f (fraction)	c_t (1/bar)	h (m)
	100	0.20	5×10^{-4}	20
Fluids	μ (cp)	ρ (g/cm³)	B_o (m³/m³)	
	1	1	1	
Well	r_w (m)	S	C (m³/bar)	
	0.0785	0	0	
Test	**Type**	**Duration (s)**	**Rate (m³/d)**	
	Pumping test	10^5	50	

*Corresponding to Figures 2–5 and 7.

Figure 2 is a semi-log plot of the results showing the grid-block pressure p_0, the BHF pressure p_{wf} derived from relations 5 and 6, and the Ei (exponential integral) analytical solution. This figure points out the following facts.

- Late in the test, the Peaceman approach provides a very good BHF pressure (less than 1% of error).
- Early in the test, the pressure perturbation did not reach the neighboring blocks, and all the produced fluid came from the well-block capacity. The well-block drawdown pressure is negligible compared to the $r_0 - r_w$ steady-state correction, and the corresponding BHP pressure is higher than the analytical BHP pressure.
- In between these two periods, the neighbor blocks start providing an increasing part of the well rate, and the BHF pressure is lower than the analytical BHF pressure to compensate for the delay early in the test.

The second question addressed concerns the influence of the grid-block size on the BHF pressure and on its derivative. The previous drawdown test was simulated using a set of different grid sizes ΔL: 20 m × 20 m, 10 m × 10 m, 5 m × 5 m, and 1 m × 1 m. Figure 3, which represents a log-log plot of the BHF pressure and its derivative versus the time, shows all the results together with the Ei analytical solution. Figure 3 illustrates the following concepts.

- Decreasing the mesh size leads, as expected, to better results at an earlier time.
- An equivalent "well-bore storage effect" appears at early time when nearly all the produced fluid comes from the well-block capacity.

This phenomena leads two more questions. The first question is, is there any rule to calculate the maximum mesh size to be used in order to get clean results from any given pressure recording time?

Figure 4, which gathers results calculated using various grid-block sizes and reservoir permeability, is a log-log plot of dimensionless BHF pressure and its derivative versus dimensionless time computed using Δx^2 (relations 9 and 10). This figure shows that for $t_D > 2.6$ the derivative is flat, as expected by the analytical solution of the fluid flow equation. This time, which was determined as the one from which the apparent well-block storage ends, leads to the relation

$$t \geq 2.6 \Delta L^2 \times \frac{\phi \mu c_t}{k} \tag{13}$$

The second question is, is it possible to use the mesh storage to simulate well bore storage whenever the numerical simulation program does not account for such a condition?

Figure 5 is the derivative plot of the PIE well test interpretation program (PIE, 1994), which has been used to interpret the results of the $\Delta L = 20$ m grid-block simulation. For such a grid the well-block capacity is equal to $\Delta x \times \Delta y \times \Delta z \times \phi \times c_t$, that is 0.8 m³/bar, and the interpretation leads to a close 0.737 m³/bar; however, the shape of the pressure derivative shows a huge discrepancy between the behavior of a true well bore storage (solid line, Figure 5) and the apparent well-block storage ("x" line, Figure 5), pointing out that the latter can be used to simulate the former.

Derivation of the Transient Numerical Productivity Index

Three different attempts have been made to correct the early time problem coming from the use of the NPI steady-state grid-block correction. Reference Case 1 (described in Table 1) has been used to check for the validity of these attempts. A one-layer system made of square (20 m × 20 m) grid-blocks has been selected for all numerical simulations. The well bore storage and skin are both equal to 0.

The Peaceman Ei Correction

The first attempt to correct the problem occurring from the use of the NPI steady-state grid-block used an unsteady-state radial flow equation between r_0 and r_w to compute the NPI as follows

$$NPI = \frac{4\pi \times \Delta z_i k}{Ei\left(-\frac{r_0^2}{4kt}\right) - Ei\left(-\frac{r_w^2}{4kt}\right)} \tag{14}$$

The corresponding BHF pressure and its derivative, which are shown in Figure 7, are far better than the conventional pressures. The pressure is relatively correct compared to the analytical Ei solution. The early part of the apparent well-block storage was removed

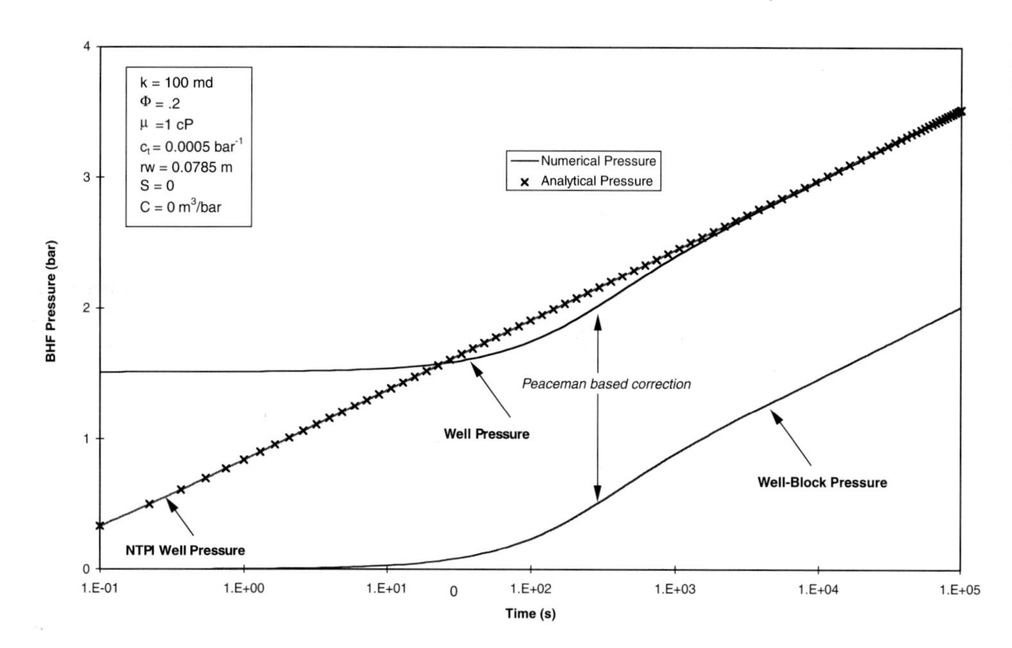

Figure 2. Conventional BHF (bottom-hole flowing) pressure computation (Reference Case 1, 10m x 10m grid-blocks).

from the derivative, which still shows a "wave" before the stabilization.

The Peaceman Explicit Transient Correction

The main drawback of the first attempt seems to come from the use of a constant equivalent radius r_o. We then extended the Peaceman transient analysis of r_0, using relation 8, far more toward the early time than Peaceman did. Figure 6 shows that the ratio $r_0/\Delta x$ first grows rapidly up to a maximum of 0.3 for $t_D = 0.01$, and then decreases down to the 0.198 limit value.

The corresponding BHF pressure and its derivative are referred to as $r_{0(t)}$ explicit in Figure 7. Compared to the analytical Ei solution, the pressure is not as good as that achieved by the Peaceman Ei, the derivative being far better than the previous pressure.

The Full TNPI Correction

The previous correction is based on a pseudo-steady-state approximation of the transient flow equation corresponding to a pseudo-steady-state flow regime around the well. The approximation allows the explicit derivation of r_0.

To be more accurate we used the full transient Ei solution. The r_0 radius was obtained using a Newton method of solution. Figure 6 shows that the corresponding radius is close to the previous one when the dimensionless time t_D becomes greater than 0.2. For $t_D < 0.2$, the two radii are slightly different. The BHF pressure

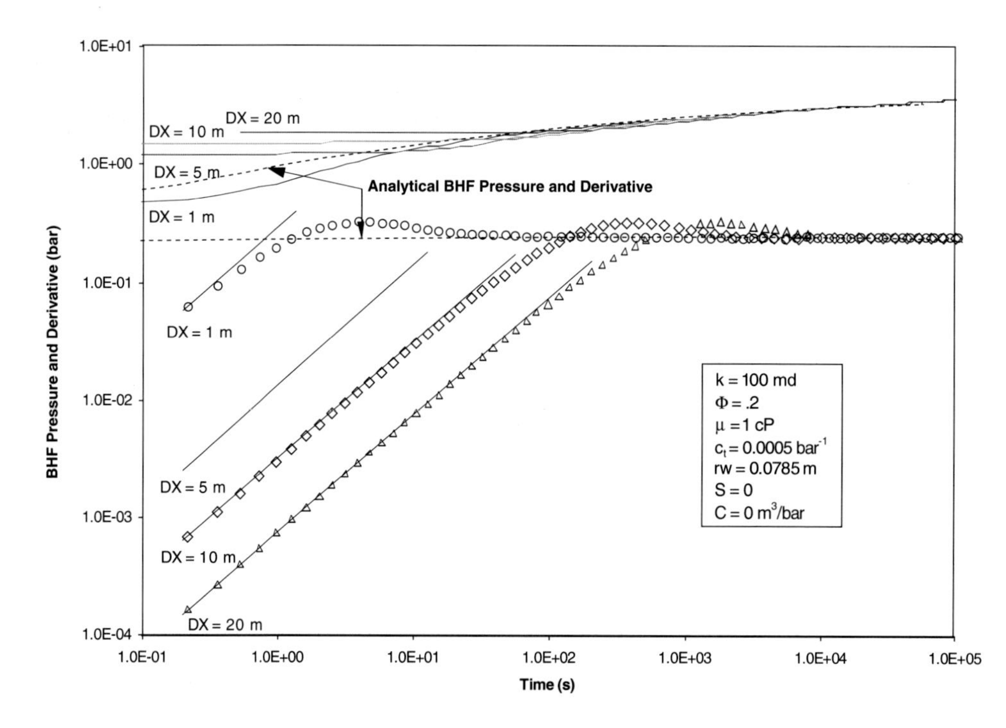

Figure 3. Conventional BHF (bottom-hole flowing) pressure and derivative plot computation (Reference Case 1)

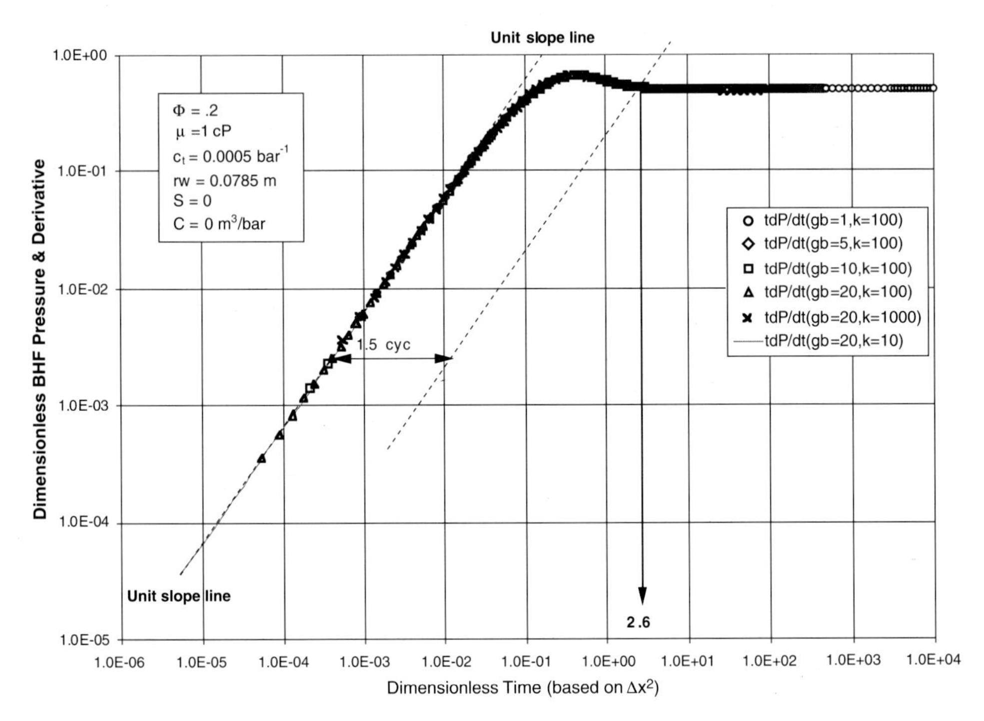

Figure 4. Well-block size effects on conventional BHF (bottom-hole flowing) derivative computation (Reference Case 1).

and derivative computed using this new radius are referred to as TNPI in Figure 7. They both correspond well with the analytical solution.

Validation of the TNPI Correction

Three different reference cases (Tables 1–4) were used to check for the validity of the full TNPI proposed correction.

The first validation was used to check for the accuracy of the proposed TNPI correction in the case of partial well penetration in a homogeneous reservoir. The last two validations were devoted to multiple rate tests, such as drill-stem tests (DST) and buildup tests. As demonstrated earlier, the correction works properly when the well rate is constant, but it vanishes when the well is shut down, as does

Table 2. Reference Case 3, Characteristics.*

Reservoir (eight layers, infinite)	k_h (md)	k_v (md)	ϕ (fraction)	c_t (1/bar)	h (m)
All layers	100	10	0.20	5×10^{-4}	10
Fluids	**μ (cp)**		**ρ (g/cm³)**	**B_o (m³/m³)**	
	1		1	1	
Well	**Skin**				
Layer 1	0		r_w = 0.0785 m		
Layers 2–8	Closed		S = 0		
			C = 0 m³/bar		
Test	**Type**		**Duration (s)**	**Rate (m³/d)**	
	Pumping test		5×10^5	50	

*Corresponding to Figure 8.

Table 3. Reference Case 4, Characteristics.*

Reservoir (three layers, infinite, ct = 5×10^{-4} bar)	k_h (md)	k_v (md)	ϕ (fraction)	h (m)
Layer 1	10	0.1	0.08	2
Layer 2	1000	1	0.25	2
Layer 3	30	0.3	0.12	2
Fluids	**μ (cp)**	**ρ (g/cm³)**	**B_o (m³/m³)**	
	1	1	1	
Well	**Skin**			
Layer 1	Closed	r_w = 0.0785 m		
Layer 2	0	S = 0		
Layer 3	Closed	C = 0 m³/bar		
Test	**Duration (hr)**	**Rate (m³/d)**		
Flow period 1	7	10		
Fow period 2	7	30		
Flow period 3	7	50		
Flow period 4	112	0 buildup		

*Corresponding to Figure 9.

Table 4. Reference Case 5, Characteristics.*

Reservoir (three layers, infinite, ct = 5 × 10^{-4} bar)				
	k_h (md)	k_v (md)	φ (fraction)	h (m)
Layer 1	10	0.1	0.20	2
Layer 2	1000	1	0.05	2
Layer 3	30	0.3	0.20	2
Fluids	μ (cp)	ρ (g/cm³)	B_o (m³/m³)	
	1	1	1	
Well	**Skin**			
Layer 1	0	r_w = 0.0785 m		
Layer 2	0	$S = 0$		
Layer 3	0	C = 0 m³/bar (Fig. 10)		
		C = 0.1 m³/bar (Fig. 11)		
Test	**Duration (hr)**	**Rate (m³/d)**		
Flow period 1	7	10		
Fow period 2	7	30		
Flow period 3	7	50		
Flow period 4	112	0 buildup		

*Corresponding to Figures 10 and 11.

the conventional Peaceman approach. To bypass this limitation we used a constant rate simulation to derive a "kernel" pressure, which was then used as the reference pressure response for a superposition (rate convolution) program. By doing so, any kind of multiple rate test can be simulated, even if it includes shutdown periods. This method can be applied only when the permeability, porosity, and compressibility are not pressure or fluid-saturation dependent; i.e. when equation 1 is linear (necessary validity condition of the superposition theorem).

Reference Case 3: Homogeneous Reservoir, Draw-Down Test, Partial Penetration

A 10-m-thick homogeneous reservoir is produced at constant rate by a well that is perforated in the top 2 m. The grid-block system is made of square (20 m × 20 m) grids 2 m thick; the numerical model is made of 10 layers. Table 3 shows all the characteristics of this reference case.

Figure 8 gives the results of the simulation. The computed BHF pressure, as well as its derivative, is close to the analytical pressure computed by the analytical well test interpretation program PIE down to 0.2 s. The derivative is far better than the derivative computed with the conventional Peaceman correction.

Reference Cases 4: Three-Layer Reservoir, DST, Partial Penetration

Reference Case 4 is a three-layer reservoir for which the characteristics are shown in Table 3. There are three layers in the numerical model (one per reservoir layer), which are made of square grid-blocks (20 m × 20 m × 2 m). The purpose of this case was to check for the validity of the superposition method proposed to simulate buildup flow periods of a DST. The simulated DST is made of three flow periods of 7 hr followed by a buildup of more than 100 hr (Table 4). The

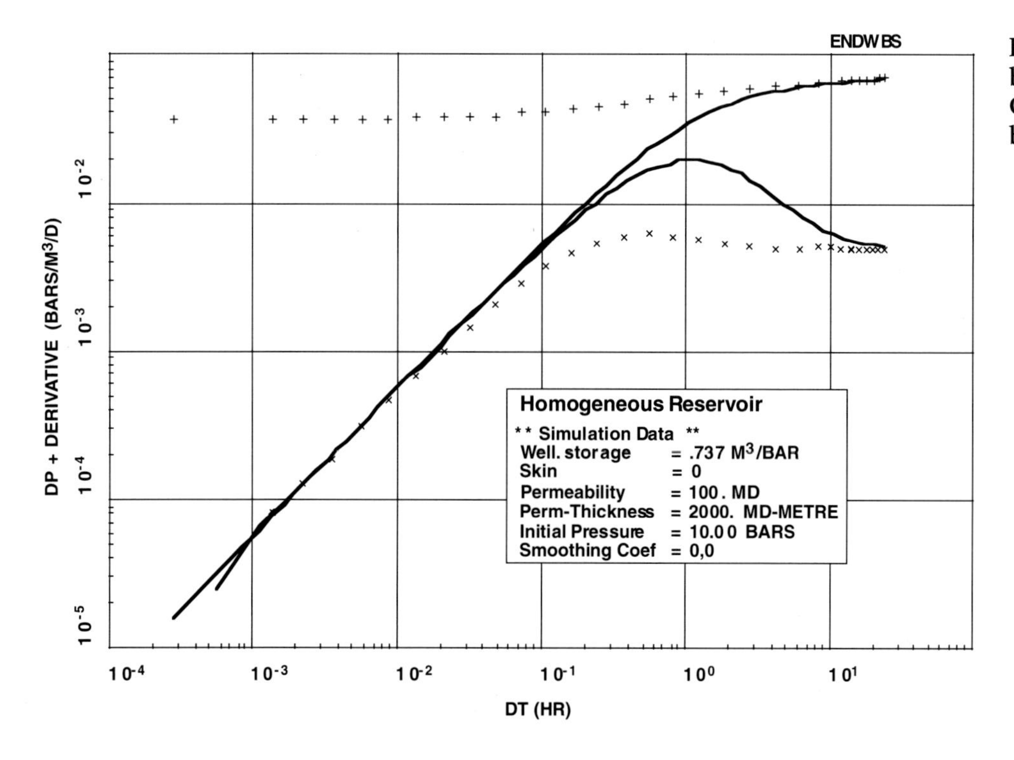

Figure 5. Apparent well-block storage (Reference Case 1, 20 m × 20 m grid-blocks).

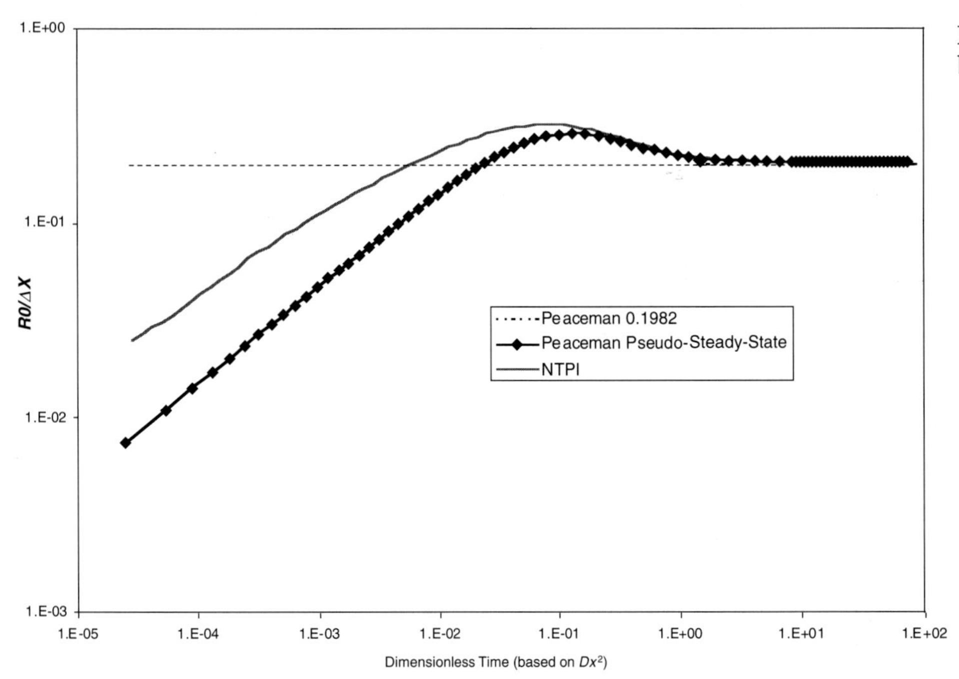

Figure 6. Equivalent well-block radius versus time.

well is perforated only in the middle layer of the reservoir. The results of the simulated buildup, which are represented in Figure 9, correspond well with results given by the analytical program. The improvement of the conventional Peaceman correction is very significant in the first 1000 s of the buildup.

Reference Case 5: Three- Layer Reservoir, DST, Full Penetration

Reference Case 5 is also a three-layer reservoir for which the characteristics are shown in Table 4. The grid-block system is the same as the one used in Reference Case 4. The purpose of this case was to check for the validity of the superposition method with and without the presence of a well bore storage for a full penetration well. The two simulated buildups are represented in Figures 10 and 11. They both correspond well with the analytical program results. When there is no well bore storage, the improvement of the proposed correction is also very significant, as previously observed with Reference Case 4.

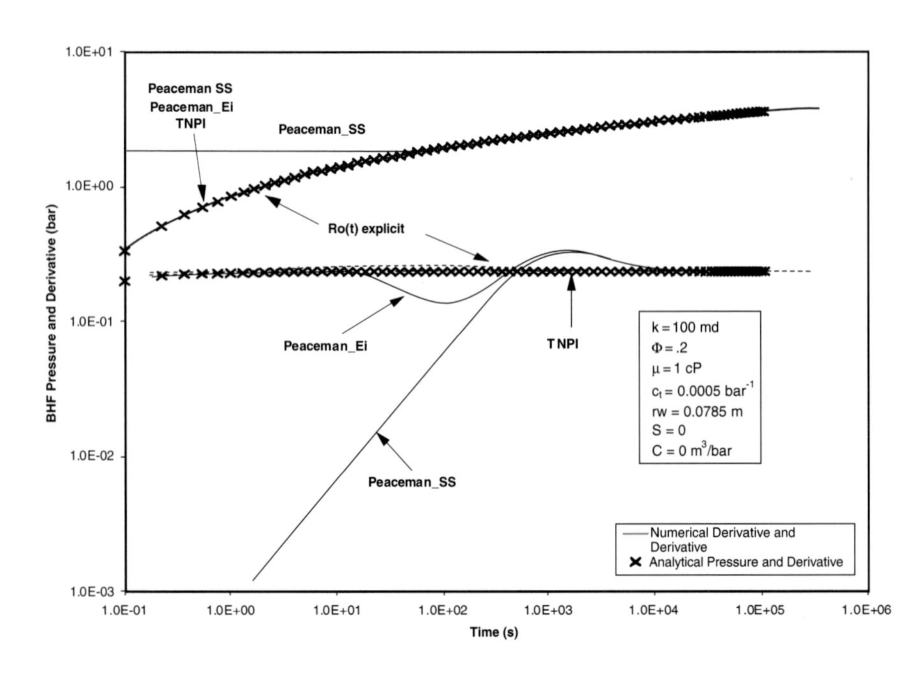

Figure 7. Summary of NPI (numerical productivity index) transient corrections (Reference Case 1, 10 m × 10 m grid-blocks).

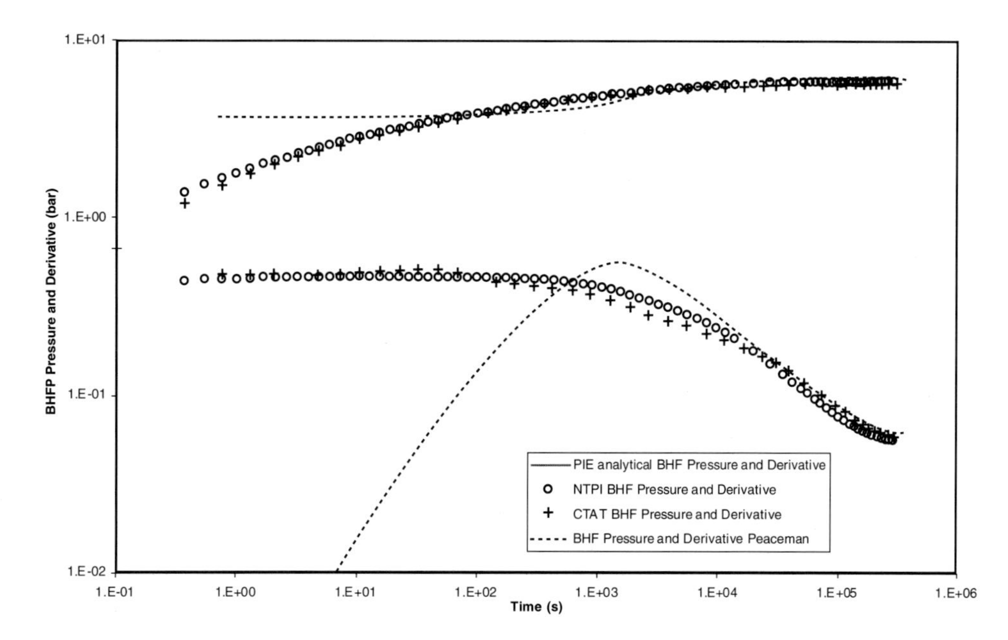

Figure 8. Homogeneous reservoir. Partial penetration (Reference Case 3, 20 m × 20 m grid-blocks).

CORRECTED TRANSMISSIBILITY AND ACCUMULATION TERMS (CTAT)

Derivation of the Corrected Transmisibility and Accumulation Terms

The transmissibility multiplier, which is based on a steady-state flow regime, was also improved to account for the short time period when the well flow rate is essentially produced from the well-block by the volume storage; that is, by the term

$$c_t\phi\int_{v_0}\frac{\partial p}{\partial t}dV \tag{15}$$

However, because p_0 is expected to be the pressure corresponding to the radius r_0, the volume storage variation will be approximated by

$$c_t\phi V_0\frac{\partial p_0}{\partial t}=-c_t\phi V_0\frac{q\mu}{4\pi kh}\frac{e^{-\frac{r_0^2}{4kt}}}{t} \tag{16}$$

where the well-block pressure is assumed to be equal to

$$p_0=p_i+\frac{q\mu}{4\pi kh}Ei\left(-\frac{r_0^2}{4kt}\right) \tag{17}$$

so that the derivative

$$\frac{\partial p_0}{\partial t}=-\frac{q\mu}{4\pi kh}\frac{e^{-\frac{r_0^2}{4kt}}}{t} \tag{18}$$

is independent of the integration variables on V_0. This approximation is not accurate enough for the short time period. To improve this calculation, we introduce the accumulation correction coefficient for volume storage by determining the coefficient β, such that

$$-\beta c_t\phi V_0\frac{q\mu}{4\pi kh}\frac{e^{-\frac{r_0^2}{4kt}}}{t}=c_t\phi\int_{v_0}\frac{\partial p}{\partial t}dV=$$
$$-c_t\phi\frac{q\mu}{4\pi kh}\int_{v_0}\frac{e^{-\frac{r^2}{4kt}}}{t}dr \tag{19}$$

The result is as follows

$$\beta=\frac{1}{V_0}\int_{v_0}e^{-\frac{\left(r^2-r_2^0\right)}{4kt}}dr \tag{20}$$

Validation of the CTAT Correction

Reference Case 2: Homogeneous Limited Reservoir, Draw-Down Test, Full Penetration

The validity of the full CTAT correction was first checked with Reference Case 2 (this case is the same as Reference Case 1, but it is for a limited reservoir, as shown on Tables 1 and 2). Figure 12 shows that although the early time improvement obtained with the CTAT correction is not as good as the full TNPI correction, it is still better than the pure Peaceman Ei correction, and it is close to the one obtained with the Peaceman explicit transient correction.

Reference Case 3: Homogeneous Reservoir, Draw-Down Test, Partial Penetration

Reference Case 3 simulation confirms the previous conclusions for early in the test, as illustrated on Figure 12: TNPI gives better results than CTAT. On the contrary, the CTAT correction works better than the TNPI correction for late in the test. This differ-

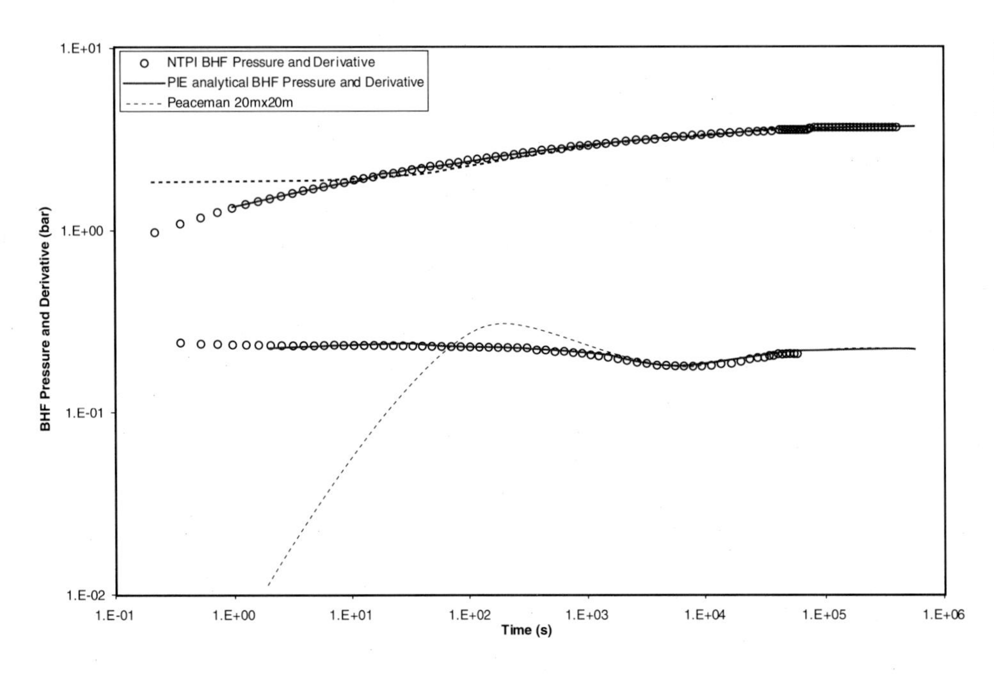

Figure 9. Three layers. Partial penetration (Reference Case 4. 20 m × 20 m grid-blocks).

ence is the result of the use of the transmissivity multipliers, which have proven to be more accurate than the conventional Peaceman approach in steady-state flow conditions (i.e., for late test times), as demonstrated by Ding (1995).

CONCLUSIONS

The Peaceman approach can be used for numerical well test simulation as long as the following relation between the petrophysical properties, the time, and the grid-block size is fulfilled

$$t \geq 2.6 \Delta L^2 \times \frac{\phi \mu c_t}{k}$$

This relation is true if and only if the well is located at the center of a local 5 × 5 square grid system within which the reservoir is isotropic and homogeneous.

For the simulation of early in the test, two approaches have been proposed.

- The transient numerical productivity index (TNPI) correction, which has been validated on several reference cases and has proven to give far better results than the conventional Peaceman approach.
- The correction transmissibility and accumulation terms (CTAT) is not as good as the TNPI approach, but is still far better than the conventional Peaceman approach; however, the use of transmissibility multipliers gives a better simulation of late test times.

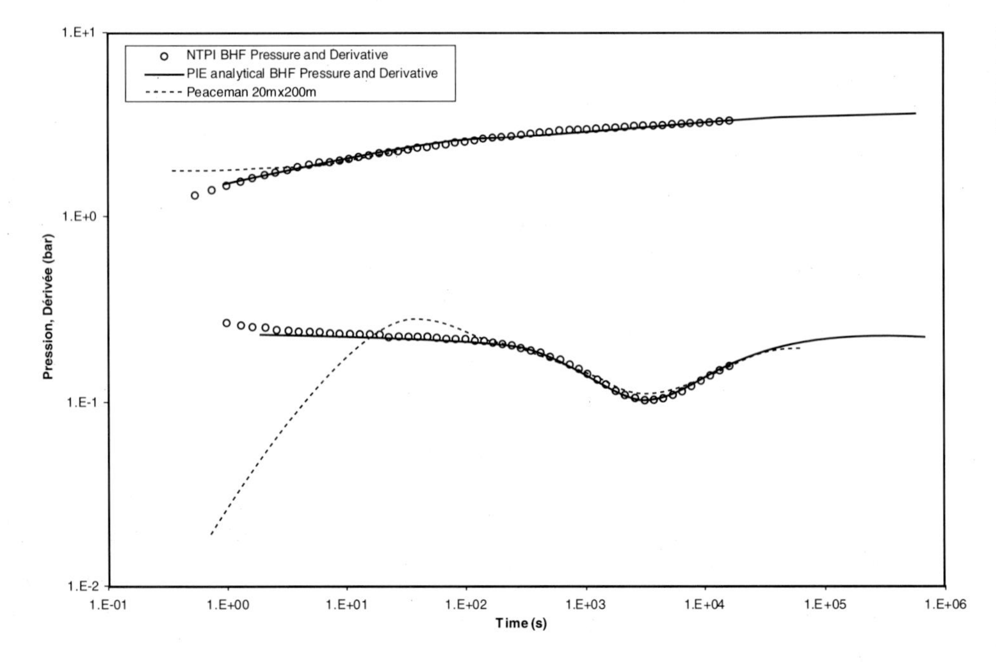

Figure 10. Three layers. Full penetration. No wellbore storage. (Reference Case 5. 20 m × 20 m grid-blocks).

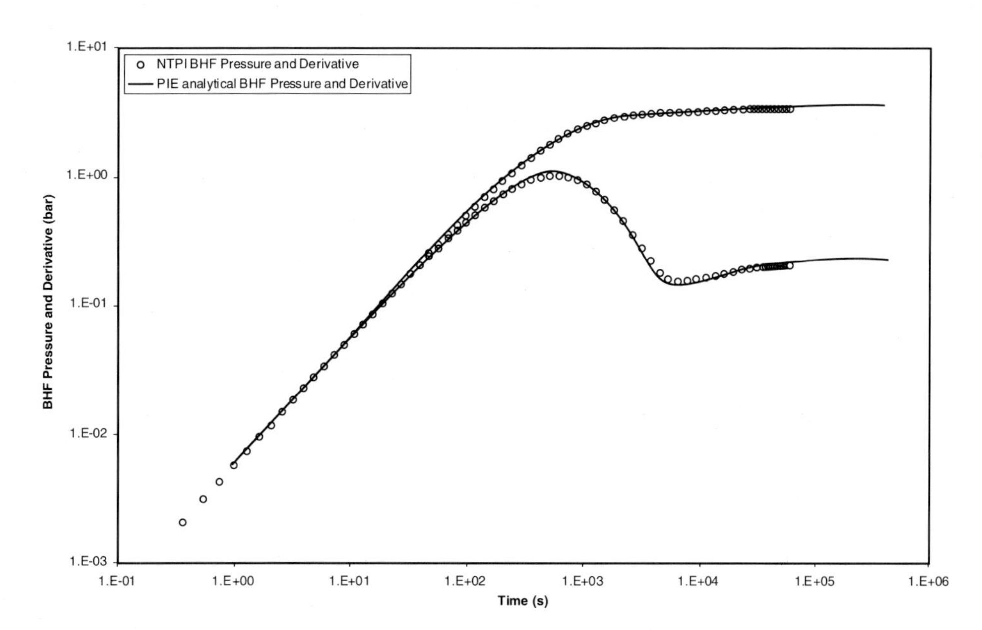

Figure 11. Three layers. Full penetration. wellbore storage (Reference Case 5. 20 m × 20 m grid-blocks).

For both approaches, the simulation of multiple rate and buildup tests must be done using the rate convolution of a constant rate simulation. This approach is valid only when the permeability, porosity, and compressibility are not pressure or fluid-saturation dependent.

This work indicates that the combination of the TNPI approach and the transmissibility multipliers would probably give both better early and late times than the previous approaches for the numerical simulation of well tests.

NOMENCLATURE

c_t = total compressibility, Lt^2/m, 1/bars
C = well bore storage, m^3/bar
$\Delta x, \Delta y, \Delta z$ = grid-block dimension, L, m
h = reservoir thickness, L, m
k = permeability, L^2, md
NPI = numerical productivity index, L^4t/m
P = pressure, m/Lt^2, bar
q = sink term, L^3/t
Q = well flow rate, L^3/t, m^3/d
r_w = well radius, L, cm
r = Peaceman equivalent radius, L, cm
S = well skin
t = time, t
T = transmissibitity, L^2/t
V_i = grid-block volume, L^3, m^3
β = accumulation multiplier
ϕ = porosity, fraction
μ = fluid viscosity, m/Lt, cp
ρ = fluid density, m/L^3, g/cm^3

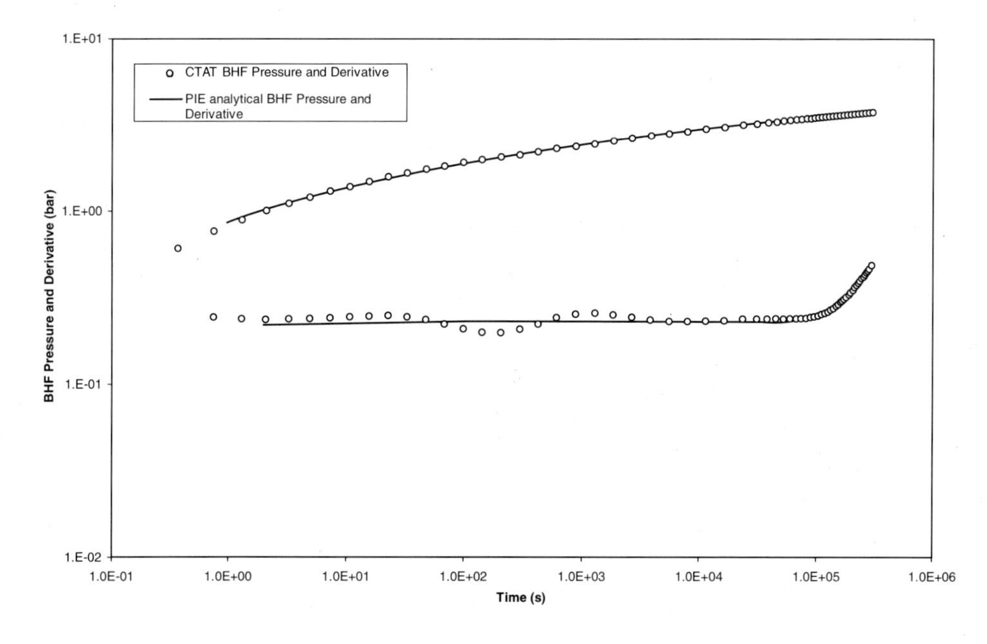

Figure 12. Homogeneous reservoir. CTAT correction (Reference Case 2. 20 m × 20 m grid-blocks).

ACKNOWLEDGMENTS

This work has been done within the framework of an ARTEP project sponsored by ELF, the Insitute Français du Pétrole, and TOTAL. The authors would like to thank both ARTEP and the Insitute Français du Pétrole for their permission to publish this work. They would also like to thank L. Piacentino for his help.

REFERENCES CITED

BLANC, G., et al., Numerical Well Tests Simulations in Heterogeneous Reservoirs. Poster presented at AAPG, Nice 95.

BLANC, G., et al, 1996, Building Geostatistical Models Constrained by Dynamic Data." SPE paper 35478 presented at the SPE/NPF European Conference held in Stavanger, Norway, 16-17 April 1996, p. 19.

DING Y., RENARD, G., and WEILL, L., Representation of Wells in Numerical Reservoir Simulation, paper SPE 29123, presented at the 13th SPE Symposium on Reservoir Simulation, San Antonio, Feb. *12-15,* 1995, p. 303–314.

DING, Y., and RENARD, G., A New Representation of Wells in Numerical Reservoir Simulation, SPERE, May 1994, p. 140.

DING, Y., A Generalized 3D Well Model for Reservoir Simulation, paper SPE 30724, presented at the SPE Annual Technical Conference held in Dallas, U.S.A., Oct. 22-25, 1995, p. 227.

EARLOUGHER, R. C., Jr., 1977. Advances in Well Test Analysis. Monograph Volume 5 of the Henry L. Doherty Series. Published by Society of Petroleum Engineers of AIME, New York, Dallas.

HORNE, R., 1997, Modern Well Test Analysis: A Computer-Aided Approach, Second Edition, Petroway Inc., Palo Alto, CA.

LEE, J., 1982, Well Testing. SPE Textbook Series, vol. 1. Published by Society of Petroleum Engineers of AIME, New York, Dallas.

PEACEMAN, D. W., 1978, Interpretation of Well-Block Pressures in Numerical Reservoir Simulation: Transactions of AIME, 1978, p. 253.

PEACEMAN, D. W., 1983, Interpretation of Well-Block Pressures in Numerical Reservoir Simulation with Non Square Grid-Blocks and Anisotropic Permeability. SPEJ (June), p. 531.

PEACEMAN, D. W., 1990, Interpretation of Well-Block Pressures in Numerical Reservoir Simulation, part 3—Off Center and Multiple Wells Within a Well Block, SPERE, May 1990, p. 1990.

PEACEMAN, D. W., 1995, A New Method for Representing Multiple Wells with Arbitrary Rates in Numerical Reservoir Simulation, paper SPE 29120, presented at the 13th SPE Symposium on Reservoir Simulation, San Antonio, Feb. 12-15, 1995, p. 257–272.

PIE—Well Test Analysis, User's Manual, S.M.C. Limited, June 1994.

SPE Reprint Series No. 9, 1967, Pressure Analysis Methods. Published by Society of Petroleum Engineers, Dallas, Texas.

SPE Reprint Series No. 14, 1980, Pressure Transient Testing Methods. Published by Society of Petroleum Engineers, Dallas, Texas.

THEIS, C. V., 1935, The relationship Between the Lowering of the Piezometric Surface and the Rate and Duration of Discharge Well Using Ground-Water Storage. Trans. Amer. Geophysical Union, vol. 16, p. 515-524.

Spinler, E. A., and B. A. Baldwin, A direct method for determining complete positive and negative capillary pressure curves for reservoir rock using the centrifuge, 1999, *in* R. Schatzinger and J. Jordan, eds., Reservoir Characterization-Recent Advances, AAPG Memoir 71, p. 175–178.

Chapter 12

A Direct Method for Determining Complete Positive and Negative Capillary Pressure Curves for Reservoir Rock Using the Centrifuge

E. A. Spinler
B. A. Baldwin
Phillips Petroleum Company
Bartlesville, Oklahoma, U.S.A.

ABSTRACT

A method is developed for direct experimental determination of capillary pressure curves from saturation distributions produced while centrifuging fluids in a rock plug. A free-water level is positioned along the length of the plugs to enable simultaneous determination of both positive and negative capillary pressures. Octadecane, as the oil phase, is solidified by temperature reduction while centrifuging to prevent fluid redistribution upon removal from the centrifuge. The water saturation is then measured via magnetic resonance imaging. The saturation profile within the plug and the calculation of pressures for each point of the saturation profile allow for a complete capillary pressure curve to be determined from one experiment. Centrifuging under oil with a free-water level into a 100% water saturated plug results in the development of a primary drainage capillary pressure curve. Centrifuging similarly at initial water saturation in the plug results in the development of an imbibition capillary pressure curve. Examples of these measurements are presented for Berea sandstone and chalk rocks.

BACKGROUND

Knowledge of capillary pressure for each specific rock/oil/water combination is important for predicting potential hydrocarbon recovery from a reservoir. Capillary pressure is a measure of the interaction between fluids and the rock pore surface. The strength of the interaction varies with the fluid saturations, the interfacial tension between the fluids, the pore structure, and the wettability of the pore surfaces. A measure of the capillary pressure can be calculated from the force exerted by the density difference between retained fluids at each height above or below the free-water level. Knowing the fluid saturations as a function of the height above and below the free-water level permits the determination of a capillary pressure curve for any given reservoir rock/fluid system; however, in practice this has been difficult, of uncertain accuracy, or very time-consuming to experimentally determine.

Capillary pressure curves are typically determined by either mercury intrusion, porous plate/membrane, or centrifuge methods. Mercury intrusion, although rapid, provides questionable results due to the use of mercury in a vacuum to mimic water/oil behavior. Its methodology limits the technique to primary drainage and possible positive imbibition capillary pressures. Porous plate/membrane methods can generate all the capillary pressure curves, but to reach an apparent equilibrium saturation can take days to months per pressure point for 5 to 8 data points. This delays the availability of the results and limits the number of tests that can be made. The lengthy experimental time also increases the chance for mechanical failure. The centrifuge method is normally used only to determine drainage or negative imbibition curves. It takes days to months to complete a test. The limitation of most centrifuge methods is that they only provide an indirect, or assumed/calculated, measure of saturation at the inlet face of a rock plug based on the amount of fluid expelled from the rock. Numerous methods over the past 50 yr have been proposed for approximating the inlet saturation from centrifuge effluent volumes, but in every case, the model chosen influences the results.

A capillary pressure curve that is determined from the measured oil/water saturation profile under a known pressure gradient as measured from a free-water level would ideally mimic the reservoir. The methodology as proposed herein provides positive and negative capillary pressures for both drainage and imbibition. Furthermore, the method appears to be accurate, rapid, and robust.

METHODOLOGY AND RESULTS

The method under development directly measures the oil/water saturation distribution in a rock plug that has been established via a known fluid differential pressure by centrifuging. Direct measurements of the saturations are possible by a number of different techniques, but for this study magnetic resonance imaging was used because it provides high spatial resolution. Previously, Baldwin and Spinler (1996) had discussed some of the different methods by which magnetic resonance images of the saturations within rock plugs could be measured. All saturations for this work were determined by imaging the liquid water phase while the oil phase was solid and does not image. The saturation profile was established and locked in place by cooling while in the centrifuge. The measurement of the complete oil/water saturation profile determined with more than a hundred measurements within a rock plug has an advantage over other typical techniques that measure saturations one point at a time.

An additional advantage is that both the positive and negative portions of the capillary pressure curve can be measured simultaneously by positioning the free-water level along the length of the plug, as shown in Figure 1. Capillary pressure is zero at the free-water level. Drainage or imbibition is achieved by centrifuging the plug under hydrocarbon or water when starting the plug at an appropriate saturation state. The sequence of the centrifuge steps is important because it determines the direction of fluid flow and possible hysteresis. Capillary pressure as a function of position in the plug and

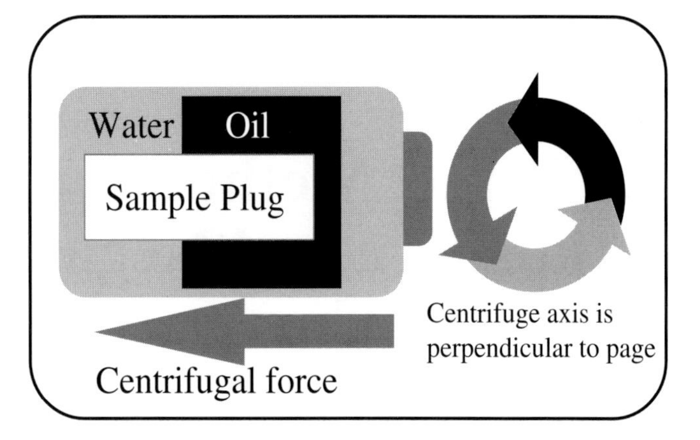

Figure 1. Schematic of plug with free-water level and large cell volume in centrifuge.

the achievable pressure range is calculated from the speed of the centrifuge, the height above the free-water level, the fluid density difference, and the sample length.

Water and octadecane, which has a melting point of about 27°C, were used. Other hydrocarbons that are solid at the ambient temperature in the magnetic resonance imager would also work. When one phase is frozen, the relaxation time in the magnetic resonance imager is so short that it is not imaged. Thus, in this system only the water was imaged. Because the hydrocarbon was frozen in place while centrifuging, redistribution of the water in the plug when handled at normal ambient temperature will not occur. The small volume change created by the contraction of the hydrocarbon during freezing can be easily corrected. One disadvantage is that the water as the wetting phase had a relaxation time that was dependent on the size of the pore spaces it occupied. In chalks with a relatively narrow pore size distribution, this was not a problem; however, sandstones with a broad pore size distribution had relaxation times for the lower water saturations that were significantly shorter than those for high water saturation that consequently affected the image intensity. A correction (Baldwin and Yananashi, 1991) was calculated for this effect in the sandstones by using T1 mapping where the relaxation time was measured on a pixel-by-pixel basis and extrapolated back to zero relaxation time to minimize the relaxation effect on intensity. This procedure produced a porosity map where intensity was proportional to the water saturation.

Laboratory experiments for this study consisted of preparing both a Berea Sandstone plug and a reservoir chalk plug. These plugs were selected because they have a great disparity in rock properties (Table 1), and it was intended to evaluate the methodology for both. The reservoir chalk plug was previously cleaned by extraction with mild solvents. No cleaning was necessary for the Berea plug.

Each plug was initially saturated with 100% of the water phase. The experimental work with each plug was done separately, but the steps were basically the same. Once placed in their centrifuge holders, water and octadecane (warmed to slightly above its melting point)

Table 1. Petrophysical Properties

Plug #	Type	Length (cm)	Diameter (cm)	Porosity (fraction)	Permeability (md)
bc	berea	4.68	2.54	0.21	~600
1052b	chalk	3.94	2.54	0.33	~5

were added. No oil imbibition was expected to occur, nor was it observed, since both samples were water-wet. The final free-water level height was adjusted by adding the appropriate amounts of each fluid. This position was expected to change because one fluid will be expelled from the rock during the test, while the other fluid will enter. Movement of the free-water level was minimized by having the volume of bulk fluids around the plugs significantly greater than the pore volume of the rock plugs as indicated in Figure 1. Centrifuging at a temperature above the melting point of octadecane was conducted for an appropriate amount of time, 1-day intervals for the Berea and 3-day intervals for the chalk. At the end of the centrifuging time, but while still centrifuging, the sample was cooled to solidify the oil but not the water. When the oil phase in the sample was frozen approximately 4 hours, it was removed from the centrifuge and imaged for saturation. The centrifuge procedure was repeated until the saturation profiles in the plugs were observed to be stabilized. The height of the free-water level was locked in place by the frozen oil phase and becomes part of the image. Distances from the free-water level for each pixel of the intensity image were determined, and local pressures were calculated from standard centrifuge equations. Local pressures determine the capillary pressure at any point on the image from the free-water level. Capillary pressure curves as measured by this method for primary drainage of Berea sandstone and reservoir chalk are shown in Figures 2 and 3. The scatter in capillary pressure data for the Berea reflects the less uniform porosity distribution found in the sandstone as compared to the chalk. Note, however, that for these tests the image represented only a 4-mm slice through each plug. Averaging or imaging more of the plug by magnetic resonance imaging can further smooth the capillary pressure curves.

Following the determination of the primary drainage capillary pressure curve, the plugs were centrifuged under 100% oil phase and inverted as necessary to drive the plugs to relatively uniform initial water saturation. The plug at initial water saturation was then centrifuged with a free-water level near the center of the plug. Again centrifuging of the plugs continued until the saturation profiles were observed to be stable. The resultant primary imbibition capillary pressure curves are illustrated in Figures 4 and 5 for the Berea Sandstone plug and the reservoir chalk core plug, respectively.

Centrifuging under 100% oil phase established a relatively uniform residual oil saturation before the development of the secondary drainage capillary pressure curves. This can be repeated, as well as the imbibition step, to obtain a set of hysteresis capillary pressure

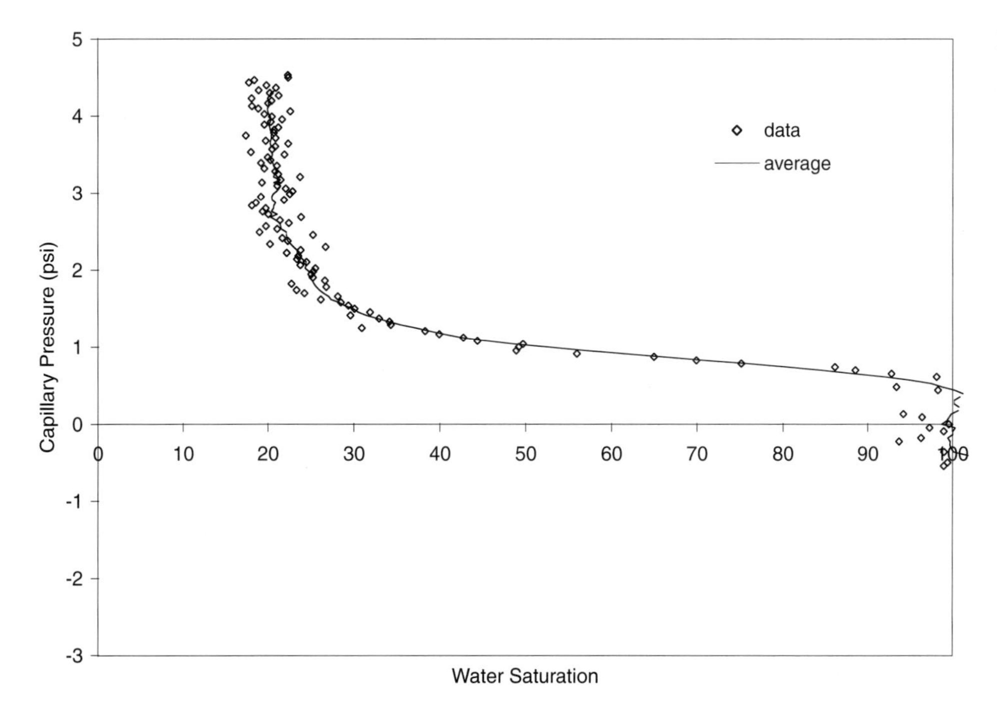

Figure 2. Primary drainage capillary pressure curve for Berea plug bc.

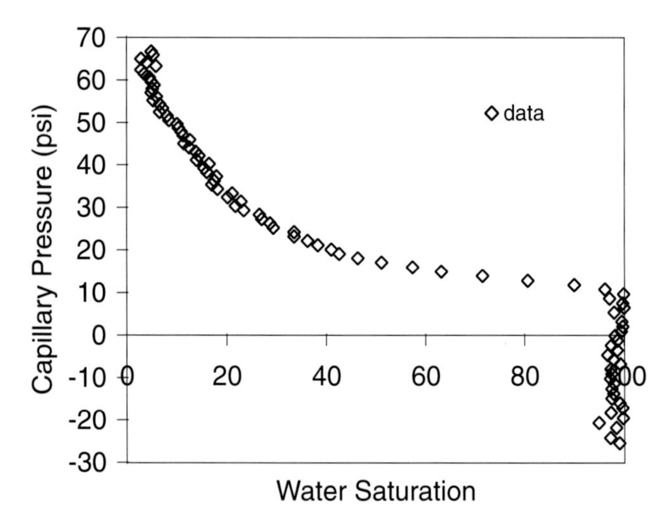

Figure 3. Primary drainage capillary pressure curve for chalk plug 1052b.

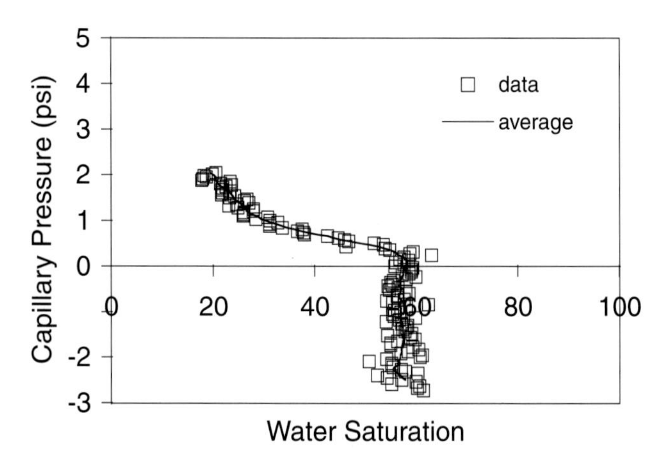

Figure 4. Imbibition capillary pressure curve for Berea plug bc.

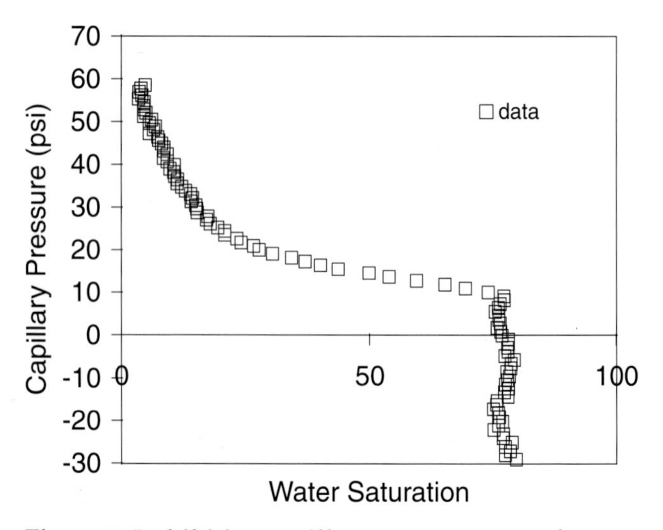

Figure 5. Imbibition capillary pressure curve for chalk plug 1052b.

curves. Secondary drainage curves for these plugs will be developed in future papers on this methodology.

The time needed to complete each step of the above test depended on how rapidly the fluids reached their equilibrium saturation in the centrifuge. This was largely determined by the permeability of the rock. This was especially true for low-permeability chalk, which had to be centrifuged up to 12 days to obtain and verify a stabilized saturation profile, compared to up to 2 days for the Berea Sandstone. The total experimental time to generate equilibrium primary drainage and primary imbibition capillary pressure curves in this study took approximately 1 week for the Berea Sandstone plug and approximately 1 month for the low-permeability chalk plug.

CONCLUSIONS

(1) Direct measurement of the saturation distribution in centrifuged rock plugs by magnetic resonance imaging or other means eliminates the need for assumed or indirect determination of fluid saturations.

(2) Placement of the free-water level along the length of rock plugs allows for the simultaneous development of both the positive and negative capillary pressures for drainage and imbibition.

(3) Freezing of the oil phase during centrifuging allows for handling and determination of the water saturation in rock plugs without fluid redistribution.

(4) The combination of centrifuging for pressure and direct measurement of the resulting saturation distribution can provide a rapid, accurate, and robust laboratory method for determining capillary pressure curves in porous media.

ACKNOWLEDGMENTS

The authors thank Phillips Petroleum Company for permission to publish this extended abstract and thank D. M. Chancellor for his mechanical assistance in building equipment for this work, J. C. Stevens for performing the centrifuging, and R. L. King for assistance with magnetic resonance imaging.

REFERENCES CITED

Baldwin, B. A., and Yananashi, W. S., Sept.–Oct. 1991, Capillary-pressure determinations from NMR images of centrifuged core plugs: Berea Sandstone: The Log Analyst, p. 550–556.

Baldwin, B. A., and Spinler, E. A., 1996, A direct method for simultaneously determining positive and negative capillary pressure curves in reservoir rock: 4th International Symposium on Evaluation of Reservoir Wettability and Its Effect on Oil Recovery Proceedings, p. 347–353.

Kasap, E., M. Altunbay, and D. Georgi, Flow units from integrated WFT and NMR data, 1999, *in* R. Schatzinger and J. Jordan, eds., Reservoir Characterization-Recent Advances, AAPG Memoir 71, p. 179–190.

Chapter 13

Flow Units from Integrated WFT and NMR Data

Ekrem Kasap[1]
Mehmet Altunbay[2]
Dan Georgi[2]
Western Atlas Logging Services[2]
Houston, Texas, U.S.A

ABSTRACT

Reliable and continuous permeability profiles are vital as both hard and soft data required for delineating reservoir architecture. These profiles can improve the vertical resolution of seismic data, well-to-well stratigraphic correlations, and kriging between the well locations. In conditional simulations, permeability profiles are imposed as the conditioning data. Variograms, covariance functions, and other geostatistical indicators are more reliable when based on good-quality permeability data.

Nuclear magnetic resonance (NMR) logging and wireline formation tests (WFTs) separately generate a wealth of information, and their synthesis extends the value of this information further by providing continuous and accurate permeability profiles without increasing the cost. NMR and WFT data present a unique combination because WFTs provide discrete, in situ permeability based on fluid flow, and NMR responds to the fluids in the pore space and yields effective porosity, pore-size distribution, bound and moveable fluid saturations, and permeability. The NMR permeability is derived from the T_2 distribution data. Several equations have been proposed to transform T_2 data to permeability. Regardless of the transform model used, the NMR-derived permeabilities depend on interpretation parameters that may be rock specific.

The objective of this study is to integrate WFT permeabilities with NMR-derived, T_2 distribution-based permeabilities and thereby arrive at core-quality, continuously measured permeability profiles. We outlined the procedures to integrate NMR and WFT data and applied the procedure to a field case. Finally, this study advocates the use of hydraulic unit concepts to extend the WFT-NMR–derived, core-quality permeabilities to uncored intervals or uncored wells.

[1] Now with GeoQuest Reservoir Technologies–Schlumberger.

[2] Now Baker Atlas.

INTRODUCTION

Over the years traditional reservoir characterization techniques have evolved into a multidisciplinary process. Geostatistical techniques provide a framework for integrating all available engineering and geoscience data over multiple scales while honoring the petrophysical well data and geological and sedimentological models.

Well logs provide direct measures of reservoir formation properties and are a vital source for hard and soft data required for delineating reservoir architecture. Logs are used to improve the vertical resolution of seismic data (Araktingi et al., 1993; Yang et al., 1995; Yang and Gao, 1995; Schultz et al., 1994, Hand et al., 1994), well-to-well stratigraphic correlation (Johann et al., 1996; Reedy and Pepper, 1996), and kriging between the well locations. In conditional simulations, well-log porosity profiles are imposed as the conditioning data and the sand-to-shale ratios from well logs are maintained throughout the relevant vertical cross sections. Well-log–measured heterogeneity in a vertical cross section is used to generate variograms, covariance functions, and other geostatistical indicators within a single well or between well pairs. Corbett et al. (1995) incorporated wireline formation pressure test data into a reservoir characterization study to identify connectivity between the adjacent zones.

Recently, great effort has been dedicated to the integration of well logs into seismic data (Araktingi et al., 1993; Yang et al., 1995; Yang and Gao, 1995; Schultz et al., 1994; Hand et al., 1994). Conventionally, seismic data have been used to delineate reservoir structure (lateral extent), but the vertical resolution is limited. Yang et al. (1995) pointed out that generally it is difficult to identify beds thinner than 25 ft (7.6 m) from 3-D seismic signals. Although seismic data are routinely and effectively used to estimate the structure of the reservoir bodies, this data typically plays no role in the essential task of estimating rock properties, which are measured or computed from well logs. In the presence of 3-D seismic surveys and logged wells, the simultaneous analysis of seismic-attribute data and borehole data often leads to better estimates of property distributions in comparison with estimates generated from either seismic or well data alone (Araktingi et al., 1993; Yang et al., 1995; Yang and Gao, 1995; Schultz et al., 1994, Hand et al., 1994; Johann et al., 1996; Reedy and Pepper, 1996; Corbett et al., 1995).

Porosity and permeability are the key reservoir properties needed for proper reservoir description and simulation. Porosity is routinely available from wireline log data and, in conjunction with 3-D seismic data, can be used to generate excellent reservoir porosity models. Until recently, cores were the only source of permeability data measured on a small enough scale to provide critical reservoir heterogeneity data. Usually permeability is derived from core-based permeability-porosity cross-plots; however, such an approach is flawed, as the porosity-permeability correlations are often poor and core data may be sparse and expensive to obtain.

Newly developed nuclear magnetic resonance (NMR) logging techniques do provide vertically continuous pore-size distributions, but do not measure permeability directly. Transient pressure tests are the only direct, in situ measurement that can yield formation permeability. Well test volume of investigation is on the largest possible scale; however, all vertical variations in permeability are averaged. Wireline formation tests (WFTs) access a limited volume of the reservoir and are well suited for characterizing reservoir heterogeneities. They also measure on a scale that closely matches the measurement scale of NMR logs.

The objective of this study is to integrate WFT-calculated permeabilities with NMR-derived permeabilities and thereby arrive at core-quality, continuously measured permeability profiles without increasing the cost. NMR and WFT data present a unique combination because WFTs provide discrete, in situ permeability measured through fluid flow, and NMR logs provide continuous T_2 distributions. This study outlines a procedure to integrate NMR and WFT data and illustrates its application to a field study. The study extends the hydraulic unit concepts of Amaefule et al. (1993) to the WFT-NMR–derived, core-quality permeabilities to uncored intervals or uncored wells.

WIRELINE FORMATION TESTS

When a WFT probe is set against the formation and a measured volume of fluid is withdrawn, a wireline formation test is initiated. Pressure changes are continuously recorded throughout the test. Decreasing pressure is sometimes followed by a stabilized flow period when the drawdown period is extended and permeability of the formation is high enough to supply fluid at a rate equal to that required by the pump withdrawal. The buildup period starts when the fluid withdrawal stops and ends when the system pressure reaches the formation fluid pressure.

Figure 1 shows a WFT test repeated three times. This particular test was performed at a depth of 8775 ft (2676 m). Approximately thirty tests at ten depths were performed. Permeability, undisturbed formation

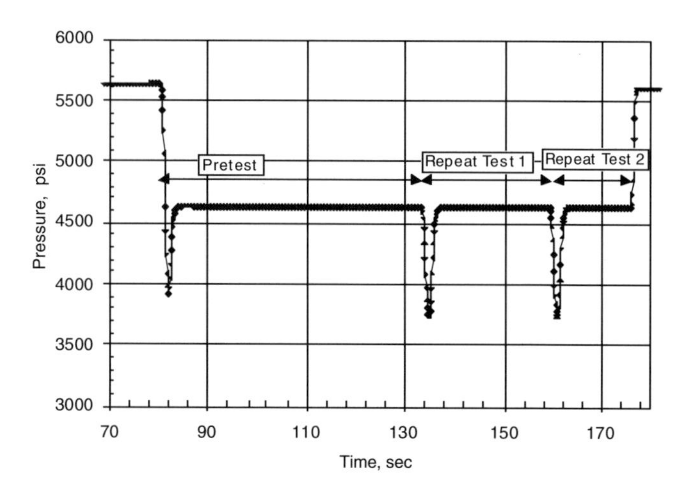

Figure 1—Pressure versus time from a WFT showing a pretest and two repeated tests.

pressure (P^*), and fluid compressibility can be extracted from transient-measured pressures and drawdown rates. Shown in Figure 2 are the calculated P^* values at ten vertical depths, which shows an excellent correlation between the depth and formation pressure, indicating a constant pressure gradient. The calculated slope of –3.62 translates to an in situ API gravity of 90.8, which may be indicative of solution gas in the oil under the reservoir conditions. Alternatively, the reservoir oil might also contain large quantities of light hydrocarbon components.

In addition to P^*, WFTs can provide a direct measure of near-wellbore permeability. Although WFTs are short (seconds to a minute), tests cannot feasibly be conducted continuously because the time to set the tool and the time for the pressure to stabilize before ending the test are too long to permit "continuous" permeability profiling. Figure 3 shows ten permeability values calculated at ten vertical depths. The figure indicates that ten measurements of permeability are not sufficient to

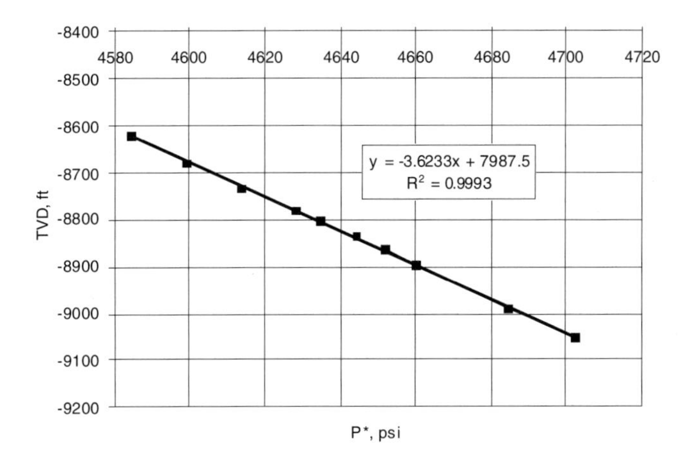

Figure 2—Depth versus formation pressure obtained from a WFT. The calculated pressure gradient corresponds to 90 API gravity formation fluid. TVD = total vertical depth.

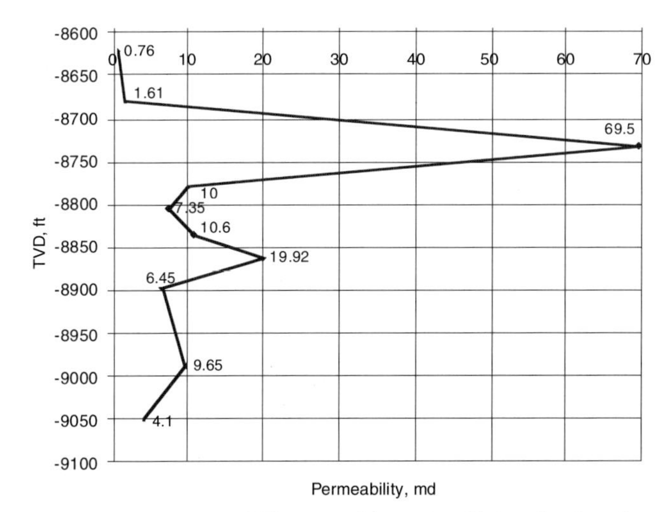

Figure 3—Permeability profile at Well A, obtained with repeated WFTs.

delineate reservoir heterogeneity, although ten pressure values are quite satisfactory for calculating reservoir fluid pressure gradient and in situ API gravity; however, even though WFT-measured permeabilities are too sparse to produce a continuous profile, they are currently the most reliable in situ measured permeability obtainable on a small enough scale to permit the identification of local heterogeneity.

In the following section, we discuss the calculation of P^* and permeability from WFTs. New concepts and analysis techniques, combined with 3-D numerical studies, have recently been reported in the literature (Kasap et al., 1996a, b; Huang et al., 1996; Samaha et al., 1996; Waid et al., 1991; Proett and Chin, 1996), and 3-D numerical simulations have also contributed to the diagnosis of WFT-related problems and the improved analysis of WFT data. We first review the formation rate analysis (FRA) technique, a new technique that combines drawdown and buildup data analysis.

Formation Rate Analysis

Kasap (1998) proposed the FRA technique for the entire WFT to obtain permeability and P*. The FRA technique combines the drawdown and buildup data into a single plot where both the drawdown and buildup periods are well represented by straight lines with identical slopes. The geometric factor (G_o) concept (Goggin et al., 1988) is the foundation of the FRA technique, from which we obtain both P* and permeability. G_o is calculated for modified hemispherical-flow geometry, which does not require a cavity in the formation nor a probe seal of infinite extent. In Darcy's equation with this modified hemispherical geometry for isothermal, steady-state flow of a liquid, the dimensionless G_o replaces the factor 2π that would apply for true semi-infinite, hemispherical-flow geometry.

During the drawdown period, fluid will be withdrawn at a rate of q_{dd}. The accumulation rate, q_{ac}, is the difference between the formation flow rate, q_f, expressed with Darcy's equation, and the drawdown rate of the pump. The pressure-transient theory for slightly compressible liquids can be derived from the definition of isothermal compressibility:

$$q_{ac} = C_t V_{sys} \frac{\partial P(t)}{\partial t} \qquad (1)$$

The rate q_{ac}, for all practical purposes, is equal to the difference between q_f, the volumetric rate flowing through the rock adjacent to the packer probe (Darcy flux), and q_{dd}, the drawdown rate of the pump. Thus, we can rewrite the Darcy equation and account for the WFT drawdown, stabilized flow, and buildup data:

$$P(t) + P^* - \left(\frac{\mu}{kG_o r_i}\right)\left(C_t V_{sys} \frac{\partial P(t)}{\partial t} + q_{dd}\right) \qquad (2)$$

Note that the terms within the last parentheses in equation 2 correspond to the accumulation and drawdown rates, respectively. The two terms act against each other, and their difference, in essence, is the flow rate from the formation. Equation 2 provides a means to estimate formation permeability and P^* from drawdown and buildup data. A plot of $P(t)$ versus formation rate, which is given in equation 2 as

$$q_f = \left(C_t V_{sys} \frac{\partial P(t)}{\partial t} + q_{dd} \right) \quad (3)$$

should approach to a straight line with a negative slope and intercept P^* at the $P(t)$ axis. The absolute value of the slope, m, is

$$m = \frac{\mu}{k G_o r_i}$$

and permeability is calculated from

$$k = \frac{\mu}{G_o r_i m} \quad (4)$$

We note that equation 2 does not require a constant drawdown rate as long as q_{dd} is continuously recorded.

Theoretically, both drawdown and buildup should have the same slope and should go through the same straight line; however, during the early stages of the drawdown period when the flow is being established, and during the early stages of the buildup period when the flow is adjusting to the changing flow conditions, the actual pressure versus formation rate plot may deviate from the assumed pseudo-steady-state solution.

Field Tests

We illustrated FRA with two field examples collected with the reservoir characterization instrument (WFT). Test A6-4 was chosen because it shows a stabilized flow during the drawdown period, and test A7-2 was chosen because it is representative of many wireline formation tests.

Figure 4 shows pressure versus formation rate and its best-fit straight line. The slope of the straight line is directly related to permeability, and the intercept with the vertical axis is P^*. When a probe radius of 2.25 cm, a viscosity of 0.75 cp, and G_o of 4.27 are used, a permeability of 70.1 md is calculated.

This calculation can be performed in real time with every new pressure value measured by using weighted averages of formation rate. A plot of such a calculation is shown in Figure 5. Figure 5 indicates that calculated permeability stabilizes within a very short period of time and remains constant for the drawdown and buildup periods at around 69.5 md. Figure 5 also shows the corresponding formation rate, calculated by using equation 2, and indicates that during the test formation rate nearly stabilizes due to high permeability and low drawdown rate.

Formation rate analysis is also applied to test A7-2, (the first repeat test, Figure 1). The FRA plot, shown in

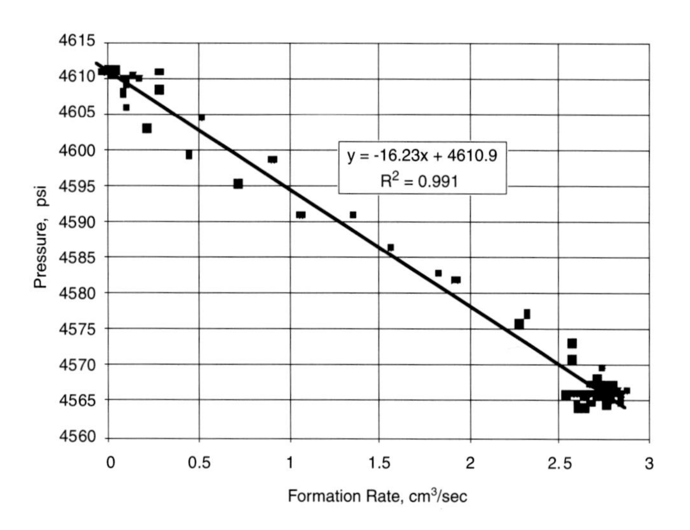

Figure 4—Pressure versus formation rate plot for test A6-4 and its best-fit straight line. Permeability of 70.1 md is calculated from the slope.

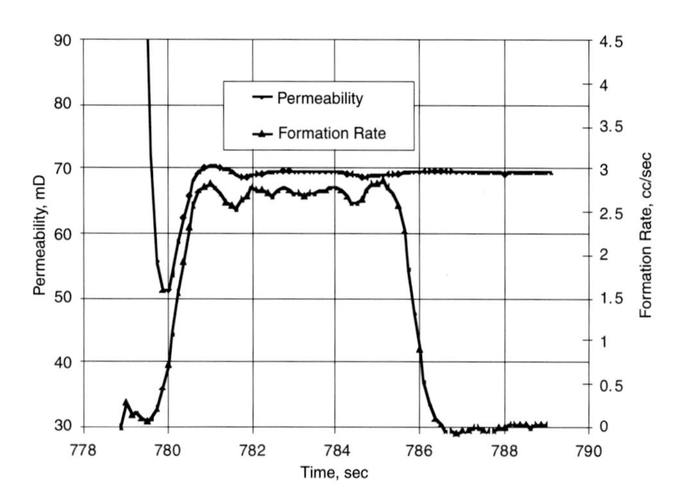

Figure 5—Continuously calculated formation rate and permeability from FRA technique for test A6-4.

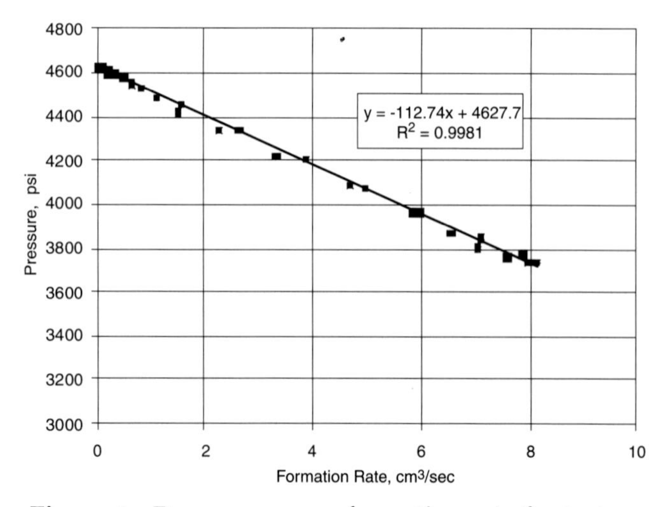

Figure 6—Pressure versus formation rate for test A7-2 and its best-fit straight line. Permeability of 10.1 md is calculated from the slope.

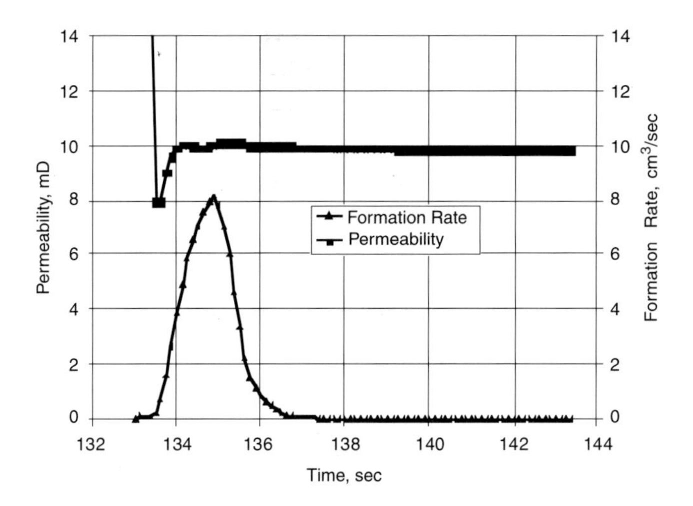

Figure 7—Continuously calculated formation rate and permeability from FRA technique for test A6-4.

Figure 6, indicates very little scatter in the data. Using the same viscosity, probe radius, and G_o as in the A6-4 test, the calculated permeability and P^*, based on the best-fit straight line, are 10.1 md and 4627.7 psi, respectively. For the formation rate calculations a system volume of 267 cm^3 and fluid compressibility of 6.8 × 10^{-6}/psi were used. Figure 7 shows the continuously calculated permeability and flow rate. The figure indicates that for this test the formation rate did not stabilize; nevertheless, the calculated permeability is not affected and reaches a constant value immediately after the start of the test and remains constant at that value throughout the drawdown and buildup periods.

Numerical Simulations

In this section, we use numerical simulations to verify the FRA results. A 3-D finite-element scheme applied to the diffusivity equation in Cartesian coordinates is used to calculate transient pressures due to the flow in the near-wellbore region of the formation. The details of the simulator are given in previous publications (Kasap et al., 1996a, b; Huang et al., 1996; Samaha et al., 1996).

Table 1. Numerical Simulation Parameters.

Parameter	Value
Flow line volume (cm^3)	267.1
Wellbore radius (cm)	15
Outer boundary, r_e (cm)	230
Probe radius (cm)	1.27
Pay thickness (cm)	242.25
Duration of drawdown (s)	5
Porosity	0.15
Fluid viscosity (cp)	0.75
Reservoir pressure, P^* (atm)	315.116
Total compressibility (/atm)	1×10^{-4}

Figure 8 shows the comparison of pressure versus time plot from the field data, and the numerical simulation data for the A6-4 test. We used measured drawdown rates at the pump and the information given in Table 1 for this simulation. The figure shows that the permeability of 69.5 md used in the simulation (FRA permeability is 70.1 md) satisfactorily matches the pressure response of the formation. The match between the field data and the numerical simulation data is excellent.

Numerical simulations were also conducted for the A7-2 test (Figure 1, the first repeat test) by using the FRA-obtained permeability and the continuously measured drawdown pump rate. Simulation results and field data are compared in Figure 9. The results indicate that the simulations run with the FRA permeability match the field data well .

These field test studies confirm that FRA analysis correctly predicts formation permeability and undisturbed formation pressure. WFTs can very accurately provide fluid type, density, and location, and formation permeability and pressure.

NMR LOGGING

The earliest industry workers (Brown and Gamson, 1959; Timur, 1969; Seevers, 1966) recognized the potential for NMR measurements to provide porosity, pore-size distribution, and permeability. Initial

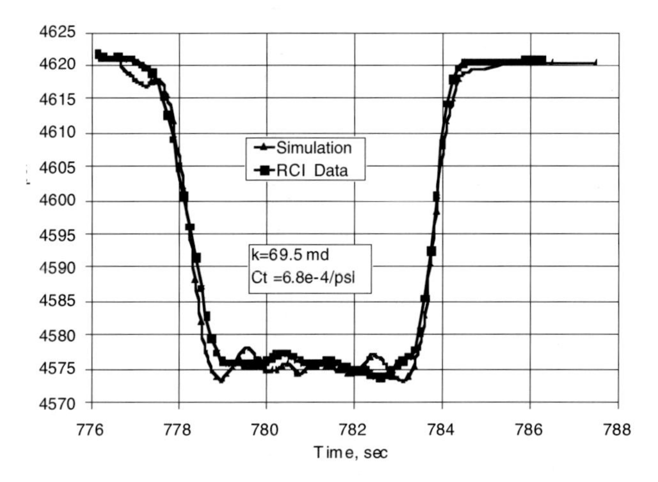

Figure 8—Comparison of the field-measured transient pressures with the numerical simulation pressures for test A6-4.

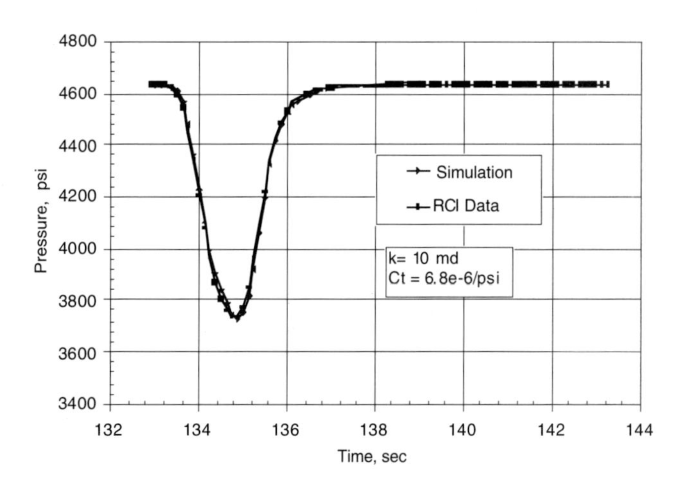

Figure 9—Comparison of the field-measured transient pressures with the numerical simulation pressures for test A7-2.

attempts to make downhole NMR measurements proved cumbersome; however, in the early 1990s a new magnet configuration (Miller et al., 1990) greatly simplified the measurement and significantly increased the data quality. Today in many areas of the world NMR logs are run routinely as part of the open-hole formation evaluation suite.

NMR logs provide the petrophysicist two types of data:

- Mineralogy-independent porosity
- T_2 distributions

The NMR measurement essentially counts the number of hydrogen protons aligned with an external magnetic field, and then monitors the dephasing of the protons' spins with time. The number of protons initially aligned with the external magnetic field is proportional to the porosity, whereas the change in amplitude of subsequent echoes, known as the transverse relaxation, is characterized by a quasi-exponential decay with a characteristic time, T_2 (Figure 10). The initial NMR signal amplitude depends only on the hydrogen proton density in the pore space because the measurement is insensitive to the hydrogen in the rock matrix (e.g., OH^- groups in clays) and is, thus, a mineralogy-independent measure of porosity.

To measure T_2 decay, a series of evenly spaced radio frequency (RF) pulses are used to rephase the transverse magnetization. These RF pulses essentially recall the NMR signal, which is then referred to as an echo. In a gradient magnetic field or in a heterogeneous magnetic field the apparent T_2 decay rate depends on the inter-echo spacing (*TE*); for liquids, diffusion effects are minimal for commonly used inter-echo times (e.g., 0.6 and 1.2 ms); however, they can be important for very light hydrocarbon liquids and gasses (Akkurt et al., 1995).

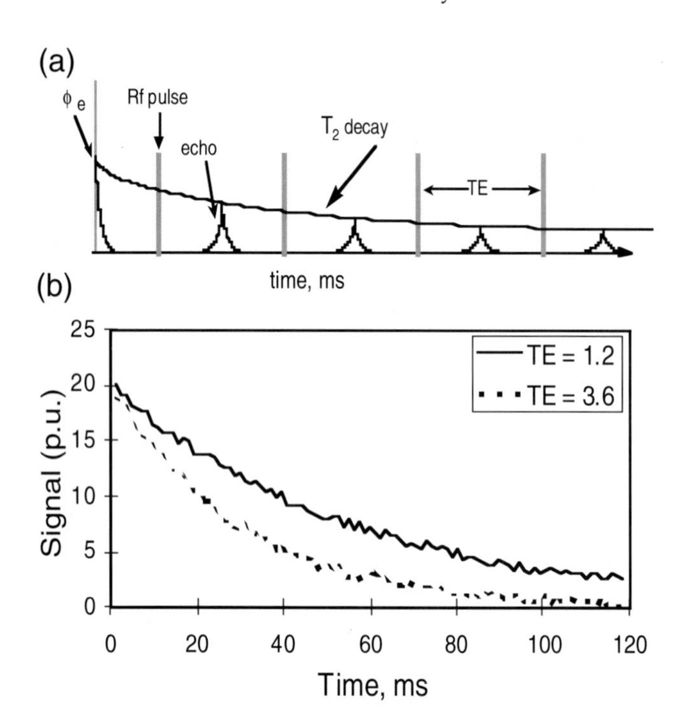

Figure 10—(a) Schematic of NMR data acquisition. (b) Typical NMR echo data.

The dominant factor on the T_2 relaxation process is controlled by nuclear magnetic interactions occurring on the pore wall. Including diffusion effects, the T_2 relaxation is well described by

$$\frac{1}{T_2} = \frac{1}{T_{2bulk}} + \frac{\lambda}{T_{2surf.}} \frac{S}{V} + \frac{1}{T_{2D}} \tag{5}$$

where T_{2bulk} = relaxation time of the bulk fluids in the pore space; $T_{2surf.}$ = relaxation time of the fluid in a few molecular layers next to the pore surface; λ = thickness of the surface fluid layer; S, V = pore surface and pore volume, respectively; T_{2D} = relaxation time due to diffusion.

If diffusion effects are negligible, and because the bulk T_2 for water is on the order of seconds, while the observed T_2 is only a few hundred milliseconds or less, then the above can be approximated by

$$\frac{1}{T_2} = \rho_s \frac{S}{V} \tag{6}$$

where ρ_s = surface relaxivity (0.003–0.03 cm/s) for clastics, and ρ_s = surface relaxivity (< 0.003 cm/s) for carbonates.

In water-filled pore systems, T_2 is a measure of the S/V ratio. Large pores will have a small surface-area-to-pore-volume ratio and long T_2s, and small pores will have a large surface-area-to-pore-volume ratio and will exhibit short T_2s. Rocks with a spectrum of pore sizes give rise to a corresponding spectrum of T_2 relaxation times (Figure 11).

The NMR echo data from the logging tool are converted to a T_2 spectrum that is closely related to the pore-size distribution for a single-phase fluid. The T_2 spectrum, derived from the echo data, measures the

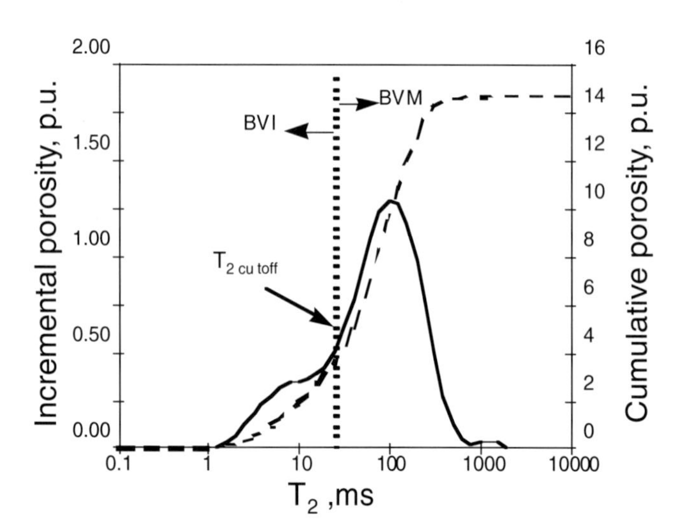

Figure 11—T_2 spectrum with indicated T_2 cutoff.

porosity fraction associated with each T_2. The integral of the T_2 spectrum (the sum of the porosities at each T_2) is the NMR porosity; however, the real benefit from NMR logging derives not just from the NMR porosity, but from the partitioning of the T_2 spectrum into fast decaying fluids (bound and irreducible water) and free fluids (Straley et al., 1991). Generally, clay-bound water relaxes in a few milliseconds or less (Prammer et al., 1996); thus, if we log with a short enough *TE* (~0.6 ms) and faithfully record the early echoes, the integral from 0 to 4 ms will be a measure of the clay-bound water. For clastics, the integral of the T_2 spectrum from 4 to ~32 ms is a measure of the irreducible water and the remainder (e.g., 32 ms to ∞) is the moveable fluid fraction. The T_2 time separating the moveable and irreducible fluids is referred to as the *T*2 cutoff time.

In clastic systems, the NMR T_2 data have been used successfully to estimate permeability. Several models are currently used to derive permeability from T_2 spectra. Timur (1969) recognized the relationship between moveable fluids and permeability.

$$k = 0.136 \frac{\phi^{4.4}}{S_{wir}^{\ 2}} \tag{7}$$

Kenyon et al. (1988) and others have related permeability to the mean T_2. This approach is formulated as

$$k_{NMR} = C \cdot T_{2gm}^{\ 2} \cdot \phi^4 \tag{8}$$

The constant *C* may need to be adjusted for different reservoirs and formations.

Coates et al. (1991) put forward a permeability model that incorporates both the moveable and bound fluid fractions and does not explicitly depend on the mean T_2.

$$k = \left(\frac{100\phi}{C}\right)^4 \left(\frac{S_{moveable}}{S_{wir}}\right)^2 \tag{9}$$

where $S_{moveable}$ = moveable fluid saturation = $(1 - S_{wir})$, S_{wir} = bulk volume irreducible fluid saturation, and *C* = formation-specific constant.

This is particularly advantageous in the presence of hydrocarbons. In a water-wet system, if liquid hydrocarbons are present then equation 5 does not simplify to equation 6, because the hydrocarbons are shielded from the rock surface and do not undergo surface relaxation. If a rock contains both water and oil, the pore space occupied by hydrocarbons will no longer exhibit the T_2 relaxation associated with the *S*/*V* ratio of these pores, and the mean T_2 will be anomalously large compared to the mean T_2 for a water-saturated rock. Thus, equation 8 will overestimate the permeability in hydrocarbon zones, whereas permeability calculated with the Coates equation, equation 9, will be relatively unaffected by hydrocarbons.

WFT CALIBRATION OF NMR PERMEABILITY

The NMR signal is obtained from the fluids in the pores of a porous media. To interpret this signal, a good understanding of the relationship among the signal, the rock matrix, and the fluids is required. An accurate interpretation of the NMR data also requires calibration of the processing parameters (e.g., T_2 cutoff time, coefficients, and exponents in the permeability model). Where available, core data are the natural source for the calibration data; however, advances in logging technology, specifically in WFT tool design, have moderated this requirement and provided an alternative solution to interpreting calibration for most fields where cores are unavailable or too expensive to obtain.

In this study, WFT-generated permeability values are used to calibrate the coefficient in the Coates equation. Plotted in Figure 12 are the WFT permeabilities and the NMR permeabilities for different values of *C*. To determine the best Coates constant, we have summed the absolute value of the difference between the WFT and NMR permeabilities and plotted the sum versus the *C* value (Figure 13). The optimum *C* for this formation is obtained from the minimum at 14.5. Calibration of the coefficient in the Coates bound-water permeability model via WFT data makes it possible to translate the NMR log responses into an accurate permeability profile for the formation despite the lack of core data.

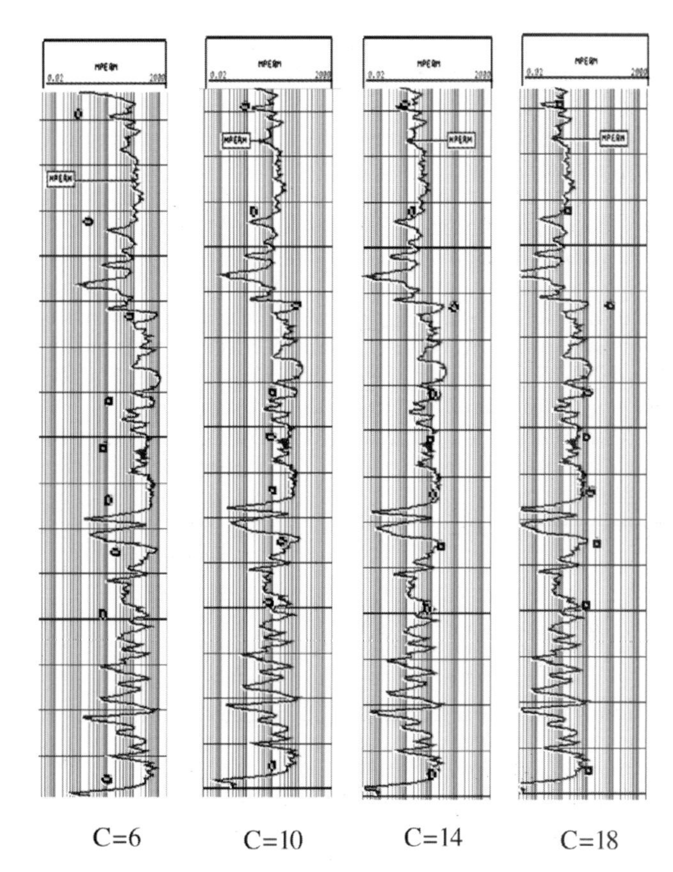

Figure 12— Calibration of *C* based on WFT permeabilities.

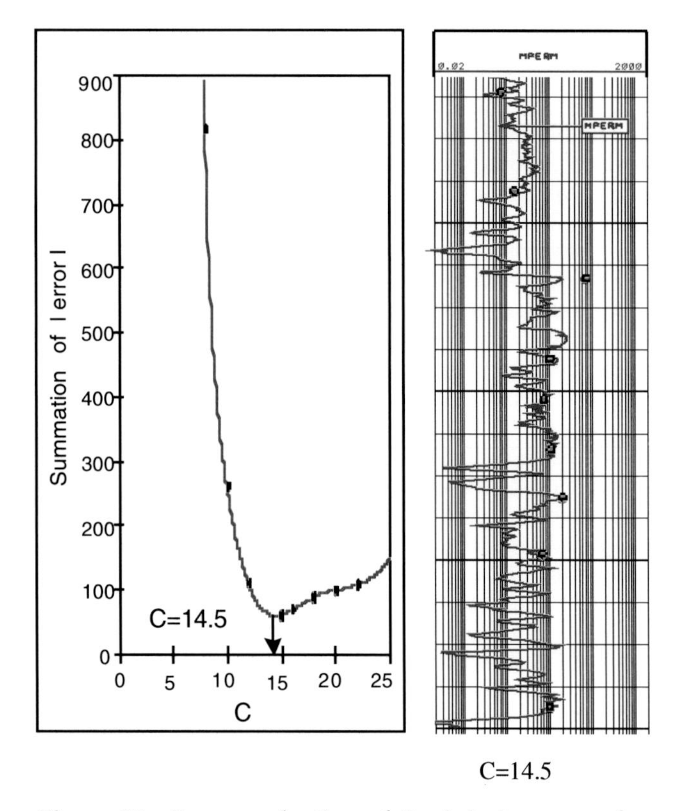

Figure 13—Proper selection of *C* minimizes error for permeability estimation.

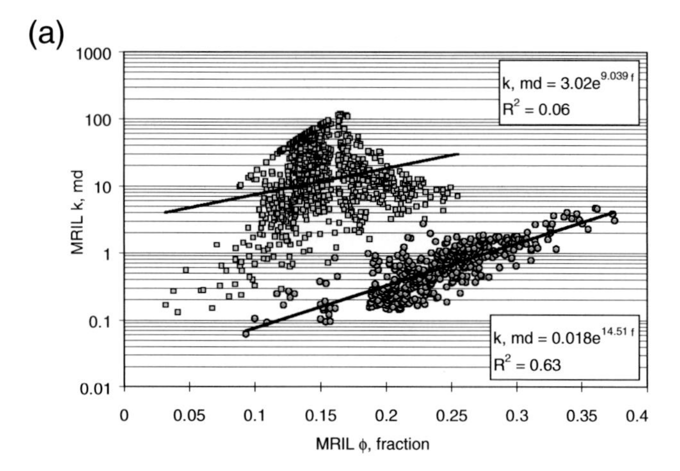

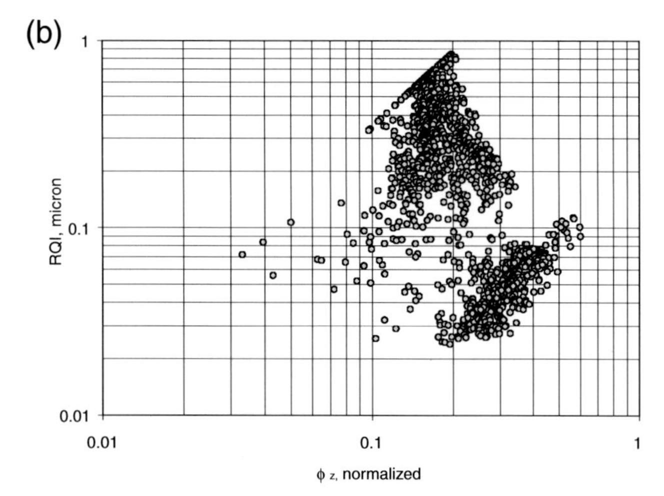

Figure 14—(a) Permeability versus porosity relationship. (b) Reservoir quality index versus normalized porosity.

DETERMINATION OF FLOW UNITS

A flow unit is a volume of reservoir rock within which geological and petrophysical properties that affect the fluid flow are internally consistent and predictably different from properties of surrounding rock volumes. Flow units are extensively used in reservoir characterization studies and in numerical simulations for performance predictions and reserve estimations. Two key elements for deriving the flow-unit profiles are accurate permeability and effective porosity profiles. WFT-calibrated NMR permeability and NMR porosity meet this requirement and, hence, enable the flow unitization process.

Permeability and porosity data from a control well or interval are processed to determine RQI, the reservoir quality index (Amaefule et al., 1993), as shown in Figure 14.

$$RQI = 0.0314\sqrt{\frac{k}{\phi}} \tag{10}$$

Then a transformation of RQI to hydraulic units is performed by employing the hydraulic unitization technique as outlined below.

Traditional plots of permeability versus. porosity (Figure 14a) have little petrophysical and no mathematical justification. Usually, log-normally distributed permeability is plotted against normally distributed porosity in a semi-logarithmic format to reduce the scatter. The underlying assumption for this plot is the capillary-tube model from which the Kozeny-Carman equation stems.

$$k = \frac{1}{8}\phi\frac{r^2}{\tau^2} \tag{11}$$

When capillary-tube geometry is assumed, the r^2 term is replaced with an expression for the specific surface area, S_{gv}:

$$k = \frac{\phi_e^3}{(1-\phi_e)^2}\frac{1}{F_s\tau^2 S_{gv}^2} \tag{12}$$

where k = WFT-calibrated NMR permeability (cm^2), ϕ_e = effective porosity (NMR MPHI), τ = tortuosity, F_s = shape factor, and S_{gv} = specific surface area (cm^{-1}).

This equation can be rewritten as

$$k = \frac{\phi_z}{K_z S_{gv}^2} \tag{13}$$

where ϕ_z = normalized ϕ ratio $\phi^2/(1-\phi)^2$, and K_z = Kozeny-Carman constant, $F_s\tau^2$.

In reality, tortuosity and shape factor vary from one sample to another reflecting formation heterogeneity; therefore, the traditional permeability versus porosity cross-plots show scatter, even in semi-logarithmic space; therefore, any transform equation derived from this approach (via statistical best-fit line) is prone to have unacceptable error. Assuming that the Kozeny-Carman constant is a single "universal" constant has proven unacceptable. The hydraulic unitization concept assigns a different constant for each hydraulic (flow) unit to account for variations in the pore-space attributes. To apply the hydraulic unitization methodology, we apply the following steps:

(1) Algebraic manipulation of the Kozeny-Carman equation yields the following when both sides of the original equation are divided by porosity and raised to the power of 0.5:

$$\sqrt{\frac{k}{\phi_e}} = \frac{1}{S_{gv}K_z^{0.5}}\frac{\phi_e}{(1-\phi_e)} \tag{14}$$

(2) Taking logarithms of both sides of the above equation results in a straight-line relationship:

$$\log\left[\sqrt{\frac{k}{\phi_e}}\right] = \log\frac{1}{S_{gv}K_z^{0.5}} + \log\frac{\phi_e}{(1-\phi_e)} \qquad y = b + mx \tag{15}$$

(3) The left side of equation 15 (with the proper conversion factor for permeability in cm^2 to μm^2) yields the RQI (Figure 14b).

The relationship of K_z and S_{gv} to permeability and porosity remains constant for data points with similar pore-space attributes; therefore, when RQI is plotted versus normalized porosity in logarithmic space, data fall on a straight line with a 45° slope. Where this straight line intersects the y-axis ($x = 1$) the intercept is equal to $\log(1/S_{gv}K_z^{0.5})$. This intercept is called the flow zone indicator (FZI). Samples with similar pore-space attributes exhibit similar y-intercepts. They remain within a confidence envelope (set by the required data accuracy) around a 45° straight line. Pore-space attributes control intrinsic flow properties; samples with similar y-intercepts belong to the same hydraulic unit. Any large deviation from this straight line indicates the existence of a separate hydraulic unit for which the relationship among K_z, S_{gv}, porosity, and permeability is different.

To identify the correct number of hydraulic units (i.e., to draw the correct number of straight lines with the prescribed 45° slope through the data), cluster analysis is used (Figure 15). The analysis is based on deciding the optimum center for each data cloud and isolating that set of data with a cluster identification

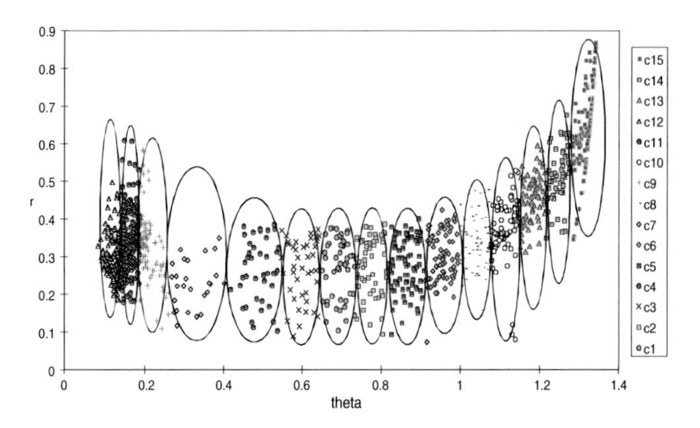

Figure 15—Cluster analysis maximizes between-cluster distance and minimizes within-cluster variations.

tag. The clustering algorithm used for the WFT-calibrated NMR data maximizes the between-cluster distance and minimizes the within-cluster variations. It should be noted that cluster analysis introduces a bias to the determination because the number of clusters should be known prior to clustering. This bias is later eliminated from the determination of flow units via sensitivity analysis.

SENSITIVITY ANALYSIS

Independent input variables used in the flow-unit formulation have associated uncertainties and known inaccuracies. Because of these uncertainties, adjoining hydraulic units may not be statistically independent. If the computed uncertainties for adjoining hydraulic units overlap, the two units are merged.

In this study, error margins are calculated for RQI and FZI based on the rms (root mean square) equation and recommended API practices for permeability and porosity determinations. The following rms equation applies, if

$$y = f(x_1, x_2, x_3, \ldots, x_n)$$

then, the error in y-function can be calculated:

$$\Delta Y = \pm\begin{pmatrix}\left[(\partial Y/\partial x_1)\Delta x_1\right]^2 + \ldots \\ +\left[(\partial Y/\partial x_n)\Delta x_n\right]^2\end{pmatrix}^{0.5} \tag{16}$$

where $\partial Y/\partial x_i$ are partial derivatives of the function with respect to its independent variables. Δx_i are uncertainties associated with independent variables.

With equation 16, an error margin is calculated for each cluster. These margins are used to establish whether clusters are truly independent. Then clusters with similar means are merged into one unique flow unit (Figures 16 and 17). Redundant clusters are eliminated. Clusters with more than 20% variation are isolated and then re-analyzed to refine the clustering. Thus, the bias introduced initially into the analysis is

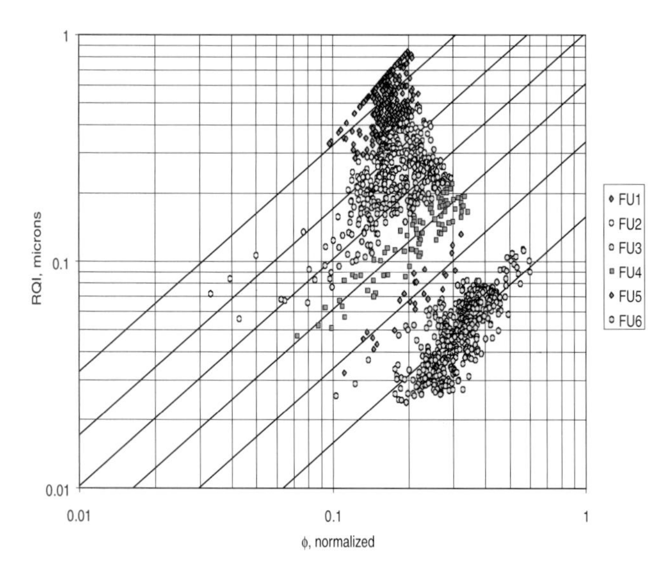

Figure 16—Sensitivity analysis of the clustering results produces the unique classification of flow (hydraulic) units.

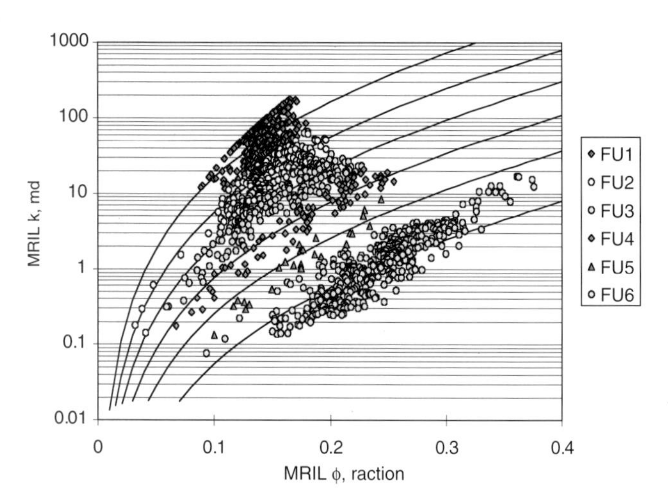

Figure 17—Traditional permeability versus porosity plot for the six unique flow units.

eliminated. At this stage of unitization, flow units are plotted in depth format with available wireline logs (Figure 18).

Figure 18 shows a composite plot of the conventional open-hole logs: gamma ray, caliper, deep and medium resistivity, and NMR logs. In addition, the flow-unit profile (the fifth track) determined from the NMR-log-derived permeability profile, and the effective porosity (MPHI) is displayed. The objective of deriving flow-unit profile is to provide a means of extracting permeability and transform equations of bulk volume irreducible (BVI) based on the conventional open hole logs. These equations are to be used for the neighboring wells in the same geological and sedimentary settings to generate continuous permeability and BVI profiles. The number of lateral divisions occupied on the fifth track (increasing to the right) at any depth refers to the flow-unit number.

The sixth track shows a reasonable agreement between the NMR-derived permeability and the pressure transient permeability (WFT). This agreement is reached by calibrating the NMR processing parameters. The seventh track shows the reservoir quality index [$RQI = 0.0314 \times (k/\phi)^{0.5}$] that is analogous to mean hydraulic radius for each depth level. The overall shape (lower resolution) of the RQI curve is in agreement with facies profile of the interval, but yields a more detailed hydraulic transmissivity picture. The eighth track displays the calculated Q_v (the cation exchange capacity per unit pore volume of the rock) from the Hill-Shirley-Kline model (Hill et al., 1956).

CONCLUSIONS

In this study, we used WFT pressure transient permeabilities to calibrate NMR-derived permeabilities. The constant in the Coates equation was changed from a generic 10 to the reservoir-specific 14.5. We then used the hydraulic units concept to classify the reservoir into six distinct flow units based on the NMR effective porosity and WFT-calibrated NMR permeabilities

We conclude from this work that

(1) FRA technique applied to WFT data provides robust local reservoir pressure and permeability.
(2) WFT permeabilities are ideally suited for calibrating NMR permeabilities.
(3) WFT-calibrated, NMR permeability profiles are an excellent source of reservoir heterogeneity data.
(4) NMR effective porosity and WFT-calibrated NMR permeability data can provide core-quality in situ data for reservoir flow-unit studies.

NOMENCLATURE

C_t	compressibility factor, 1/psi
F_s	shape factor
G_o	geometric factor
k	permeability, md
K_z	Kozeny-Carman constant
P	pressure, psi
q	volumetric flow rate, cm^3/s
Q_v	cation exchange capacity per unit rock volume
r_i	packer-probe radius, cm
S	pore surface area, cm^2
S_{gv}	specific surface area, cm^{-1}
t	time, s
T_{2bulk}	bulk fluid relaxation time, ms
$T_{2surf.}$	surface relaxation time, ms
T_{2D}	relaxation time due to diffusion, ms
V	pore volume, cm^3
V_{sys}	system volume, cm^3
ϕ	porosity, fraction
ϕ_e	effective porosity, fraction
ϕ_z	normalized ϕ ratio
λ	thickness of the surface fluid layer
μ	viscosity of fluid, cp
ρ_s	surface relaxivity, cm/s
τ	tortuosity

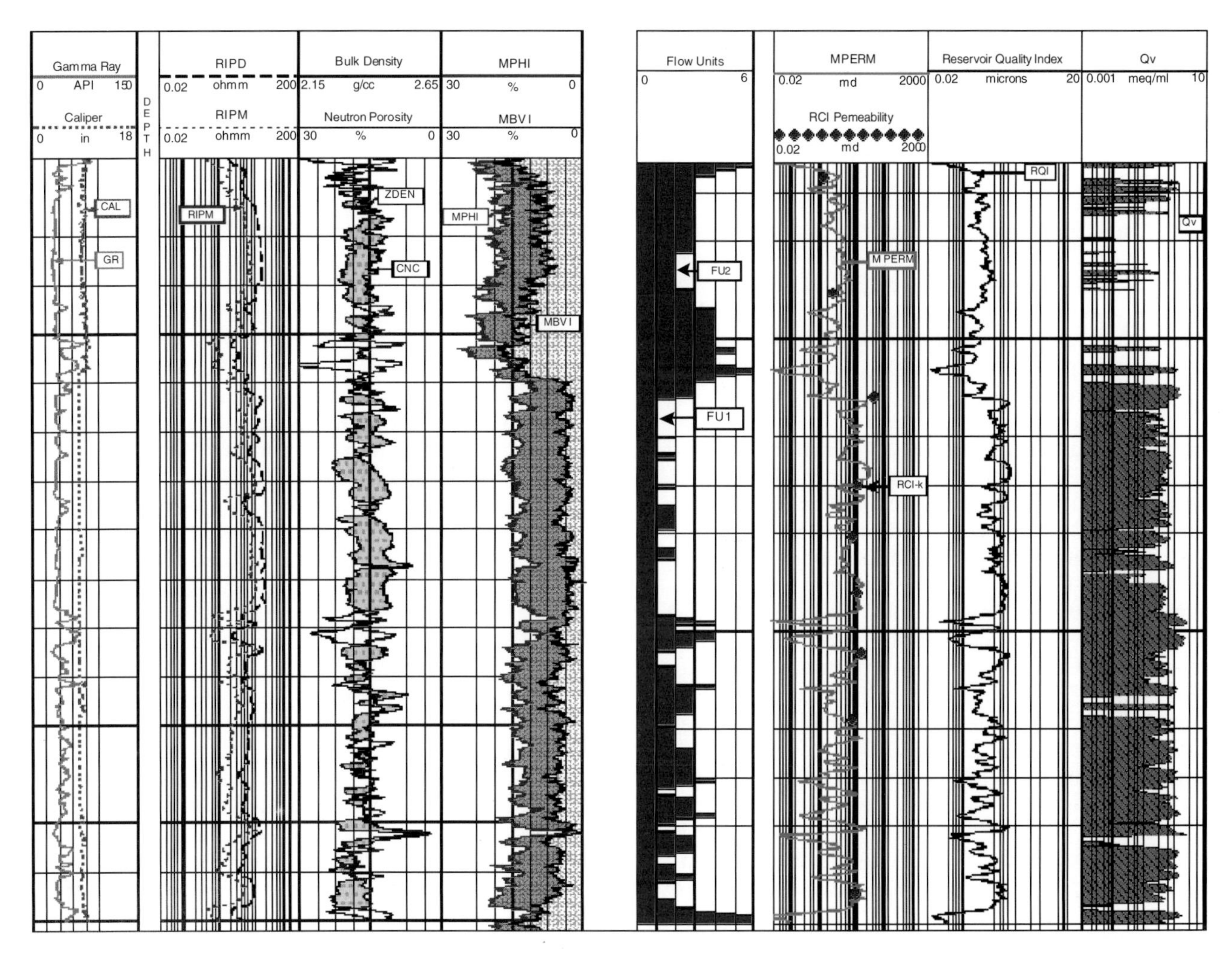

Figure 18—WFT-calibrated NMR permeability and hydraulic unitization with conventional log data enable the derivation of permeability transform equations for the neighboring cased wells.

Subscripts

ac	accumulation
dd	drawdown
f	formation

ACKNOWLEDGMENTS

We acknowledge the assistance of J. Michaels and M. Moody with data collection and preparation, and D. Oliver and T. Jbeili for editorial assistance. We thank Western Atlas Logging Services for permission to publish this paper.

REFERENCES CITED

Akkurt, R., Vinegar, H. J., Tutunjian, P. N., and Guillory, A. J., NMR Logging of Natural Gas Reservoirs, Paper N, Presented at the 36th Annual Logging Symposium, SPWLA, Paris, France, 1995, p. 1–5.

Amaefule, J. O., Altunbay, M., Tiab, D., Kersey, D. G., and Keelan, D. K., Enhanced Reservoir Description: Using Core and Log Data to Identify Hydraulic (Flow) Units and Predict Permeability in Uncored Intervals/Wells, SPE 26436, 68th Annual SPE Conference, Oct. 1993, p. 1–8.

Araktingi, U. G., Bashore, W. M., Tran, T. T. B., and Hewett, T. A., Integration of Seismic and Well Log Data in Reservoir Modeling, Reservoir Characterization III, Edited by B. Linville, T. E. Burchfield, and T. C. Wesson, PennWell Publishing Co., 1993, p. 515-554.

Brown, R. J. S., and Gamson, B. W., Nuclear Magnetism Logging, SPE 1305-G, Dallas, Tx, October 4-7, 1959.

Coates, G. R., Miller, M., Gillen, M., and Henderson, G., An Investigation of a New Magnetic Resonance Imaging Log, SPWLA Convention in Midland, TX, June 18, 1991.

Corbett, C., Solomon, G. J., Kartikay, S., Ujang, S., and Ariffin, T., Application of Seismic-Guided Reservoir Property Mapping to the Dulang West Field, Offshore Peninsula Malaysia, SPE 30568, presented at the SPE Annual Technical Conference and Exhibition, 22–25 October, Dallas, Texas, 1995, p. 381–389.

Goggin, D. J., Trasher, R. L., and Lake, L. W., A Theoretical and Experimental Analysis of Minipermeameter Response Including Gas Slippage and High Velocity Flow Effects, In Situ 12, Nos. 1 and 2 (1988), p. 79-116.

Hand, J. L., Moritz, A. L., Jr., Yang, C-T., and Chopra, A. K., Geostatistical Integration of Geological, Petrophysical, and Outcrop Data for Evaluation of Gravity Drainage Infill Drilling at Prudhoe Bay, SPE 28396, presented at the SPE Annual Technical Conference and Exhibition, 25–28 September, New Orleans, LA, 1994, p. 347–358.

Hill, H. J., and Milburu, J. D., Effect of Clay and Water Salinity on Electrochemical Behavior of Reservoir Rocks, Trans., AIME (1956), vol. 207, p. 65–72.

Huang, K., Samaha, A., and Kasap, E., Dimensionless Parameters for Interpretation of WFT Data: Simulations and Experiments, paper presented at the 1996 SPWLA Annual Logging Symposium, New Orleans, LA, June 16–19, Paper Y.

Johann, P., Fournier, F., Souza, O., Eschard, R., and Beucher, H., 3-D Stochastic Reservoir Modeling Constrained by Well and Seismic Data on a Turbidite Field, SPE 36501, presented at the SPE Annual Technical Conference and Exhibition, 6–9 October, Denver, CO, 1996, p. 51–66.

Kasap, E., Huang, K., Shwe, T., and Georgi D., Robust and Simple Graphical Solution for Wireline Formation Tests: Combined Drawdown and Buildup Analysis, SPE Paper 36525, presented at the SPE Annual Technical Conference and Exhibition, Denver, Colorado, 1996a, Vol. Ω, p. 343–357.

Kasap, E., Georgi, D., Micheals, J., and Shwe, T., A New Simplified, Unified Technique for the Analysis of Wireline Formation Test Data, paper presented at the SPWLA Annual Logging Symposium, New Orleans, LA, June 16–19, 1996b Paper AA.

Kasap, E., 1998, Fluid Flow Analysis Method for Wireline Formation Testing Tools, United States Patent # 5,708,204, January 13.

Kenyon, W. E., Day, P. I., Straley, C., and Willemsen, J. F., A Three-Part Study of NMR Longitudinal Relaxation Properties of Water-Saturated Sandstone, SPE Formation Evaluation, September 1988.

Kubica, P., Statistical Tests of Permeability Estimates Based on NMR Measurements, SPWLA 36th Annual Logging Convention, Paper VVV, June 26–29, 1995.

Miller, M. N., Paltiel, M. E., Gillen, J. Granot, and J. C. Bouton, Spin Echo Magnetic Resonance Logging: Porosity and Free Fluid Index Determination, SPE 20561, 65th Annual Tech. Conf., New Orleans, Sep. 23–26, 1990.

Prammer, M. G., Drack, E. D., Bouton, J. C., Gardner, J. S., and Coates, G. R., Measurements of Clay-Bound Water and Total Porosity by Magnetic Resonance Logging; SPE 36522, presented at the SPE Annual Technical Conference and Exhibition, 6–9 October, Denver, CO, 1996, p. 311–320.

Proett, M. A., and Chin, W. C., Supercharge Pressure Compensation Using a New Wireline Testing Method and Newly Developed Early Time Spherical Flow Model, SPE 36524, presented at the 1996 SPE Annual Technical Conference and Exhibition, Denver, CO, Oct. 6–9, p. 329–342.

Reedy, G. K., and Pepper, C. F., Analysis of Finely Laminated Deep Marine Turbidites: Integration of Core and Log Data Yields a Novel Interpretation Model, SPE 36506, presented at the SPE Annual Technical Conference and Exhibition, 6–9 October, Denver, CO, 1996, p. 119–127.

Samaha, A., Huang, K., Kasap, E., Shwe, T., and Georgi, D., Near Wellbore Permeability and Damage Measurements: Experiments and Numerical Simulations for Interpretation of WFT Data, SPE 35150, presented at the SPE 1996 International Symposium on Formation Damage Control in Lafayette, Louisiana, U.S.A. 14–15 February.

Schultz, P. S., Ronen, S., Hattori, M., Mantran, P., Hoskins, J., and Crobett, J., Seismic-Guided Estimation of Reservoir Properties, SPE 28386, presented at the SPE Annual Technical Conference and Exhibition, 25–28 September, New Orleans, LA, 1994, p. 235–250.

Seevers, D. O., A Nuclear Magnetic Method for Determining the Permeability of Sandstones, Trans. SPWLA, v. 6, Sec. L, 1966.

Starley, C., Morris, C. E., Kenyon, W. E., and Howard, J. J., NMR in Partially Saturated Rocks: Laboratory Insights on Free Fluid Index and Comparison With Borehole Logs, Paper CC, presented at the 32nd Annual Logging Symposium, SPWLA, Midland, TX, June 16–19, 1991.

Timur, A., Pulsed Nuclear Magnetic Resonance Studies of Porosity, Movable Fluid, and Permeability of Sandstones, Trans. AIME (1969), v. 246, p. 775.

Waid, M. C., Proett, M. A., Chen, C. C., and Ford, W. T., Improved Models for Interpreting the Pressure Response of Formation Testers, SPE 22754, presented at the 1991 SPE Annual Technical Conference and Exhibition, Dallas, TX, Oct. 6–9, p. 889–904.

Yang, A-P., and Gao, Y., Reservoir Characterization by Integrating Well Data and Seismic Attributes, SPE 30563, presented at the SPE Annual Technical Conference and Exhibition, 22–25 October, Dallas, Texas, 1995, p. 337–341.

Yang, C. T., Chopra, A. K., J. Chu, Huang, X., and Kelkar, M. G., Integrated Geostatistical Reservoir Description Using Petrophysical, Geological and Seismic Data for Yachheng 13-1 Gas Field, SPE 30566, presented at the SPE Annual Technical Conference and Exhibition, 22–25 October, Dallas, Texas, 1995, p. 357–372.

Ali, M., et al., Extracting maximum petrophysical and geological information from a limited reservoir database, 1999, *in* R. Schatzinger and J. Jordan, eds., Reservoir Characterization-Recent Advances, AAPG Memoir 71, p. 191–208.

Chapter 14

Extracting Maximum Petrophysical and Geological Information From a Limited Reservoir Database

Maqsood Ali
Schlumberger-GeoQuest
Denver, Colorado, U.S.A.

Adwait Chawathé
Chevron Research Tech Company
La Habra, California, U.S.A.

Ahmed Ouenes
Reservoir Characterization Research and Consulting
Denver, Colorado, U.S.A.

Martha Cather
William Weiss
New Mexico Petroleum Recovery Research Center
Socorro, New Mexico, U.S.A.

ABSTRACT

The characterization of old fields lacking sufficient core and log data is a challenging task. This paper describes a methodology that uses both new and conventional tools to build a reliable reservoir model for the Sulimar Queen field. At the fine scale, permeability measured on a fine grid with a minipermeameter was used in conjunction with the petrographic data collected on multiple thin sections. The use of regression analysis and a newly developed fuzzy logic algorithm led to the identification of key petrographic elements that control permeability. At the log scale, old gamma ray logs were first rescaled/calibrated throughout the entire field for consistency and reliability using only four modern logs. Using data from one cored well and the rescaled gamma ray logs, correlations among core porosity, permeability, total water content, and gamma ray were developed to complete the small-scale characterization. At the reservoir scale, outcrop data and the rescaled gamma logs were used to define the reservoir structure over an area of 10 mi^2 (26 km^2) where only 36 wells were available. Given the structure, the rescaled gamma ray logs were used to build the reservoir volume by identifying the flow units and their continuity. Finally, history-matching results constrained to the primary production were used to estimate the dynamic reservoir properties, such as relative permeabilities, to complete the characterization. The obtained reservoir model was tested by forecasting the waterflood performance and was in good agreement with the actual performance.

INTRODUCTION

In an ideal situation, the data required for detailed reservoir characterization range from regional to field scale to small scale, and consist of lithological description, petrophysical properties, fluid properties, well-test, production, and pressure data. Optimal data density is also crucial in addition to the variety and quality of data, and is governed by the type and scales of heterogeneity. Adequate data density is needed so that statistically significant conclusions can be drawn; however, in actual reservoir characterization problems neither all types of data nor the appropriate data density may be available.

A large number of oil and gas fields developed prior to the 1970s lack modern logs and sufficient core material. The economic constraints and risks involved prohibit both small and large operators from spending large sums of money on drilling new wells, coring, logging, and running special well-test operations in these old fields. The characterization of such fields using available information is a challenging task, and no proven methodology is available in the industry. Also, the production data available from these fields are sometimes not reliable. Under these circumstances the simulation/history-matching process may result in petrophysical properties (porosity, permeability, relative permeabilities, etc.) far removed from reality.

This paper describes the methodology employed for the reservoir characterization of an old field, the Sulimar Queen. This field is deficient in core material, modern logs, and complete production data. The Sulimar Queen field is located in Chaves County, southeast New Mexico (Figure 1). The reservoir is present in the upper Queen Formation, called the Shattuck Member. The internally complex Shattuck Member was deposited in a mixed lagoonal-sabkha-eolian environment, and can be successfully exploited only if the distributions of individual porosity and permeability units within the reservoir are clearly understood.

For the characterization of the field, the only control points available consisted of one core, core reports from two wells, and modern log suites from four wells. From the rest of the field, only old gamma ray and neutron logs were available. Neutron logs, because of their wide scale ranges and anomalous high and low porosity values, were discarded and were not used in this study (Ali et al., 1996). Two additional cores were obtained from the South Lucky Lake and Queen fields, and outcrop studies were conducted to supplement the database (Figure 1). Cores and logs from the adjacent fields were collected to capture the regional trend, so that the relationships developed in the Sulimar Queen field could be verified.

SMALL-SCALE HETEROGENEITY STUDY

The first objective was to analyze the scale of heterogeneity in the vertical direction from the cores available from the Sulimar Queen (one core), South Lucky Lake, and Queen fields. Permeability measurements were made using a computer-controlled minipermeameter (Suboor and Heller, 1995), and approximately 5000 permeability measurements were made. Permeability measurements were made on a square grid with an interval of 12.5 mm, which created five vertical permeability profiles along the length of the core. The averages of the five profiles for the three cores are shown in Figure 2 (also refer to Figure 3). The permeability ranged from less than 0.1 md to 500 md, and the scale of permeability heterogeneity was the same for three cores (Figure 2). This similarity suggested that the vertical permeability heterogeneity in the Sulimar Queen and adjacent fields may be caused by the same depositional and diagenetic events. This regional trend provided the first confirmation of the assumption that the data from the adjacent fields may be used.

The producing zone of the Shattuck Member consists of greenish brown, greenish gray, gray, brown, red, very fine grained sandstone. It is bounded at the top and the bottom by anhydrite (Figure 2). The permeability heterogeneity is caused by minor changes in the depositional environment. Minor changes in the lithology are expressed by the changes in the permeability (Figure 2). Permeability measurements made on a smaller scale (millimeter to centimeter scale) also improved the lithologic description of the core. On the basis of permeability distribution, the reservoir was divided into two separate layers, an upper high-permeability layer (Zone 1) with one low-permeability subzone (Subzone A), and a lower tighter layer (Zone 2), as illustrated in Figure 2.

Improving the Quantity and Quality of Petrographic Data

An appropriate sample density is necessary to ascertain the control of petrographic elements, such as porosity types, pore morphologies, mineralogy, textures, and type, amount, and distribution of clay and cement, in creating permeability heterogeneity. Because only one core was available from the Sulimar Queen field, it was impossible to collect enough samples for

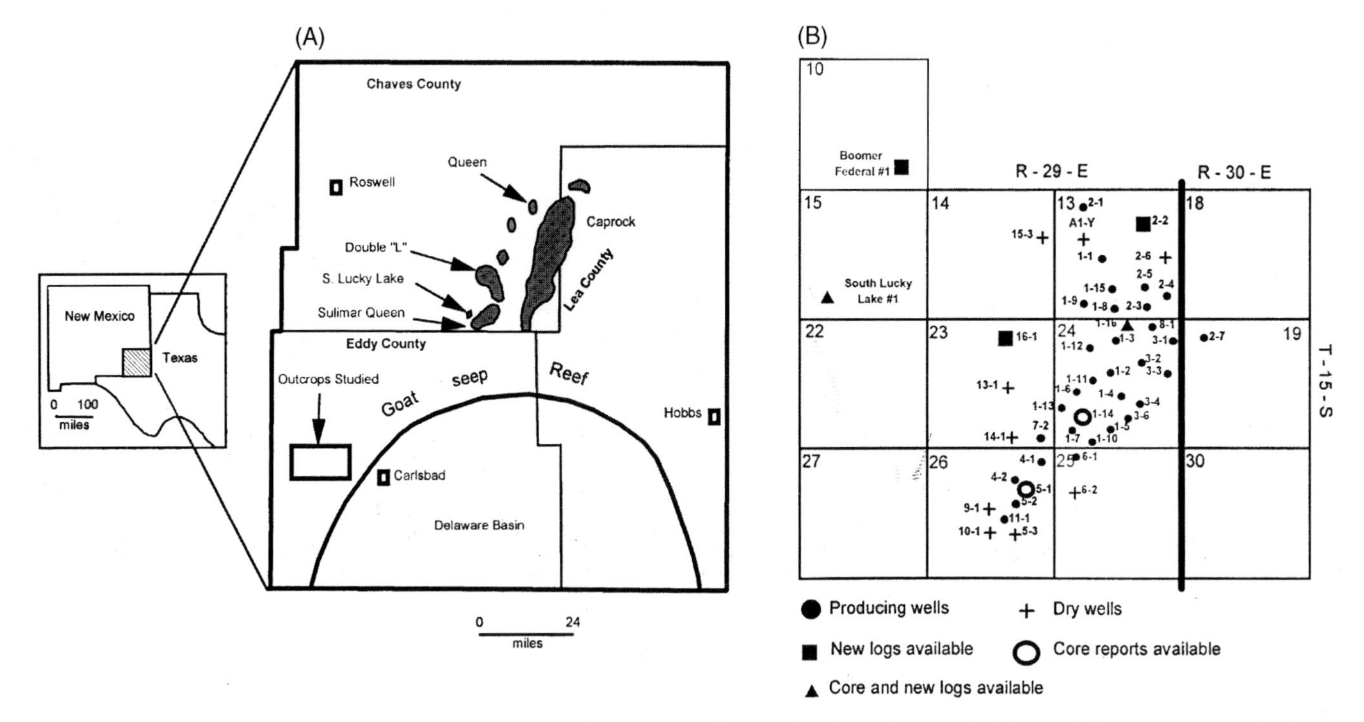

Figure 1. Locations of the Sulimar Queen, South Lucky Lake, and other Queen fields and the outcrops relative to the Goat Seep reef. Distribution of the wells in the Sulimar Queen field is also shown.

the petrographic analysis without destroying the core. Conventionally, relationships between permeability and petrographic elements are determined by using core plugs and thin sections, but there is a large difference in their respective volumes of investigation. Permeability varies from point to point within most core plugs; therefore, a thin section prepared from the edge of that core plug may not contain the petrographic elements that are representative of the core plug's permeability. Consequently, such correlations may be misleading. Because of its destructive nature and the size, core plug analysis also restricts the detailed investigation of permeability heterogeneities on a small scale, especially for sedimentary structures with dimensions less than 2.5 cm. In the Shattuck Member, permeability heterogeneities exist on a small scale (Figure 2), and the whole range of permeability distribution cannot be sampled using core plugs. A new methodology was developed by using a fine-scale grid (on which permeability measurements were made with a minipermeameter) based on the premises that (1) the area of investigation of the minipermeameter probe tip is small and can be examined in thin sections, and (2) each thin section contains many permeability points (Figure 3). The advantage of collecting a large set of data points is that detailed statistical analyses can be done and the effects of each petrographic element on permeability can be determined. This also helps assess the effects of diagenesis and porosity evolution on permeability more accurately.

According to Goggin et al. (1988), the effective radius and depth of investigation of a minipermeameter probe tip is four times the internal radius of the probe tip; however, it was found during calibration and permeability measurements with the minipermeameter that the area immediately under and around a probe tip exerts the main control over the permeability. After careful examination, it was decided that for a probe tip with an inner radius of 0.125 in (3.175 mm), the area of the thin section to be analyzed should have a diameter of 0.4 in (10 mm). Depending on the number of permeability points located on each thin section, the thin section was divided into that many equal parts, and petrographic data were collected from each part separately (Figure 3). With this new method, data equivalent to seven thin sections on the average were collected from one thin section, which would not have been possible using conventional core plugs. The minipermeameter, in addition to providing large amounts of permeability data and resolving small-scale heterogeneity, also provided the opportunity to collect more petrographic data using few thin sections. Because of the similarity of the areas of investigation of both the thin sections and the minipermeameter, the correlations established between permeability and petrographic elements were also more accurate than the correlations obtained from conventional core plugs.

Petrographic Analysis

Thirty-eight thin sections were made from the cores and a total of 267 data points were obtained. The petrographic elements collected for each permeability point were total porosity, total secondary porosity, secondary intergranular porosity, microporosity, intraconstituent porosity, moldic porosity, microfractures, quartz, feldspar, rock fragments, anhydrite, dolomite, detrital clay, dead oil, pore size, grain size, grain sorting, pore

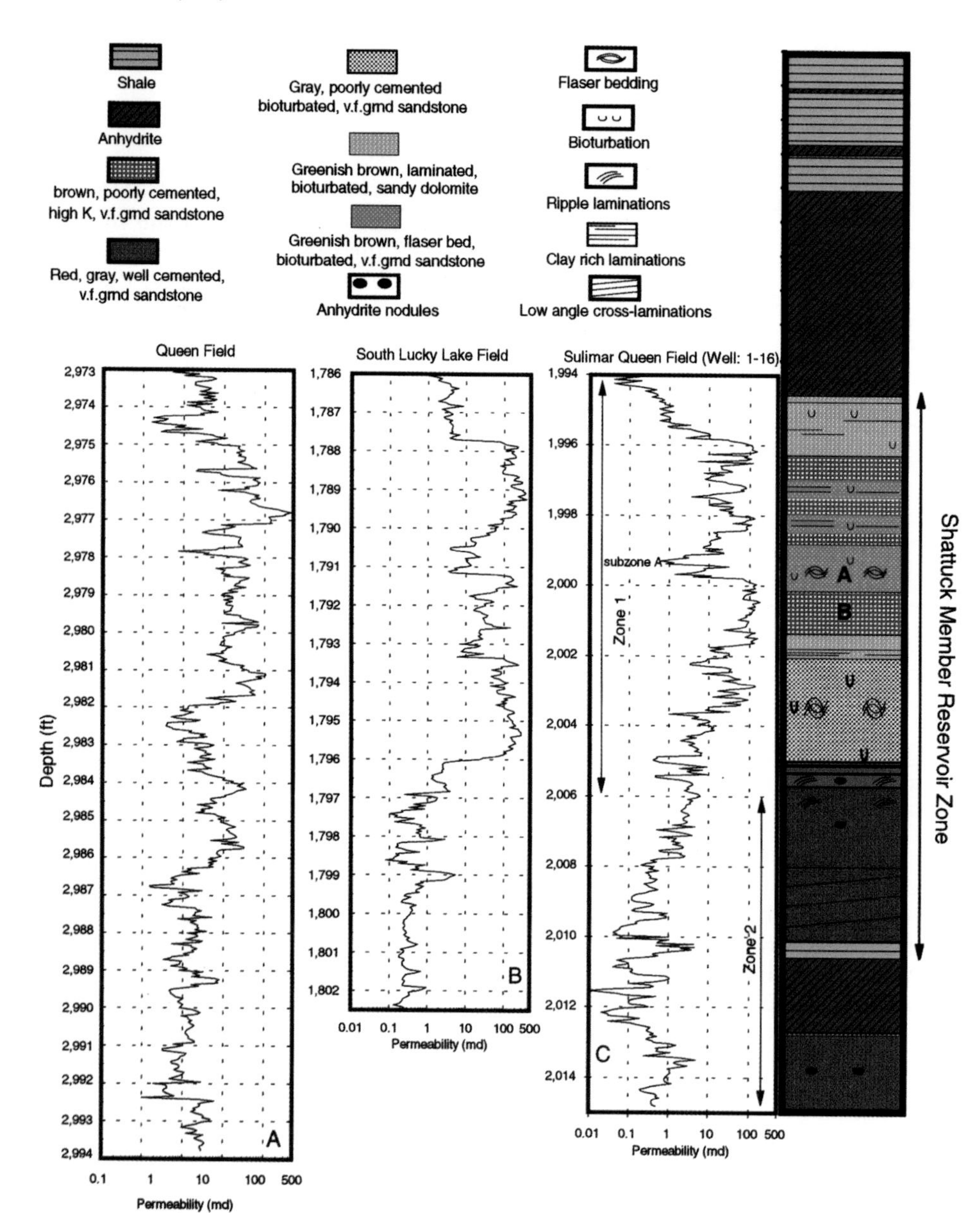

Figure 2. Permeability distribution in the cores from (A) Queen, (B) South Lucky Lake, and (C) Sulimar Queen fields. Core description from the Sulimar Queen field is also shown alongside the permeability distribution. Note the similar permeability distribution, especially in (B) South Lucky Lake and (C) Sulimar Queen fields.

sorting, and pore interconnection. Detailed petrographic analysis also helped determine the depositional environment, paragenetic sequence, and porosity evolution. A generalized paragenetic sequence for the Shattuck Member is shown in Figure 4.

Petrographic analysis was done to establish the ranking of each petrographic element in controlling the permeability. In addition to simple regression analysis (Figure 5; Table 1), a newly developed fuzzy logic algorithm was also used to determine the importance (ranking) of each petrographic element controlling the permeability (Figure 6; Table 2). Depending on the ranking, the permeability may be estimated by using only the most important petrographic elements. In old fields, where the samples are scarce or not suited for making permeability measurements, thin section analysis may be used to estimate the permeability. The results of the fuzzy logic algorithm were in excellent agreement with conventional regression analysis. The advantage of using this new algorithm lies in the speed of analysis.

Petrographic analysis also indicated that quartz (60–87%) and feldspar (10–34%) are the dominant detrital grains, whereas anhydrite (0–35%) and dolomite (0–50%) are the dominant cements. Both quartz and feldspar overgrowths are ubiquitous throughout the reservoir, but never exceeded 2%. The presence of dust rims and hematite between the grain and overgrowth helped to identify the overgrowths. Most of the feldspar grains are altered (sericitized), and the majority of intraconstituent porosity is present among the partially dissolved feldspar grains, especially in Zone 1. Although alteration of feldspar has been considered to reflect climatic conditions at the time or site of deposition of feldspar grains, chemical alteration of feldspar grains can take place as easily and rapidly during diagenesis, as well as during surface weathering (Blatt, 1982). In the Shattuck Member the zones that are well cemented by early poikilotopic anhydrite do not show considerable alteration of feldspar, whereas the zones that escaped early cementation or were not extensively cemented

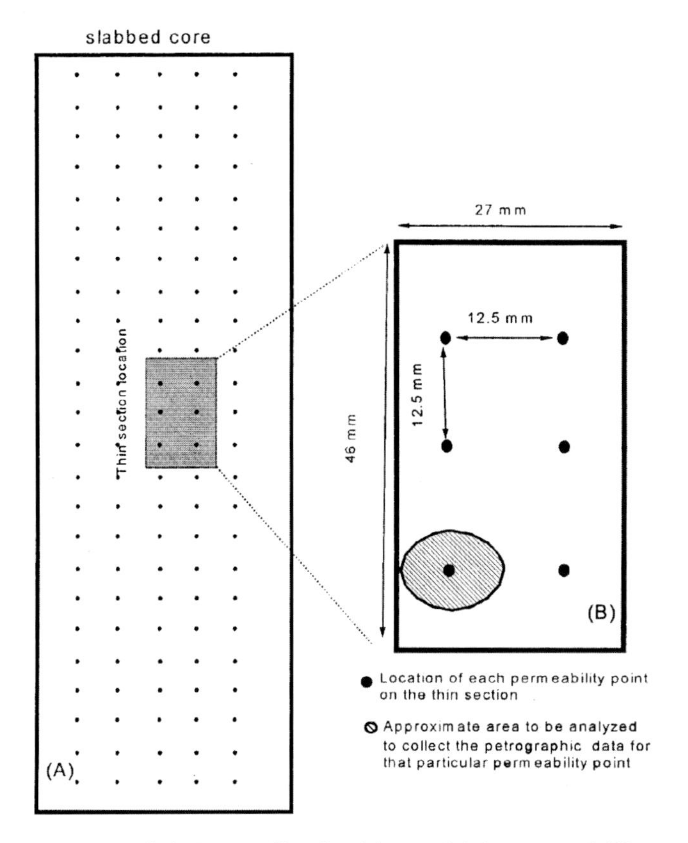

Figure 3. (A) Generalized grid on which permeability measurements were made. Five vertical permeability profiles were generated by this grid. The averages of the five profiles in three cores are shown in Figure 2. Thin section location is also shown. (B) Distribution of permeability points on the thin section.

have sericitized feldspar (Ali, 1997). Quartz and feldspar grains are also coated with clay and hematite. These clay rims consist of authigenic chlorite (Ali, 1997). Hematite is more conspicuous in the low permeability zone (Zone 2), where it coats the grain and also stains the anhydrite (poikilotopic) cement.

Both quartz and feldspar ranked in the middle with respect to controlling the permeability based on the fact that as their percentage increases, the percentage of anhydrite and dolomite cement decreases (Ali, 1997). The rock fragments consist of clay, dolomite (dolomicrite), and metamorphic rock fragments (Zone 1). Rock fragments do not exert much control over the permeability because of their low percentage (less than 2%) and fine sand size. Only when clay fragments form the pseudomatrix as they get squeezed between the harder grains are both porosity and permeability reduced. Dolomicrite might have originally been lime mud that was diagenetically altered to dolomicrite. Dolomicrite could be identified easily due to its exclusive presence in the primary pores. Dolomicrite is present in substantial amounts only in the thinly laminated, poorly sorted silty sandstone zones not thicker than 2 in (5 cm) (Zone 1), which is why clay does not affect the permeability and is close to the bottom of the conventional ranking (Table 1).

Cement is present in the form of anhydrite and dolomite. Anhydrite is present in three morphologies: (1) fine crystalline nodules, mainly in the upper silty-shaly zone at the top of the Shattuck Member and in small amounts in Zone 2 (Figure 2), (2) coarse, pore-filling crystals, and (3) large poikilotopic crystals surrounding several grains. It is the poikilotopic

Table 1. Summary of the Petrographic Elements and Their Relationship (R^2) with Permeability and Total Porosity.*

Petrographic Elements	Amount (Ranges)	Correlation (R^2) with Permeability	Correlation (R^2) with Total Porosity	Rank Based on R^2 with Permeability
Secondary Intergranular Porosity	0–27%	0.75	0.91	1
Total Secondary Porosity	0–27%	0.71	0.97	2
Total Porosity	0–27%	0.70	–	3
Pore Size (μm)	0–70	0.70	0.81	4
Quartz	0–75%	0.47	0.37	5
Feldspar	0–26%	0.43	0.33	6
Dolomite	0–55%	–0.36	–0.26	7
Grain Size (μm)	0–120	0.30	0.25	8
Intraconstituent Porosity	0–3.5%	0.21	0.36	9
Primary Porosity	0–4%	0.17	0.20	10
Moldic Porosity	0–3%	0.13	0.14	11
Anhydrite	0–40%	–0.12	–0.07	12
Dead Oil	0–11%	0.11	0.09	13
Microporosity	0–5.5%	–0.10	–0.03	14
Clay	0–21%	–0.06	–0.13	15
Rock Fragments	0–9%	0.01	–0.06	16

*The ranking shows the importance of that petrographic element in controlling the permeability as determined by conventional regression analysis.

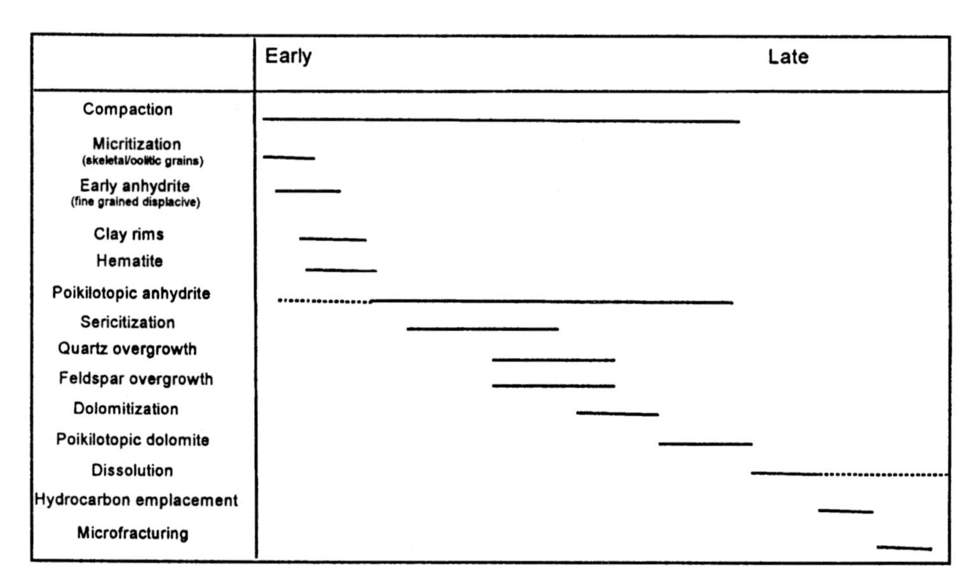

Figure 4. Generalized paragenetic sequence for the Shattuck Member in the Sulimar Queen and adjacent fields.

morphology that has affected the permeability by completely plugging the pores and pore throats, especially in Zone 2 (Ali, 1997). Dolomite is present in two morphologies: (1) micritic dolomite, probably formed by the dolomitization of the carbonate mud, and (2) large poikilotopic patches. The presence of the inclusions of dolomicrite and anhydrite in the poikilotopic dolomite suggests that poikilotopic dolomite post-dates both of them. Dolomite and anhydrite are distributed in the form of irregular patches (2–4 mm in diameter) throughout the reservoir zone. This patchy distribution is probably due to the heterogeneous dissolution pattern. Both dolomite and anhydrite have also replaced the detrital grains (Ali, 1997). Based on the correlation coefficient and fuzzy logic, dolomite seems to be more important than anhydrite in controlling permeability. The dead oil present reduced the porosity and the interconnection between the pores and, therefore, its effect on the permeability was also acknowledged.

Among the porosity, the secondary intergranular porosity is the most dominant type and exerts the most influence on the permeability (Table 1; Figure 5K). As dissolution increases, secondary intergranular porosity increases, and the interconnection between the pores becomes good. As the amount of secondary porosity increases, so does the pore size. This explains why the pore size is very important in controlling the permeability (Figure 5M). The relationship between the grain size and permeability is not very good (Table 1; Figure 5N). The larger size grains are present among the well-sorted part of the reservoir (Zone 2) and were affected by early anhydrite cementation, and almost all the original porosity was lost. The moderately sorted lagoonal sandstones (Zone 1) were not affected by early cementation, and enough primary porosity was available for the fluids responsible for dissolution to move freely through the system and produce secondary porosity.

The Fuzzy Logic Algorithm

As indicated earlier, a new fuzzy logic algorithm was applied to determine the ranking of the petrographic elements with respect to permeability. The following paragraphs briefly describe the application of the fuzzy logic algorithm, and Table 2 shows the comparison between the conventional ranking and the fuzzy logic ranking.

All the petrophysical attributes examined under the thin section contribute to permeability. Although this is true, we are also aware that each attribute alters permeability in a unique manner. It is difficult to accurately quantify the non-linear interaction between each of the petrophysical parameters with permeability. The fact that permeability is affected differently by each attribute brings out the question of orthogonality of the petrophysical attributes with respect to each other. Truly dependent attributes should alter permeability in a similar fashion.

Facing the dilemma of not knowing the explicit relationship between the petrophysical attributes and permeability, the question of the most significant attributes that contribute to permeability was resolved using the fuzzy logic algorithm. This data-directed algorithm compares the effect of each individual input parameter (the petrophysical measurement) on the output (the minipermeameter permeability). Briefly, the algorithm achieves this comparison by building fuzzy membership functions for each of the input parameters. The fuzzy membership functions are then defuzzified using the centroid defuzzification rule to plot fuzzy curves. The range of each of these fuzzy curves on the ordinate reflects the effect of each input parameter on the output (Figure 6). Details of the fuzzy logic algorithm are beyond the scope of this paper, and the interested reader is referred to Lin (1994). Our objective in this study was to rank the petrophysical attributes in a descending order of their influence on the permeability of the sample.

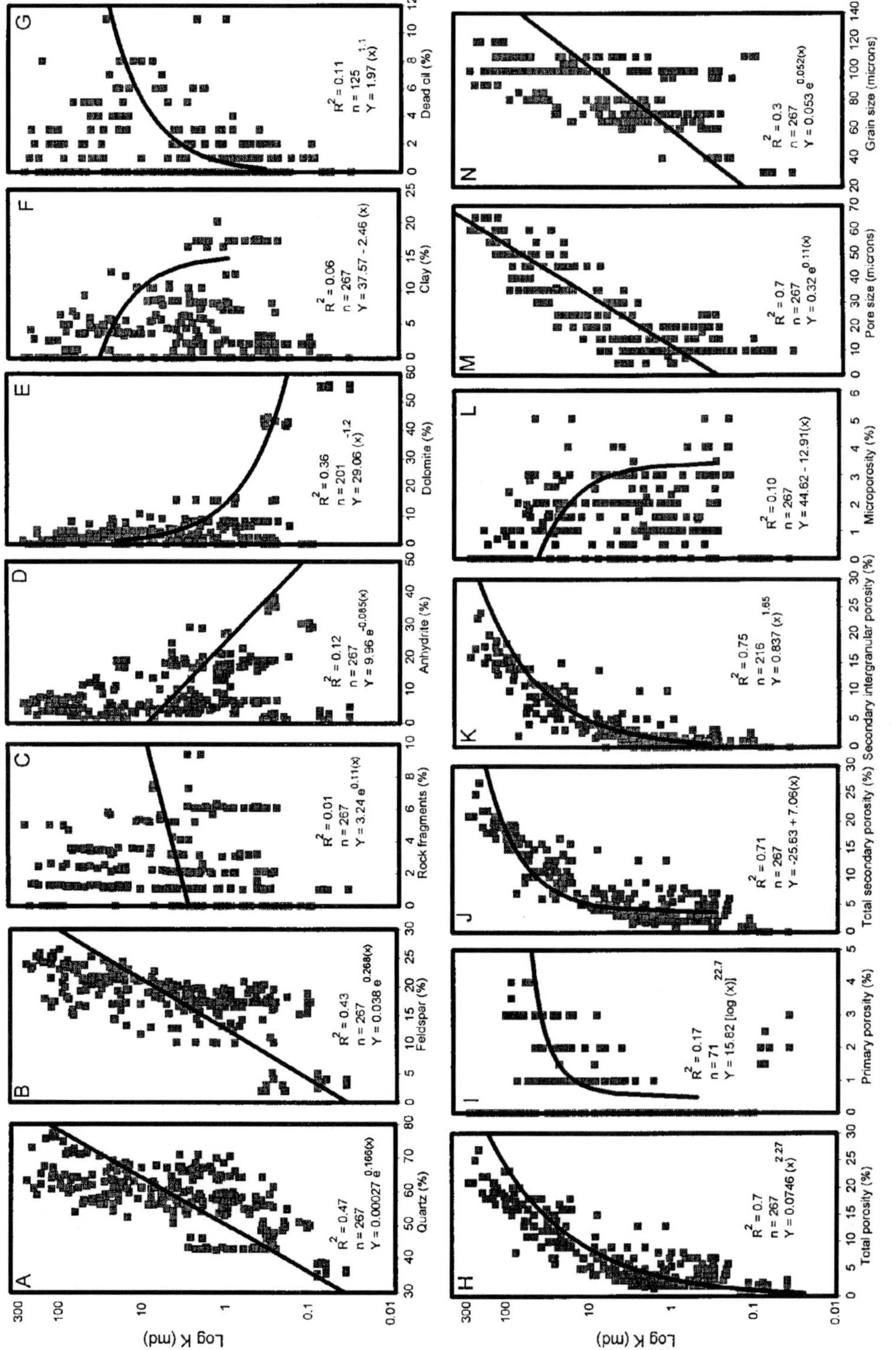

Figure 5. Relationships between permeability and different petrographic elements in the Shattuck Member, Sulimar Queen and South Lucky Lake fields.

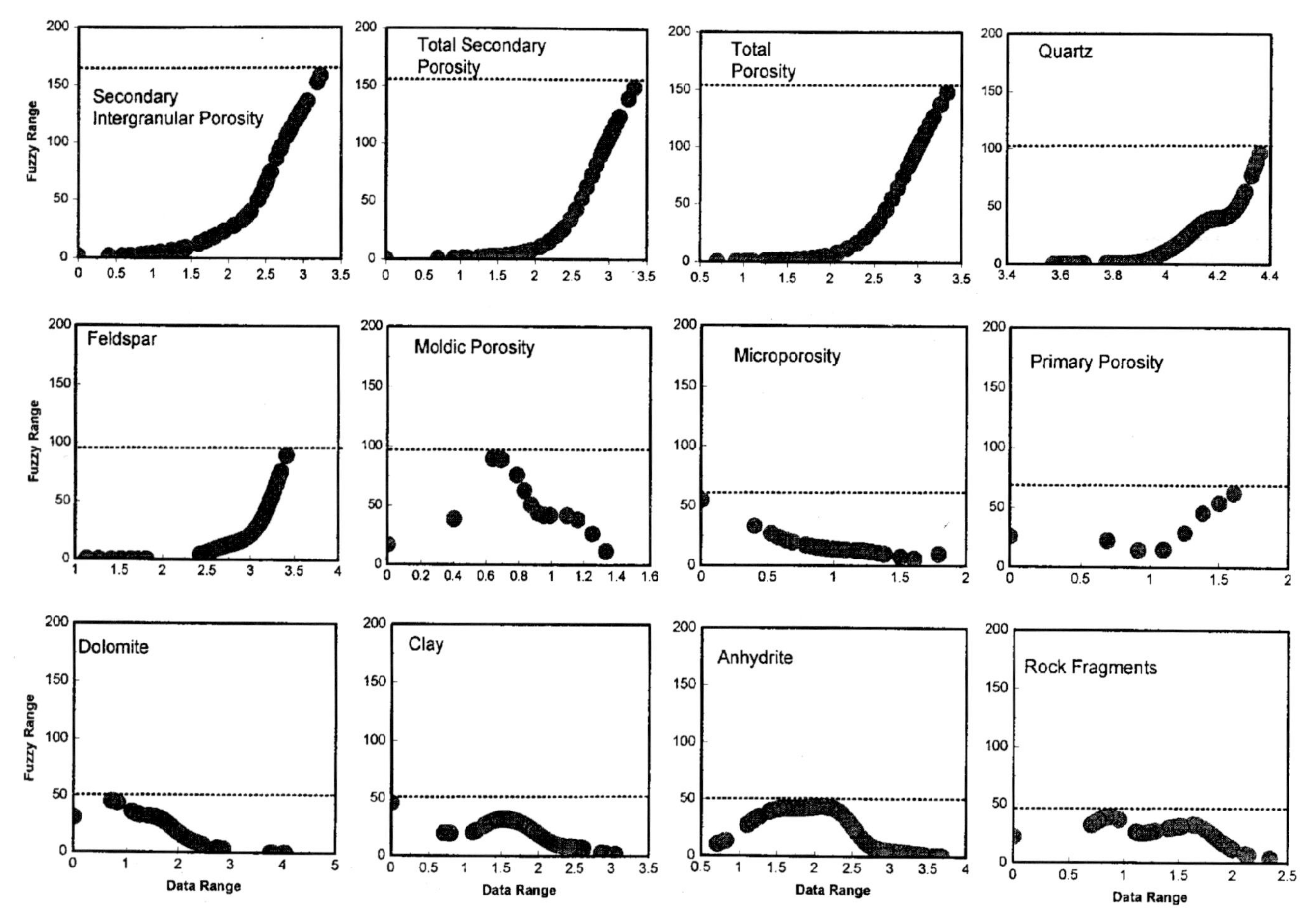

Figure 6. Fuzzy curves for various petrographic elements in the Shattuck Member. The importance of each element in controlling the permeability is decided by the fuzzy range (*y*-axis) of its fuzzy curve.

Figure 6 shows the fuzzy curves resulting from the fuzzy logic method. The first row represents the petrophysical parameters that have the largest influence on the permeability. This influence is evident from the range of each fuzzy curve on the corresponding ordinate. For example, the fuzzy curve for total porosity exhibits a range of approximately 160, whereas feldspar and anhydrite exhibit ranges of approximately 90 and 50, respectively. Here, total porosity has greater influence than feldspar, which, in turn, affects permeability more than anhydrite.

Table 2. Comparison of Conventional and Fuzzy Logic-Based Ranking of Petrographic Elements.

Petrographic Element	Ranking Due to Regression	Ranking Due to Fuzzy Logic
Secondary Intergranular Porosity	1	1
Total Secondary Porosity	2	2
Total Porosity	3	3
Quartz	4	4
Feldspar	5	5
Dolomite	6	7
Anhydrite	7	9
Microporosity	8	6
Clay	9	8
Rock Fragments	10	10

Prior to generating the fuzzy curves for these petrophysical attributes, histograms were plotted for each attribute. The histograms ensure that the data have approximately Gaussian distribution. Heavily skewed data result in erroneous results from the fuzzy logic algorithm. Histograms for moldic porosity and intraconstituent porosity were skewed even after normalization, and hence the fuzzy logic results pertaining to these parameters were considered inconclusive. These parameters were eliminated from analysis. Table 2 compares the ranking obtained from the fuzzy logic algorithm with conventional regression analysis. It is evident from Table 2 that the most important petrographic elements (ranking 1 to 5) controlling permeability are accurately predicted by the new algorithm. From Table 2, we can also see that the fuzzy logic algorithm ranked microporosity in sixth place. From conventional ranking, we see that microporosity was ranked eighth. In the case of Sulimar Queen, we know that microporosity should affect permeability more than dolomite, which appears to be in agreement with the fuzzy logic prediction.

It is important to realize that the conventional ranking technique compares the best regression models for different petrographic parameters. These regression models do not have the same polynomial order, and hence the comparison is not entirely equitable. The fuzzy logic algorithm compares all the parameters on the same basis and hence is superior in that aspect.

From the petrographic analysis, we concluded that Zone 1 is the main reservoir because it was not affected by anhydrite cementation. Zone 1 also indicated the effects of dissolution and thus has the highest porosity and permeability. This detailed petrographic analysis also helped to understand the subtle changes in the depositional environment and the development of the depositional model.

LOG ANALYSIS

As mentioned previously, only four modern log suites were available, and the rest of the field contained only old gamma ray and neutron logs. Due to the absence of other logs (e.g., resistivity, sonic, density, etc.) and the unreliability of the old neutron logs, correlations of core porosity, permeability, water saturation, and petrographic elements had to be made with gamma ray logs to predict these properties in the uncored wells. Both the old and new logs were digitized on 0.5 ft (0.15 m) intervals.

Developing Relationships Among Gamma Ray, Porosity, Total Water Content, and Permeability

In the Sulimar Queen and adjacent fields, a negative correlation was observed between old gamma ray (API) and neutron (API) values. This correlation implied that porosity increases with the increase in gamma (Ali et al., 1996). Even though old neutron and gamma ray logs were not reliable, this behavior still helped us to understand the overall relationship between gamma ray values and porosity. A positive correlation was also observed between core porosity and gamma ray in the Sulimar Queen field (Figure 7A), as well as in the Queen field (Figure 7B), which is located 20 mi (32 km) northeast of Sulimar Queen field. Similar relationships between gamma ray and porosity in the Sulimar Queen and adjacent fields suggested a possible regional trend. This increased our confidence in using the correlation between core porosity and gamma ray, developed in one well (well 1-16) to be used for predicting porosity throughout the field (Figure 7A). The correlation may be expressed as

$$\phi = 0.334e^{0.0526(\gamma)} \tag{1}$$

where ϕ is the porosity in percent and γ represents the gamma ray value in the API units.

This relationship was good only for the sand portion of the Shattuck Member as determined from the core description. In the Shattuck zone, the gamma ray values were not controlled by the amount of clay, but by the amount of potassium feldspar (confirmed by petrographic and x-ray diffraction analysis) and uranium (confirmed by the spectral gamma ray log from one well in the Queen field) (Figure 8A). Unfortunately, the spectral gamma ray log was not available from well 1-16, in which core porosity, water saturation, and permeability data were available (Figure 8C). This uranium was assumed to be present in the formation water since gamma ray values are proportional to the total water content (W_{TC}) (Figure 9A). Similar relationships were also observed in the Queen (Figure 9B) and adjacent fields, again implying a regional trend. These trends emphasized that the relationship developed in only one well (1-16) could be applied with reasonable confidence throughout the field (Ali et al., 1996).

The total water content represents the percentage of the total rock volume occupied by the water. The conventional way of representing the water saturation (S_w)

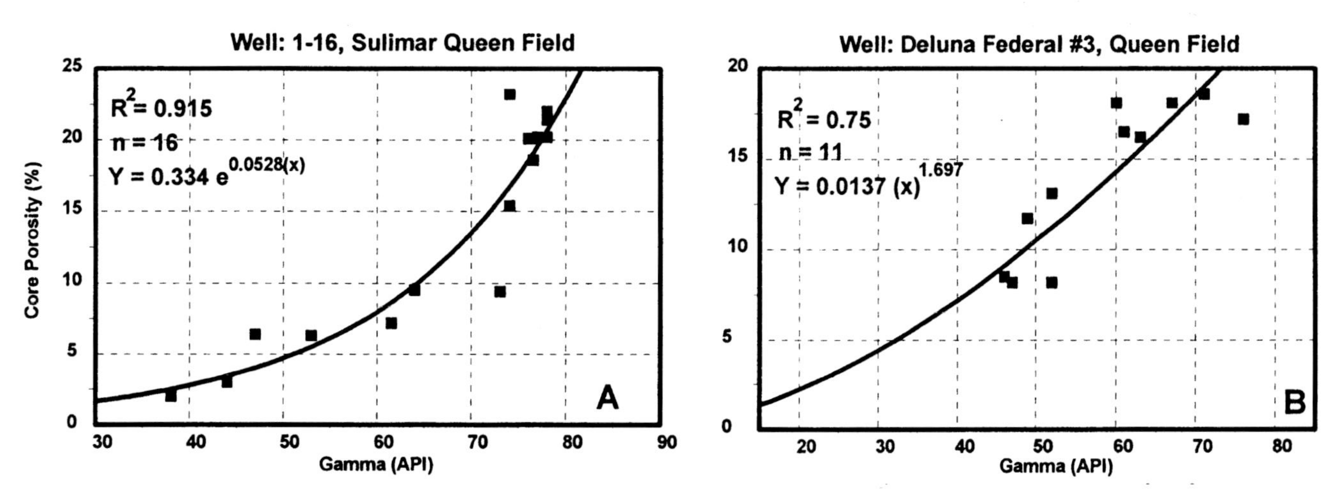

Figure 7. Plots showing the relationship between gamma ray values and core porosity in the (A) Sulimar Queen and (B) Queen fields.

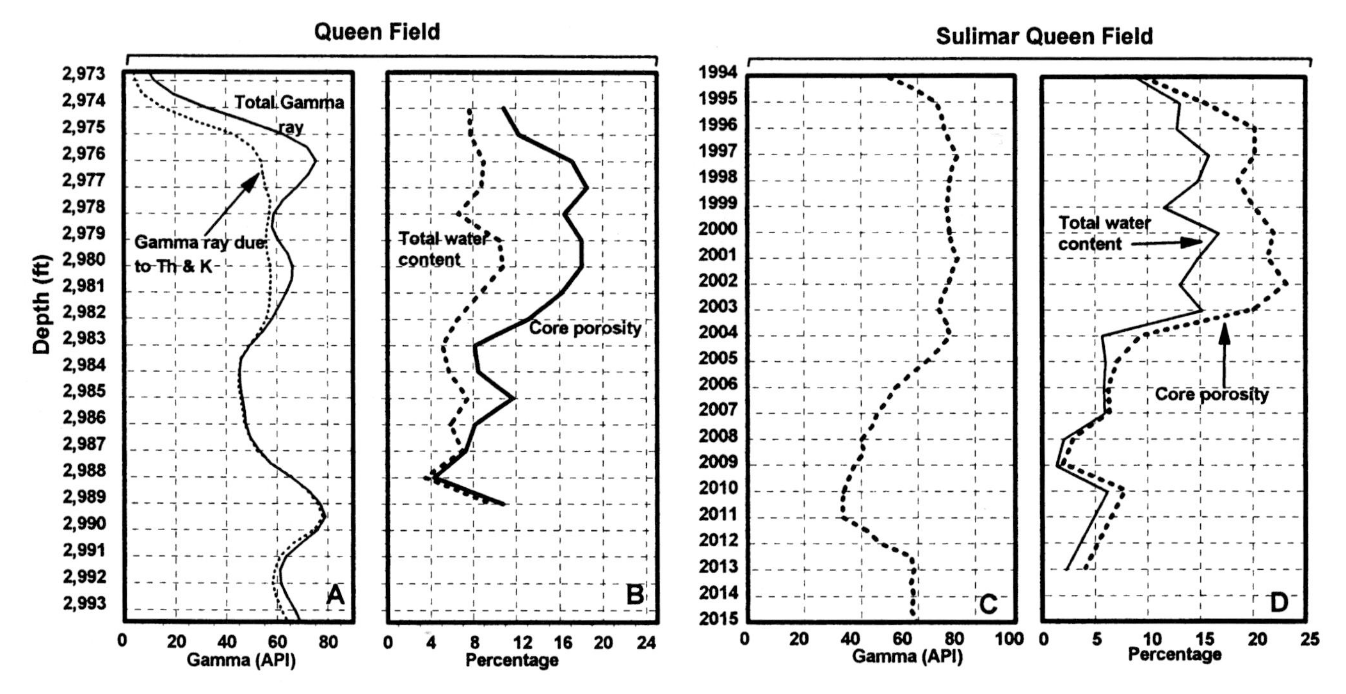

Figure 8. Plots showing the distribution of (A) spectral gamma ray and (B) porosity and total water content in the Queen field. Also shown are the (C) gamma ray log and porosity and (D) total water content in the Sulimar Queen field. As the porosity and total water content increase, so does the gamma ray value.

as the percentage of porosity occupied by water was not used because of its inadequacy in conveying the amount of water present. For example, a 5% porosity rock with water saturation of 90% contains less volume of water than 20% porosity rock with only 35% water saturation. If we use the conventional representation of the water saturation, the relationship between gamma ray response and volume of water could not be identified correctly. We suggest using total water content (W_{TC}) instead of conventional water saturation (S_w) in order to understand the relationship between the gamma ray response and the control of radioactive dissolved species in the water. The following relationship developed in well 1-16 was then used with caution to predict the total water content (W_{TC}) (Figure 9A):

$$W_{TC} = 4.610^{-5}\gamma^{2.9} \tag{2}$$

Here, W_{TC} is the total water content in percentage, and γ the gamma ray value in API units. Once the total water content is calculated, water saturation with a particular porosity can be calculated using:

$$S_W = \left(W_{TC}/\phi\right)\times 100 \tag{3}$$

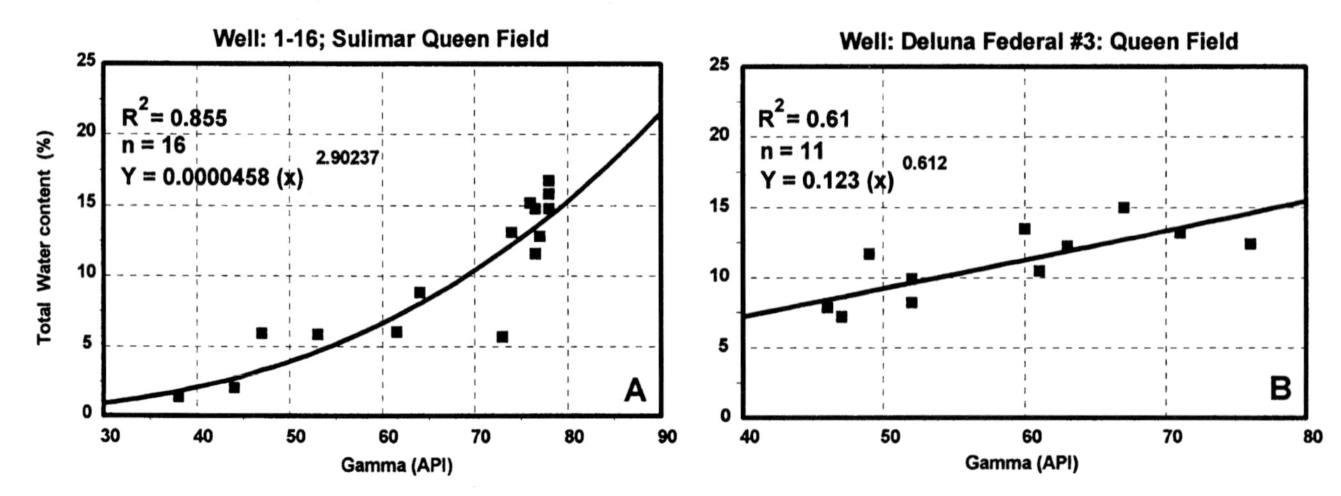

Figure 9. Plots showing the relationships between total water content and gamma ray values in the (A) Sulimar Queen and (B) Queen fields.

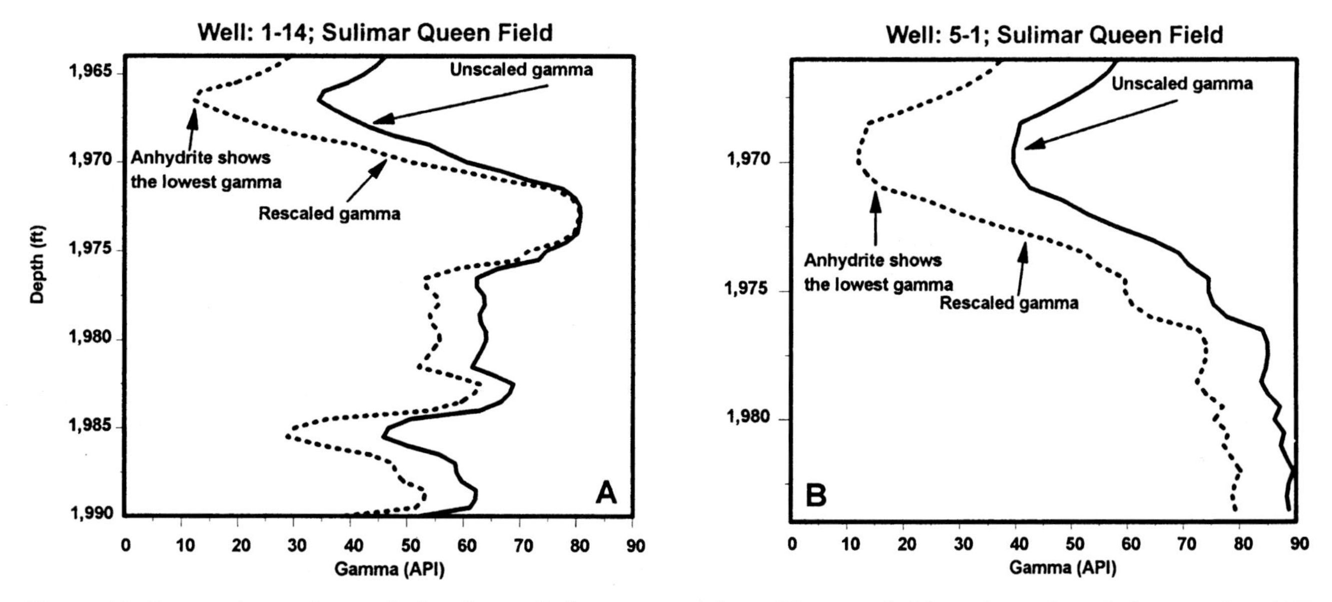

Figure 10. Comparison of unscaled and rescaled gamma ray logs. The rescaled logs have the minimum of 12 API and maximum of 80 API values. Note the recognition of anhydritic portions in the rescaled gamma ray logs that were not easily identifiable in unscaled logs.

Improving the Reliability of Old Gamma Ray Logs

The old gamma ray logs had varying scale ranges; therefore, a comparison of gamma ray API values from different wells was meaningless (Ali et al., 1996). Also, porosity and total water content obtained from the unscaled old gamma ray logs using the relationships developed in well 1-16 (equations 1 and 2) were erroneous. Because of the different scale ranges of the old gamma ray logs, the extrapolation beyond the range of original data values (well 1-16) yielded wrong results. In order to make old gamma ray logs useful and reliable, they had to be rescaled to the same datum so that the information provided by each well became comparable.

The old gamma ray logs were rescaled using the methodology proposed by Barrett (1994). In this methodology, modern logs are needed to determine the representative value ranges and the average high and low values in the zone of interest in that area. The average high (80.5 API units) and low (12.0 API units) gamma ray values were determined from the four modern logs available in the field. Using these average values, all the old gamma ray logs were

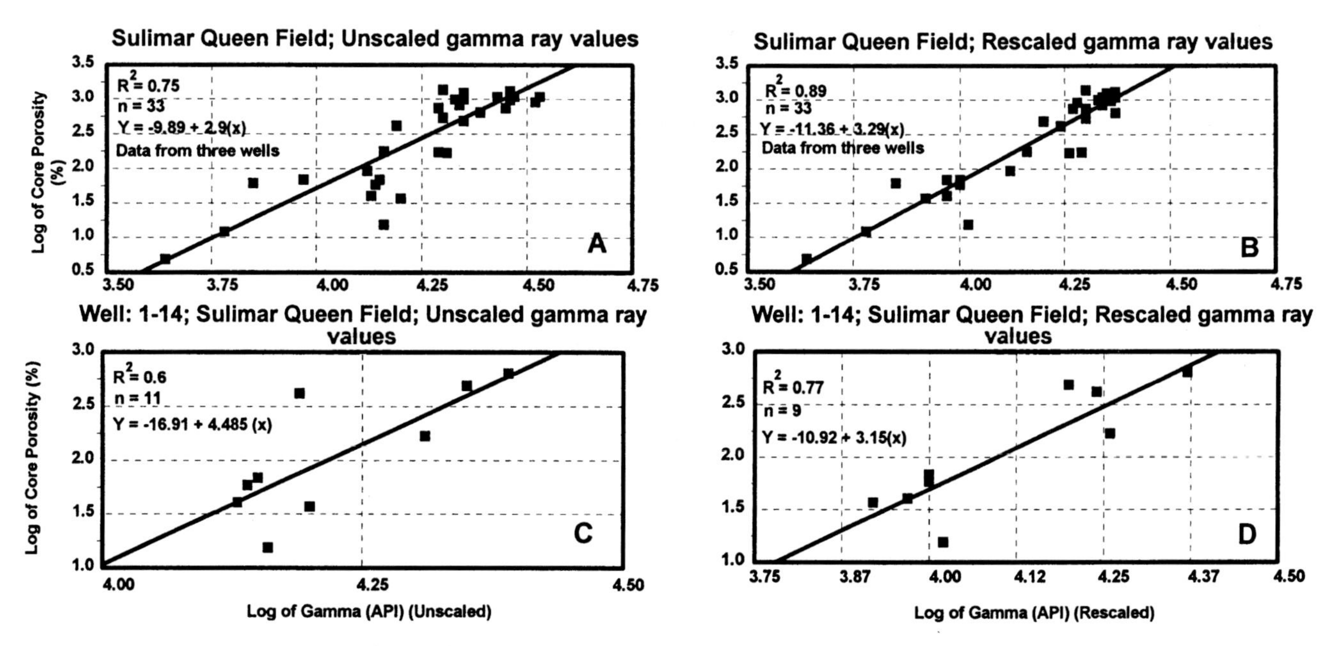

Figure 11. Plots showing the effects of rescaling on the correlation between core porosity and gamma ray values. Note the improvement of correlation from (A) to (B) and (C) to (D).

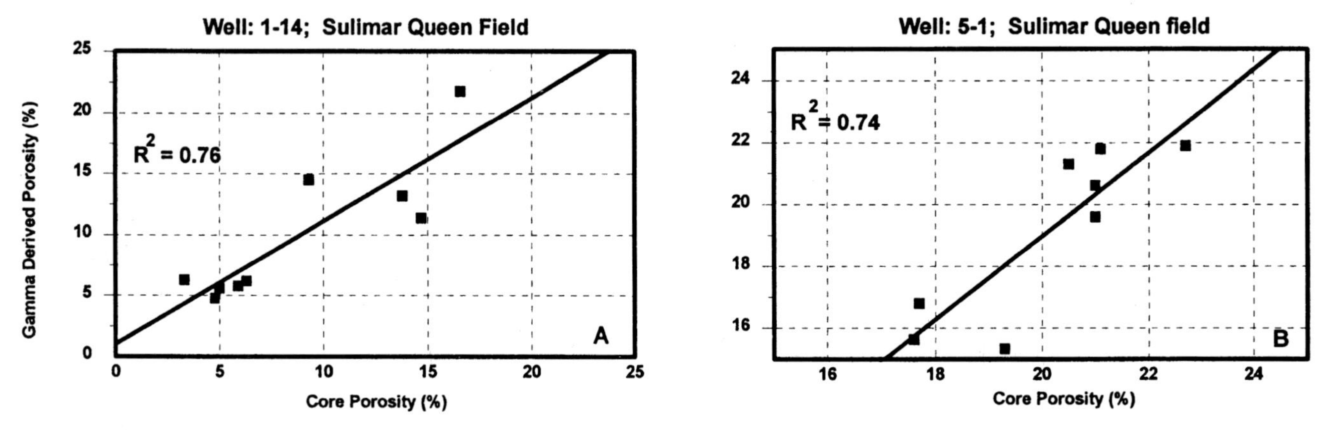

Figure 12. Comparison of core porosity and the porosity predicted using rescaled gamma ray logs in the Sulimar Queen field.

rescaled. Examples of the rescaled logs from well 1-14 and 5-1 are shown in Figure 10. The increase in the reliability of the old gamma ray logs after rescaling is also evident from the improved correlation coefficient between the core porosity and the gamma ray values (Figure 11). The gamma ray-derived porosity and total water content using equations 1 and 2 were compared with the core porosity and total water content in wells 1-14 and 5-1 for reliability (Figures 12, 13). Porosity calculated from the rescaled gamma ray logs also captured the presence of individual high and low porosity zones which were not distinguishable in the porosity obtained from unscaled gamma ray logs (Figure 14).

Air permeability and porosity also indicated a positive relationship in the Sulimar Queen field (Figure 15A). Similar relationships were also observed in the adjacent fields (Ali et al., 1996). A positive correlation was also observed between gamma ray-derived porosity and permeability (Figure 15B). The relationship between gamma ray-derived porosity and permeability appears to be more reliable because it covers the whole permeability spectrum. When permeability was predicted using gamma ray values and relationship in Figure 15A, the highest predictable permeability was only 46 md, whereas when gamma ray derived porosity and the relationship in Figure 15B was used, the highest permeability value predicted was 116 md, which is close to the minipermeameter-measured permeability. Unfortunately, due to a lack of core material in Sulimar Queen field, the reliability of the permeability prediction could not be confirmed.

DEPOSITIONAL ENVIRONMENT

Two opposing interpretations for the deposition of the Queen Formation are available in the literature. Boyd (1958), Pray (1977), and Sarg (1977) proposed an aqueous "all wet" depositional model. They attribute the deposition in a subsiding lagoon based on the lack of evidence for extensive subaerial exposure in the Queen or the adjacent formations in the Guadalupe Mountains in Texas and New Mexico. On the other hand, Silver and Todd (1969), Mazzullo et

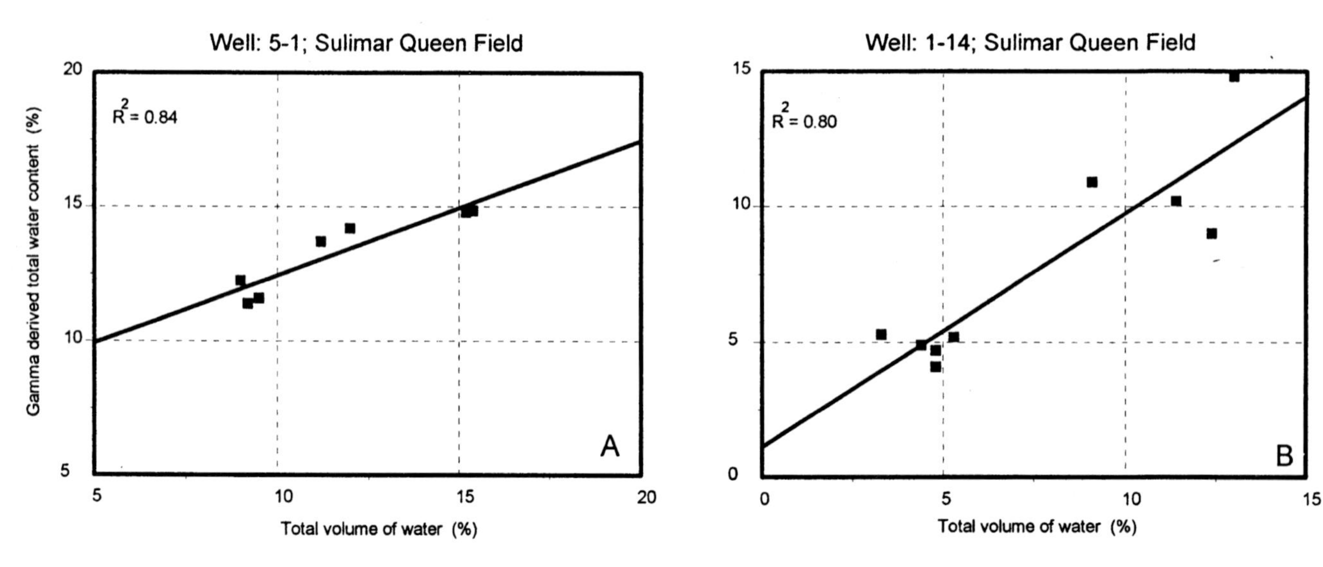

Figure 13. Comparison of core derived and rescaled gamma ray log derived total water content in the Sulimar Queen field. Note a strong correlation in both wells.

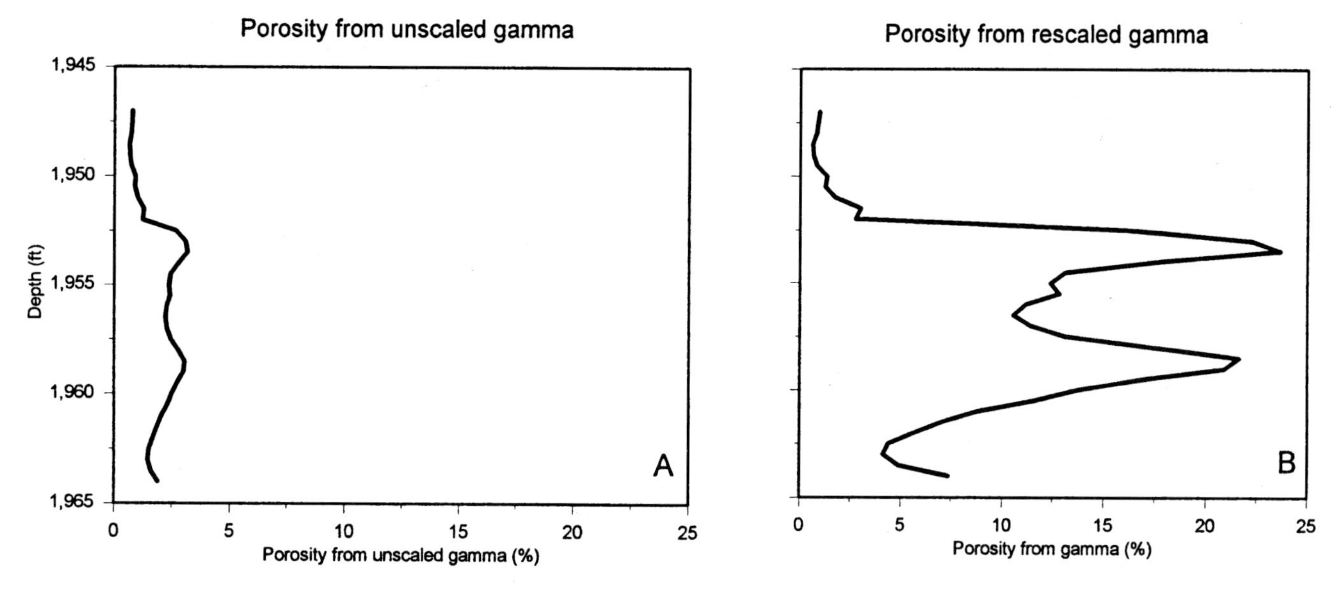

Figure 14. Comparison of porosity distribution as calculated from (A) unscaled and (B) rescaled gamma ray logs in well 5-2. High and low porosity zones are clearly visible in the porosity derived from (B) the rescaled gamma ray log.

al. (1984), Mazzullo and Hendrick (1985), and Malisce and Mazzullo (1990) suggested an eolian/continental environment. During the Guadalupian, the northwest shelf was a flat, slowly subsiding platform to the north of the Delaware Basin (Figure 1). During Queen deposition, a broad lagoon was present between carbonate sand shoals and the land–sea margin (Ball et al., 1971).

Our depositional model was constructed after assimilating all the information from the cores, petrographic analysis, logs, lithologic description of the cores and chip analysis (Haynes, 1978), and the outcrop data. In the Sulimar Queen and the South Lucky Lake fields, the Shattuck Member was deposited in a mixed coastal sabkha, shallow lagoon (with variable energy and depth), and eolian environment [dry eolian sand sheets of Malisce and Mazzullo (1990)] (Figure 16). These environments were established based on the variations in grain size and sorting, poorly developed sedimentary features (flaser bedding, ripple laminations, cross-laminations, bioturbation, and haloturbation), and the distribution of overlying and underlying anhydrite (Figure 16).

Zone 2 constitutes the bottom 10 ft (3 m) of the Shattuck Member and consists of gray and red sandstones and anhydrite (Figure 16). Massive, gray, moderately to well sorted, very fine grained arkosic sandstone is present toward the lower part of the Shattuck Member (2013–23015 ft; 614–706 m). Gray sandstone exhibits extensive haloturbation due to the growth of displacive anhydrite nodules and contains very thin discontinuous clay laminations. The gray sandstone is

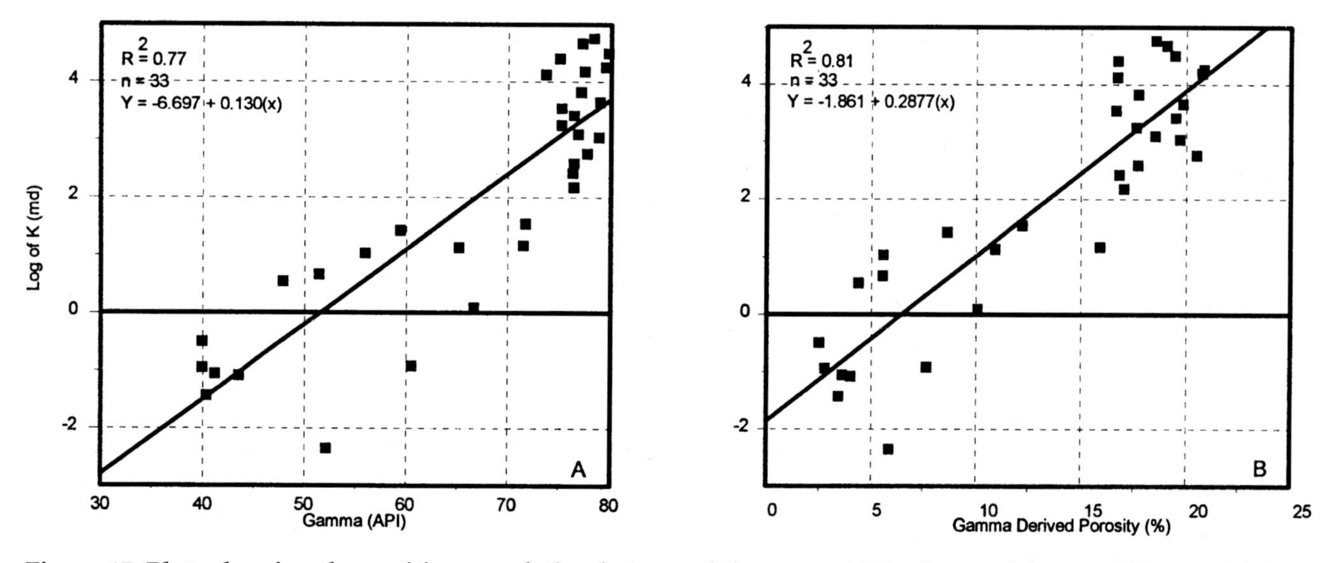

Figure 15. Plots showing the positive correlation between (A) gamma API values and permeability, and (B) gamma derived porosity and permeability in well 1-16 in the Sulimar Queen field.

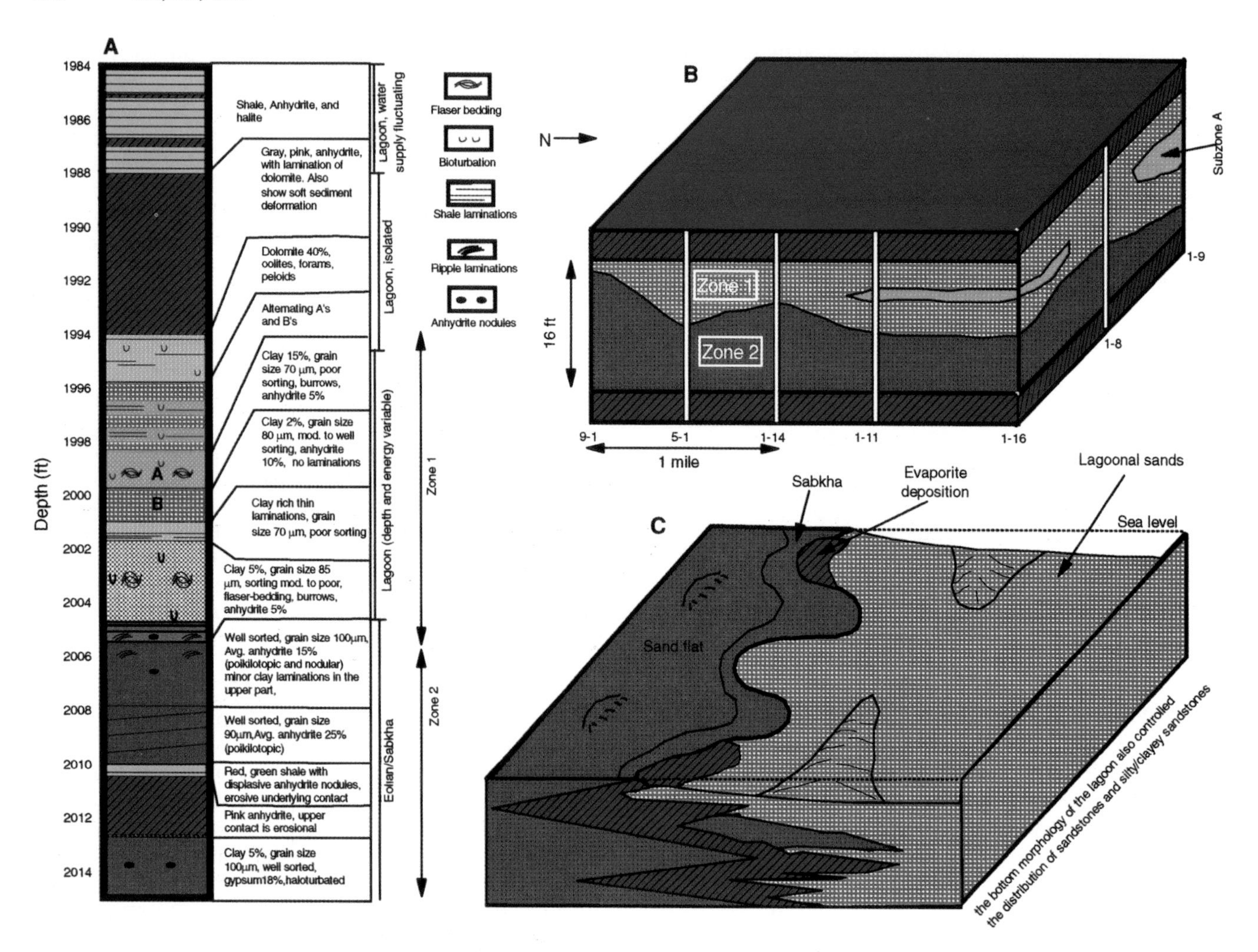

Figure 16. (A) Core description along with inferred depositional environments. (B) Block diagram of the Sulimar Queen field showing the distribution of Zone 1, Zone 2, and low-permeability Subzone A. (C) Generalized depositional model for the Shattuck Member in the Sulimar Queen and adjacent fields. The interfingering of different environments created the present distribution of the Shattuck Member.

overlain by a dominantly nodular anhydrite bed (Figure 16). Nodular anhydrite forms when gypsum precipitates displacively within the sediments. With increasing concentration of pore fluids across the sabkha, gypsum crystals are replaced by a fine mush of equant and lath-shaped anhydrite crystals (Tucker, 1981). Continued precipitation of anhydrite results in closely packed nodules with host sediments restricted to thin stringers. The nodular texture produced is referred to as chicken-wire anhydrite (Tucker, 1981). Nodular anhydrite forms above the water table in the capillary zone (Warren, 1991). According to Kendall and Warren (1988), the nodular anhydrite usually indicates deposition in a sabkha.

Both the lower and upper contacts of the nodular anhydrite present in Zone 2 are erosive (Figure 16). According to Warren (1991), erosion surface in the sandstone with nodular anhydrite and the capping nodular anhydrite are the most important features of a sabkha sequence, as observed in Abu Dhabi. Because of the nodular morphology of anhydrite beds, the presence of nodular anhydrite in the underlying gray sandstone, and the presence of erosional surfaces, a sabkha depositional environment is assigned to the gray sandstone and nodular anhydrite.

Overlying the anhydrite is a 5 ft (1.5 m) thick sequence of red and gray, well-sorted sandstone. The red sandstone grades upward into gray sandstone. The red color is mainly due to hematite (Fe_2O_3), whereas the gray color could be due to the presence of organic matter or pyrite (FeS_2) (Berner, 1971). We assume that the gray coloration is due to the presence of pyrite because there is no mention in the literature of the presence of a considerable amount of organic matter, which can cause gray coloration alone. If this is the case, the upward change from red to gray color probably represents a change from an oxidizing to a slightly reducing condition. These sandstones exhibit weak low-angle cross-stratification in the red sandstone and haloturbation in the gray sandstone (Figure 16). We assume that the cross-laminated and well-sorted red sandstone in Zone 2 was deposited in an eolian (sand flat) environment. In the Shattuck Member, low-angle cross-bedded sandstones deposited in sand flats were also reported by Malisce and Mazzullo (1990). The presence of haloturbation (anhydrite nodules) in the gray sandstone (similar to bottom gray sandstone) indicates its deposition in a sabkha environment. The distribution of Zone 2 was traced throughout the field using the rescaled gamma ray logs. Zone 2 thickens and becomes dominant in the western part of the field.

Zone 1 is the main reservoir zone. Zone 1 consists of the upper 12 ft (3.6 m) of the Shattuck Member, and consists of greenish gray and brown sandstones (Figure 16). The sandstones are composed of moderately to poorly sorted, very fine to fine-grained arkosic sandstones. These sandstones contain parallel to subparallel, continuous and discontinuous silt laminations, flaser bedding, wavy bedding, minor cross-laminations, and bioturbation. Nodular anhydrite is completely absent in Zone 1.

The upper part of Zone 1 shows a gradual vertical change from anhydrite-cemented to dolomite-cemented sandstones to sandy dolomite. Zone 1 grades upward through sandy dolomite into laminated anhydrite. This transition was observed in the cores from Sulimar Queen and adjacent fields. The gradation of these sandstones vertically into bedded evaporites, especially laminated anhydrite, indicates deposition in a lagoon or a large salina (Kerr and Thomson, 1963). The laminated morphology and the regional extent of the overlying anhydrite point toward deposition in a lagoon/large salina instead of local ponds.

The distribution of sediments and sedimentary structures within a lagoon is controlled mainly by hydrographic conditions and availability of sediments (Reineck and Singh, 1973). Sands can be brought into the lagoon by several means—storms, tidal channels, rivers, or by wind blowing from land (desert) toward the lagoon. The alternating clay-rich beds, which contain thin laminations of clayey-silty sandstone, with the featureless poorly to moderately sorted sandstone beds in Zone 1 (Figure 16) suggest a minor fluctuation either in the energy of the lagoon or in the supply of sand from the land. The sands probably were supplied to the lagoon by the winds blowing from the desert as well as by occasional rainstorms.

Presence of flaser bedding, wavy laminations, horizontal laminations, bioturbation, minor ripple laminations, trace amounts of well-rounded and frosted grains, and absence of nodular anhydrite indicate that much of the deposition of these sandstones took place in a lagoonal/salina environment.

The Shattuck Member present in the Sulimar Queen field was deposited in the part of the lagoon present toward the land as inferred from the location of the Goat Seep reef relative to the Sulimar Queen field. We assume that diurnal tides were not responsible for the formation of flaser beds. Flaser beds probably were formed due to the wind-induced tides. In shallow waters, the sediment surface may be strongly influenced by waves actively driven by winds (forced waves) (Collins and Thompson, 1982), which can cause the formation of ripples and flaser bedding. We suggest that strong wind-induced tides created the flaser bedding, and weaker wind-induced tides formed the continuous and discontinuous horizontal and wavy laminations which are frequently present in Zone 1. Elliot (1986) reported the formation of silt/sand laminae in Laguna Madre due to slight wave agitation.

The lack of evidence of subaerial exposure and sedimentary structures typical of sabkha, fluvial, or eolian environments also supports the deposition of sandstones in Zone 1 in a lagoonal environment. The absence of fossils in the sandstones may be attributed to hypersaline conditions.

We conclude that the Shattuck Member in the Sulimar Queen and adjacent fields was deposited in an imperceptible merging of lagoonal, sabkha, and eolian environments that produced the widespread sheetlike geometry. Eolian sands blew into the lagoon when winds blew seaward. Sea level fluctuations were probably small, but they caused the interfingering of quartz sands (Zone 1 and Zone 2) of different environments.

NUMERICAL SIMULATION OF THE SULIMAR QUEEN

The overall objective of this project was to characterize the Sulimar Queen from a limited data set. The confirmation of successful characterization is accomplished through a reliable reservoir model that honors most static (petrophysical and geological) and dynamic (production and pressure) data, and results in a reasonably accurate forecast of the reservoir performance.

The old gamma ray and neutron logs (unscaled) were useful only for detecting the radioactive Shattuck sand, and were treated as perforating logs. Rescaling the logs provided improved correlations to predict porosity in the pay zone. The process of rescaling also improved the log signatures for detecting the anhydrite sequences bounding the pay sand, which helped in delineating the sand thickness and the top of the structure for the 36 wells present in the field. Once the top of the structure was accurately identified in all the wells, a simple geostatistical mapping algorithm was applied to obtain a smooth fieldwide top of the structure. Using geologic input, this structure was modified manually (Figure 17). The initial porosity field distribution was also estimated using the same geostatistical mapping technique. The permeability control points in the field were obtained from inverse drill-stem tests, which were performed at many wells. Using an analytical k-ϕ relationship for each layer, derived from well 1-16, the initial permeability map was obtained.

The rescaling of the logs indicated the presence of a thin, tight zone that separated Zones 1 and 2. The outcrop histograms also suggested three distinct lithological units (Martin et al., 1996). The proposed geologic model and some of the well logs indicated that this zone may not be continuous throughout the field. Because the tight zone was thin and assumed to be

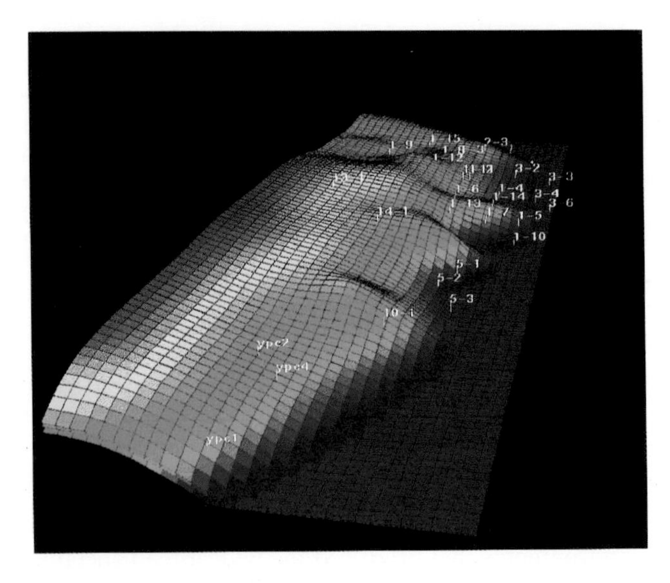

Figure 17. Top of the Sulimar Queen structure predicted using a simple geostatistical method manually adjusted to geologic input.

discontinuous, the final simulation model comprised two layers only: the upper simulation flow layer representing the zone with high permeability and the lower flow layer for the low-permeability zone. The thin, tight zone was combined with the lower layer.

The PVT analysis performed on oil samples collected from an adjacent field was used as initial PVT data in the simulator. These PVT data were unable to explain the large gas production in the initial producing life of the field because the reservoir was initially assumed to be undersaturated. Careful analysis of the structure and production (gas:oil ratio) profile indicated the presence of a gas cap. This was later confirmed independently (W. Lemay, 1996, personal communication). Because the only accurate measurement in the field was the oil production, this was used as an inner boundary condition constraint in an automatic history-matching algorithm developed at the Petroleum Recovery Research Center. Based on this constraint, a good history match was obtained for the primary production. The waterflood forecast obtained by using the reservoir model derived after primary production history match is shown in Figure 18.

We considered the Sulimar Queen to be successfully characterized based on the detecting and locating the previously unconfirmed gas cap, combined with a consistent geologic description and reasonable simulation forecasts.

CONCLUSIONS

In this study, new techniques were applied for improving the quantity and quality of data for better reservoir characterization of old fields.

(1) The minipermeameter is useful in capturing the small-scale heterogeneities. Permeability measurements made on fine grids also improve the lithologic description of the cores.

(2) Combination of minipermeameter and petrographic analysis allows the collection of a large amount of petrographic data using fewer thin sections. The control of different petrographic elements on permeability can be easily and accurately determined. Correlations developed between the permeability and petrographic elements are also reliable.

(3) The newly developed fuzzy logic algorithm is a fast, unbiased, and reliable method for establishing the importance of each petrographic element on permeability.

(4) Rescaling of the old logs improves their reliability. In the Shattuck Member, porosity, permeability, and water saturation could be estimated using the gamma ray logs.

(5) Maximum geological and petrophysical information was gathered from the limited database by combining small-, medium-, and large-scale data obtained from minipermeameter, cores, petrographic

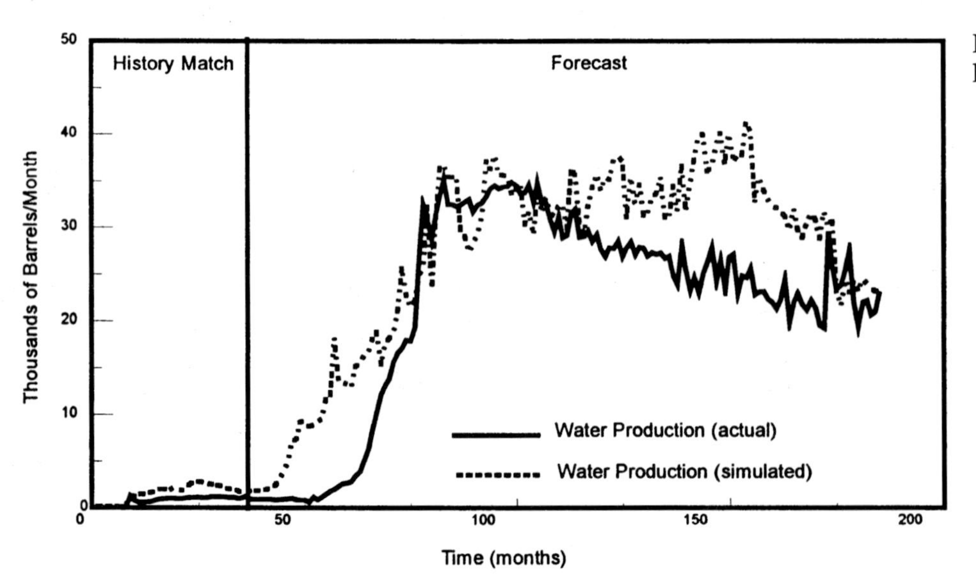

Figure 18. Waterflood history match and forecast.

analysis, and logs. The distribution of the different zones was mapped successfully in the field, and a depositional model was developed.

(6) Using these data, a successful characterization of the Sulimar Queen was obtained and confirmed through a good history match of the primary production and a reasonably accurate forecast of the waterflood performance.

ACKNOWLEDGMENTS

This research was funded by the Department of Energy (contract no. DE-AC22-93BC14893), State of New Mexico, Landmark Graphics Corporation (Halliburton), Texaco, and Yates Petroleum. The authors greatly appreciate the financial support.

REFERENCES CITED

Ali, M., A. Chawathé, A. Ouenes, and W. Weiss, 1996, Improved log analysis for the characterization of the Sulimar Queen field, southeast New Mexico: PRRC Report No. 96-33, 32 p.

Ali, M., 1997, Improved geological characterization of old hydrocarbon fields with sparse control points: A case study from the Sulimar Queen Field, southeast New Mexico: Ph.D. Dissertation, New Mexico Institute of Mining and Technology, Socorro, New Mexico, 410 p.

Ball, S. M., J.W. Roberts, J.A. Norton, and W.D. Pollard, 1971, Queen Formation (Guadalupian, Permian) outcrops of Eddy County, New Mexico, and their bearing on recently proposed depositional models: American Association of Petroleum Geologists Bulletin, v. 55, p. 1348-1355.

Barrett, G. D., 1994, Use of computers to perform old log analysis: SPE Computer Applications, August, p. 9-12.

Berner, A. R., 1971, Principles of chemical sedimentology: McGraw-Hill Book Company, New York, 240 p.

Blatt, H., 1982, Sedimentary petrology: W. H. Freeman and Company, New York, 564 p.

Boyd, D. W., 1958, Permian sedimentary facies, central Guadalupe Mountains, New Mexico: New Mexico Bureau of Mines and Mineral Resources Bulletin, No. 40, 100 p.

Collins, D. J., and B. D. Thompson, 1982, Sedimentary structures: George Allen and Unwin, London, 194 p.

Elliot, T., 1986, Siliciclastic shorelines: in ed., Reading, H. G., Sedimentary Environments and Facies, Blackwell Scientific Publications, p. 155-189.

Goggin, D. J., R.L. Thrasher, and L.W. Lake, 1988, A theoretical and experimental analysis of minipermeameter response including gas slippage and high velocity flow effects: In Situ, v. 12, p. 79-116.

Haynes, L. C., 1978, Sandstone diagenesis and development of secondary porosity, Shattuck Member, Queen Formation, Chaves County, New Mexico: M.A. thesis the University of Texas at Austin, 148 p.

Kendall, C. G. St. C., and K. J. Warren, 1988, Peritidal evaporites and their sedimentary assemblages: in Schreiber, C. B., ed., Evaporites and hydrocarbons: Columbia University Press, New York, p. 66-138.

Kerr, D.S., and A. Thomson, 1963, Origin of nodular and bedded anhydrite in Permian shelf sediments, Texas and New Mexico: American Association of Petroleum Geologists Bulletin, v. 47, p. 1726-1732.

Lin, Y., 1994, Input identification of Modeling with fuzzy and neural systems: Ph.D. dissertation, New Mexico Institute of Mining and Technology, Socorro, New Mexico, 89 p.

Malisce, A., and J. Mazzullo, 1990, Reservoir properties of the desert Shattuck Member, Caprock field, New Mexico: *in* eds., Barwis, J. H., McPherson, J. G., and Studlick, R. J., Sandstone Petroleum Reservoirs: Springer-Verlag, New York, p. 133-152.

Martin, D. F., J. Buckley, W. Weiss, and A. Ouenes, 1996, Integration of advanced geoscience and engineering techniques to quantify interwell heterogeneity: Quarterly Technical Report submitted to DOE under contract no. DE-AC22-93BC14893, 24 p.

Mazzullo, J. M., M. Williams, and S.J. Mazzullo, 1984, The Queen Formation of Millard Field, Pecos County, Texas: its lithologic characteristics, environment of deposition, and reservoir petrophysics: Transactions, southwest section American Association of Petroleum Geologists Publication 84-78, p. 103–110.

Mazzullo, S. J., and C.L. Hendrick, 1985, Road log and locality guide, lithofacies, stratigraphy, and depositional models of the back-reef Guadalupian section (Queen, Seven Rivers, Yates, and Tansill formations), SEPM Field Guidebook, p. 1-30.

Pray, L. C., 1977, The all wet, constant sea level hypothesis of Upper Guadalupian, shelf and shelf edge strata, Guadalupe Mountains, New Mexico and Texas: West Texas Geological Society Special Publication No. 77-16, 443 p.

Reineck, H. E., and I. B. Singh, 1973, Depositional sedimentary environments: Springer-Verlag, Berlin, 439 p.

Sarg, J. F., 1977, Sedimentology of the carbonate-evaporite facies transition of the Seven Rivers Formation (Guadalupian, Permian) in southeast New Mexico: SEPM Permian Basin Section Special Publication, No. 77-16, p. 451-478.

Silver, B. A., and R.G. Todd, 1969, Permian cyclic strata, northern Midland and Delaware basins, west Texas and southeastern New Mexico: American Association of Petroleum Geologists Bulletin, v. 53, p. 2223-2251.

Suboor, M. A., and J.P. Heller, 1995, Minipermeameter characteristics critical to its use: In Situ, v. 19, no. 3, p. 225-248.

Tucker, E. M., 1981, Sedimentary petrology, An introduction: John Wiley & Sons, New York, 252 p.

Warren, K. J., 1991, Sulfate dominated sea-marginal and platform evaporite settings: sabkhas and salinas, mudflats and salterns: Melvin, L. J. ed., Evaporites, petroleum and mineral resources: Developments in Sedimentology, v. 50, Elsevier, New York, p. 69-188.

Refunjol, B.T., L.W. Lake, Reservoir characterization based on tracer response and rank analysis of production and injection rates, 1999, *in* R. Schatzinger and J. Jordan, eds., Reservoir Characterization-Recent Advances, AAPG Memoir 71, p. 209–218.

Chapter 15

Reservoir Characterization Based on Tracer Response and Rank Analysis of Production and Injection Rates

Belkis T. Refunjol
Lagoven, S.A., PDVSA
Venezuela

Larry W. Lake
Center for Petroleum and Geosystems Engineering
The University of Texas at Austin
Austin, Texas, U.S.A.

ABSTRACT

This paper presents the results of a practical technique to determine preferential flow trends in a reservoir. The technique is a combination of reservoir geology, tracer data, and Spearman rank correlation coefficient analysis of injection/production rate data. The Spearman analysis, in particular, will prove to be important because it appears to be insightful and uses data that are prevalent when other data are nonexistent. The technique is applied to the North Buck Draw field, Campbell County, Wyoming.

This work provides guidelines to assess information about reservoir continuity in interwell regions from widely available measurements of production and injection rates at existing wells. When successfully applied, the information gained can contribute to both the daily reservoir management and the future design, control, and interpretation of subsequent projects in the reservoir, without the need for additional data. As with other techniques, however, the method gives the most confidence when corroborated by other procedures.

INTRODUCTION

Appropriate representation of reservoir heterogeneity is important to successful modeling of past and future production performance. Because of this, description and quantification of reservoir heterogeneity have achieved significant progress in recent years. Among the new developments are depositional and stochastic models of reservoir heterogeneities.

Generating good stochastic or geological models requires a lot of data. These modeling approaches are difficult, if not impossible, in reservoirs where production records are the only data available. Moreover, many geological models do not account for the fluid processes taking place in the reservoir. The technique proposed here relies entirely on fluid data by using only production records. In this trial use of the technique, we also use tracer data and results of a geologic study to validate the model.

The following sections contain the results from the tracer-response analysis and from the nonparametric statistical analysis of the production and injection rate ranks. The analysis is extended to investigate the effect of lag time on well-to-well correlations. The use of lag time analysis is, as far as we can tell, new to the application of Spearman rank analysis to production data. We conclude with a reservoir description based on rank correlation coefficient and gas tracer analysis.

RESERVOIR OVERVIEW

The study area is the North Buck Draw field, located in Campbell County, Wyoming, approximately 60 miles northeast of Casper (Figure 1). The field produces from the Lower Cretaceous Fall River Formation in the west-central portion of the Powder River Basin. The limits of the Buck Draw field and the delineation of the Fall River channel are shown on a structural map in Figure 1.

General Geology

Several studies have been made of the Buck Draw field stratigraphy and depositional patterns (Hawkins and Formhals, 1985; Rasmussen et al., 1985; Sellars and Hawkins, 1992). A recent interpretation (Gardner et al., 1994) based on outcrop studies revealed a complex valley-fill architecture and reservoir compartmentalization. The valley-fill system is interpreted to be composed of multilateral and vertical channel facies that change in character with geographic and stratigraphic position within a valley.

The main hydrocarbon-trapping mechanism at Buck Draw is the meandering of the channel. The reservoir is more shaly in the northwest portion of the field, with thin, low-permeability sands. This change in lithology is interpreted to form a barrier to hydrocarbon migration in the northwesterly direction. Other barriers are the shaly facies of the Fall River that form the bottom seal, and the overlying shale of the lower Skull Creek (Sellars and Hawkins, 1992).

Reservoir Development

North Buck Draw field porosity and permeability are areally and vertically heterogeneous. Both properties appear to be greater along the middle-stream axis of the channel, with expected preferred fluid-flow direction along this axis. Generally, reservoir sand quality declines in a direction normal to the channel axis.

The reservoir fluid is a near-volatile oil; fluid properties fall between those of black and volatile oils. The fluid meets the majority of volatile-oil criteria, including large oil formation volume factors and solution gas-oil ratios. The bubblepoint pressure is 4680 psia, and the reservoir fluid is a single-phase, low-viscosity (0.12 cp) fluid above this pressure. There is no free water in the reservoir.

North Buck Draw commercial production began in June 1983. In 1988 a pressure maintenance project was initiated by injecting gas. Radioactive tracers were injected into the reservoir, and their occurrence was monitored at the producing wells from February 1989 until March 1993. These tracer data should provide helpful information for the reservoir description.

TRACER BEHAVIOR AND RESERVOIR DESCRIPTION

A total of seven injectors were tagged with one of six gas tracers in a three-stage program. The gas tracers used to distinguish breakthrough from the different injectors at each producer were tritium (HT), krypton-85 (Kr-85), tritiated methane (CH3T), tritiated ethane (C2H5T), tritiated propane (C3H7T), and sulfur hexafluoride (SF6). The tracer program was carried out over a 4-yr period, long enough to use the tracer-response data to define fluid movement for investigation of reservoir heterogeneity.

During the tracer project, samples to determine tracer concentration were taken from all of the producers in the field. Figure 2 shows the areal distribution of the tracer response in each producer. Injector-producer communication is good for wells on the west side of the channel (injectors 11-18, 12-7, and 14-18). There was apparently no communication in producers from injectors 22-17, 22-20, 23-31, and 23-8 on the east side of the channel.

Tracer or tagged gas production at producers in the west, some at early and some at late breakthrough times, indicated no sealing geological barriers are present. Conversely, the lack of response to tracers injected in wells on the east side of the reservoir indicates no communication or, at least, less communication between these wells and the rest of the reservoir.

The apparent absence of multiphase flow is ideal for the use of tracer-response curves for calculating gas distribution and swept volumes based on the procedures commonly used in waterflood tracing. The assumptions made were that partitioning into oil would be minimal at reservoir conditions, and that since the majority of tracers used in this project were light components, they would partition exclusively into the gas phase in the field test separators where samples were gathered for analysis.

Calculations were performed for each pair of wells that had a detectable tracer response. As summarized in Table 1, the tracer recoveries were extremely small, less than 9% after 4 yr of production; however, there is a good chance that at least 50% of the wells showing tracer production were sampled after breakthrough had occurred. A large volume of tracer could have been produced before measurement started. We made no further attempt to identify the causes for the low tracer recovery obtained. The qualitative results of the tracing program were helpful in identifying directional flow trends when analyzed together with the Spearman rank correlation results as follows.

NONPARAMETRIC STATISTICS IN RESERVOIR DESCRIPTION

Parametric statistics involve the calculation of a statistic from a sample and the comparison of this statistic with a population parameter. If the statistic corresponds to a highly unlikely value of the parameter, the

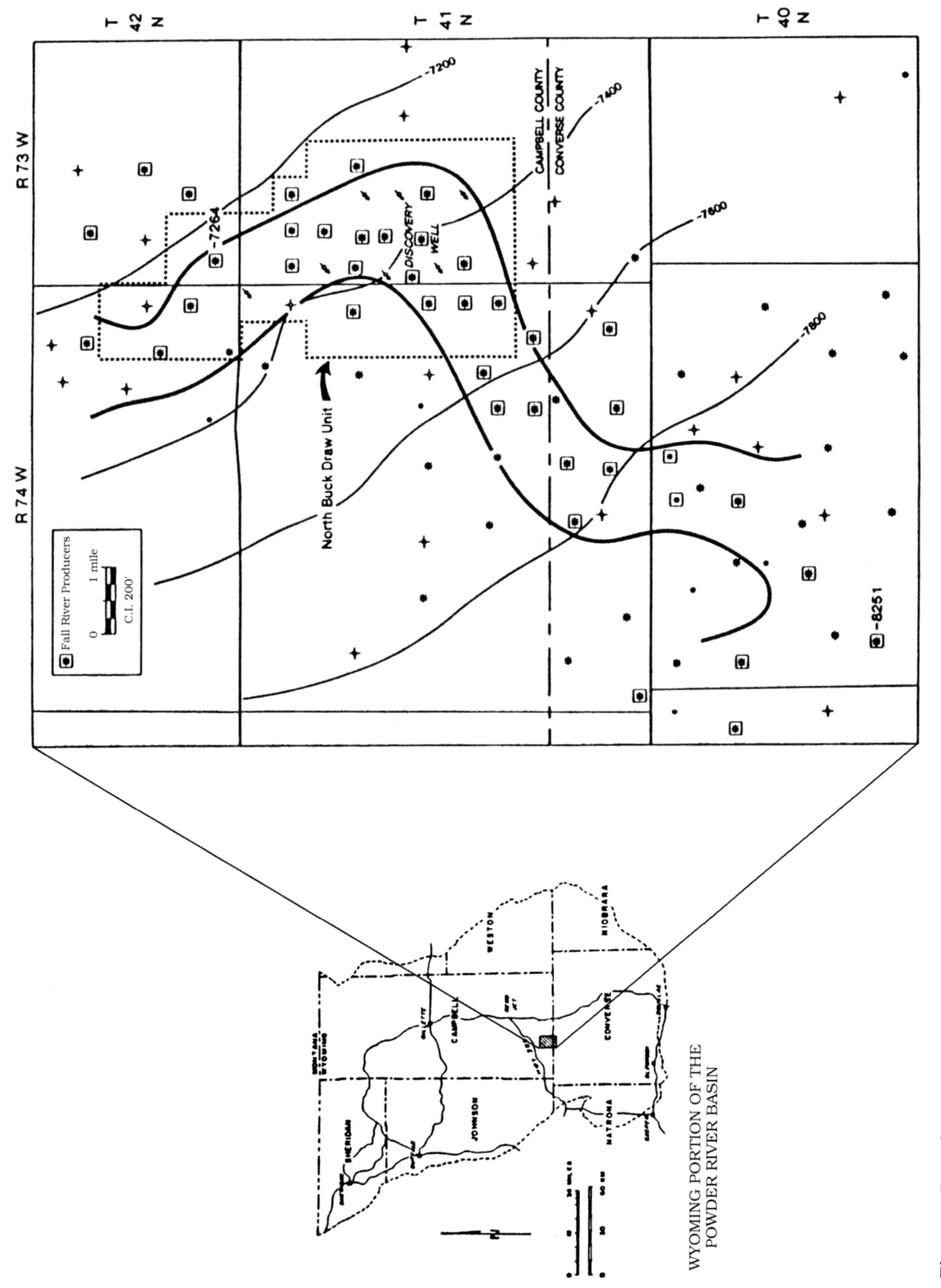

Figure 1. Location map and limits of North Buck Draw field.

Table 1. Tracer Recoveries.

Tracer	Injected into Well No.	Recovery (%)
CH3T	11–18	9.0
C2H5T	12–7	8.6
C2H5T	14–18	6.2
Kr-85	23–8	9.2

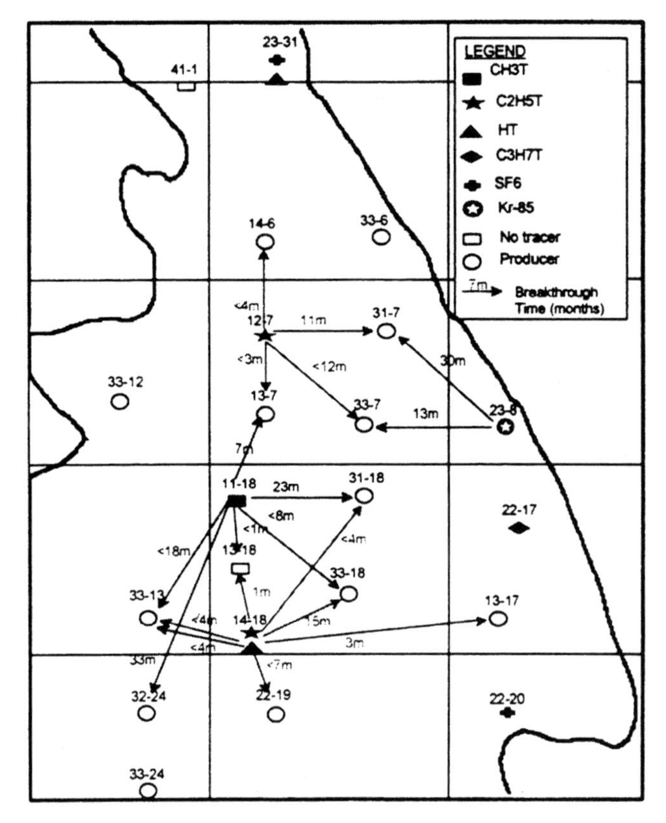

Figure 2. Tracer response pattern map. CH3T = tritiated methane, C2H5T = tritiated ethane, HT = tritium, C3H7T = tritiated propane, SF6 = sulfur hexafluoride, Kr-85 = krypton-85.

sample is assumed not to have come from the population described by the parameter (Volk, 1969). Parametric statistical tests often use the mean and variance of a distribution, which require operations on the original numerical values. Tests not employing the original data directly are called nonparametric methods; they generally depend on ranking (that is, a set of observations arranged in order of size) rather than using their actual numerical values.

Nonparametric statistics do not require conditions on the parameters of the population from which the samples are drawn. They do not involve any assumptions about the distribution of the population; they are parameter free. Certain assumptions are associated with most nonparametric statistical tests, such as that the observations are independent and that the variable under study has underlying continuity, but these assumptions are fewer and less restrictive than those associated with parametric tests (Siegel, 1956). For measuring the degree of correlation, correlation coefficients such as the Spearman or the Kendall coefficients are appropriate for ordinal scales.

The Spearman rank correlation coefficient (r_s) is a function of the sum of the squares of the difference of the two rankings for each observation and the number of observations.

$$r_s = 1 - \frac{6}{n\left(n^2 - 1\right)} \sum_{i=1}^{n} d_i^2$$

where r_s = Spearman rank correlation coefficient, d_i = difference between the rankings of the ith observations, and n = number of observations. Because r_s is a correlation coefficient, it has the property that $-1 \le r_s \le 1$. If there is perfect positive correlation, all the differences will be zero and $r_s = 1$. If there is perfect negative correlation, in which the low-ranking observation in one classification corresponds to the high-ranking observation in the other, the term $6\Sigma d^2$ will be equal to $2n(n^2 - 1)$, and the correlation coefficient will be equal to –1 (Volk, 1969). If the two ranking sets are independent, r_s will be zero.

The Spearman rank correlation coefficient is a quick, simple, and powerful test of the existence of the association between variables, regardless of the population distribution from which the samples are drawn. If all the assumptions of the parametric statistical model are met in the data, the nonparametric statistical tests are wasteful of data. The degree of wastefulness is expressed by the power-efficiency of the nonparametric test (Siegel, 1956). The efficiency of the Spearman rank correlation when compared with the parametric correlation, the Pearson r, is about 91%.

Method Used

Rank correlations are more suitable than parametric correlations when data are not normally distributed. There is no reason for production and injection rates at Buck Draw to be normally distributed, so the parametric statistics should not be used. All rates from production and injection histories in the North Buck Draw field show fluctuations about some average value, and it is the relationship between these fluctuations that should reflect reservoir heterogeneities and preferential flow trends. None of the rates at Buck Draw were normally distributed.

The Spearman rank correlation coefficient, therefore, provides an ideal tool for investigating the correlation between production and injection rates in the reservoir. Data used for the analysis were the monthly production and injection rates from wells in the North Buck Draw field. The Spearman rank correlation coefficient was calculated between flow rates of injector/producer pairs of wells to establish dominant communication trends in the reservoir.

The calculations were based on the total (oil and gas) monthly fluid production and injection rates. The rates were converted to ranks, and the Spearman rank correlation coefficient was calculated for pairs composed of each injection well and all its adjacent production wells.

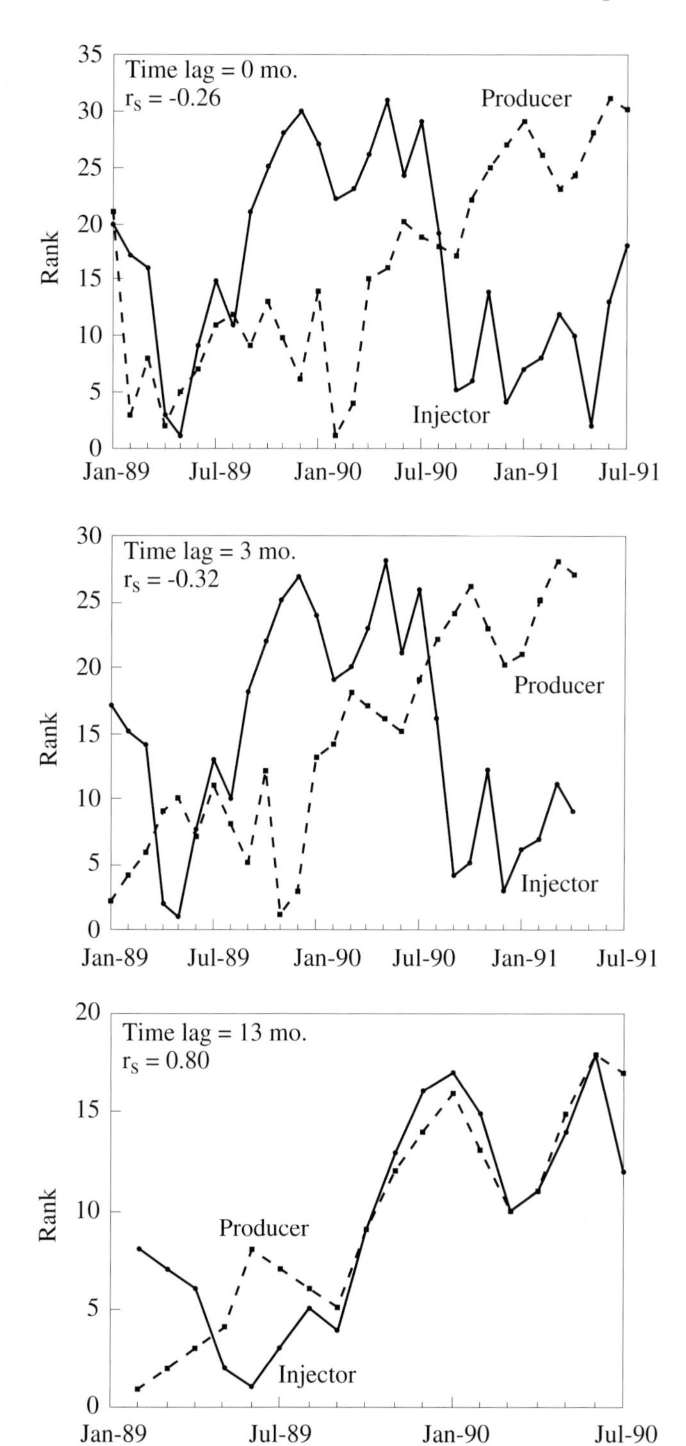

Figure 3. Spearman ranks between producer 33-7 and injector 11-16 vs. time for three time lags.

The upper plot in Figure 3 shows the well rate ranks for two typical wells, injector 11-18 and producer 33-7, in the Buck Draw field as a function of time.

An earlier application of the Spearman method is the Heffer et al. (1995) study, which attempted to corroborate fracture orientations determined by other techniques. In that application, correlations were sought at zero time lag, a reasonable assumption since that work was intended to determine variations in fracture density whose pressure fluctuations are carried primarily through the rock. Because we are dealing with fluctuations in fluid rates, the assumption of instantaneous response is inappropriate here because fluids are more compressible; consequently, we investigated the effect of nonzero time lags between injector-producer pairs.

Any response at a producer must occur at some time after the stimulus in the injector. The procedure employed for each pair of wells was first to convert production and injection rates to ranks, and then to calculate the rank correlation coefficient at zero lag time. From this, a series of correlation coefficients were calculated by shifting the time for the producers starting at 1 month up to a shifted time no greater than half the total number of months in the period analyzed. The two lower plots in Figure 3 show how the well rank histories change with lag time. In the case shown, the maximum r_s occurs at a time lag of 13 months. Note how shifting the lag time causes data to be discarded for both the injector and the producer.

The observed correlation coefficient at each lag time was plotted for every possible adjoining pair of wells in the field. Plots in Figure 4 present the correlation of the rates from injectors to the rates of surrounding producer wells. The results exhibit some negative r_{smax} values, which were interpreted as negative correlations. These negative correlations, which physically represent a decrease in injection rate with a subsequent increase in the production rate of the nearby well, could be explained by the influence of a third well. More analysis should be done to understand these negative correlations. See Refunjol (1996) for a possible way to account for them.

We also tried giving every injector with every producer in the reservoir (96 pairs). This method gave a few well pairs that showed significant r_{smax} at multiple pattern spacing. Similar to the negative correlations mentioned above, this must be because of multiple well effects; consequently, we limited subsequent study to adjacent well pairs (40 pairs). Heffer et al. (1995) also mention such large correlation at a distance.

Preferential Flow Direction Based on r_{smax}

One of the main objectives of this work was to infer preferential flow directions by the information provided from the rank correlation coefficient technique. To do this, the r_{smax} value was analyzed against information about the producer-injector distance in each pair, grouping the pairs based on the spatial orientation of the well pairs. Graphs were constructed for pairs of wells in five different groups: pairs oriented into 30° bins as shown in Figure 5. We constructed a histogram based on the orientation of the well pairs with maximum correlation (r_{smax}). Using 30° angle classes, Figure 5 shows a higher frequency of correlation in pairs with an orientation between N30E and N60E. This is in accordance with the southwest-northeast permeability trend observed from the geologic studies.

Along a given direction, the value of r_{smax} decreases with increasing distance in all the data analyzed, especially in the N30E to N60E direction group (Figure 6). This figure is actually an autocorrelogram that could

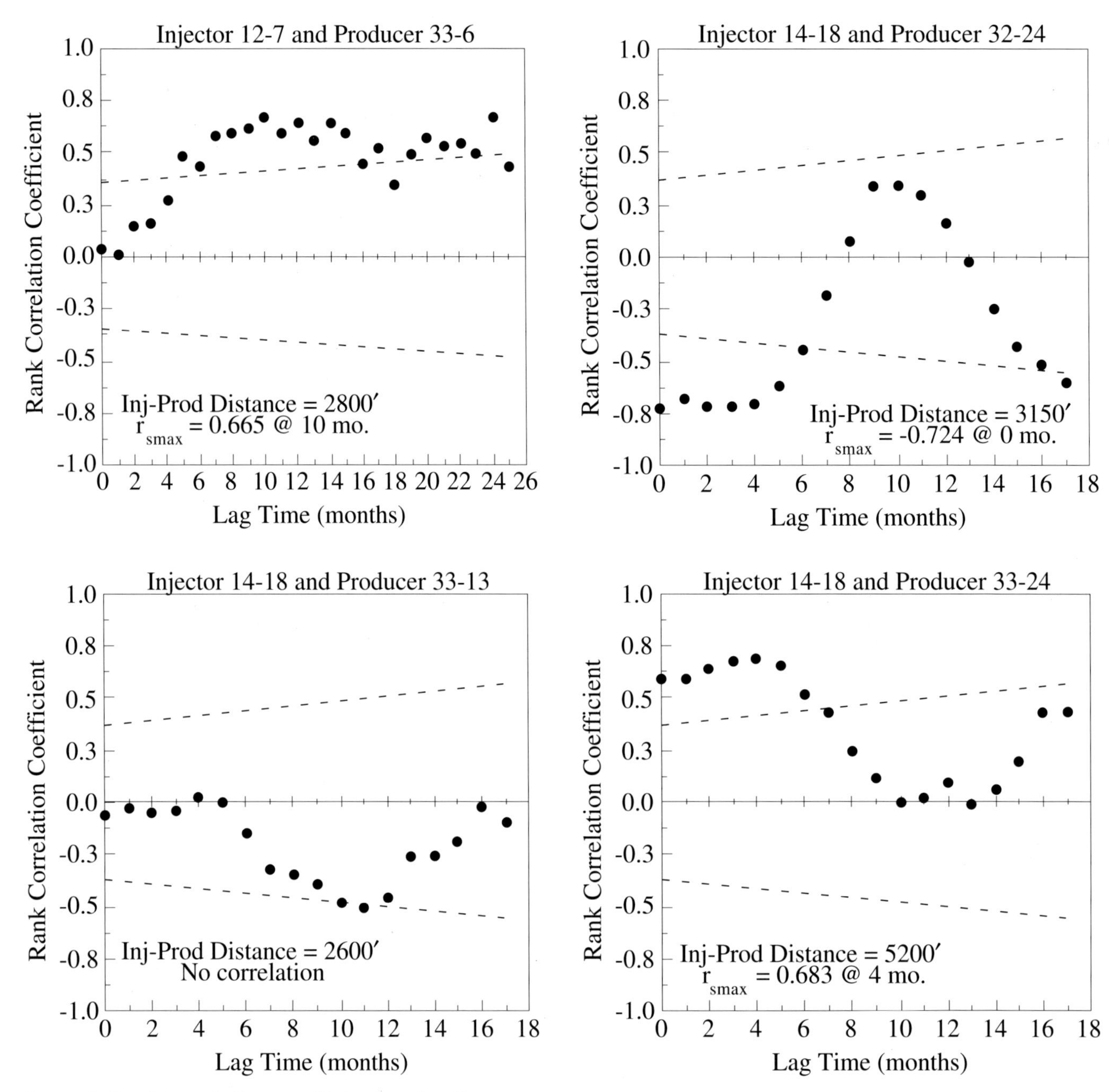

Figure 4. Rank correlation coefficient vs. lag time.

be interpreted with conventional models, such as the exponential or spherical (Jensen et al., 1996); we show a least-square straight-line fit in Figure 6. Whatever the model being used, there is clearly very long-range autocorrelation in the N30E to N60E direction, well in excess of 10,000 ft (3050 m).

Another way to illustrate directional trends is to map r_{smax}. We did this by linearly interpolating along a line from an injector to a producer between the calculated value of r_{smax} and $r_{smax} = 1$. This assumes that perfect correlation, $r_{smax} = 1$, occurs when the injector and producer are coincident. After this interpolation, we contoured the interpolated values around each injector. Figures 7–10 display these maps for injectors 11-18, 12-7, 13-18, and 14-18, respectively. All four maps exhibit a clear pattern of southwest-northeast preferential flow trend as expected. These wells are located on the west flank of the channel. The four injectors to the east of the channel showed no preferential flow trends (22-20) or even no correlation at all (22-17, 23-8, and 23-31).

Injector 22-20 presented significant rank correlation coefficients when related to the four surrounding wells, but with no clear preferential direction of correlation. In the analysis of each of these four pairs of wells, we observed that the producers were closer to and more influenced by other injectors than to well 22-20, suggesting no reliable information.

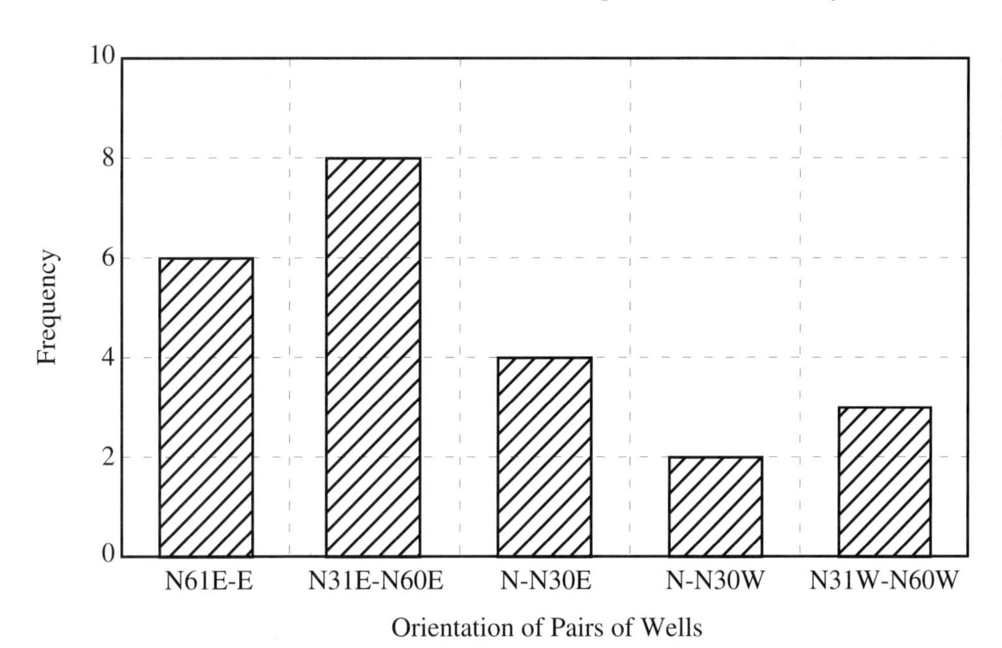

Figure 5. Histogram for preferential direction from maximum rank correlation in all wells.

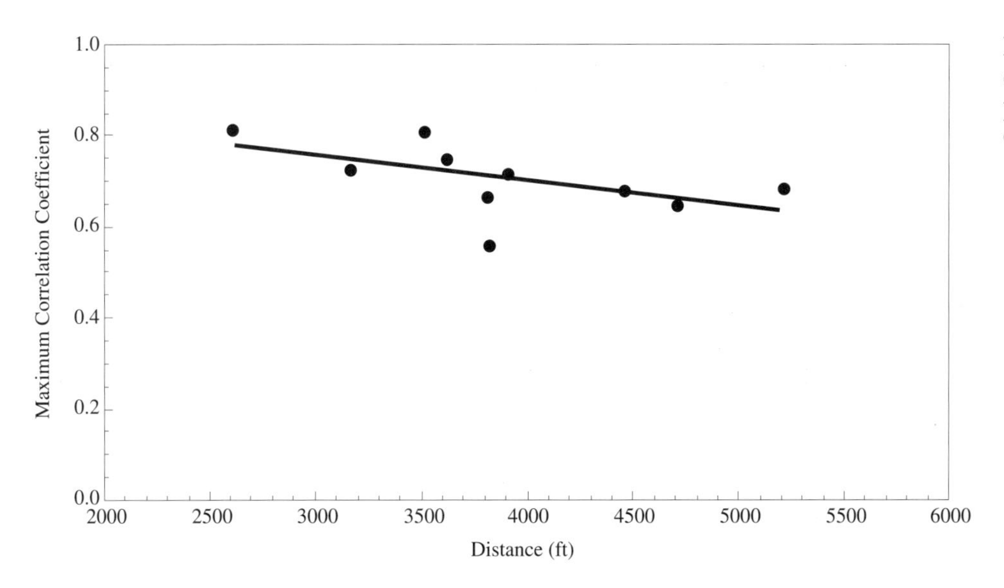

Figure 6. Rank correlation coefficient vs. distance for pairs on the N30E to N60E orientation.

SUMMARY OF RESERVOIR DESCRIPTION AND CONCLUSIONS

With this research project, a practical and theoretically based technique was developed to determine preferential flow trends in a reservoir by integrating tracer response, the results of Spearman rank correlation coefficient applied to reservoir geology, and reservoir geology. Most importantly, the work's basic information is tracer data and the widely available measurements of production and injection rates. This makes the technique suitable for implementation in all types of reservoirs, including ones with scarce information.

The Spearman rank correlation coefficient method appears to be successful, as measured by consistency with tracer breakthrough and geologic inference. The integrated reservoir description is summarized as follows:

- A southwest-northeast preferential flow trend is indicated.
- The middle-stream axis of the channel contains high-quality reservoir sand, with preferential flow direction along this axis.
- Communication is good between injectors on the west flank of the channel with the rest of the reservoir, especially in the southwest-northeast direction. Injectors located on the east flank of the channel are not effectively connected to the rest of the reservoir.

Although we have had some success in the description, significant questions must be answered before we

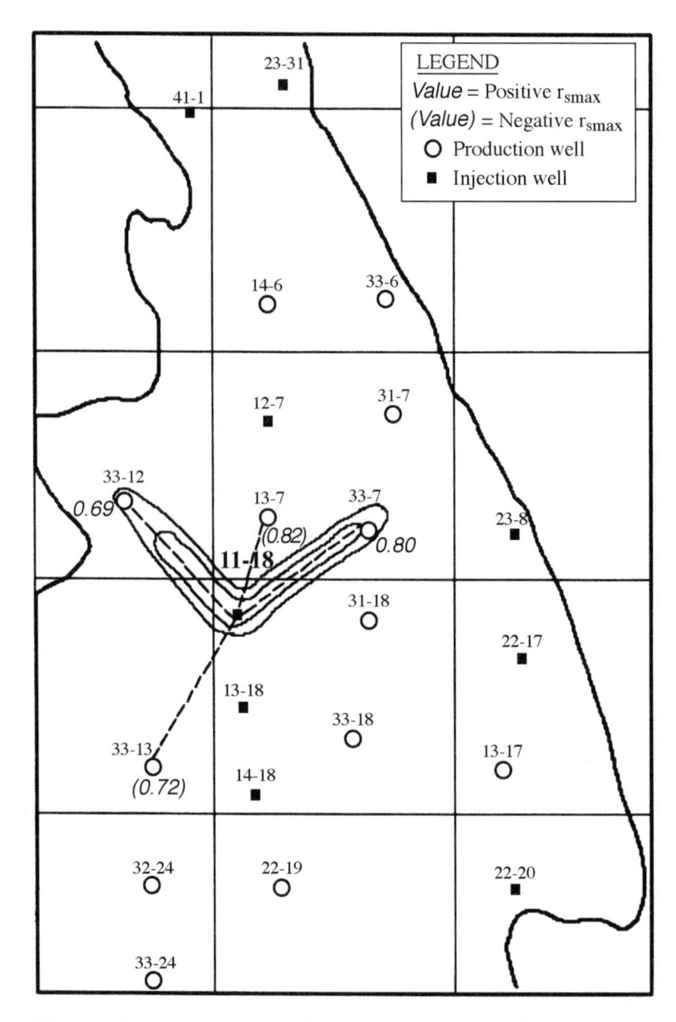

Figure 7. Iso-r_{smax} map for injector 11-18. Contour intervals are 0.1 starting with r_{smax} = 1 at the central injection well.

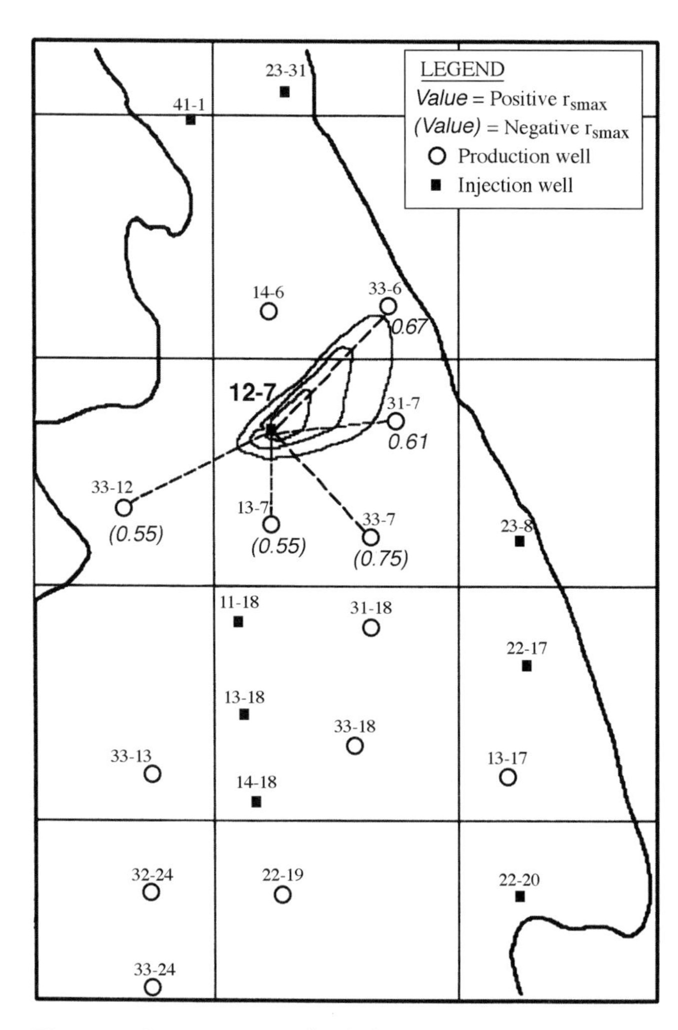

Figure 8. Iso-r_{smax} map for injector 12-7. Contour intervals are 0.1 starting with r_{smax} = 1 at the central injection well.

can claim that the Spearman technique, by itself, is a useful characterization technique. These considerations include the role of fluid compressibilities in the correlations, accounting for multiple-well correlation, and explanations for correlations at a distance; however, the Spearman technique uses data that are available in virtually all mature fields, and data that can be acquired with little additional expense. These advantages, plus the encouragement from this work, give ample justification for more research.

ACKNOWLEDGMENTS

The authors acknowledge the data provided by Kerr-McGee, through the Deltas Industrial Affiliate Project at the Bureau of Economic Geology, and the permission to publish this paper. Thanks go to Christopher White for his valuable suggestions and to Drs. Mike King and Kes Heffer for their discussions on applications of the Spearman rank correlation technique. Thanks are also due to Lagoven, S.A. for permission for preparation and publication of this paper. Larry W. Lake holds the W.A. (Monty) Moncrief Centennial Endowed Chair at The University of Texas.

REFERENCES CITED

Gardner, M.H., W. Dharmasamadhi, B.J. Willis, S.P. Dutton, Q. Fang, S. Kattah, J. Yeh, and F. Wang, October 1994, Reservoir characterization of Buck Draw field: Bureau of Economic Geology, prepared for the Deltas Industrial Associates field trip manual, 29 p.

Hawkins, C.M., and S. Formhals, 1985, Geology and engineering aspects of Buck Draw field, Campbell and Converse counties, Wyoming: Wyoming Geological Association 36th Annual Field Conference guidebook, p. 33-45.

Heffer, K.J., R.J. Fox, and C.A. McGill, October 1995, Novel techniques show links between reservoir flow directionality, earth stress, fault structure and geomechanical changes in mature waterfloods:

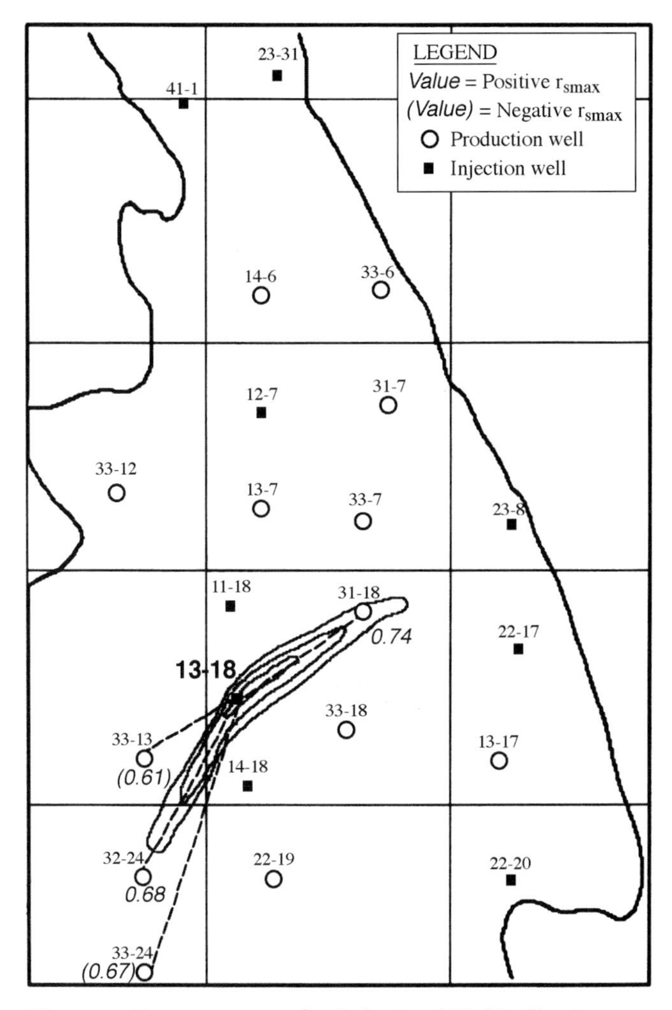

Figure 9. Iso-r_{smax} map for injector 113-18. Contour intervals are 0.1 starting with r_{smax} = 1 at the central injection well.

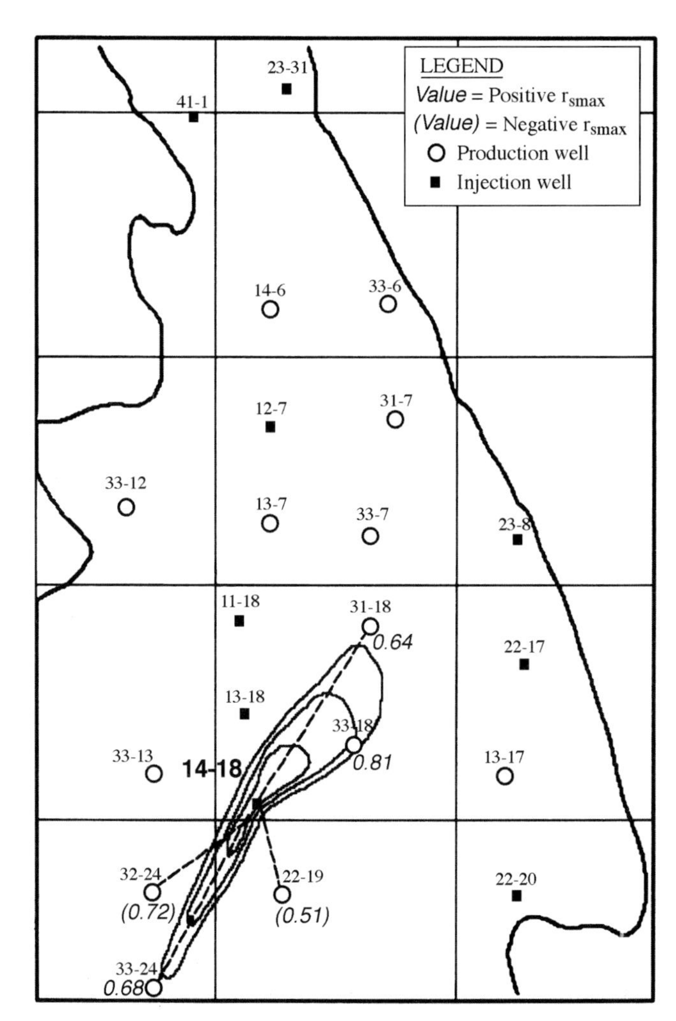

Figure 10. Iso-r_{smax} map for injector 14-18. Contour intervals are 0.1 starting with r_{smax} = 1 at the central injection well.

paper SPE 30711 presented at the Annual Technical Conference and Exhibition of the Society of Petroleum Engineers held in Dallas, Texas, p. 77–87.

Jensen, J.L., L.W. Lake, P.W.M. Corbett, and D.J. Goggin, 1996, Statistics for petroleum engineers and geoscientists: Prentice Hall PTR, Upper Saddle River, New Jersey, 390 p.

Mirzadjanzade, A.Kh., I.M. Ametov, A.A. Bokserman, and V.P. Filippov, October 1993, New perspective trends of research in oil and gas recovery: paper presented at the 7th European IOR Symposium held in Moscow, Russia.

Rasmussen, D.L., C.L. Jump, and K.A. Wallace, 1985, Deltaic system in the early Cretaceous Fall River formation, Southern Powder River basin, Wyoming: Wyoming Geological Association 36th Annual Field Conference guidebook, p. 91–111.

Refunjol, B.T., 1996, Reservoir characterization of North Buck Draw field based on tracer response and production/injection analysis: M.S. thesis, The University of Texas at Austin, 124 leaves.

Sellars, R., and C.M. Hawkins, 1992, Geology and stratigraphic aspects of Buck Draw field, Powder River basin, Wyoming: Wyoming Geological Association 43rd Annual Field Conference guidebook, p. 97–110.

Siegel, S., 1956, Nonparametric statistics for the behavioral sciences: McGraw-Hill Book Company, New York City, 312 p.

Volk, W., 1969, Applied statistics for engineers: McGraw-Hill Book Company, New York City, 354 p.

Zemel, B., 1995, Tracers in the oil field: Elsevier Science B. V., Amsterdam, 487 p.

Section V

Fracture Analysis

Guo, G., S. A. George, and R. P. Lindsey, Statistical analysis of surface lineaments and fractures for characterizing naturally fractured reservoirs, 1999, *in* R. Schatzinger and J. Jordan, eds., Reservoir Characterization-Recent Advances, AAPG Memoir 71, p. 221–250.

Chapter 16

Statistical Analysis of Surface Lineaments and Fractures for Characterizing Naturally Fractured Reservoirs

Genliang Guo
Stephen A. George
BDM Petroleum Technologies
Bartlesville, Oklahoma, U.S.A.

Rhonda P. Lindsey
DOE National Petroleum Technology Office
Tulsa, Oklahoma, U.S.A.

ABSTRACT

Thirty-six sets of surface lineaments and fractures mapped from satellite images and aerial photos from parts of the Mid-continent and Colorado Plateau regions were collected, digitized, and statistically analyzed to obtain the probability distribution functions of natural fractures for characterizing naturally fractured reservoirs. The orientations and lengths of the surface linear features were calculated using the digitized coordinates of the two end points of each individual linear feature. The spacing data of the surface linear features within an individual set were obtained using a new analytical sampling technique that involves overlapping a set of uniform imaginary scanlines orthogonally on top of an individual fracture set and calculating the distance between two adjacent intersection points along each scanline. Statistical analyses were then performed to find the best-fit probability distribution functions for the orientation, length, and spacing of each data set. Twenty-five hypothesized probability distribution functions were used to fit each data set. A chi-square goodness-of-fit value was considered the best-fit distribution.

The orientations of surface linear features were best-fitted by triangular, normal, or logistic distributions; the lengths were best-fitted by PearsonVI, PearsonV, lognormal2, or extreme-value distributions; and the spacing data were best-fitted by lognormal2, PearsonVI, or lognormal distributions. These probability functions can be used to stochastically characterize naturally fractured reservoirs.

INTRODUCTION

The characterization of naturally fractured reservoirs has challenged geologists and engineers for decades (Evans, 1981). From a reservoir engineering point of view, the objective of characterizing a naturally fractured reservoir is to estimate the flow and storage capacity, both of which require a quantitative description of natural fracture characteristics. Unfortunately, subsurface natural fracture data are rarely available and very expensive to obtain. The scarcity of subsurface natural fracture data and our inability to cost-effectively acquire sufficient amounts of them have prevented us from obtaining a realistic and quantitative description of subsurface fracture systems.

One technique for characterizing a naturally fractured reservoir, which may overcome the limitation in the availability of subsurface natural fracture data, is to stochastically simulate the subsurface fracture systems and numerically calculate their permeability tensor distributions (Guo and Evans, 1994). The reliability of this technique, however, rests on the validity of the probability distribution functions of natural fracture characteristics used in the stochastic simulations. Many attempts have been made to use surface fracture data mapped from outcrop studies to derive the probability distributions of natural fracture characteristics.

Studies show that natural fracture orientations have a normal or Arnold's hemispherical normal distribution for two-dimensional data and a Fisher-Von Mises distribution for three-dimensional data (Long et al., 1982; Dverstorp and Andersson, 1989; Cacas et al., 1990; Belfield and Sovich, 1995). Fracture lengths have been found by various investigators to have a lognormal, exponential, or power-law distribution (Baecher et al., 1977; Long et al., 1982; Baecher, 1983; Rouleau and Gale, 1985; Dverstorp et al., 1989; Cacas et al., 1990; Heffer and Bevan, 1990). The distribution functions that provide best fits to many fracture aperture data include normal and lognormal distributions (Snow, 1970; Long et al., 1982; Belfield et al., 1995). As the most extensively investigated characteristic of natural fractures, fracture spacing is found to have an exponential, lognormal, Weibull, Gamma, or power-law distribution (Baecher et al., 1977; Long et al., 1982; Baecher, 1983; Roulean and Gale, 1985; Heffer and Bevan, 1990; Lorentz and Hill, 1991; Loosveld and Franssen, 1992; Belfield et al., 1995). These distribution functions were usually identified by best-fitting measured data to a few hypothesized distributions.

In this paper, we present the results of a statistical analysis of natural fracture characteristics using a large number of surface lineaments and fractures in the Mid-continent and Colorado Plateau regions collected and digitized from previous studies reported in the literature. Twenty-five different probability distribution functions (see Appendix 1) are used to best-fit the orientation, length, and spacing of each of the 36 fracture and lineament sets identified. A chi-square goodness-of-fit test is employed to identify the best fits. The abundance of surface lineaments and fractures and the list of 25 hypothesized distribution functions used in the statistical analysis will inevitably result in more reliable probability distributions for describing natural fracture characteristics.

STATISTICAL ANALYSIS PROCEDURE

The basic data used in this study were obtained by digitizing the two end points of the surface linear features mapped from satellite images and aerial photos of various parts of the Mid-continent and Colorado Plateau regions. Due to the nature of the data, only orientation, length, and spacing were analyzed. No data were available for aperture analysis. The following procedure was used for the statistical analysis of surface lineament and fracture characteristics.

- A rose diagram analysis for identifying the number of subsets in a surface lineament or fracture system.
- A filtering analysis for partitioning a surface fracture system into subsets.
- A sampling analysis for collecting surface fracture spacing data using a new analytical technique.
- A best-fit analysis for obtaining distribution functions for surface lineament and fracture orientation, length, and spacing.

In the following discussion, the surface fractures in Osage County, Oklahoma, are used to demonstrate the procedure for statistical analysis of surface fracture characteristics.

Shown in Figure 1 are the surface fractures in Osage County. They were mapped from satellite images and aerial photos (Guo and Carroll, 1995). A rose diagram for these surface fractures was generated (see Figure 2). This rose diagram shows that the surface fractures primarily consist of two sets trending northeast and

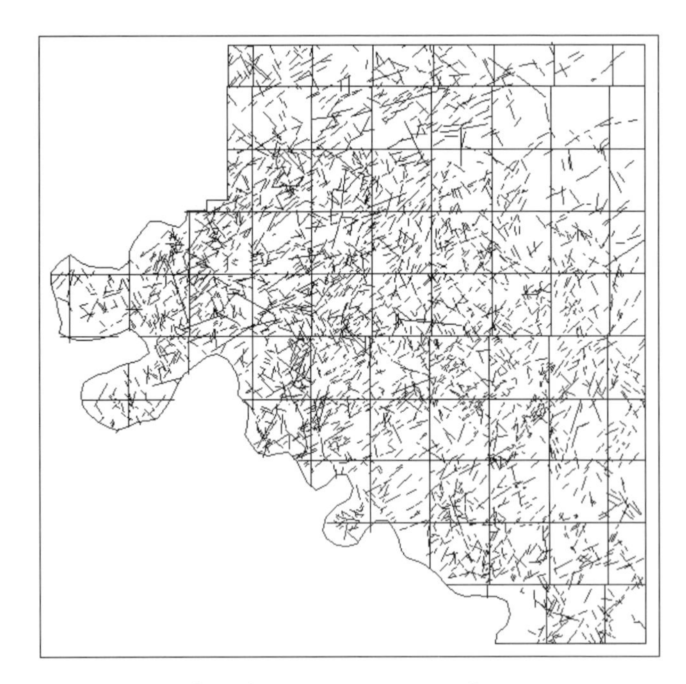

Figure 1. Surface fractures in Osage County, Oklahoma.

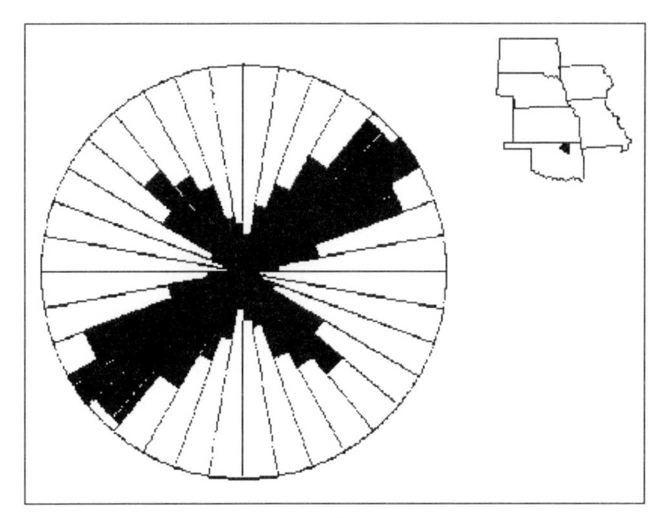

Figure 2. A rose diagram of the surface fractures in Osage County, Oklahoma.

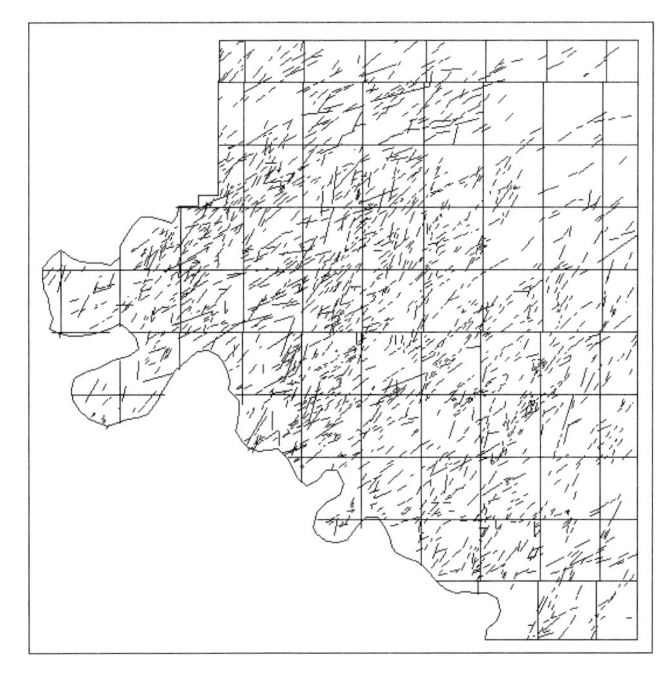

Figure 3. Northeast-trending surface fractures in Osage County, Oklahoma.

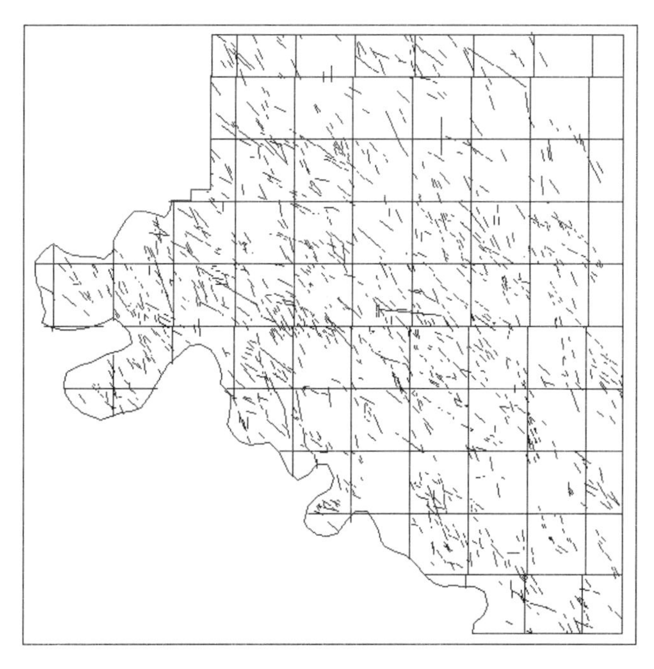

Figure 4. Northwest-trending surface fractures in Osage County, Oklahoma.

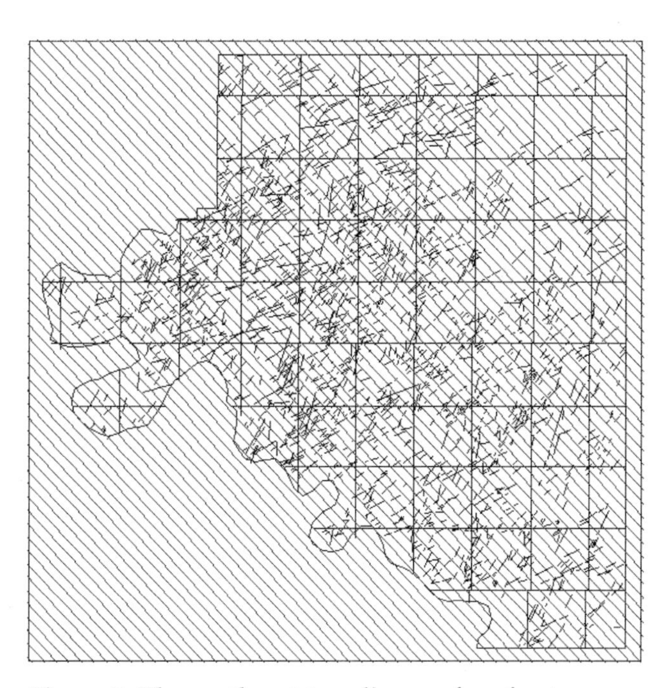

Figure 5. The northeast-trending surface fractures and a set of imaginary scanlines in Osage County, Oklahoma.

northwest. Figures 3 and 4 show the northeast- and northwest-trending surface fractures, respectively, after they are separated through a filtering analysis based on their orientations.

Within each subset, the orientation and length data are calculated using the digitized coordinates of the two end points for each surface fracture. The spacing data are obtained using an analytical sampling technique. The technique involves overlapping a set of uniform imaginary scanlines on top of an individual fracture set (see Figures 5 and 6). The scanlines are oriented perpendicular to the average orientation of the fracture set. The intersection points along each imaginary scanline are then identified and sorted. The spacing data are obtained by calculating the distance between two adjacent intersection points along a scanline. After orientation, length, and spacing data are obtained for each fracture set, they are fitted to 25 hypothesized probability distributions (see Appendix 1). A chi-square value is calculated for each fit. The smaller the chi-square value, the better the fit. The distribution with the smallest chi-square value is considered as the best-fit distribution to the data.

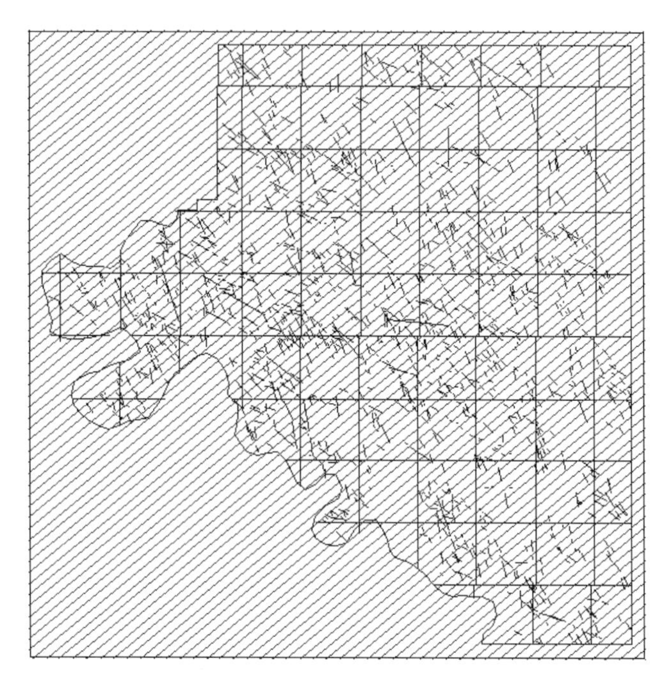

Figure 6. The northwest-trending surface fractures and a set of imaginary scanlines in Osage County, Oklahoma.

Table 1. Best-Fit Results for the Orientation Data of the Northeast-Trending Surface Fractures in Osage County, Oklahoma.

Function	Chi–Square
Triang(1.47,37.38,95.63)	54.09
Logistic(41.68,12.71)	114.27
Weibull(2.07,52.72)	141.82
Normal(43.59,19.80)	147.40
Rayleigh(33.85)	177.53
Beta(2.01,2.14) * 90.00	294.61
ExtremeValue(32.83,16.88)	349.13
Erlang(3.00,15.89)	505.66
PearsonVI(2.27,0.39,12.40)	938.31
Expon(61.11)	1820.47
Gamma(3.76,11.61)	1912.22
Geomet(2.24e–2)	2004.13
Erf(1.90e–2)	5646.72
NegBin(6.00,0.12)	7071.78
LogLogistic(–21.12,1.00e–4,0.11)	1.73E+05
Lognorm(45.86,31.18)	2.29E+05
Lognorm2(3.64,0.62)	2.29E+05
Student's T(2.00)	4.95E+06
Chisq(24.00)	6.87E+08
PearsonV(1.66,45.27)	6.43E+09
Pareto(1.00,0.00)	9.47E+09
Poisson(38.66)	1.70E+14
Binomial(95.00,0.45)	1.09E+22
InverseGaussian(43.59,4.02e+2)	1.00E+34
HyperGeo(91.00,90.00,1.81e+2)	1.00E+34

Table 1 summarizes the best-fit results for the orientation data of the northeast-trending fracture set. A chi-square value is calculated for each of the 25 fitting distributions. One can see from Table 1 that a triangular distribution, TRIANG(a = –1.47, b = 37.38, c = 95.63) (see Appendix 1), appears to provide the best fit to the data. Figure 7 shows the histogram and best-fit distribution for the orientation data of the northeast-trending surface fractures.

Similarly, Figure 8 shows the histogram and best-fit distribution for the orientation data of the northwest-trending surface fractures. The orientations of this set of surface fractures were best-fitted by a logistic distribution, LOGISTIC(α = 128, β = 12.01) (see Appendix 1).

As shown in Figure 9, the length data of the northeast-trending surface fractures were best-fitted by a PearsonVI distribution, PearsonVI(α_1 = 11.35, α_2 = 4.54, β = 1480) (see Appendix 1). Figure 10 shows the histogram and best-fit length distribution for the northwest-trending surface fractures. A PearsonVI distribution, PearsonVI(α_1 = 8.95, α_2 = 3.95, β = 1500), provides the best fit.

The spacing data of the northeast-trending surface fractures, were best-fitted by a lognormal2 distribution, LOGNORM2(μ = 8.21, σ = 1.11) (see Appendix 1). The histogram, best-fit distribution, and statistical parameters of the northeast-trending surface fracture spacing data are shown in Figure 11. Similar results

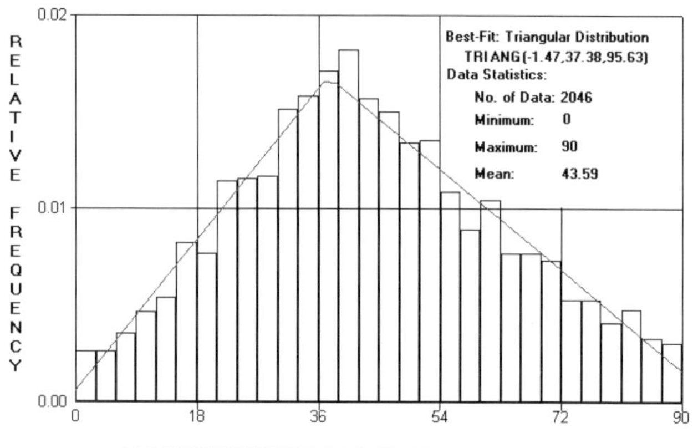

Figure 7. The histogram and best-fit orientation distribution of the northeast-trending surface fractures in Osage County, Oklahoma.

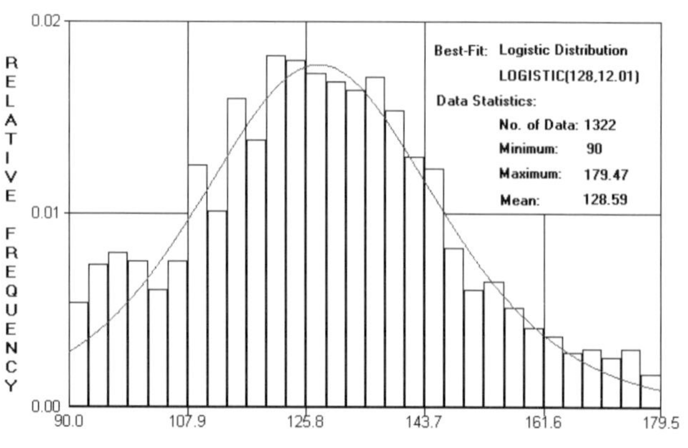

Figure 8. The histogram and best-fit orientation distribution of the northwest-trending surface fractures in Osage County, Oklahoma.

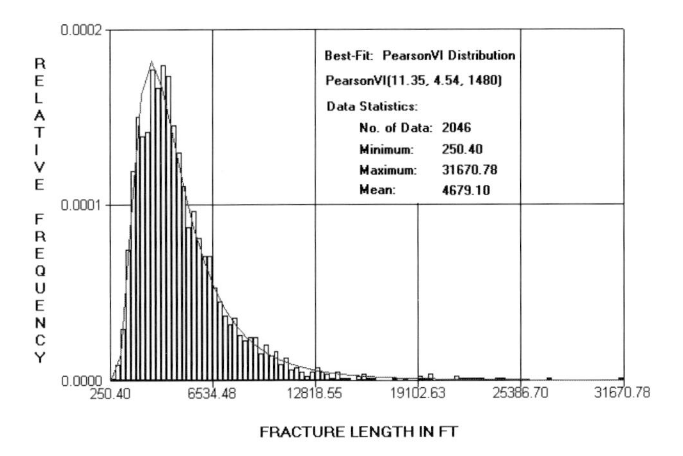

Figure 9. The histogram and best-fit length distribution of the northeast-trending surface fractures in Osage County, Oklahoma.

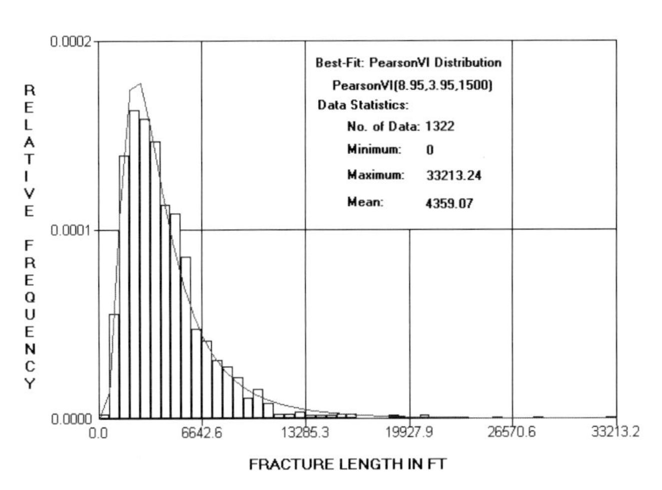

Figure 10. The histogram and best-fit length distribution of the northwest-trending surface fractures in Osage County, Oklahoma.

were also obtained for the spacing data of the northwest-trending surface fractures, as shown in Figure 12. A lognormal2 distribution, LOGNORM2(μ = 8.64, σ = 1.18), is the best-fit distribution.

STATISTICAL ANALYSIS OF SURFACE LINEAMENTS AND FRACTURES IN NORTHEASTERN ARIZONA

A photogeological interpretation study was conducted in the Cameron Basin and the Black Mesa Basin areas in northeastern Arizona (Guo et al., 1995). Over 50,000 surface fractures were mapped from aerial photos and subsequently digitized. In addition, large-scale surface lineaments were collected and digitized in the general northeastern quarter of Arizona (Lepley, 1977). Using the statistical analysis procedure discussed in the previous section, the characteristics of the surface lineaments and fractures in northeastern Arizona were statistically analyzed.

Surface Lineaments in Northeastern Arizona

Figure 13 is a rose diagram for the surface lineaments in northeastern Arizona. The surface lineaments apparently can be partitioned into four subsets, including two major sets trending northeast and northwest and two minor sets trending east and north. Figures 14–17 show the histograms, best-fit orientation distributions, and sample data statistics for the four sets of the surface lineaments. A normal distribution provides the best fit to the orientation data of the east-trending surface lineaments, whereas those of the northeast-, north-, and northwest-trending surface lineaments were best-fitted by three triangular distributions.

The results of a statistical analysis on the corresponding length data are given in Figures 18–21. The length data of the east- and north-trending surface lineaments were best-fitted by two PearsonV distributions; those of the northeast- and northwest-trending surface lineaments were best-fitted by two PearsonVI distributions.

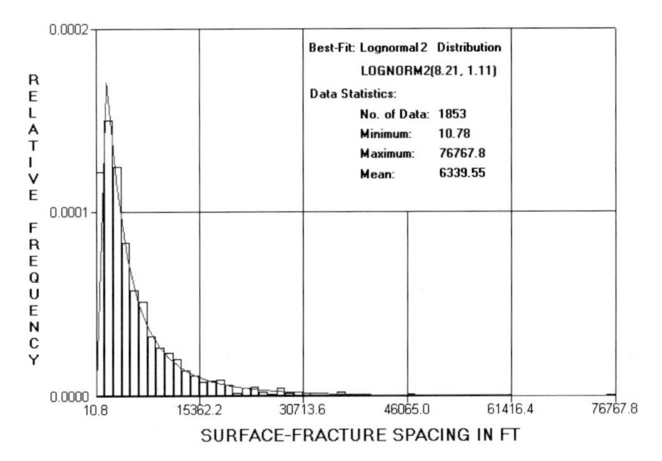

Figure 11. The histogram and best-fit spacing distribution of the northeast-trending surface fractures in Osage County, Oklahoma.

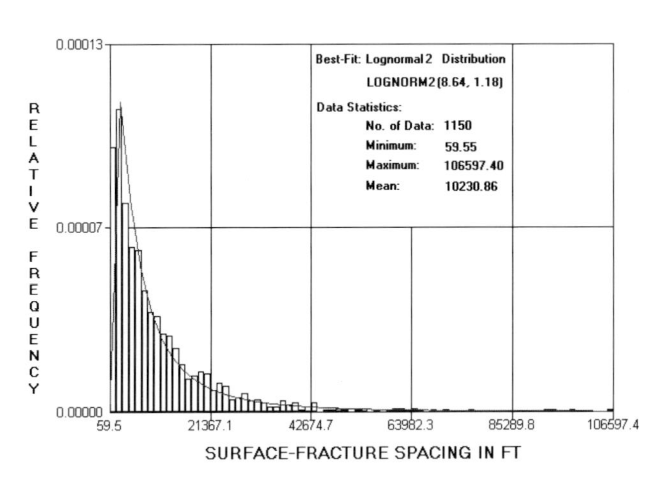

Figure 12. The histogram and best-fit spacing distribution of the northwest-trending surface fractures in Osage County, Oklahoma.

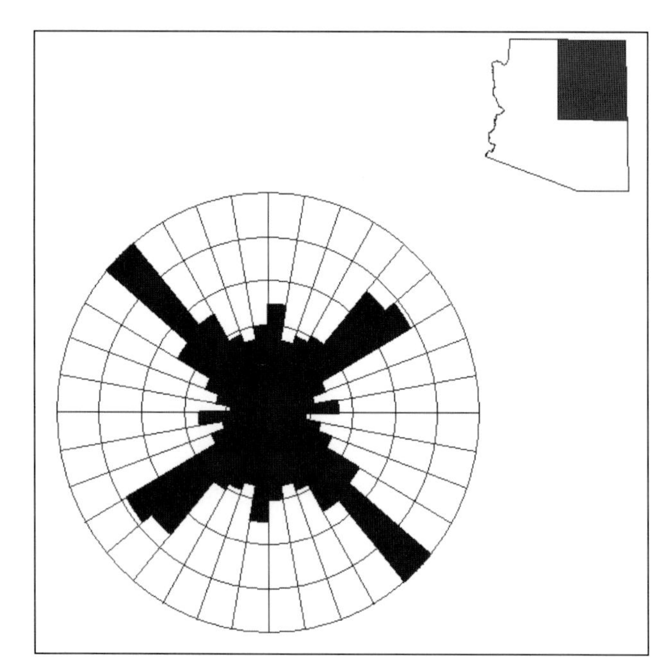

Figure 13. A rose diagram of the surface lineaments in northeastern Arizona.

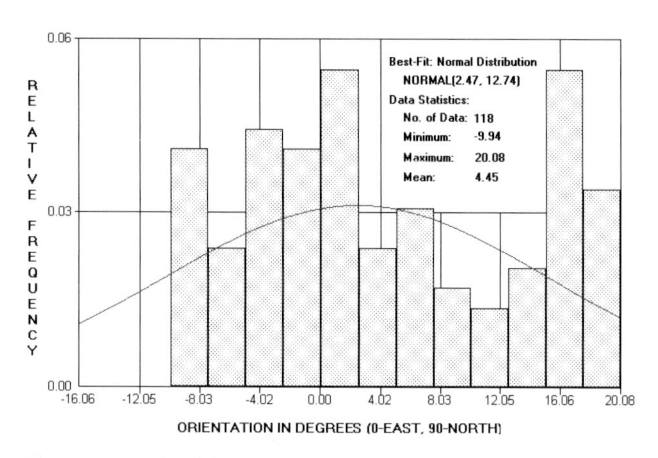

Figure 14. The histogram and best-fit orientation distribution of the east-trending surface lineaments in northeastern Arizona.

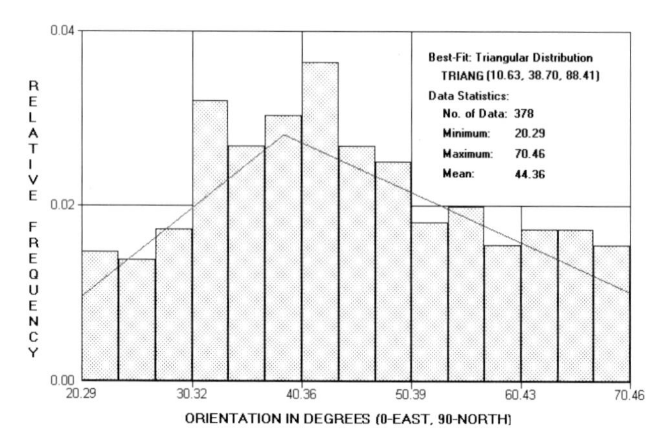

Figure 15. The histogram and best-fit orientation distribution of the northeast-trending surface lineaments in northeastern Arizona.

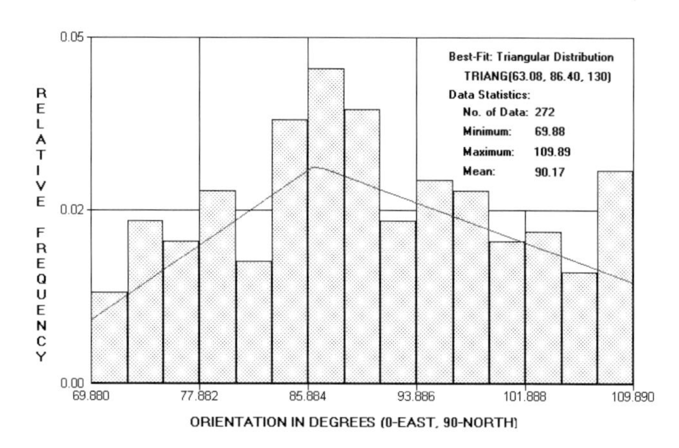

Figure 16. The histogram and best-fit orientation distribution of the north-trending surface lineaments in northeastern Arizona.

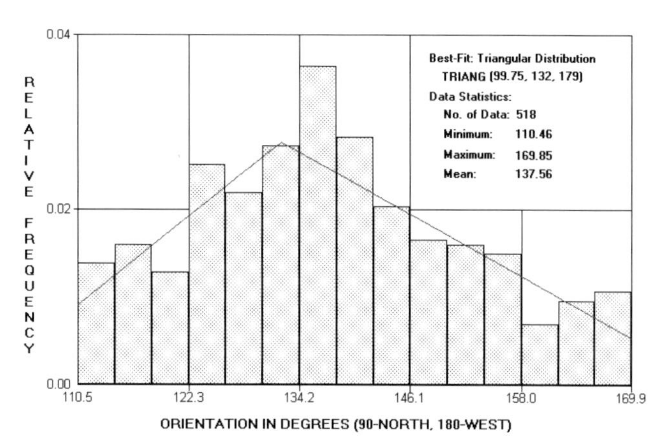

Figure 17. The histogram and best-fit orientation distribution of the northwest-trending surface lineaments in northeastern Arizona.

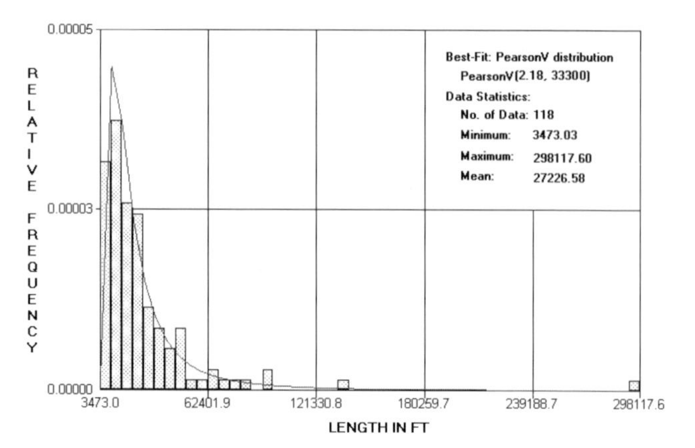

Figure 18. The histogram and best-fit length distribution of the east-trending surface lineaments in northeastern Arizona.

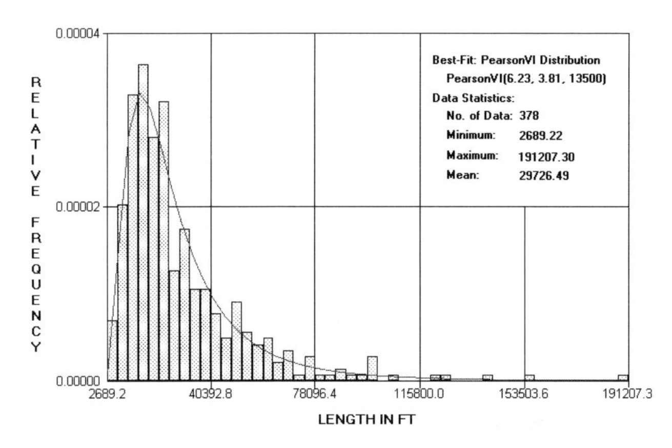

Figure 19. The histogram and best-fit length distribution of the northeast-trending surface lineaments in northeastern Arizona.

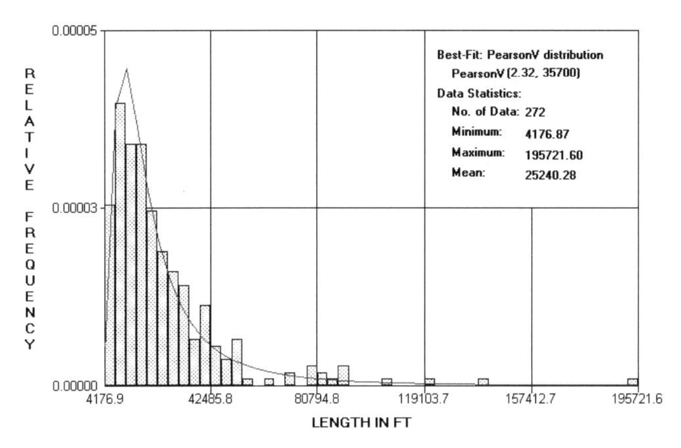

Figure 20. The histogram and best-fit length distribution of the north-trending surface lineaments in northeastern Arizona.

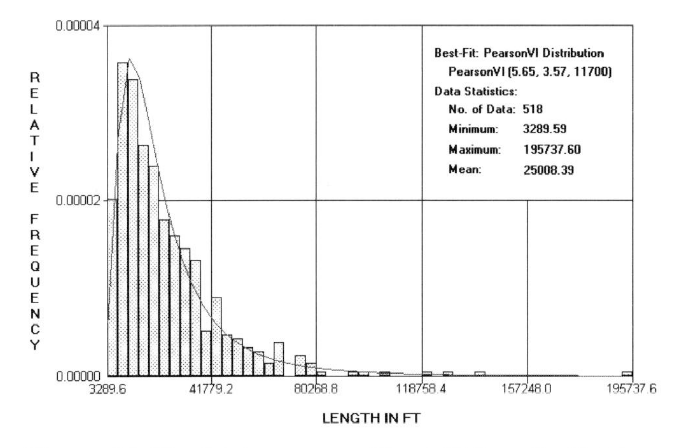

Figure 21. The histogram and best-fit length distribution of the northwest-trending surface lineaments in northeastern Arizona.

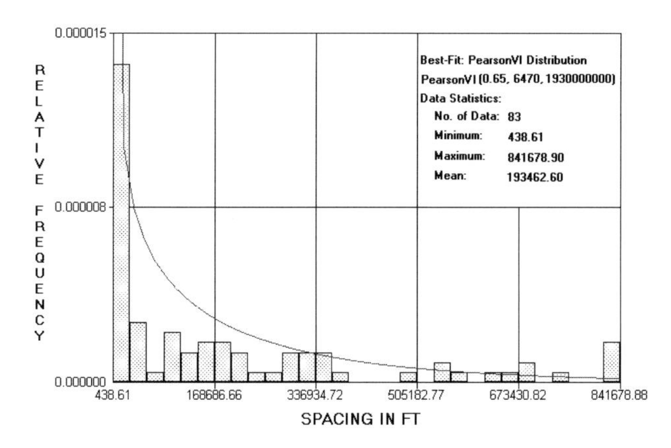

Figure 22. The histogram and best-fit spacing distribution of the east-trending surface lineaments in northeastern Arizona.

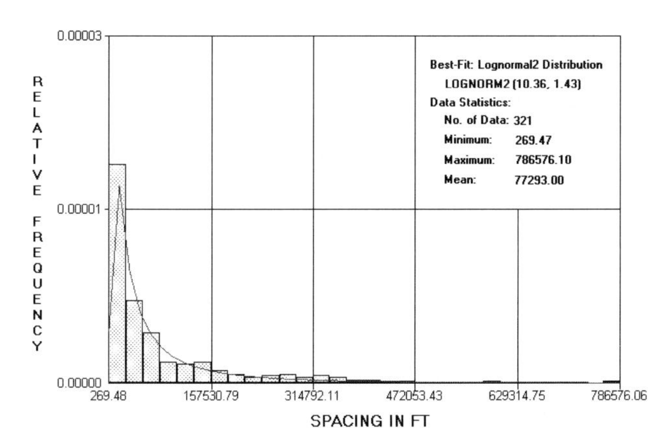

Figure 23. The histogram and best-fit spacing distribution of the northeast-trending surface lineaments in northeastern Arizona.

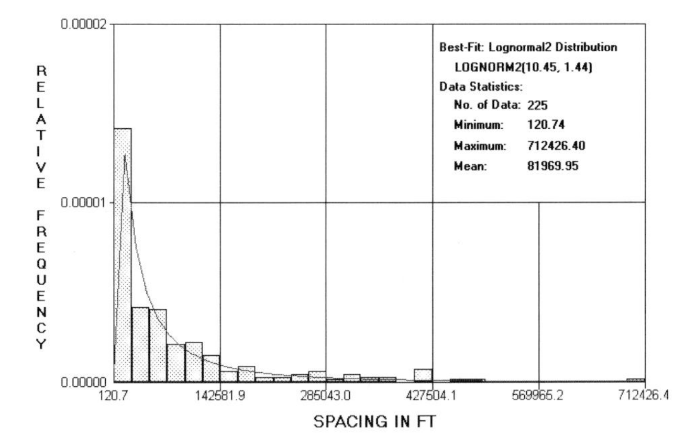

Figure 24. The histogram and best-fit spacing distribution of the north-trending surface lineaments in northeastern Arizona.

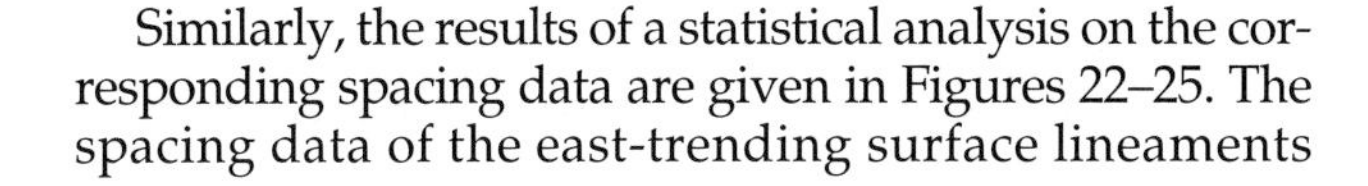

Similarly, the results of a statistical analysis on the corresponding spacing data are given in Figures 22–25. The spacing data of the east-trending surface lineaments were best-fitted by a PearsonVI distribution; those of the northeast-, north-, and northwest-trending surface lineaments were best-fitted by three lognormal2 distributions.

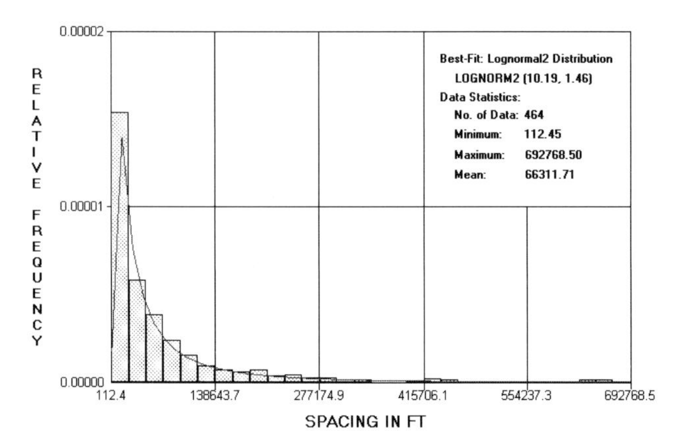

Figure 25. The histogram and best-fit spacing distribution of the northwest-trending surface lineaments in northeastern Arizona.

Surface Fractures in the Cameron Area in Northeastern Arizona

Over 6100 surface fractures were mapped from aerial photos in the Cameron area in northeastern Arizona (Guo et al., 1995). Figure 26 is a rose diagram for those surface fractures. From this rose diagram, three subsets were identified, including two major sets trending northeast and north-northeast and a minor set trending east. The histograms, best-fit orientation distributions, and sample data statistics for the three sets of the surface fractures are shown in Figures 27–29. The orientation data of the northeast-trending surface fracture set were best-fitted by a normal distribution, whereas those of the other two fracture sets were best-fitted by two triangular distributions.

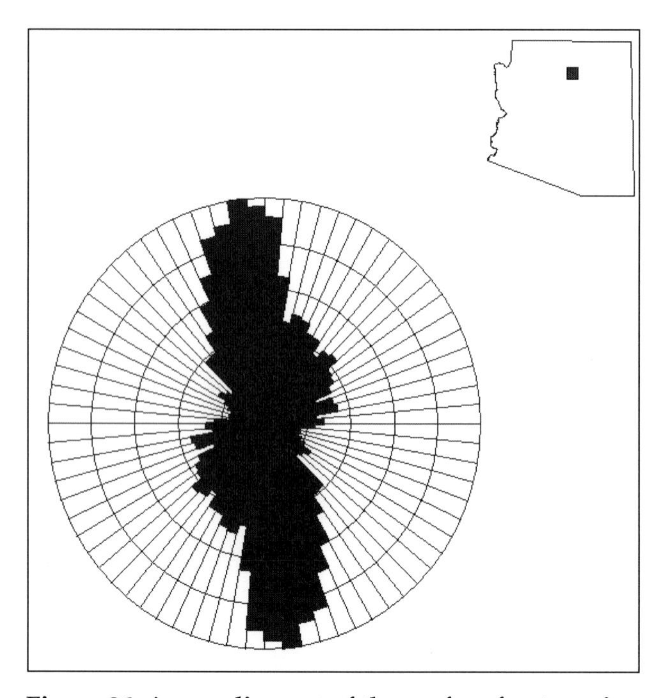

Figure 26. A rose diagram of the surface fractures in the Cameron area in northeastern Arizona.

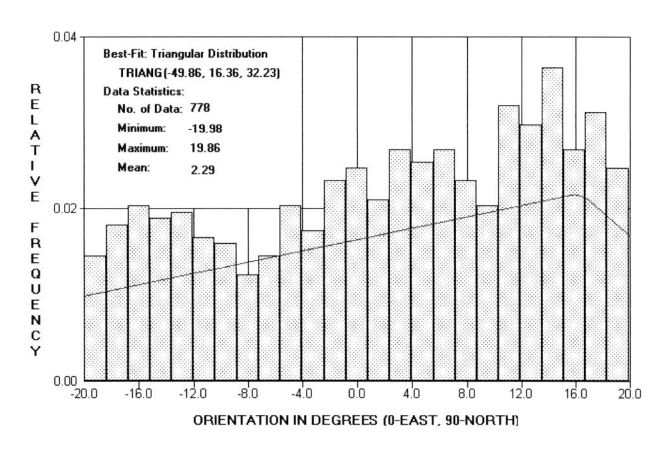

Figure 27. The histogram and best-fit orientation distribution of the east-trending surface fractures in the Cameron area.

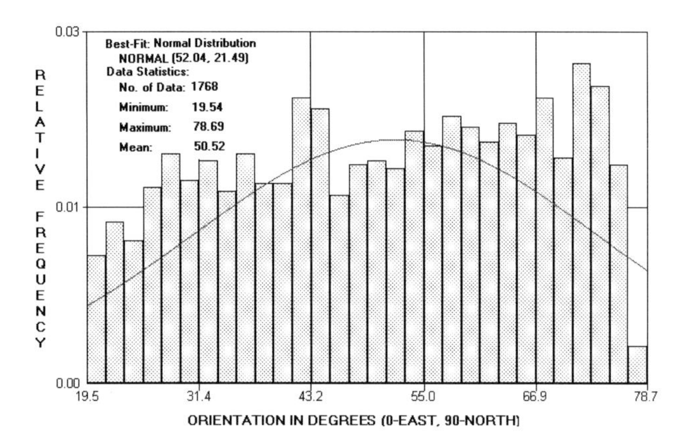

Figure 28. The histogram and best-fit orientation distribution of the northeast-trending surface fractures in the Cameron area.

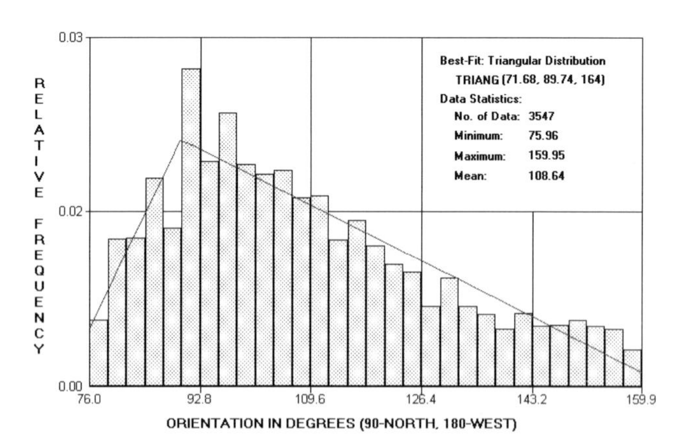

Figure 29. The histogram and best-fit orientation distribution of the north-northwest–trending surface fractures in the Cameron area.

Figures 30–32 show the results of a statistical analysis on the corresponding three sets of surface fracture length data in the Cameron area. A PearsonVI distribution appears to provide the best fit to the length data of the

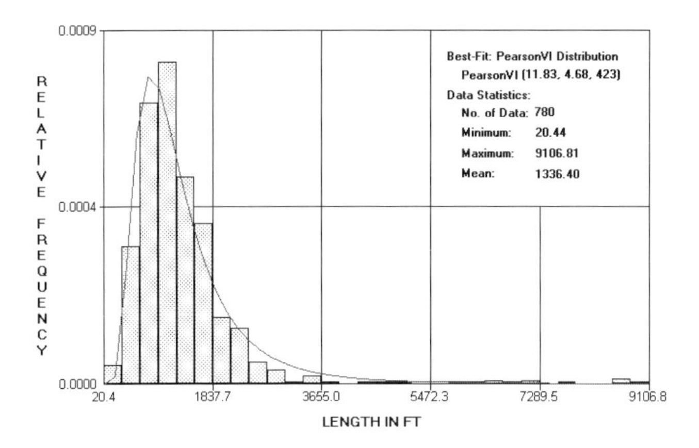

Figure 30. The histogram and best-fit length distribution of the east-trending surface fractures in the Cameron area.

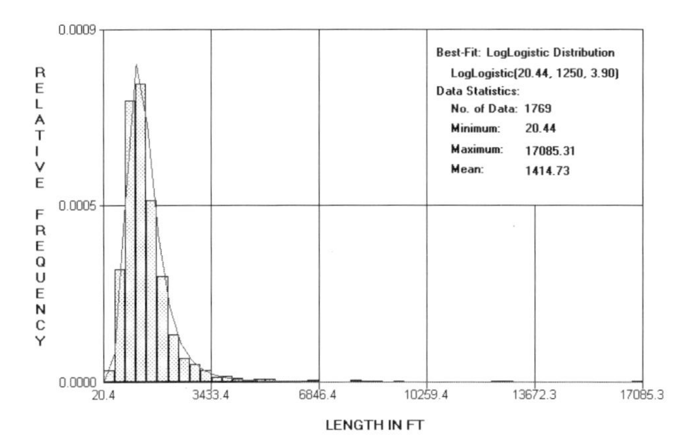

Figure 31. The histogram and best-fit length distribution of the northeast-trending surface fractures in the Cameron area.

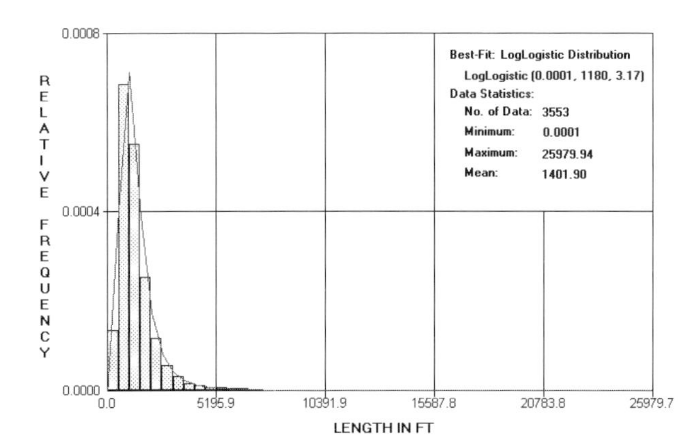

Figure 32. The histogram and best-fit length distribution of the north-northwest–trending surface fractures in the Cameron area.

east-trending fracture set; two loglogistic distributions provide the best fits to the other two fracture sets.

Similarly, Figures 33–35 show the histograms, best-fit distributions, and sample data statistics for

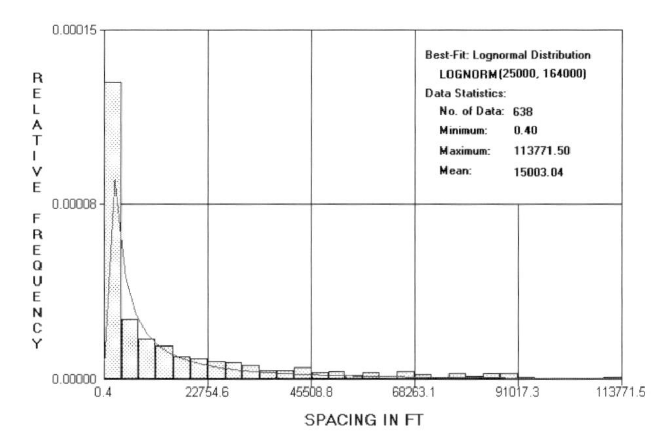

Figure 33. The histogram and best-fit spacing distribution of the east-trending surface fractures in the Cameron area.

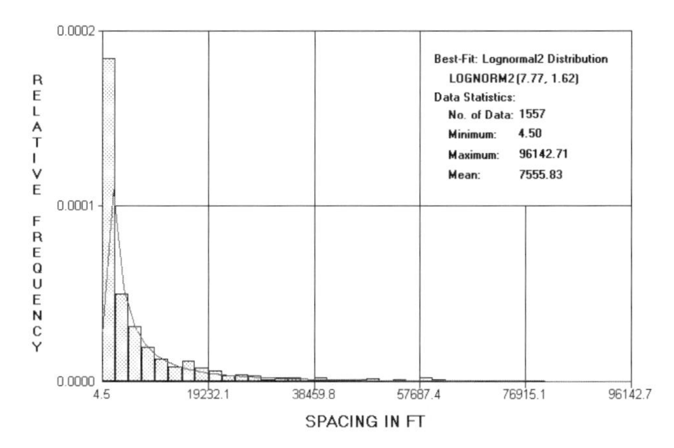

Figure 34. The histogram and best-fit spacing distribution of the northeast-trending surface fractures in the Cameron area.

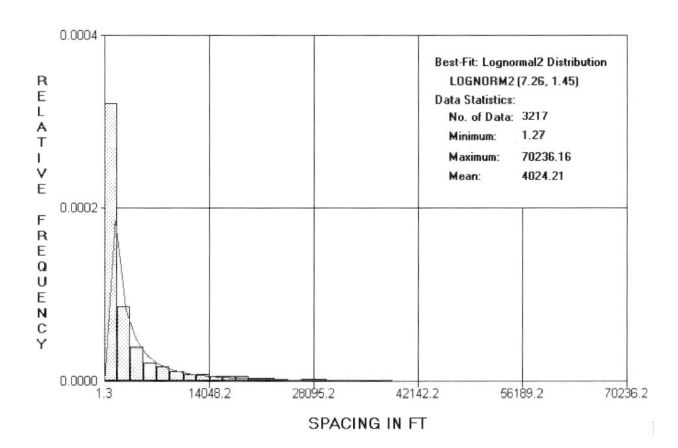

Figure 35. The histogram and best-fit spacing distribution of the northwest-trending surface fractures in the Cameron area.

the spacing data of the three sets of surface fractures in the Cameron area. A lognormal distribution provides the best fit to the spacing data of the east-trending surface fracture set. Those of the northeast- and north-northeast–trending surface fractures were best fitted by two lognormal2 distributions.

Surface Fractures in the Black Mesa Basin Area in Northeastern Arizona

Over 44,000 surface fractures were mapped from aerial photos of the Black Mesa Basin area in northeastern Arizona (Guo and Carroll, 1995). Figure 36 shows a rose diagram of these surface fractures. Four subsets were identified from Figure 36 trending east, northeast, north-northeast, and northwest. A statistical analysis was performed on the orientations of these four sets of surface fractures. The results are given in Figures 37–40. The orientation data of the east- and northeast-trending

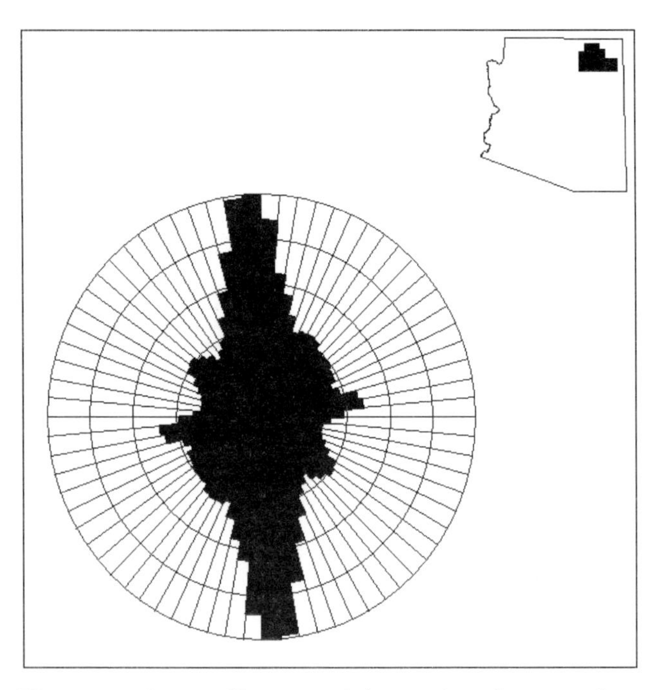

Figure 36. A rose diagram of the surface fractures in the Black Mesa Basin area in northeastern Arizona.

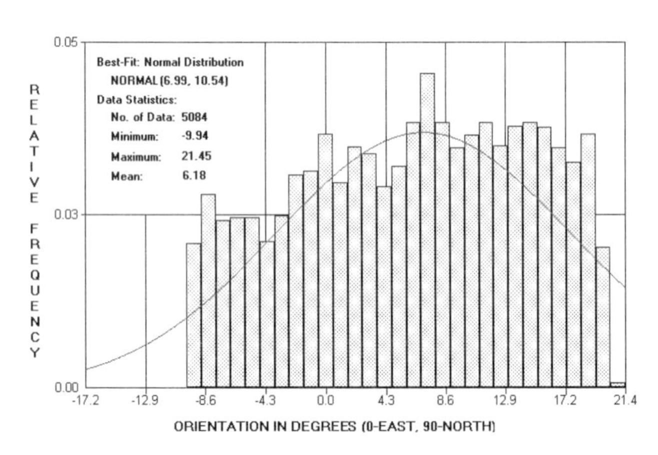

Figure 37. The histogram and best-fit orientation distribution of the east-trending surface fractures in the Black Mesa Basin area.

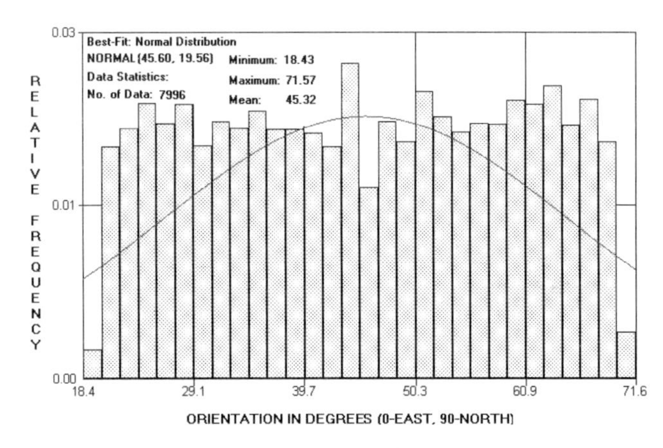

Figure 38. The histogram and best-fit orientation distribution of the northeast-trending surface fractures in the Black Mesa Basin area.

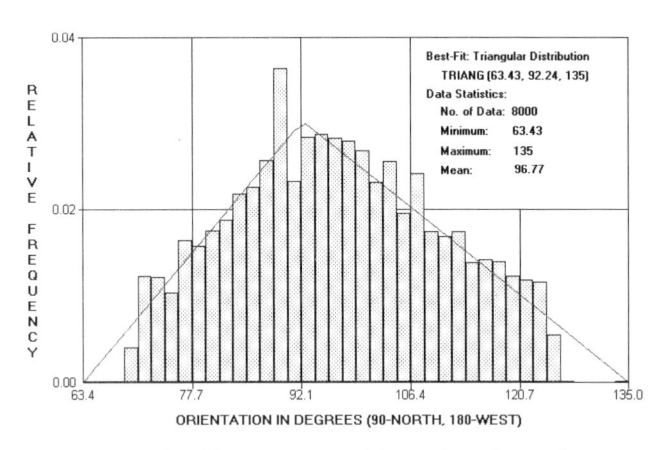

Figure 39. The histogram and best-fit orientation distribution of the north-northwest–trending surface fractures in the Black Mesa Basin area.

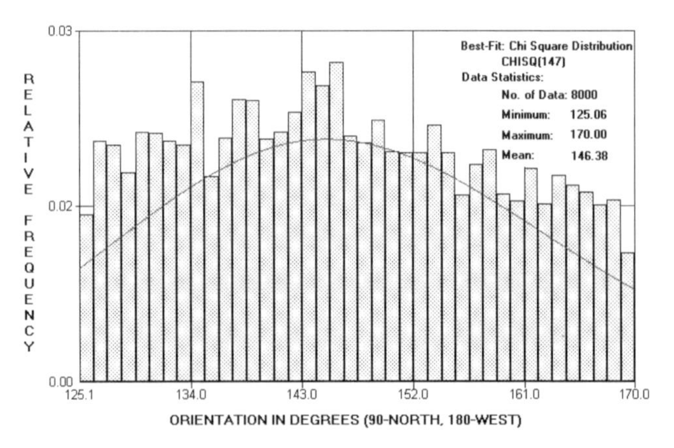

Figure 40. The histogram and best-fit orientation distribution of the northwest-trending surface fractures in the Black Mesa Basin area.

fractures were best-fitted by two normal distributions. Those of the north-northwest–trending fractures were best-fitted by a triangular distribution, whereas those of the northwest-trending fractures were best-fitted by a chi-square distribution.

Similarly, the length data of the four sets of surface fractures were also statistically analyzed. The results are shown in Figures 41–44. It appears that the length data of the east- and northwest-trending surface fractures in this area were best-fitted by two PearsonVI distributions. Those of the northeast-trending surface fractures were best-fitted by an extreme-value distribution, whereas those of the north-northwest–trending surface fractures were best-fitted by a lognormal2 distribution.

Figures 45–48 show the results of a statistical analysis on the spacing data of the four sets of surface fractures in the Black Mesa Basin area. A lognormal distribution appears to provide the best fit to the spacing data of the northwest-trending surface fractures, but those of the other three sets were best-fitted by lognormal2 distributions.

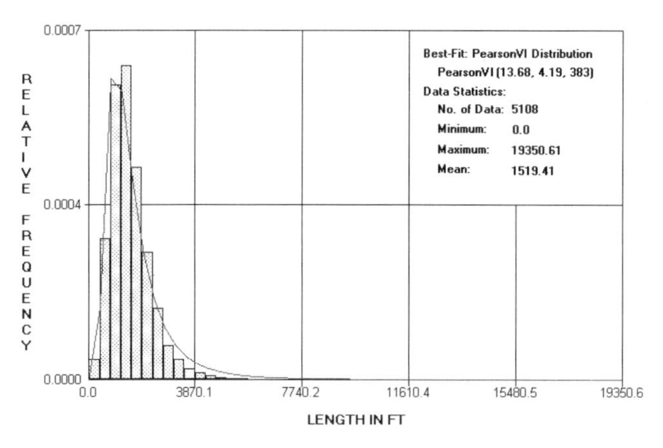

Figure 41. The histogram and best-fit length distribution of the east-trending surface fractures in the Black Mesa Basin area.

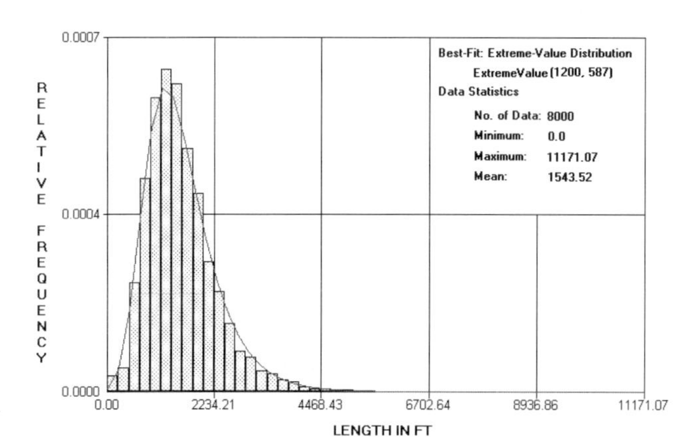

Figure 42. The histogram and best-fit length distribution of the northeast-trending surface fractures in the Black Mesa Basin area.

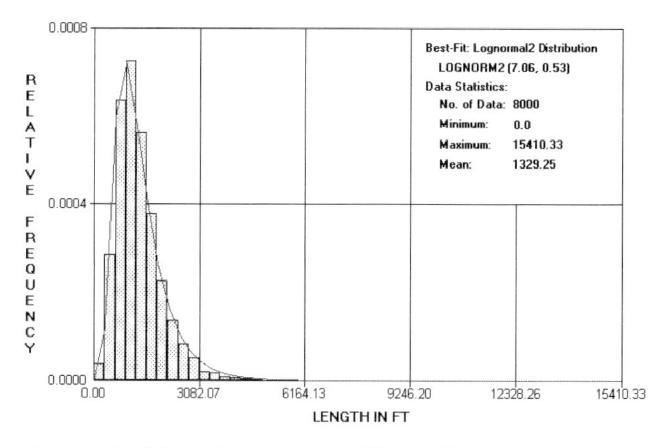

Figure 43. The histogram and best-fit length distribution of the north-northwest–trending surface fractures in the Black Mesa Basin area.

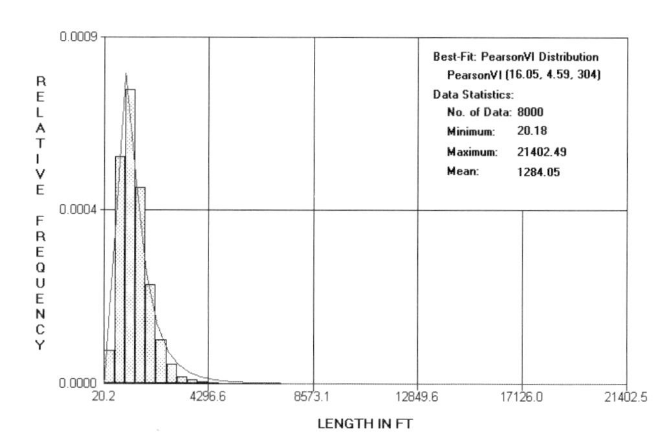

Figure 44. The histogram and best-fit length distribution of the northwest-trending surface fractures in the Black Mesa Basin area.

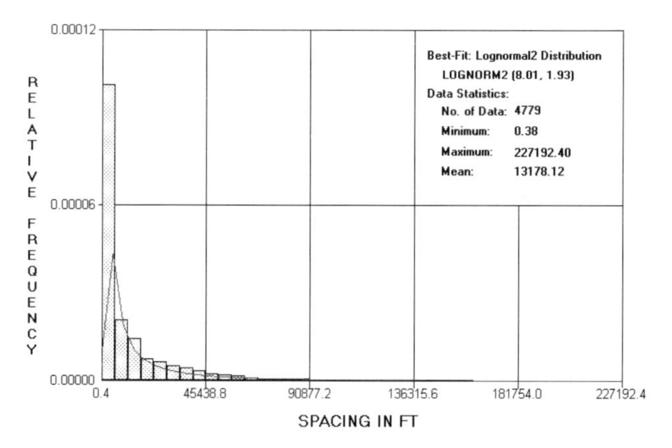

Figure 45. The histogram and best-fit spacing distribution of the east-trending surface fractures in the Black Mesa Basin area.

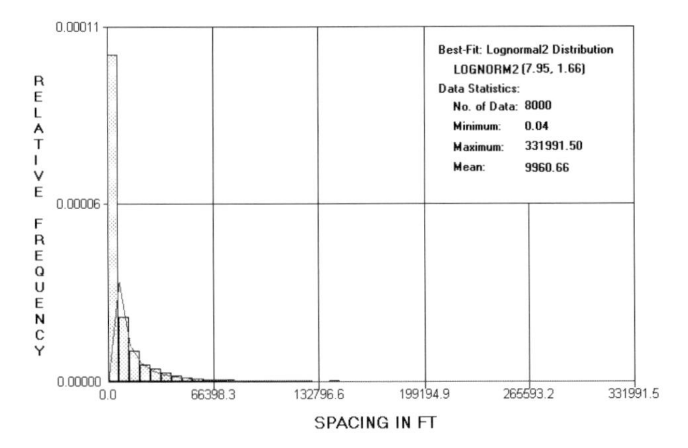

Figure 46. The histogram and best-fit spacing distribution of the northeast-trending surface fractures in the Black Mesa Basin area.

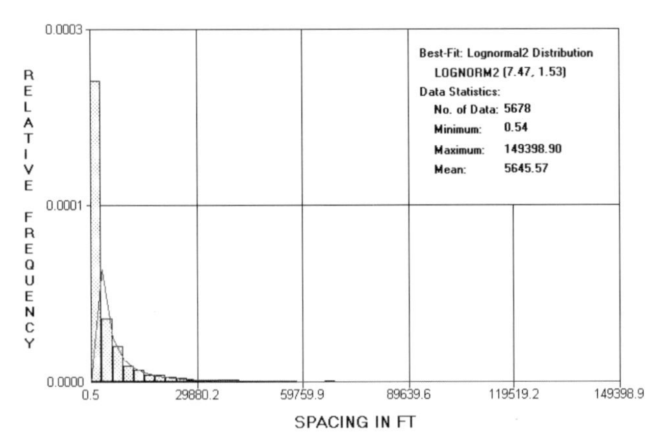

Figure 47. The histogram and best-fit spacing distribution of the north-northwest–trending surface fractures in the Black Mesa Basin area.

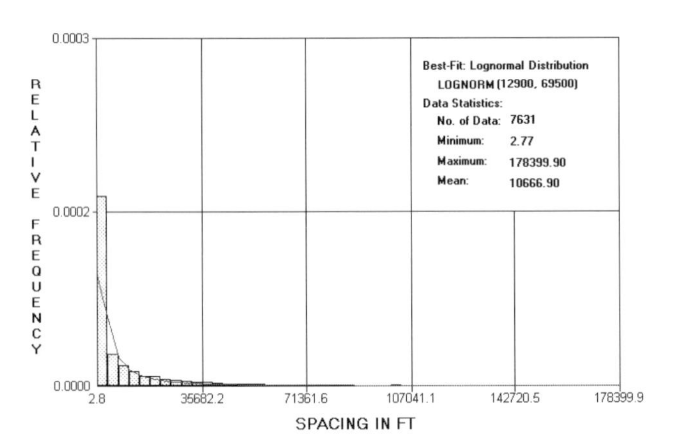

Figure 48. The histogram and best-fit spacing distribution of the northwest-trending surface fractures in the Black Mesa Basin area.

STATISTICAL ANALYSIS OF SURFACE LINEAMENTS AND FRACTURES IN THE MID-CONTINENT REGION

A large number of surface lineaments and fractures also were collected and digitized from many previous studies of various parts of the Mid-continent region (Guo and George, 1996). Using the procedure discussed earlier, best-fit statistical distribution functions were obtained for the surface lineament and fracture characteristics in this region through extensive best-fitting analyses.

Surface Lineaments in Northeastern Oklahoma

Figure 49 is a rose diagram for the surface lineaments in northeastern Oklahoma (Burchett et al., 1983). From this rose diagram, one can see that the surface lineaments in this area can be partitioned into five subsets oriented east-northeast, north-northeast, north, north-northwest, and northwest. Figures 50–54 show the histograms, best-fit distributions,

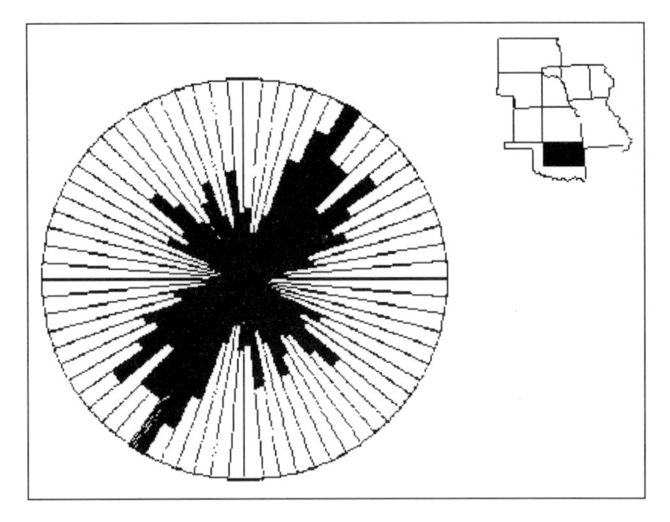

Figure 49. A rose diagram of the surface lineaments in northeastern Oklahoma.

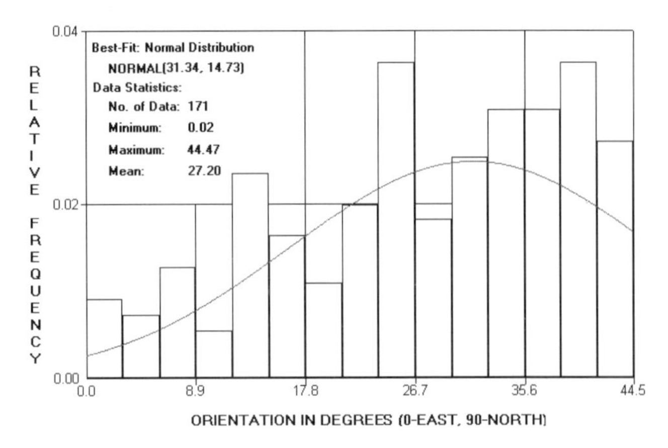

Figure 50. The histogram and best-fit orientation distribution of the east-northeast–trending surface lineaments in northeastern Oklahoma.

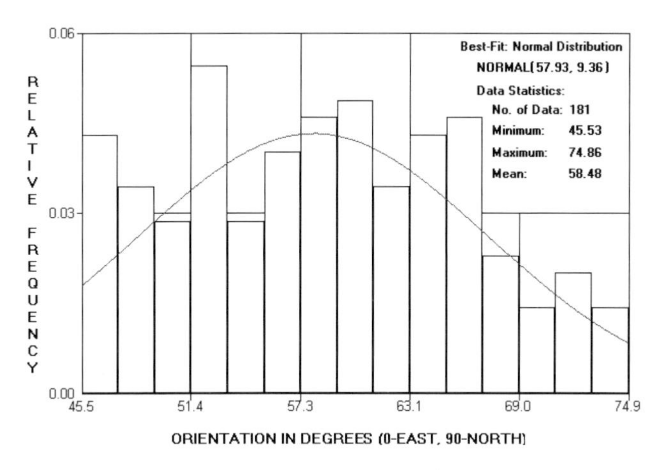

Figure 51. The histogram and best-fit orientation distribution of the north-northeast–trending surface lineaments in northeastern Oklahoma.

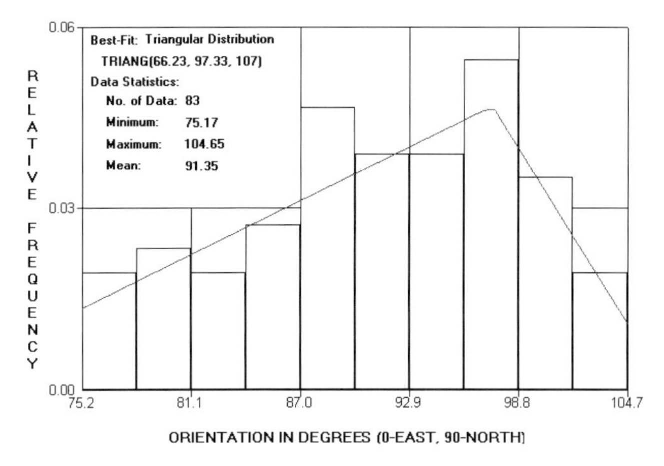

Figure 52. The histogram and best-fit orientation distribution of the north-trending surface lineaments in northeastern Oklahoma.

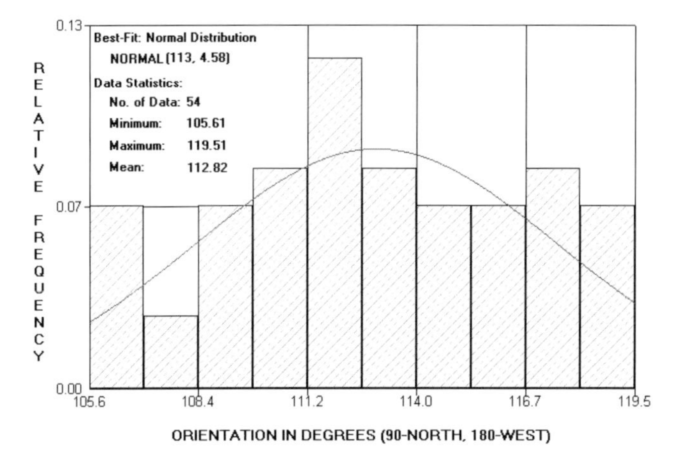

Figure 53. The histogram and best-fit orientation distribution of the north-northwest–trending surface lineaments in northeastern Oklahoma.

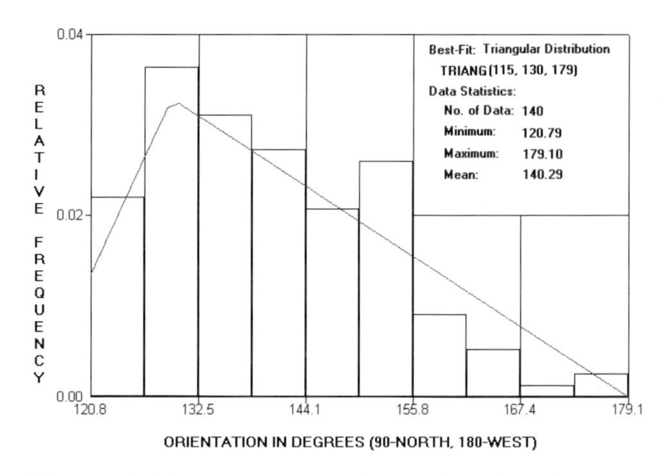

Figure 54. The histogram and best-fit orientation distribution of the northwest-trending surface lineaments in northeastern Oklahoma.

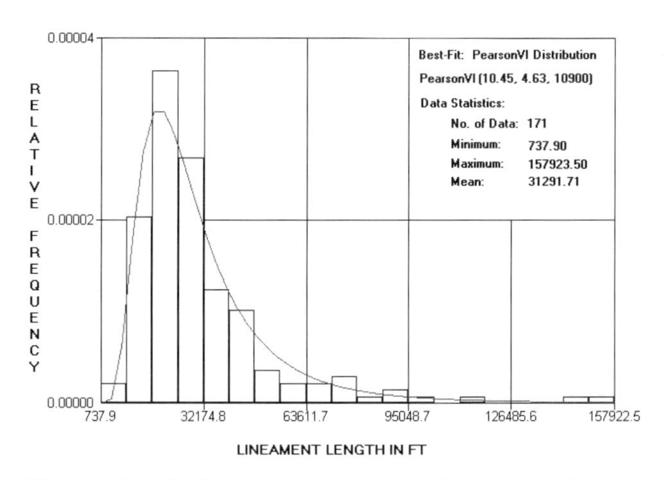

Figure 55. The histogram and best-fit length distribution of the east-northeast–trending surface lineaments in northeastern Oklahoma.

and sample data statistics for the five sets of the surface lineament orientation data. The orientations of the east-northeast–, north-northeast–, and north-northwest–trending surface lineaments in this area were best-fitted by normal distributions. Those of the north- and northwest-trending surface lineaments were best-fitted by triangular distributions.

The histograms, best-fit distributions, and sample data statistics of the corresponding lengths for the five sets of surface lineaments in the northeastern Oklahoma are shown in Figures 55–59. It appears that PearsonV or PearsonVI distributions provide the best fits to the five sets of the lineament length data.

The spacing data for the surface lineaments in northeastern Oklahoma were also analyzed. The results are shown in Figures 60–64. Inverse Gaussian, lognormal, lognormal2, and PearsonVI distributions appear to provide best fits to the five sets of the spacing data.

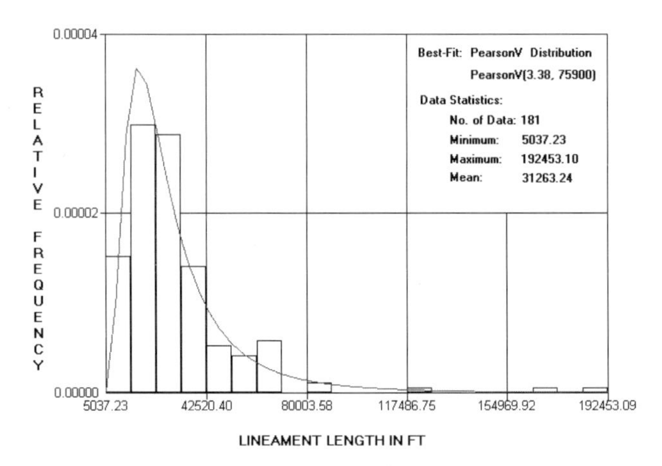

Figure 56. The histogram and best-fit length distribution of the north-northeast–trending surface lineaments in northeastern Oklahoma.

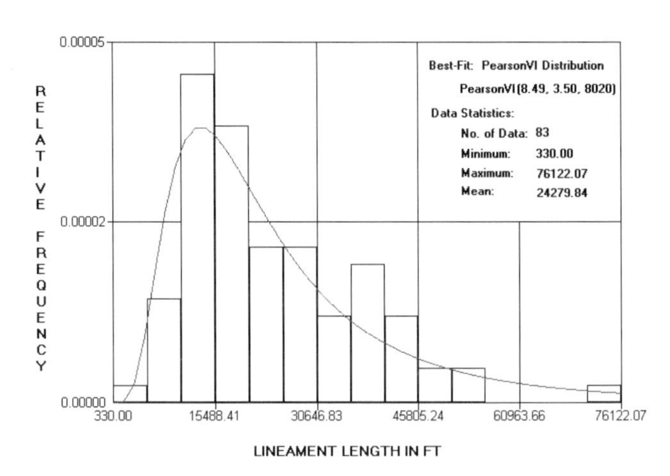

Figure 57. The histogram and best-fit length distribution of the north-trending surface lineaments in northeastern Oklahoma.

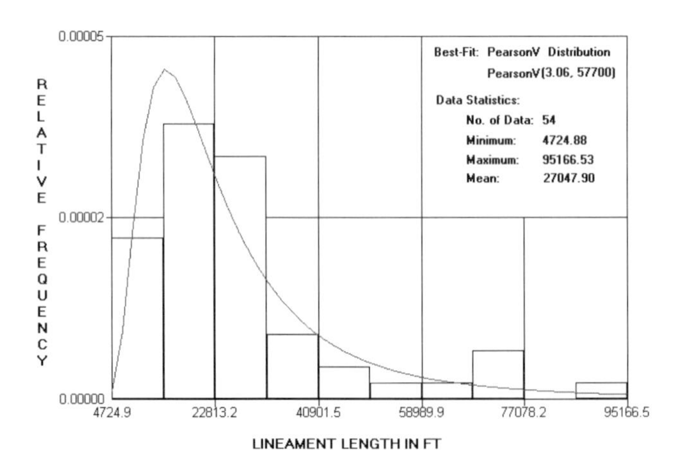

Figure 58. The histogram and best-fit length distribution of the north-northwest–trending surface lineaments in northeastern Oklahoma.

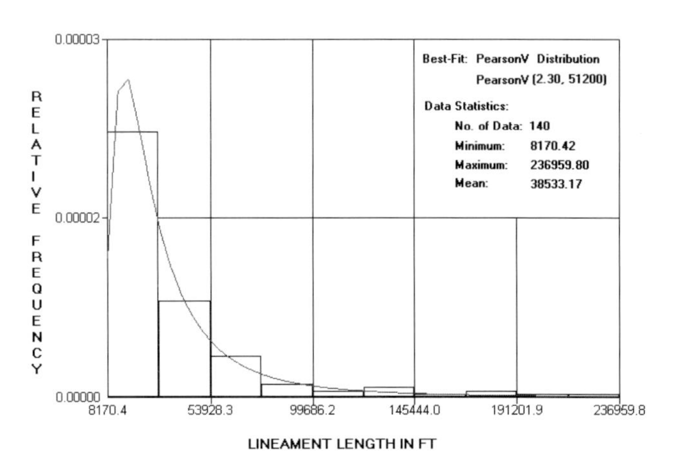

Figure 59. The histogram and best-fit length distribution of the northwest-trending surface lineaments in northeastern Oklahoma.

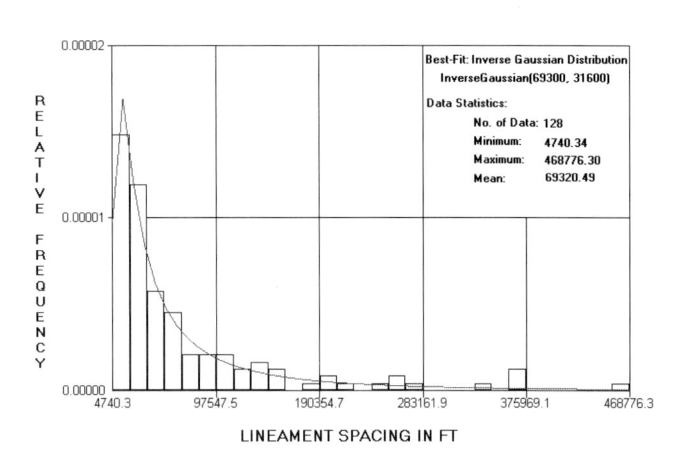

Figure 60. The histogram and best-fit spacing distribution of the east-northeast–trending surface lineaments in northeastern Oklahoma.

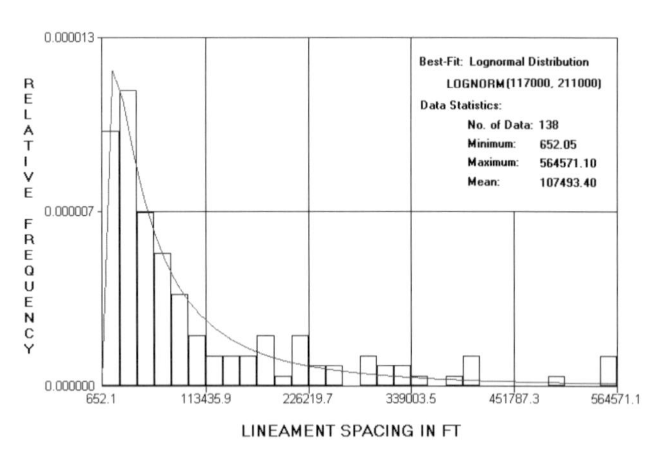

Figure 61. The histogram and best-fit spacing distribution of the north-northeast–trending surface lineaments in northeastern Oklahoma.

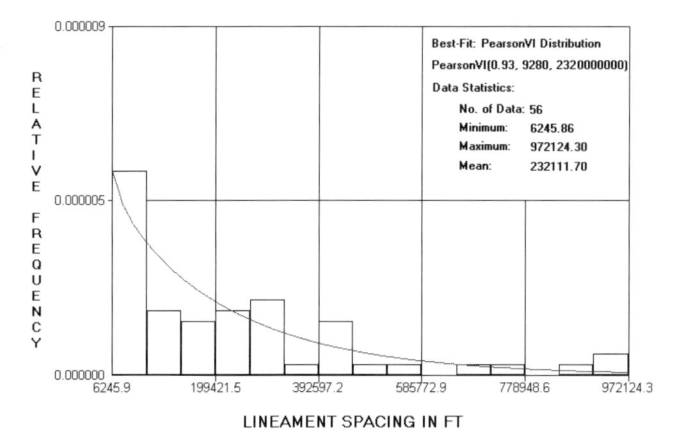

Figure 62. The histogram and best-fit spacing distribution of the north-trending surface lineaments in northeastern Oklahoma.

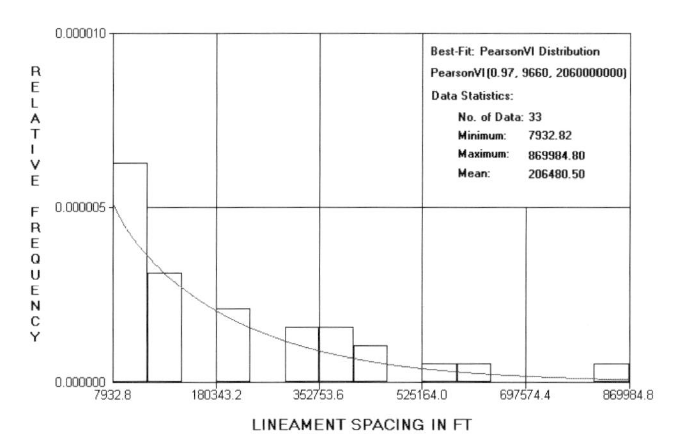

Figure 63. The histogram and best-fit spacing distribution of the north-northwest–trending surface lineaments in northeastern Oklahoma.

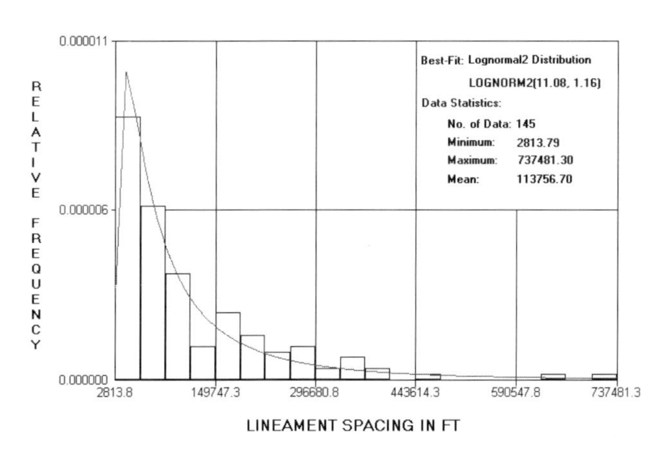

Figure 64. The histogram and best-fit spacing distribution of the northwest-trending surface lineaments in northeastern Oklahoma.

Surface Lineaments in Eastern Kansas

Figure 65 shows a rose diagram of the surface lineaments in eastern Kansas (Burchett et al., 1983). Three major subsets can be identified from this diagram. They are oriented in the northeast, north-northwest, and northwest directions. Figures 66–68 show the histograms, best-fit distributions, and sample data statistics of the orientation data. A triangular, a normal, and a chi-square distribution provide the best-fit probability distribution functions for the orientation data of the surface lineaments in eastern Kansas.

The corresponding length data were also analyzed. The results are shown in Figures 69–71. One can see that the lengths of the surface lineaments in eastern Kansas are best-fitted by lognormal, PearsonVI, and PearsonV distributions.

The results of a statistical analysis on the corresponding spacing data are shown in Figures 72–74. Lognormal2 and PearsonVI distributions were found to provide the best fits to the three sets of spacing data for the surface lineaments.

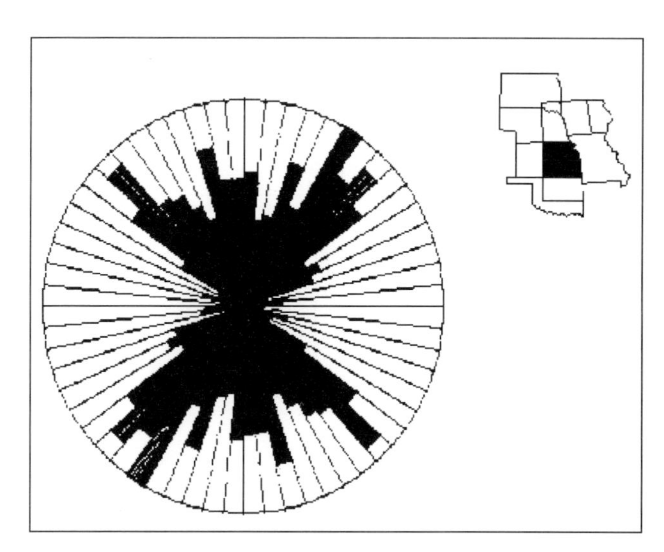

Figure 65. A rose diagram of the surface lineaments in eastern Kansas.

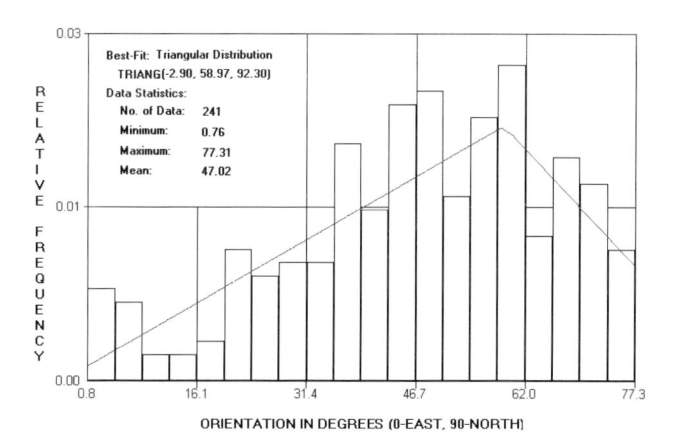

Figure 66. The histogram and best-fit orientation distribution of the northeast-trending surface lineaments in eastern Kansas.

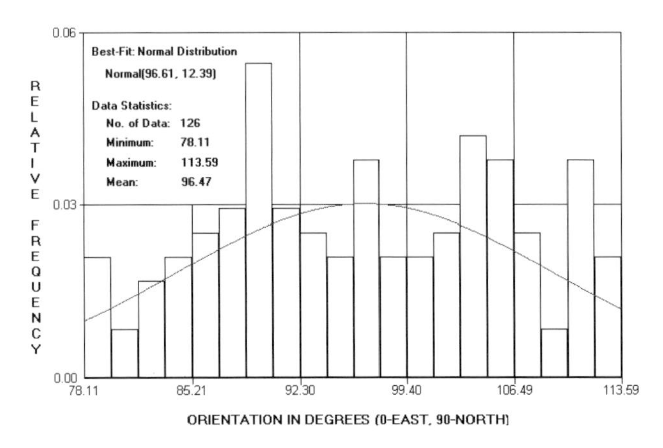

Figure 67. The histogram and best-fit orientation distribution of the north-northwest–trending surface lineaments in eastern Kansas.

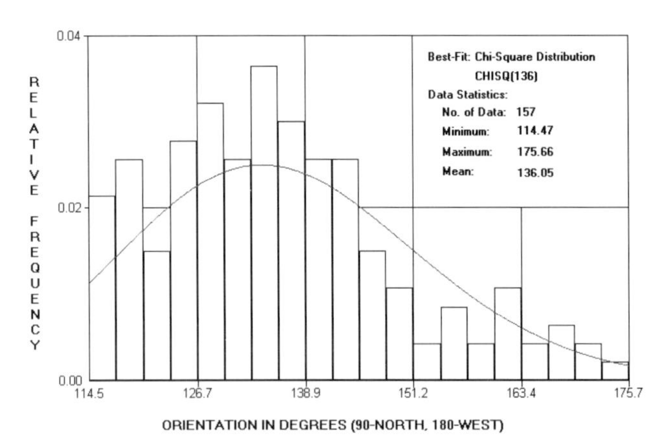

Figure 68. The histogram and best-fit orientation distribution of the northwest-trending surface lineaments in eastern Kansas.

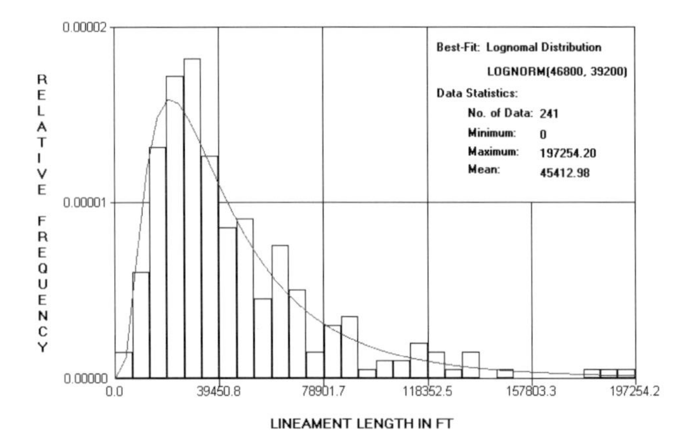

Figure 69. The histogram and best-fit length distribution of the northeast-trending surface lineaments in eastern Kansas.

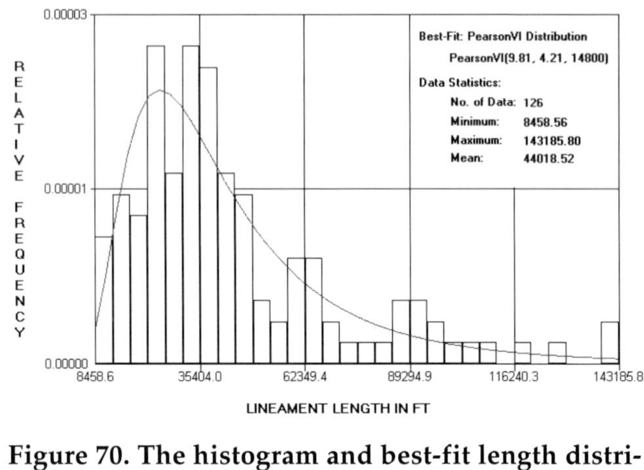

Figure 70. The histogram and best-fit length distribution of the north-northwest–trending surface lineaments in eastern Kansas.

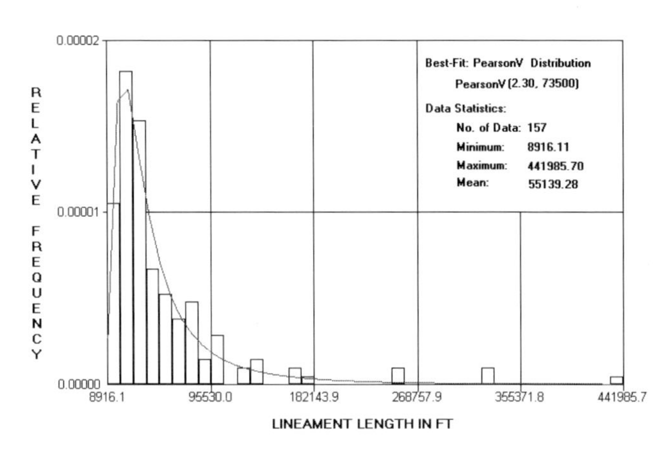

Figure 71. The histogram and best-fit length distribution of the northwest-trending surface lineaments in eastern Kansas.

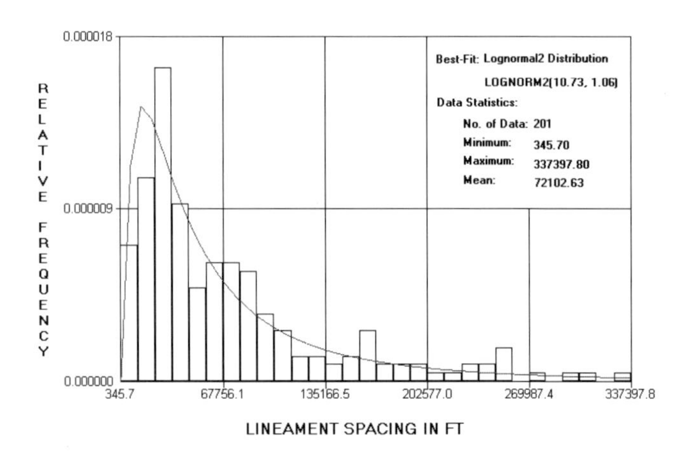

Figure 72. The histogram and best-fit spacing distribution of the northeast-trending surface lineaments in eastern Kansas.

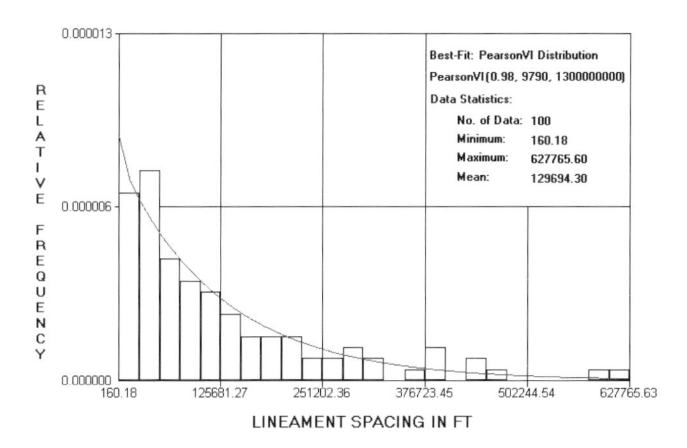

Figure 73. The histogram and best-fit spacing distribution of the north-northwest–trending surface lineaments in eastern Kansas.

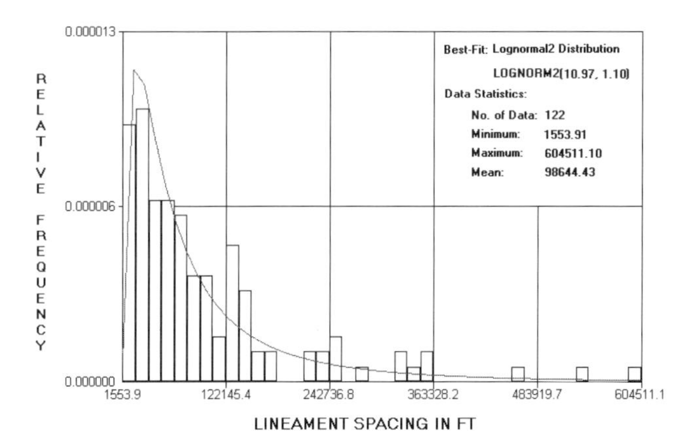

Figure 74. The histogram and best-fit spacing distribution of the northwest-trending surface lineaments in eastern Kansas.

Surface Lineaments in Eastern Nebraska and Western Iowa

Figure 75 is a rose diagram of the surface lineaments in eastern Nebraska and western Iowa (Burchett et al., 1983). Clearly, the lineaments can be partitioned into three sets, oriented northeast, north, and northwest. A statistical analysis of the lineament orientations shows that triangular distributions provide the best fits to the three sets of surface lineaments in the area. Figures 76–78 show the histograms, best-fit distributions, and sample data statistics.

Similarly, a statistical analysis was also performed on the corresponding three sets of length and spacing data. The results for the length data analysis are given in Figures 79–81. PearsonVI distributions appear to provide the best fits to the length data. The results for the spacing data analysis are shown in Figures 82–84. An inverse Gaussian, a PearsonVI, and a lognormal distribution are the best-fitted spacing distributions for the surface lineaments in eastern Nebraska and western Iowa.

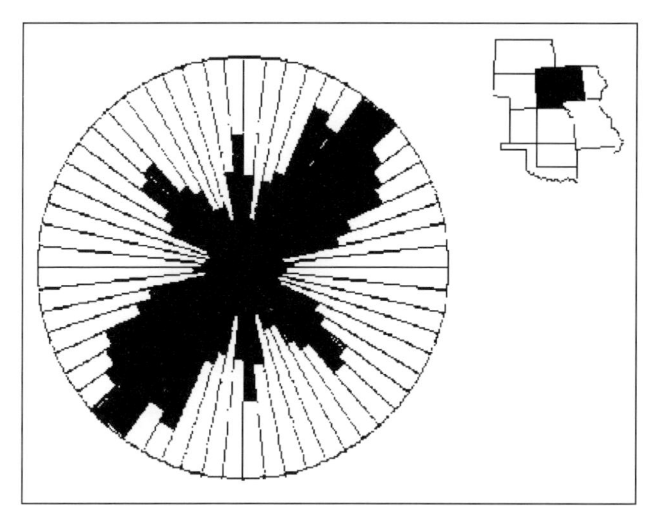

Figure 75. A rose diagram of the surface lineaments in eastern Nebraska and western Iowa.

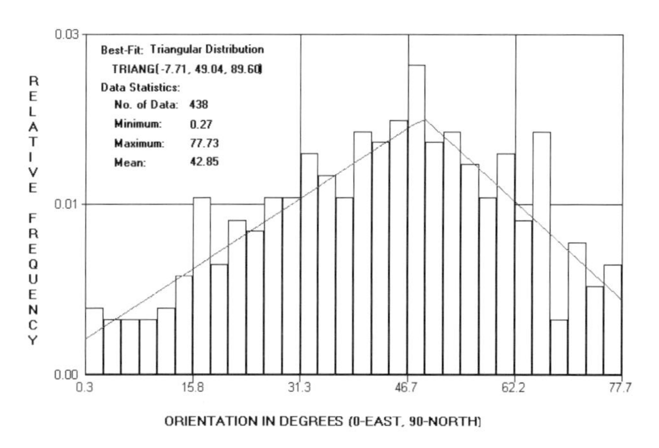

Figure 76. The histogram and best-fit orientation distribution of the northeast-trending surface lineaments in eastern Nebraska and western Iowa.

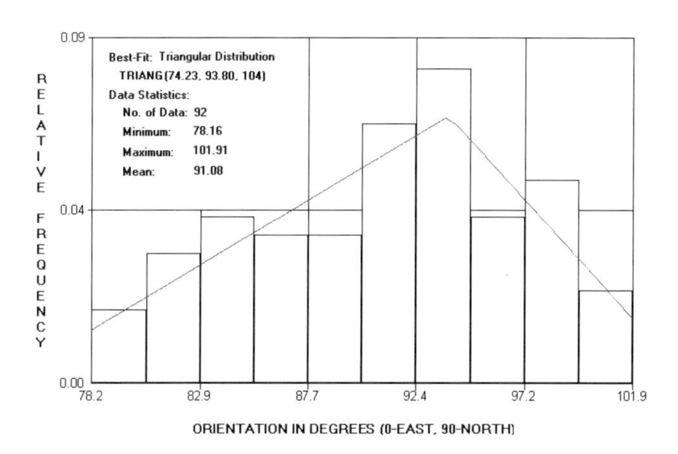

Figure 77. The histogram and best-fit orientation distribution of the north-trending surface lineaments in eastern Nebraska and western Iowa.

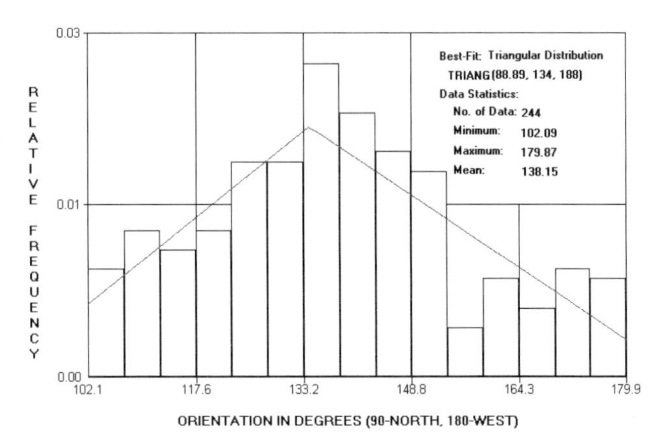

Figure 78. The histogram and best-fit orientation distribution of the northwest-trending surface lineaments in eastern Nebraska and western Iowa.

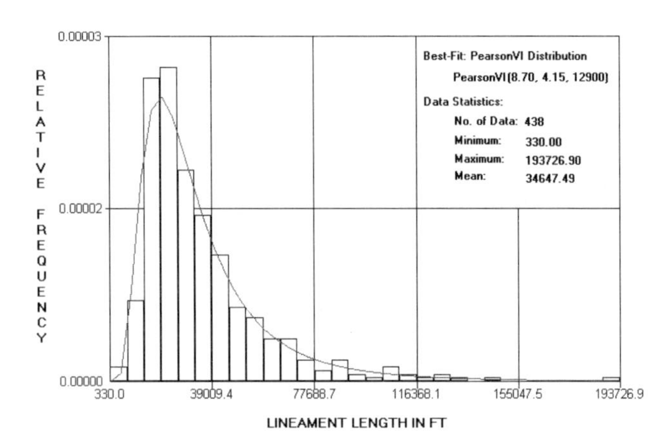

Figure 79. The histogram and best-fit length distribution of the northeast-trending surface lineaments in eastern Nebraska and western Iowa.

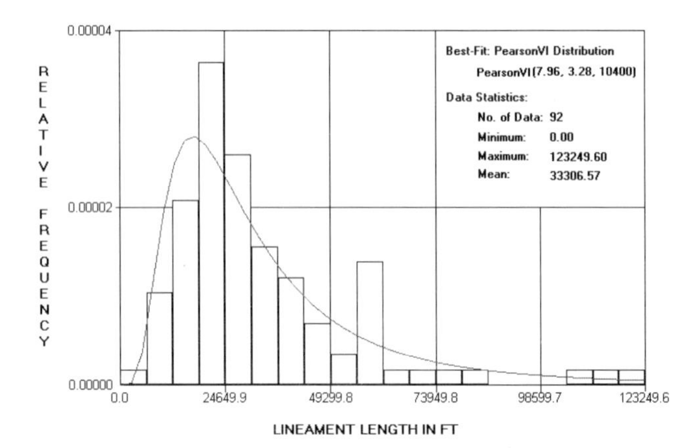

Figure 80. The histogram and best-fit length distribution of the north-trending surface lineaments in eastern Nebraska and western Iowa.

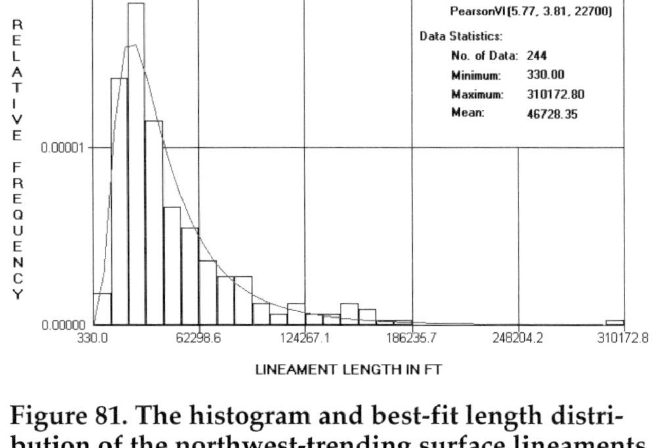

Figure 81. The histogram and best-fit length distribution of the northwest-trending surface lineaments in eastern Nebraska and western Iowa.

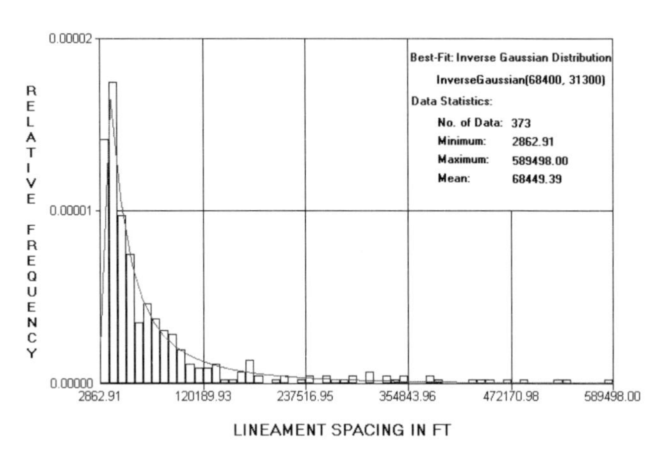

Figure 82. The histogram and best-fit spacing distribution of the northeast-trending surface lineaments in eastern Nebraska and western Iowa.

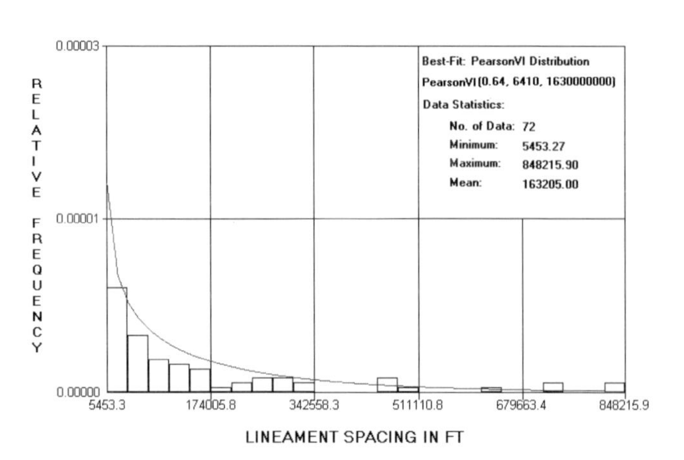

Figure 83. The histogram and best-fit spacing distribution of the north-trending surface lineaments in eastern Nebraska and western Iowa.

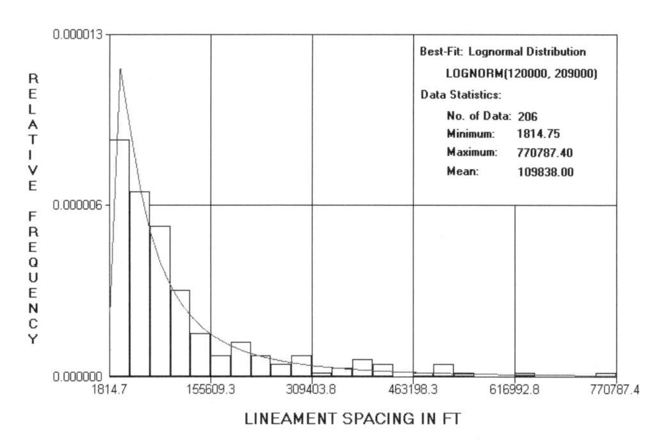

Figure 84. The histogram and best-fit spacing distribution of the northwest-trending surface lineaments in eastern Nebraska and western Iowa.

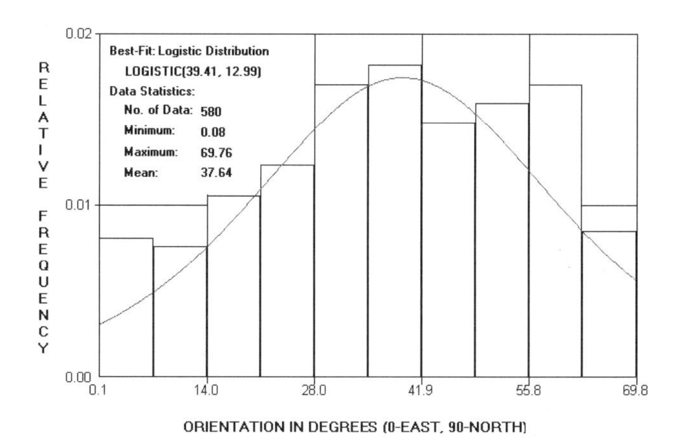

Figure 86. The histogram and best-fit orientation distribution of the northeast-trending surface lineaments and fractures in western Oklahoma.

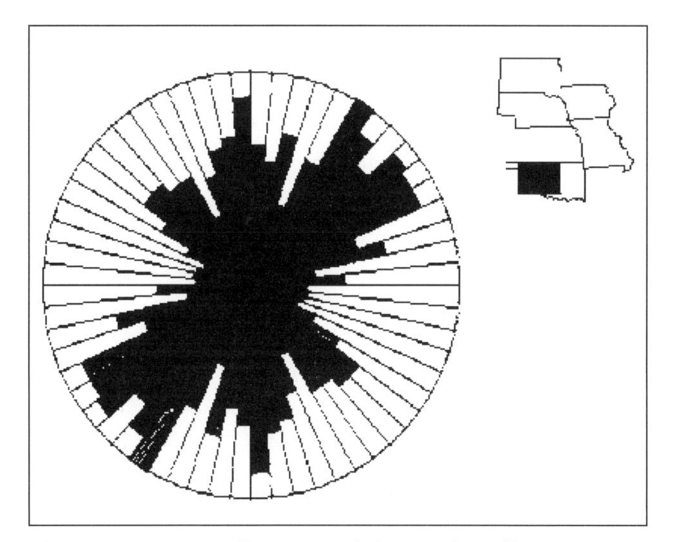

Figure 85. A rose diagram of the surface lineaments and fractures in western Oklahoma.

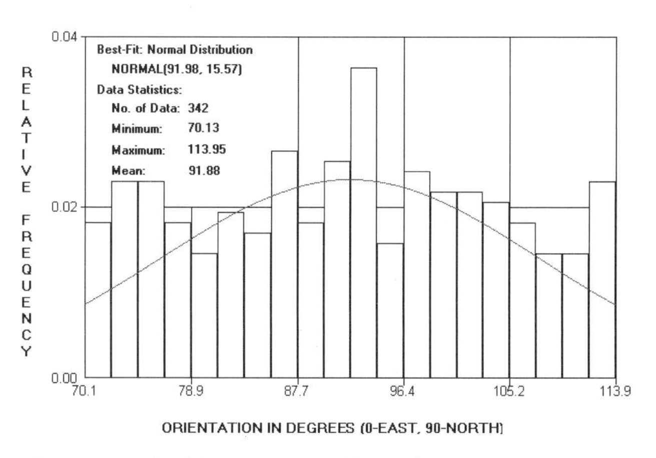

Figure 87. The histogram and best-fit orientation distribution of the north-trending surface lineaments and fractures in western Oklahoma.

Surface Lineaments and Fractures in Western Oklahoma

Figure 85 shows a rose diagram of the surface lineaments and fractures in western Oklahoma (Collins et al., 1974). From this figure, the surface lineaments and fractures can apparently be partitioned into three subsets trending northeast, north, and northwest. Figures 86–88 show the histograms, best-fit distributions, and sample data statistics for the three sets of orientation data. A logistic, a normal, and a triangular distribution provide the best fits to the orientation data.

The results of a statistical analysis on the three sets of length data are given in Figures 89–91. A lognormal, a lognormal2, and a PearsonV distribution appear to provide the best fits to the length data. The corresponding three sets of spacing data were sampled and statistically analyzed. The results are shown in Figures 92–94. The spacing data were best-fitted by three lognormal2 distributions.

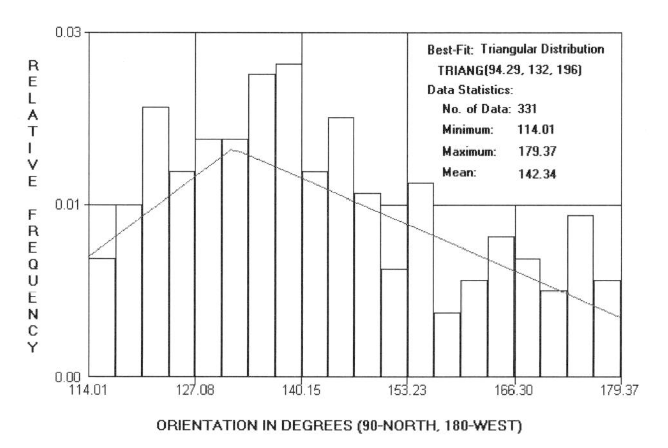

Figure 88. The histogram and best-fit orientation distribution of the northwest-trending surface lineaments and fractures in western Oklahoma.

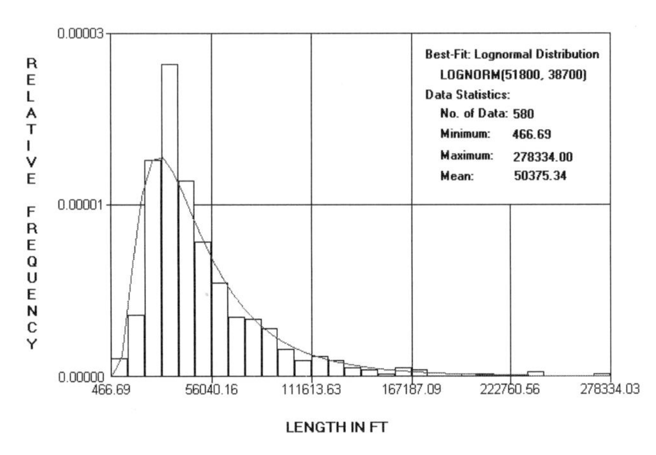

Figure 89. The histogram and best-fit length distribution of the northeast-trending surface lineaments and fractures in western Oklahoma.

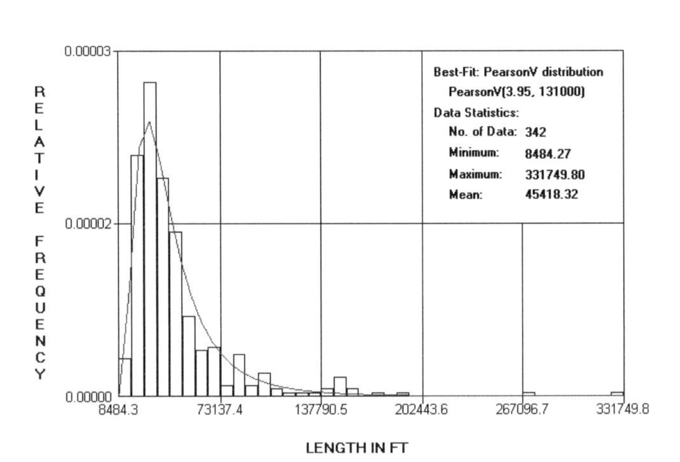

Figure 90. The histogram and best-fit length distribution of the north-trending surface lineaments and fractures in western Oklahoma.

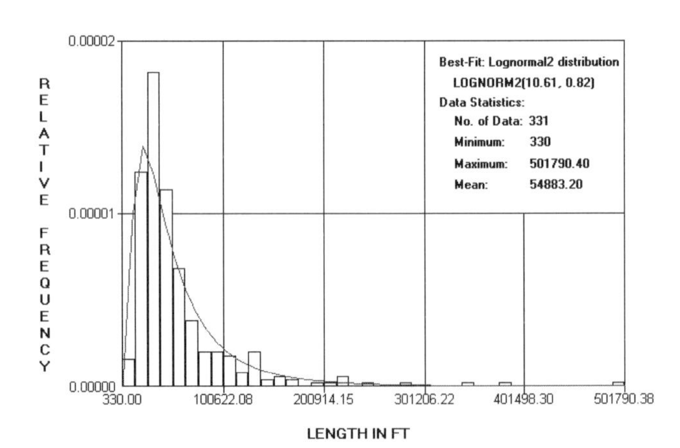

Figure 91. The histogram and best-fit length distribution of the northwest-trending surface lineaments and fractures in western Oklahoma.

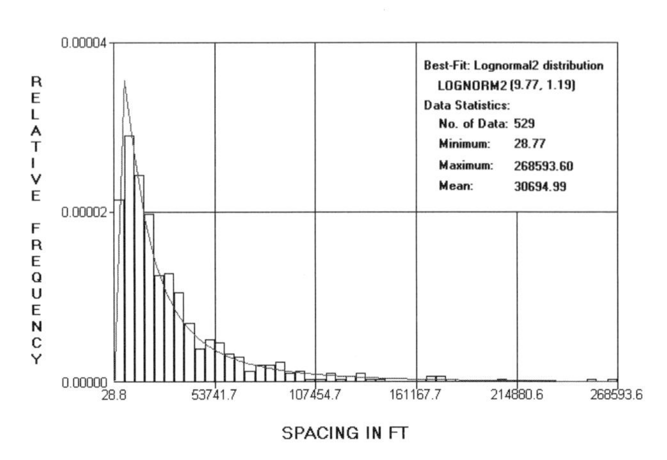

Figure 92. The histogram and best-fit spacing distribution of the northeast-trending surface lineaments and fractures in western Oklahoma.

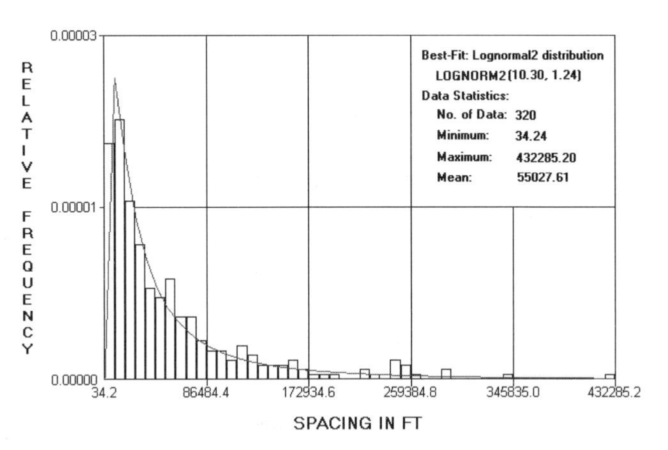

Figure 93. The histogram and best-fit spacing distribution of the north-trending surface lineaments and fractures in western Oklahoma.

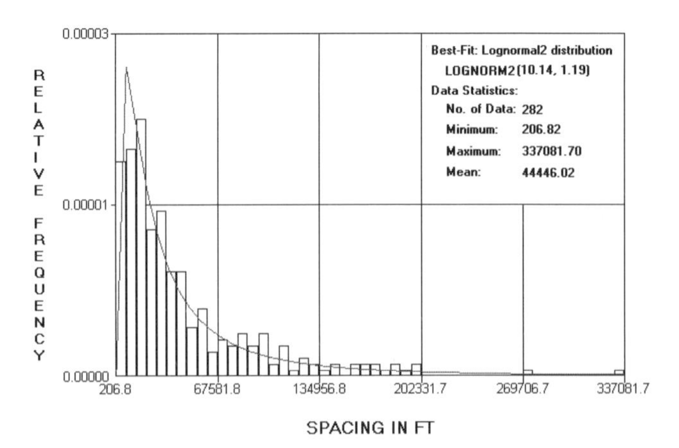

Figure 94. The histogram and best-fit spacing distribution of the northwest-trending surface lineaments and fractures in western Oklahoma.

Surface Lineaments and Fractures in Western Kansas

Figure 95 shows a rose diagram of the surface lineaments and fractures in western Kansas (Cooley, 1984). The surface linear features in this area can be divided into two subsets trending northeast and northwest, although there appears to be an insignificant set trending north. Upon being partitioned into two sets, the orientation data were statistically analyzed. The best-fitted distributions are a triangular distribution for the northeast-trending set and a logistic distribution for the northwest-trending set. The histograms, best-fit distributions, and sample data statistics for the orientation data are shown in Figures 96 and 97.

Similarly, the corresponding length and spacing data are also analyzed. The results for the two sets of the length data are shown in Figures 98 and 99. PearsonVI distributions appear to provide the best fits to the length data. The results for the spacing data are shown in Figures 100 and 101. The spacing data are also best-fitted by PearsonVI distributions.

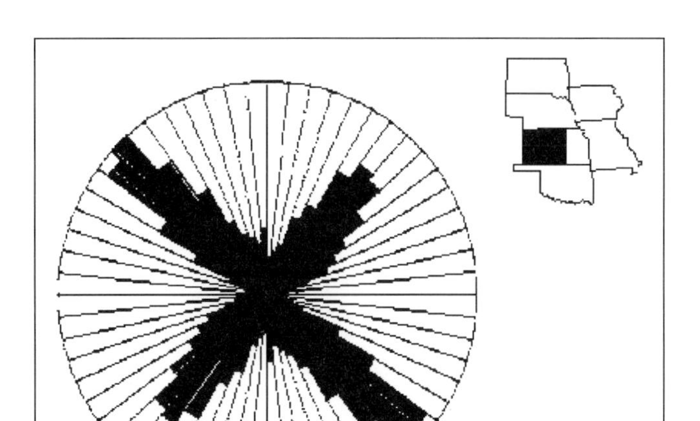

Figure 95. A rose diagram of the surface lineaments and fractures in western Kansas.

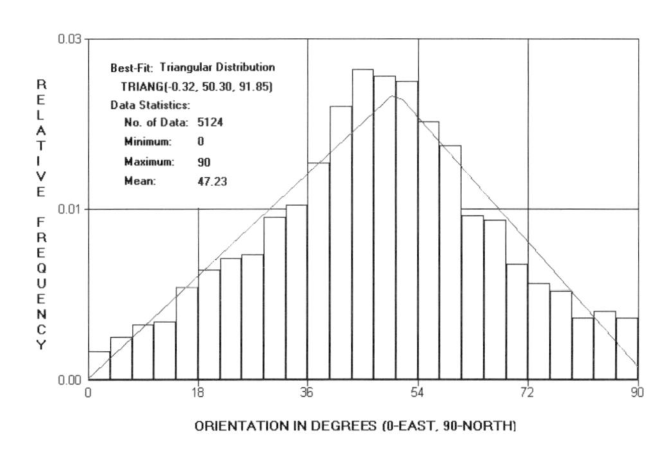

Figure 96. The histogram and best-fit orientation distribution of the northeast-trending surface lineaments and fractures in western Kansas.

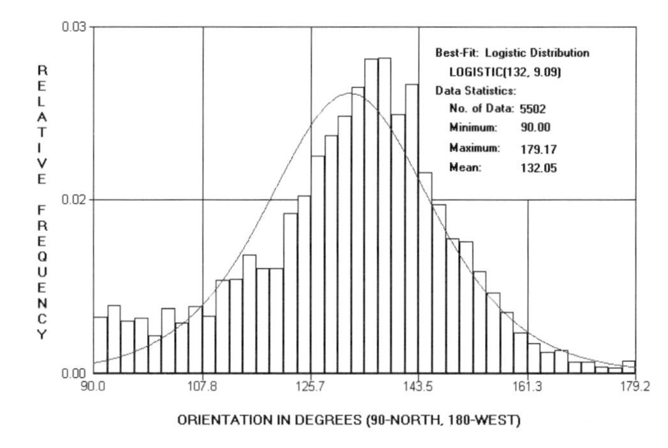

Figure 97. The histogram and best-fit orientation distribution of the northwest-trending surface lineaments and fractures in western Kansas.

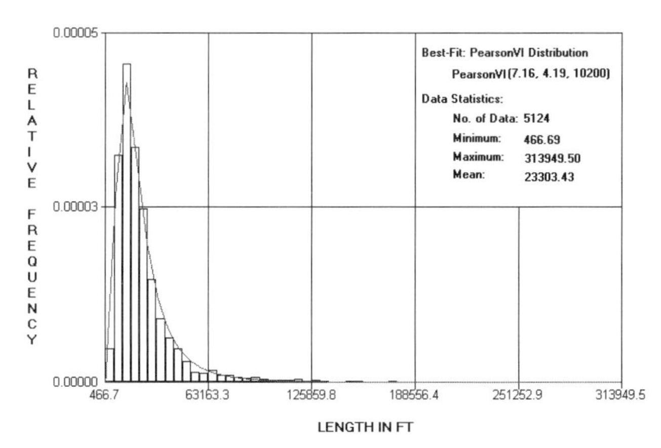

Figure 98. The histogram and best-fit length distribution of the northeast-trending surface lineaments and fractures in western Kansas.

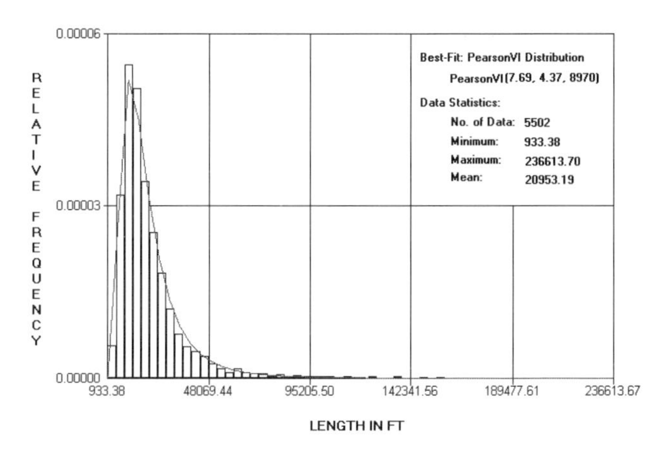

Figure 99. The histogram and best-fit length distribution of the northwest-trending surface lineaments and fractures in western Kansas.

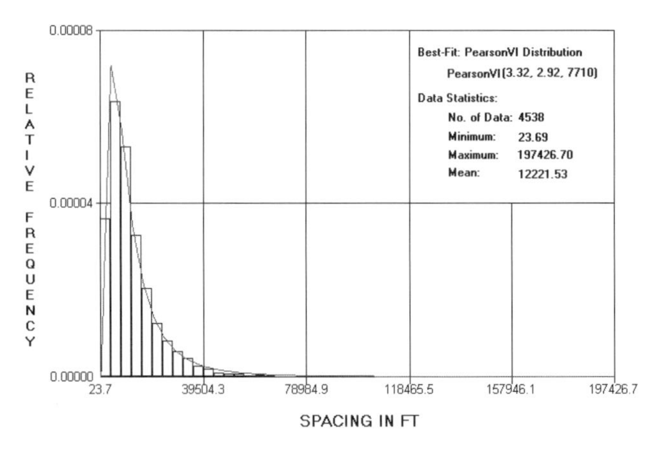

Figure 100. The histogram and best-fit spacing distribution of the northeast-trending surface lineaments and fractures in western Kansas.

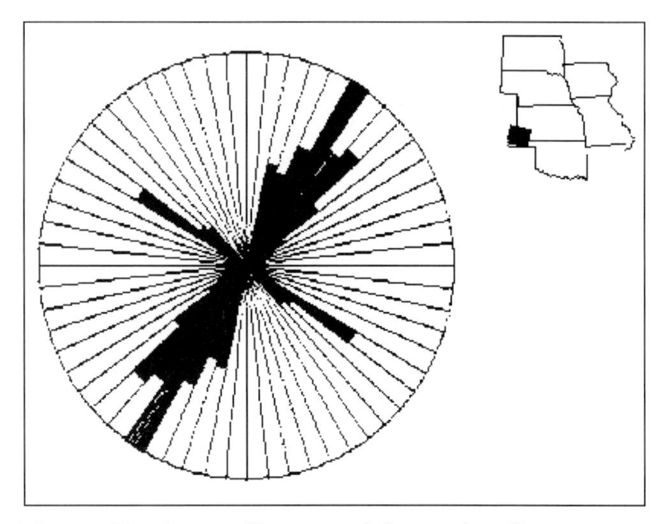

Figure 102. A rose diagram of the surface lineaments and fractures in the western Hugoton gas field.

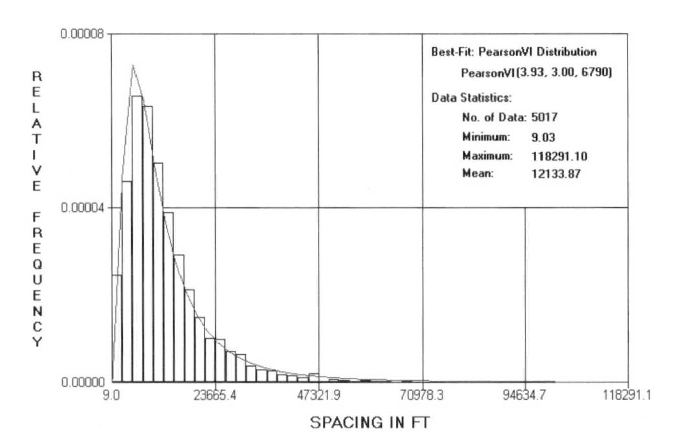

Figure 101. The histogram and best-fit spacing distribution of the northeast-trending surface lineaments and fractures in western Kansas.

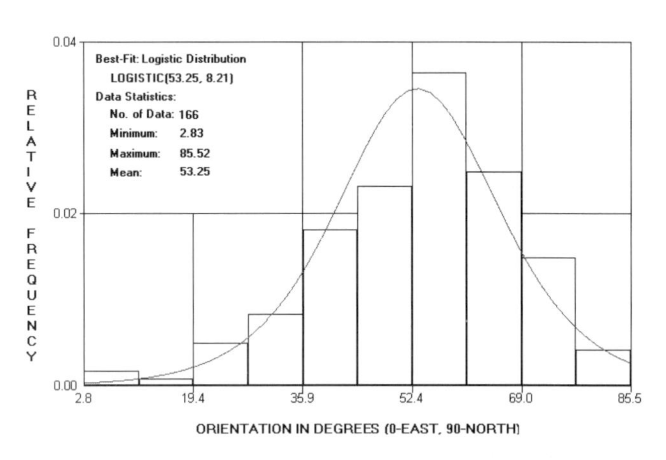

Figure 103. The histogram and best-fit orientation distribution of the northeast-trending surface lineaments and fractures in the western Hugoton gas field.

Surface Lineaments and Fractures in the Hugoton Gas Field

Figure 102 is a rose diagram of the surface lineaments and fractures in the western Hugoton gas field in southwestern Kansas and the Oklahoma Panhandle (Swanson and Shannon, 1990). It is clear from this rose diagram that the surface linear features in this area can be partitioned into two subsets trending northeast and northwest. Figures 103 and 104 show the histograms, best-fit distributions, and sample data statistics for the two sets of orientation data. A logistic and a triangular distribution appear to provide the best fits to the orientation data.

Similar analyses were also performed on the corresponding length and spacing data. The results for the length data are given in Figures 105 and 106. Lognormal2 and PearsonVI distributions were found to provide the best fits to the two sets of length data. The results for the spacing data are shown in Figures 107 and 108. The best-fitted distributions are a lognormal2

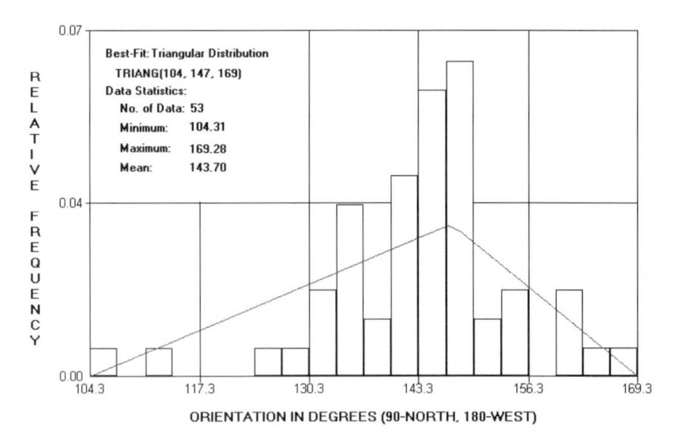

Figure 104. The histogram and best-fit orientation distribution of the northwest-trending surface lineaments and fractures in the western Hugoton gas field.

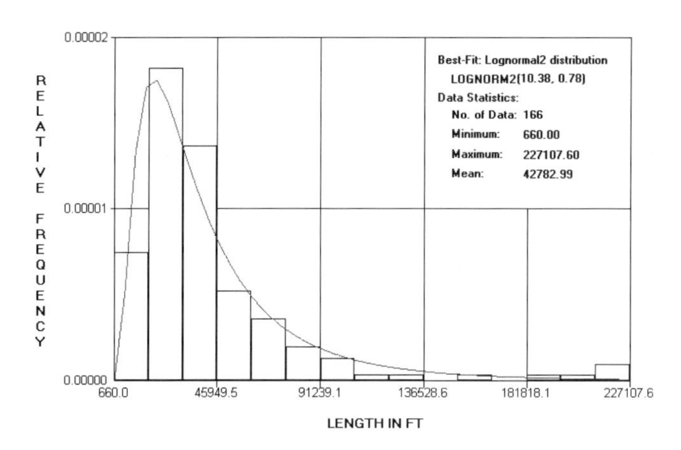

Figure 105. The histogram and best-fit length distribution of the northeast-trending surface lineaments and fractures in the western Hugoton gas field.

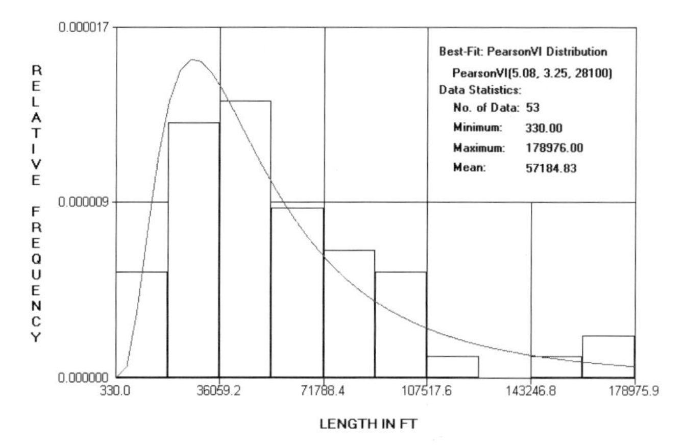

Figure 106. The histogram and best-fit length distribution of the northwest-trending surface lineaments and fractures in the western Hugoton gas field.

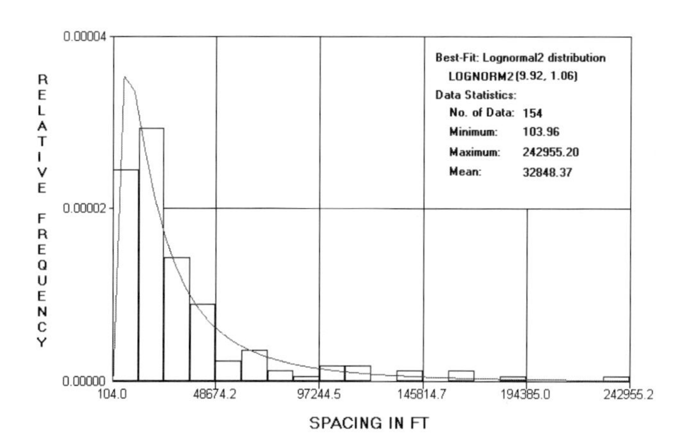

Figure 107. The histogram and best-fit spacing distribution of the northeast-trending surface lineaments and fractures in the western Hugoton gas field.

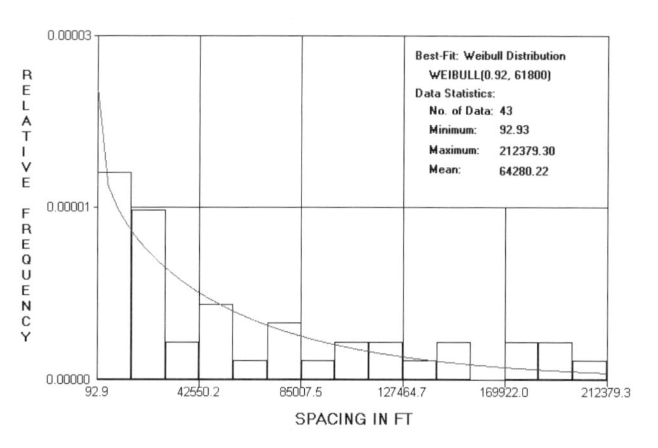

Figure 108. The histogram and best-fit spacing distribution of the northwest-trending surface lineaments and fractures in the western Hugoton gas field.

distribution for the spacing data of the northeast-trending linear features and a Weibull distribution for the spacing data of the northwest-trending linear features.

Surface Fractures in the Northern Forest City Basin

Figure 109 shows a rose diagram of the surface fractures in northern Forest City Basin in southwestern Iowa (Herman et al., 1986). The surface fractures in this area appear to consist of two subsets trending east and north-northeast. The results of a statistical analysis on the orientation data are given in Figures 110 and 111. A triangular and a normal distribution provide the best fits to the two sets of orientation data.

Figures 112 and 113 show the histograms, best-fit distributions, and sample data statistics for the corresponding length data. A logistic and an extreme-value distribution give the best fits to the two sets of length data.

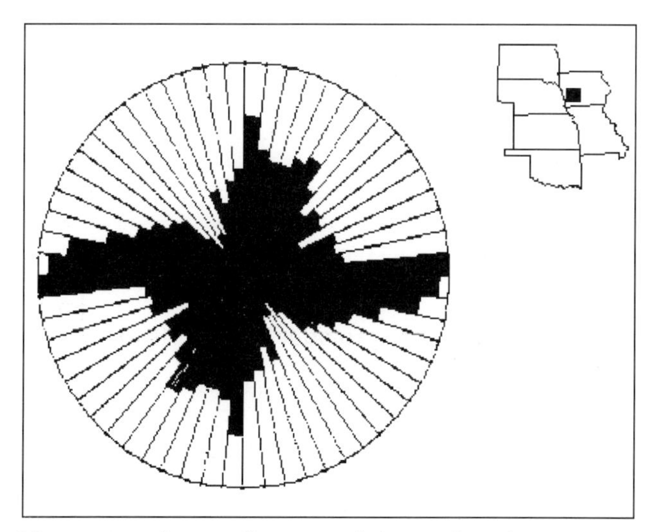

Figure 109. A rose diagram of the surface fractures in northern Forest City Basin.

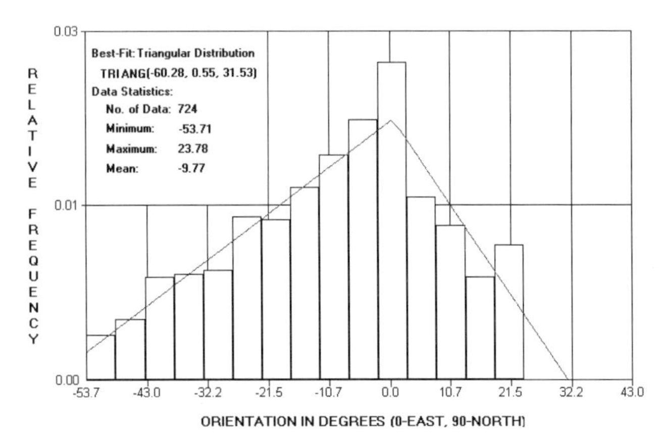

Figure 110. The histogram and best-fit orientation distribution of the east-trending surface fractures in northern Forest City Basin.

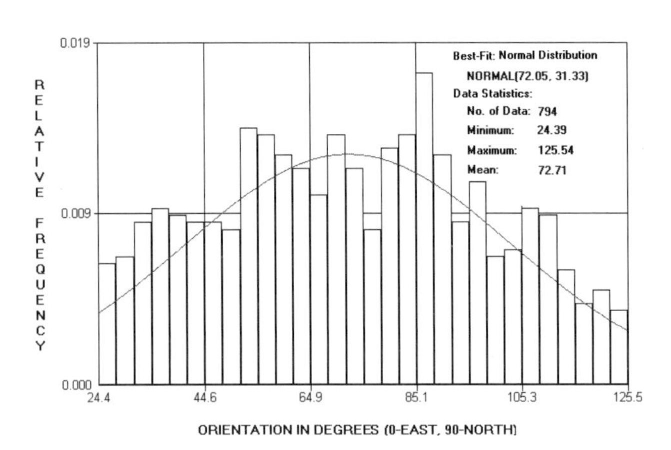

Figure 111. The histogram and best-fit orientation distribution of the north-northeast–trending surface fractures in northern Forest City Basin.

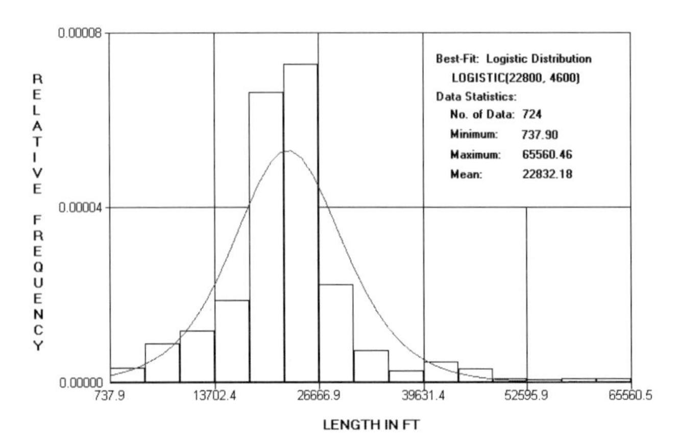

Figure 112. The histogram and best-fit length distribution of the east-trending surface fractures in northern Forest City Basin.

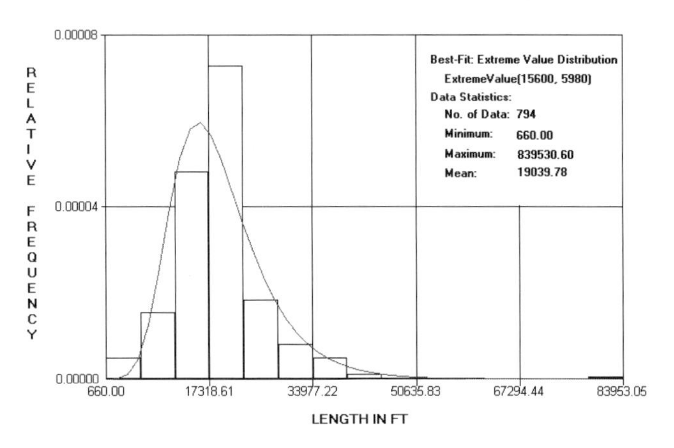

Figure 113. The histogram and best-fit length distribution of the north-northeast–trending surface fractures in northern Forest City Basin.

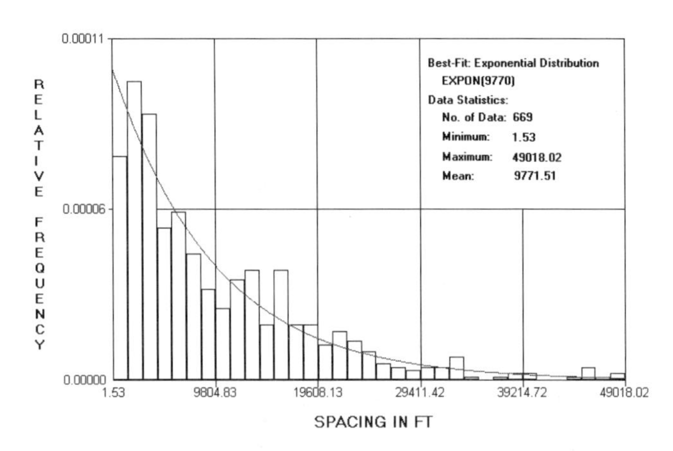

Figure 114. The histogram and best-fit spacing distribution of the east-trending surface fractures in northern Forest City Basin.

Similarly, the results for an analysis of the corresponding spacing data are shown in Figures 114 and 115. The best-fitted distributions for the two sets of spacing data are an exponential distribution for the east-trending surface fractures and a lognormal2 distribution for the north-northeast–trending surface fractures.

Surface Lineaments and Fractures in Northeastern Iowa

Figure 116 shows a rose diagram of the surface lineaments and fractures in northeastern Iowa (Chen, 1992). Three subsets can be identified trending east, northeast, and northwest. Figures 117–119 show the histograms, the best-fit distributions, and the sample data statistics for the three sets of orientation data. A PearsonV and two triangular distributions provide the best fits to the three sets of orientation data.

Similarly, Figures 120–122 show the results of a statistical analysis on the corresponding length data. The

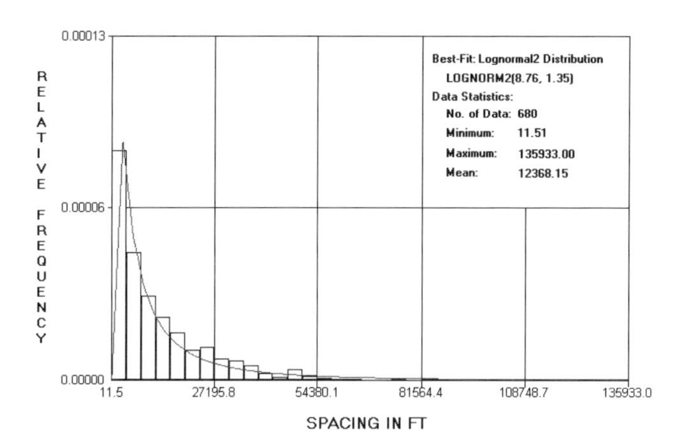

Figure 115. The histogram and best-fit spacing distribution of the north-northeast–trending surface fractures in northern Forest City Basin.

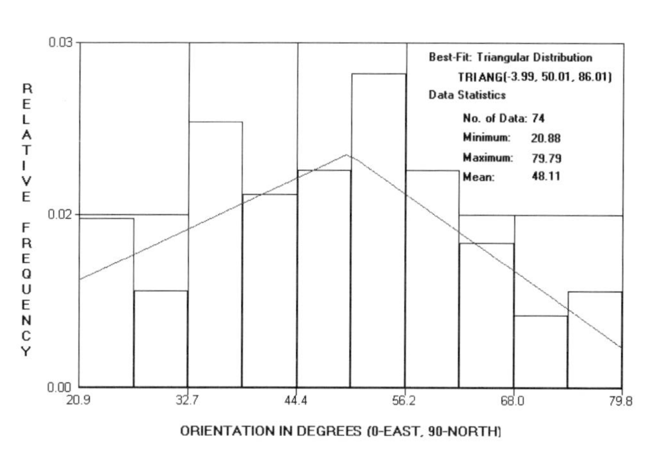

Figure 118. The histogram and best-fit orientation distribution of the northeast-trending surface fractures in northeastern Iowa.

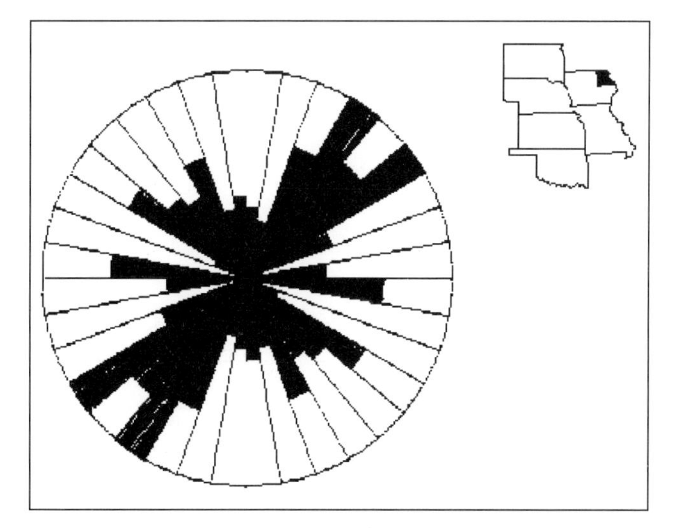

Figure 116. A rose diagram of the surface fractures in northeastern Iowa.

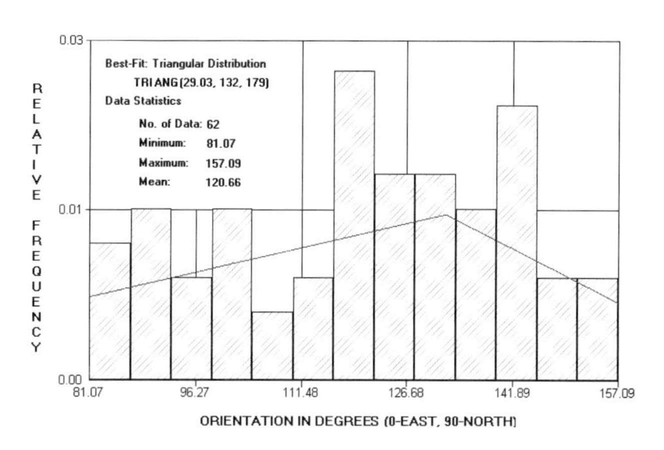

Figure 119. The histogram and best-fit orientation distribution of the northwest-trending surface fractures in northeastern Iowa.

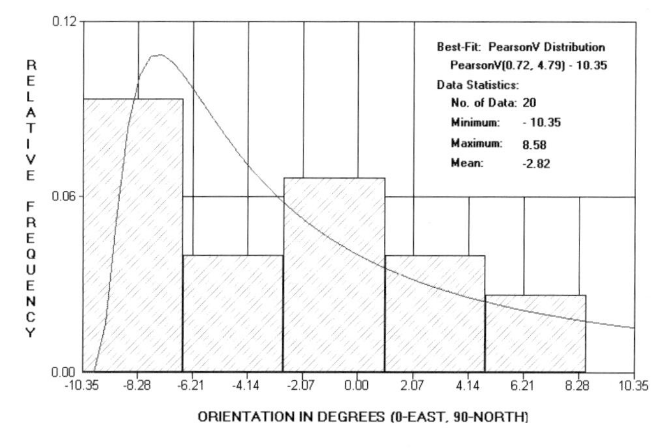

Figure 117. The histogram and best-fit orientation distribution of the east-trending surface fractures in northeastern Iowa.

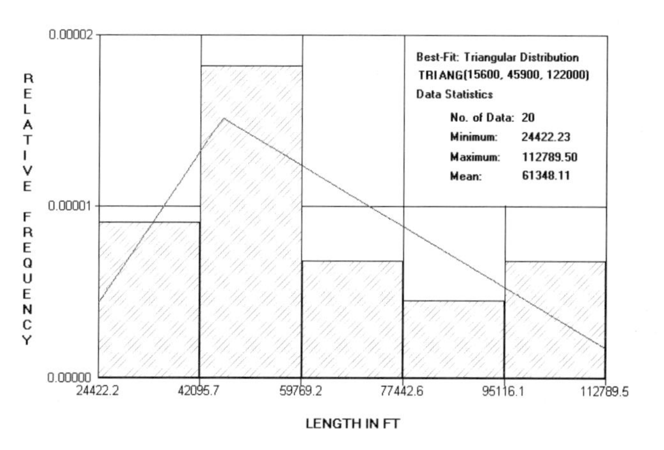

Figure 120. The histogram and best-fit length distribution of the east-trending surface fractures in northeastern Iowa.

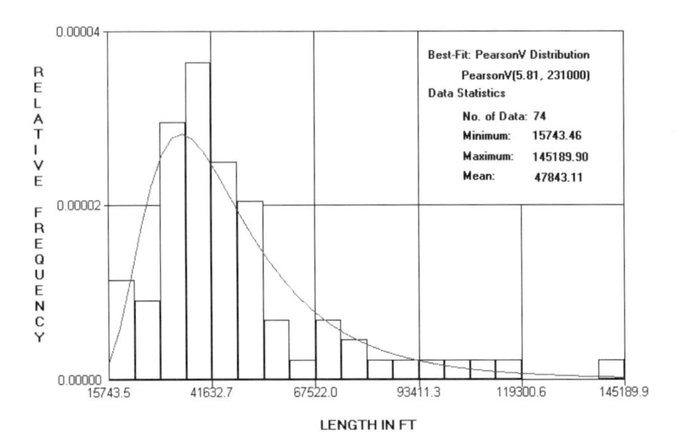

Figure 121. The histogram and best-fit length distribution of the northeast-trending surface fractures in northeastern Iowa.

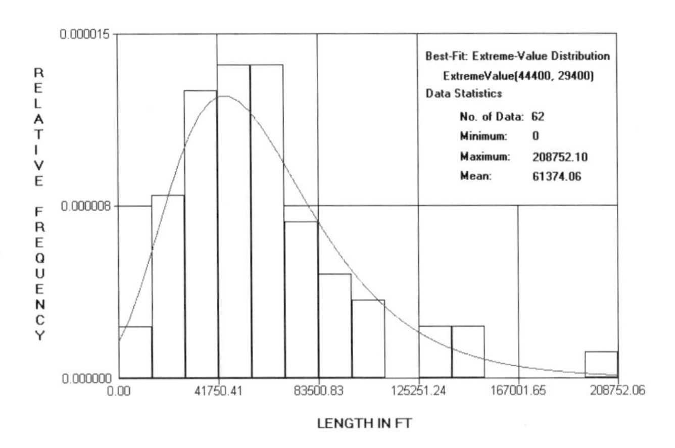

Figure 122. The histogram and best-fit length distribution of the northwest-trending surface fractures in northeastern Iowa.

best-fitted distributions are a triangular distribution for the lengths of the east-trending linear features, a PearsonV distribution for the lengths of the northeast-trending linear features, and an extreme-value distribution for the lengths of the northwest-trending linear features.

The results of a statistical analysis on the corresponding spacing data are shown in Figures 123–125. A Weibull, a PearsonV, and a PearsonVI distribution provide the best fits to the three sets of spacing data for the surface lineaments and fractures in northeastern Iowa.

DISCUSSIONS ON THE STATISTICAL DISTRIBUTIONS OF SURFACE LINEAMENTS AND FRACTURES

The surface lineaments and fractures mapped from nine areas in the Mid-continent region and three areas in northeastern Arizona were collected, digitized, and

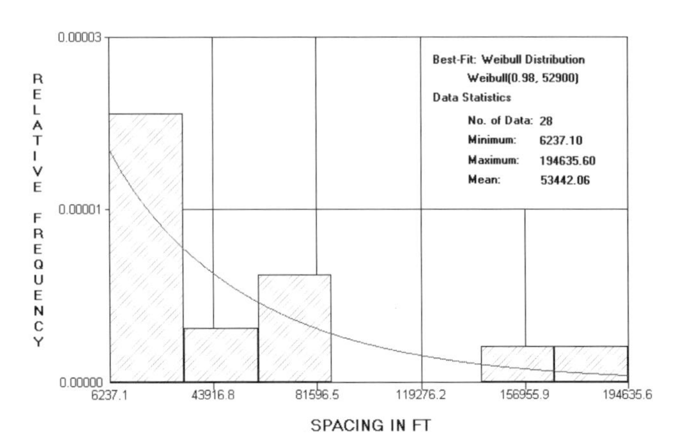

Figure 123. The histogram and best-fit spacing distribution of the east-trending surface fractures in northeastern Iowa.

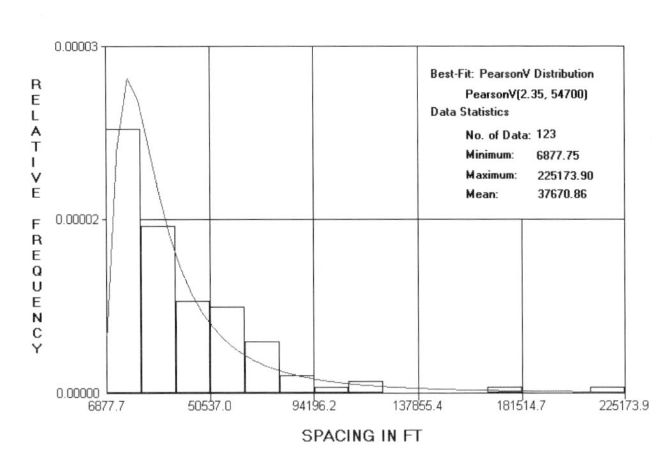

Figure 124. The histogram and best-fit spacing distribution of the northeast-trending surface fractures in northeastern Iowa.

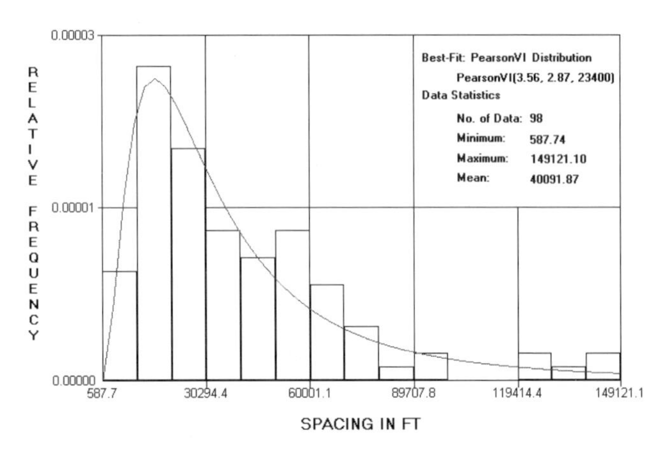

Figure 125. The histogram and best-fit spacing distribution of the northwest-trending surface fractures in northeastern Iowa.

statistically analyzed in this paper. In all, 36 sets were identified based on their orientations. The best-fit distributions were obtained for the orientation, length, and spacing of each set.

Figure 126 shows the best-fit distributions and their frequencies for the orientations of the 36 sets of surface lineaments and fractures. Triangular, normal, and logistic distributions provided the best fits to 33 of the 36 sets of orientation data. The best-fit distributions and their frequencies for the corresponding 36 sets of length data are shown in Figure 127. PearsonVI, PearsonV, lognormal, and extreme-value distributions provided the best fits to 30 of the 36 sets of length data. Similarly, the best-fit distributions and their frequencies for the corresponding 36 sets of spacing data are given in Figure 128. Lognormal2, PearsonVI, and lognormal distributions provided the best fits to 30 of the 36 sets of spacing data.

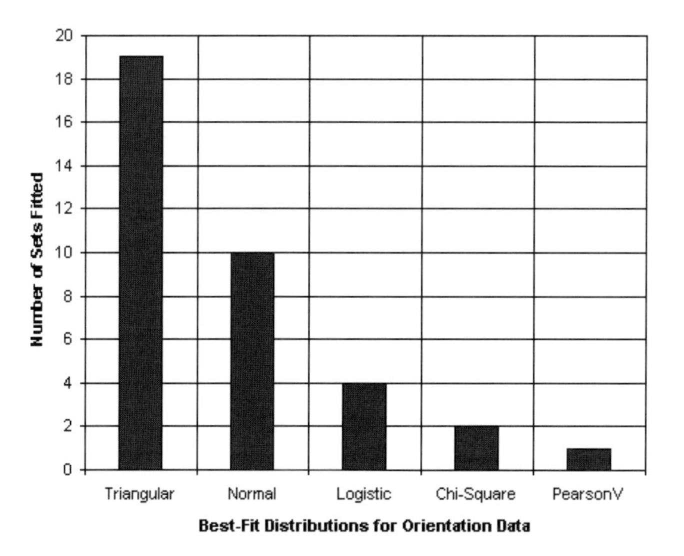

Figure 126. The best-fit distributions for the orientations of 36 sets of surface lineaments and fractures based on chi-square goodness-of-fit tests.

Comparing the results of this study to those of the previous studies reviewed earlier in this paper, one can notice both consistencies to a certain degree and significant discrepancies between them. Many probability distribution functions identified in the previous studies, such as normal distribution for orientation data, lognormal distribution for length data, and lognormal and Weibull distributions for spacing data, also provided best fits to some of the 36 sets of surface lineaments and fractures; however, these distribution functions fit only a few of the 36 sets of data. Differing from previous studies, this study shows triangular distributions are the most significant type of probability functions for describing natural fracture orientation data, followed by normal and logistic distributions (see Figure 126). PearsonVI distributions are the most significant type of probability functions for describing natural fracture length data, followed by PearsonV, lognormal2, and extreme-value distributions (see Figure 127). Natural fracture spacing data, as shown in Figure 128, appear to be best characterized by lognormal2 distribution, followed by PearsonVI and lognormal distributions.

The probability distribution functions identified in this study for describing natural fracture characteristics were obtained through an extensive statistical analysis of a large number of surface lineaments and fractures, using 25 different hypothesized probability functions during best-fitting. We believe they provide realistic descriptions of natural fracture orientation, length, and spacing; however, these distribution functions must be conditioned using any available subsurface fracture data before they can be used for stochastic simulations of natural fractures in a naturally fractured reservoir.

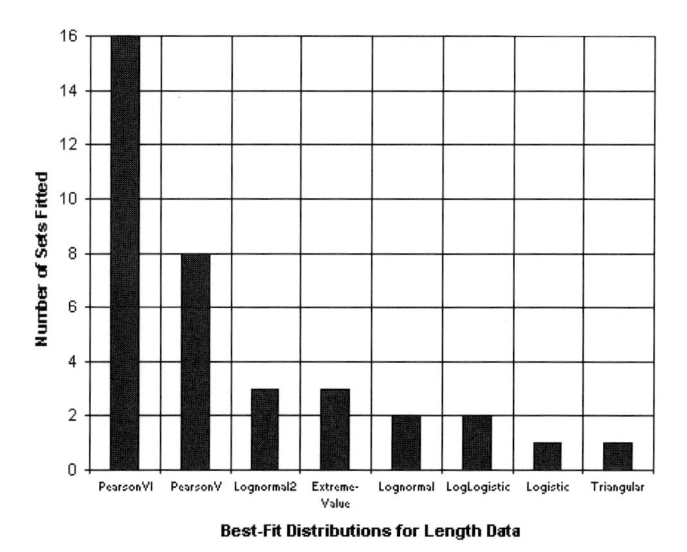

Figure 127. The best-fit distributions for the lengths of 36 sets of surface lineaments and fractures based on chi-square goodness-of-fit tests.

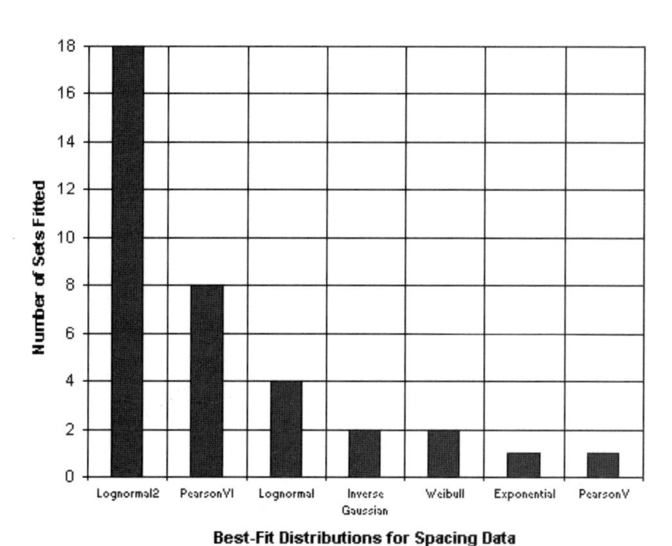

Figure 128. The best-fit distributions for the spacing of 36 sets of surface lineaments and fractures based on chi-square goodness-of-fit tests.

CONCLUSIONS

In this paper, we presented the results of an extensive statistical analysis of natural fracture characteristics using a huge number of surface lineaments and fractures mapped in the Mid-continent and Colorado Plateau regions. The following conclusions are obtained.

Natural fracture orientations are best described by triangular and normal distributions, followed by logistic, chi-square, and PearsonV distributions. Triangular and normal distributions, by far, are the most significant distributions for characterizing natural fracture orientation data.

Natural fracture lengths are best described by PearsonVI and PearsonV distributions. Other favorable distribution functions for characterizing natural fracture length data include extreme-value, lognormal2, lognormal, loglogistic, logistic, and triangular distributions.

Natural fracture spacing data are best described statistically by lognormal2 and PearsonVI distributions, followed by lognormal, inverse Gaussian, Weibull, exponential, and PearsonV distributions.

The probability distribution functions identified for characterizing natural fracture characteristics using surface lineaments and fractures must be calibrated using any available subsurface fracture data before they can be used for stochastic simulations of subsurface natural fracture systems in a naturally fractured reservoir.

APPENDIX 1. PROBABILITY DENSITY FUNCTIONS

$$BETA(\alpha_1,\alpha_2): f(x) = \frac{x^{\alpha_1-1}(1-x)^{\alpha_2-1}}{B(\alpha_1,\alpha_2)}$$

where $B(x_1,x_2) = \int_0^1 t^{x_1-1}(1-t)^{x_2-1}dt$

$$BINOMIAL(n,p): f(x) = \binom{n}{x} p^x (1-p)^{n-x}$$

$$CHISQ(v): f(x) = \frac{x^{\frac{v}{2}-1} e^{\frac{-x}{2}}}{2^{\frac{v}{2}}\Gamma\left(\frac{v}{2}\right)}$$

$$ERF(h): f(x) = \frac{h}{\sqrt{\pi}} e^{-h^2x^2}$$

$$ERLANG(m,\beta): f(x) = \frac{\beta^{-m}x^{m-1}e^{\frac{-x}{\beta}}}{\Gamma(m)}$$

$$EXPON(\beta): f(x) = \frac{e^{\frac{-x}{\beta}}}{\beta}$$

$ExtremeValue(a,b)$:

$$f(x) = \left(\frac{1}{b}\right)\exp\left(-\frac{x-a}{b}\right)\exp\left[-\exp\left(-\frac{x-a}{b}\right)\right]$$

$$GAMMA(\alpha,\beta): f(x) = \frac{\beta^{-\alpha}x^{\alpha-1}e^{\frac{-x}{\beta}}}{\Gamma(\alpha)}$$

$$GEOMET(p): f(x) = p(1-p)^x$$

$$HYPERGEO(n,D,M): f(x) = \frac{\binom{D}{x}\binom{M-D}{n-x}}{\binom{M}{n}}$$

$InverseGaussian(\lambda,\mu): f(x) =$

$$\left(\frac{\lambda}{2\pi x^3}\right)^{\frac{1}{2}} \exp\left[\frac{-\lambda(x-\mu)^2}{2\mu^2 x}\right]$$

$$LOGISTIC(\alpha,\beta): f(x) = \frac{z}{\beta(1+z)^2}$$

where $z = \exp\left[\frac{-(x-\alpha)}{\beta}\right]$

$$LOGLOGISTIC(\alpha,\beta,\gamma): f(x) = \frac{\alpha\left(\frac{x-\gamma}{\beta}\right)^{\alpha-1}}{\beta\left[1+\left(\frac{x-\gamma}{\beta}\right)^{-\alpha}\right]^2}$$

$$LOGNORM(\mu,\sigma): f(x) = \frac{1}{x\sqrt{2\pi\sigma_1^2}}\exp\frac{-(\ln x-\mu_1)^2}{2\sigma_1^2}$$

where $\mu_1 = \ln\left(\frac{\mu^2}{\sqrt{\sigma^2+\mu^2}}\right)$

and $\sigma_1 = \sqrt{\ln\left(\frac{\sigma^2+\mu^2}{\mu^2}\right)}$

$$LOGNORM2(\mu,\sigma): f(x) = \frac{1}{x\sqrt{2\pi\sigma^2}} \exp \frac{-(\ln x - \mu)^2}{2\sigma^2}$$

$$NEGBIN(s,p): f(x) = \binom{s+x-1}{x} p^s (1-p)^x$$

$$NORMAL(\mu\sigma): f(x) = \frac{1}{\sqrt{2\pi\sigma^2}} e^{\frac{-(x-\mu)^2}{2\sigma^2}}$$

$$PARETO(\theta,a): f(x) = \frac{\theta a^\theta}{x^{\theta+1}}$$

$$PearsonV(\alpha,\beta): f(x) = \frac{x^{-(\alpha+1)} \exp\left(-\frac{\beta}{x}\right)}{\beta^{-\alpha}\Gamma(\alpha)}$$

$$PearsonVI(\alpha_1,\alpha_2,\beta): f(x) = \frac{\left(\frac{x}{\beta}\right)^{\alpha_1 - 1}}{\beta B(\alpha_1,\alpha_2)\left(1+\frac{x}{\beta}\right)^{\alpha_1+\alpha_2}}$$

$$POISSON(\lambda): f(x) = \frac{e^{-\lambda}\lambda^x}{x!}$$

$$RAYLEIGH(b): f(x) = \frac{x}{b^2} \exp\left(-\frac{x^2}{2b^2}\right)$$

$$Student'sT(u): f(x) = \frac{\Gamma\left(\frac{\upsilon+1}{2}\right)}{(\pi\upsilon)^{\frac{1}{2}}\Gamma\left(\frac{\upsilon}{2}\right)\left[1+\frac{x^2}{\upsilon}\right]^{\frac{\upsilon+1}{2}}}$$

$$TRIANG(a,b,c): f(x) = \begin{cases} \frac{2(x-a)}{(b-a)(c-a)} & if \quad a \le x \le b \\ \frac{2(c-x)}{(c-a)(c-b)} & if \quad b \le x \le c \end{cases}$$

$$WEIBULL(\alpha,\beta): f(x) = \alpha\beta^{-\alpha}x^{\alpha-1}e^{-\left(x/\beta\right)^\alpha}$$

REFERENCES CITED

Baecher, G.B., Lanney, N.A., and Einstein, H.H., 1977. Statistical Description of Rock Properties and Sampling, presented at the 18th U.S. Symp. on Rock Mechanics, Energy Resources and Excavation Technology, 1977, 5C1-1.

Baecher, G.B., 1983. Statistical Analysis of Rock Mass Fracturing, Math. Geol. (Vol. 15, No. 2, 1983) 329.

Belfield, W.C., and Sovich, J.P., 1995. Fracture Statistics from Horizontal Wellbores, JCPT (Vol. 34, No. 6, June 1995), p. 47–50.

Burchett, R.R., Luza, K.V., Van Eck, O.J., and Wilson, F.W., 1983. Seismicity and Tectonic Relationships of the Nemaha Uplift and Midcontinent Geophysical Anomaly, Division of Health, Siting and Waste Management, Office of Nuclear Regulatory Commission, 33 pp.

Cacas, M. C., et al., 1990. Modeling Fracture Flow with a Stochastic Discrete Fracture Network: Calibration and Validation 1. The Flow Model, Water Resources Research (Vol. 26, No. 3, March 1990), p. 479–489.

Chen, X., 1992. Applications of Remote Sensing and GIS Techniques for Environmental Geologic Investigation, Northeastern Iowa, Ph.D. thesis, University of Iowa, 171 pp.

Collins, R.J., McCown, F.P., Stonis, L.P., Petzel, G.J., and Everett, J.R., 1974. An Evaluation of ERTS Data for the Purposes of Petroleum Exploration, final report prepared for Goddard Space Flight Center.

Cooley, M.E., 1984. Linear Features Determined from Landsat Imagery in Western Kansas, U.S. Geological Survey Open-File Report p. 84-241.

Dverstorp, B., and Andersson, J., 1989. Application of the Discrete Fracture Network Concept with Field Data: Possibilities of Model Calibration and Validation, Water Resources Research (Vol. 25, No. 3, March 1989), p. 540–550.

Evans, R.D., 1981. A Proposed Model for Multiphase Flow Through Naturally Fractured Reservoirs, SPE 9940 presented at the 1981 California Regional Meeting, Bakersfield, California, March 25–26.

Guo, G., and Carroll, H.B., 1995. A New Methodology for Oil and Gas Exploration Using Remote Sensing Data and Surface Fracture Analysis, NIPER/BDM-0163, Aug. 1995, 83 pp.

Guo, G., and George, S.A., 1996. An Analysis of Surface and Subsurface Lineaments and Fractures for Oil and Gas Exploration in the Midcontinent Region, NIPER/BDM-0223, March 1996, 36 pp.

Guo, G., and Evans, R.D., 1994. Geologic and Stochastic Characterization of Naturally Fractured Reservoirs, SPE 27025 presented at the 1994 SPE III Latin American & Caribbean Petroleum Engineering Conference, Buenos Aires, Argentina, April 27–29, p. 1179–1191.

Heffer, K.J., and Bevan, T.G., 1990. Scaling Relationships in Natural Fractures: Data, Theory, and Application, SPE 20981 presented at the 1990 European Petroleum Conference held in Hague, Netherlands, Oct. 22–24, p. 367–376.

Herman, J., Etzler, P.J., Wilson, M.L., and Vincent, R.K., 1986. Remote Sensing Study of the Mid-continent Geophysical Anomaly in Iowa, Paper presented at the Society of Mining Engineers Fall Meeting, St. Louis, Missouri, Sep. 7–10, 1986.

Lepley, L.K., 1977. Landsat Lineament Map of Arizona with Emphasis on Quaternary Fractures, University of Arizona Bureau of Geology and Mineral Technology Open-File Report p. 77–2.

Long, J.C.S., Remer, J.S., Wilson, C.R., and Witherspoon, P.A., 1982. Porous Media Equivalents for Networks of Discontinuous Fractures, Water Resources Research (Vol. 18, No. 3, June 1982), p. 645–685.

Loosveld, R.J.H., and Franssen, R.C.M.W., 1992. Extensional vs. Shear Fractures: Implications for Reservoir Characterization, SPE 25017 presented at the 1992 SPE European Petroleum Conference held in Cannes, France, Nov. 16–18, p. 23–30.

Lorentz, J.C., and Hill, R.E., 1991. Subsurface Fracture Spacing: Comparison of Inferences from Slant/Horizontal Core and Vertical Core in Mesaverde Reservoirs, SPE 21877 presented at the 1991 Rocky Mountain Regional Meeting and Low-Permeability Reservoirs Symposium held in Denver, April 15–17, p. 705–716.

Rouleau, A., and Gale, J.E., 1985. Statistical Characterization of the Fracture System in the Stripa Granite, Sweden, Int. J. Rock Mech. Min. Sci. & Geomech. Abstr. (Vol. 22, No. 6, 1985) p. 353.

Snow, D.T., 1970. The Frequency and Apertures of Fractures in Rock, Int. J. Rock Mech. Min. Sci. (Vol. 7, 1970), p. 23–40.

Swanson, D.C., and Shannon, P.J., 1990. Landsat Interpretation Useful in Hugoton Gas Field, World Oil (January 1990), p. 108–111

Parra, J. O., H. A. Collier, B. G. Angstman, Feasibility of detecting seismic waves between wells at the fractured Twin Creek reservoir, Utah-Wyoming overthrust belt, 1999, *in* R. Schatzinger and J. Jordan, eds., Reservoir Characterization-Recent Advances, AAPG Memoir 71, p. 251–262.

Chapter 17

Feasibility of Detecting Seismic Waves Between Wells at the Fractured Twin Creek Reservoir, Utah-Wyoming Overthrust Belt

Jorge O. Parra
Southwest Research Institute
San Antonio, Texas, U.S.A.

Hughbert A. Collier
Tarleton State University
Stephenville, Texas, U.S.A.

Burke G. Angstman
HS Resources
Denver, Colorado, U.S.A.

ABSTRACT

In this paper, we present the feasibility of using seismic measurement techniques to map the fracture zones between wells spaced 800 m apart at a depth of 3000 m. The wells are in the fractured Twin Creek reservoir in the Utah-Wyoming Overthrust Belt. In particular, we want to demonstrate that fracture zones cannot be resolved by surface seismic measurements alone. Instead, it is more appropriate to use high-resolution crosswell seismic data. Surface seismic data244 integrated with well logs from the Lodgepole field are used to delineate the members of the Twin Creek carbonate reservoir. Petrophysical analysis provides the rock physical properties and thickness of the Leeds Creek, Watton Canyon, Boundary Ridge, and Rich members of the Twin Creek Formation. Surface seismic and horizontal well information delineates a fracture zone in the Watton Canyon Member. The result is a 12-layer model describing the fracture zone, with petrophysical parameters for each geological unit in the reservoir. We pre-sent the feasibility of transmitting seismic waves between two wells in the Twin Creek reservoir at a distance up to 800 m using synthetic interwell seismic data. We also show the geology, petrophysics, and migrated seismic data, which are used to describe the fractured zone, as well as the members used to produce the model for planning interwell seismic measurements.

INTRODUCTION

In low-porosity, low-permeability formations, natural fractures are the primary source of permeability, controlling both production and injection of fluids. Open fractures do not contribute significantly to total porosity, but do provide an increased drainage network for the matrix porosity. An important approach to characterizing the fracture orientation and fracture permeability of reservoirs is one based upon the effects of such conditions on the propagation of acoustic and seismic waves in the rock.

To better understand the influence of fractures on the production of oil and gas, core analyses and a detailed logging program are usually required. The general objectives of such a program are first to identify fractures, second to determine fracture orientation, and third to predict fracture influence on the production of individual wells.

An emerging technology called cross dipole acoustic logging can provide detailed information on the anisotropy associated with the presence of vertical fractures. This method, based on the detection of split flexure modes, has recently been developed by Schlumberger (Esmersoy et al., 1994) and Amoco (Muller et al., 1994) and is designed to determine the orientation of vertical fractures and microcracks and to differentiate variations in horizontal stresses caused by azimuthal anisotropy. At the interwell scale, high-quality reflection imaging is required to predict the presence and orientation of a fracture system. The reflection data can be obtained by processing multicomponent seismic data (Tatham and McCormack, 1991); however, to resolve fracture zones in heterogeneous reservoirs at depths of about 3000 m, high-resolution crosswell seismic measurements are more appropriate.

Because well separation in the Twin Creek reservoir is about 800 m, careful planning of interwell seismic measurements is desirable. Such planning requires knowledge of the lithology, structural features, and rock physical properties of the formation. The boundaries between the formations can be determined using well logs and surface seismic data, and the rock physical parameters can be determined from acoustic and density logs.

We used computer models of the Twin Creek reservoir to predict seismic signatures recorded at the crosswell and surface seismic scales. Well logs and thin sections of cuttings were integrated with 2-D seismic sections to produce geologic cross sections, determine petrophysical parameters, and define reservoir geometries needed for constructing computer models. Finally, we conducted model studies for planning high-resolution interwell seismic experiments to map the fracture zones in the Twin Creek Formation at the Lodgepole field (Parra et al., 1997).

GEOLOGY AND PETROPHYSICS OF THE STUDY AREA

Location

Lodgepole field is the southernmost in a series of oil fields in the Overthrust Belt of southwestern Wyoming and neighboring Utah that produce from the Jurassic Twin Creek Formation. Union Pacific Resources (UPRC) has three fields in the play in Summit County, Utah: Lodgepole, Elkhorn, and Pineview. These three fields are shown in Figure 1.

Six horizontal and 12 vertical boreholes have been drilled in Lodgepole field. The field is a depletion

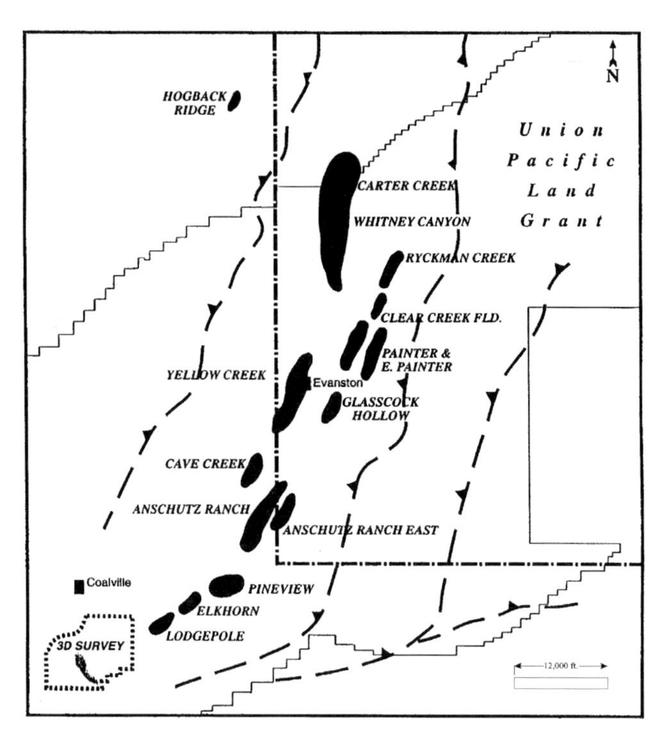

Figure 1. Regional map showing location of the Utah-Wyoming Overthrust Belt oil fields.

drive reservoir, but most of the wells have significant water production. The well location map (see Figure 2) shows the vertical and horizontal well paths. In particular, the map shows the horizontal path for the 34-1H Judd well, seismic line CREA-25K, and the orientation of the 34-2 to 34-3 cross section.

Geology

The Twin Creek Formation is approximately 1500 ft (457 m) thick and is divided into seven members. The following brief description of each member of the Twin Creek Formation is taken from Bruce (1988):

Gypsum Springs. Approximately 50 ft (15 m) of sabkha evaporites, red beds, and minor carbonates. It serves as a detachment surface between the Nugget and Twin Creek formations in the thrusting process. It is also a barrier to hydrocarbon formations in most places.

Sliderock. Approximately 90 ft (27 m) of micritic limestone with thin beds of oolitic grainstone and thin shaly zones. No primary porosity has been preserved, but calcite-filled fractures are present.

Rich. Approximately 250 ft (76 m) of argillaceous limestone that cleans upward. The upper 20–50 ft (6–15 m) has some intergranular porosity and is often dolomitized.

Boundary Ridge. Approximately 50 ft (15 m) of red siltstones and claystones with a sabkha character. It is a good marker bed in the Twin Creek Formation.

Watton Canyon. Approximately 250 ft (76 m) of limestone with thin, tightly cemented oolitic zones. The limestones of the Watton Canyon Member are thinner bedded and more terrigenous than the Rich Member. An interval approximately 20 ft (6 m) thick near the

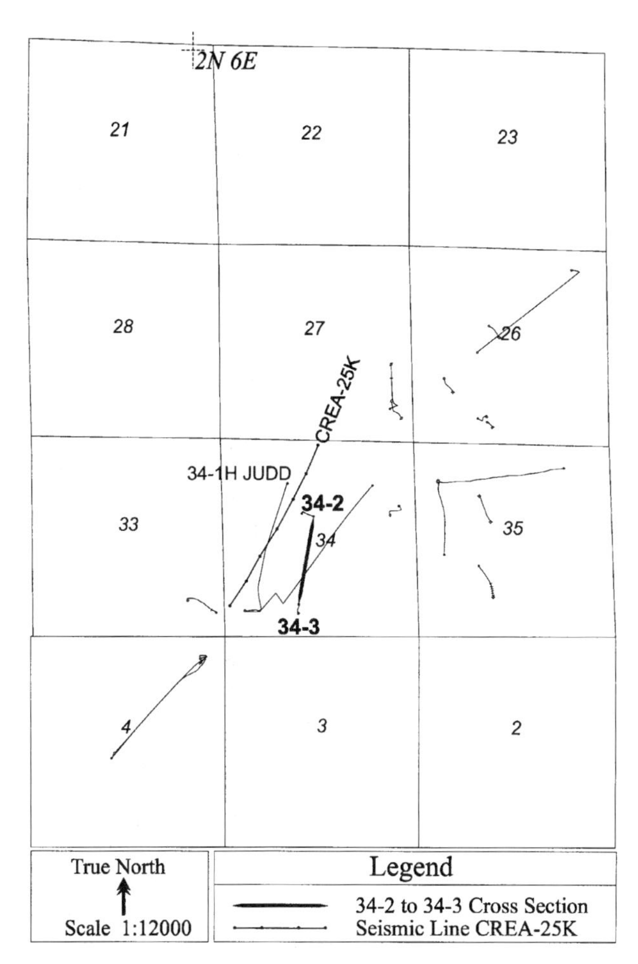

Figure 2. Map of well locations in the Lodgepole field. Labeled are the horizontal path of well 34-1H Judd, the trace of a stratigraphic cross section between wells 34-2 and 34-3, and the trace of a portion of seismic line CREA-25K.

base of the Watton Canyon Member is the primary target for most horizontal wells, including the 34-1H Judd.

Leeds Creek. Approximately 300 ft (91 m) of interbedded argillaceous and relatively clean limestones. A highly radioactive unit near the top of the member is a devitrified tuff.

Giraffe Creek. Approximately 450 ft (137 m) of micritic and oolitic carbonates. It becomes sandy toward the top.

The basal Preuss silt, a 30–100 ft (9–30 m) interval of tightly cemented quartz siltstone, overlies the Twin Creek Formation. The Preuss salt overlies the siltstone. It varies from 20 to 300 ft (6 to 91 m) in thickness. The Twin Creek Formation overlies the Nugget formation, which is eolian sandstone. The Nugget is one of the primary hydrocarbon producing reservoirs in the Overthrust Belt.

Petrophysical cross sections were constructed with logs from the vertical wells at the Lodgepole field. The cross sections center on the Watton Canyon and Rich members, the most fractured members of the Twin Creek Formation. Figure 3 is a stratigraphic cross section between wells 34-2 and 34-3. The fractured interval within the Watton Canyon is a dolomitic limestone, and here sonic velocity ranges from 18,500 to 20,000 ft/s (5640 to 6096 m/s) and bulk density ranges from 2.6 to 2.75 g/cm^3. The upper boundary does not appear to have a significant change in sonic velocity, but the lower boundary is a shale with a sonic velocity of 16,000–17,000 ft/s (4877–5182 m/s).

For most of the fractured interval in the Rich Member, sonic velocities range from 18,800 to 21,000 ft/s (5730–6400 m/s) and bulk density ranges from 2.65 to 2.8 g/cm^3. The interval is bounded above by a shale with a sonic velocity of 16,000 ft/s (4877 m/s) and below by a shaly limestone with a velocity of approximately 18,000 ft/s (5486 m/s). The upper 15 ft (4.5 m) of the fractured interval contains a significant percentage of dolomite, which may explain, at least in part, the higher sonic velocities. The fractured interval in the Rich Member has more shale than the fractured interval in the Watton Canyon Member.

INTEGRATION OF SURFACE SEISMIC AND WELL LOG DATA

Seismic line CREA-25K was selected because it follows the horizontal path of well 34-1H Judd. The paths of this well and well 34-2 were plotted in the migrated seismic section displayed in Figure 4. A synthetic seismogram was produced using the compressional wave velocity log and the density log from well 34-2. The synthetic seismogram was superimposed on the migrated seismic line together with the geological boundaries defined by the well logs (see Figure 4). The Watton Canyon Member is the most prominent geologic unit in the seismic section. Its upper and lower boundaries correlate with the sonic, density, and gamma logs, as well as the seismic events on the seismic section. In fact, the synthetic seismogram fits reasonably well at the position of Judd 34-1H in "time-depth" on the seismic section. For example, the top of the Watton Canyon Member is at 1600 ms, and the top of the Rich Member is at 1630 ms.

In general, the petrophysical boundaries defined by the well logs correlate with the seismic events in the seismic section. The seismic signatures associated with the Watton Canyon Member suggest lateral velocity changes between wells 34-1H and 34-2. There is a change in the seismic signature associated with this region between the trace at position 2190 and well 34-1H. This event is probably an indication of faulting.

INTERWELL SEISMIC MODELING

Computer Models

To evaluate transmission and detection of seismic waves in the Twin Creek Formation of the Utah-Wyoming Overthrust Belt, computer models were constructed using rock physical properties from the Lodgepole field. We selected petrophysical parameters from a cross section including the Watton Canyon, Boundary Ridge, and Rich members between wells 34-2 and 34-3. The P-wave and S-wave velocities, densities, and thicknesses from these two wells were used to construct a 12-layer model to produce particle velocity

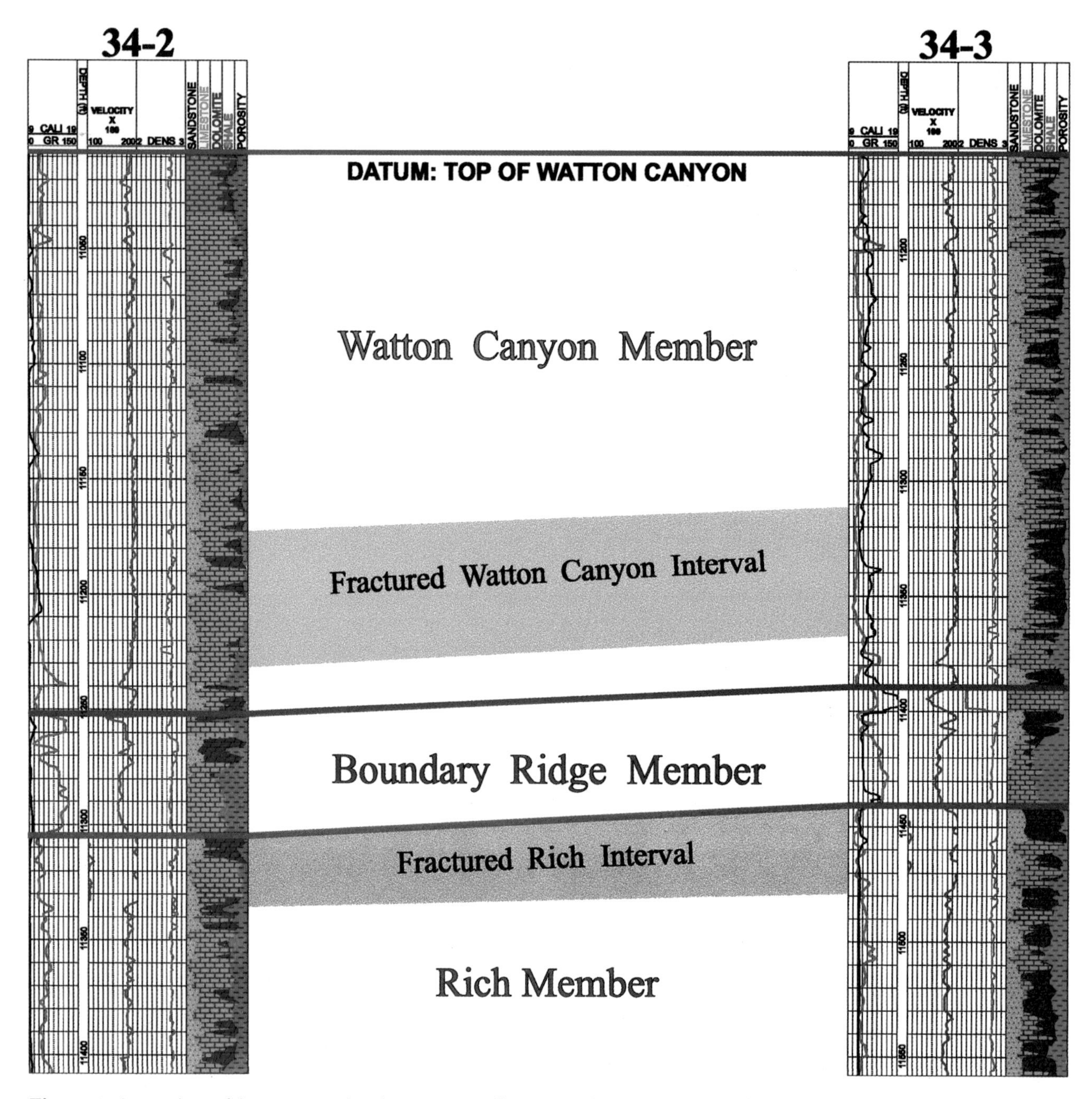

Figure 3. A stratigraphic cross section between wells 34-2 and 34-3. Horizontal wells in the Lodgepole field have been drilled in the Watton Canyon and Rich members. Units of measurement for the log curves are caliper (CALI) in inches, gamma ray (GR) in API units, velocity in ft/s, and bulk density (DENS) in g/cm^3. Lithology is scaled in percent.

and pressure seismograms to evaluate if the seismic energy in the formations would reach the desired interwell distance of about 0.5 mi (800 m) in the Lodgepole field. The P-wave and S-wave velocities of the 12-layer model are shown in Figure 5.

The seismic responses were calculated using a Ricker wavelet as a source pulse function for a pressure source placed at 31.5 m and 55 m below the top layer interface of the Watton Canyon Member and an array of 60 detectors placed at 800 m from the source well. Synthetic seismograms and spectral density plots of interwell seismic signatures were produced by taking into account the source signal strength of 10 cm^3 and receiver noise levels using a buried point source mechanism (i.e., a point source in a fluid-filled borehole that is represented by an equivalent force system of a monopole and a dipole). The interwell seismic responses were simulated using the modified Clenshaw-Curties quadrature method given by Parra et al. (1993). The spectral density plots and amplitude-depth distribution curves were produced to characterize interwell seismic signatures and to evaluate the propagation characteristics of seismic waves in the Twin Creek Formation for planning interwell seismic experiments.

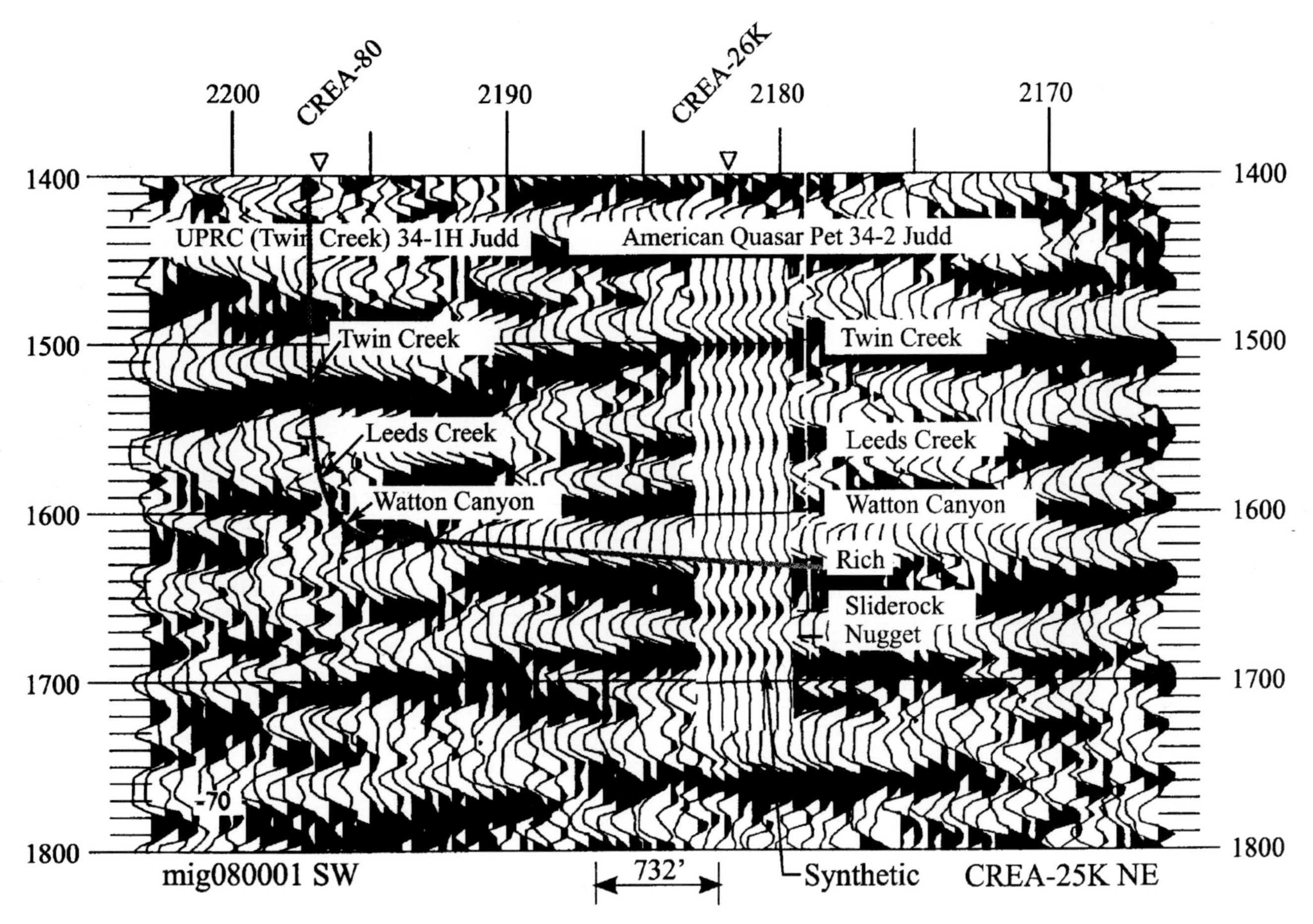

Figure 4. Migrated seismic section CREA-25K along horizontal well path of well 34-1H Judd with synthetic seismogram inserted at well location 34-2 Judd.

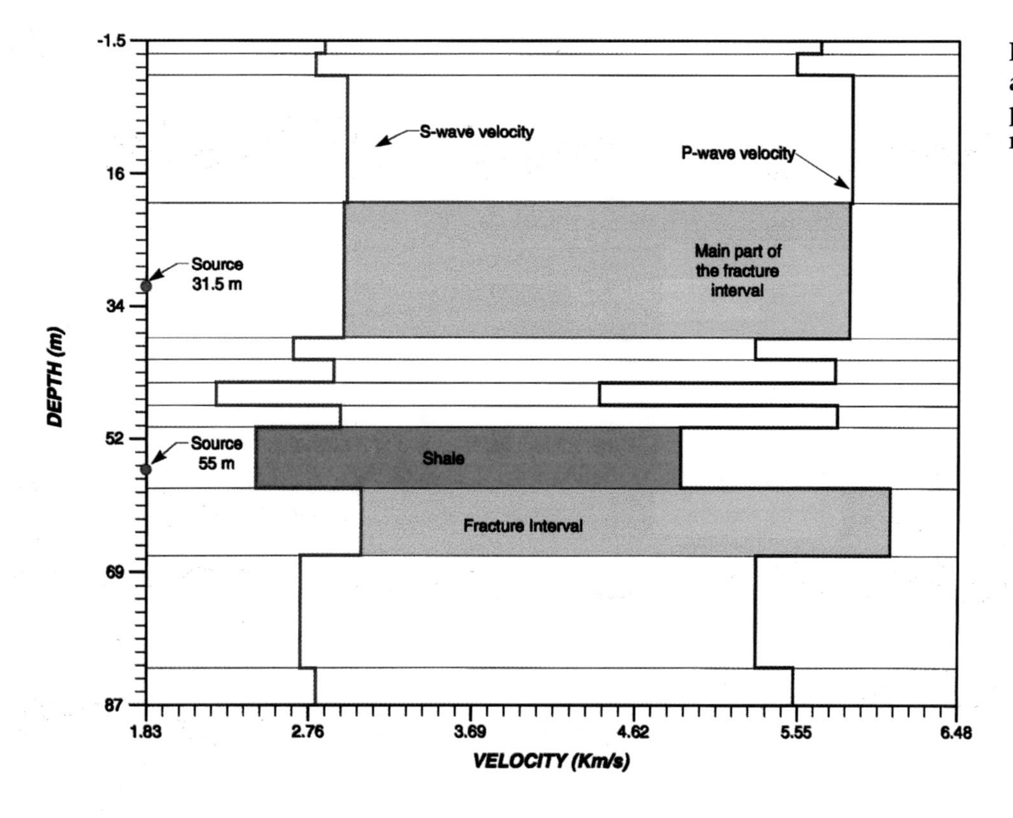

Figure 5. Compressional and shear wave velocity parameters for the 12-layer model

Synthetic Seismograms and Amplitude Depth Distribution Curves

The full waveform particle velocity seismograms given in Figure 6a and 6b show strong P headwave events and reflections. The vertical component particle velocity captures most of the main boundaries, such as the Watton Canyon Member fracture/heterogeneous zone and the shale/lower fracture interval. The heterogeneous zone and the shale unit are part of the Boundary Ridge Member, where trapped waves are observed in the vertical particle velocity seismogram. In the horizontal particle velocity seismogram, it is difficult to correlate reflections with formation interfaces; however, amplitude-depth distribution curves of the horizontal component show a better correlation with the layer interfaces, in particular with the contact between the main fracture interval in the Watton Canyon Member and the heterogeneous zone (formed by units having different rock physical properties) above the shale unit (see Figure 7). These plots also show that most of the energy appears to be distributed in the main fracture interval at higher frequencies. This suggests that a source having a center frequency of about 400–500 Hz will be appropriate to resolve the main features in the Watton Canyon formation at a well separation of 800 m.

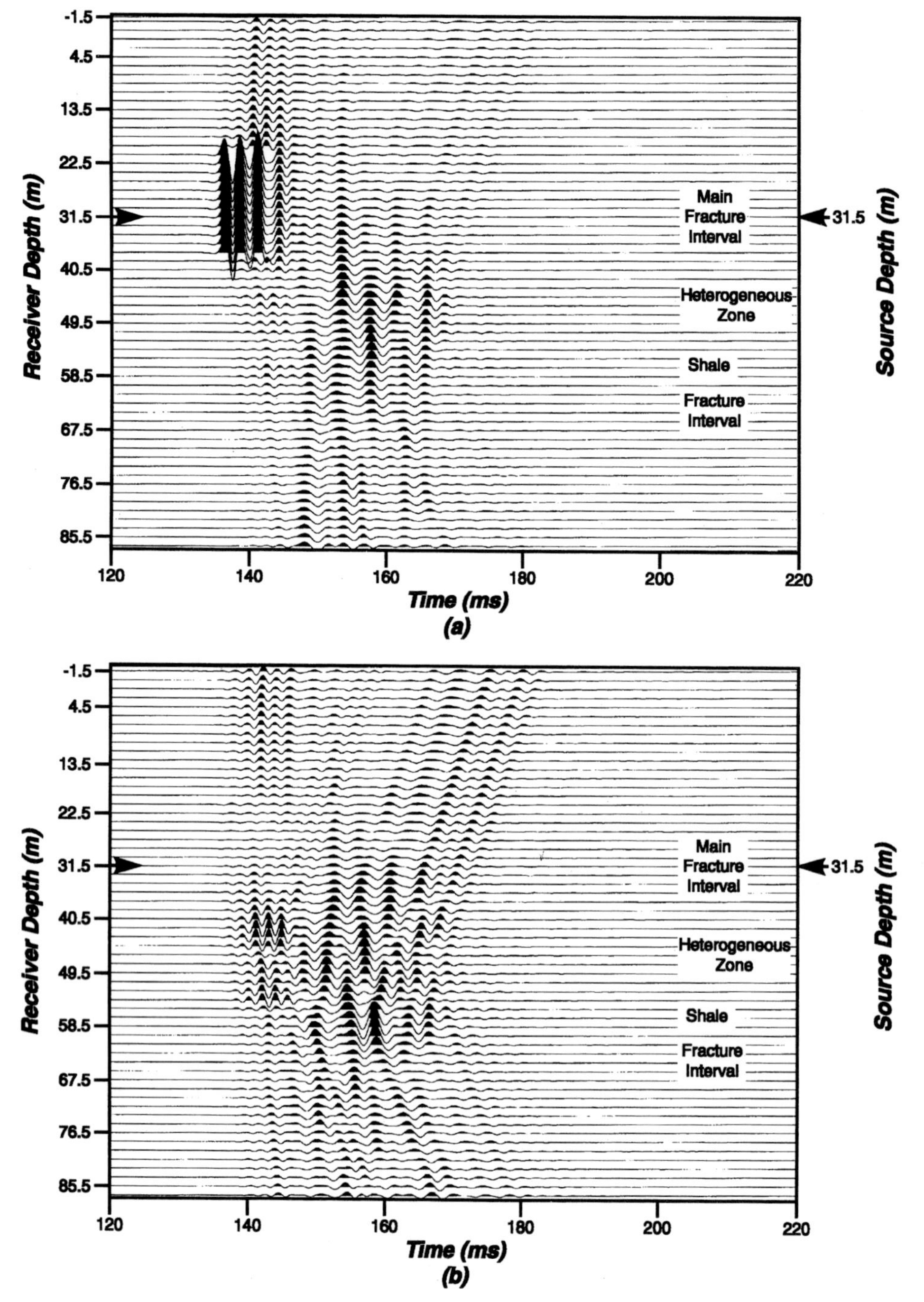

Figure 6. Common source (a) horizontal and (b) vertical particle velocity seismograms with source at a depth of 31.5 m below top layer interface. Well separation of 800 m.

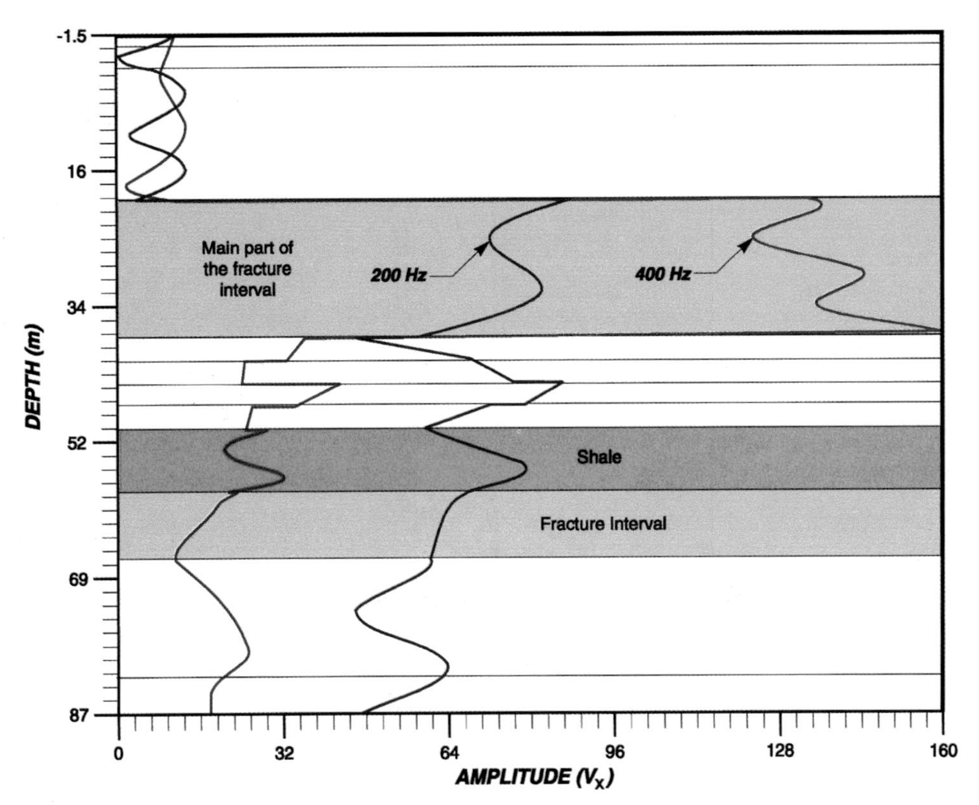

Figure 7. Amplitude-depth distribution of horizontal particle velocity produced by a source at a depth of 31.5 m below top layer interface.

The full waveform synthetic seismograms given in Figure 8a and 8b show direct waves, head waves, trapped waves (leaky modes), reflections, and normal modes (Rayleigh-type waves) for a source placed in the Boundary Ridge Member at 55 m below the top layer interface. The horizontal component particle velocity and the pressure seismograms show strong head waves and trapped waves. In particular, the horizontal component particle velocity has head wave events followed by direct events, reflection, and leaky modes. This seismogram captures the boundary between the main fractured interval (top) and the heterogeneous zone (bottom), which is formed by four petrophysical units having different rock physical properties. Also, the interface between the shale and lower fractured zone can be detected by measuring the horizontal particle velocity. Similarly, the vertical component seismogram captures the same interfaces (Figure 9); however, it also shows Rayleigh-type waves (normal modes) associated with the shale waveguide at about the source position of 55 m. The modes are excited in the shale, as well as in the heterogeneous zone below the main fracture interval. These results suggest that guided waves can be used to map the continuity of shales that are permeability barriers in the Twin Creek reservoir.

Amplitude-depth distribution curves of the horizontal and vertical particle velocities were produced to evaluate the propagation frequency for resolving the features of interest in the formation at a well separation of 800 m when a source is placed in the low-velocity shale. A center frequency of 200 Hz can excite Rayleigh-type waves in the shale waveguide, as well as waves in the low-velocity units within the heterogeneous zone (see Figure 10). On the other hand, a center frequency of about 400 Hz will resolve geological units of about 2–3 m thick.

Spectral Density for a Source 55 m Below the Top Layer Interface

Seismic traces were selected for the pressure source at 55 m and detectors at 55.5 m and 61.5 m below top layer interface to produce spectral density plots as shown in Figures 11 and 12. These plots were compared with ambient seismic noise levels having rms values of 1 mPa and 100 mPa, respectively (Parra et al., 1998). The spectral density signatures of the pressure are below the specified ambient seismic noise level of 1 mPa and the electronic noise level of the OAS hydrophone (A. L. Kurkjian, 1994, personal communication). Alternatively, the computed spectral density of vertical and horizontal particle velocity components is above the ambient seismic noise level in quiet and noisy environments. The spectral amplitude of the horizontal component is about 25 dB above the specified rms velocity noise level of 6.1 μcm/s, which is based on the rms pressure noise level of 100 mPa, and the spectral amplitude of the vertical component is about 20 dB above the noise level of 6.1 μcm/s. In this case, when the source is placed in the waveguide the spectral density is greater than when the source is placed in the

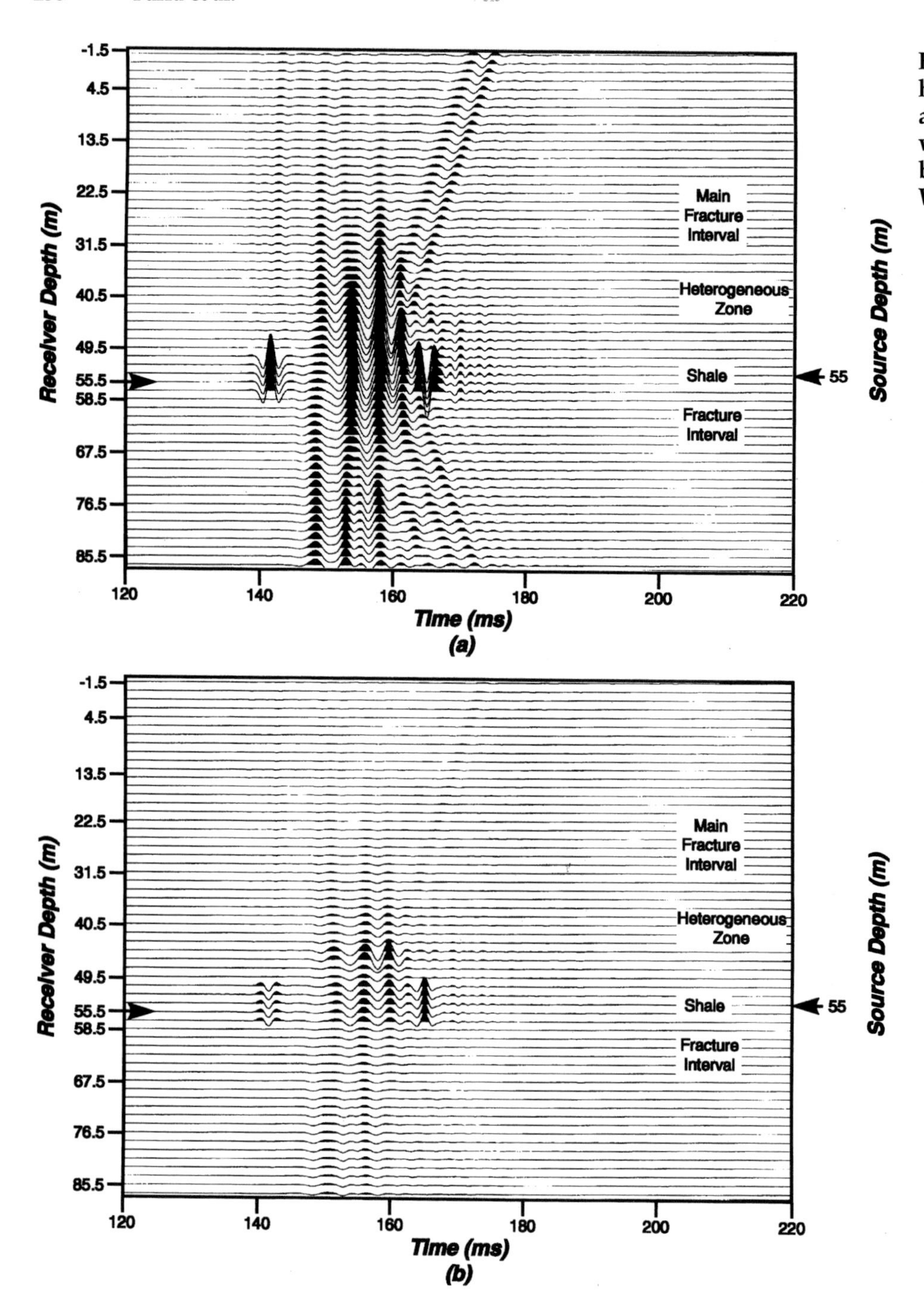

Figure 8. Common source (a) horizontal particle velocity and (b) pressure seismograms with source at a depth of 55 m below top layer interface. Well separation of 800 m.

main fracture interval of the Watton Canyon Member. The main reason for this difference is that energy can be trapped in the waveguide, and it can travel for long distances in the low-velocity zone. On the other hand, energy in high-velocity formations will leak as the wave propagates between wells at large well separations.

DISCUSSIONS AND CONCLUSIONS

Correlation of the lithology, the well logs, and the migrated seismic data from the Lodgepole field served to identify seismic events associated with geological units of interest. The surface seismic delineated the major geological boundaries between members of the Twin Creek Formation. The fracture zones were not resolved directly by surface measurement techniques. The surface seismic section showed a reflection at about 1630 ms (at the bottom of the Watton Canyon Member) that was interpreted as a boundary surface between the Watton Canyon Member and the Rich Member. In this case, the fracture zones (in the Watton Canyon and Rich Members) and the Boundary Ridge Member were interpreted to be part of this boundary surface.

To resolve the petrophysical units at the scale of the fracture zones and the Boundary Ridge Member,

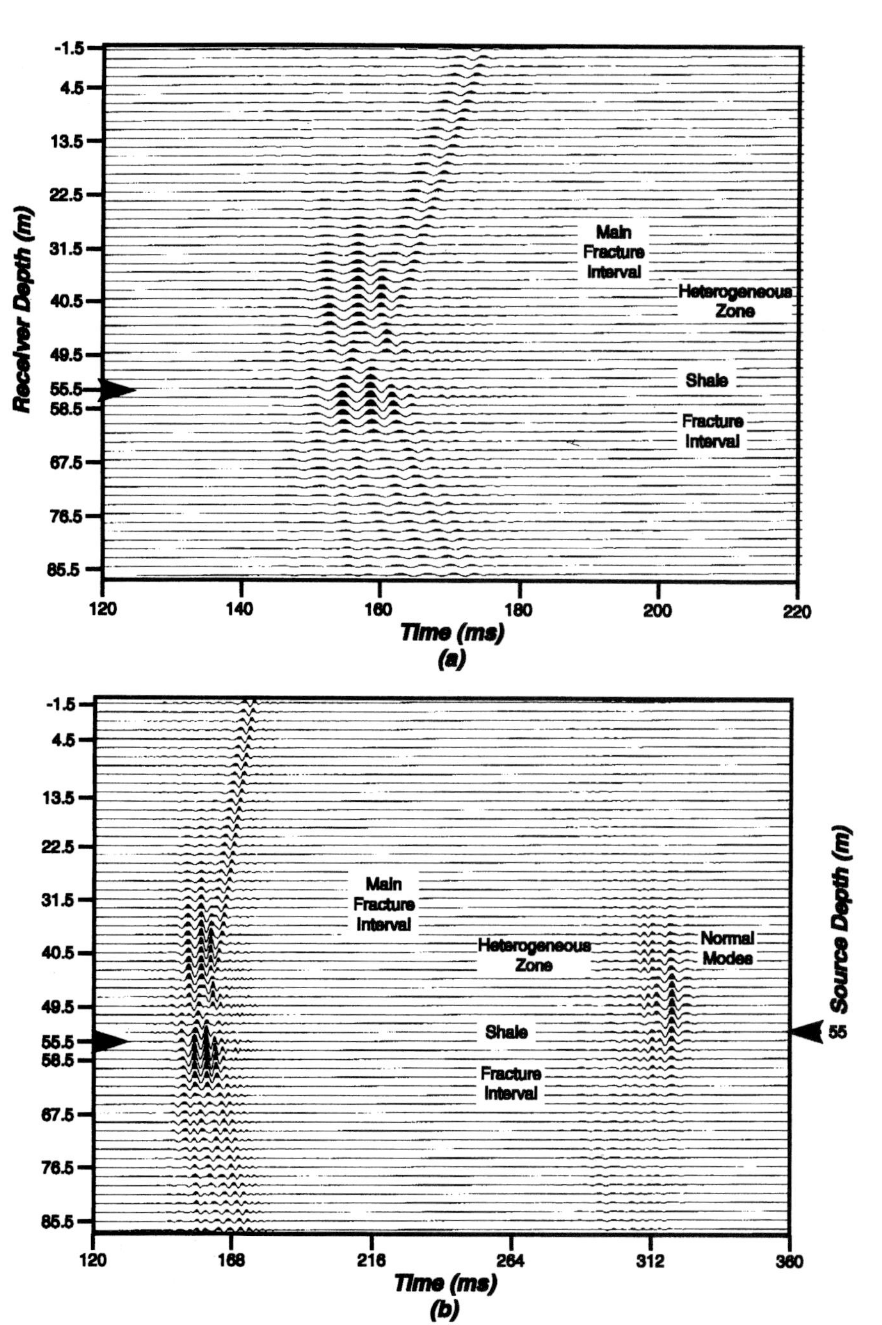

Figure 9. Common source vertical particle velocity seismograms with source at a depth of 55 m below top layer interface for a well separation of 800 m for (a) 100-ms time window and (b) 200-ms time window.

crosswell seismic measurements are more appropriate. As a result, we conducted a model study for planning high-resolution interwell seismic experiments to map or delineate the fracture zones of interest in Lodgepole field. The models demonstrated that a source having a center frequency of 400–500 Hz can resolve the features of interest in Lodgepole between wells 800 m apart. We expect that if we can transmit energy in the frequency range of 1000 Hz, we have the potential to map the fracture zones in the Watton Canyon Member and in the top of the Rich Member, as well as to properly identify the boundaries that were not resolved by the surface seismic. The surface seismic did not resolve the fracture zones and the heterogeneous Boundary Ridge Member, which is formed by four units with different petrophysical properties and a low-velocity shale with a thickness of approximately 9 m.

The computer model study suggests that high-resolution crosswell measurements are necessary to map the fracture zones in Lodgepole field, in particular in the Watton Canyon Member and the upper part of the Rich Member. The model results also show that it is feasible to detect seismic waves propagating in the Twin Creek Formation above an rms velocity noise level of 6.1 μcm/s by measuring the vertical particle motion and using a broadband seismic source operating in the frequency range of 0.3–1 kHz at well separations exceeding 800 m. In addition, the spacing between detectors should be about 1.5 m. We expect that traveltime tomography and

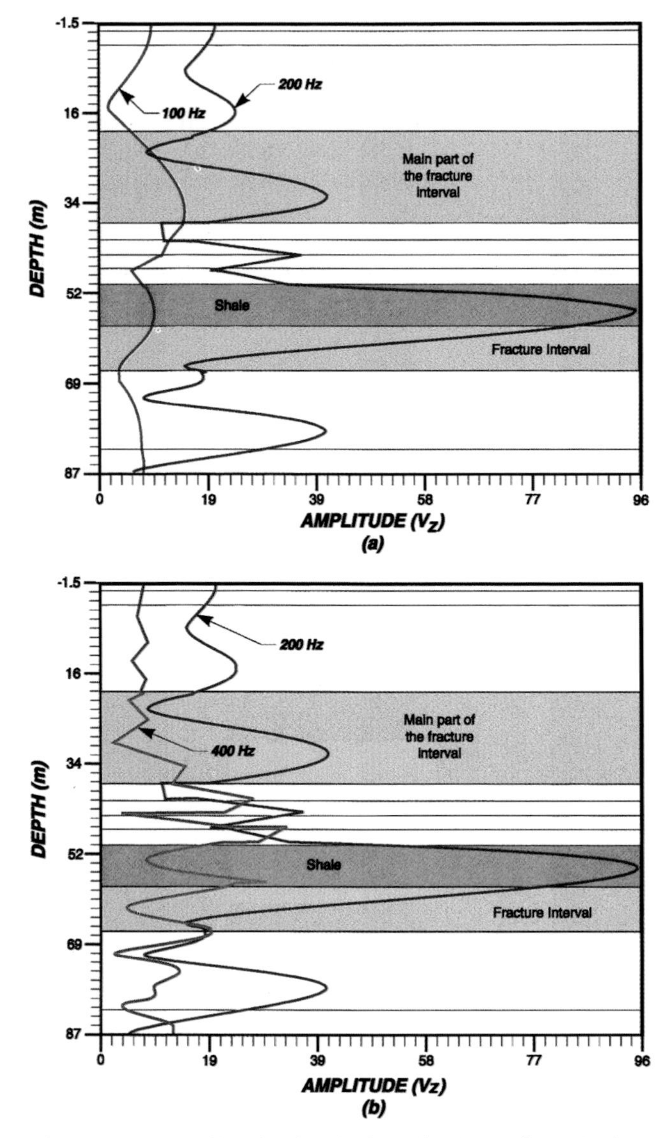

Figure 10. Amplitude-depth distribution of vertical particle velocity produced by a source at a depth of 55 m below top layer interface for (a) amplitudes of 100 and 200 Hz and (b) amplitudes of 200 and 400 Hz.

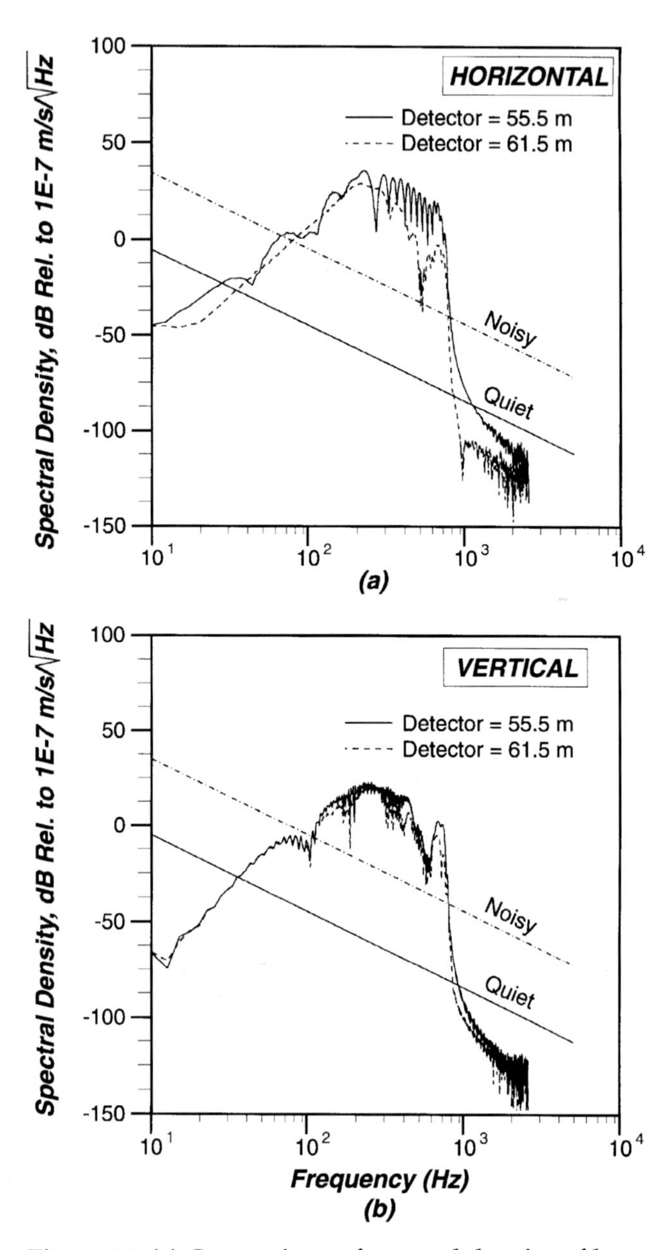

Figure 11. (a) Comparison of spectral density of horizontal and vertical (b) particle velocity with seismic noise levels in quiet and noisy environments for a source at 55 m below upper layer interface. Well separation of 800 m.

reflection imaging can be used to map the fracture boundary zones. The trapped energy and normal modes excited in waveguides can be useful information to determine the degree of continuity of zones of interest, such as permeability barriers, layer thickness variations, inhomogeneities, and boundary surfaces (Parra et al., 1996). Polarization diagrams from three-component seismic data can be used to evaluate the anisotropy associated with the presence of vertical fractures in the formations.

ACKNOWLEDGMENTS

This work was supported by DOE contract No. DE-AC22-94DC91008 through the BDM-Oklahoma, Inc., subcontract G4S51731. The assistance of Dr. Min Tham is gratefully acknowledged. We thank Union Pacific Resources, in particular James Peters, Gerry Wilbourn, and Jerry Wristers, for their contribution of Twin Creek Reservoir data. Neuralog software was used to digitize the logs, and Terrasciences was used for the log analysis.

REFERENCES CITED

Bruce, C.L., 1988, Jurassic Twin Creek Formation: a fractured limestone reservoir in the Overthrust Belt, Wyoming and Utah, Rocky Mountain Association of Geologists, p. 105-120.

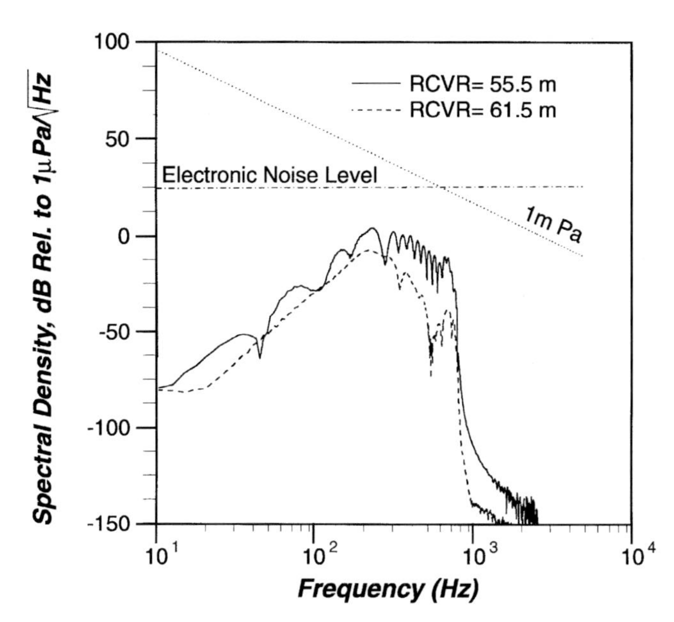

Figure 12. Comparison of spectral density of pressure with electronic and seismic noise levels in quiet and noisy environments for a source at 55 m below upper layer interface. Well separation of 800 m.

Esmersoy, C., Koster, W., Williams, M., Boyd, A., and Kane, M., 1994, Dipole shear anisotropy logging, 64th Annual International Meeting, Society of Exploration Geophysics, p. 1139-1142.

Muller, M., Boyd, A., and Esmersoy, C., 1994, Case studies of the dipole shear anisotropy logs, 64th Annual International Meeting, Society of Exploration Geophysics, p. 1143-1146.

Parra, J.O., Sturdivant, V.R., and Xu, P.-C., 1993, Interwell seismic transmission and reflections through a dipping low-velocity layer, Journal of the Acoustical Society of America, v. 93, p. 1954-1969.

Parra, J.O., Zook, B.J., and Collier, H.A., 1996, Interwell seismic logging for formation continuity at the Gypsy test site, Oklahoma, Journal of Applied Geophysics, v. 35, p. 45-62.

Parra, J.O., Collier, H.A., Datta-Gupta, A., Owen, T.O., Peddibhotla, S., Xu, P.-C., and Zook, B.J., 1997, Characterization of fracture reservoirs using static and dynamic data: from sonic and 3-D seismic to permeability distribution, Annual Report, U.S. Department of Energy, Contract No. DE-AC22-94DC91008, Subcontract No. G4S51731, and SwRI Project 15-7939, 1848, 184 p.

Parra, J.O., Zook, B.J., Xu, P.-C., and Brown, R.L., 1998, Transmission and detection of guided seismic waves in attenuating media, Geophysics, v. 63, p. 1190–1199.

Tatham, R.H., and McCormack, M.D., 1991, Multicomponent seismology in petroleum exploration (E.B. Neitzel and D.F. Winterstein, eds.), Society of Exploration Geophysicists, 248 p.

Section VI
Upscaling and Simulation

Ringrose, P., et al., The Ardross Reservoir gridblock analog: sedimentology, statistical representivity, and flow upscaling, 1999, *in* R. Schatzinger and J. Jordan, eds., Reservoir Characterization-Recent Advances, AAPG Memoir 71, p. 265–276.

Chapter 18

The Ardross Reservoir Gridblock Analog: Sedimentology, Statistical Representivity, and Flow Upscaling

Philip Ringrose[1]
Gillian Pickup
Jerry Jensen[2]
Margaret Forrester[3]
Heriot-Watt University
Edinburgh, Scotland, U.K.

ABSTRACT

We have used a reservoir gridblock-size outcrop (10 × 100 m) of fluvio-deltaic sandstones to evaluate the importance of internal heterogeneity for a hypothetical waterflood displacement process. Using a dataset based on probe permeameter measurements taken from two vertical transects representing "wells" (5 cm sampling) and one "core" sample (exhaustive 2-mm-spaced sampling), we evaluate the permeability variability at different lengthscales, the correlation characteristics (structure of the variogram function), and importance of volume and data support. We then relate these statistical measures to the sedimentology.

We show how the sediment architecture influences the effective tensor permeability at the lamina and bed scales, and then calculate the effective relative permeability functions for a waterflood. We compare the degree of oil recovery from the formation: (1) using averaged borehole data and no geological structure, and (2) modeling the sediment architecture of the interwell volume using mixed stochastic/deterministic methods.

We find that the sediment architecture has an important effect on flow performance, mainly due to bed-scale capillary trapping and a consequent

[1]Now at Statoil Research Centre, Rotvoll, Trondheim, Norway.
[2]Now at Petroleum Engineering Department, Texas A&M University, College Station, TX 77843–3116, U.S.A.
[3]Now at Edinburgh Petroleum Systems Ltd., Research Park, Riccarton, Edinburgh, Scotland, U.K.

reduction in the effective oil mobility. The predicted oil recovery differs by 18% when these small-scale effects are included in the model. Traditional reservoir engineering methods using average permeability values only prove acceptable in high-permeability and low-heterogeneity zones. The main outstanding challenge, represented by this illustration of sub-gridblock scale heterogeneity, is how to capture the relevant geological structure along with the inherent geo-statistical variability. An approach to this problem is proposed.

INTRODUCTION

The effects of sedimentary architecture on flow in petroleum reservoirs have been widely studied (e.g., Weber, 1982; van de Graaff and Ealey, 1989; Corbett et al., 1992; Jones et al., 1993; Kjønsvik et al., 1994; Saad et al. 1995). In most of these studies, the importance of considering the sedimentary architecture at a lengthscale relevant to the fluid flow problem (e.g., connectivity between wells versus sweep efficiency within flow units) has been stressed. But very often, the uncertainties associated with permeability estimation, interwell variability, and relevance to effective flow behavior at different lengthscales are undifferentiated. Thus, a general flow estimation problem with large uncertainties emerges. In this paper, we illustrate how sedimentary variability at different scales can be separated out into lithology-dependent components in order to provide a better means of accurately judging effective flow properties of a reservoir unit. We base the illustration on a study of an outcrop the size of a typical reservoir gridblock.

The Ardross Cliff, an outcrop of Lower Carboniferous deltaic and fluvial sandstones, lies on the south coast of Fife, in eastern Scotland. This outcrop was chosen because it contains a variety of sediment architecture types that illustrate the heterogeneity that typically occurs at the sub-gridblock scale within reservoir simulation models. The main exposure is a 10 × 100 m cliff comprising two sandstone units, the Upper Ardross Castle Sand Unit and the Lower Ardross Castle Sand Unit, separated by a prominent coal stratum (Figures 1 and 2). The lower unit is underlain by heterolithic sand and shale units, the upper unit is overlain by a limestone bed, and both are assumed to be no-flow boundaries. The lower unit is characterized by ripple lamination, manifested in parts by well-preserved climbing ripple sets and related to the building out of a delta front. The upper unit contains abundant trough cross-bedding formed within a fluvio-deltaic channel.

Analysis of Permeability Measurements

Two types of probe permeameter measurements have been taken in the Ardross units. First, two vertical transects of 5-cm-spaced measurements were taken in the field using a steady-state field permeameter calibrated against known samples. The two transects are 70 m apart and represent "wells" at either end of the "gridblock" analog. The probe tip used had a 4 mm inner radius and a 24 mm outer radius (implying a hemispheric volume of investigation of about 2 cm diameter). Surface preparation comprised chipping rock edges to expose a fresh surface 1–2 cm beneath the natural surface. This avoided most of the effects of the weathered crust, although no systematic study of surface effects was conducted. Second, measurements were made in the laboratory using a pressure-decay permeameter on a block sampled from the logged section, representing a "whole core" sample for determination of relative permeability and capillary pressure curves. Exhaustive 2-mm-spaced permeability profiles were acquired from a slab cut from this sample. The probe tip used had a 3 mm inner radius and a 6 mm outer

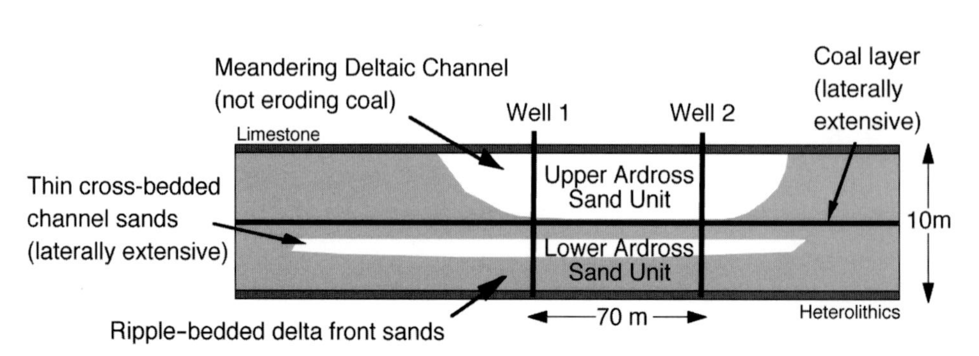

Figure 1. Sketch of the Ardross outcrop showing the sampled "wells" in relation to the main stratigraphic units.

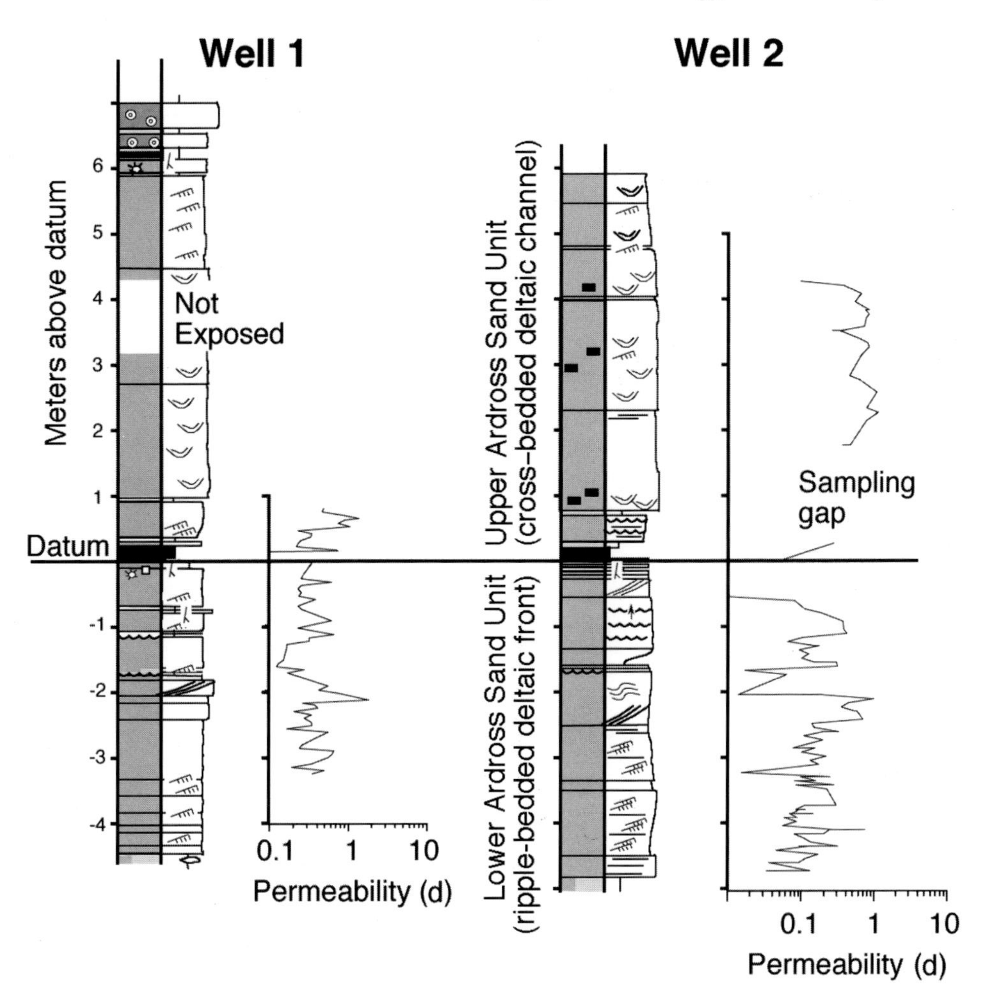

Figure 2. Stratigraphic logs and permeability data for the two wells sampled at either end of the outcrop. (The coefficient of variation, Cv, is defined as the standard deviation divided by the arithmetic mean).

	Both Units	Lower Unit	Upper Unit
Well 1	N = 57	N = 43	N = 14
Arith Av	433.41	391.19	571.35
St. Dev.	313.40	290.98	353.64
Cv	0.72	0.74	0.62
Well 2	N = 115	N = 85	N = 30
Arith Av	290.15	194.68	563.84
St. Dev.	272.08	189.82	288.84
Cv	0.94	0.97	0.51
Both Wells	N = 172	N = 128	N = 44
Arith Av	341.09	266.01	566.34
St. Dev.	294.71	249.36	307.89
Cv	0.86	0.94	0.54

radius (implying a hemispheric volume of investigation of about 1.5 cm diameter). These data thus represent higher resolution and more accurate measurements (as the pressure-decay device has a higher dynamic range). The calibration established for the field permeameter appears to give a consistent dataset, with no major systematic error.

Permeability data indicate that the upper sand unit has high permeabilities, averaging 566 md, whereas the lower unit is less permeable, with an average of 266 md (Figure 2). The main features of permeability variation in the two wells (e.g., high permeability layer at 2.0 m below datum) can be correlated between wells.

Permeability histograms indicate that the whole formation has an approximately log-normal distribution spanning over 3 orders of magnitude (1 md to 1.4 d) with a mode at 100 md (Figure 3A). The arithmetic averages for the two units are statistically different at the 95% level; furthermore, the high-end tail to the permeability distribution for the Lower Sand Unit (i.e., Lower Ardross Castle Sand Unit) is associated with a

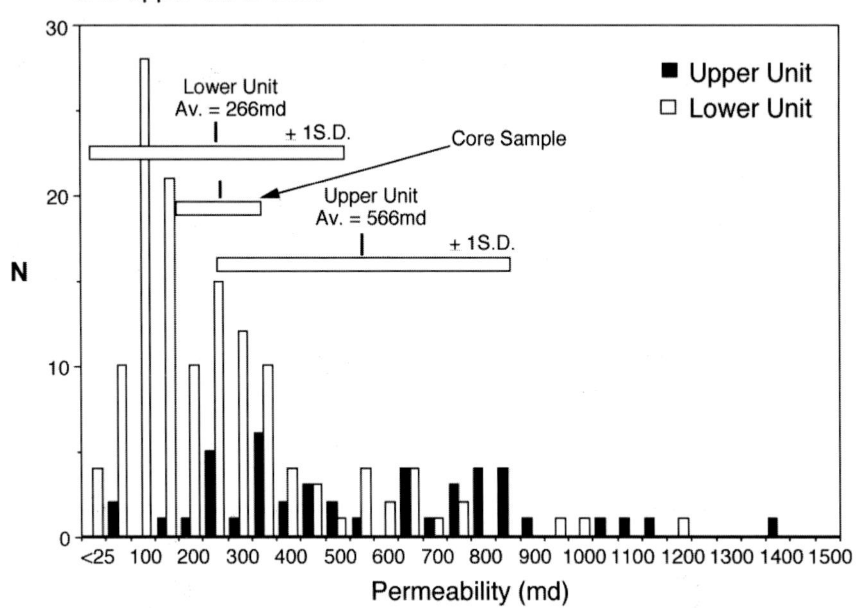

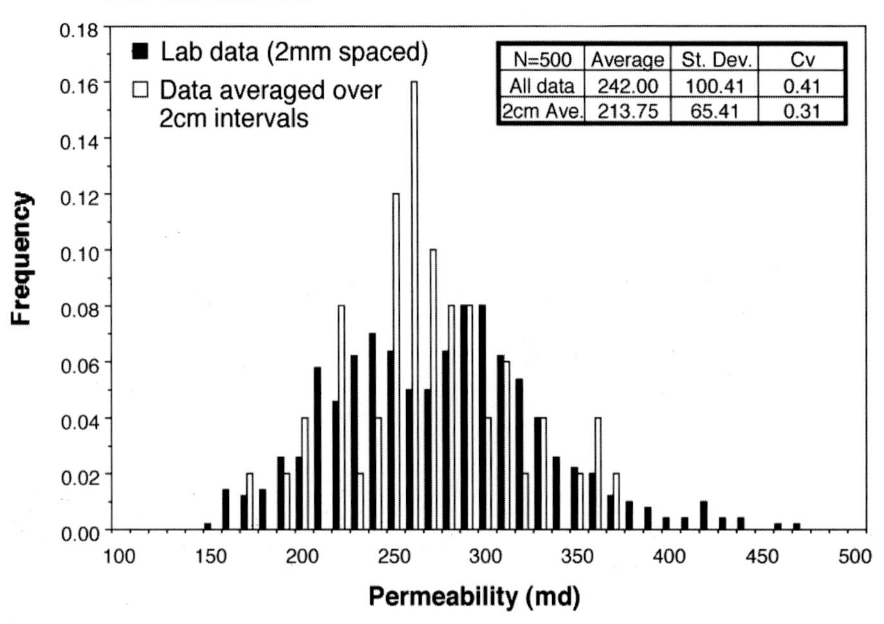

N=500	Average	St. Dev.	Cv
All data	242.00	100.41	0.41
2cm Ave.	213.75	65.41	0.31

Figure 3. Permeability histograms for the two sand units and the lab sample.

thin cross-bedded unit (at around –2.0 m on well 1 and at –2.3 m on well 2, Figure 2), and the lower values in the Upper Sand Unit (i.e., Upper Ardross Castle Sand Unit) correspond with ripple bedding at the base (between 0.0 and 1.0 m, Figure 2). Thus, the field permeability data can be separated into two distinct groups on the basis of lithology.

At the core scale, one can resolve the sedimentological components (lamina and beds), whose effects are aggregated in the larger scale dataset. Bimodality is evident in the histogram of core-scale data (Figure 3B). Permeability profiles collected on the core show a marked oscillatory pattern as the sand-rich and mica-rich laminae alternate (Figure 4), and these laminae are clearly seen on the core surface. The lamina spacing is around 2 cm. Despite a drift in the precise location and permeability of these laminae as successive permeability profiles are acquired (five profiles are shown in Figure 5), the lamina-scale influence is persistent and evident in the bimodal permeability histogram (Figure 4). The permeability contrast between sand-rich and mica-rich laminae varies between about 2:1 and 5:1. In a similar manner, the upper cross-bedded sand unit reveals lamina- and bed-scale patterns in permeability. These patterns form the basis for bed-scale permeability models for each lithofacies in the following section.

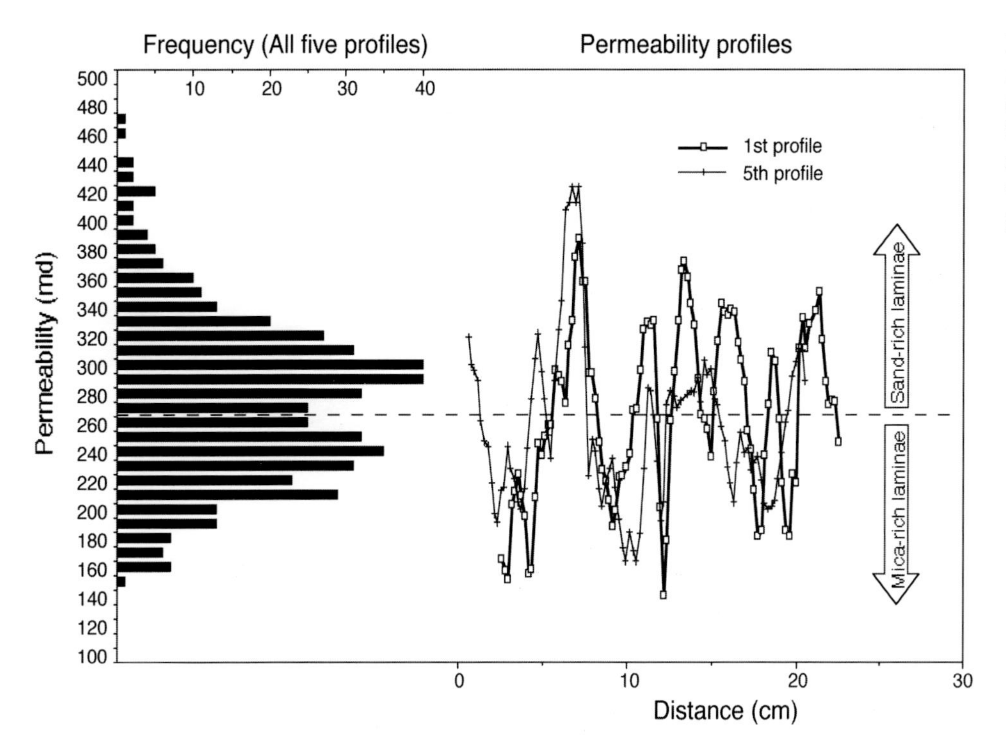

Figure 4. Permeability profiles from the core data revealing the presence of the alternating sand-rich and mica-rich laminae within the ripple-bedded facies.

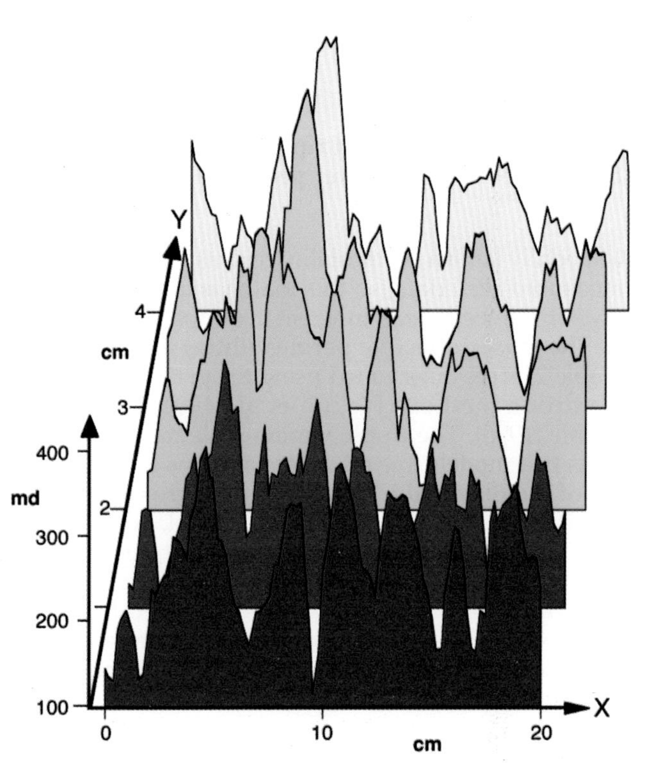

Figure 5. Graphical illustration of the five permeability profiles (each profile 1 cm apart) showing the degree of persistence of the sand-rich and mica-rich lamination.

Thus, the high resolution lab-scale permeability data clearly show the presence and persistence of lamina architecture. This structural component of the permeability data is easily lost or masked by measurements at larger scales. For example, if the 2-mm-spaced data are averaged over 2-cm intervals, the resultant probability density function (Figure 3B) loses the bimodality and has a much reduced standard deviation. The decrease in sample variance (σ^2) as a function of an increase in the volume of support (V/v) is well known (e.g., Isaaks and Srivastava, 1989) and is represented by the variance reduction factor, $f = \sigma^2(V)/\sigma^2(v)$. In this case, $f = 0.42$ (i.e., only 42% of the sample variance is retained for V/v = 1000). f is also influenced by the correlation structure of the data, and can be estimated if the correlation parameters are known. It is easy to overlook this factor and to assume that measurements taken at a particular scale (e.g., core plugs) adequately capture the true population variance. What is needed is a clear recognition of the inherent sedimentological architecture in relation to the available measurements.

Semivariogram analysis of the 5-cm-spaced well data (Figure 6) reveals the presence of holes at lags of around 0.3 m, 0.9 m, and 1.2 m. By inspection of the outcrop, we relate this evidence of cyclicity to bed-scale repetitions in the ripple-bedded lithofacies. The large nugget reflects the undersampling of lamina-scale heterogeneity by the 5-cm-spaced data. The Upper Sand Unit does not display strong vertical cyclicity, but an average cross-bed thickness can be identified. These lengthscales of cyclicity are used as a guide to defining the bed-scale grid dimensions used in the flow modeling. They also define the volume scale at which the observed variability in bed-scale permeability is captured as an effective flow property; however, once again, this structural aspect of the data can be easily lost in larger scale measurements. Semivariograms are also poor at detecting this type of repetition if there are fluctuations in the signal wavelength (i.e., height of the bed), as discussed recently by Jensen et al. (1996).

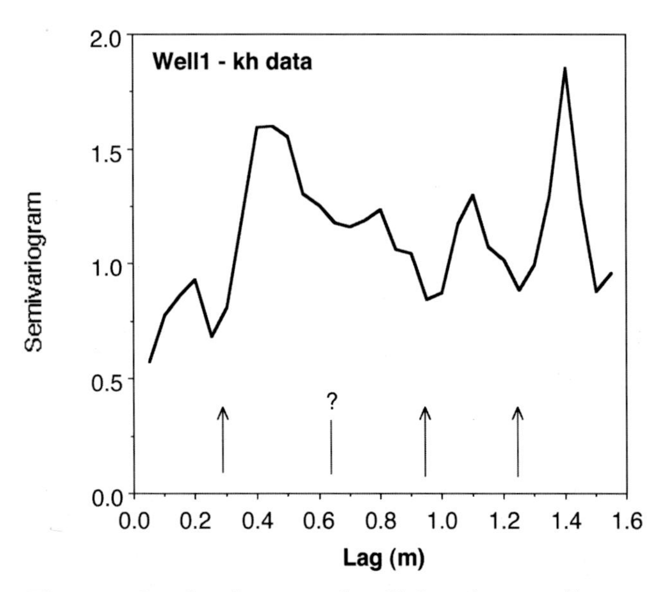

Figure 6. Semivariogram of well data from well 1 showing cyclicity with a 0.3-m wavelength.

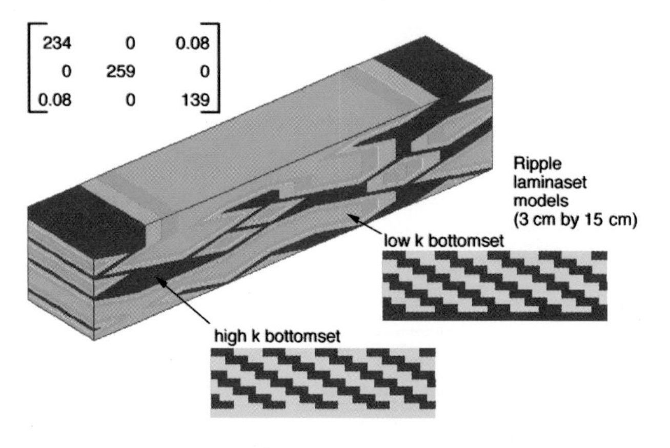

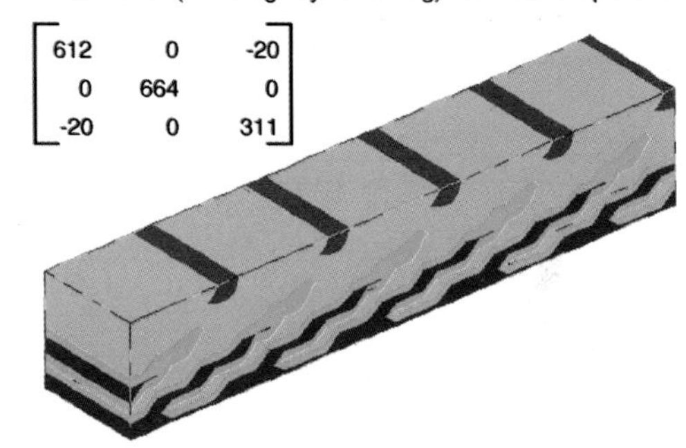

Figure 7. Flow models of ripple and cross-bed facies. Bluer tones indicate lower permeabilities.

Single-Phase Permeability Estimation

Using the field and lab permeability data, we constructed lamina-scale permeability models for the ripple-laminated Lower Sand Unit and the cross-bedded Upper Sand Unit. These deterministic templates were designed to portray the characteristic permeability structure of each bed type (i.e., permeability contrast, lamina spacing, lamina shape, lamina continuity, and internal grading). Variations in these characteristics, such as drift of the average between the two wells and vertical variations within the same facies, were treated by imposing stochastic variations and interwell trends on these templates. The models are illustrated in Figure 7. Two end-member cases were considered for the ripple model by allocating higher or lower permeability values for the bottomset to reflect observed variations. These end-members also represent well-connected and poorly connected extremes as far as flow is concerned. The lamina-scale ripple models were then scaled up and assembled into stochastic bed-scale templates to capture the range of variability observed in the wells.

A comparison of model statistics with well data shows a reasonable match (Table 1). This could be improved on by detailed adjustment of the model values; however, our aim has been to honor the sedimentary architecture (especially lamina permeability contrasts) as much as the measured well data, so this match is quite acceptable given the uncertainties inherent in sampling.

The tensor effective permeabilities for the models (Table 2) were determined using the periodic boundary condition method (Pickup et al., 1995; Pickup and Sorbie, 1996). The ripple lamina-scale models and the cross-bed model have off-diagonal terms that are about 4% of the diagonal term, due to their cross-laminated architecture. The bed-scale ripple model, however, has negligible off-diagonal terms because, at this scale, the effects of smaller scale cross-lamination are masked by the effects of the approximately layered bed-scale architecture. The tensor permeability values also differ significantly from the arithmetic (k_a) and harmonic averages (k_h), which one might have used to estimate

Table 1. Comparison of Permeability Statistics for Lithofacies Models and Well Data.

Lithofacies Model	Arithmetic Average (st. dev.) of Model	Arithmetic Average (st. dev.) of Corresponding Well Data
Cross-bed model (for both wells)	664 (356)	566 (308)
Ripple model, well 1, high-k bottomset	308 (211)	329 (153)
Ripple model, well 1, low-k bottomset	213 (158)	
Ripple model, well 2, high-k bottomset	154 (106)	153 (97)
Ripple model, well 2, low-k bottomset	107 (79)	

Table 2. Permeability Tensors for the Ripple and Cross-Bed Lithofacies Models (Well 1).

Key to tensor permeability terms		
k_{xx}	k_{xy}	k_{xz}
k_{yz}	k_{yy}	k_{yz}
k_{zx}	k_{zy}	k_{zz}
Lamina-scale ripple model (low-k bottomset)		
192	0	−9
0	213	0
−9	0	124
Lamina-scale ripple model (high-k bottomset)		
289	0	−11
0	308	0
−11	0	162
Bed-scale ripple model		
234	0	0.08
0	259	0
0.08	0	139
Bed-scale cross-bed model		
612	0	−20
0	664	0
−20	0	311

effective permeability. The cross-bed model has k_{xx} = 0.92 k_a. The actual size of the off-diagonal term depends on the cross-bed geometry and lamina permeability contrast. For example, an idealized cross-bed unit with layers of 10:1 permeability contrast, an angle of 26.565 (arctan 0.5), and no bottomset would have k_{xx} = 0.866 k_a (a reduction of about 13%). Thus, the lithofacies model geometries impose a small but significant control on the effective single-phase permeability, with small but potentially flow-significant off-diagonal terms; however, the geometrical control is much more important in determining the multiphase flow behavior.

Waterflood Scale-Up of the Lower Ardross Sand Unit

The lithofacies models of the Ardross Cliff have been evaluated in terms of their likely impact on a waterflood displacement, using the geo-pseudo method (Corbett et al., 1992; Ringrose et al., 1993; Pickup et al., 1994). This approach attempts to capture the impact of small-scale permeability architecture on multiphase flow using some form of pseudofunction numerical scheme. For this illustration we use the Kyte and Berry (1975) method. We have evaluated the potential errors implicit to the pseudofunction scheme by comparison with other methods (e.g., P_c-equilibrium steady state). The differences can be significant, but do not alter the overall conclusion, which is more influenced by the choice of rock and fluid properties than by the numerical scheme.

In the case of the cross-bed model, one scale-up step is sufficient to define pseudo relative permeability functions for the gridblock model (0.3-m-high gridcells); however, in the case of the ripple-bedded model, pseudofunctions were first defined at the laminaset scale (3-cm-high model) for the high bottomset and low bottomset models. These pseudos were then applied to the stochastic bed-scale model (Figure 7) to define pseudofunctions for the gridblock model (0.3-m-high gridcells). All models are 2-D vertical sections. Simulations at all scales are done at a flow rate of 0.2 m/day, and assume a water-wet system with an endpoint mobility ratio of 1.76. The capillary pressure, P_c, and relative permeability functions, k_{ro} and k_{rw}, assumed for input at the lamina scale (implicitly the "pore-scale" multiphase flow properties) were

$$P_c \text{ [bars]} = 3.0\, S_{wn}^{-2/3} (\phi/_k)^{0.5}$$
$$k_{ro} = 0.85\,(1 - S_{wn})^3$$
$$k_{rw} = 0.3\,(S_{wn})^3$$
$$S_{wir} = 0.6 - 0.165 \text{ Log } k$$
$$S_{wor} = 0.7$$

with S_{wir} = irreducible water saturation, S_{wor} = wetting phase saturation corresponding to the residual oil saturation, and S_{wn} = effective wetting phase saturation (i.e., normalized in the range S_{wir} S_{wor}).

Details of the assumptions implied by these functions are given in Ringrose et al. (1993).

Results for these models are given in Table 3. As a measure of the significance of these calculated recovery factors, the lithofacies-based model of the Lower Ardross Sand Unit has been compared with a model using 30-cm layers defined using 30-cm averages of the

Table 3. Waterflood Recovery (Percent of Original Oil in Place) After Injection of 1 Pore Volume for Different Models of the Ardross Cliff.

	Small Scale (3 cm High)	Bed Scale (30 cm High)	Field Scale
Upper Ardross Sand Unit			
Curved graded cross-bed model		50.19	54.34
Lower Ardross Sand Unit			
Ripples: Low-k bottomset	38.00		
Ripples: High-k bottomset	44.54		
Ripple bed-scale model with stochastically distributed high and low bottomsets		41.36	44.29
Model using 30 cm average permeability data in place of lithofacies models			52.52
Difference in recovery if small-scale sediment architecture is ignored (Lower Sand Unit)		+18.6%	

permeability data from well 2. This case represents a common approach of upscaling only the permeability data (usually by averaging) and assuming the core-scale relative permeability functions apply directly to the upscaled gridcell. The difference in recovery is 18%. This appreciable difference can be attributed mainly to the process of capillary oil retention within the small-scale cross-lamina and cross-bed architecture of the two lithofacies present in this section. In other work, Huang et al. (1995) and Honarpour et al. (1995) have conducted core-scale laboratory waterfloods to demonstrate that these small-scale capillary trapping phenomena do indeed occur. We are therefore confident that this degree of systematic difference in recovery (~20%) between models that capture and ignore the effects of lamina architecture is reasonable. In fact, because lamina-scale permeability contrasts may be much greater than those observed here (e.g., Weber, 1982), the recovery difference could be much larger.

Figure 8 compares the oil production and watercut for the geopseudo model of the Lower Ardross Sand Unit with the 30-cm average model. Not only is the recovery much poorer for the geopseudo model, but the watercut has earlier breakthrough and a steeper rise. Clearly, the effects of sediment architecture in this formation could have important economic significance if this formation were an oil reservoir. A cross-sectional model based on 30-cm averages (1-ft spacing) would overestimate recovery and field performance considerably. In practice, averaging as a basis for upscaling is often done at a much larger scale (~10 m), and so the errors may be even larger.

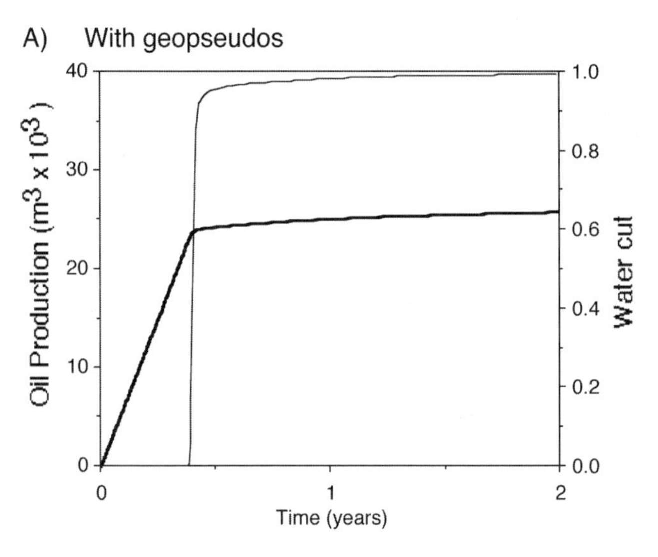

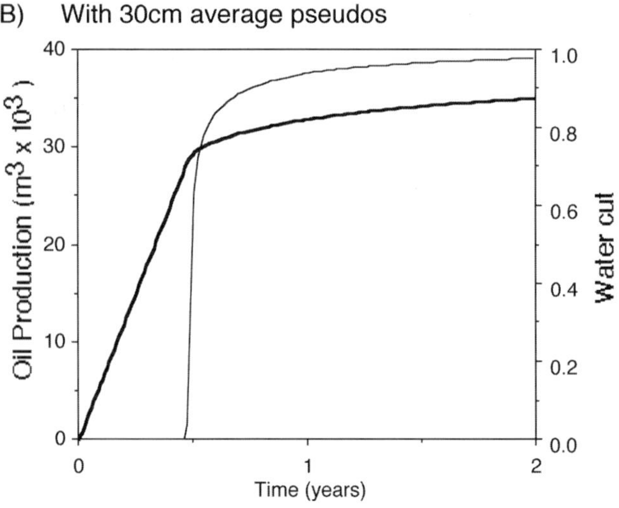

Figure 8. Simulated oil production and watercut for a cross-sectional model of the Lower Ardross Sand Unit: (A) using geopseudo upscaling and (B) using pseudos of 30-cm averaged well data.

DISCUSSION

This study of the sedimentary architecture, permeability data, and flow property calculation serves to illustrate the potential effects of internal rock structure and variability on flow at the scale of a single reservoir model gridcell. Other studies have evaluated the impact of sedimentary architecture on waterflood over a wider range of scales (e.g., Corbett et al., 1992; Jones et al., 1993; Kjønsvik et al., 1994; Ciammetti et al., 1995; Saad et al., 1995). The general conclusion that can be drawn from these integrated studies is that small-scale heterogeneity can be important for multiphase flow, but exactly how much depends on the specific details of the sedimentary and reservoir architecture, and the displacement process being considered. One must consider the problem on a case-by-case basis. We have, however, developed some simple guidelines for assessing the potential influence of small-scale sedimentary structure on a waterflood:

(1) Are immiscible fluids flowing (rates <1m/day)?
(2) Are significant small-scale heterogeneities present? Specifically

Small-Scale Heterogeneity Questions	Criterion
Is the permeability contrast	greater than 5:1 ?
Is the layer thickness pertaining to this contrast	less than 20 cm ?
Is the mean permeability	less than 500 md ?

If these criteria are all satisfied, then capillary/heterogeneity effects should be evaluated.

A further criterion involves the relative importance of small-scale and large-scale reservoir architecture. Even if the small-scale structure is important, it may be that the large-scale reservoir connectivity issues are still the dominant uncertainty. Using the simple classification of reservoir heterogeneity proposed by Weber and van Guens (1990), we infer the following guidelines on this aspect of the problem:

- Layer cake reservoirs—small-scale structure will usually have primary importance.
- Jigsaw puzzle reservoir—small-scale structure may be important.
- Labyrinth reservoir—small-scale structure will usually be of secondary importance.

Once the potential importance of small-scale heterogeneities has been considered and a decision to proceed with a detailed evaluation has been made, we are then faced with detailed questions about how to go about the study. We have attempted to demonstrate an approach for doing this with the gridblock analog described here. Figure 9 summarizes the procedure we have used. The underlying problem is how to handle the uncertainties inherent in an incomplete dataset (the sampling and estimation problem) along with the scale-dependent flow-structure interactions. We advocate the importance of referring both the parameter estimation problem and the flow upscaling problem to the sedimentological lengthscales inherent in the reservoir system. The procedures for doing this include the following.

(1) Considering the sample sufficiency of the well data. An important guide is the permeability C_v (Jensen et al., 1997).
(2) Evaluate the scales at which permeability variability is expressed. In this example, we compared an exhaustive permeability dataset acquired from one core sample with the well-scale permeability data to establish the degree of small-scale variability and importantly the typical lamina permeability contrasts. Wireline microscanner measurements provide a useful tool for estimating small-scale variability, permeability contrasts, and small-scale sedimentary architecture. The variance reduction factor, f, is a simple and useful measure of the degree to which the sample at a particular volume scale underestimates the population variance.
(3) Use correlation measures (e.g., semivariogram of well data) to identify inherent sedimentary lengthscales. In this example, we identified bed cyclicity at the 30-cm lengthscale and used this to guide the definition of the small-scale gridcells.
(4) Apply a knowledge of sedimentology to define likely bed-scale architectural patterns. In this example, the ripple and cross-bed architectures were found to have a significant effect on remaining oil. Where aspects of the architecture are uncertain, stochastic approaches can be used.
(5) Use a flow upscaling framework based on the sedimentary hierarchy of lengthscales. In this example, we used two different schemes to define upscaled flow models for the two lithofacies present in the analog (i.e., ripple bedding and cross-bedding).

In unpublished studies for oil companies we have applied this approach to 3-D reservoir studies. This work shows that the approach is practicable within a realistic field development program and can have a significant impact on field performance predictions (in these cases the geologically based upscaling gave up to 29% difference in recovery); however, more automated methods are needed in order to perform this type of study on a more routine basis.

The difficulties in actually detecting and modeling geological structure in the subsurface should not be underestimated (Jensen et al., 1996); however, a greater emphasis on the smaller scale aspects of reservoir architecture is warranted in view of their potentially significant impact on flow. Conventional geostatistical approaches (modeling the reservoir at a single scale using correlation functions) may fail to account for the reservoir architecture that actually affects flow.

Finally, there is considerable uncertainty about the choice of multiphase flow function to use as input to the scale-up procedure. We have not considered this problem here, but have argued elsewhere (Ringrose et al., 1996) that the lithofacies control is also vital when considering the choice of special core analysis sample, the laboratory method, and the procedure for applying

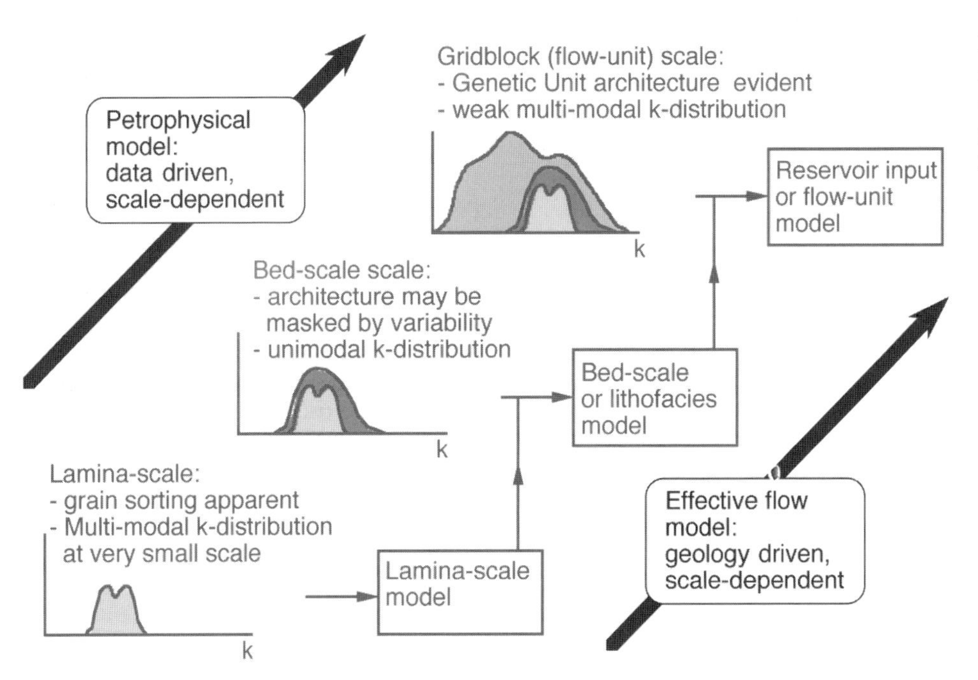

Figure 9. Illustration of the integrated approach to flow upscaling, accounting for sedimentology and statistical representivity.

it to the simulator model. We have also tested our upscaling approach against a whole-core experiment to lend support to our choice of multiphase flow functions for input at the lamina scale (Huang et al., 1995).

SUMMARY

The Ardross analog provides an illustration of flow-heterogeneity interactions at the reservoir gridblock scale. Although the analog is a specific example from a deltaic sequence, the sediment structures involved (ripple and cross-bed lamination) occur quite widely in many other sediment environments. Our studies of effective single-phase permeability and two-phase waterflood of the cliff have demonstrated some very significant effects of sediment structure:

(1) The tensor effective permeability k_{xx} term can be 10% less than the arithmetic average.
(2) Waterflood oil recovery could be 20% lower than that estimated from arithmetic averages of 30-cm (1-ft) intervals.

The architectural characteristics of any particular reservoir unit can be established using well data, provided that the data sufficiency is evaluated and the true lengthscales and patterns of permeability variability are identified. The basic tools for doing this involve the comparison of data from different lengthscales, different sedimentological models and the use of correlation measures. We have shown how spatial and scale-dependent variability can be captured using mixed deterministic/stochastic models that incorporate the sediment structure as well as known spatial variability. This small-scale sedimentary architecture is particularly important when assessing waterflood behavior.

ACKNOWLEDGMENTS

This work has been funded by sponsors of the Reservoir Heterogeneity Project: Amerada Hess, British Gas, Chevron, Conoco, Deminex, Elf, Esso, Fina, Mobil, Pan Canadian, Petrobras, Phillips, Shell, Statoil, Talisman, and the U.K. Department of Trade and Industry. We thank our colleagues in the Reservoir Description Group for advice and help, especially Patrick Corbett and Ken Sorbie. Yaduo Huang kindly collected and provided the lab-sample permeability data. John Underhill introduced us to the sedimentology of this outcrop.

REFERENCES CITED

Ciammetti, G., Ringrose, P. S., Good, T. R., Lewis, J. M. L., and Sorbie, K. S., 1995. Waterflood recovery and fluid flow upscaling in a shallow marine and fluvial sandstone sequence. SPE 30783, presented at the SPE Annual Technical Conference and Exhibition, Dallas, USA, 22–25 Oct., 1995, p. 845–858.

Corbett, P. W. M., Ringrose, P. S., Jensen, J. L., and Sorbie, K. S., 1992. Laminated clastic reservoirs—the interplay of capillary pressure and sedimentary architecture. SPE Paper 24699, presented at the SPE Annual Technical Conference, Washington, 4–7 October, 1992, p. 365–376.

Honarpour, M. M. Cullick, A. S., Saad, N., and Humphreys, N. V., 1995. Effect of rock heterogeneity on relative permeability: implications for scale-up. Journal of Petroleum Technology, November 1995, p. 980–986.

Huang, Y., Ringrose, P. S., and Sorbie, K. S., 1995. Capillary Trapping Mechanisms in Water-wet Laminated Rock. SPE Reservoir Engineering, November 1995, p. 287–292.

Isaaks, E. H., and Srivastava, R. M., 1989. Applied Geostatistics, Oxford University Press, New York, 561 pp.

Jones, A., Doyle, J., Jacobsen, T., and Kjønsvik, D., 1993. Which sub-seismic heterogeneities influence waterflood performance? A case study of a low net-to-gross fluvial reservoir. Proceedings of the 7th European IOR Symposium, Moscow, Russia, 27–29 October, 1993, p. 35–46.

Jensen, J. L., Corbett, P. W. M., Pickup, G. E., and Ringrose, P. S., 1996. Permeability semivariograms, geological structure and flow performance. Mathematical Geology, 28 (4), p. 419–435.

Jensen, J. L., Lake, L., Corbett, P. W. M., and Goggin, D. J., 1997. Statistics for Petroleum Engineers and Geoscientists. Prentice Hall PTR, New Jersey, 390 p.

Kjønsvik, D., Doyle, J., Jacobsen, T., and Jones, A., 1994. The effects of sedimentary heterogeneities on production from a shallow marine reservoir—What really matters? SPE paper 28445, presented at the European Petroleum Conference, London, 25–27 October 1994, p. 27–40.

Kyte, J. R., and Berry, D. W., 1975. New Pseudo Functions to Control Numerical Dispersion. Soc. Pet. Eng. J., August 1975, p. 269–275.

Pickup, G. E., Ringrose, P. S., Forrester, M. M., Jensen, J. L., and Sorbie, K. S., 1994. The Geopseudo Atlas: Geologically-Based Upscaling of Multiphase Flow. SPE Paper 27565 given at the SPE European Petroleum Conf., Aberdeen, 15–17 March 1994, p. 277–289.

Pickup, G. E., Ringrose, P. S., Corbett, P.W.M., Jensen, J. L., and Sorbie, K. S., 1995. Geology, geometry and effective flow. Petroleum Geoscience, 1(1), p. 37–42.

Pickup, G. E., and Sorbie, K. S., 1996. The Scaleup of Two-Phase Flow in Porous Media Using Phase Permeability Tensors. Society of Petroleum Engineers Journal, December 1996, p. 369–381.

Ringrose, P. S., Sorbie, K. S., Corbett, P. W. M., and Jensen, J. L., 1993. Immiscible flow behaviour in laminated and cross-bedded sandstones. Journal of Petroleum Science and Engineering, 9, p. 103–124.

Ringrose, P. S., Jensen, J. L., and Sorbie, K. S., 1996. Use of geology in the interpretation of core-scale relative permeability data. SPE Formation Evaluation, Sept. 1996, p. 171–176.

Saad, N., Cullick, A. S., and Honarpour, M. M., 1995. Effective relative permeability in scaleup and simulation. Paper SPE 29592, presented at the 1995 SPE Joint Rocky Mountain Regional/Low-Permeability Reservoirs Symposium, Denver, CO, March 20–22, p. 451–464.
van de Graaff, W. J. E., and Ealey, P. J., 1989. Geological modelling for simulation studies. American Association of Petroleum Geologists Bulletin, 73(11), p. 1436–1444.
Weber, K. J., 1982. Influence of common sedimentary structures on fluid flow in reservoir models. Journal of Petroleum Technology, March 1982, p. 665–672.
Weber, K. J., and van Geuns, L. C., 1990. Framework for constructing clastic reservoir simulation models. Journal of Petroleum Technology, October 1990, p. 1248–1297.

Lemouzy, P., Evaluation of multiple geostatistical models using scaling-up with coarse grids: a practical study, 1999, *in* R. Schatzinger and J. Jordan, eds., Reservoir Characterization-Recent Advances, AAPG Memoir 71, p. 277–292.

Chapter 19

Evaluation of Multiple Geostatistical Models Using Scaling-Up with Coarse Grids: A Practical Study

P. Lemouzy
Institut Français du Pétrole
Elf/IFP Helios Group
Pau, France

ABSTRACT

In the field delineation phase, the uncertainty in hydrocarbon reservoir descriptions is large. In order to quickly examine the impact of this uncertainty on production performance, it is necessary to evaluate a large number of descriptions in relation to possible production methods (well spacing, injection rate, etc.). The method of using coarse scaled-up models was first proposed by Ballin et al. Unlike other methods (connectivity analysis, tracer simulations), it considers parameters such as thermodynamic behavior of the fluids, well management, etc.

This paper presents a detailed review of scaling-up issues, along with applications of the coarse scaling-up method to various water-injection cases, as well as to a depletion case of an oil reservoir in the presence of aquifer coning.

The need for correct scaling-up of wellbore and near-wellbore parameters is emphasized and is far more important than correct scaling-up far from wells. I present methods to accurately represent fluid volumes in coarse models. I propose simple methods to scale-up the relative permeabilities, and methods to efficiently correct for numerical artifacts.

I obtained good results for water injection. The coarse scaling-up method allows the performance of sensitivity analyses on model parameters in a much lower computer time than comprehensive simulations. Models representing extreme behaviors can be easily distinguished.

For the depletion of an oil reservoir in the presence of aquifer coning, however, the method is not as promising. It is my opinion that further research is required for scaling-up close to the wells.

INTRODUCTION

The Need for Fast Evaluation of Multiple Geostatistical Realizations

During the delineation phase of reservoir development, when only a few wells have been drilled, a lot of uncertainty remains in the reservoir description. The sources of uncertainty include the

- Geometry of the reservoir, which affects the calculation of hydrocarbon volume in place;
- Amount of reservoir facies (which can be expressed as a net:gross ratio N:G), which similarly affects the calculation of volume in place;
- Dimensions of geological bodies (i.e., channels, lobes) and their internal heterogeneities, which control reservoir connectivity, and thus the recovery; and
- Petrophysical characteristics of the facies.

The major problem is to estimate the oil in place, the expected production profiles, and the uncertainties of these estimations. If the economic value of the expected production profiles is below the acceptable threshold of the operator, further development will be canceled or postponed. When the uncertainty is too large, additional data will be required (seismic, drilling) until a reasonable degree of certainty is reached, allowing for development. Determining the uncertainty is certainly more critical offshore than onshore.

To estimate the uncertainty, two approaches are possible:

- One can determine scenarios (e.g., minimum, mode, maximum, or proven, probable, possible, etc.) for each factor participating in oil in place or in production profile computations (e.g., reservoir top depth map, porosity, N:G ratio) and combine them to obtain scenarios for the studied value. This approach can go from simple products of scenario values to more sophisticated Monte Carlo computations, but it neglects possible interaction between factors.
- From a few wells, geological analysis can provide very realistic reservoir geological models. This knowledge can be expressed in terms of geostatistical parameters, used to build up very detailed reservoir models in lithofacies or petrophysics. The overall uncertainty is determined by the examination of the production behavior of a large number of realizations, allowing uncertainty to interact between petrophysics, geometry, etc. The issue studied in the present paper is the evaluation of the production behavior of a large number of realizations.

Previous Works

Guérillot and Morelon (1992) proposed a simplified flow simulator to sort several realizations. They used a single pressure field calculated once, and they computed the saturation evolution. This method is limited to cases when pressure has a limited impact on flow; e.g., for waterflood at a pressure above oil saturation pressure. It is exact for unit mobility ratio and negligible gravity and capillary effects, and inexact otherwise; nevertheless, it yields a satisfactory ranking of geostatistical realizations with respect to some production parameters (e.g., water breakthrough time) in its domain of validity.

Ballin et al. (1992) used tracer simulations to rank realizations. Similar to the previous method, it requires only a single calculation of the pressure field. Both methods have the same limitations. Saad et al. (1996) confirmed that both methods are nearly equivalent with respect to cumulative oil production or water breakthrough time ranks.

Ballin et al. (1993) investigated how coarse scaled-up model simulations can achieve a good ranking of multiple realizations. They proposed a methodology to approximate the cumulative distribution function (*cdf*) of production data that would be obtained through extensive simulations of fine models. Deutsch and Srinivasan (1996) investigated statistical criteria and the use of coarse scaled-up models, and concluded that the latter method is a valuable one.

The emergence of streamline-based computations (e.g., Thiele et al., 1996) is noteworthy. The method is similar to that of Guérillot and Morelon (1992), as it uses a pressure field calculated once or a limited number of times, but it propagates one-dimensional (1-D) flow solutions along streamlines, making the method faster and more general. Simplified simulations or streamline computations should present the same limitations concerning gravity and capillary effects, or strongly pressure dependent effects (e.g., gas liberation from oil).

The coarse scaling-up approach is appealing because it is not restricted to a limited physical domain, as are the previous methods. It can handle cases with strong effects of pressure (e.g., three-phases flow, compositional); moreover, sophisticated management of wells and reservoirs can be reproduced thanks to the possibilities of commercial simulators (e.g., a control of well flow-rate to limit the gas:oil ratio (GOR) can be handled automatically, a work-over to plug a water-producing layer can be simulated).

SCALING-UP METHODS

General

Scaling-up must be performed considering the actual operation of a reservoir simulator. First, the simulator processes data describing the reservoir (such as geometrical, petrophysical, pressure-volume-temperature (PVT) relationships), considered as properties attributed to homogeneous grid blocks. The output of this preprocessing or initialization phase is a network composed of nodes (to which are attributed some grid-block data such as pore volume, pressure, and saturation) and links (to which a transmissivity is attributed). Flow equations are then solved for this network. This is not unfamiliar for people who used analog models made of resistors before the emergence of numerical models.

When scaling-up flow models, it is better to define directly the components of the scaled-up network, rather than the scaled-up grid-block properties, which are changed in the preprocessing phase. The errors of the preprocessing of scaled-up grid-block properties come from the

- Geometric approximations;
- Hypothesis of homogeneity of grid blocks; and
- Averaging rule of grid blocks permeability used for transmissivity calculations.

Another point to emphasize is that the scaling-up problem in question is a pseudoization (calculation of properties depending on local flow pattern and boundary conditions), rather than a homogenization (calculation of effective characteristics of an equivalent homogeneous medium).

Permeability

I determined directly scaled-up transmissivity, rather than scaling-up block permeability and computing the transmissivity afterward. This approach was proposed by Gomez-Hernandez and Journel (1990), who scaled-up interblock permeabilities. As the geometry of their model was regular, the geometric term within transmissivity (the ratio S:L, S being the surface of the interface between two blocks, and L the distance between block centers) was constant; therefore, scaling-up either interblock permeability or transmissivity was equivalent.

An explanation of the interest of direct scaling-up of transmissivities may be found in Romeu and Noetinger (1995). They investigated how to determine numerically the equivalent permeability of a block consisting of several sub-blocks with permeabilities obeying a spatially correlated distribution. They showed that the accuracy of a numerical determination depends on the ratio of the size of the sub-block to the variogram range, and on the numerical method used. The larger the sub-block size, the larger the error in the equivalent numerical permeability. They demonstrated that a special finite-difference (FD) method, termed "direct," consisting of using permeability directly within the transmissivity term in place of the harmonic average of adjacent block permeabilities, yields equivalent numerical permeability closer to the value that would be obtained with an exact numerical method.

Computing directly scaled-up transmissivity is an application of this direct method. Scaling-up implies that the ratio of block size to variogram range will be much larger in the coarse model, leading to large errors when using traditional FD methods. Determining transmissivity directly reduces the error.

The scaled-up transmissivities are computed by solving numerically on the fine grid the steady-state single-phase flow problem for the model composed of the fine blocks belonging to the coarse block under consideration and to the existing neighboring coarse blocks (up to 27 coarse blocks). Constant potential conditions are applied on two opposite faces of the model, and no-flow conditions on the remaining faces. One flow problem must be solved for each principal direction X, Y and Z. Processing the results of the flow models (potential and flow rate) gives the interface transmissivities T. I have to sum up the flow rates q_i across each fine block interface composing a coarse block interface, and to calculate the average potential Φ for each coarse block.

$$Q = \sum q_i \qquad T_{IJ} = \frac{\mu Q}{\overline{\Phi}_J - \overline{\Phi}_I} \tag{1}$$

The potentials are averaged by considering an arithmetic average of fine block potentials weighted by pore volume ($V\,\phi$). This is obtained assuming that total compressibility is constant over fine blocks. Detailed derivation and discussion are presented in Appendix A.

$$\overline{\Phi} = \frac{\sum V_i \phi_i \Phi_i}{\sum V_i \phi_i} \tag{2}$$

Initialization

Initialization of flow models consists of computing, for each block, its initial saturation and pressure, and attributing functions of pressure (PVT) or saturation (relative permeability (K_r) and capillary pressure (Pc)). This requires the knowledge of the gravity center depths of blocks. In current use of FD numerical simulators, this gravity center depth is calculated from the geometrical description of the block, assuming the block is homogeneous. Thus, the gravity center coincides with the geometric center of the block. When the coarse block is made of sub-blocks of different properties, these two points are not identical. When density of fluid is constant, the calculation detailed in Appendix B leads to

$$\overline{Z} \sum V_i \phi_i = \sum V_i \phi_i Z_i \tag{3}$$

The pore volume is also needed for the simulation. It is computed from fine-scale models with equations A3 and A4.

Initial pressure is determined in the same way for coarse and fine blocks: given the initial reservoir pressure known at datum, PVT, and block depth, the pressure is computed during the preprocessing phase of simulations. The same PVT is assumed to apply to the two scales. The question of whether to perform some kind of scaling-up for PVT may be raised, as shown by the next equation expressing equivalence of oil-in-place volume at standard conditions for fine and coarse models:

$$\frac{\overline{\phi}\,\overline{V}\,\overline{S_0}}{\overline{B_0}(\overline{P})} \sum \frac{\phi_i V_i S_{0i}}{B_0(P_i)} \tag{4}$$

At initial equilibrium, pressure depends only on depth, and it is reasonable to keep the same PVT for coarse and fine blocks, as long as the dimensions of the coarse blocks are not too large.

To compute initial saturations, flow simulators consider gravity-capillarity equilibrium, assuming that during migration, hydrocarbon displaced water by a drainage process. The saturation computation requires the definition of drainage capillary pressure (*Pc*) curves and of fluid contacts, which define the null *Pc* depth. Two methods have been tested:

- Scaled-up drainage *Pc* curves may be determined, as explained in Appendix C. Using these curves for the coarse model initialization ensures that gravity-capillary equilibrium is honored.
- Average saturations of coarse blocks resulting from the initialization of the fine model (equation C4) can be directly entered into the coarse model, bypassing the initialization step. In that case, if scaled-up *Pc* curves are not provided to the coarse model (e.g., assuming that *Pc* is null everywhere in the reservoir), gravity-capillary equilibrium will not be respected.

Well Numerical Representation

Wells are represented in numerical models by special nodes and links of the numerical network. Where the well pressure is computed, a new node is connected to the perforated reservoir block. The link between this node and the reservoir node representing the perforated block is characterized by a production index (*PI*) and by relative permeability functions, often those of the perforated block for a producing well.

The radial flow pattern near wells is different from the nearly linear behavior far away from wells: therefore, scaling-up near wells is not the same as far from wells. It is of prime importance to correctly scale-up the well functions (*PI* and K_r). The bottom-hole pressure (BHP) is the most frequent pressure data available (directly or from well-head pressure through a friction loss equation), and it must be represented accurately. Moreover, the production operations (i. e., production lifting and injection) provide constraints for BHP and well flow rate. Their impact on the reservoir behavior can only be assessed if well functions are carefully computed.

Classically, numerical *PI* is computed with the Peaceman (1978) equation. This equation has been established for centered vertical wells in a regular grid and homogeneous reservoir, far from reservoir limits and from other wells. Ding (1995) has proposed to process the results of a fine-scale simulation of steady-state single-phase flow around wells to compute scaled-up *PI* and scaled-up transmissivities of interfaces of the perforated coarse blocks. This method takes into account within the accuracy of the fine-scale simulation:

- The true position of the well in the coarse block (not always centered);
- The deviation of the well from the vertical;
- A partial perforation of the coarse block;
- The effect of reservoir limits close to the well;
- The interference with another well (provided the flow rates of wells keep the same ratio during simulation); and
- The heterogeneity in the vicinity of the well.

Practically, the steady-state single-phase simulation is performed on a fine-scale model including the studied well and the neighboring wells. In that model, the Peaceman equation is used. Well flow rates are set according to realistic values. A steady-state regime may be established when producing and injecting wells are present. For depletion, a pseudo–steady-state regime may be simulated. From simulation results, the well flow rates *qi* of perforated fine blocks composing a perforated coarse block are summed up to provide its flow rate *Q*. Average potentials of the perforated coarse block Φ_b and of surrounding blocks are obtained by a pore volume weighted average (equation 2). Scaled-up transmissivities near wells are calculated with equation 1. Finally, the scaled-up *PI* is simply obtained from well potential Φ_w:

$$\overline{PI} = \frac{\mu Q}{\Phi_b - \Phi_w} \tag{5}$$

K_r is needed to calculate the phase productions into the well. Due to the radial flow around the well, it is clear that saturation within fine blocks close to the well influences very much the overall mobility around the well. The saturation detail is lost in the coarse block, where only average saturation is considered; therefore, specific well K_r should be built up. Furthermore, PVT must not be forgotten; one can imagine a case where only a few fine blocks close to the well have pressures below oil saturation pressure. The presence of gas has an impact on K_r, and thus on phase flows. But no gas appears in the coarse block if its average pressure is above the saturation pressure; therefore, significant differences between fine and coarse models could occur. These PVT and K_r points have not been tackled in the present work, but I think they deserve further attention.

Relative Permeability

Relative permeability (K_r) may be scaled-up in two steps:

(1) A physical scaling-up gives the coarse-scale K_r.
(2) A correction of numerical artifacts is applied.

Physical Scaling-Up

Scaled-up K_r depends on relative magnitude of gravity, capillary, and viscous forces. When one force dominates, scaling-up may be performed quickly, as the saturation distribution at fine scale may be obtained without any flood simulation. In this case applying permeability scaling-up techniques to phase permeability provides scaled-up K_r; otherwise, the current practice is to use fine-scale flow simulations and to process their results (e.g., Kyte and Berry, 1975). This is obviously impractical when evaluating multiple realizations in the lowest computer time possible.

This is why I used an approximate but fast method. K_r curves are split into two sets of data:

- The set of end points (saturation and K_r).
- The set of K_r shapes, obtained by normalizing the curves in the [0, 1] range for both saturation and K_r axes with linear transformations.

K_r shapes may be preserved through scales for some physical situations (e.g., Kossak et al., 1989). I kept the fine-scale K_r shapes at the coarse scale, and I only scaled-up end points. This choice is erroneous in general cases, but I assumed that ranking capability of scaled-up models is not affected.

In the presence of lithotypes exhibiting more than one K_r shape, I selected the shape of the most permeable lithotype, as it will dominate the flow capacity.

For saturation end points, I recalculated effective residual saturation, considering the initial saturation status in the fine model. When the aquifer crosses a coarse block, the fine blocks below the contact show a null oil saturation. Thus, residual oil saturation for these blocks is set to zero. Once corrected, end-point saturations are scaled-up by a pore volume weighted average (equation C4).

For K_r end points, I considered permeability scaling-up methods applied to phase permeability ($K\,K_r$). Algebraic methods are preferable, being faster than numerical ones. Among them, the power average method is useful, provided the exponent ω characteristic of the reservoir is known. The scaled-up K_r is obtained knowing the proportions of lithotypes x_i:

$$\overline{K}_r = \frac{\left[\sum x_i \left(K_i K_r^i\right)^{\omega}\right]^{\frac{1}{\omega}}}{\left[\sum x_i K_i^{\omega}\right]^{\frac{1}{\omega}}} \qquad (6)$$

The determination of ω requires a calibration step. As I did not perform this step, I assumed ω = 1. This exponent corresponds to a layered reservoir that is obviously often far from the actual configuration. When a lithotype permeability is much larger than others, the scaled-up K_r end point is the same as the end point of this lithotype. Thus, I also tested the use of a proportion weighted average of K_r, which gives a lower value:

$$\overline{K}_r = \sum x_i K_r^i \qquad (7)$$

Correction of Numerical Artifacts

I must correct for two artifacts.

- The K_r relationships coming from laboratory measurements express the dependence of K_r on saturation computed on a null volume support. Such K_r relationships can be used either for a continuous problem (e.g., Buckley and Leverett, 1942) or for steady-state multiphase flow FD simulations, as the saturation is constant in each grid block. For the usual FD simulations of transient multiphase displacement, K_r depends on average saturation computed in the grid blocks. Thus, the K_r relationships must be corrected to account for the change of support in the saturation calculation and to correctly reproduce the multiphase fluxes at block interfaces.
- Traditional FD methods are known to spread the displacing fluid fronts. The analysis of this artifact is well known in 1-D displacements (Lantz, 1971).

I can correct numerical artifacts in 1-D by computing pseudo-K_r. I obtain these values for pseudo-K_r by applying the Kyte and Berry (1975) method to pressures and flow rates given by a two-phase displacement simulation performed in a finely gridded homogeneous 1-D medium. I can compute pseudo-K_r for different aggregation ratios of fine blocks into coarse blocks. Pseudo-K_r depends on several factors: viscosity ratio, distance to injection block, and capillary number; nevertheless, a single pseudo-K_r computed for the actual aggregation ratio at half the mean distance between injector and producer for the viscosity, capillary, and velocity mean conditions dominating in the reservoir enables a reasonable reproduction of the fine-scale 1-D simulation with the coarser grid.

It is thus attractive to use the pseudo-K_r computed in 1-D to correct the numerical artifacts in 2-D (two-dimensional) or 3-D (three-dimensional) cases. As vertical velocity is usually low in reservoirs, I assume that correction in 3-D can be achieved by a 2-D correction obtained by considering an areal flow.

The aggregation ratio to compute 1-D pseudo-K_r must be chosen. To do this, I performed 2-D areal two-phase displacement simulations on a finely gridded homogeneous medium. I considered a flood direction either parallel or diagonal to the grid, and either favorable or unfavorable mobility ratios. I calculated pseudo-K_r in 1-D for different aggregation ratios, and looked at the best choice of this ratio to reproduce the results of fine 2-D simulations on coarse 2-D grids.

From these simulations, the following conclusions may be drawn:

- When the mobility ratio is unfavorable, flood results of fine and coarse grids are almost identical. In this case, FD method appears to be insensitive to the grid size and to the flow direction.
- For a favorable mobility ratio, the best aggregation ratio for floods parallel to a grid direction is the actual aggregation ratio; e.g., if fine blocks are

grouped into 5 × 5 to give a coarse block, a satisfactory pseudo-K_r is obtained from 1-D fine simulation with an aggregation ratio of 5:1.
- For a favorable mobility ratio, the best aggregation ratio for floods in the direction of a diagonal to the grid is half the actual aggregation ratio. For a 4 × 4 grouping of fine blocks, a pseudo-K_r from 1-D fine simulation with an aggregation ratio of 2:1 is efficient to reproduce the results of the fine 2-D model on the coarse 2-D model.

PRESENTATION OF CASE STUDIES

First Case: Waterflooding a Reservoir Without Active Aquifer

This case, C1, deals with a turbiditic reservoir with two drilled wells. The study was aimed at determining the sensitivity of production forecasts to the uncertainty on geostatistical parameters. Four lithotypes were described.

- Two types of channels (facies 1 and 2), of different petrophysical characteristics.
- A less permeable lithotype (facies 3), corresponding to levees.
- A shaly lithotype (facies 4), filling the remaining volume of the valley in which the turbiditic system was deposited.

A portion of this reservoir was represented by a rectangular box, elongated along the principal axis of the system. Geological uncertainty was thought to be important for two factors:

- The proportions of the facies. Due to relations between the proportions of the facies, the proportion of channel 1 was the single factor considered.
- The dimensions of the reservoir bodies (facies 1–3). These dimensions are indirectly modeled by the variogram ranges (spherical model). The variogram range of channel 1 in the *X* direction perpendicular to the principal direction of the system (*Y*) is the single factor considered.

The possible value ranges of these factors was determined by a geologist. Applying experimental design techniques, five sets of geostatistical data (Table 1) were compiled.

Table 1. Definition of Geostatistical Sets, Case C1.

Set	Proportions	Variogram Range (m)
S_0	0.15	100
S_1	0.10	50
S_2	0.20	50
S_3	0.20	200
S_4	0.10	200

Table 2. Grid Description, Case, C1.

	Block Number/Aggregation Ratio			
Grid	X	Y	Z	Total
Fine	41	51	20	41,820
Coarse 1	8/~5:1	10/~5:1	4/~5:1	320/~130:1
Coarse 2	10/~4:1	13/~4:1	5/~4:1	650/~64:1

- The first set S_0 corresponded to what the geologist thought was the most probable. It is the central point of the experimental plan, used to validate the model derived from the others sets.
- The four others sets S_1–S_4 were built according to a complete factorial plan, each factor being fixed at one of the two limits of its range.

The parameter sets were used as input of the SISIMPDF routine of GSLIB (Deutsch and Journel, 1992) to obtain several facies simulations for each set. The chosen grid is shown in Table 2. The block dimensions were 25 m in *X*, 60 m in *Y*, and 1 m in *Z*. Each facies was given constant petrophysical data (Table 3). K_r end points and shapes were different for each lithotype. Because the aquifer is far from this part of the reservoir, *Pc* was neglected (*Pc* = 0 during simulations) and initial water saturation was at its irreducible value.

A waterflood with the two existing vertical wells and two new wells was simulated (Figure 1). A perforation method for new wells was determined: the blocks containing channel facies were systematically perforated. To maximize the impact of varying geostatistical parameters, high liquid production rates were imposed at producing wells. Injection rates were set to balance the total production, as long as injection pressure did not reach an upper limit. This led to an unbalanced waterflood, as the injecting rate was not always able to balance production. Thus, depletion below bubble-point pressure occurred near producing wells, and the production gas:oil ratio (GOR) increased. The GOR was controlled by decreasing the liquid rate when an upper GOR limit was reached. Results were recorded at two different times: at 500 days, shortly after water breakthrough, and at 2000 days, at which time the water-cut reached values close to 80%.

Second Case: Depleting or Waterflooding a Reservoir With Active Aquifer

The reservoir was an anticline structure, with a bed thickness of 40 m and an oil column of about 60 m above an active aquifer. Two wells were drilled, and the study was aimed at determining the uncertainty of production either under balanced waterflood or under depletion. Forty-nine geostatistical simulations of horizontal permeability conditioned to the two existing wells were obtained in a grid presented in Table 4. The block dimensions were 50 m × 50 m × 2 m. Porosity was attributed through a correlation

Table 3. Petrophysical Characteristics of Facies, Case C1.

Facies	Permeability Horizontal (md)	K_v/K_h	Porosity	S_{wi}	S_{or}	K_{rwmax}
1	3000	0.1	0.30	0.15	0.20	0.40
2	300	0.01	0.26	0.30	0.17	0.20
3	25	0.001	0.06	0.35	0.18	0.15
4	1	0.00001	0.022	0.45	0.15	0.11

Table 4. Grid Description, Case C2.

Grid	Block Number/Aggregation Ratio X	Y	Z	Total
Fine	20	25	20	10,000
Coarse 1	5/4:1	5/5:1	2/10:1	50/200:1
Coarse 2	5/4:1	5/5:1	4/5:1	100/100:1

between logarithm of horizontal permeability and porosity. A vertical to horizontal permeability ratio was assigned according to the level of horizontal permeability. The same K_r curves applied on all blocks.

For the waterflood case, two producing wells (P_1 and P_2) and four injecting wells (I_1 to I_4) were considered (Figure 2). P_1 and I_4 were the drilled wells, and they were perforated in the same way for all realizations. The other wells were perforated differently depending on the realization. Liquid rates were fixed at producing wells. The sum of injection rates compensated the total fluid production of producers. As the total rate was moderate, no imbalance occurred during simulations, and reservoir pressure was above saturation pressure. The production results were recorded at two times: first when reservoir water-cut was at 5%, and second at the end of production (20 yr).

For the depletion case, seven producing wells (P_1 to P_7) were implemented in the oil zone (Figure 3). P_1 and P_2 were the two drilled wells, always perforated in the same manner for all realizations. The other wells were perforated differently depending on the realization. The production results are presented at end of production (20 yr).

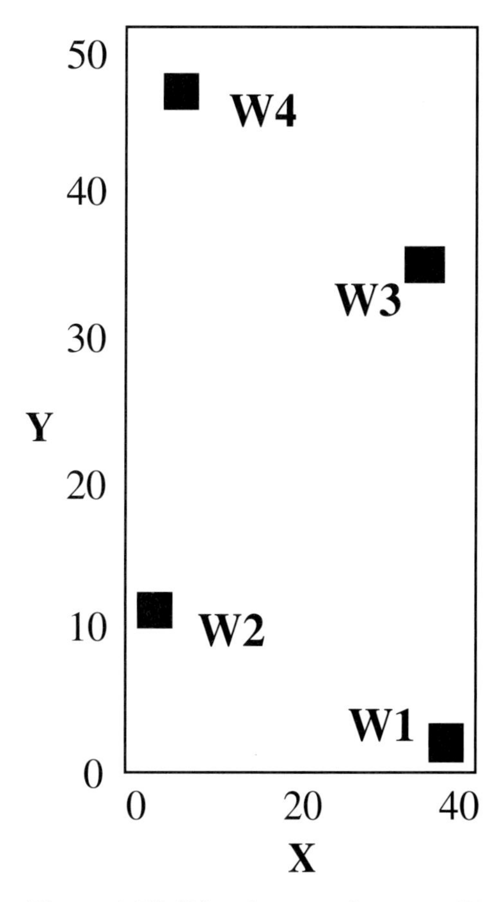

Figure 1. Well implementation, case C1.

ANALYSIS OF COMPARISON TESTS BETWEEN FINE AND COARSE SIMULATIONS

General

To test the value of scaling-up to rank multiple realizations, a comparison between simulated productions results (termed X) of fine and scaled-up models is performed. This yields the error δX:

$$\delta X = X_f - X_u$$

δX is composed of two parts: its mean is the systematic error of using scaled-up models, and is meaningless with respect to ranking. The important part of δX is its standard deviation, which gives the magnitude of random-like fluctuations around the mean: in the event of this standard deviation being null, one would obtain a perfect ranking of X, even if δX were far from being null. As the value of δX in itself is not sufficient to compare the ranking capability for different production results X_i, I propose to look at:

$$\varepsilon_1^i = \frac{\sigma(\delta X^i)}{\mu(X_f^i)} \text{ and } \varepsilon_2^i = \frac{\sigma(\delta X^i)}{\sigma(X_f^i)}$$

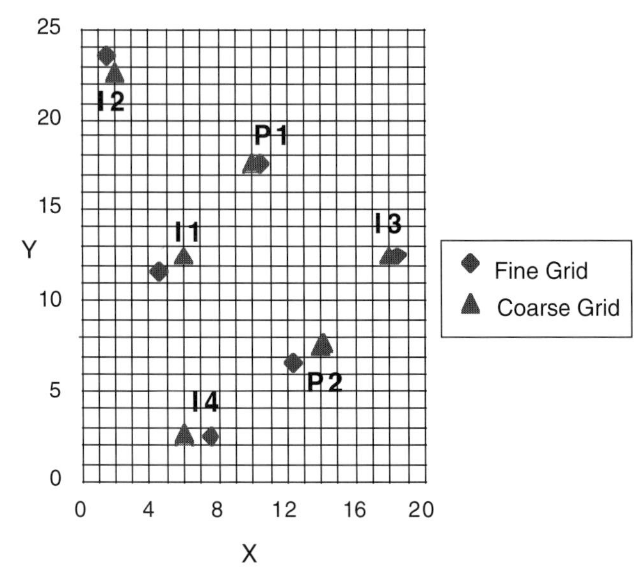

Figure 2. Well implementation and grids, case C2 waterflood.

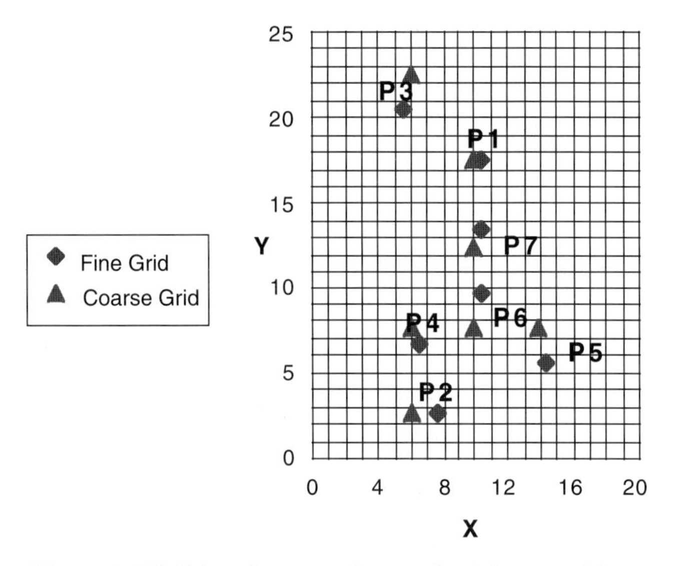

Figure 3. Well implementation and grids, case C2 depletion.

ε_1 can be viewed as the normalized accuracy of the "random" part of the error. ε_2 represents the ranking capability: the smaller this figure is, the better the ranking. A good ranking capability can be obtained either with small "random" errors or because of a large variability of *X*. Conversely, if *X* has a small variability, ranking will be difficult, but in this case, uncertainty is not of much concern.

Another way to analyze ranking capability is to compute linear correlation coefficients. Exact scaling-up methods would provide a perfect linear relationship: $X_u = X_f$. Because scaling-up is not exact, one should expect a departure from this linear behavior, reasonably approximated by a linear correlation. Correlation coefficients between fine and scaled-up simulation results will thus be calculated for both production parameters (ρ_X) and their ranks (ρ_r).

The simulated production data examined are of two types:

- Cumulative data: produced volumes of oil (N_p), gas (G_p), water (W_p), and oil recovery factor. These data are smooth functions of time. The produced cumulative volumes are crucial for economic calculations (e.g., net present value NPV). Normalized data such as recovery factor are important from a technical point of view to compare performances of different reservoirs.
- Instantaneous rates or rate ratios: oil flow rate (Q_o), gas-oil ratio (GOR), and water-cut (F_w). These data may vary a lot versus time and are the most difficult to reproduce through scaled-up models.

Tests of Case C1

The tests (w1–w5) differ according to the following options (see Table 5):

- Coarse grid used.
- Method to scale up well functions. (1) A simple method is used to compute scaled-up *PI*. The Peaceman equation is applied to perforated coarse blocks, attributing a permeability which is the arithmetic average of the perforated fine blocks. No correction of transmissivities close to wells is done. (2) Only *PI* is scaled-up (equation 5), without any correction of transmissivities close to wells. (3) *PI* and near-well transmissivities are scaled-up from fine-scale well steady-state simulations.
- Aggregation ratio for computing corrected K_r shapes.
- Processing of end points, either (1) with equation 6 ($\omega = 1$) or (2) with equation 7.

Tests of Case C2, Waterflood

The tests (w1–w6) differ according to the following options (see Table 6):

- Coarse grid used.
- Aggregation ratio for computing corrected K_r shapes.

Table 5. Test Specifications, Case C1.

Test	Grid	Wells	Correction of K_r Shape	Correction of K_r E. P.*
w1	CG1	1	1:1	1
w2	CG1	2	5:1	1
w3	CG1	3	5:1	1
w4	CG2	3	5:1	1
w5	CG1	3	5:1	2

*E.P. = end points.

Table 6. Test Specifications, Case C2 Waterflood.

Test	Grid	Wells	K_r Shape	Saturation Initialization	Gravity Correction
w1	CG1	1	1:1	1	No
w2	CG1	1	5:1	1	No
w3	CG1	1	2:1	2	No
w4	CG1	1	2:1	1	No
w5	CG2	1	2:1	1	No
w6	CG1	1	2:1	1	Yes

- Initialization of saturation (1) with a transfer of saturation from fine-scale model initialization, and a null *Pc* considered during waterflood, or (2) by using scaled-up drainage Pc curves, used also during the imbibition phase.
- Using a correction of gravity term close to producing wells. As these wells are perforated close to the aquifer, water coning or cusping occurs during production. To accurately describe the variation of the potential in the oil phase close to the well, a fine grid is necessary. When the grid is coarse, the driving potential difference that builds up the water cone is underestimated. This can be tackled in our numerical simulator by increasing the driving gravity term $[g(\rho_w - \rho_o)]$ applied in the blocks close to the well. A constant multiplying factor has been used in all realizations for all producing wells.

Tests of Case C2, Depletion

The tests (d1 to d3) differ according to the following options (see Table 7):

- Method to scale up well functions: (1) only *PI* is scaled-up (equation 5), without any correction of transmissivities close to wells, or (2) near-well transmissivities are scaled-up from fine-scale well steady-state simulations.
- Using a correction of gravity term close to producing wells, as for the waterflood case.

The finest coarse grid (CG2) was always used.

Influence of Correction of K_r for Numerical Artifacts

This influence is shown when comparing the following waterflood tests of case 2 (Table 8):

- w1: K_r curves are not corrected for numerical dispersion.

Table 7. Test Specification, Case C2 Depletion.

Test	Grid	Wells	Gravity Correction
d1	CG2	1	No
d2	CG2	2	No
d3	CG2	1	Yes

- w2: K_r curves are corrected with an aggregation ratio 5:1.
- w4: K_r curves are corrected with an aggregation ratio 2:1.

Test w4 exhibits the best ranking performance for both intermediate and final production data. This observation confirms the conclusion of the 2-D numerical dispersion investigation. The aggregation factor for computing corrected K_r curve (2:1) is close to half the 1-D block aggregation factor (5:1).

Influence of K_r Physical Scaling-Up

The influence of K_r physical scaling-up is shown when looking at tests w3 and w5 of case 1 (Table 9). The option to compute K_r end point in w5 gives better overall results for intermediate and final times. This shows that scaling-up K_r is important to improve ranking. Methods used here are simplistic, and better ones should be developed. Li et al. (1995) proposed a promising method that seems accurate, though not expensive in terms of computer time.

Influence of Scaling-Up Well Parameters

This influence is shown when comparing the following depletion tests of case 2 (Table 10):

- d2: Only *PI* is scaled-up.
- d1: *PI* and near-well transmissivities are scaled-up.

The results of test d2 are clearly worse than those of test d1, and the necessity of accurately computing near-well transmissivities is clear.

Influence of Saturation Initialization

This influence is shown when comparing the following waterflood tests of case 2 (Table 8):

- w3: Coarse models are initialized with scaled-up *Pc* drainage curves.
- w4: Coarse models are initialized with initial saturation computed on the fine-scale models.

These tests do not show significant differences, whether in correlation (ρ) or accuracy (ε) coefficients. Thus, the simplest method is preferable in practical cases.

Table 8. Case C2 Waterflood Production Results.

		End of Production (20 yr)						$F_w = 5\%$						
	Data	w1	w2	w3	w4	w5	w6	w1	w2	w3	w4	w5	w6	Mean
ρ_x	N_p	0.93	0.96	0.95	0.95	0.94	0.95	0.63	0.63	0.65	0.64	0.76	0.71	0.81
	Q_o	0.45	0.42	0.47	0.47	0.58	0.51	0.05	0.09	0.02	0.01	0.04	0.11	0.27
	G_p	0.93	0.96	0.95	0.95	0.93	0.95	0.64	0.63	0.65	0.64	0.76	0.71	0.81
	W_p	0.93	0.94	0.95	0.95	0.94	0.95	0.23	0.27	0.15	0.20	0.08	0.02	0.55
	F_w	0.45	0.19	0.48	0.28	0.57	0.50	0.08	0.04	0.10	0.06	0.08	0.07	0.24
	Recov.	0.55	0.59	0.57	0.58	0.74	0.55	0.51	0.49	0.52	0.51	0.69	0.61	0.58
	Mean	0.71	0.68	0.73	0.70	0.78	0.74	0.36	0.36	0.35	0.34	0.40	0.37	
ρ_r	N_p	0.87	0.93	0.91	0.90	0.90	0.92	0.76	0.78	0.77	0.79	0.71	0.79	0.84
	Q_o	0.41	0.45	0.46	0.45	0.53	0.49	0.12	0.03	0.22	0.24	0.00	0.07	0.29
	G_p	0.87	0.93	0.91	0.90	0.91	0.92	0.76	0.78	0.77	0.79	0.71	0.80	0.84
	W_p	0.87	0.94	0.91	0.93	0.90	0.92	0.09	0.07	0.11	0.02	0.05	0.02	0.49
	F_w	0.43	0.41	0.51	0.40	0.57	0.50	0.11	0.05	0.20	0.10	0.05	0.17	0.29
	Recov.	0.31	0.35	0.33	0.32	0.63	0.37	0.64	0.64	0.65	0.67	0.64	0.70	0.52
	Mean	0.63	0.67	0.67	0.65	0.74	0.68	0.41	0.39	0.45	0.43	0.36	0.42	
ε_1 (%)	N_p	5.3	3.4	3.9	2.5	2.7	3.1	12.8	36.5	18.0	23.3	15.1	18.9	12.1
	Q_o	13.5	20.0	13.7	11.8	12.8	10.5	746	0.3	0.4	0.3	0.3	0.5	69.1
	G_p	4.2	4.8	3.9	2.6	2.3	2.6	13.1	38.4	18.0	24.5	15.7	18.9	12.4
	W_p	6.1	4.1	4.5	2.7	3.1	3.6	16.8	40.8	62.3	35.3	39.4	48.6	22.3
	F_w	1.4	2.1	1.4	1.3	1.3	1.1	51.3	5.8	6.2	5.7	5.9	9.8	7.8
	Recov.	5.3	3.5	3.9	2.5	2.7	3.1	15.3	37.1	18.2	23.8	15.3	19.6	12.5
	Mean	6.0	6.3	5.2	3.9	4.1	4.0	143	26.5	20.5	18.8	15.3	19.4	
ε_2 (%)	N_p	33.8	29.7	28.0	22.5	24.7	24.2	46.2	70.8	42.1	67.9	49.6	72.3	42.6
	Q_o	63.2	81.6	66.3	63.7	64.3	54.2	2460	106	226	189	109	186	306
	G_p	30.0	31.9	28.0	25.6	21.9	21.7	47.3	70.5	42.1	68.8	51.2	72.3	42.6
	W_p	33.9	32.8	28.0	22.1	24.6	24.2	84.9	88.7	80.6	60.8	75.1	114	55.8
	F_w	62.7	93.7	65.1	78.8	68.4	54.7	235	84.3	186	157	90.9	168	112
	Recov.	69.2	70.2	57.5	50.8	55.2	50.9	53.6	83.4	48.2	79.2	56.2	83.6	63.2
	Mean	48.8	56.7	45.5	43.9	43.2	38.3	488	83.9	104	104	72.1	116	

Influence of Grid Size

This influence is shown when comparing

- Tests w4 and w5 of case 2 (Table 8).
- Tests w3 and w4 of case 1 (Table 9).

Correlation coefficients are improved at final times for the finest grids. They are also improved near breakthrough (BT) for case C1. For case C2 near BT, the accuracy (ε_1) is improved, although the correlation coefficient is not.

Coning

Water coning occurs in case C2, especially for depletion. It can be better represented either with a finer grid (test w4, Table 8), or by the trick on gravity coefficient (test w5, Table 8; test d3, Table 10). The results show effectively an improvement of correlation coefficients when these options are chosen, emphasizing the importance of correctly scaling-up near wells.

The depletion case is demonstrative; if the aquifer was not present, the depletion would be completely controlled by the pore volume and well *PI*, which are correctly scaled-up by the methods used in this study. Coning, which occurs in the vicinity of wells, deteriorates dramatically the results of scaled-up models. Our present methods of K_r scaling-up have to be improved and extended. I feel that the well PVT and K_r issues touched on in the discussion about scaling-up methods are critical.

Additional Results

The scaling-up method appears to be effective for waterflood cases, whether an aquifer is present or whether injection balances production. It is possible to pick up realizations exhibiting extreme behaviors (Figures 4–7). Cumulative production data appear easier to reproduce through scaled-up models. Oil recovery is less successfully reproduced (Figure 8). As shown by error analysis, this comes from the small variability of this parameter, although recovery is obtained with the same precision as cumulative oil production. Instantaneous data as flow rates or rate ratios are less easily captured, but yearly flow rates used in NPV calculations can be satisfactorily derived from cumulative data. Depletion in the presence of an active aquifer cannot be accurately simulated with the present scaling-up methods (Figure 9), making this approach ineffective.

Table 9. Case C1 Waterflood Production Results.

	Data	2000 Days w1	w2	w3	w4	w5	500 Days w1	w2	w3	w4	w5	Mean
ρ_x	N_p	0.98	0.99	0.98	0.98	0.98	0.93	0.95	0.92	0.93	0.95	0.96
	Q_o	0.28	0.07	0.44	0.25	0.75	0.97	0.96	0.95	0.95	0.96	0.66
	G_p	0.96	0.96	0.93	0.96	0.96	0.80	0.90	0.82	0.84	0.87	0.90
	GOR	0.28	0.65	0.17	0.31	0.67	0.91	0.94	0.92	0.90	0.92	0.67
	W_p	0.68	0.74	0.66	0.84	0.88	0.73	0.24	0.17	0.50	0.16	0.56
	F_w	0.50	0.49	0.24	0.49	0.90	0.44	0.11	0.37	0.61	0.08	0.42
	Recov.	0.56	0.76	0.60	0.65	0.79	0.74	0.89	0.82	0.83	0.86	0.75
	Mean	0.61	0.67	0.57	0.64	0.85	0.79	0.71	0.71	0.80	0.68	
ρ_r	N_p	0.97	0.98	0.97	0.97	0.98	0.85	0.90	0.84	0.87	0.91	0.92
	Q_o	0.38	0.38	0.55	0.36	0.77	0.86	0.85	0.79	0.81	0.88	0.66
	G_p	0.94	0.95	0.92	0.94	0.96	0.73	0.80	0.76	0.74	0.80	0.85
	GOR	0.41	0.64	0.38	0.42	0.70	0.89	0.90	0.82	0.86	0.87	0.69
	W_p	0.77	0.78	0.75	0.86	0.90	0.77	0.57	0.49	0.60	0.46	0.70
	F_w	0.72	0.70	0.62	0.72	0.87	0.84	0.63	0.60	0.71	0.60	0.70
	Recov.	0.61	0.70	0.59	0.61	0.79	0.70	0.89	0.83	0.83	0.85	0.74
	Mean	0.69	0.73	0.68	0.70	0.85	0.81	0.79	0.73	0.77	0.77	
ε_1 (%)	N_p	9.2	5.0	6.2	6.1	4.2	7.6	5.1	6.9	6.3	6.2	6.3
	Q_o	33.5	37.6	41.6	37.4	23.1	6.3	6.4	8.5	7.2	6.7	20.8
	G_p	8.4	5.9	7.7	6.0	5.8	10.7	6.8	8.9	8.1	8.1	7.7
	GOR	27.3	20.9	30.1	24.5	18.9	7.8	5.0	5.8	6.0	6.8	15.3
	W_p	64.7	43.1	55.0	43.9	23.9	157	87.6	89.3	71.1	89.4	72.5
	F_w	38.0	31.6	38.4	30.6	15.4	102	97.3	88.4	74.5	93.4	61.0
	Recov.	9.3	5.1	6.4	6.3	4.5	9.9	6.2	8.5	7.7	7.5	7.1
	Mean	27.2	21.3	26.5	22.1	13.7	43.0	30.6	30.9	25.8	31.1	
ε_2 (%)	N_p	19.4	12.5	15.6	16.1	13.0	40.2	32.1	41.7	39.4	36.1	26.6
	Q_o	82.4	100	80.8	85.5	49.0	23.5	24.9	27.9	26.1	23.8	52.4
	G_p	24.8	19.3	26.9	21.6	19.0	56.3	38.3	48.4	47.9	44.9	34.7
	GOR	91.8	74.2	128	95.6	62.8	41.4	29.5	31.9	37.5	35.0	62.7
	W_p	56.0	55.8	58.2	45.1	39.0	139	92.4	94.5	80.0	94.6	75.4
	F_w	73.8	75.7	88.8	76.8	38.7	120	99.9	94.5	76.2	99.3	84.4
	Recov.	47.7	38.0	47.7	46.3	41.3	69.4	45.3	62.3	55.7	52.1	50.6
	Mean	56.6	53.6	63.7	55.3	37.5	69.9	51.8	57.3	51.8	55.1	

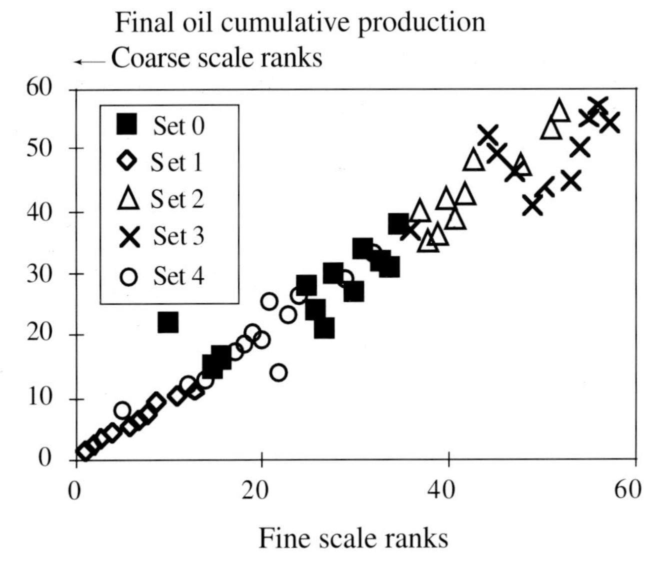

Figure 4. Ranking of final oil production, case C1, test w5.

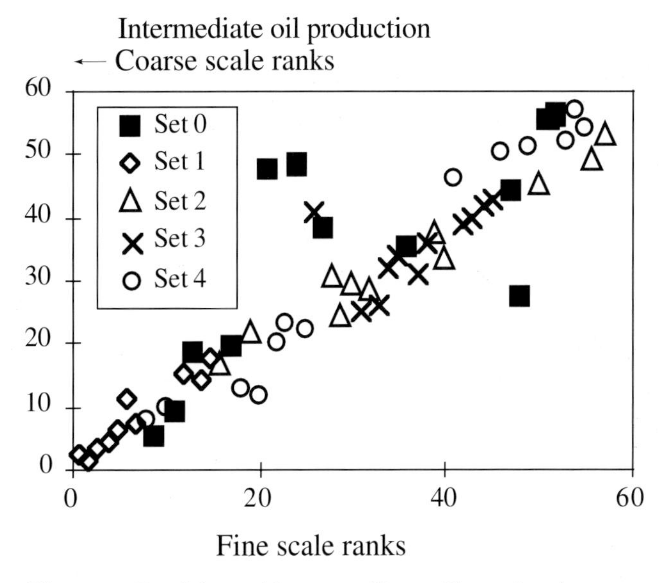

Figure 5. Ranking of intermediate oil production, case C1, test w5.

Table 10. Case C2 Depletion Production Results.

	Data	End of Production (20 yr) d1	d2	d3	Mean
ρ_x	N_p	0.57	0.22	0.57	0.45
	Q_o	0.10	0.00	0.28	0.13
	G_p	0.63	0.29	0.63	0.52
	GOR	0.09	0.00	0.12	0.07
	W_p	0.54	0.19	0.47	0.40
	F_w	0.04	0.00	0.13	0.06
	Recov.	0.37	0.02	0.36	0.25
	Mean	0.33	0.10	0.37	
ρ_r	N_p	0.56	0.23	0.55	0.45
	Q_o	0.26	0.20	0.15	0.20
	G_p	0.61	0.26	0.65	0.51
	GOR	0.23	0.40	0.20	0.27
	W_p	0.49	0.13	0.51	0.38
	F_w	0.13	0.27	0.13	0.18
	Recov.	0.38	0.09	0.34	0.27
	Mean	0.38	0.23	0.36	
ε_1 (%)	N_p	36.3	77.8	29.5	47.9
	Q_o	86.0	100	83.3	89.8
	G_p	9.0	72.3	8.6	30.0
	GOR	60.3	100	64.5	74.9
	W_p	34.5	94.2	34.0	54.2
	F_w	129	100	171	133
	Recov.	36.1	77.8	29.4	47.8
	Mean	55.8	88.9	60.1	
ε_2 (%)	N_p	103	97.5	102	101
	Q_o	98.5	100	82.5	93.7
	G_p	70.7	95.8	70.5	79.0
	GOR	97.5	100	96.6	98.0
	W_p	75.4	99.6	68.2	81.1
	F_w	103	100	113	106
	Recov.	115	102	114	110
	Mean	94.9	99.2	92.4	

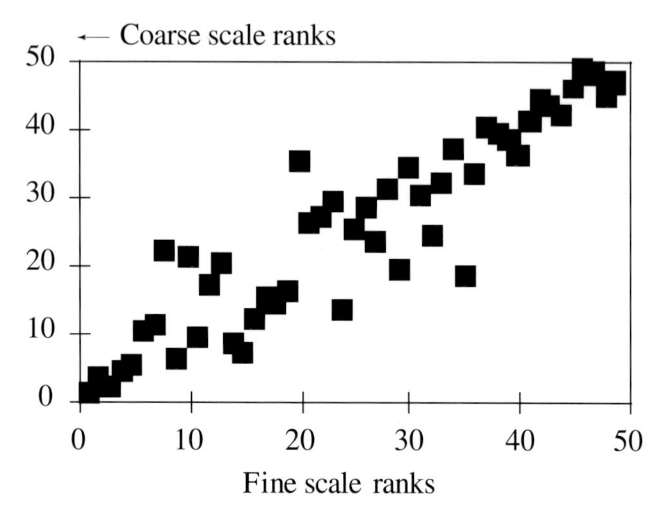

Figure 6. Ranking of final oil production, case C2, test w6.

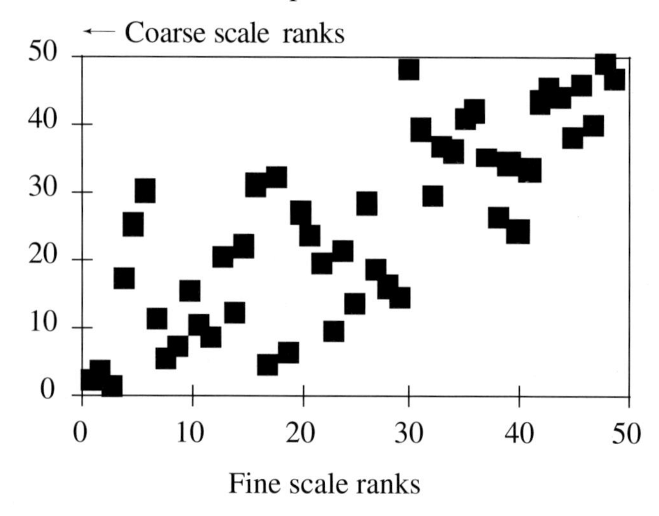

Figure 7. Ranking of intermediate oil production, case C2, test w6.

Figure 10 represents the mean final oil production computed for each geostatistical set of case 1. The mean values of fine-scale models are satisfactorily approximated by scaled-up models. In the preliminary phase of a sensitivity study, when the most important factors are not known, it is interesting to use scaled-up models to investigate the relative importance of factors, and then to use fine models in a second phase focused on the controlling factors.

Figure 11 depicts how the variation coefficients of several simulated productions parameters are approximated with scaled-up models; their relative magnitude is preserved, thereby allowing the focus to be on the uncertainty of the most variable ones.

Table 11. Comparison of Computational Times for Fine and Coarse Models.

Task	Computer Time
Fine model simulation	100
Computation of pore volume, depth, saturation	0.24
Computation of transmissivities	1.19
Well model for PI computation	0.40
Scaled-up flow simulation	0.79
Total	2.62
Ratio scaled-up model:fine model	1:38

COMPARISON OF COMPUTATIONAL COSTS

The mean performances for case C1, compared to a mean computer time of 100 for the comprehensive simulation, are shown in Table 11.

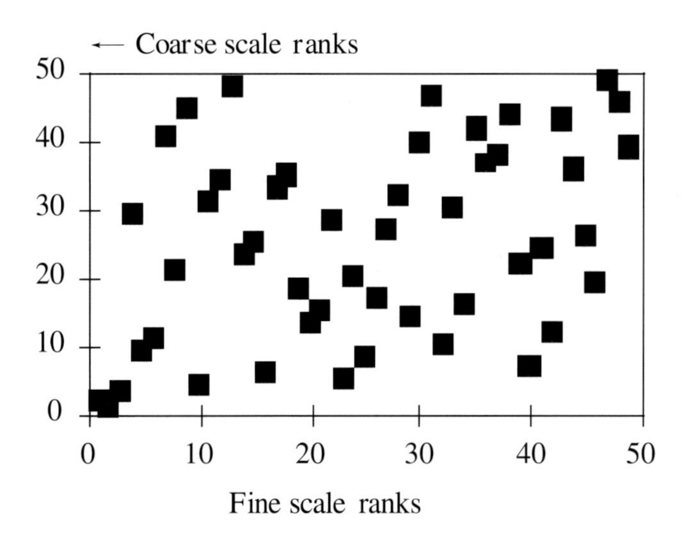

Figure 8. Ranking of final oil recovery, case C2, test w6.

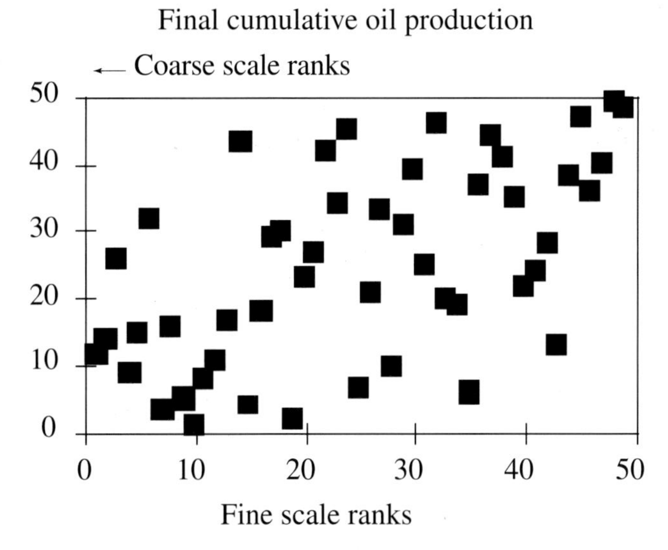

Figure 9. Ranking of final oil production, case C2, test d1.

Under some conditions (moderate permeability anisotropy and variance of the permeability field), the numerical solutions used for transmissivities and for well representations could be replaced by algebraic solutions, leading to smaller ratios of total time for scaled-up simulation to fine-scale simulation time. With the present options, smaller ratios would also be obtained for larger fine-scale models because the cost of scaling-up varies linearly with the number of coarse grid blocks, whereas the cost of simulations is of higher order.

CONCLUSIONS

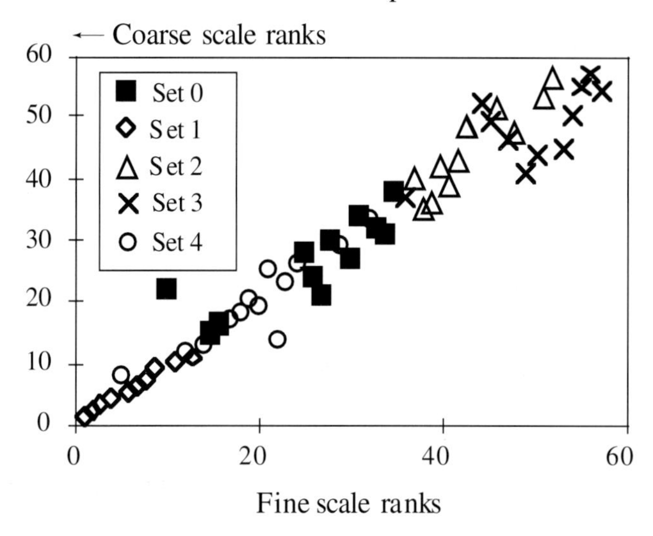

Figure 10. Comparison of experimental planning results on fine and coarse models, case 1, test w5.

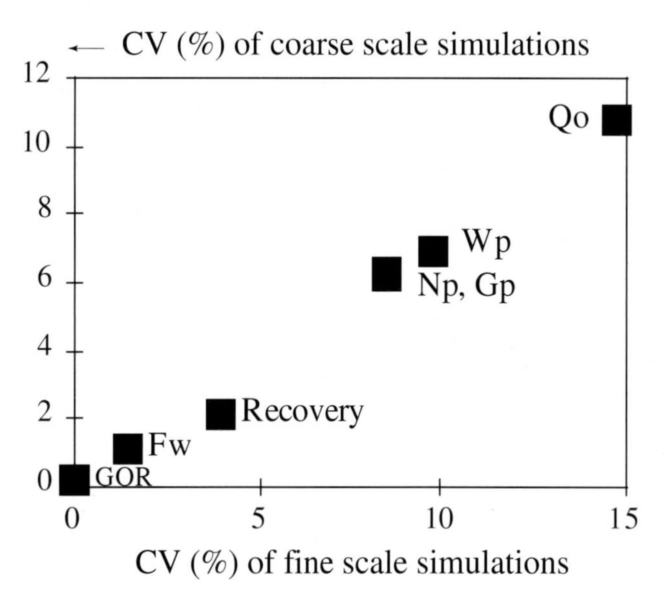

Figure 11. Comparison of coefficients of variation for fine and coarse models.

- For waterflood cases, I have successfully used scaling-up to rank multiple geostatistical realizations. Scaling-up with a large aggregation ratio is thus a practical method, which dramatically speeds up sensitivity studies, otherwise precluded by business constraints.
- When depletion is controlled by aquifer coning, the scaling-up techniques I used failed to provide a correct ranking of the realizations. Scaling-up close to the well appears to be a critical point. A practical recommendation is to use fine grid blocks around wells when scaling-up from a geological model to a flow model.

- Scaling-up of relative permeability is also critical. I used oversimplified K_r scaling-up methods for the sake of rapidity, and further progress is necessary.
- When working with a regular coarse grid, correction of numerical artifacts is necessary. A practical method consisting of computing a correction for 1-D flow and applying it in 3-D worked well. This correction depends on the respective directions of flow and grid.

NOMENCLATURE

c^t = total compressibility, $m^{-1}Lt^2$, $m^3/m^3/Pa$
K = permeability, L^2, m^2
K_r = relative permeability, adimensional
$n_1:n_2$ = aggregation ratio, n_1 fine blocks grouped into n_2 coarse blocks
P = pressure, $mL^{-1}t^{-2}$, Pa
Pc = capillary pressure, $mL^{-1}t^{-2}$, Pa
q = flow rate, L^3t^{-1}, m^3/s
S = saturation, L^3/L^3
t = time, t, s
V = volume, L^3, m^2
$\overline{X}$ = scaled-up or averaged property **X**
x = proportion, adimensional
Z = depth, L, m
ϕ = porosity, L^3/L^3
Φ = potential, $mL^{-1}t^{-2}$, Pa
μ = viscosity, $mL^{-1}t^{-1}$, Pa.s
ρ = density, mL^{-3}, kg/m^3
$\mu(X)$ = mean of X
$\sigma(X)$ = standard deviation of X

Subscripts

f = fine
i = index
o = oil
u = scaled-up
w = water

Tables

N_p = cumulative oil production, L^3, m^3
W_p = cumulative water production, L^3, m^3
G_p = cumulative gas production, L^3, m^3
Q_o = oil flow rate, L^3t^{-1}, m^3/d
F_w = water-cut, L^3/L^3
GOR = gas-oil ratio, L^3/L^3
Recov. = recovery factor, L^3/L^3
S_{wi} = irreducible water saturation, L^3/L^3
S_{or} = residual oil saturation, L^3/L^3

APPENDIX

Appendix A: Computation of Average Potential

In the transient regime of single-phase flow, the discretized equation of conservation of volume (assuming that the density of the fluid is constant) is written as follows for either fine or coarse blocks:

$$\frac{V\phi c^t\left(\Phi^{n+1}-\Phi^n\right)}{\Delta t}=\sum_{i\in I} q_i \qquad \text{(A1)}$$

where I = set of interfaces of the block.

Because the sum of all flow rates across interfaces of fine blocks is equal to the sum of flow rates across interfaces of the coarse block, you have

$$\frac{\overline{V}\overline{\phi}\overline{c^t}\left(\overline{\Phi}^{n+1}-\overline{\Phi}^n\right)}{\Delta t}=\sum_{i\in B}\frac{V_i\phi_i c_i^t\left(\Phi_i^{n+1}-\Phi_i^n\right)}{\Delta t} \qquad \text{(A2)}$$

where B = set of fine blocks.

Total volume and pore volume of the coarse block are computed thanks to simple equivalence equations:

$$\overline{V}=\sum_{i\in B} V_i \qquad \text{(A3)}$$

and

$$\overline{V}\overline{\phi}=\sum_{i\in B} V_i\phi_i \qquad \text{(A4)}$$

Equation A2 shows that averaging of compressibility and potential are not independent. One possibility is to compute average potential with pore volume weighting.

$$\overline{V}\overline{\phi}\overline{\Phi}=\sum_{i\in B} V_i\phi_i\Phi \qquad \text{(A5)}$$

It follows that average compressibility is computed according to

$$\overline{c^t}=\sum_{i\in B} V_i\phi_i\Delta\Phi_i=\sum_{i\in B} V_i\phi_i c_i^t\Delta\Phi_i \qquad \text{(A6)}$$

Equation A6 means that average compressibility varies with time, as boundary conditions on the coarse block evolve in transient regime.

A second possibility is to compute the average compressibility as follows:

$$\overline{V}\overline{\phi}\overline{c^t}=\sum_{i\in B} V_i\phi_i c_i^t \qquad \text{(A7)}$$

The average potential resulting from compressibility averaging by A7 is given by the following equation:

$$\overline{V}\overline{\phi}\overline{c^t}\overline{\Phi}=\sum_{i\in B} V_i\phi_i c_i^t\Phi_i \qquad \text{(A8)}$$

In the special case where total compressibility is constant, the average compressibility is equal to the small-scale compressibility for both choices, and the potential is simply averaged with pore volume weighting. This occurs when the total compressibility is composed of a negligible rock term and a prominent constant fluid term. For cases where small-scale compressibility varies, further work is needed to determine which formulation is the best.

Appendix B: Computation of Gravity Center Depth

From the equivalence of gravitational potential between fine and coarse scale in a single-phase case I obtain the following equation:

$$\overline{\rho}\overline{V}\overline{\phi}g\overline{Z} = \sum \rho_i V_i \phi_i g Z_i \qquad \text{(B1)}$$

From equivalence of mass you have:

$$\overline{\rho}\overline{V}\overline{\phi} = \sum \rho_i V_i \phi_i \qquad \text{(B2)}$$

And finally

$$\overline{Z} = \frac{\sum \rho_i V_i \phi_i Z_i}{\sum \rho_i V_i \phi_i} \qquad \text{(B3)}$$

When the density *r* is constant, it is a pore volume weighted average of block center depths:

$$\overline{Z} = \frac{\sum V_i \phi_i Z_i}{\sum V_i \phi_i} \qquad \text{(B4)}$$

Appendix C: Computation of Scaled-Up Drainage Capillary Pressure Curve

This is going to be done for a water-oil case. Before production starts, the reservoir is in equilibrium between capillary and gravity forces, resulting in a saturation vertical profile honoring the drainage *Pc* curve. The *Pc* in a fine block *i* may be expressed versus the small-scale *Pc* at the gravity center of the coarse block:

$$Pc_i = Pc(\overline{Z}) + \varepsilon g(\rho_w - \rho_o)(\overline{Z} - Z_i) \qquad \text{(C1)}$$

$\varepsilon = \pm 1$ depending on vertical axis orientation.

The average *Pc* in the coarse block is obtained thanks to a volume average:

$$\overline{Pc} = \frac{\sum Pc_i \phi_i V_i}{\sum \phi_i V_i} = Pc(\overline{Z}) + \varepsilon g(\rho_w - \rho_o)\frac{\sum \phi_i V_i(\overline{Z} - Z_i)}{\sum \phi_i V_i} \qquad \text{(C2)}$$

From equation B4, it follows that

$$\overline{Pc} = Pc(\overline{Z}) \qquad \text{(C3)}$$

The appropriate choice of the gravity center of the coarse block leads to this simple relation, allowing us to build up an scaled-up drainage curve for a coarse block:

(1) Choose a value for

$$\overline{Pc} = Pc(\overline{Z})$$

(2) Compute Pc_i for each fine block (C1), and the corresponding water saturation S_{iw} from the drainage *Pc* curve.
(3) Average water saturation for the coarse block:

$$\overline{S_w} = \frac{\sum V_i \phi_i S_w^i}{\sum V_i \phi_i} \qquad \text{(C4)}$$

(4) Join the new point (S_w,Pc) to the scaled-up drainage *Pc* curve, and iterate the sequence.

The fine-scale *Pc* curve is characterized by a maximum S_w for which $Pc = Pc_{min}$, and a minimum S_w for which $Pc = Pc_{max}$. This curve has to be extended outside this *Pc* range as follows:

$$Pc > Pc_{max} \Rightarrow S_w = S_{w\,min};\ Pc > Pc_{min} \Rightarrow S_w = S_{w\,max}$$

To completely define the scaled-up *Pc* curve, the value picked up in step 1 must vary in the range between

$$Pc_{min} - \varepsilon g(\rho_w - \rho_o)(\overline{Z} - z_{top})$$

and

$$Pc_{max} - \varepsilon g(\rho_w - \rho_o)(\overline{Z} - z_{bot})$$

where z_{top} and z_{bot} are the depths of the uppermost and lowermost fine blocks, respectively, inside the coarse block. This results in an scaled-up drainage *Pc* curve with a range of negative *Pc* values.

This method may be applied to homogeneous blocks to initialize their saturation according to the capillary vertical profile, and to cases of pure gravity segregation when the capillary transition zone has a negligible thickness (the drainage *Pc* curve is then defined by

$$Pc \geq 0 \Rightarrow S_w = S_{w\,min};\ Pc < 0 \Rightarrow S_w = S_{w\,max}).$$

ACKNOWLEDGMENTS

The author would like to thank the managements of

the Institut Français du Pétrole and Elf for permission to publish this work performed within the Elf/IFP Helios Joint Research Group.

REFERENCES CITED

Ballin, P. R., A. G. Journel, and K. Aziz, 1992, Prediction of uncertainty in reservoir performance forecast, Journal of Canadian Petroleum Technology, v. 31, no. 4, p. 52–62.

Ballin, P. R., K. Aziz, and A. G. Journel, 1993, Quantifying the impact of geological uncertainty on reservoir performing forecasts, Paper SPE 25238 in Proceedings of the 12th SPE Symposium on Reservoir Simulation held in New Orleans, LA, USA, p. 47–57.

Buckley, S. E., and M. C. Leverett, 1942, Mechanism of fluid displacement in sands, in Transactions of SPE of AIME, v. 146, p. 107–116.

Deutsch, C. V., and A. G. Journel, 1992, GSLIB, Geostatistical software library and user's guide, Oxford University Press, editor, p. 167–170.

Deutsch, C., and S. Srinivasan, 1996, Ranking stochastic reservoir models, Paper SPE/DOE 35411 in Proceedings of the 10th Symposium on Improved Oil Recovery, Tulsa, OK, USA, v. 2, p. 105–113.

Ding Y., 1995, Scaling-up in the vicinity of wells in heterogeneous fields, paper SPE 29137, in Proceedings of 13th SPE Symposium on Reservoir Simulation, San Antonio, p. 441–451.

Gomez-Hernandez, J. J., and A. G. Journel, 1990, Stochastic characterization of grid-block permeabilities: from point values to block tensors, in D. Guérillot and O. Guillon, eds., Second European Conference on the Mathematics of Oil Recovery: Paris, Editions Technip, p. 83–90.

Guérillot, D. R., and I. F. Morelon, 1992, Sorting equiprobable geostatistical images by simplified flow calculations, Paper SPE 24891 in Proceedings of the 67th Annual Technical Conference and Exhibition of the SPE held in Washington, D.C., USA, v. Σ, p. 327–342.

Kossak, C. A., J. O. Aasen, and S. T. Opdal, 1989, Scaling-up laboratory relative permeabilities and rock heterogeneities with pseudo functions for field simulations, Paper SPE 18436 in proceedings of SPE symposium on Reservoir Simulation held in Houston, TX, USA, p. 367–390.

Kyte, J. R., and D. W. Berry, 1975, New Pseudo Functions to Control Numerical Dispersion, in SPE Journal (August), p. 269–275.

Lantz, R. B., 1971, Quantitative evaluation of numerical diffusion (Truncation error), in Transactions of SPE of AIME, v. 251-II, p. 315–320.

Li, D., A. S. Cullick, and L. W. Lake, 1995, Scale-up of reservoir models' relative permeability using a global method, Paper SPE 29872 in Proceedings of 9th Middle East Oil Show and Conference, Bahrain. v. 2, p. 341–354.

Peaceman, D. W., 1978, Interpretation of well-block pressures in numerical reservoir simulation, in Transactions of SPE of AIME, v. 265, p. 183–194.

Romeu R. K., and B. Noetinger, 1995, Calculation of internodal transmissibilities in finite-difference models of flow in heterogeneous media, Paper 94WR02422, Water Resources Research, v. 31, no. 4, p. 943–959.

Saad, N., V. Maroongroge, and C. T. Kalkomey, 1996, Ranking geostatistical models using tracer production data: Paper SPE 35494 in Proceedings of European 3-D Reservoir Modelling Conference, Stavanger, Norway, p. 131–142.

Thiele, M. R., R. P. Baticky, M. J. Blunt, and F. M. Orr, 1996, Simulating flow in heterogeneous systems using streamtubes and streamlines, SPE Reservoir Engineering, v. 11, no. 1, p. 5–12.

Urgelli, D., Y. Ding, An efficient permeability scaling-up technique applied to discretized flow equations, 1999, *in* R. Schatzinger and J. Jordan, eds., Reservoir Characterization-Recent Advances, AAPG Memoir 71, p. 293–304.

Chapter 20

An Efficient Permeability Scaling-Up Technique Applied to Discretized Flow Equations

Denise Urgelli
Yu Ding
Institut Français du Pétrole
Rueil Malmaison, France

ABSTRACT

Grid-block permeability scaling-up for numerical reservoir simulations has been discussed for a long time in the literature. It is now recognized that a full permeability tensor is needed to get an accurate reservoir description at large scale; however, two major difficulties are encountered: (1) grid-block permeability cannot be properly defined because it depends on boundary conditions and (2) discretization of flow equations with a full permeability tensor is not straightforward, and little work has been done on this subject.

In this paper, we propose a new method that allows us to get around both difficulties. As the two major problems are closely related, a global approach will preserve the accuracy. In the proposed method, the permeability up-scaling technique is integrated in the discretized numerical scheme for flow simulation. The permeability is scaled-up via the transmissibility term, in accordance with the fluid flow calculation in the numerical scheme. A finite-volume scheme is particularly studied, and the transmissibility scaling-up technique for this scheme is presented.

Two numerical examples are tested for flow simulation. This new method is compared with two published numerical schemes for full permeability tensor discretization where the full permeability tensor is scaled-up through various techniques. Comparing the results with fine grid simulations shows that the new method is more accurate and more efficient.

INTRODUCTION

Reservoir heterogeneity can be described using geostatistical models, but these models generate the heterogeneity on a million fine grid blocks, which leads to prohibitive computational costs for reservoir simulations. In order to reduce the number of grid blocks, averaging techniques are needed to scale-up the fine-scale permeabilities to the larger scales appropriate for flow simulations and engineering calculations.

Generally, when the principal flow direction is not parallel to the grid-block axes, the effective permeability must be represented as a full tensor to capture the flow properties of the system; however, this scaling-up procedure leads to two difficulties: (1) grid-block permeability cannot be properly defined because it depends on boundary conditions and (2) discretization of flow equations with a full permeability tensor is not straightforward. One of the solutions is using a global approach to improve the accuracy.

Nowadays, more and more geoscientists recommend the use of internodal permeabilities calculated at the center of the interface and not at the center of the block. This calculation is equivalent to the calculation of transmissibilities, once the numerical scheme is fixed. Scaling-up of the transmissibility is more accurate than scaling-up of the permeability because the transmissibility is calculated directly with the fine-scale permeability; furthermore, reservoir simulators are incorporating the transmissibilities from each block to each of its neighbors as input.

The purpose of this paper is to present a new procedure for computing the equivalent transmissibility in the discretized flow equations. An application of this technique to a finite-volume–type numerical scheme is detailed, and we also give a non-exhaustive review of scaling-up of full permeability tensor and discuss the different methods for the computation of transmissibilities. Then, we present a new procedure to scale-up the transmissibility for a finite-volume scheme. Finally, we compare the results of the transmissibility scaling-up and the permeability scaling-up by two examples.

REVIEW OF SCALING-UP TECHNIQUE FOR SINGLE-PHASE FLOW

Permeability Scaling-Up

The determination of the effective permeability in grid blocks has been discussed for a long time in the literature (Warren and Price, 1961). It is now recognized that a full-block permeability tensor (Bamberger, 1977; Guerillot, 1988; Guerillot et al., 1989; Samier, 1990, Durlofsky, 1991; Wang et al., 1992; Gautier and Noetinger, 1994; Pickup and Sorbie, 1994a, b;) is needed to get an accurate reservoir description. In this section, several methods currently used are reviewed, and the advantages and disadvantages of each are analyzed.

Bamberger (1977) proposed to define the equivalent block permeability using an energy criterion. The equivalent permeability is determined using a bilinear form of the variational formula associated to the elliptic operator $A = -div[k\nabla]$ and the bilinear form associated to the elliptic operator $\tilde{A} = -div[K_{eq}\nabla]$. This method is based on the fact that the work realized by the viscous force in the heterogeneous media and in the equivalent homogeneous media must be equal. The energy criterion presents an immediate physical sense. This method is also studied by Njifenjou (1993, 1994). This kind of method can be applied to complex geometries.

Another numerical approach to obtain a full tensor consists of using linear boundary conditions (Guerillot, 1988; Guerillot et al., 1989). A pressure gradient is imposed in the flow direction and a linear pressure in the two other opposite faces. This method allows us to construct a nonsymmetrical permeability tensor. This type of condition is equivalent to putting the block in a uniform large-scale flow. This method takes into account the cross-flow term, but it may lead to overestimating the grid-block permeability in some cases (Galli et al., 1996).

The most successful approach to the computation of equivalent permeability tensor consists of using a periodic boundary condition (Durlofsky, 1991). The permeability tensor is determined using a pressure gradient in the x-direction to obtain K_{eq}^{xy} and K_{eq}^{xy} and a pressure gradient in the y-direction to obtain K_{eq}^{yx} and K_{eq}^{yy}. This method is exact for infinitely repeated permeability structures. These boundary conditions give a symmetric tensor. Periodic boundary conditions were also adopted by Pickup and Sorbie (1994a, b) for their studies, which demonstrate that the periodic boundary conditions are the most robust.

The periodic boundary condition was also applied by Gautier and Noetinger (1994) to the renormalization scaling-up technique (tensorial renormalization). The renormalization method was introduced by King (1989). The scaled-up permeability is calculated by a series of successive aggregations. King computed the permeability of an elementary cell of four meshes (in two dimensions) with a finite difference approximation and with standard no-flow boundary conditions. This method can be improved by the simplified renormalization where the general idea is to group the cells only two by two and to calculate the permeability of the cell pairs by applying the harmonic mean if the cells are in series, or the arithmetic mean if the cells are in parallel (Le Loc'h, 1987; Kruel Romeu, 1994); however, the effective permeability is diagonal. If the nondiagonal terms are necessary, it is important to use the tensorial renormalization.

Transmissibility Scaling-Up

The scaled-up block permeability will be integrated in the transmissibility terms in a numerical scheme for flow simulation on coarse grid blocks; however, it is difficult to determine accurately the transmissibility from the block permeability. Usually, the harmonic average of the block permeability is used for the transmissibility calculation (Kruel Romeu and Noetinger, 1995), but using the harmonic average might cause a loss of precision for 2-D (two-dimensional) problems. So, several authors proposed to directly scale-up the transmissibility from the fine grid heterogeneity (White and Horne, 1987; Palagi, 1992; Palagi et al., 1993; Kruel Romeu and Noetinger, 1995; Peaceman, 1996).

White and Horne (1987) seem to be the first authors who considered the scaling-up of transmissibilities between grid blocks rather than the equivalent permeabilities of a grid block. They scaled-up the transmissibility on a global domain, which included all the fine grid blocks. Several boundary conditions are imposed on the whole domain to evaluate the transmissibility tensor between the connected blocks. The well conditions are particularly considered as one of the boundary conditions. Then the pressures and fluxes from fine scale simulations are averaged and summed to obtain pressures and fluxes on a coarse scale. Using these pressures and fluxes on a coarse scale and the least square method, they determined macro-transmissibilities. Because the whole simulation domain is considered with different boundary conditions, the linear system will be solved several times on fine grid scale with all grid blocks, which might be very expensive in terms of computing time.

Palagi (1992) and Palagi et al. (1993) presented a scaling-up procedure for the permeability at the interface of Voronoi grid blocks. They used the power law average to calculate the homogenized permeability. The optimum value for the power law coefficient ω is obtained by the minimum between the fine grid results and coarse grid results for various value of ω. This requires solving several times the coarse grid simulation, which might be very much time consuming.

Kruel Romeu and Noetinger (1995) presented a numerical method with impermeable boundary conditions on rectangular grid blocks. The impermeable conditions consist of imposing a pressure drop between two opposite faces and a no-flow boundary on the remaining two faces. The Kruel Romeu and Noetinger's method is equivalent to the calculation of the permeability in a block placed between the two blocks where the transmissibility is calculated. This method was applied by Lemouzy et al. (1995) in a field study.

More recently, Peaceman (1996) proposed a transmissibility scaling-up method on corner point geometry. For each coarse grid block, six half-block transmissibilities are scaled-up. In a first step, Peaceman represented each half-block as a rectangular block. To calculate the transmissibility on this rectangular block, he proposed a direct method that consists of using the impermeable boundary conditions to calculate the flow through the block by solving sets of different equations for the pressure. Then he compared this direct method with the renormalization. He found that the renormalization is less accurate and less easy to implement than the direct method.

It is clear that using transmissibility scaling-up can give more accurate results than using permeability scaling-up; however, White and Horne's (1987) method needs to consider a global domain, which is computer time consuming. The other methods, which calculate the transmissibility locally, are limited to a finite-difference type numerical scheme and do not fully incorporate the full tensor property. In this paper, we present a transmissibility scaling-up technique related only to the local grid block for a finite-volume type method with a full tensor discretization.

SCALING-UP OF TRANSMISSIBILITY FOR A FINITE-VOLUME SCHEME

Finite-Volume Scheme

Consider a two-phase flow problem in 2-D:

$$\frac{\partial}{\partial t}\left(\frac{\Phi S_l}{B_l}\right) = \nabla(\lambda_l k \nabla p) - Q_l$$

where λ = ω or o indicates the phase, λ is the mobility, and k is a full permeability tensor given by

$$k = \begin{pmatrix} k^{xx} & k^{xy} \\ k^{yx} & k^{yy} \end{pmatrix}$$

One of the difficulties for the numerical method is the discretization of the elliptic operator $\nabla(\lambda k \nabla p)$ with a full permeability tensor. We present a finite-volume–type method for the discretization of this operator. Integrating on a considered grid block (i,j) and applying the Gauss divergence theorem (Figure 1), we have

$$-\int_\Omega \nabla(\lambda k \nabla p) d\Omega = \sum_{m=1}^{4} \int_{\Gamma_m} \lambda k \nabla \vec{p}.\vec{n}.d\gamma = \sum_{m=1}^{4} Fm$$

Without loss of generality, we will study the fluid-flow calculation through the interface Γ_1:

$$F_1 = \int_{\Gamma_1} \lambda k \nabla \vec{p}.\vec{n}.d\gamma =$$

$$\lambda_{i+1/2,j} \int_{Y_1}^{Y_2} \left(k^{xx}\frac{\partial p}{\partial x} + k^{xy}\frac{\partial p}{\partial y} \right) dy =$$

$$\lambda_{i+1/2,j}\left(F_{1a} + F_{1b}\right)$$

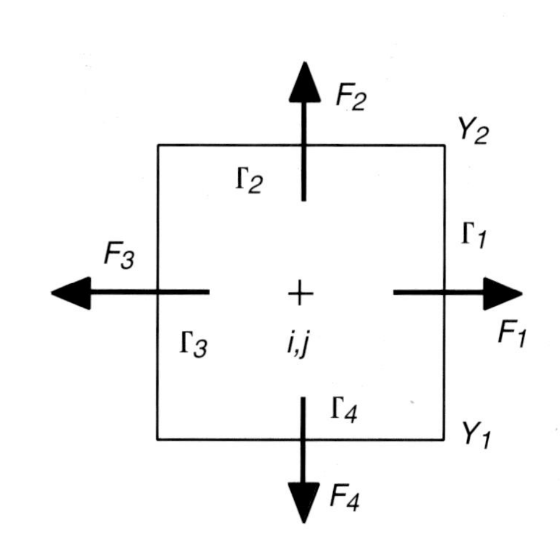

Figure 1. Finite-volume scheme (on coarse grid block).

where

$$F_{1a} = -\int_{Y_1}^{Y_2} k^{xx} \frac{\partial p}{\partial x} d$$

represents the fluid flow through the interface Γ_1 by a force in x-direction and

$$F_{1b} = \int_{Y_1}^{Y_2} k^{xy} \frac{\partial p}{\partial y} dy$$

represents the fluid flow through this interface by a force in y-direction. The last representation is also called the cross flow.

For instance, for an uniform grid with $\Delta x = Cte$ and $\Delta y = Cte$ for the sake of simplicity, these two terms can be approximated by

$$F_{1a} = -\frac{\Delta y}{\Delta x} k^{ax}_{i+1/2,j}\left(p_{i+1,j} - p_{i,j}\right)$$

and

$$F_{1b} = \frac{-1}{4} k^{xy}_{i+1/2,j}\left(p_{i,j+1} + p_{i+1,j+1} - p_{i,j-1} - p_{i+1,j-1}\right)$$

Finally, the fluid flow through the interface Γ_1 is calculated by

$$F_1 = -\lambda_{i+1/2,j}\left[T^{xx}_{i+1/2,j}\left(p_{i+1,j} - p_{i,j}\right) + \frac{1}{4} T^{xy}_{i+1/2,j}\left(p_{i,j+1} + p_{i+1,j+1} - p_{i,j-1} - p_{i+1,j-1}\right)\right]$$

From this discretization, we observe that

$$T^{xx}_{i+1/2,j} = \frac{\Delta y}{\Delta x} k^{xx}_{i+1/2,j}$$

represents the transmissibility (permeability) between the blocks (i,j) and $(i+1,j)$ and

$$T^{xy}_{i+1/2,j} = k^{xy}_{i+1/2,j}$$

represents the cross-transmissibility (permeability) between these two blocks; however, one of the difficulties is the determination of the transmissibility (permeability) at the interface, if the permeability is known on the grid block. In homogeneous media, these terms are well defined; however, in heterogeneous media, the interface permeability $K^{xx}_{i+1/2,j}$ or $K^{xy}_{i+1/2,j}$ is usually unknown. To get an accurate transmissibility formula, one of the possibilities is to calculate these terms directly from the scaling-up procedure.

For the fluid flow at the interface Γ_2, it can be calculated by

$$F_2 = -\lambda_{i,j+1/2}\left[T^{yy}_{i,j+1/2}\left(p_{i,j+1} - p_{i,j}\right) + \frac{1}{4} T^{yx}_{i,j+1/2}\left(p_{i+1,j} + p_{i+1,j+1} - p_{i-1,j+1} - p_{i-1,j}\right)\right]$$

where $T^{yy}_{i,j+1/2}$ represents the transmissibility between the blocks (i,j) and $(i,j+1)$, and $T^{yx}_{i,j+1/2}$ represents the cross transmissibility between these two blocks.

A complete discretization formula is given in Appendix 1.

Transmissibility Scaling-Up on a Shifted Block

A transmissibility scaling-up method was proposed by White and Horne (1987) for this kind of scheme; however, their method needs to solve several linear systems in the whole domain with millions of fine grid blocks, so it is very time consuming. In this section, we propose to scale-up the transmissibility that depends only on the local heterogeneity. The principle of the proposed technique is similar to the calculation of a full-block permeability tensor, but the considered block is not the coarse grid block. The block is shifted. To calculate the transmissibility related to the flow in the x-direction, a shifted block is used, as shown in Figure 2a. With a given scaling-up procedure, a permeability tensor

$$k_{i+1/2,j} = \begin{pmatrix} k^{xx}_{i+1/2,j} & k^{xy}_{i+1/2,j} \\ k^{yx}_{i+1/2,j} & k^{yy}_{i+1/2,j} \end{pmatrix}$$

can be determined on this block, where $K^{xx}_{i+1/2,j}$ represents the permeability on the x-direction, and $K^{xy}_{i+1/2,j}$ represents the cross-permeability on this direction. From this formula, we can determine the transmissibility terms for the flow in the x-direction for the finite-volume scheme with

$$T^{xx}_{i+1/2,j} = \frac{\Delta y}{\Delta x} k^{xx}_{i+1/2,j}$$

and

$$T^{xy}_{i+1/2,j} = k^{xy}_{i+1/2,j}$$

In the same way, we can determine the transmissibility terms related to the flow in the y-direction. Calculating the permeability tensor

$$k_{i,j+1/2} = \begin{pmatrix} k^{xx}_{i,j+1/2} & k^{xy}_{i,j+1/2} \\ k^{yx}_{i,j+1/2} & k^{yy}_{i,j+1/2} \end{pmatrix}$$

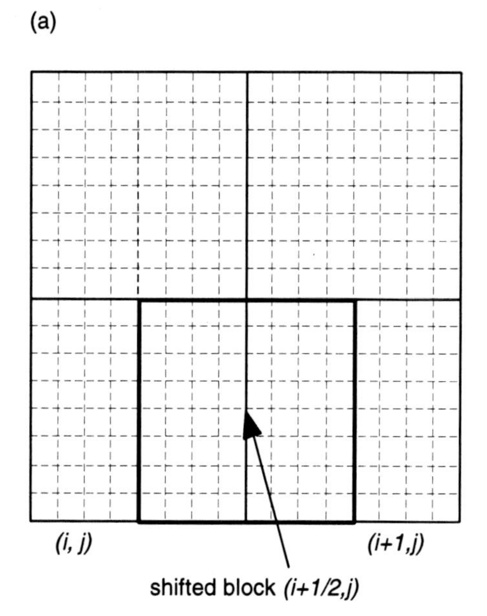

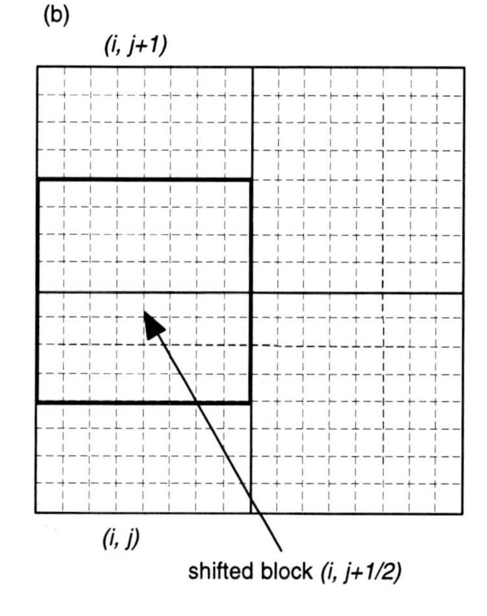

Figure 2. Transmissibility scaling-up on shifted blocks. (a) Scaling-up in the *x*-direction. (b) Scaling-up in the *y*-direction.

on the shifted block in the y-direction (Figure 2b), the transmissibility terms are

$$T^{yy}_{i,j+1/2} = \frac{\Delta x}{\Delta y} k^{yy}_{i,j+1/2}$$

and

$$T^{yx}_{i,j+1/2} = k^{yx}_{i,j+1/2}$$

In the examples presented below for the transmissibility scaling-up procedure, we use the periodic boundary conditions that were successfully used for block permeability scaling-up. In fact, on the shifted block, which can be refined by the user for a good affectation of fine scale heterogeneities, a pressure gradient is imposed on the faces perpendicular to the flow direction, and the flow rates in the same direction are equal. On the other faces, pressures and flow rates on opposite boundaries are equal. This can be expressed as follows (Figure 1) for pressure gradient in x-direction:

$$\begin{cases} p(x,y=0) = p(x,y=L) \text{ on } \Gamma_2 \text{ and } \Gamma_4 \\ q(x,y=0) = p(x,y=L) \text{ on } \Gamma_2 \text{ and } \Gamma_4 \\ p(x=0,y) = p(x=L,y) - Cte \text{ on } \Gamma_3 \text{ and } \Gamma_1 \\ q(x=0,y) = q(x=L,y) \text{ on } \Gamma_3 \text{ and } \Gamma_1 \end{cases}$$

The pressure and, consequently, the flow rate on each refined block are calculated using a 5-point scheme and an iterative solver. Imposing a pressure gradient Δp_x in the x-direction for a shifted block as shown in Figure 2a, the transmissibility (permeability) $T^{xx}_{i,j+1/2}$; can be determined by

$$k^{xx}_{i+1/2,j} = \frac{1}{Nx_r} \sum_{i_1=1}^{Nx_r} \frac{Nx_r \Delta x_r \sum_{j_1=1}^{Ny_r} q^{xx}_{i_1,j_1}}{\Delta p_x Ny_r \Delta y_r} = \frac{\Delta x_r \sum_{i_1=1}^{Nx_r} \sum_{j_1=1}^{Ny_r} q^{xx}_{i_1,j_1}}{\Delta p_x Ny_r \Delta y_r}$$

Imposing a pressure gradient Δp_y in the y-direction for the same block, we can determine the cross-transmissivity (permeability) term $T^{xy}_{i+1/2,j}$:

$$k^{xy}_{i+1/2,j} = \frac{1}{Nx_r} \sum_{i_1=1}^{Nx_r} \frac{\sum_{j_1=1}^{Ny_r} q^{xx}_{i_1,j_1}}{\Delta p_y} = \frac{\sum_{i_1=1}^{Nx_r} \sum_{j_1=1}^{Ny_r} q^{xx}_{i_1,j_1}}{Nx_r \Delta p_y}$$

where Nx_r, Ny_r = number of the refined blocks in the x-, and y-directions, respectively; Δx_r, Δy_r = size of the refined blocks in the x-, and y-directions, respectively.

To calculate $K^{yy}_{i,j+1/2}$, and $K^{yx}_{i,j+1/2}$, we use the same procedure for the shifted block as shown in Figure 2b.

It must be mentioned that the scaled-up transmissibility might strongly depend on the boundary condition as in the case of scaling-up the block permeability; however, the transmissibility scaling-up can be considered as an improvement with respect to the block permeability scaling-up because the transmissibility will not be determined again from the scaled-up block permeability. Moreover, we think that the effect of boundary condition might be less sensitive at the field scale, and the flow simulation at the reservoir scale might be approximated by considering two kinds of flow pattern, as discussed by Ding (1995).

NUMERICAL RESULTS

Two numerical examples are presented to compare this new transmissibility scaling-up method and two published permeability scaling-up methods. For the transmissibility scaling-up procedure, we use the periodic boundary condition (Durlofsky, 1991). For permeability scaling-up, we use the periodic boundary condition (Durlofsky, 1991) and the optimized linear boundary condition (Bamberger, 1977). For the permeability scaling-up, the transmissibilities are calculated using a harmonic average for block permeability, as presented in Appendix 1.

The general data are given in Table 1. To validate the proposed method, the flow is simulated using the fine-scale isotropic permeability distribution. The fine-scale simulation results are assumed to be the reference solution.

Example 1

The permeability distribution is given in Figure 3. This heterogeneity field is well known and was studied by several authors (Samier 1990; White and Horne, 1987; Wang et al., 1992). They showed the necessity of using a full permeability tensor. Here, we will compare the transmissibility scaling-up method and some permeability scaling-up methods.

Two wells, one injector and one producer, are respectively placed at the left-bottom corner and at the right-top corner of the mesh (see Figure 3). Water is injected into the oil reservoir with a fixed flow rate. The water-cut profiles are presented in Figure 4, and the saturation distribution maps are presented in Figure 5. All the coarse grid simulation results are very similar; however, we can observe that scaling-up of transmissibility, which uses periodic boundary conditions, gives slightly better results than the scaling-up of permeability. In this case, the improvement using transmissibility scaling-up is not significant; however, it proves that scaling-up of transmissibility is a good initiative.

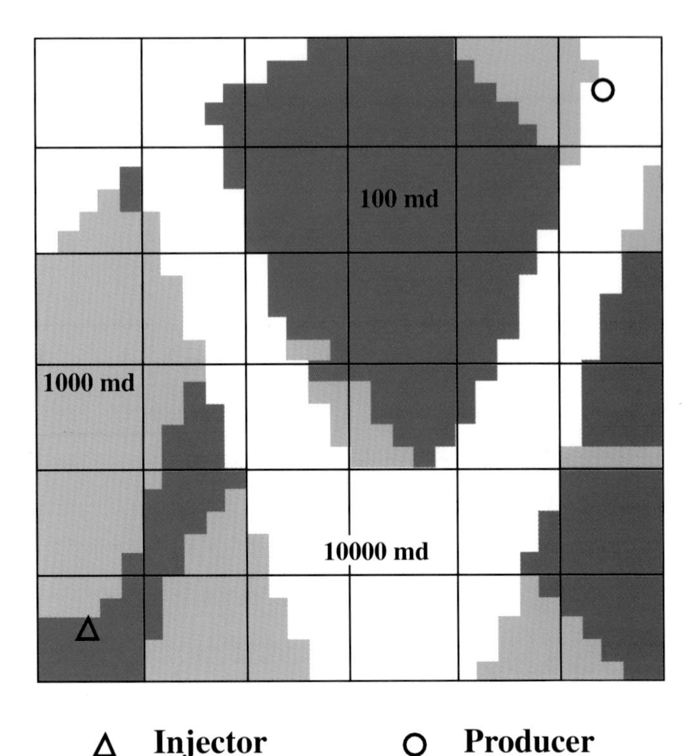

Figure 3. Heterogeneous field in Example 1.

Table 1. Simulation Input Data.

Parameter	Input Data (S.I. unit)
Injection rate	0.00081 m^3/s
Production well pressure	7×10^6 Pa
Oil viscosity	0.001 Pa.s
Water viscosity	0.001 Pa.s
Initial/irreducible water saturation	0.2
Residual oil saturation	0.2
Porosity	0.2
Thickness	1 m
Areal dimensions	300×300 m for example 1 990×990 m for example 2
Initial pressure	85×10^5 Pa

Example 2

A field with parallel barriers of permeability is used as shown in Figure 6 with a strong permeability contrast. The fine grid has 99×99 grid blocks, and the coarse grid has 11×11 grid blocks. Water is injected into an oil reservoir. In this example, we have one injector and three producers. For the fine grid simulation, it is clear that the #3 producer produces the most because it is connected to the injector by a permeable layer, and the #2 producer produces less because it is separated by several "impermeable" zones. For coarse grid simulation, if the permeability is scaled-up on grid blocks, the phenomenon resulting from anisotropy might be homogenized in the transmissibility calculation. Figure 7 shows the water-cut. Even though using the transmissibility scaling-up gives the results far from the fine grid simulation, they are much better than using the permeability scaling-up method. Figure 8 and Figure 9 represent the saturation distribution and the pressure distribution, respectively, for the fine grid and the coarse grid simulations with different scaling-up methods. It is clear that using transmissibility scaling-up gives the best result, which represents the same trend as fine grid simulation. The improvement of using transmissibility scaling-up is significant.

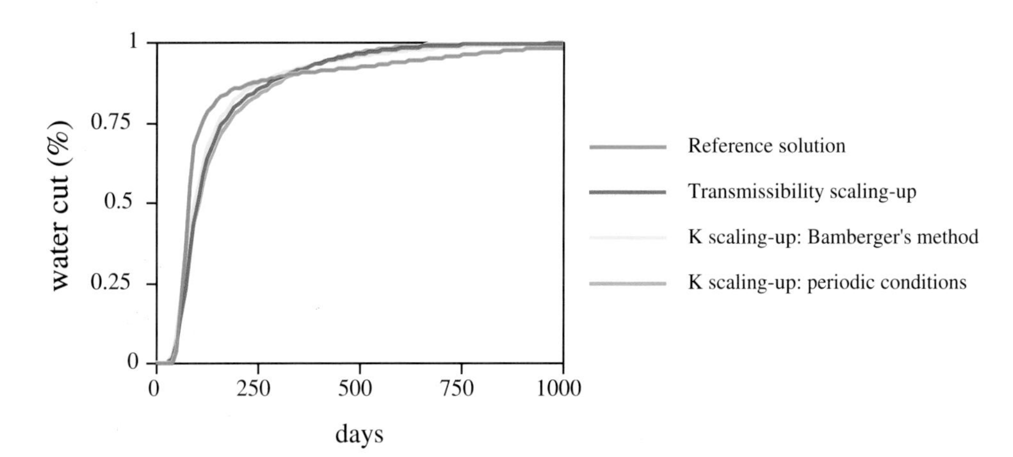

Figure 4. Water cut for production well in Example 1. Bamberger's method from Bamberger (1977).

CONCLUSION

The scaling-up of transmissibility and permeability is discussed. Scaling-up of transmissibility is more accurate than scaling-up of permeability. A new transmissibility scaling-up procedure, which uses shifted blocks to calculate locally the transmissibility, is proposed for a finite-volume type numerical scheme with a full permeability tensor.

Some numerical examples are presented. The results show that using the transmissibility scaling-up improves the coarse grid simulation accuracy. In some cases, like a stratified field, this improvement is significant.

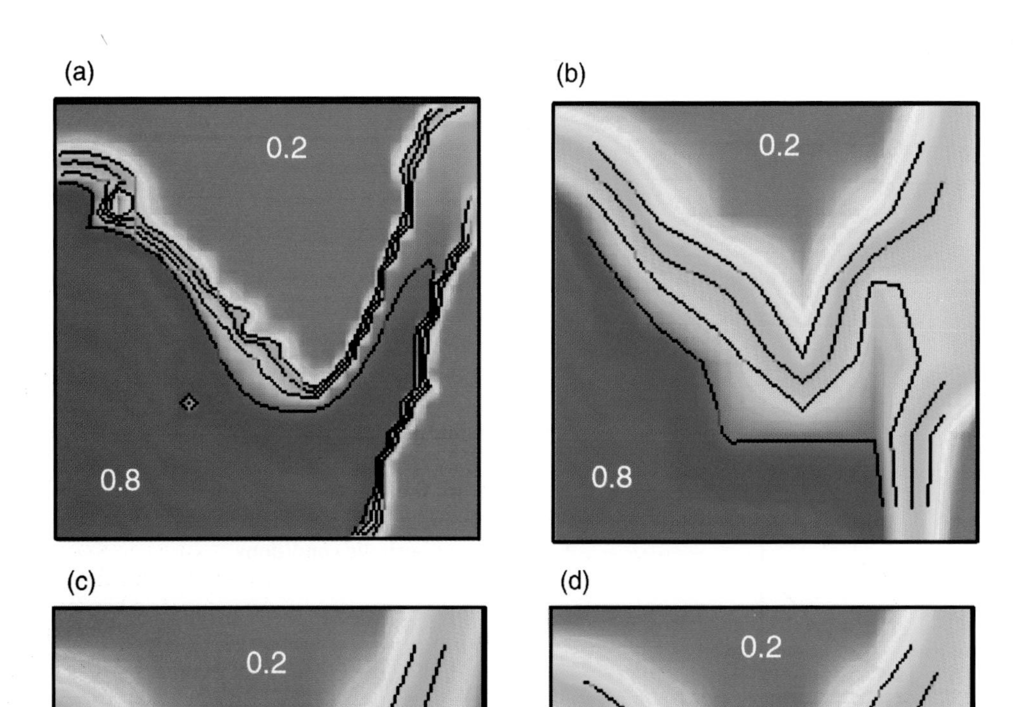

Figure 5. Water saturation distribution for Example 1. (a) Fine grid simulation. (b) Coarse grid simulation with transmissibility scaling-up. (c) Coarse grid simulation with permeability scaling-up using Bamberger's (1977) method. (d) Coarse grid simulation with permeability scaling-up using periodic boundary conditions.

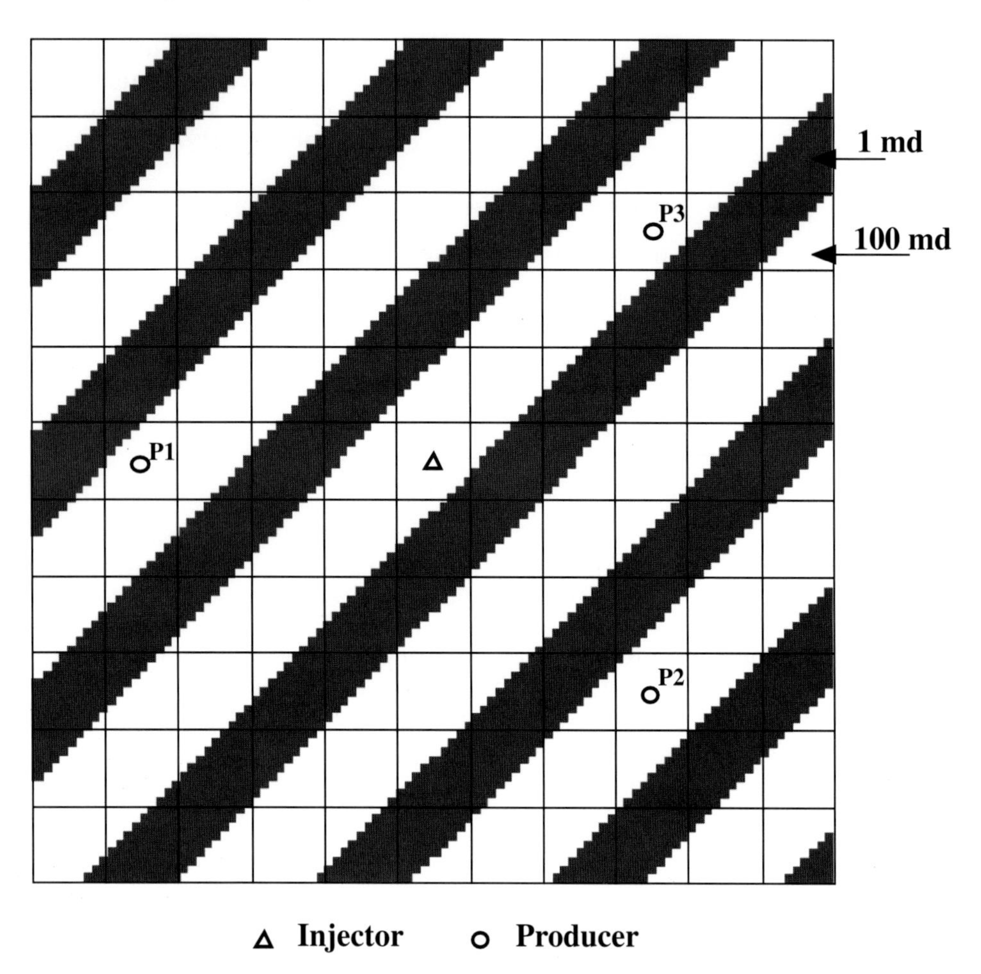

Figure 6. Heterogeneous field in Example 2.

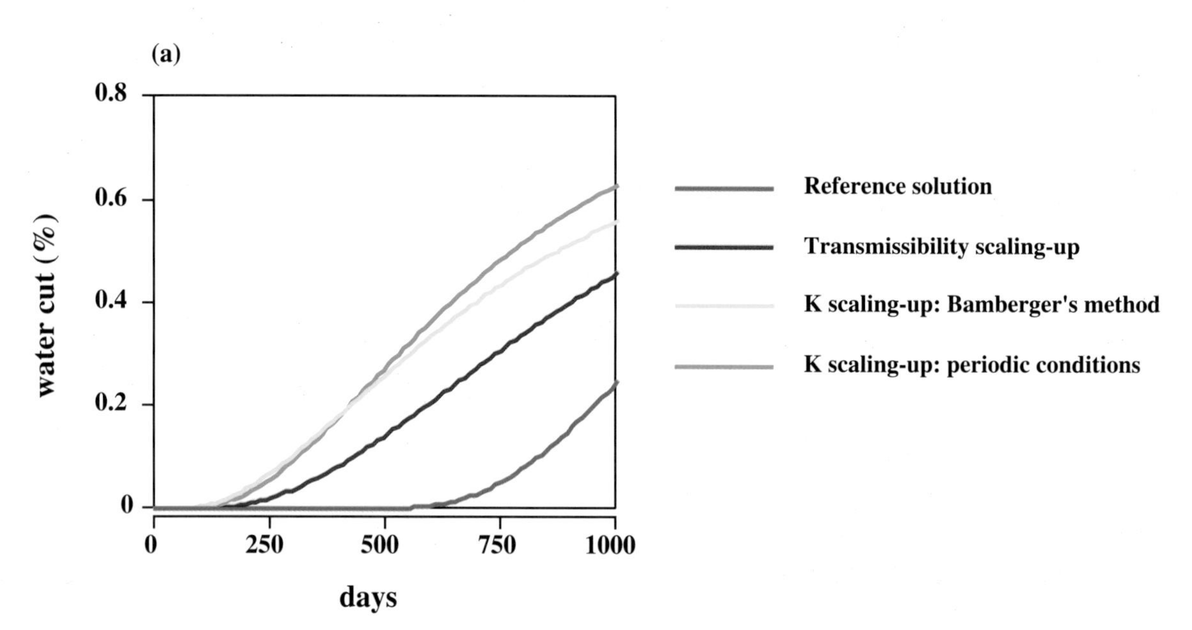

Figure 7. Water cut curves in Example 2. (a) Water cut for the production well P1. Bamberger's method from Bamberger (1977). (b) Water cut for the production well P2. (c) Water cut for the production well P3.

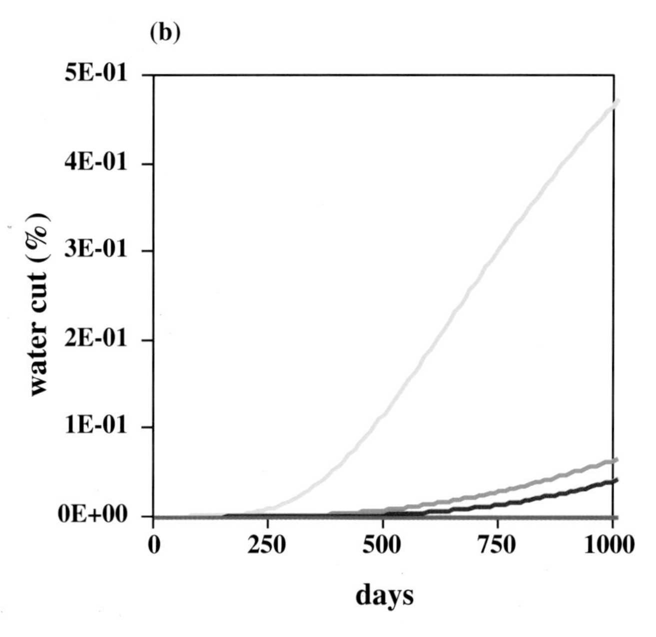

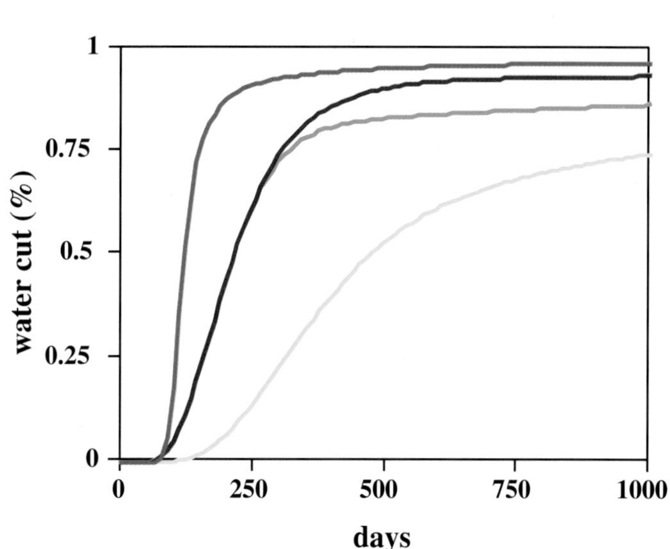

Figure 7 (continued).

APPENDIX 1

Discretization of Elliptic Operator with a Full Permeability Tensor

According to the discussion in the section on the finite-volume scheme, the elliptic operator can be discretized as

$$\begin{aligned}\Delta x\Delta y div(k\nabla p) = {} & \lambda_{i+1/2,j}\Big[T^{xx}_{i+1/2,j}\left(p_{i+1,j}-p_{i,j}\right)+\\ & \frac{1}{4}T^{xy}_{i+1/2,j}\left(p_{i,j+1}+p_{i+1,j+1}-p_{i,j-1}-p_{i+1,j-1}\right)\Big]+\\ & \lambda_{i,j+1/2}\Big[T^{yy}_{i,j+1/2}\left(p_{i,j+1}-p_{i,j}\right)+\\ & \frac{1}{4}T^{yx}_{i,j+1/2}\left(p_{i+1,j}+p_{i+1,j+1}-p_{i-1,j+1}-p_{i-1,j}\right)\Big]+\\ & \lambda_{i-1/2,j}\Big[T^{xx}_{i-1/2,j}\left(p_{i-1,j}-p_{i,j}\right)-\\ & \frac{1}{4}T^{xy}_{i-1/2,j}\left(p_{i,j+1}+p_{i-1,j+1}-p_{i,j-1}-p_{i-1,j-1}\right)\Big]+\\ & \lambda_{i,j-1/2}\Big[T^{yy}_{i,j-1/2}\left(p_{i,j-1}-p_{i,j}\right)-\\ & \frac{1}{4}T^{yx}_{i,j-1/2}\left(p_{i+1,j}+p_{i+1,j-1}-p_{i-1,j-1}-p_{i-1,j}\right)\Big]\end{aligned}$$

In this formula, λ represents the mobility. It takes an upstream value according to the flow direction at the interface. For example,

$$\lambda_{i+1/2,j}=\begin{cases}\lambda_{i+1,j} & \text{if} \quad \delta_{i+1/2,j\geq 0}\\ \lambda_{i,j} & \text{if} \quad \delta_{i+1/2,j\leq 0}\end{cases}$$

with

$$\begin{aligned}\delta_{i+1/2,j} = {} & T^{xx}_{i+1/2,j}\left(p_{i+1,j}-p_{i,j}\right)+\\ & \frac{1}{4}T^{xy}_{i+1/2,j}\left(p_{i,j+1}+p_{i+1,j+1}-p_{i,j-1}-p_{i+1,j-1}\right)\end{aligned}$$

This scheme is conservative, because the fluid flux through an interface is always the same for the discretization of the block in both sides; i.e.,

$$\delta_{i+1/2,j}=\delta_{i\to i+1,j}=-\delta_{i+1\to i,j}$$

In the section on scaling-up of transmissibility for a finite-volume scheme, we present a scaling-up procedure to calculate these transmissibility terms; however, if the block permeability is scaled-up, these transmissibility terms should be calculated depending on the block permeability value. For this case, we propose to use the harmonic average for the calculation, which is calculated by

$$T^{xx}_{i+1/2,j}=\frac{\Delta y}{\Delta x}\frac{2k^{xx}_{i+1,j}k^{xx}_{i,j}}{k^{xx}_{i,j}+k^{xx}_{i+1,j}}$$

for the cross-term $T^{xy}_{i+1/2,j}$, which is calculated by

$$T^{xy}_{i+1/2,j}=\frac{1}{2}\left[\frac{4k^{xy}_{i,j+1}k^{xy}_{i,j-1}k^{xy}_{i,j}}{k^{xy}_{i,j+1}k^{xy}_{i,j}+2k^{xy}_{i,j+1}k^{xy}_{i,j-1}+k^{xy}_{i,j}k^{xy}_{i,j-1}}+\frac{4k^{xy}_{i+1,j+1}k^{xy}_{i+1,j-1}k^{xy}_{i+1,j}}{k^{xy}_{i+1,j+1}k^{xy}_{i+1,j}+2k^{xy}_{i+1,j+1}k^{xy}_{i+1,j-1}+k^{xy}_{i+1,j}k^{xy}_{i+1,j-1}}\right]$$

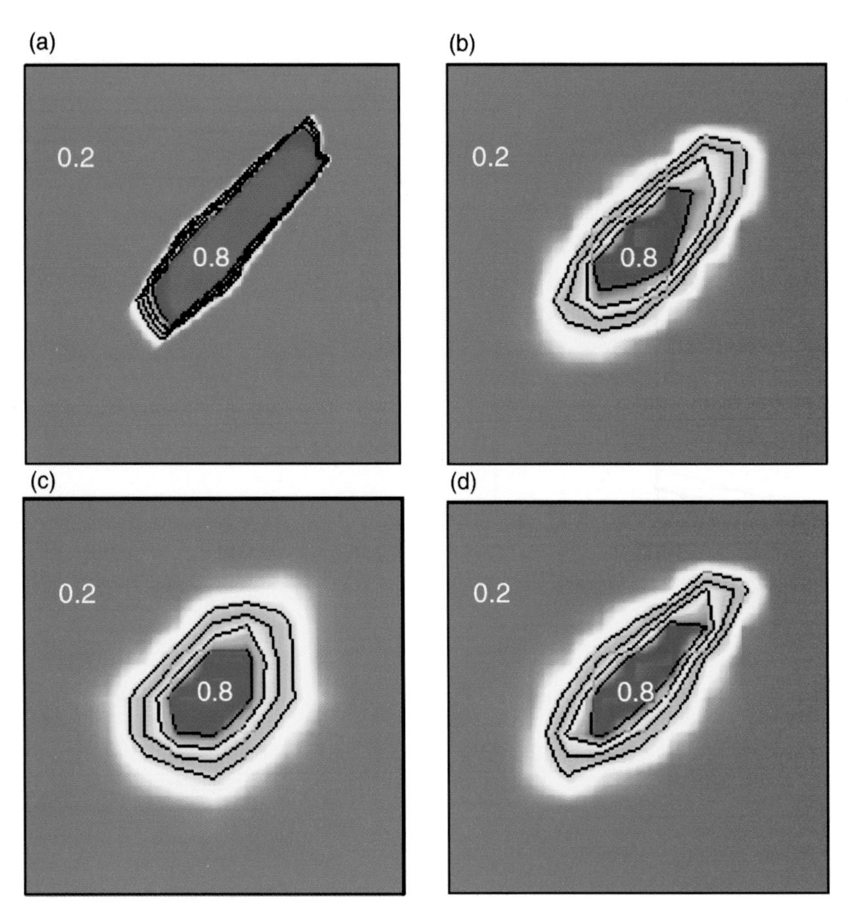

Figure 8. Water saturation distribution for Example 2. (a) Fine grid simulation. (b) Coarse grid simulation with transmissibility scaling-up. (c) Coarse grid simulation with permeability scaling-up using Bamberger's method from Bamberger (1977). (d) Coarse grid simulation with permeability scaling-up using periodic boundary conditions.

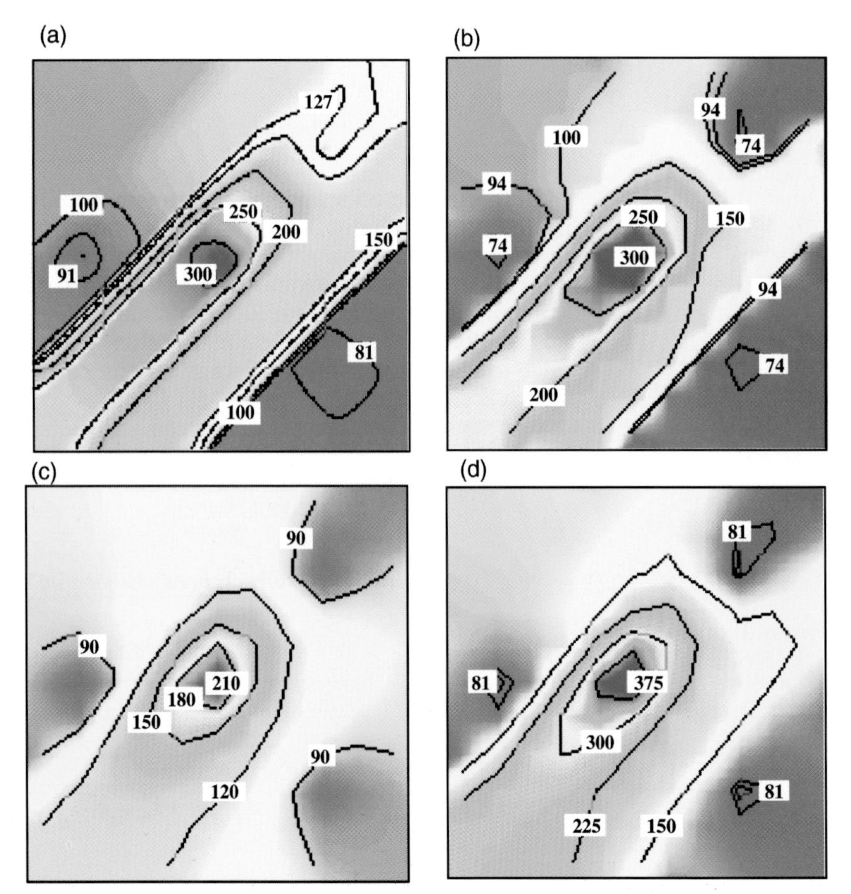

Figure 9. Pressure distribution for Example 2. (a) Fine grid simulation. (b) Coarse grid simulation with transmissibility scaling-up. (c) Coarse grid simulation with permeability scaling-up using Bamberger's method from Bamberger (1977). (d) Coarse grid simulation with permeability scaling-up using periodic boundary conditions.

This formula is used in the examples for the comparison with the block permeability scaling-up.

It cannot be written in the general standard nine-point–scheme form (Yanosik and McCracken, 1979); however, for the single-phase flow modeling, the discretization formula can be written in a standard form, but it is a nonsymmetrical scheme:

$$\int_{\partial\Gamma} k\nabla\vec{p}.\vec{n}.d\gamma \approx$$

$$\left[T^{xx}_{i+1/2,j}+\frac{T^{yx}_{i,j+1/2}}{4}-\frac{T^{yx}_{i,j-1/2}}{4}\right]\left(p_{i+1,j}-p_{i,j}\right)+$$

$$\left[T^{xx}_{i-1/2,j}-\frac{T^{yx}_{i,j+1/2}}{4}+\frac{T^{yx}_{i,j-1/2}}{4}\right]\left(p_{i-1,j}-p_{i,j}\right)+$$

$$\left[T^{yy}_{i,j+1/2}+\frac{T^{xy}_{i+1/2,j}}{4}-\frac{T^{xy}_{i-1/2,j}}{4}\right]\left(p_{i,j+1}-p_{i,j}\right)+$$

$$\left[T^{yy}_{i,j-1/2}-\frac{T^{xy}_{i+1/2,j}}{4}+\frac{T^{xy}_{i-1/2,j}}{4}\right]\left(p_{i,j-1}-p_{i,j}\right)+$$

$$\left[\frac{T^{xy}_{i+1/2,j}}{4}+\frac{T^{yx}_{i,j+1/2}}{4}\right]\left(p_{i+1,j+1}-p_{i,j}\right)+$$

$$\left[\frac{T^{xy}_{i+1/2,j}}{4}-\frac{T^{yx}_{i,j-1/2}}{4}\right]\left(p_{i+1,j-1}-p_{i,j}\right)+$$

$$\left[\frac{T^{xy}_{i-1/2,j}}{4}-\frac{T^{yx}_{i,j+1/2}}{4}\right]\left(p_{i-1,j+1}-p_{i,j}\right)+$$

$$\left[\frac{T^{xy}_{i-1/2,j}}{4}+\frac{T^{yx}_{i,j-1/2}}{4}\right]\left(p_{i-1,j-1}-p_{i,j}\right)$$

NOMENCLATURE

A = elliptic operator
B = formation volume factor
F = flow term
K_{eq} = effective permeability tensor
L = length of the shifted block
Nx_r, Ny_r = number of the refined blocks in the x-, y-direction, respectively
T = internodal transmissibility
k = permeability tensor of fine grid block
n = normal direction
p = pressure
Q = well flow rate
u = Darcy filtration velocity
Δx, Δy = size of the coarse grid block
Δx_r, Δy_r = size of the refined block
μ = fluid viscosity
λ = mobility
Φ = porosity
δ = fluid flux
Ω = domain
$\partial\Omega$ = Γ = domain boundary
ω = power law coefficient.

Subscript

i,j = indices of blocks in x-, y-direction, respectively
o = oil phase
w = water phase
x,y = spatial directions

ACKNOWLEDGMENTS

We would like to thank P. Lemonnier and E. Delamaide for their encouragement and critical comment during the writing of this paper.

REFERENCES CITED

Bamberger, A., 1977, Approximation des coefficients d'opérateurs elliptiques stables pour la G-convergence, Rapport du Centre de Mathématiques Appliquées, École Polytechnique.

Ding, Y., 1995, Scaling-up in the Vicinity of Wells in Heterogeneous Fields, SPE 29137, presented at the 13th SPE Symposium on Reservoir Simulation, 12-15 Feb., San Antonio, Texas, p. 441–451.

Durlofsky, L. J., 1991, Numerical calculation of equivalent grid block permeability tensors for heterogeneous porous media, Water Resources Research, v. 27 (5), p. 699-708.

Galli, A., Goblet, P., Griffin, D., Ledoux, E., Le Loc'h, G., Mackay, R., and Renard, P., 1996, Quick upscaling of flow and transport related parameters, Technical Report, Ecole des Mines de Paris.

Gautier, Y., and Noetinger, B., 1994, Preferential flow paths detection for heterogeneous reservoirs using a new renormalization technique, 4th European Conference on the Mathematics of Oil Recovery, 7-10 June, Roros, Norway.

Guerillot, D., 1988, Composition des perméabilités absolues aux échelles intermédiaires pour les gisements d'hydrocarbures hétérogènes; application à un affleurement du Yorkshire; rapport d'avancement, IFP Report 36738.

Guerillot, D., Rudkiewicz, J. L., Ravenne, Ch., Renard, G., and Galli, A., 1989, An integrated model for computer aided reservoir description: from outcrop study to fluid flow simulations, presented at the IOR Symposium, Budapest, Hungary.

King, P. R., 1989, The use of renormalization for calculating effective permeability, Transport in Porous Media, v. 4, p. 37-58.

Kruel Romeu, R., 1994, Écoulement en milieux hétérogènes: prise de moyenne de perméabilités en régimes permanent et transitoire, Ph.D. thesis, Université Paris VI, France.

Kruel Romeu, R., and Noetinger, B., 1995, Calculation of internodal transmissivities in finite difference models of flow in heterogeneous media, Water Resources Research, v. 31, p. 943-959.

Le Loc'h, G., 1987, Etude de la decomposition des

perméabilités par des méthodes variationnelles, Ph.D. thesis, Ecole Normale Superieure des Mines de Paris, France.

Lemouzy, P. M., Parpant, J., Eschard, R., Bacchiana, C., Morelon, I., and Smart, B., 1995, Successful History Matching of Chaunoy Field Reservoir Behavior Using Geostatistical Modelling, SPE 30707 presented at the SPE ATC&E, 22-25 Oct., Dallas, Texas, p. 23–38.

Njifenjou, A., 1993, Éléments finis mixtes hybrides duaux et homogénéisation des paramètres pétrophysiques; application à l'étude numérique de l'écoulement en milieux poreux, Ph.D. thesis, Université Paris VI, France.

Njifenjou, A., 1994, Expression en termes d'énergié pour la perméabilité absolue effective, IFP Review, v. 49 (4), p. 345-358.

Palagi, C., 1992, Generation and application of Voronoi grid to model flow in heterogeneous reservoirs, Ph. D. thesis, Stanford University, California.

Palagi, C. L., Ballin, P. R., and Aziz, K., 1993, The modelling of flow in the heterogeneous reservoirs with Voronoi grid, SPE 25259 presented at the SPE Symposium on Reservoir Simulation, 28 Feb.-3 March, New Orleans, Louisiana, p 291–300.

Peaceman, D. W., 1996, Effective transmissibilities of a gridblock by upscaling; why use renormalization?, SPE 36722 presented at the SPE Annual Technical Conference and Exhibition, 6-9 Oct., Denver, Colorado, p. 417–430.

Pickup, G. E., and Sorbie, K. S.,1994, Development and application of a new two phase scale up method based on tensor permeabilities, SPE 28586 presented at the SPE Annual Technical Conference and Exhibition, 25-28 Sept., New Orleans, Louisiana, p. 217–230.

Pickup, G. E., and Sorbie, K. S., 1994, The scale-up of two phase flow using permeability tensors, 4th European Conference on the Mathematics of Oil Recovery, 7-10 June, Roros, Norway.

Samier, P., 1990, A finite element method for calculating transmissibilities in N-point difference equations using a non diagonal permeability tensor, 2nd European Conference on the Mathematics of Oil Recovery, 11–14 Sept., Arles, France, p. 121–130.

Wang, J. T., Daltaban, T. S., and Archer, J. S., 1992, The use of permeability tensor in modelling heterogeneous and fractured flow media, SPE 24503, p. 54–64.

Warren, J. E., and Price, H. S., 1961, Flow in heterogeneous porous media, SPEJ Sept., p. 153-169.

White, C. D., and Horne, R. N., 1987, Computing absolute transmissibility in the presence of fine scale heterogeneity, SPE 16011 presented at 9th Symposium on Reservoir Simulation, 1-4 Feb., San Antonio, Texas, p. 209–220.

Yanosik, J. L., and McCracken, T. A., 1979, A nine-point finite difference reservoir simulator for realistic prediction of adverse mobility ratio displacements, SPEJ Aug., p. 253-262.

Chang, Y.C., V. Mani, K.K. Mohanty, Effect of wettability on scale-up of multiphase flow from core-scale to reservoir fine-gird scale, 1999, *in* R. Schatzinger and J. Jordan, eds., Reservoir Characterization-Recent Advances, AAPG Memoir 71, p. 305–318

Chapter 21

Effect of Wettability on Scale-Up of Multiphase Flow From Core-Scale to Reservoir Fine-Grid Scale

Y. C. Chang
V. Mani
K. K. Mohanty
Department of Chemical Engineering
University of Houston
Houston, Texas, U.S.A.

ABSTRACT

Reservoir rocks modeled in typical field-scale simulation grids are internally heterogeneous. The objective of this work is to study how rock wettability affects the scale-up of multiphase flow properties from core-scale to fine-grid reservoir simulation scale (~10 ft × 10 ft × 5 ft; 3 m × 3 m × 1.5 m). Upscaling from fine-grid reservoir simulation scale to coarse-grid simulation scale is not addressed here. Heterogeneity is modeled as a correlated random field, parameterized in terms of its variance and two-point variogram. Variogram models of both finite (spherical) and infinite (fractal) correlation lengths are included as special cases. Local core-scale porosity, permeability, capillary pressure, relative permeability, and initial water saturation are assumed to be correlated. Water injection is simulated; effective flow properties and flow equations are calculated.

For strongly water-wet media, capillarity has a stabilizing/homogenizing effect on multiphase flow. For a permeability field with a small variance and a small correlation length, effective relative permeability can be described by capillary equilibrium models. At higher variance and moderate correlation length, the average flow can be described by a dynamic relative permeability. As the oil wettability increases, the capillary stabilizing effect decreases and deviation from this average flow increases. For fractal fields with large variance in permeability, effective relative permeability is not adequate in describing the flow.

INTRODUCTION

Heterogeneities exist at several length scales in naturally occurring porous media. Variations in properties such as permeability and porosity can occur at the core scale, stratum scale, bedding scale, well-log scale, and interwell scale (Worthington, 1991). Reservoir heterogeneity at different scales has different impacts on ultimate oil recovery in displacement processes. In the last decade, much progress has been made on the characterization of heterogeneities in oil reservoirs and the integration of these heterogeneities into flow calculations (Wolcott and Chopra, 1993). Geostatistical techniques have been developed for the estimation of the heterogeneities (Journel, 1990). Emerging parallel computers and new numerical techniques are increasing the power of reservoir simulation (Bhogeswara and Killough, 1993; Thiele et al., 1994); however, typical reservoir-scale simulation models still do not have the resolution to incorporate fine-scale heterogeneities directly. These models rely on the use of effective properties to represent the effect of small-scale heterogeneities in large-scale numerical grids (Edwards and Christie, 1993).

Upscaling techniques have been developed to estimate effective flow properties in heterogeneous media (Warren and Price, 1961; Gelhar, 1984; King, 1989; Glimm et al., 1993; Durlofsky et al., 1995). The techniques developed for single-phase flow are accurate and range from simple statistical estimates to detailed numerical simulations. The upscaling of multiphase flow has proven to be difficult. Multiphase flow in porous media is governed by functions such as relative permeability (k_r) and capillary pressure (P_c). Barker and Thibeau (1997) have reviewed the pseudo-relative permeability method and point out its lack of generality. The capillary-equilibrium method has been professed by many (Smith, 1991), but is applicable in water-wet rocks at small scales (e.g., laboratory scale) at low flow rates. The homogenization method proposed by Quintard and Whitaker (1990) is valid for small correlation lengths. The stochastic method proposed by Gelhar (1984) and Butts (1991) is valid for small permeability variations. The effective relative permeability approach has been proposed (Muggeridge, 1991; Hewett and Behrens, 1991) to take into account the dynamic effects of flow. It can be calculated by many methods: fine-grid simulation (Muggeridge, 1991), space renormalization (King et al., 1993), or heuristic methods (Li et al., 1996).

The applicability of the effective relative permeability approach for core-to-numerical grid (in typical pattern simulations) scale-up was tested for water-wet media by Chang and Mohanty (1997). This issue involves two questions. First, can flow at a larger scale be adequately described by the multiphase Darcy's law involving an effective relative permeability (and an effective capillary pressure)? Second, if Darcy's law is adequate, what is the effective permeability (and effective capillary pressure) as a function of intrinsic relative permeability and heterogeneity? Chang and Mohanty (1997) found that in water-wet rocks the average flow can be described by an effective relative permeability if the permeability variance and correlation length are not very large. Such a formulation is not valid if the permeability variance is large and the spatial correlation is fractal. Capillary pressure plays an important role in stabilizing waterflood displacement fronts in water-wet media.

Many reservoirs are not water wet. The objective of this work is to study how rock wettability affects the scale-up of multiphase flow properties from core scale (~2 in; 5 cm) to pattern simulation grid scale (~10 ft; 3 m). Reservoir models need another level of upscaling to coarse-grid simulation scale (Li et al., 1996; Barker and Thibeau, 1996), which is not addressed here. Two key simplifying assumptions are used in this paper: the domain is limited to a two-dimensional flow region and effects of gravity are neglected. The scale-up of flow in only the horizontal direction is considered in several permeability fields. The anisotropic nature of the scaled-up relative permeability is outside the scope of this paper. Also, capillary pressure scale-up is not addressed here. In the following section, the methodology used to represent reservoir wettability, heterogeneity, and flow is summarized. The results are described in the section after the methodology section.

METHODOLOGY

Heterogeneity is modeled here by a spatially correlated random permeability field parameterized in terms of its variance and two-point variogram, similar to our earlier work (Chang and Mohanty, 1997). Local porosity, permeability, capillary pressure, relative permeability, and initial water saturation are assumed to be correlated to one another. High-resolution reservoir simulation of water/oil displacement is conducted to understand the effect of heterogeneity on detailed multiphase flow. Saturation fronts and pressure drops are monitored during the simulation. The JBN method (Johnson et al., 1959) is used to compute the effective relative permeabilities from simulation data.

The computational flow field is taken to be 16 ft × 8 ft (5 m × 2 m), about the size of a typical grid block in a reservoir pattern simulator. This system is modeled by a 100 × 50 grid of uniform size. The individual grid blocks are 0.16 ft × 0.16 ft (0.05 m × 0.05 m), about the size of typical laboratory systems for which relative permeability and capillary pressure functions are measured. Each grid block is assumed to be homogeneous with prescribed intrinsic multiphase flow functions. Darcy's law for multiphase flow and capillary pressure-saturation relations are presumed to be valid at the scale of each grid block.

The permeability heterogeneity is specified by a probability distribution function and two-point variogram. The probability distribution function for permeability is assumed to be log-normal, with an average of 100 md and a standard deviation of $\sigma_{\log k}$. $\sigma_{\log k}$ is varied from 0.2 to 0.8. Two types of spatial variograms are considered: spherical and fractal. The ratio of the correlation length to the system dimension in the systems with a spherical variogram is varied from 0.05 to 0.2. Because the overall system dimension modeled is 16 ft × 8 ft (5 m × 2 m), the correlation length in the horizontal direction is twice that in the vertical direction and the flow field is anisotropic; however, only the overall flow in the horizontal direction is studied in this work. The fractal systems are modeled by fractional Brownian motion (*fBm*) (Peitgen and Saupe, 1988). These models represent rocks with long-range correlations and often have been used to characterize reservoir rocks (Hewett, 1986; Painter, 1995). Fractional Brownian motion models are defined by a parameter, the Hurst exponent (H), which describes the roughness of the function at small scales. Rough, anti-persistent trends are observed for fields with $H < 0.5$. Smoother, persistent trends are observed for $H > 0.5$. Representative values of 0.2 and 0.87 for the Hurst exponent are considered in this work. In addition to correlations in the permeability field, the porosity distribution is correlated using

$$K = a \cdot 10^{b\phi}$$

where a and b are set to 0.001 and 25, respectively (Beier and Hardy, 1993). Rock wettability has a strong effect on the initial oil saturation, residual oil saturation, relative permeability, and capillary pressure functions. These effects are discussed in the following sections for strongly water-wet, mixed-wet, and oil-wet systems. Model parameters used in the simulations are listed in Table 1.

Water-Wet System

The water-wet media were modeled similar to our earlier work (Chang and Mohanty, 1997). The irreducible water saturation (S_{wr}) is assumed to be a constant for all grid blocks and equal to 0.2. The intrinsic relative permeability function is given by the Corey model (Honarpour et al., 1982) as

$$k_{ro} = k_{ro}^{0}\left(\frac{S_o - S_{or}}{1 - S_{or} - S_{wr}}\right)^{n_o}$$

and

$$k_{rw} = k_{rw}^{0}\left(\frac{S_w - S_{wr}}{1 - S_{or} - S_{wr}}\right)^{n_w}$$

The residual oil saturation (S_{or}) and initial oil saturation (S_{oi}), for water-wet media, can be related using the correlation (Land, 1968)

$$S_{or} = \frac{S_{oi}}{1 + c \cdot S_{oi}}$$

Table 1. Model Parameters.

Water-Wet System

Parameter	Value	Parameter	Value
k_{ro}^0	1	k_{rw}^0	0.16
n_o	2	n_w	3
c	1.5	σ	30
Γ (Drainage)	1.537	Γ (Imbibition)	0.461
β (Drainage)	2	β (Imbibition)	2.76
S_{wr}	0.2		

Mixed-Wet System

Parameter	Value	Parameter	Value
m	7.5	$I_{w\text{-}o}^*$	2
a	0.3	b	2
I_{w-o}^{min}	0.25	S_{or}^{min}	0.15
ε	0.05	p_c^{init}	13 psi
S_{wcp}	0.1	λ_p	0.5
S_{ocm}	0	λ_m	0.5
S_{wcm}	0.2	Γ	3

Oil-Wet System

Parameter	Value	Parameter	Value
k_{ro}^0	1	k_{rw}^0	0.5
n_o	4	n_w	2
σ	30	Γ	0.745
β	0.7	S_{oc}	0
S_{wc}	0.2		

The initial oil saturation depends on the initial capillary pressure of the system. The permeability, porosity, and primary drainage capillary pressure functions are correlated using the *J*-function (Lake, 1989). Because the *J*-function depends on only the effective saturation (S), the capillary pressure function can be expressed in terms of permeability, porosity, and effective saturation and is given by

$$P_c^{Dr}(S) = \Gamma\sigma(\phi/K)^{1/2} S^{-1/\beta}$$

where Γ and β are constants and S is given by

$$S = \frac{S_w - S_{wr}}{1 - S_{wr}}$$

The imbibition capillary pressure, P_c^{Im}, is given by a similar equation but with different constants, Γ and β. The effective saturation, S, for imbibition is defined as

$$S = \frac{S_w - S_{wr}}{1 - S_{or} - S_{wr}}$$

Capillary pressure hysteresis and relative permeability hysteresis are assumed to follow the model developed by Killough (1976).

Mixed-Wet System

For mixed-wet media, the wettability is heterogeneous and depends strongly on the initial water saturation. Jadhunandan and Morrow (1995) studied the effect of the initial water saturation on wettability and the effect of wettability on oil recovery for an oil/brine/rock system. They found that the initial water saturation increases as the wettability changes from oil wet to water wet. The wettability index I_{w-o} obtained from Amott tests is linearly related to initial water saturation. The relation can be described by

$$I_{w-o} = S_{wi} \cdot m + I_{w-o}^*$$

The slope (m) and the intercept (I_{w-o}^*) depend on the crude oil and the brine composition. They also demonstrated that the residual oil saturation decreases as the wettability changes from strongly water wet to a neutral wettability, and then increases as wettability changes to strongly oil wet. Residual oil saturation, S_{or}, and wettability index, I_{w-o}, can be expressed by

$$S_{or} = a\left(\left|I_{w-o} - I_{w-o}^{min}\right|\right)^b - S_{or}^{min}$$

From the two preceding equations, the initial water saturation and the residual oil saturation can be explicitly related by

$$S_{or} = a\left(\left|S_{wi} \cdot m + I_{w-o}^* - I_{w-o}^{min}\right|\right)^b - S_{or}^{min}$$

Depending on the initial water saturation, we classify the grid blocks into three categories: strongly water wet ($S_{wi} \geq 0.4$); mixed wet ($0.2 < S_{wi} < 0.4$); and strongly oil wet ($S_{wi} \leq 0.2$). For the grid blocks that are classified as strongly water wet, S_{or} and S_{wi} can be related by Land's correlation (Land, 1968). Figure 1 shows the relation between S_{wi} and S_{or} in mixed-wet systems.

Relative permeability for mixed-wet systems is again given by the Corey model, but the parameters k_{rw}^0, k_{rw}^0, n_w, and n_o are assumed to depend on the rock wettability. For mixed-wet media, the parameters are interpolated between the values for strong water wettability and strong oil wettability, according to the initial water saturation. The parameters used in the relative permeability model for mixed-wet systems in grids of different wettability are shown in Table 2. The calculated relative permeability curves for the different regions (of different wettability and hence at different initial water saturations) are shown in Figure 2.

The imbibition capillary pressure function also depends on wettability. For water-wet grids ($S_{wi} \geq 0.4$), capillary pressure is positive at all saturations, whereas capillary pressure is negative at all saturations for oil-wet grids ($S_{wi} \leq 0.2$). For mixed-wet grids ($0.2 < S_{wi} < 0.4$), capillary pressure is positive at low water saturations and negative at higher saturations. The imbibition capillary pressure for mixed-wet grids is described by

$$P_c = P_c^{ww} - F_{mw}\left(P_c^{ww} - P_c^{ow}\right)$$

where

$$F_{mw} = \begin{cases} 1 & \text{for } S_{wi} \leq S_{wi}^{ow} \\ \dfrac{\dfrac{1}{\left(S_w - S_{wi}\right)\cdot\left(S_{wi}^{ww} - S_{wi}\right) + \varepsilon} - \dfrac{1}{\varepsilon}}{\dfrac{1}{\left(S_{wi}^{ww} - S_{wi}^{ow}\right)\cdot\left(S_w - S_{wi}^{ow}\right) + \varepsilon} - \dfrac{1}{\varepsilon}} & \text{for } S_{wo}^{ow} < S_{wi} < S_{wi}^{ww} \\ 0 & \text{for } S_{wi} \geq S_{wi}^{ww} \end{cases}$$

and S_{wi}^{ww}, the critical initial water saturation for water-wet media, is set to 0.4; S_{wi}^{ow}, the critical initial water saturation for oil-wet media, is equal to 0.2; ε is a parameter. The

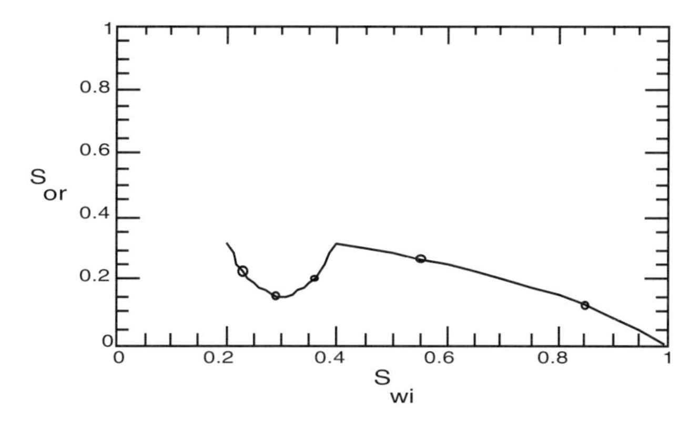

Figure 1. Relation between S_{oi} (= 1 – S_{wi}) and S_{or} in mixed-wet media.

water-wet capillary pressure, P_c^{ww}, and the oil-wet capillary pressure, P_c^{ow}, in the above equation are given by

$$P_c^{ww} = P_c^{init} \cdot \left(\frac{S_w - S_{wcp}}{S_{wi} - S_{wcp}} \right)^{-1/\lambda_p}$$

and

$$P_c^{ow} = -\Gamma \cdot \left(\frac{S_o - S_{ocm}}{1 - S_{ocm} - S_{wcm}} \right)^{-1/\lambda_m}$$

where P_c^{init} is the equilibrium capillary pressure at S_{wi}; λ_p, λ_m, S_{wcp}, S_{wcm}, S_{ocm}, and Γ are parameters (see Table 1). Figure 3 shows the effect of the initial water saturation on the capillary pressure functions.

Oil-Wet System

The intrinsic relative permeability and capillary pressure of the oil-wet system are represented by functions similar to those of the water-wet system. The relative permeabilities are described by the Corey model, but with parameters listed in Table 1. Capillary pressure in oil-wet systems is defined by

$$P_c(S) = \Gamma \sigma (\phi/K)^{1/2} S^{1/\beta}$$

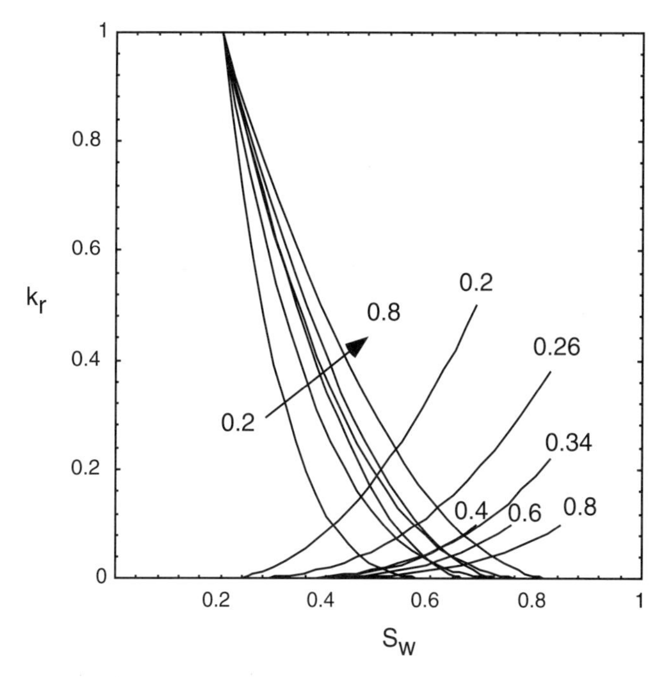

Figure 2. Relative permeability at different initial water saturation in mixed-wet media.

where Γ and β are constants and S is given by

$$S = \frac{S_o - S_{oc}}{1 - S_{oc} - S_{wc}}$$

For simplicity, no hysteresis in relative permeability and capillary pressure was assumed for oil-wet media.

Waterflood Simulation

Two-dimensional, fine-grid waterflood simulations were conducted to investigate the impact of heterogeneity on two-phase flow. A two-phase, immiscible, third-order, finite difference simulator with flux limiters was used. The numerical scheme is implicit in pressure and explicit in saturation. Water was injected at the left boundary at a specified total rate. The pressure at the right boundary was specified. The top and bottom were no-flow boundaries. Gravitational effects were neglected. The following parameters affect

Table 2. Relative Permeability Model Parameters for Mixed-Wet Media.

	Oil-Wet Grid	Mixed-Wet Grid	Water-Wet Grid
S_{wi}	≤ 0.2	0.2~0.4	≥0.4
S_{or}	Land correlation	Jadhunandan and Morrow (1995)	Land correlation
k^0_{rw}	0.5	Interpolation	0.1
k^0_{ro}	1.0	1.0	1.0
n_w	2.0	Interpolation	4.0
n_o	4.0	Interpolation	2.0

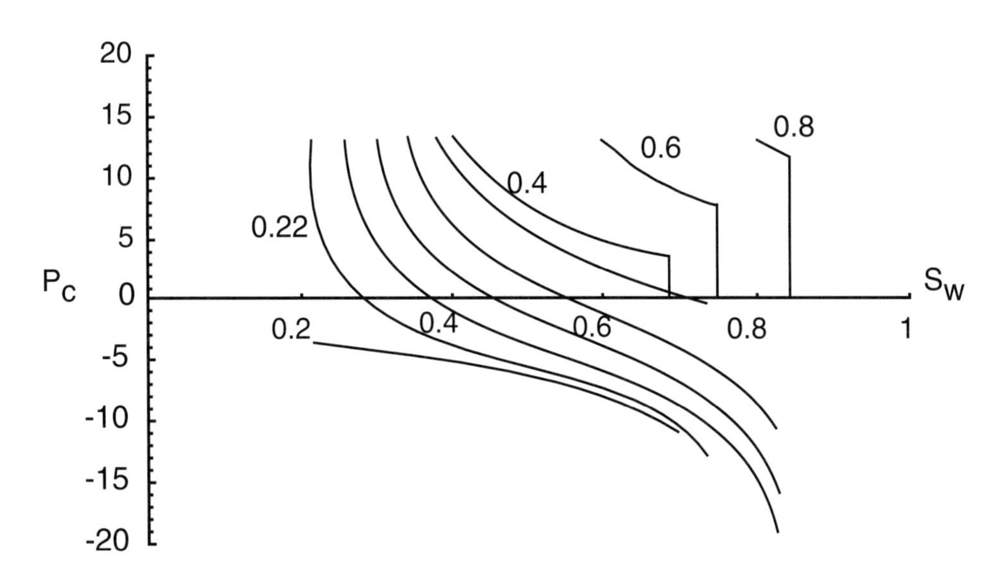

Figure 3. Capillary pressure in psi at different initial water saturation in mixed-wet media.

waterflood, but were kept constant in this study at the following values: $\Delta\rho = 0$, $\mu_o = 10$ cp, $\mu_w = 1$ cp, velocity $= 0.49 \times 10^{-2}$ cm/s, $N_{gv} = 0$, $N_{cv} \approx 3$, $R_l = 2$. The effect of these parameters is discussed elsewhere (Li et al., 1996; Chang and Mohanty, 1997).

From computed 2-D (two-dimensional) saturation distributions, cross-sectionally averaged 1-D (one-dimensional) saturation fronts are calculated and monitored. The objective in scale-up is to relate the flux of a certain phase to its average saturation around any location and thus predict approximate 1-D saturation distributions at any time. In this work, we determine whether the relative permeability formulations can be used to estimate these average saturations and the effective relative permeabilities. For homogeneous media and 1-D flow, Darcy's law leads to the traditional fractional flow theory (Lake, 1989). When the capillary pressure term is negligible, fractional flow theory dictates that each saturation, S_w, has its own constant characteristic velocity given by df_w/dS_w. Even when the capillary pressure term is not negligible, the characteristic velocity of each saturation depends only slightly on

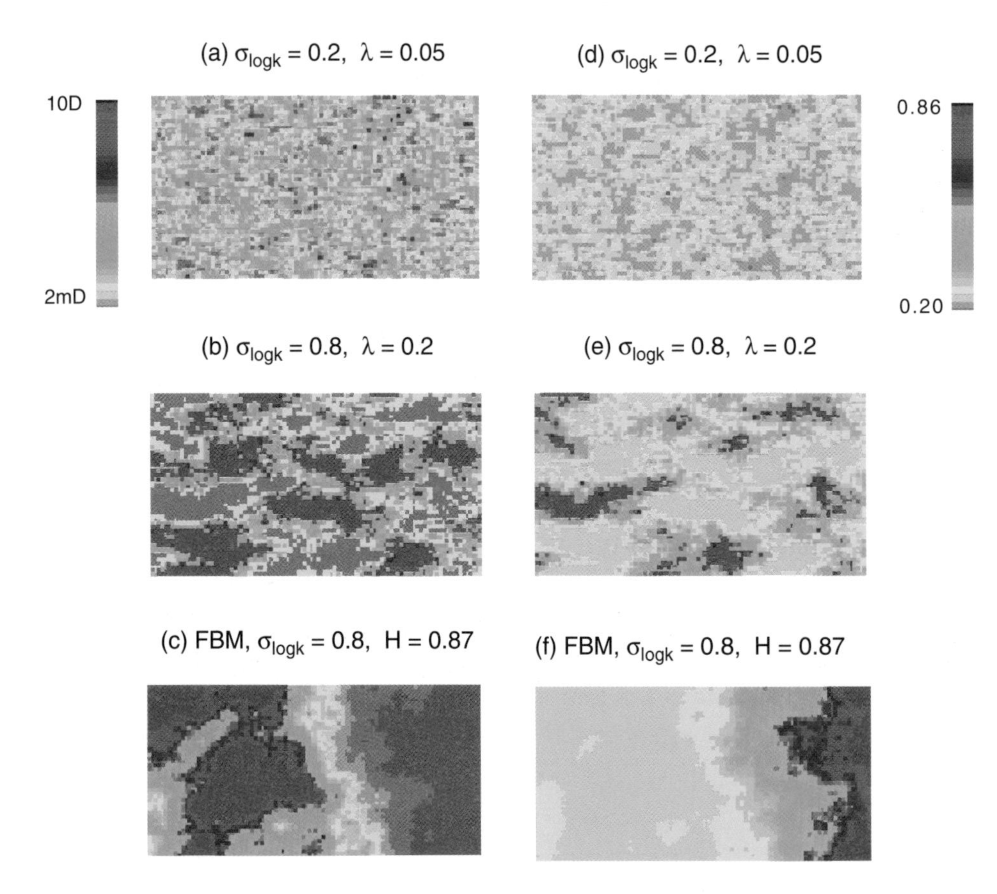

Figure 4. Permeability and S_{wi} distribution.

time. If a heterogeneous system can be averaged to a 1-D homogeneous system, the plot of S_w vs. its characteristic velocity (x/t) must be independent of time. Only then can Darcy's law with relative permeability be used to describe multiphase flow in a large-scale heterogeneous system and appropriate effective relative permeability functions be identified. We will call such systems "k_r-formulation adequate." If, however, the characteristic velocity of each saturation depends strongly on time (i.e., the plots of S_w vs. velocity (x/t) do not fall on a single line), the system will be identified as "k_r-formulation inadequate." These systems cannot be described by a traditional relative permeability at the large scale. New flow equations need to be developed for effective multiphase flow of these systems.

Effective relative permeabilities were calculated by the JBN method (Johnson et al., 1959) by monitoring the effluent fractional flow and pressure drop during waterflood simulations. This method is commonly used to determine relative permeabilities from coreflood experiments.

RESULTS

Three typical heterogeneous permeability fields are illustrated in Figure 4. Figure 4a shows the permeability distribution for the case of a small standard deviation and a small correlation length ($\sigma_{\log k}$ = 0.2 and λ = 0.05). Figure 4b shows the permeability distribution for the case of a large standard deviation and a large

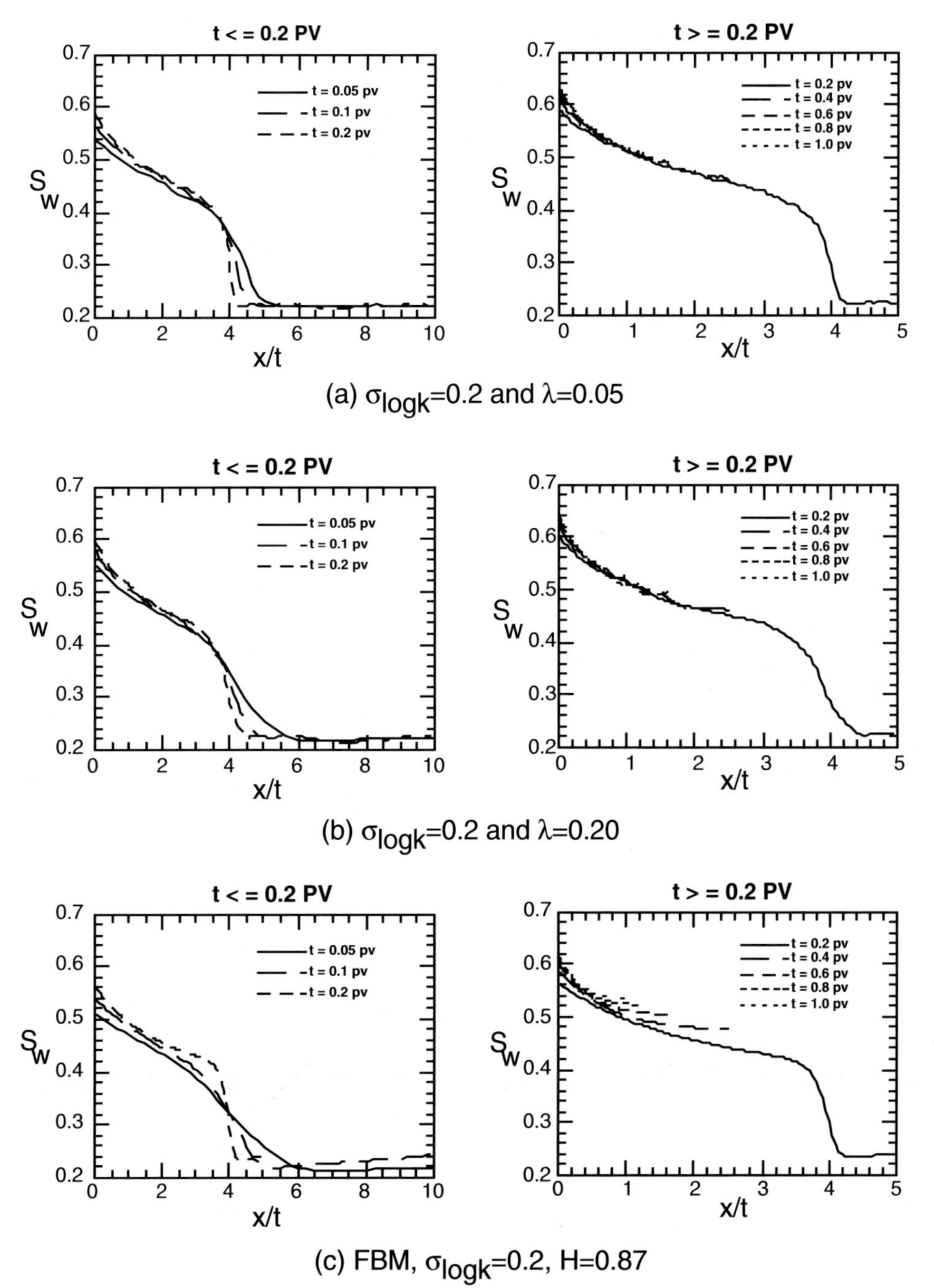

Figure 5. Saturation-velocity profiles at $\sigma_{\log k}$ = 0.2 for mixed-wet system.

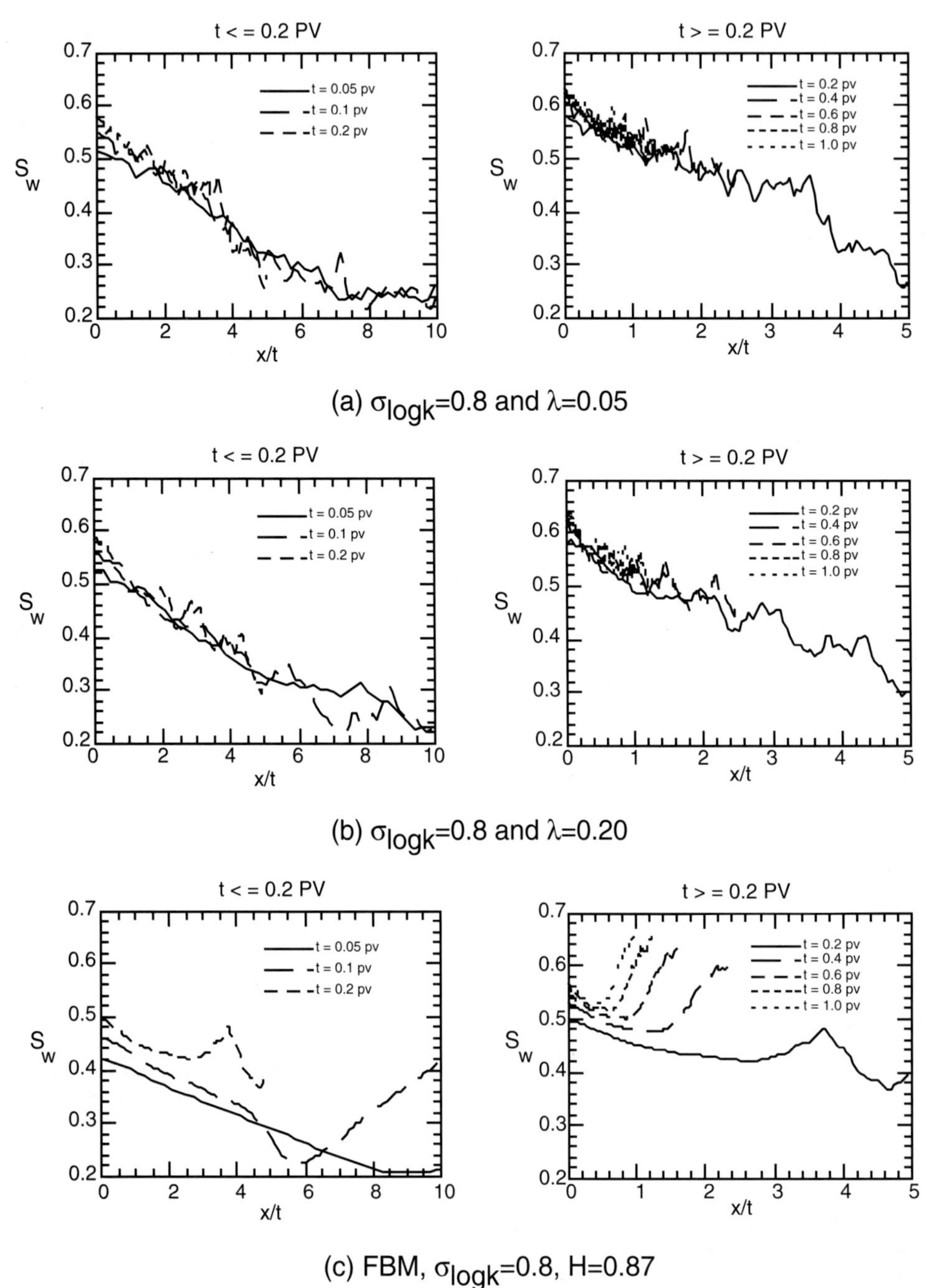

Figure 6. Saturation-velocity profiles at $\sigma_{\log k}$ = 0.8 for mixed-wet system.

correlation length ($\sigma_{\log k}$ = 0.8 and λ = 0.2). Figure 4c shows the permeability distribution for the case of a fractal model with a large standard deviation and large Hurst exponent ($\sigma_{\log k}$ = 0.8 and H = 0.87). As discussed earlier, three kinds of wettability scenarios are studied: completely water wet, mixed wet, and oil wet. In the mixed-wet scenario, the grids with $S_{wi} > 0.4$ (black in Figure 4d–f) are water wet, those with $0.2 < S_{wi} < 0.4$ are mixed wet, and those with $S_{wi} \leq 0.2$ are oil wet. The distribution of the water-wet region depends on the correlation length of the permeability field.

The results for strongly water-wet media have been presented elsewhere (Chang and Mohanty, 1997). In this work, we extend the earlier results to include mixed-wet and oil-wet media. For the mixed-wet media, the wettability is heterogeneous, and both capillary pressure and relative permeability depend on initial water saturation as described earlier. Figure 5a and b shows saturation-velocity profiles for mixed-wet systems at $\sigma_{\log k}$ = 0.2 with λ = 0.05 and λ = 0.20, respectively. As discussed previously, these profiles were obtained by averaging the simulated 2-D saturation profile in the vertical direction and plotting them against the ratio of the distance over time. The shock fronts are more dispersive and the fluctuations of characteristic velocity are larger than those for water-wet media with similar heterogeneity. This is due to wettability heterogeneity and more oil-wet characteristics. Saturation-velocity curves at large times fall on top of each other, so k_r-formulation is adequate for small standard deviation and finite correlation length cases. For the fractal field (H = 0.87) as shown in

Figure 5c, the saturation-velocity curves do not overlay each other, so k_r-formulation is inadequate for this case.

Figures 6a–c show saturation profiles for mixed-wet systems at $\sigma_{\log k} = 0.8$ with different correlation lengths. In an earlier work (Chang and Mohanty, 1997), we observed that water-wet systems with a high $\sigma_{\log k}$ exhibit fluctuations about the average velocity. The saturation-velocity profiles shown in Figures 6a–c are similar, except that the fluctuations are larger in the mixed-wet systems. For $\lambda = 0.05$ and 0.20, the average velocity is a function of saturation alone; thus k_r-formulation is adequate at large $\sigma_{\log k}$. Fractal fields with a Hurst exponent of 0.2 (not shown in the figure) display a behavior similar to spherically correlated systems with $\lambda = 0.2$, namely, k_r-formulation is adequate for small $\sigma_{\log k}$ and the average flow behavior is modeled adequately at large $\sigma_{\log k}$. For the fractal field with $H = 0.87$, shown in Figure 6c, the saturation velocity profile does not fluctuate around an average. The average characteristic velocity is not a function of saturation alone, thus k_r-formulation is not adequate for fractal fields with large Hurst exponents and large $\sigma_{\log k}$.

Figure 7 illustrates the effective relative permeabilities at $\sigma_{\log k} = 0.2$. The effective relative permeabilities at small permeability deviation with $\lambda = 0.05$ and $\lambda = 0.2$ are almost identical, whereas the effective relative permeabilities for the *fBm* fractal field ($H = 0.87$) are a little lower than those for small correlation ones. Because the relative permeability depends on initial water saturation, the intrinsic relative permeability chosen was based on the initial average saturation of the whole medium. All the effective relative permeabilities at this small permeability heterogeneity are comparable to intrinsic relative permeability. As permeability variance increases, the deviation between effective relative permeability and intrinsic relative permeability increases as shown in Figure 8, especially for water relative permeability. All effective relative permeabilities to water at large $\sigma_{\log k}$ are higher than the intrinsic relative permeability.

Figure 9 shows the saturation-velocity profiles for oil-wet systems at $\sigma_{\log k} = 0.8$. The saturation-velocity profiles for oil-wet systems at $\sigma_{\log k} = 0.2$ are very similar to those of the mixed-wet systems. The fluctuations of saturation-velocity at large $\sigma_{\log k}$ and finite λ are much larger than those for mixed-wet and water-wet systems; however, the characteristic velocities after breakthrough for the *fBm* fractal field are smoother and closer to each other at large times when compared to those for mixed-wet and water-wet media. This difference occurs because of the constant initial water saturation assumed in this oil-wet system. The situation is similar to a stratified layer system with constant initial water saturation, and the saturation fronts move at the same velocity in each layer; therefore, the 1-D cross-sectionally averaged saturations are smoother than those of nonuniform initial water saturations.

Figure 10 shows the effective relative permeabilities at $\sigma_{\log k} = 0.2$. The effective relative permeabilities are close to the intrinsic relative permeability when the permeability variance is small; however, at large permeability variance, the effective relative permeabilities for various correlation lengths are quite different, as illustrated in Figure 11. Intrinsic oil relative permeability overestimates all the oil effective relative permeabilities, while intrinsic water relative permeability underestimates the water effective relative permeabilities at low water saturation. The water effective relative permeability at $\sigma_{\log k} = 0.8$ and $\lambda = 0.2$ is higher than the other two effective relative permeabilities with distinct λ, whereas oil effective relative permeability is lower than the other permeabilities. The flow in this large permeability variance and large correlation length media has the earliest breakthrough. The effect of permeability field correlation

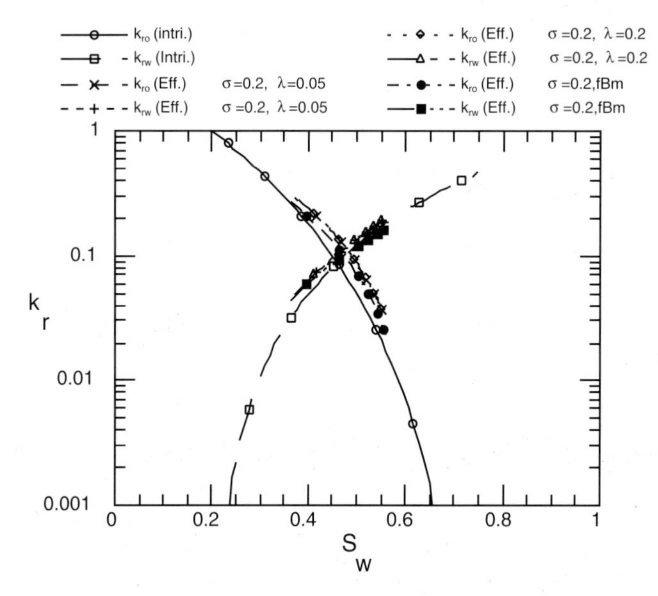

Figure 7. Relative permeability at $\sigma_{\log k} = 0.2$ for mixed-wet system.

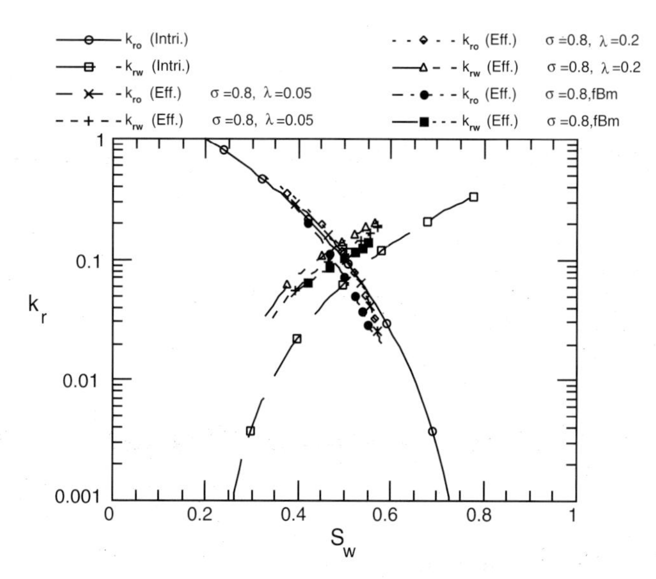

Figure 8. Relative permeability at $\sigma_{\log k} = 0.8$ for mixed-wet system.

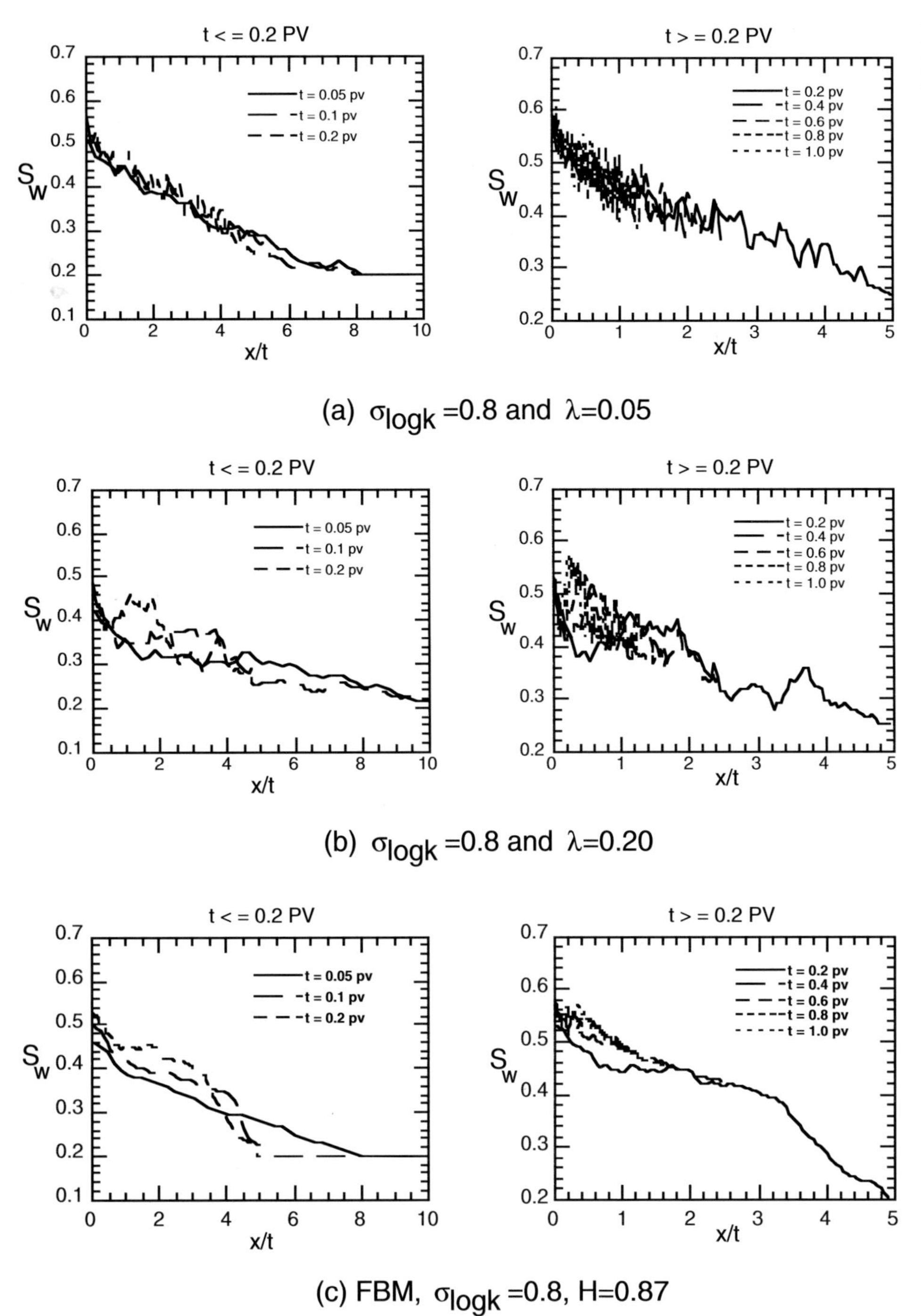

Figure 9. Saturation-velocity profiles at $\sigma_{\log k} = 0.8$ for oil-wet system.

length on effective relative permeability increases as the medium become more oil wet.

From these simulation results, we find that permeability variance has more of an impact on the fluctuation of saturation velocity than correlation length for media with spherically correlated permeability fields. To quantify the effect of the permeability variance on the fluctuation of saturation velocity, the average saturation velocity was first determined based on the saturation velocity profiles after breakthrough using a weighted least square method. Then the difference between average saturation velocity and actual saturation velocity was calculated. Figure 12 shows the relation between the standard deviation of fluctuation of saturation-velocity, $\sigma_{f'(Sw)}$, and standard deviation of permeability field, $\sigma_{\log k}$, at $\lambda = 0.05$. The standard deviation of fluctuations of saturation-velocity for both water-wet and oil-wet systems increases almost linearly with the standard deviation in the permeability. The velocity fluctuation is larger for oil-wet systems than for water-wet systems at the same level of heterogeneity; therefore, the heterogeneity region for which the k_r-formulation is adequate is larger for the water-wet media than for the oil-wet media.

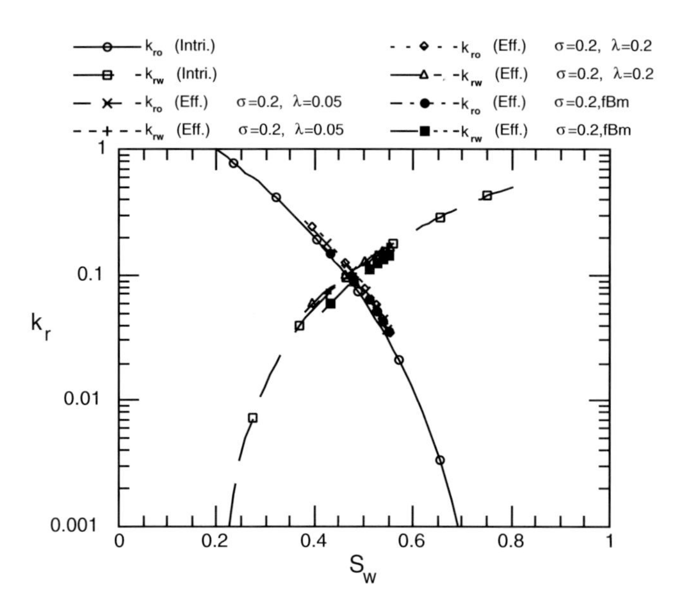

Figure 10. Relative permeability at σ_{logk} = 0.2 for oil-wet system.

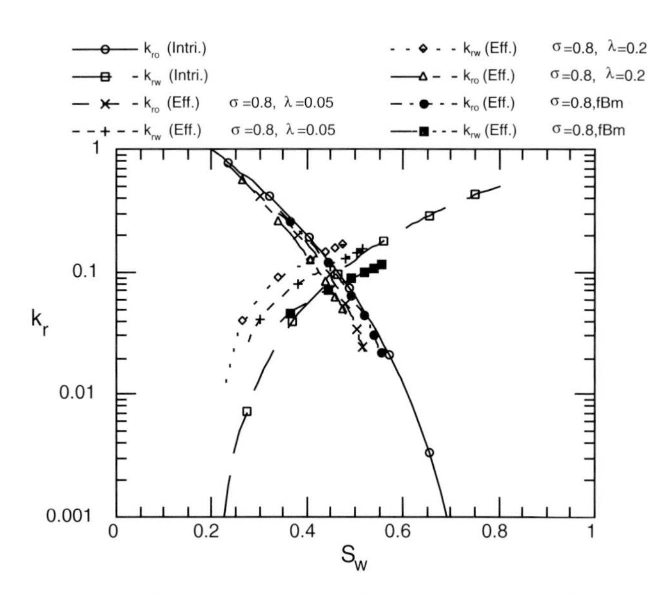

Figure 11. Relative permeability at σ_{logk} = 0.8 for oil-wet system.

In-situ water saturations provide insights into the effect of wettability on multiphase flow in heterogeneous media. Figure 13 compares the in-situ water saturations at $\sigma_{logk} = 0.8$ and $\lambda = 0.2$ for water-wet and oil-wet media. A constant initial water saturation is assumed for the oil-wet media. The right half of the figure shows the flow distribution at different injected pore volumes for the oil-wet medium. As water flows through the oil-wet medium, water first moves to high-permeability grids as illustrated in the figure at 0.05 PV, 0.1 PV, and 0.2 PV. Capillary pressure in the oil-wet medium resists the flow of water through low-permeability grids. Water can move through the low-permeability grids only

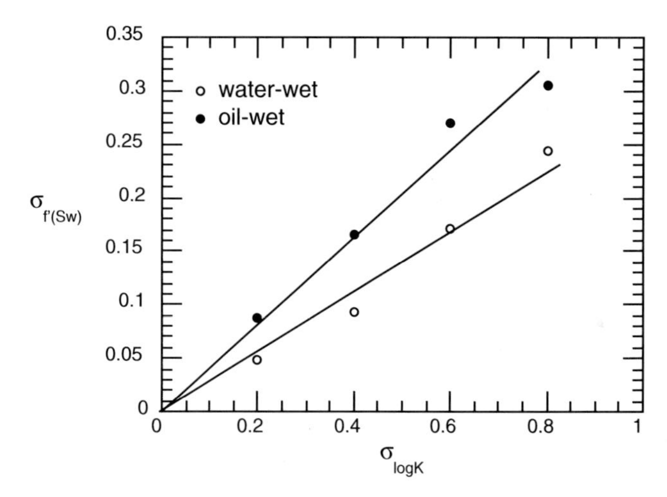

Figure 12. The effect of standard deviation of permeability on the fluctuation of saturation-velocity for λ = 0.05.

when the viscous forces overcome the capillary forces. Even at 1.0 PV (not shown), many of the low-permeability regions are bypassed. The capillary pressure term in the oil-wet medium tends to destabilize front movement. In water-wet media, capillary pressure helps imbibe water into low-permeability grids, whereas viscous pressure drop tends to move water into high-permeability grids. As shown on the left half of Figure 13, the saturation fronts in the water-wet medium are more uniform than those in the oil-wet medium. The water-wet medium is almost completely swept at 1.0 PV. The capillary pressure term in the water-wet medium helps to stabilize front movement.

CONCLUSIONS

This study has looked at relative permeability upscaling from the laboratory scale (several centimeters) to fine-grid field scale (several meters). The effects of wettability, permeability variation, and correlation length scale are studied. Only the flow in the horizontal direction is considered. Gravitational effects and three-dimensional flow are neglected.

- For strongly water-wet media, capillarity has a stabilizing/homogenizing effect on multiphase flow. As the oil wettability increases, the capillary stabilizing effect decreases.
- For small variance in permeability and small correlation length, effective relative permeability can be described by the intrinsic relative permeability. At higher variance and moderate correlation length, the average flow can be described by a dynamic relative permeability. The deviation from this average increases with increasing oil wettability.
- For fractal fields with large Hurst exponents and large variance in permeability, effective relative permeability is not adequate in describing the flow.

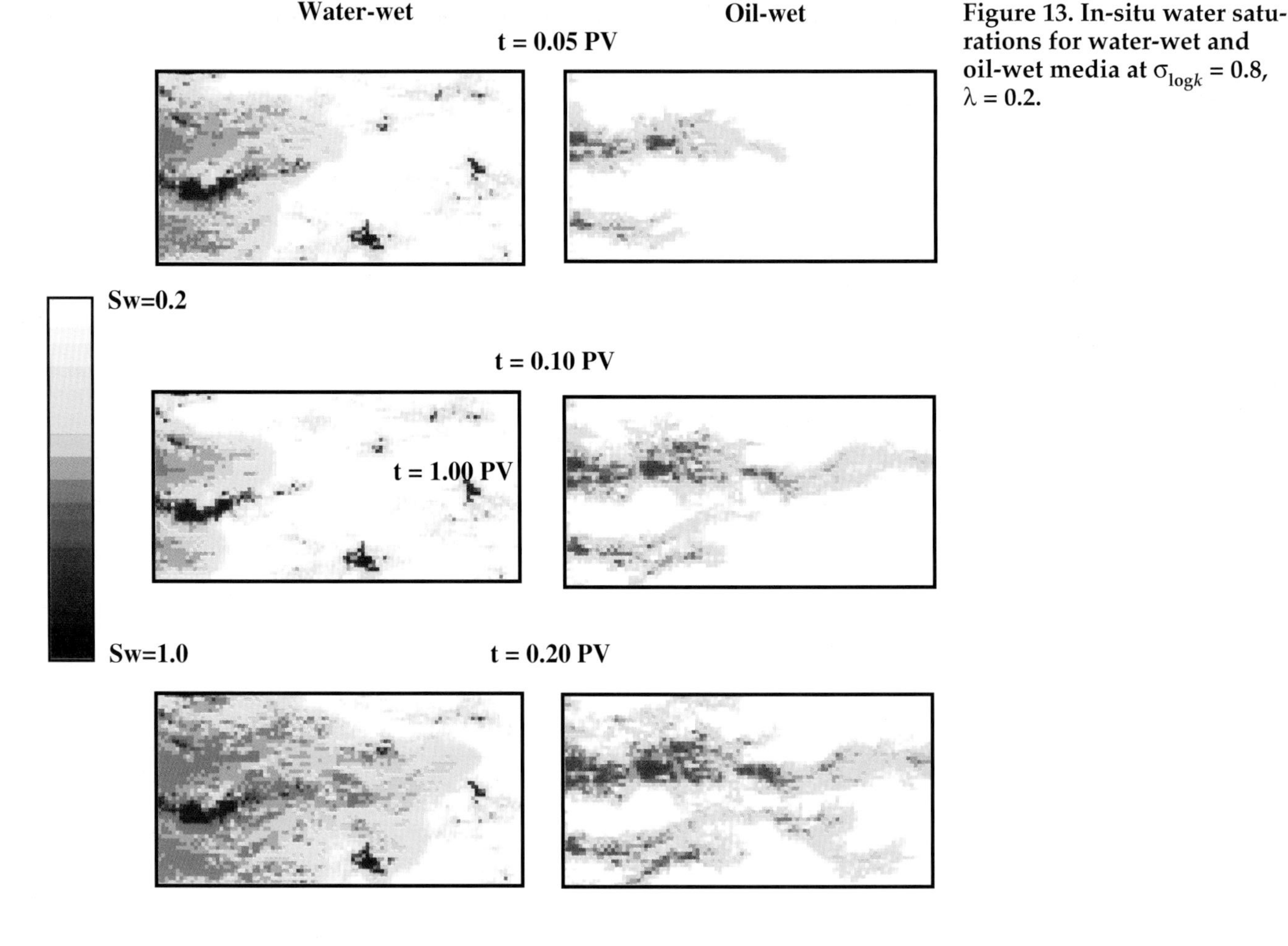

Figure 13. In-situ water saturations for water-wet and oil-wet media at $\sigma_{\log k} = 0.8$, $\lambda = 0.2$.

ACKNOWLEDGMENTS

This work was partially funded by the Energy Lab, ARCO, Chevron, and Mobil.

REFERENCES CITED

Barker, J. W., and S. Thibeau, 1997, A critical review of the use of pseudo relative permeabilities for upscaling, SPERE, p. 138–143.

Beier, R. A., and H. H. Hardy, 1993, Comparison of 2-D and 3-D fractal distribution in reservoir simulation, SPE 25236, SPE 12th Symposium on Reservoir Simulation, New Orleans, LA, Feb. 28-March 3, p. 25–34.

Bhogeswara, R., and J. E. Killough, 1993, Parallel linear solvers for reservoir simulation: A generic approach for existing and emerging computer architectures, SPE 25240, SPE 12th Symposium on Reservoir Simulation, New Orleans, LA, Feb. 28-March 3, p. 71–82.

Butts, M.B., 1991, A stochastic model for two-phase flow in heterogeneous porous media, Progress Report 73, Inst. Hydrodynamics and Hydraulic Eng., Tech. Univ. Denmark, p. 31-54.

Chang, Y.C., and K. K. Mohanty, 1997, Scale-up of two-phase flow in heterogeneous porous media, Journal of Petroleum Science and Engineering, 18, p. 21-34.

Durlofsky, L. J., R. A. Behrens, R. C. Jones, and A. Bernath, 1995, Scale-up of heterogeneous three-dimensional reservoir descriptions, SPE 30709, ATCE of SPE, Dallas, TX, Oct. 22-25, p. 53–66

Edwards, M., and M. A. Christie, 1993, Dynamically adaptive Gudonov schemes with renormalization in reservoir simulation, SPE 25268, SPE 12th Symposium on Reservoir Simulation, New Orleans, LA, Feb. 28-March 3, p. 413–422.

Gelhar, L. W., 1984, Stochastic analysis of flow in heterogeneous media, in J. Bear and M. Y. Corapcioglu, eds., Fundamentals of transport phenomena in porous media: Martinus Nijhoff Publishers, Boston, p. 673-718.

Glimm, J., W. B. Lindquist, F. Pereira, and Q. Zhang, 1993, A theory of macrodispersion for the scale-up problem, Transport in Porous Media 13, p. 97-122.

Hewett, T.A., 1986, Fractal distributions of reservoir heterogeneity and their influence on fluid transport, SPE 15386, ATCE of SPE, New Orleans, LA.

Hewett, T., and R. Behrens, 1991, Scaling laws in reservoir simulation and their use in a hybrid finite difference/streamtube approach to simulating the effects of permeability heterogeneity, in L. W. Lake, H. B. Carroll, Jr., and T. C. Wesson, eds., Reservoir characterization II: Academic Press, p. 402-441.

Honarpour, M., L. F. Koederitz, and H. A. Harvey,

1982, Empirical equations for estimating two-phase relative permeability in consolidated rock, Journal of Petroleum Technology 34, p. 2905-2908.

Jadhunandan, P. P., and N. R. Morrow, 1995, Effect of wettability on waterflood recovery for crude-oil/brine/rock systems, SPERE, p. 40-48.

Johnson, E.F., D. P. Bossler, and V. O. Naumann, 1959, Calculation of relative permeability from displacement experiments, Transactions of the AIME 216, p. 370–372.

Journel, A. G., 1990, Geostatistics for reservoir characterization, SPE 20750, ATCE of SPE, New Orleans, LA, Sep. 23–26, p. 127–136.

Killough, J. E., 1976, Reservoir simulation with history-dependent saturation function, SPE Journal 16, p. 37-48.

King, P. R., 1989, The use of renormalization for calculating effective permeability, Transport in Porous Media 4, p. 37-58.

King, P.R., A. H. Muggeridge, and W. G. Price, 1993, Renormalization calculation of immiscible flow, Transport in Porous Media 12, p. 237-260.

Lake, L. W., 1989, Enhanced oil recovery, Prentice Hall, p. 128–140.

Land, C. S., 1968, The optimum gas saturation for maximum oil recovery from displacement by water, SPE 2216, ATCE of SPE, Houston, TX, Sep. 29-Oct. 2.

Li, D., A. S. Cullick, and L. W. Lake, 1996, Scale-up of reservoir model relative permeability with a global method, SPERE, p. 149-157.

Muggeridge, A. H., 1991, Generation of effective relative permeabilities from detailed simulation of flow in heterogeneous porous media, in L. W. Lake, H. B. Carroll, Jr., and T. C. Wesson, eds., Reservoir characterization II: Academic Press, p. 197-225.

Quintard, M., and S. Whitaker, 1990, Two-phase flow in heterogeneous porous media I: The influence of large spatial and temporal gradients, Transport in Porous Media 5, p. 341-379.

Painter, S., 1995, Random fractal models of heterogeneity: The Levy-stable approach, Mathematical Geology 27, p. 813-830.

Peitgen, H.-O., and D. Saupe, 1988, The science of fractal images, Springer-Verlag, New York, p. 11–20.

Smith, E. H., 1991, The influence of small-scale heterogeneity on average relative permeability, in L. W. Lake, H. B. Carroll, Jr., and T. C. Wesson, eds., Reservoir characterization II: Academic Press, p. 52-76.

Thiele, M. R., M. J. Blunt, and F. M. Orr, 1994, A new technique for predicting flow in heterogeneous systems using streamtubes, SPE/DOE 27834, SPE/DOE 9th Symposium on Improved Oil Recovery, Tulsa, OK, p. 571–582.

Warren, J. E., and H. S. Price, 1961, Flow in heterogeneous porous media, Journal of Petroleum Technology 13, p. 153-169.

Wolcott, D. S., and A. K. Chopra, 1993, Incorporating reservoir heterogeneity with geostatistics to investigate waterflood recoveries, SPEFE, p. 26-32.

Worthington, P. F., 1991, Reservoir characterization at the mesoscopic scale, in L. W. Lake, H. B. Carroll, Jr., and T. C. Wesson, eds., Reservoir characterization II: Academic Press, p. 123-165.

Section VII
Modeling

Seifert, D., et al., Evaluation of field development plans using 3-D reservoir modeling, 1999, *in* R. Schatzinger and J. Jordan, eds., Reservoir Characterization-Recent Advances, AAPG Memoir 71, p. 321–332.

Chapter 22

Evaluation of Field Development Plans Using 3-D Reservoir Modeling

D. Seifert[1]
Heriot-Watt University
Edinburgh, Scotland

J.D.H. Newbery[2]
C. Ramsey[3]
Conoco (U.K.) Ltd.
Aberdeen, Scotland

J.J.M. Lewis[4]
Heriot-Watt University
Edinburgh, Scotland

ABSTRACT

Three-dimensional reservoir modeling has become an accepted tool in reservoir description and is used for various purposes, such as reservoir performance prediction or integration and visualization of data. In this case study, a small northern North Sea turbiditic reservoir was to be developed with a line-drive strategy utilizing a series of horizontal producer and injector pairs, oriented north-south. This development plan was to be evaluated and the expected outcome of the wells was to be assessed and risked.

Detailed analyses of core, well log, and analog data have led to the development of two geological end member scenarios, thus accounting for uncertainties associated with the geological model. Both scenarios have been modeled using the sequential indicator simulation method in a hybrid deterministic-stochastic approach. The resulting equiprobable realizations have been subjected to detailed statistical well placement optimization and analysis techniques. Based upon bivariate statistical evaluation of more than 1000 numerical well trajectories for each of the two scenarios, it was found that the inclinations and lengths of the wells had a greater impact on the wells' success, whereas the azimuth was found to have only a minor impact. After integration of these results, the actual well paths were redesigned to meet external drilling constraints, resulting in substantial reductions in drilling time and costs.

[1]Present Address: ARCO E&P Technology, Plano, Texas, U.S.A.
[2]Present Address: Enterprise Oil Ltd, London, England.
[3]Present Address: LASMO North Sea Plc, London, England.
[4]Present Address: Landmark, Houston, Texas, U.S.A.

Although three development wells drilled subsequent to this study were very successful, their outcome raises questions about the validity of the stochastic model, which is based on geological assumptions which, in turn, were derived from much fewer well data. It is clear that a better quantitative sedimentological understanding of the reservoir, specifically in the lateral dimension, would have resulted in a more reliable reservoir model.

INTRODUCTION

Journel and Alabert (1990) stated that "the goal of detailed reservoir description is to provide a numerical model of the reservoir attributes for input into various flow simulators." Today, reservoir (flow) simulations form the basis of almost all economic decisions for a field development strategy, including improved oil recovery (IOR) and enhanced oil recovery (EOR), and reservoir management (Haldorsen and van Golf-Racht, 1992). Even though flow simulation probably is still the most important application of numerical reservoir models, there are by now other very important applications, such as well placement optimization studies, the provision of a common database (forcing quantification and integration of data), and the three-dimensional (3-D) visualization of data.

This study is an example of one of these applications. This paper will show how multiple realizations of two stochastic reservoir models of a North Sea reservoir were used to assess the success of high-angle development wells.

Based on two-layer, single-phase flow simulation, a development plan was devised for this reservoir utilizing a series of horizontal injector and producer pairs, oriented north-south. The objective of this study was to develop a stochastic model of the reservoir, evaluate the planned drilling program, and transfer the technology to the operating environment.

To ensure the technology transfer, the company seconded the development geologist for the field to Heriot-Watt University for the duration of the entire study so he would gain hands-on experience with every step within the study process. The study process was building on earlier work (Hern et al., 1996; Seifert et al., 1996) and involved quantitative sedimentological analysis, stochastic reservoir modeling, and statistical well placement optimization. More specifically, the section on data acquisition and analysis explains how lithofacies were identified, genetic units were defined, and spatial statistics were derived. The section on reservoir modeling procedures briefly introduces the modeling method chosen in this study and provides further references. Also, it outlines how the modeling procedure was adapted to handle geological uncertainties and reproduce lateral deterministic trends. The section on well placement optimization procedures illustrates how a large number of numerical wells were drilled through the models and their results were analyzed with respect to identifying optimum well trajectory (analysis of results).

In this statistical well placement optimization technique, a number of static measurements are used to identify the optimum trajectory for development wells. Based on these data, actual drilling trajectories are designed by developing drilling diagrams that envelop the potential range of drilling targets (Solomon et al., 1993). During the drilling process, newly developed LWD (logging-while-drilling) tools are able to provide new data every hour; these data may be quickly integrated into a deterministic framework of the models to help steer the drill bit between the target boundaries (Bryant and Baygün, 1996).

Finally, the impact of the results on the drilling plan is presented, showing how time and money may be saved by changing the azimuth of the proposed development wells.

GEOLOGICAL OVERVIEW AND INITIAL DEVELOPMENT PLAN

The field under study is a small turbidite sandstone reservoir of late Paleocene age located in the northern North Sea. It comprises a structure, with closure in all four directions, that is relatively flat but has steep flanks along the channel margins. Prior to development, the field was delineated by four wells, all of which were comprehensively logged and cored.

A wide variety of facies have been identified in the core. These facies range from massively stacked channel sands representing high-density turbidite flows to thinly interbedded very fine sands and shales that represent channel margin deposits. Relatively thick hemipelagic shales have also been identified in the core. Over parts of the field, some of the sands and shales have moved due to postdepositional liquefaction. This movement has caused load structures, sand injection features and where body shear has occurred, significant slumping.

The initial (Phase 1) development plan has been designed to access proven, minimum reserves derived from seismic data and appraisal wells. Three horizontal production wells have been targeted at areas where the likelihood of encountering stacked submarine channel sands is highest, close to the existing appraisal wells. These areas also coincide with those parts of the field with the highest structural relief, and hence have the potential to achieve highest deliverability. Added together, the Phase 1 development areas are estimated to contain over half of the oil-in-place for the field.

In order to pursue a strategy that confirms the minimum-reserves case while providing the potential to access additional economic reserves, the original development plan included up to nine high-angle/horizontal wells, comprising four producers and five injectors. The three producers drilled as Phase 1 were to be followed by at least one more producer and up to five injectors as Phase 2.

Pressure maintenance is the biggest concern regarding recovery efficiency and the final well count after development Phase 2. Even in the Phase 1 development area where sand connectivity is assumed to be good, basal aquifer support has been considered to be a major uncertainty. As a result, the base case development strategy assumes that following an initial period of primary depletion during which formation pressure data will be obtained, a line-drive waterflood may need to be implemented to maintain reservoir pressure above the bubble point and to improve sweep efficiency. This second phase of drilling will begin when production data have been acquired to define the most appropriate water injection strategy.

DATA ACQUISITION AND ANALYSIS (RESERVOIR AND ANALOG)

To perform any modeling work, it is necessary to delineate the modeling elements and gain quantitative data on their form and spatial distribution. To acquire the appropriate data for the modeling process, a multistep approach has been employed that involves (1) identification of the lithofacies, (2) definition of the genetic units, (3) derivation of spatial statistics of the genetic units (proportion, geometry, size, orientation), and (4) construction of a conceptual model.

Identification of Lithofacies

The sedimentological logs, probe-permeametry data, petrophysical data, and core photographs were analyzed with each probe-permeametry point being given a lithofacies indicator label. The probe-permeametry data give significant control on this identification process (Hurst and Goggin, 1995). It appeared that a significant number of the sandstone flow units have an upward-increasing permeability profile. Similar profiles have been identified in the Tabernas Basin (Kleverlaan, 1994), where it may be associated with dewatering structures. On identification of each lithofacies unit in the cored section, a full core viewing was held to calibrate the results.

On completion of the lithofacies identification from the core, petrophysical descriptors were applied to each type. These descriptors, or signatures, were then applied to the remainder of the reservoir zone in the uncored sections. In practice, this was difficult to do because the log characteristics did not allow for detailed facies identification; however, the hemipelagic shales could be identified from their hot gamma ray responses.

Definition of the Genetic Units

A genetic unit is defined as a body of rock that is distinct from other bodies on the basis of geometrical, petrophysical, and spatial properties (Dreyer, 1993). This definition can be expanded to "...a genetic unit is an association of facies which are related by the same depositional process and which are distinct from other genetic units..." (C.Y. Hern and T. Good, 1996, personal communication); consequently, a genetic unit may contain several lithofacies types if it is believed they were deposited at the same time and have the same or similar spatial properties.

From core data, a total of six genetic units were interpreted (Table 1). These relate to the depositional processes involved and the grain size of the resulting deposits.

Derivation of Spatial Statistics

The genetic units (GUs) represent the basic building blocks of the reservoir model. It is therefore important to characterize the relevant GUs in terms of shape, size, and orientation within the reservoir. Of fundamental importance is the proportion of a given GU at a given location within the reservoir. Some of these data can be derived from core and well data, such as proportions and thicknesses of the relevant GUs. Regional orientations on the reservoir scale most likely will be derived from regional geological interpretations, including data from nearby fields. Dipmeter data, however, may provide an indication of (local) orientation of, for example, fluvial GUs.

Other data, such as width and approximate length of GUs, can be deduced only from outcrop analog data. It is, of course, very difficult to find a "true" analog for any reservoir (unless the very formation crops out in the "neighborhood"); however, certain measures can be used to assess the similarity of the outcrop to the reservoir at hand (C.Y. Hern and T. Good, 1996, personal communication). For example, thicknesses, aspect ratios, and proportions of individual GUs can be compared, as well as the size of the entire depositional system.

For this study, the Eocene Ainsa II channel complex in the south-central Pyrenées was chosen as an analog (Clark and Pickering, 1996). The Ainsa II is not a true analog to the field under study; however, there are similarities between them in terms of turbidite system dimensions, channels, and flow units. Integrating the quantitative sedimentological and petrophysical data from the Ainsa II and published ancient and modern data, the key genetic units could be delineated. Data gathered included average thicknesses and proportions (from wells), and aspect ratios and estimated lengths (from outcrop) of the key GUs.

Table 1. Modeling Categories, Genetic Unit Description, and Interpretation.

Category	Genetic Unit Description	Interpretation
Channel sand	Clean sandstones, often stacked	High-density turbidites
	Argillaceous sandstones, often stacked	Low-density turbidites
	Clean sandstones with mixed sandstones and shales at top	High-density turbidites topped by load structures and slumped materials
	Clean sandstones with mixed sandstones and shales at base	High-density turbidites with loaded shales and sandstones at base
Slumped	Mixed sandstones and shales, often containing clastic injections	Slump deposits and sandstone injections
Shale	Bioturbated mudstone	Hemipelagic shales

Conceptual Reservoir Model

Before any stochastic modeling is undertaken, it is essential that a conceptual geological model be developed. In this case, it was important to understand the stacking and channel fill patterns of the channels and flow units. Clark and Pickering (1996) have shown that in a typical upper fan channel fill, different stacking and channel fill patterns are to be expected. To capture this degree of complexity stochastically within one model is very difficult, and it was therefore decided to model the two endpoints of the sedimentation process separately: scour and fill and lateral accretion. The scour and fill model assumed the flow units to be horizontal, and the lateral accretion model had the units dipping at a low angle.

RESERVOIR MODELING PROCEDURES

Modeling Technique

To build a computer-based model of the reservoir, a stochastic reservoir modeling technique, called Sequential Indicator Simulation (SIS) is used. The SIS method (pixel-based stochastic simulation technique) was chosen over a Boolean method (object-based simulation technique). The main advantage of this modeling method is its capability to reproduce very complex heterogeneity patterns, allowing for different orientations, aspect ratios, and frequencies for each modeling category (discrete or continuous). This method also allows for reproduction of geological trends and facies associations through spatial cross-correlation between indicator variables. The flexibility comes from the use of one indicator variogram for each indicator variable to be modeled. Detailed discussions of the SIS technique have been provided by Deutsch and Journel (1992) and Alabert and Modot (1992). Numerous case studies have shown that SIS proves to be effective in modeling reservoir heterogeneities (Journel and Gle to be modeled. DetaAlabert and Massonnat, 1990; Journel and Alabert, 1990; Massonnat et al., 1992).

Model Setup

The aim of this study was not to produce a full-field 3-D model for reservoir simulation purposes, but to model the reservoir heterogeneity as accurately as possible to understand its impact on well placement. As a result, stochastic modeling was applied only to model discrete variables; i.e., the key genetic units.

Geological uncertainty exists at two scales. First, on a large scale, the possibility exists that there are two channel systems present within the reservoir, instead of only the one that has been found by Well A (Figure 1). To reduce this uncertainty, it was decided to focus on the

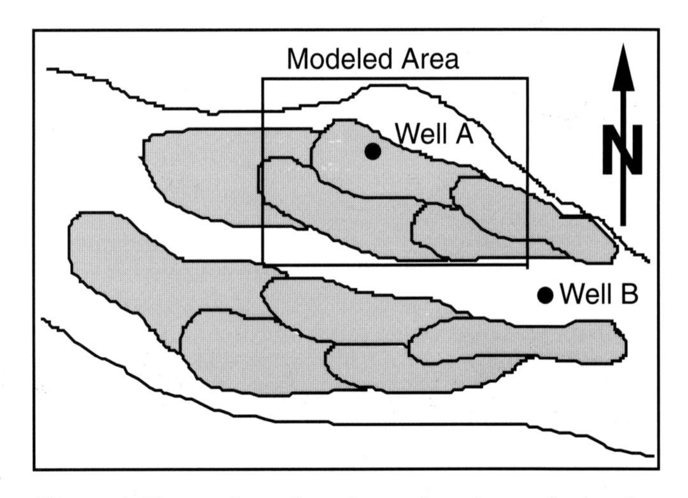

Figure 1. Two submarine channel systems, derived from seismic data, are shown as shaded. The northern channel system is confirmed by Well A, whereas the existence of the southern channel system is uncertain. The black box outlines the portion of the reservoir that has been modeled.

northern part of the field where two of the three Phase 1 wells would be drilled; however, this model also could serve as a generic model for the southern channel system, if it were present. Second, on a smaller scale, uncertainty exists whether the sands within the channel system or systems are laterally accreted or represent a scour and fill scenario (Figure 2). In order to account for this uncertainty, both possibilities, representing two sedimentological end members, have been modeled and subsequently analyzed separately.

In addition, three general assumptions were made for the modeling:

(1) All structure has been removed, resulting in an orthogonal reservoir model,
(2) No faulting is present within the modeled area, and
(3) The thickness of the reservoir model was set to 80 ft (24 m) even though the proven oil column is about 150 ft (45 m).

Assumptions (1) and (2) allow the model and subsequent well placement optimization results to be adapted quickly to fit any geophysical interpretation. This was deemed important because this reservoir is currently in its earliest stage of development, and the seismic interpretation is likely to be revised after more well data become available. Assumption (3) was made because (1) the top 20 ft (6 m) of the reservoir was classed as nonpay to avoid potential poorly connected sand bodies, which included sand injection features or thin, isolated turbidite flows, (2) the uncertainty on the geophysical top structure map was about ±20 ft (6 m), and (3) based on flow simulation studies it was decided that no producing well should go within 50 ft (15 m) of the oil-water contact, to avoid water coning.

As a result the modeled area covered a volume of 4600 × 4300 × 80 ft (1403 × 1311 × 24 m). The size of the smallest genetic unit that needed to be modeled, the "slumped" GU, determined the cell size, resulting in a total model of 632,960 cells, each of which was 50 × 50 × 1 ft (15 × 15 × 0.3 m) in size.

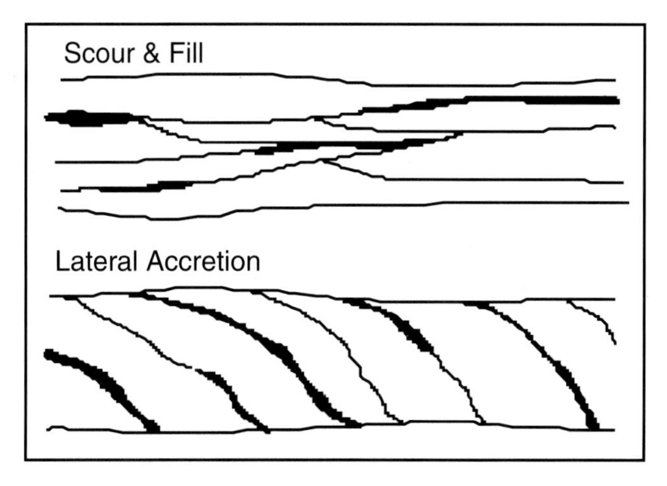

Figure 2. Smaller scale geological uncertainty resulted in two possible channel fill scenarios, scour and fill and lateral accretion. This result is due to a lack of core coverage (only one well that hit the channel system).

Sensitivity Testing

Extensive sensitivity testing is an essential first step within the reservoir modeling procedure. In sensitivity testing, it is the aim to establish the proper parameters for both, technique specific (e.g., variogram type, search ellipsoid) and geological (e.g., lateral trends, correlation lengths) parameters. Up to thirty binary, ternary, and four-component models were created for both depositional end members, using models that have a reduced total thickness of only 25 ft (7 m) (to save computation time), but full lateral dimensions.

It was not possible to model all six genetic units stochastically (Table 1) because each simulation run would take too long to complete. It was therefore necessary to group several GUs together appropriately into modeling categories. Thus, after establishing the proper variogram parameters for each GU (e.g., variogram type, correlation lengths, anisotropies) and search neighborhood (i.e., ellipsoid), it was investigated how well the heterogeneity would be represented using models with two, three, and four modeling categories.

For the binary model, pay was modeled versus nonpay. Effectively, the clean sands were modeled versus the mixed sands and shales. For the ternary model, three different groupings were investigated. The most appropriate grouping modeled the shale GU versus the slumped GU versus the clean sand GU. For the four-component model, two different groupings were investigated. The most appropriate grouping modeled the shale GU versus the slumped GU versus the clean sand GU versus the sands with loaded structures GU.

By comparing the results of the three and four-component models, it can be seen that the four-component model improved the level of heterogeneity by splitting the pay up into two different sand genetic units (clean sand GU and sands with loaded structures GU). If flow simulation had been the end product of this study, the four-component representation would have allowed for different effective properties within the pay sands and therefore yielded a more accurate description of the heterogeneity. This was, however, considered unnecessary, because only the intersection of sands with the wells was of interest, not their type. The binary model was not believed to be appropriate because it forced an amalgamation of the slumped GU and the shale GU into one modeling category. Because both have significantly different shapes and sizes, it would not have been possible to represent their presence appropriately within the channel sands. Their appropriate representation, however, is most crucial in this high net-to-gross (N/G) reservoir because these govern the potential compartmentalization of the channel sand bodies; therefore, the approach taken in the ternary model (Table 1) was chosen as the optimum way to proceed.

A further complication was introduced when trying to reproduce the lateral deterministic trends as expected within this reservoir. The model was to extend beyond the main channel system to the north and south (Figure 1). Both edges are expected to have much lower

N/G ratios than the channelized center, as derived from Well B. Second, both edges would have different proportions of each GU because the northern edge would represent the end of the turbidite system, exhibiting a very low proportion of hemipelagic shales, but a fairly high proportion of slumped material, whereas the southern edge would represent a barrier or transition zone between two possible channel systems, resulting in mixed proportions of hemipelagic shales and slumped materials. Finally, all deposits are expected to be flat lying within the edge volumes, even for the case that the channelized center would be inclined to represent the lateral accretion system. As a result, these "edge" volumes would have had to be modeled differently from the main channelized section.

The stationarity principle in stochastic modeling ensures that a certain set of parameters, represented by indicator variograms and proportion statistics for the SIS method, will be applied and honored throughout the entire model; therefore, lateral nonstationarity was modeled by splitting up the model into three sections along the north-south axis, with a center of 3000 ft (915 m), and a northern and southern edge of 650 ft (198 m) each (Figure 3). Appropriate sets of statistical parameters were defined for each section based on well data and geological assumptions with reference to the respective depositional end member to be modeled. The sections were then modeled sequentially, conditioning each section to the adjacent face of the neighboring section.

This sequential process turned out to be complex. It became important in which order the sections were modeled. In the case of modeling the edges before the center section, results looked much different from the case where the center was modeled before the edges (Figure 4). This is an effect of the conditioning process

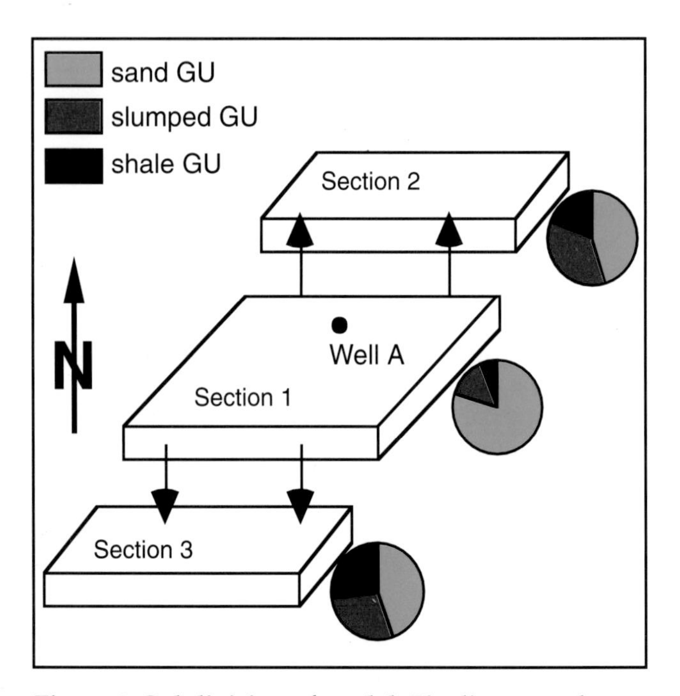

Figure 3. Subdivision of model. Pie diagrams show proportions of GUs present within sections.

(a)

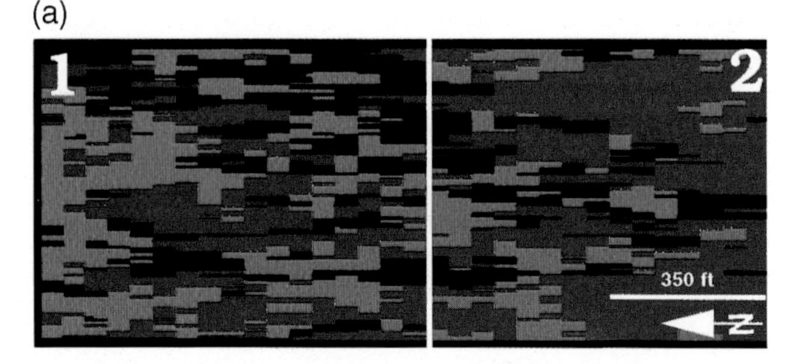

(b)

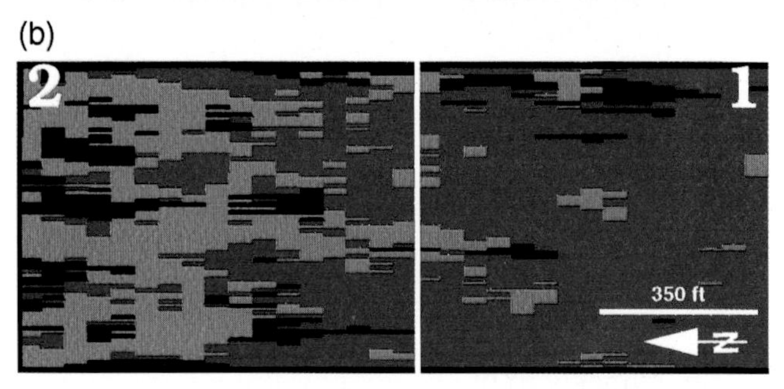

Figure 4. (a) Vertical cross-section (80 ft; 24 m). Modeling the edge first (left) and then the center (right). Only part of the center section is shown. Sand GU (red) vs. slumped GU (green) vs. shale GU (blue). (b) Vertical cross-section (80 ft; 24 m). Modeling the center first (right) and then the edge (left). Only part of the center section is shown. Sand GU (red) vs. slumped GU (green) vs. shale GU (blue).

using the adjacent face of the neighboring section. When modeling the low N/G edges first, all GUs would be distributed evenly throughout the edge sections. This would result in a significant amount of slumped and shale GUs and a relatively low amount of channel sands at the border to the center section. These would then have to be honored by the subsequently modeled center section, pushing away a lot of the channel sands from the border in the center section. This resulted in the presence of smaller, isolated channel bodies encased in slumped sands and shales throughout the edge section and the area near the boundary of the center section (Figure 4a). When modeling the center section first, all GUs would be distributed evenly, and a high proportion of channel sands would be placed at the border to the edges. These would then be honored by the edge statistics, drawing a lot of the channel sands within these sections toward the boundary to the center section. This resulted in large connected channels across the boundary, with only minor proportions of channel sands in the far areas of the edge sections (Figure 4b).

This difference illustrates how dependent the result can be on the modeling approach; however, in this case, these different results can be related to sedimentological principles. The case where the edge section is modeled first (Figure 4a) can be related to a turbidite system with vertically aggradating confined channels and well defined levees. On the other side of the levees, individual channels may develop as well (Figure 5a).

(a)

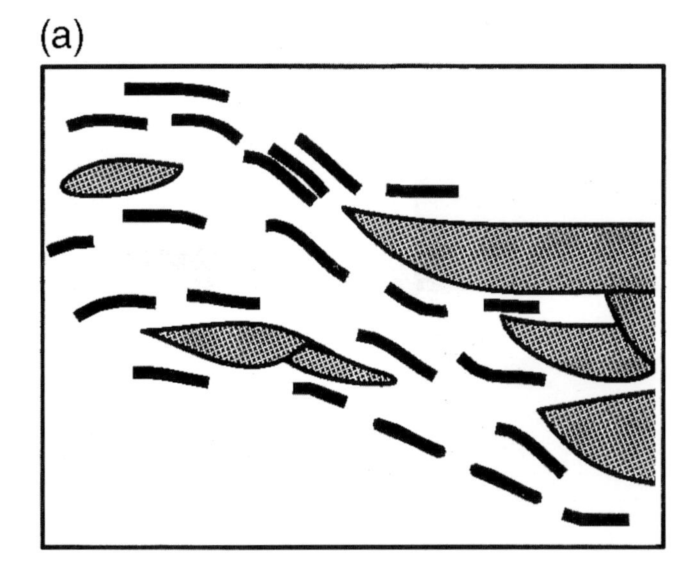

(b)

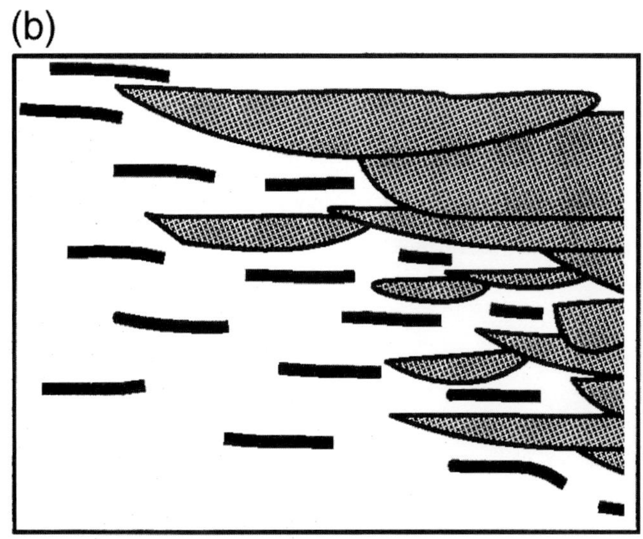

Figure 5. (a) Schematic of a turbidite system characterized by vertically aggradating confined channels and well defined levees. (b) Schematic of a turbidite system characterized by less confined channels with a component of lateral migration.

The case where the center section is modeled first (Figure 4b) can be related to a turbidite system that is characterized by less confined channels with a component of lateral migration (Figure 5b). Hence, it is of crucial importance to check the results of each modeling step with respect to the geological model of the reservoir under study. In this study, the second approach was deemed more appropriate.

Modeling and Model Descriptions

Analysis of the geological data and the sensitivity testing, resulted in two different geological scenarios, representing the scour and fill and the lateral accretion end members. Each scenario was modeled by subdivision into three sections, the first section to be simulated (center section, Figure 3) being constrained by Well A, Sections 2 and 3 (northern and southern edge sections, Figure 3) being conditioned to the adjacent faces of Section 1. For each scenario, a total of fifteen equiprobable realizations were obtained.

Figure 6 shows a vertical cross-section through the scour and fill model. The center has a very high proportion of channel sands (red) and very low proportions of slumped deposits (green) and hemipelagic shale (blue). It is obvious that size and geometry information for the sand GU are of minor effect to this model. The slumped and injected sands (green) are of much smaller size than the hemipelagic shales. The northern edge of the model, representing the edge of the turbidite system where the channels are getting thinner and other deposits increase in proportion, is characterized by a medium N/G ratio, a slightly lower proportion of slumped deposits, and a relatively low proportion of hemipelagic shale. The ratio between the slumped and the hemipelagic shale is approximately 2:1. The southern edge of the model, representing the area between two potential channel systems, is characterized by a medium N/G ratio and lower proportions of slumped deposits and hemipelagic shales (ratio of 1:1).

Figure 7 shows a vertical cross-section through the lateral accretion model. Notice the approximately 5° inclination (strong vertical exaggeration) of all genetic units in the center model. The genetic units in the edges, however, have been modeled horizontally because this was found more appropriate for the northern edge and southern transition zone.

To complete the visual representation of this reservoir, Figure 8 shows an areal cross-section through the lateral accretion model. Notice the different proportions of the GUs in the northern and southern sections. The narrow shales (blue) represent the thickness of the inclined shales; furthermore, the boundaries for the sections and the well trajectories (below) have been superimposed.

Assessment of Potential Object-Based Modeling

For the channel genetic unit, excellent geometry and size data were available from outcrop analog. This would destine this genetic unit to be best modeled using an object-based approach; however, these geometries and sizes have lost importance due to the extremely high N/G ratio within the center portion of the reservoir. Geometry and size data for the slumped and shale GUs are much more important because they govern the interconnectivity of the sands. Because their sizes and geometries are only poorly known, they are likely to be modeled best using a pixel-based approach. In conclusion, this particular reservoir is likely to be modeled optimally with a pixel-based approach such as SIS. Object-based modeling would introduce geometrical artifacts based on assumed geometries that are not known; however, in cases where N/G ratios are significantly lower, a good knowledge of the channel geometries may elevate an object-based or combined modeling approach to be the method of choice.

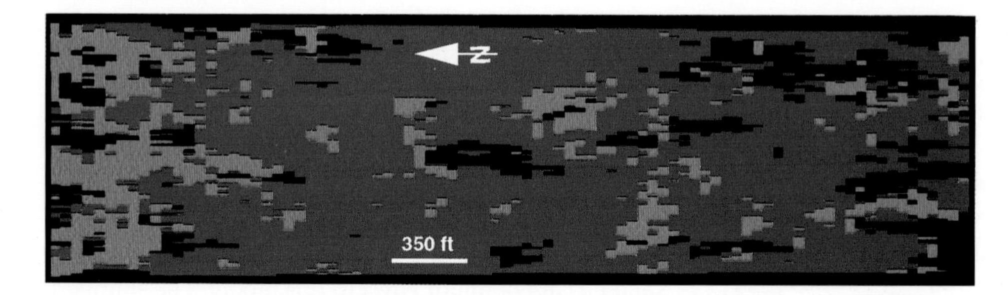

Figure 6. Vertical cross-section (80 ft; 24 m) of the scour and fill scenario. Sand GU (red) versus slumped GU (green) versus shale GU (blue).

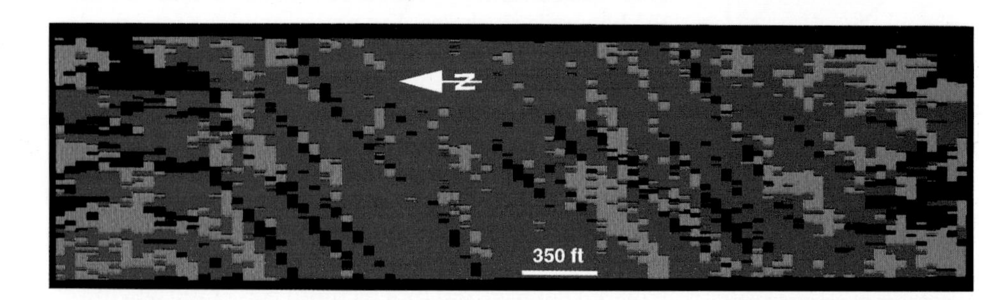

Figure 7. Vertical cross-section (80 ft; 24 m) of the lateral accretion scenario. Sand GU (red) versus slumped GU (green) versus shale GU (blue).

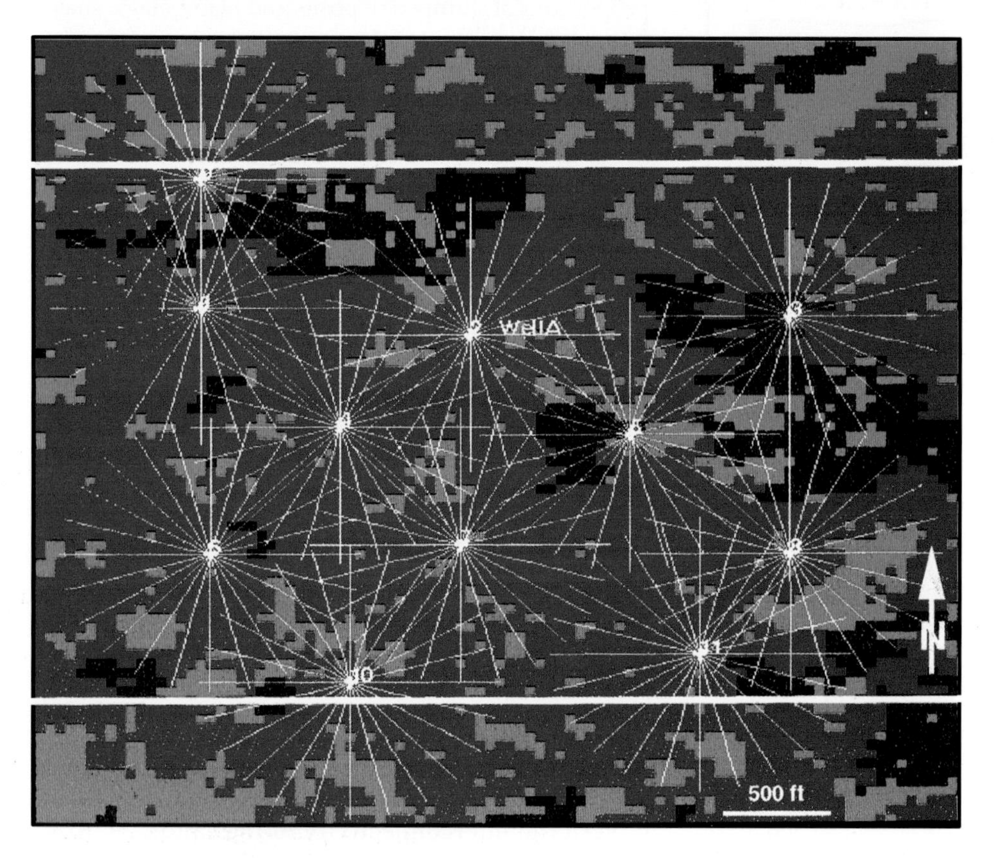

Figure 8. Areal cross-section of the lateral accretion scenario. Model size: 4600 × 4300 × 80 ft (1403 × 1311 × 24 m). Well trajectories and boundaries of sections are superimposed.

WELL PLACEMENT OPTIMIZATION PROCEDURES

The resulting fifteen realizations per geological end member have been subsequently, but independently, subjected to detailed statistical well placement optimization techniques (Seifert et al., 1996). In essence, originating from eleven cluster locations (Figure 8), a variety of linear wells have been drilled at every 15° azimuth with varying well lengths (ranging from 1000 to 2250 ft; 305 to 686 m) and inclinations (ranging from 0° to 5°). A total of 1004 numerical wells were drilled through each of the thirty realizations. Along each well trajectory, the amount of channel sand penetrated (sand GU proportions, or SGUPs) and the number of sand bodies intersected (SBIs) were extracted and analyzed. The following bivariate statistical evaluation of these data resulted in a statistical database of 60,240 data points.

ANALYSIS OF RESULTS

These 60,240 data points have been statistically analyzed to define the static success of different well trajectories. Bivariate statistical analysis (SGUPs and SBIs) of the 2 × 15 realizations results in minimum, mean, and maximum outcome values, which need to be evaluated. The mean value represents the likely outcome, and the difference between the minimum and the maximum values represents the range of the possible outcomes. The range of SGUPs and SBIs for each well trajectory can be quantified in terms of

- Location within the reservoir model,
- Azimuth angles,
- Angles of inclination, and
- Well success versus well lengths.

Because this paper intends to show the procedure rather than actual results, the presentation of the analysis will focus on the results from Cluster 4 (located in the western center of the center section) and the fifteen realizations with respect to the scour and fill scenario.

Location Within the Reservoir Model

All development wells are placed with the aim of keeping them in the area of highest N/G, which corresponds to the center section of the model; therefore, eight of the clusters have been placed all across the center section, making sure that the wells do not extend beyond the edges (Figure 8). By analyzing these clusters only, the outcomes evaluated will refer to the most likely case; however, in order to assess the impact on the wells' success in the case that the actual wells will extend beyond the first channel system into the transition zone to the south, two clusters have been placed there for analysis. Similarly, one cluster was placed in the very north to provide an outcome assessment for the case that the wells would reach the reservoir depth prematurely and drill through the edge of the turbidite system. These clusters could be analyzed together or separately, taking their locations into account.

Azimuth Angle

For each well cluster, twelve azimuth angles were analyzed which covered the whole 360° spectrum on a 15° incremental basis. Figure 9 plots the SGUPs versus azimuth for Cluster 4. All wells have the same inclination and length. As can be seen, the mean value remains almost the same throughout the data set, and the range of possible values displays only little variation. It is therefore obvious that azimuth (i.e., drilling across or along channels) will have a minor impact on the success of the wells. The reason is that because of the very high N/G ratio, channel bodies tend to lie next to each other and are not often separated by slumped material or hemipelagic shales. This is a very important result, particularly when considering that the drilling platform location is one of the external constraints on the drilling program; therefore, the azimuth of 15° was considered optimum because it allowed us to minimize the well length prior to entering the reservoir at target depth and azimuth.

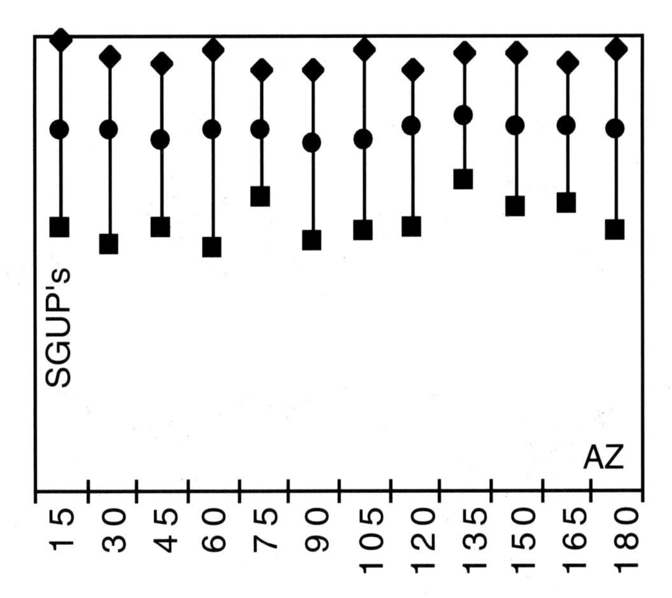

Figure 9. Cluster 4, scour and fill model, INC = 3°; well length = 1500 ft (457 m). Sand proportions versus azimuth.

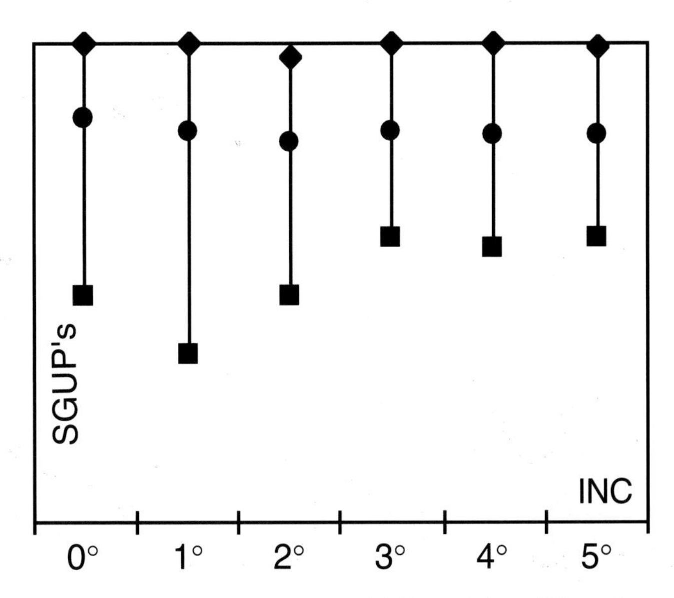

Figure 10. Cluster 4, scour and fill model, well length = 1500 ft (457 m). Sand proportions versus inclination.

Angle of Inclination

By changing the inclination angles from 0° (horizontal) to 5° (subhorizontal) in 1° increments, it is apparent in Figure 10 (for optimum azimuth) that even though the mean value does not change much, the range of outcome values reduces significantly as the inclination angle increases to about 3°. For example, the minimum value for the 3° well is more than 20% greater than for the 1° well, therefore reducing the risk of drilling a poor well.

With increasing the inclination, one can also observe that the amount of bodies intersected increases (Figure 11). Compared to the horizontal well

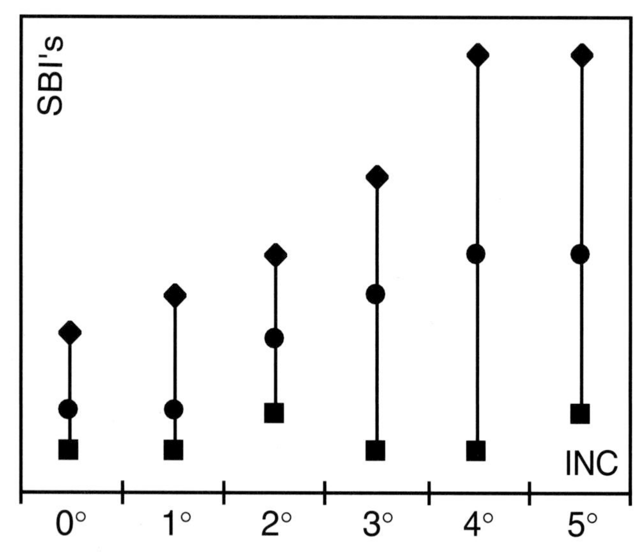

Figure 11. Cluster 4 scour and fill model, well length = 1500 ft (457 m). Sand bodies intersected versus inclination.

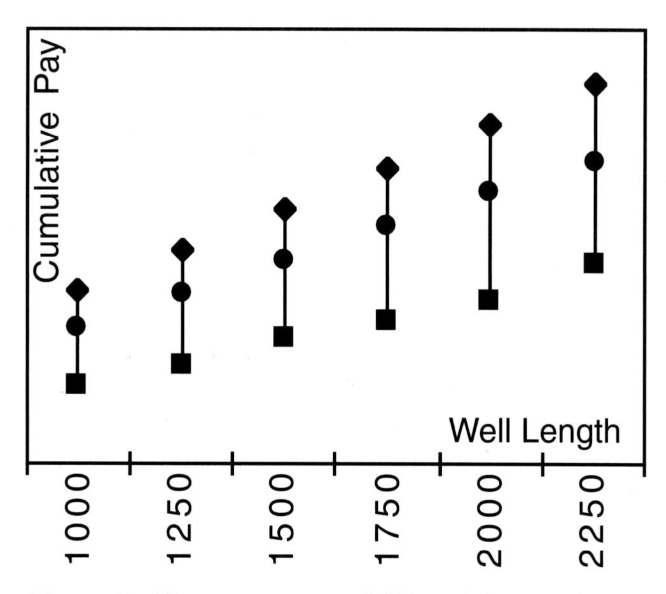

Figure 12. Cluster 4 scour and fill model. Cumulative pay versus well lengths for optimum azimuth.

of 0° inclination, the 3° inclined well almost triples the average (mean) number of sand bodies intersected, therefore significantly increasing the connected pay volume. Upon further inclination, the average (mean) values remain very similar, and only the range increases significantly; however, because of the very high N/G ratio in this reservoir, this indicator is of minor importance, especially when considering that the intervening shales may not act as barriers, thus not isolating one sand body from another.

It has been concluded that inclination of 3° is the optimum drilling angle because it allows a penetration of the target zone of up to 1500 ft (457 m). Upon further inclining the well, well length would have to be reduced in order to not extend into the 50 ft (15 m) envelope of the oil-water contact. The outcome of such a well, however, would be similar to the 3° inclined well.

Well Success as a Function of Well Length

From reservoir simulation studies, it had been concluded that a certain amount of net pay intersected by each of the development wells was required for well deliverability; therefore, six different well lengths have been investigated in this study. Figure 12 plots the cumulative pay intersected as a function of well length for the optimum azimuth with respect to the location of the drilling platform. There appears to be a linear relationship between cumulative pay and well length. Using the slope of the lines connecting the values, one can identify what well length is likely to be needed to ascertain a minimum penetration of net pay.

It is further of importance for the drilling crew how much shale (proportion and number of bodies) is likely to be encountered for a given well. Figure 13 gives an example of the likely proportions of each genetic unit to be encountered during drilling.

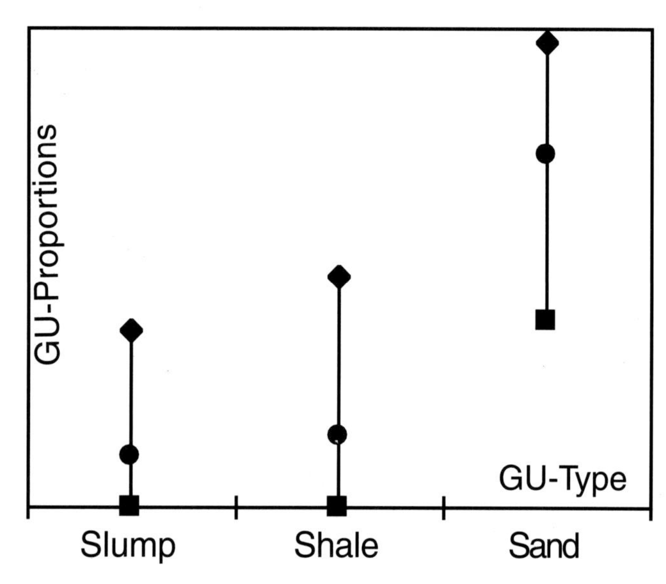

Figure 13. Cluster 4 scour and fill model, INC = 3°, well length = 1500 ft (457 m), AZ = 30°. Genetic unit proportions by genetic unit type.

General Comments

It is important to note that the results of the well placement optimization have to be analyzed by taking into account geometrical considerations when setting up the model and the well trajectories, as well as the geological assumptions that led to the stochastic models. For example, the optimum inclination of 3° is based on the assumption that a linear well entering the target zone 20 ft (6 m) below top structure is not to come within 50 ft (15 m) of the oil-water contact in a 150 ft (45 m) thick reservoir section. A well inclined at a greater angle would yield similar outcomes, but violates the geometrical constraints of the reservoir model and, therefore, is considered only suboptimum.

Second, this analysis is based upon only two static parameters, the amount (or proportion) and number of bodies of net pay intersected. Because of the high N/G ratio, the second parameter is of minor importance. In order to make these interpretation results more robust, one should engage into multivariate statistical evaluations, taking dynamic parameters into account as well. This could be done, for example, by modeling the inflow performance of the best well trajectories and relating this to different well lengths: for example, using tensors for the dynamic properties derived from lamina scale models of the genetic units, the azimuth at which the channel bodies are being intersected for drainage may become important.

Finally, taking such multivariate evaluations into account, wells could be ranked again and by assigning confidence intervals to the results, risk, and uncertainty could be assessed more accurately.

IMPACT OF RESULTS ON DRILLING PLAN

The results from the above analysis showed that azimuth would have a minor impact on the success of the well. Because the drilling platform location is one of the given constraints, a target azimuth of 15° was considered optimum because it reduces the well length prior to entering the reservoir at target depth (Figure 14) while keeping the well in the across-channel orientation. For the drilling process, this not only resulted in time savings per well, but further reduced the amount of material used (i.e., pipe, drill bit), ultimately resulting in reductions in drilling costs.

To date, three development wells (Phase 1) were drilled following the recommendations as closely as practical. The wells showed very high sand proportions as predicted; however, the wells also showed that some of the geological assumptions made prior to the modeling process were invalid. For example, the edge of the turbidite system is characterized by confined channels and well defined levees, whereas it was modeled as laterally migrating and less confined channels.

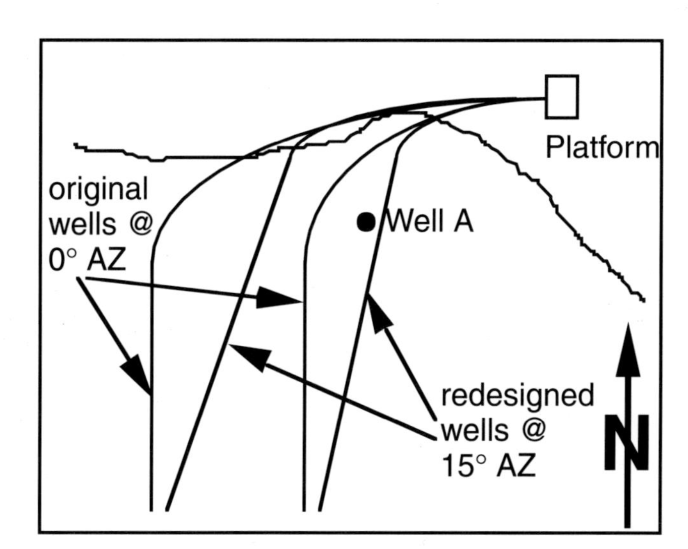

Figure 14: Original and revised drilling plan (schematic).

This proves that there is a need for much improved understanding of the sedimentological processes that generate turbidite reservoirs; furthermore, detailed quantitative sedimentological data for a wide range of submarine channel systems are needed. Such data can be used in the modeling process to investigate sedimentological uncertainties and will result in greater geological control during the modeling process and in greater confidence in the stochastic end product.

SUMMARY AND CONCLUSIONS

This paper describes the hybrid deterministic-stochastic modeling of genetic unit distributions within a deep marine clastic (turbidite) reservoir. Using Sequential Indicator Simulation (SIS), two end-member scenarios have been modeled in a part of the reservoir in order to account for uncertainties associated with the geological model. Extensive sensitivity tests were run in two dimensions and three dimensions in order to derive the appropriate modeling parameters. A pixel-based modeling technique is believed to be best suited for modeling this particular reservoir because of the lack of good geometry data on the most important elements of this reservoir, the slumped sands and the hemipelagic shales.

The objective of this study was to evaluate the existing development drilling plan, which consisted of pairs of horizontal injectors and producers, oriented parallel to the north. The resulting set of realizations for both end members were subjected to detailed well placement optimization and analysis techniques. The optimum wells were defined in terms of azimuth, inclination angle, and well length. These results had to be put into context of the structural and bedding dip, as well as existing external constraints, in order to derive the optimum drilling trajectories for the reservoir. As a result, the existing drilling program was revised, saving significant amounts of drilling time, materials, and ultimately costs.

It is acknowledged that there still exists uncertainty within several elements of this study. This includes uncertainty as to the stacking and fill pattern of the channels, the identification of petrophysical GU descriptors applied to the uncored sections of the wells, the stochastical modeling procedure and the static bivariate statistical analysis of the data; however, it is believed that these uncertainties have been dealt with appropriately considering the time constraints and objectives of the study.

Today, after drilling of three development wells, it can be concluded that azimuth, inclination, and well length have a relatively minor impact on the wells, largely due to the high N/G ratio of the reservoir. The outcomes of the (successful) wells, however, raise questions about the validity of the stochastic model, which is based on geological assumptions which, in turn, were derived from much fewer well data. It is evident that a better quantitative sedimentological understanding of this reservoir would have resulted in a more reliable reservoir model. Thus, this study should be seen as a steppingstone in the development process. The new wells provide quantitative data in lateral dimensions that, incorporated into a new modeling study, would yield a reservoir model that is much better constrained geologically, as this study could have been.

In summary, a thorough understanding of the sedimentology and the uncertainties associated with the stochastic modeling process are necessary prerequisites for developing appropriate stochastic reservoir models.

NOMENCLATURE

IOR = improved oil recovery
EOR = enhanced oil recovery
LWD = logging while drilling
SIS = Sequential Indicator Simulation
N/G = net-to-gross ratio
GU = genetic unit
SGUP = sand genetic unit proportion
SBI = sand bodies intersected

ACKNOWLEDGMENTS

The authors wish to thank C. Hern, J. Clark, K. Pickering, P. Sherwood, and D. Macklon for their helpful comments during the duration of the project. Their expertise helped ensure that the data were obtained and interpreted in a way that could readily benefit the development of the field. Furthermore, the authors wish to thank John Warrender, who critically reviewed the manuscript. Thanks are also due to the managements of Conoco (U.K.) Ltd. and LASMO North Sea Plc for permission to publish this study. Finally, the authors are grateful to Landmark Graphics Ltd. for the donation of their SGM software.

REFERENCES CITED

Alabert, F.G., and G.J. Massonnat, 1990, Heterogeneity in a Complex Turbiditic Reservoir: Stochastic Modeling of Facies and Petrophysical Variability, Paper SPE 20604 presented at the 1990 SPE Annual Technical Conference and Exhibition, New Orleans, LA, Sept. 23-26, p. 775-790.

Alabert, F.G., and V. Modot, 1992, Stochastic Models of Reservoir Heterogeneity: Impact on Connectivity and Average Permeabilities, Paper SPE 24893 presented at the 1992 SPE Annual Technical Conference and Exhibition, Washington, D.C., Oct. 4-7, p. 355-370.

Bryant, I.D., and B. Baygün, 1996, Reservoir Description for Optimal Placement of Horizontal Wells, Paper SPE 35521 presented at the 1996 SPE/NPF European 3-D Reservoir Modeling Conference, Stavanger, Norway, April 16-17, p. 301-305.

Clark, J.D., and K.T. Pickering, 1996, Submarine Channels: Processes and Architecture, Vallis Press, London, 231 p.

Deutsch, C.V., and A.G. Journel, 1992, GSLIB - Geostatistical Software Library and User's Guide, Oxford, Oxford University Press, 340 p.

Dreyer, T., 1993, Quantified Fluvial Architecture in Ephemeral Stream Deposits of the Esplugafreda Formation (Palaeocene), Tremp-Gaus Basin, Northern Spain, Alluvial Sedimentation, IAS Special Publication 17, p. 337-362.

Haldorsen, H.H., and T. van Golf-Racht, 1992, Reservoir Management into the Next Century, SEG Investigations in Geophysics 7, Reservoir Geophysics, R.E. Sheriff (ed.), Society of Exploration Geophysics, Tulsa, p. 12-24.

Hern, C.Y., J.J.M. Lewis, D. Seifert, and N.C.T. Steel, 1996, Geological Aspects of Model Construction for Well Placement Optimization in a Mixed Fluvio-Aeolian Reservoir, presented at the 1996 AAPG Annual Meeting, San Diego, CA, May 19-22, Annual Convention Program (abs.), p. A64.

Hurst, A., and D. Goggin, 1995, Probe permeametry: An overview and bibliography, AAPG Bulletin, Vol. 79, No. 3 (March), p. 463-473.

Journel, A.G., and J.J. Gle, 1995, Probe permeametry: An overview and bibliography, AAPG Bulletin, Vol. 79, No. 3 (March), p. 463-473.

Journel, A.G., and F.G. Alabert, 1990, New Method for Reservoir Mapping, Journal of Petroleum Technology, Feb., p. 212-218.

Kleverlaan, K., 1994, Architecture of a Sand-Rich Fan From The Tabernas Submarine Fan Complex, SE Spain, presented at the SEPM Gulf Coast Section 15th Annual Research Conference, Houston, TX, Dec. 4-7, p. 209-216.

Massonnat, G.J., F.G. Alabert, and C.B. Giudicelli, 1992, Anguille Marine, a Deepsea-Fan Reservoir Offshore Gabon: From Geology to Stochastic Modeling, Paper SPE 24709 presented at the 1992 SPE Annual Technical Conference and Exhibition, Washington, D.C., Oct. 4-7, p. 477-492.

Seifert, D., J.J.M. Lewis, C.Y. Hern, and N.C.T. Steel, 1996, Well Placement Optimisation and Risking Using 3-D Stochastic Reservoir Modeling Techniques Paper SPE 35520 presented at the 1996 SPE/NPF European 3-D Reservoir Modeling Conference, Stavanger, Norway, April 16-17, p. 289-300.

Solomon, S.T., K.C. Ross, R.C. Burton, and J.E. Wellborn, 1993, A Multi-Disciplined Approach to Designing Targets for Horizontal Wells, Paper SPE 25506 presented at the 1993 SPE Production Operations Symposium, Oklahoma City, OK, March 21-23, p. 53-67.

Doligez, B. et al., Integrated reservoir characterization: improvement in heterogeneous stochastic modeling by integration of additional external constraints, 1999, *in* R. Schatzinger and J. Jordan, eds., Reservoir Characterization-Recent Advances, AAPG Memoir 71, p. 333–342.

Chapter 23

Integrated Reservoir Characterization: Improvement in Heterogeneous Stochastic Modeling by Integration of Additional External Constraints

B. Doligez
Institut Français du Pétrole
Rueil Malmaison, France

H. Beucher
F. Geffroy
Centre de Géostatistique
Fontainebleau Cédex, France

R. Eschard
Institut Français du Pétrole
Rueil Malmaison, France

ABSTRACT

The classical approach to constructing reservoir models is to start with a fine-scale geological model that is densely populated with petrophysical properties. Then scaling-up techniques allow us to integrate this detailed information on a coarser grid and to obtain a reservoir model with fewer grid cells, which can be input in a fluid flow simulator.

Geostatistical modeling techniques are widely used to build the geological models before scaling-up. These methods provide possible images of the area under investigation that honor the well data and have the same variability computed from the original data. At an appraisal phase, when few data are available or when data obtained from the wells are insufficient to describe the heterogeneities and the petrophysical behavior of the field, additional constraints are needed to obtain a more realistic geological model. For example, seismic data or stratigraphic models can provide average reservoir information with an excellent areal coverage, but with a poor vertical resolution.

New advances in modeling techniques allow integration of this type of additional external information in order to constrain the simulations. In particular, two-dimensional or three-dimensional seismic derived information grids or sand-shale ratio maps coming from stratigraphic models can be used as external drifts to compute the geological image of the reservoir at the fine scale. Examples illustrate the use of these new tools, their impact on the final reservoir model, and their sensitivity to some key parameters.

INTRODUCTION

Reservoir characterization techniques involve integration of immense geoscientific information for better recovery of hydrocarbons; however, in such reservoir integrated studies, the available data are of different kinds: qualitative, quantitative, and related to different scales of measurement. One problem in reservoir studies is to be able to account for most of this available information.

Generally, qualitative structural and sedimentological models are constructed in order to give the general framework of the reservoir layering, using such methods as sequence stratigraphy or regional geological synthesis. Such models provide soft information and global trends only.

Quantitative data are also numerous and of different resolution scales. They can be classified into the following different categories.

- Well data are one-dimensional and provide detailed information about vertical variations of many reservoir properties, but are restricted to areas close to the wells. One problem is to incorporate all this information in a suitable database.
- Seismic data are generally laterally continuous and good for mapping, but with a low vertical resolution. They are not always easy to interprete in terms of geological parameters; moreover, at a reservoir scale, the precision of the picking is of primary importance for the construction of the reservoir layering. New methods based on the interpretation of seismic attributes in terms of reservoir properties using multivariate statistics, however, can give some key to correlate seismic attributes to geological parameters. The results of these analyses are average geological property maps related to the volume to be characterized, but laterally informed. Seismic also can provide information on lithological, petrophysical, or fluid variations between wells.
- Other geological "soft" data can provide geological interpretations and help to introduce some rules and constraints in the models. Also, the stratigraphic modeling gives trends of lateral facies evolution and of their lithological components in the studied area.

Geostatistical modeling techniques are widely used to build the geological models at the fine scale within predefined sedimentary units. Geostatistical tools also are well adapted to deal with data of various resolution and are efficient for their integration. Heresim® is a geostatistical tool developed by IFP (Institut Français du Pétrole) and CG (Centre de Géostatistique) that uses an algorithm based on the truncated Gaussian method (Matheron et al., 1987; Ravenne et al., 1990). New advances in modeling techniques allow us to integrate average geological parameter grids derived from seismic data or from stratigraphic models to build three-dimensional (3D) matrices of proportions that are used in the nonstationary conditional simulations.

Some examples illustrate these new methods and point out the influence of these additional external constraints on the petrophysical properties distribution in the final reservoir grid.

PRINCIPLES OF THE METHODOLOGY

Some Preliminary Recalls

The methodology implemented into Heresim has already been presented in other papers (Ravenne and Beucher, 1988; de Fouquet et al., 1988; Galli et al., 1990; Eschard et al., 1991; Doligez et al., 1992). The main purpose of this method and software is to provide to reservoir engineers with several equiprobable images of a reservoir that may be used for fluid flow simulations. Heresim is both a methodology and a geostatistical tool for reservoir characterization. The underlying philosophy is that the description of the petrophysical heterogeneities of a reservoir is directly linked to the description of the lithofacies heterogeneities. Thus, the first step in this method is to simulate the geological model as a grid informed with lithofacies, and then to compute the petrophysical parameters in each grid cell, depending on the lithofacies. The main steps of the method, illustrated in Figure 1, are the following:

- Database compilation and the construction of the conceptual geological model (Figure 1a)
- Lithofacies reservoir simulations in a high-resolution grid (Figure 1b)
- Petrophysical properties attribution on these fine-scale images (Figure 1c)

Figure 1. The main steps of the Heresim® methodology. (a) Geological conceptual model, (b) lithofacies simulation on a fine grid, and petrophysical properties computation of (c) porosities and (d) scaling-up.

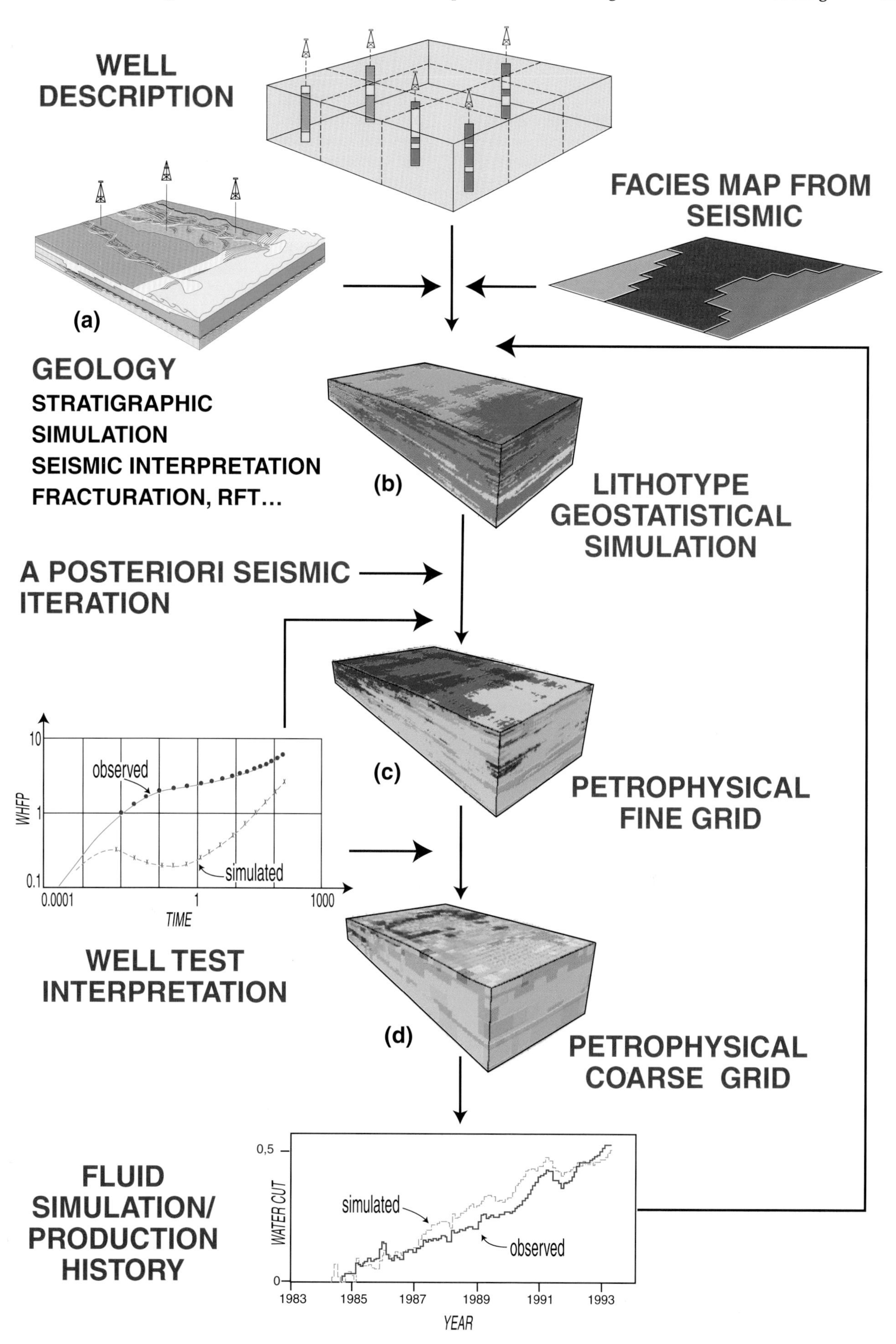

WELL DESCRIPTION
FACIES MAP FROM SEISMIC
(a)
GEOLOGY
STRATIGRAPHIC
SIMULATION
SEISMIC INTERPRETATION
FRACTURATION, RFT...
(b)
LITHOTYPE GEOSTATISTICAL SIMULATION
A POSTERIORI SEISMIC ITERATION
10
observed
WHFP
1
0.1
simulated
0.0001
1
1000
TIME
(c)
PETROPHYSICAL FINE GRID
WELL TEST INTERPRETATION
(d)
PETROPHYSICAL COARSE GRID
FLUID SIMULATION/ PRODUCTION HISTORY
0,5
WATER CUT
simulated
observed
0
1983
1985
1987
1989
1991
1993
YEAR

- Scaling-up of petrophysical properties in a coarser grid which will be the input for fluid flow simulators (Figure 1d)

The sedimentological interpretation of the well data and concepts from sequence stratigraphy allow us to define the sedimentary units of the studied reservoir, their geometries, and the chronostratigraphic horizons that can be used as reference level from a datum. In some cases, such as differential subsidence, the units also may be defined as proportional when beds are concordant with the limits of the units.

For each unit, which is simulated independently, the proportion curves and variograms are first computed to characterize the spatial variability of the sedimentary facies both vertically and horizontally, and both qualitatively and quantitatively. In particular, the vertical proportion curve represents a cumulative histogram of the proportions of the lithofacies present on the discretized wells computed at each stratigraphic level, parallel to the reference level of the considered unit (Figure 2a). This curve characterizes the vertical sequence and distribution of the lithofacies in the reservoir unit from the available well data. The horizontal proportion curve (Figure 2b) is a projection of the cumulated proportions of each facies computed well by well and projected on a horizontal line. It allows us to appreciate the degree of lateral variability of the facies within the studied unit.

The lithofacies simulations are performed using a stationary simulation of a Gaussian random function, which is truncated level by level using thresholds. These are directly linked to the proportions of the lithofacies on each level. The model uses exponential models for the variograms of the Gaussian function, which may be calibrated on the experimental variograms computed from the data on the indicator functions of the lithofacies at wells. The resulting simulations honor the well data. The global variograms and proportion curves computed from the simulated blocks are the same as those computed from the well data.

Originally, Heresim was used with a horizontal stationary hypothesis for the lithofacies distribution, which meant that the geological lateral variations are averaged across the field. In this case, the simulation algorithm used constant thresholds level by level to truncate the Gaussian function that was generated (Figure 3a). In this process, the thresholds are directly linked to the lithofacies proportions given by the vertical proportion curves. Depending on the values of the Gaussian function compared to the different thresholds, it is possible to assign a lithofacies value in each space location, as displayed on Figure 3a for one direction.

This horizontal stationary hypothesis may be valid in some cases when the facies distribution does not show

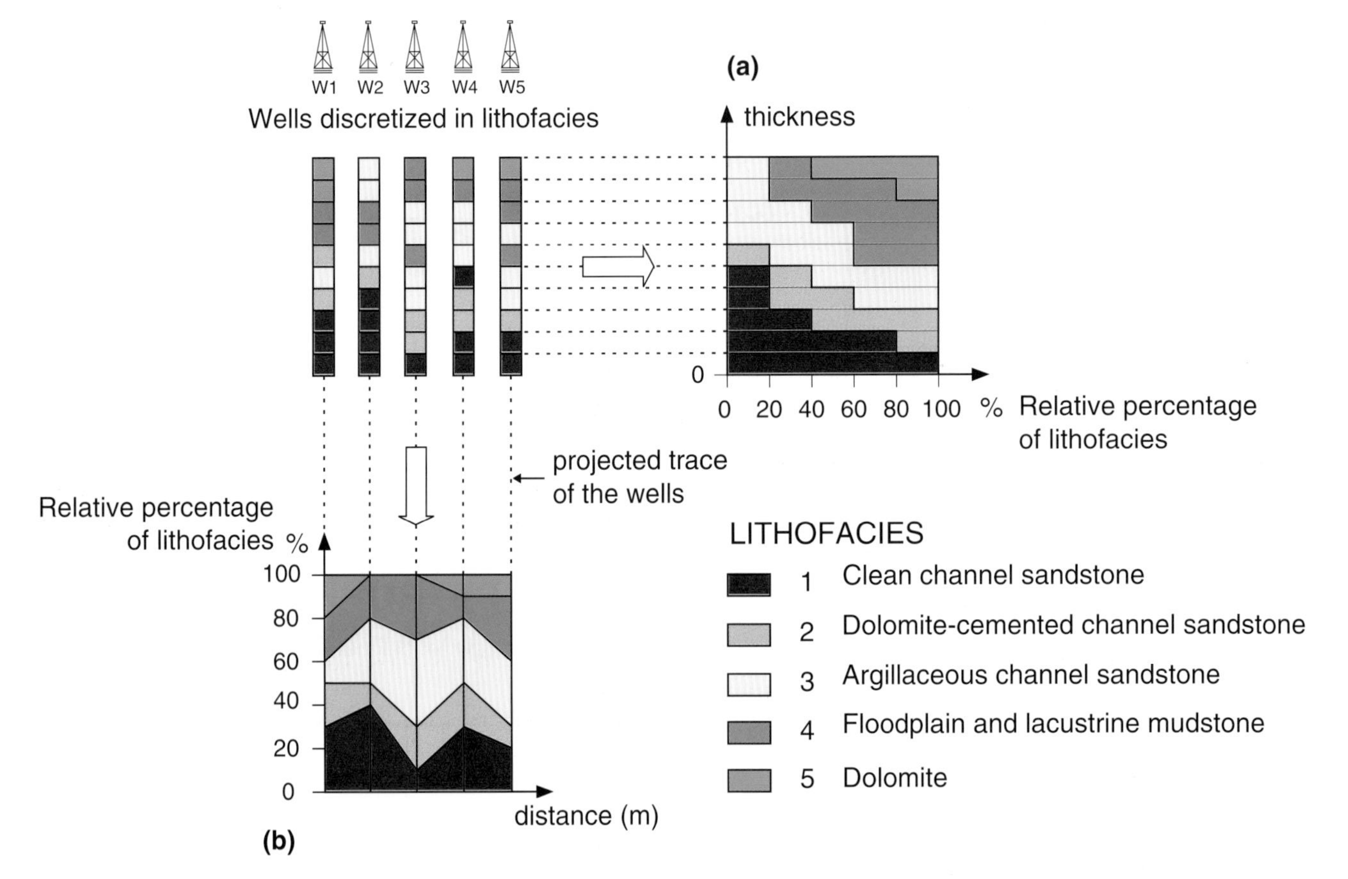

Figure 2. Principles of computation of proportion curves. (a) Vertical proportion curve showing the relative proportion of lithofacies along the horizontal plane. (b) Horizontal proportion curve showing the relative proportion of lithofacies in the wells.

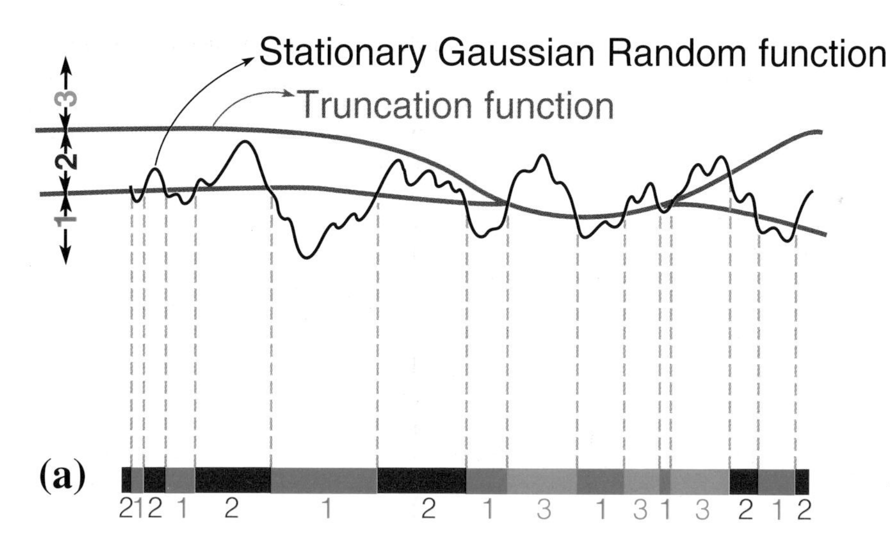

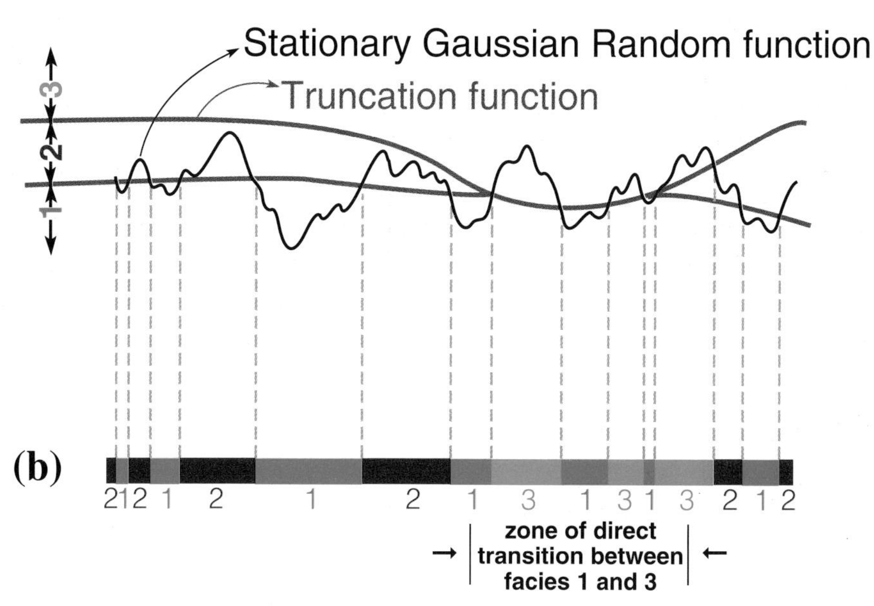

Figure 3. Truncation of the Gaussian function. (a) Examples for the stationary and the (b) nonstationary approach.

any major trend in the simulated area, but this hypothesis often is not geologically satisfactory, especially when few well controls are available or when the wells cannot be considered to be representative of the field behavior. For these reasons, a nonstationary version of the software has been developed, which allows us to take into account both the vertical and the horizontal trends of the facies repartition during the simulations. Instead of using only one vertical proportion curve that represents the main vertical facies sequences in the unit and that is linked to the truncation thresholds of the Gaussian function, the model deals in that case with a 3D proportion function that represents the mean spatial variations of the facies. The truncation thresholds of the Gaussian function are variable in space (Figure 3b), and the corresponding lithofacies values follow these variations.

How to Build a 3D Block of Proportions

Depending on the available data, several methods are possible to estimate the 3D functions of proportions in a regular block support. The target is to account for the maximum amount of data, qualitative or quantitative, from geological interpretations and seismic derived maps.

A first approach is to divide the reservoir into several zones and attribute a vertical proportion curve to each zone. The partition of the reservoir may be based on qualitative seismic interpretation using seismic facies analysis techniques. This methodology mainly uses pattern recognition techniques and statistical tools to relate each seismic trace of the field to a class while being linked to a group of wells and thus to a vertical proportion curve (Johann et al., 1996; Fournier and Derain, 1997; Fournier, 1995). The result of such an analysis is a zonation of the studied field with an associated probability of assignment of the traces to the different classes. Geological criteria also can be used to define the partition of the reservoir when a conceptual geological model allows us to assign different depositional environments and their lateral extension in different areas of the field, again associated to an uncertainty coefficient. Once the reservoir has been divided into zones, a representative vertical proportion curve is assigned to each zone, computed from groups of wells or input from analog data in order to fill the 3D block of proportions.

In other cases the 3D block of proportions can be computed directly from the well data if the wells are considered to be representative of the behavior of the reservoir, or from initial vertical proportion curves computed from groups of wells or imported from analog examples. The principle remains to compute mosaic or smooth 3D functions of lithofacies proportions on block support using more-or-less refined geostatistical methods as illustrated on the next diagram:

The estimation of the proportions from the well data or from initial and local vertical proportion curves, without additional constraints, is done using an ordinary kriging method. When additional information is available the kriging system is written with an aggregation constraint on the sum of the proportions of the facies in each cell of the 3D proportion block. For that purpose, this additional information has to be expressed as two-dimensional (2D) maps of mean lithofacies thickness or proportions derived from stratigraphic modeling techniques, for instance (Joseph et al., 1996). Statistical calibration techniques of seismic attributes allow us to convert seismic attributes into proportions or thickness of lithofacies and provide these 2D maps of average geological properties, which constrain the computation of the 3D block of proportions.

At least the different techniques may be mixed to fill the 3D block of proportions. Representative vertical proportion curves can be assigned in some areas, while in other parts the local vertical proportions are

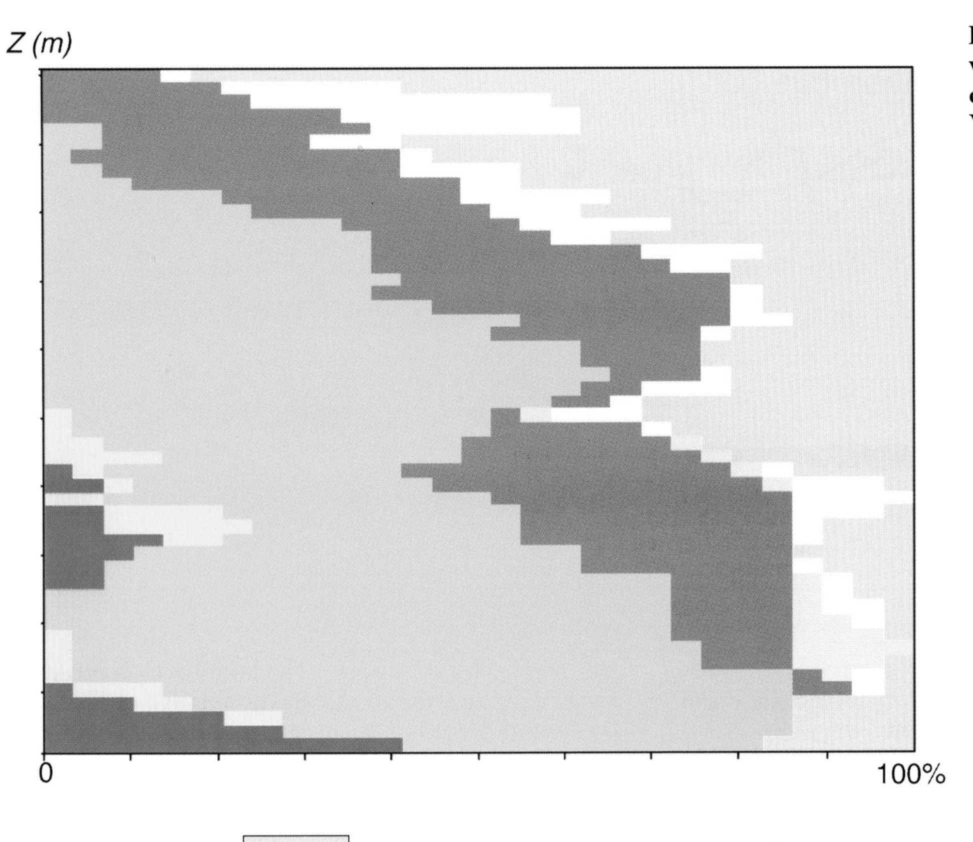

Figure 4. Experimental vertical proportion curve computed for the Mesa Verde example.

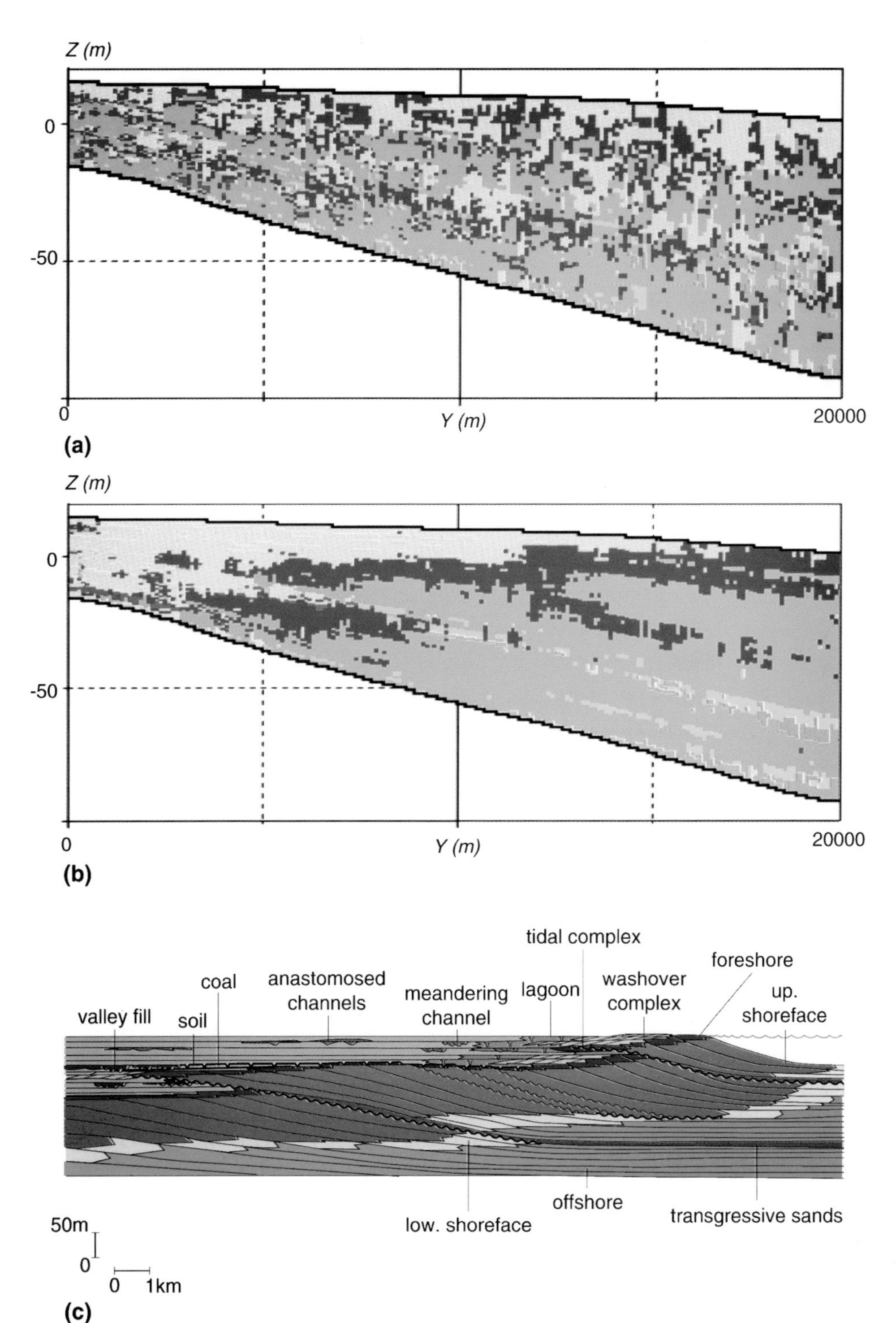

Figure 5. North-south vertical cross section as the result of (a) a stationary simulation, (b) a nonstationary simulation, and from (c) the geological model.

estimated by kriging of initial data, taking into account additional external constraints.

The choice of the methodology mainly depends on the available data (Mouliere et al., 1996). The computation by areas is specific to the cases when the information, geological and that derived from seismic, is qualitative, or when the transition between the reservoir zones is very abrupt. Generally the kriging methods and the use of the 3D functions of lithofacies proportions allow us to obtain more realistic results than those obtained from the stationary algorithm.

Simulations

The nonstationary algorithm is an extension of the stationary algorithm. In the nonstationary case, the thresholds that are used to truncate the Gaussian function are dependent on the location of the considered point of computation. In this option the transitions between lithofacies may vary and do not necessarily follow the sequential order of the facies.

Once the lithofacies grids are built, the process remains the same as in the stationary case to translate the

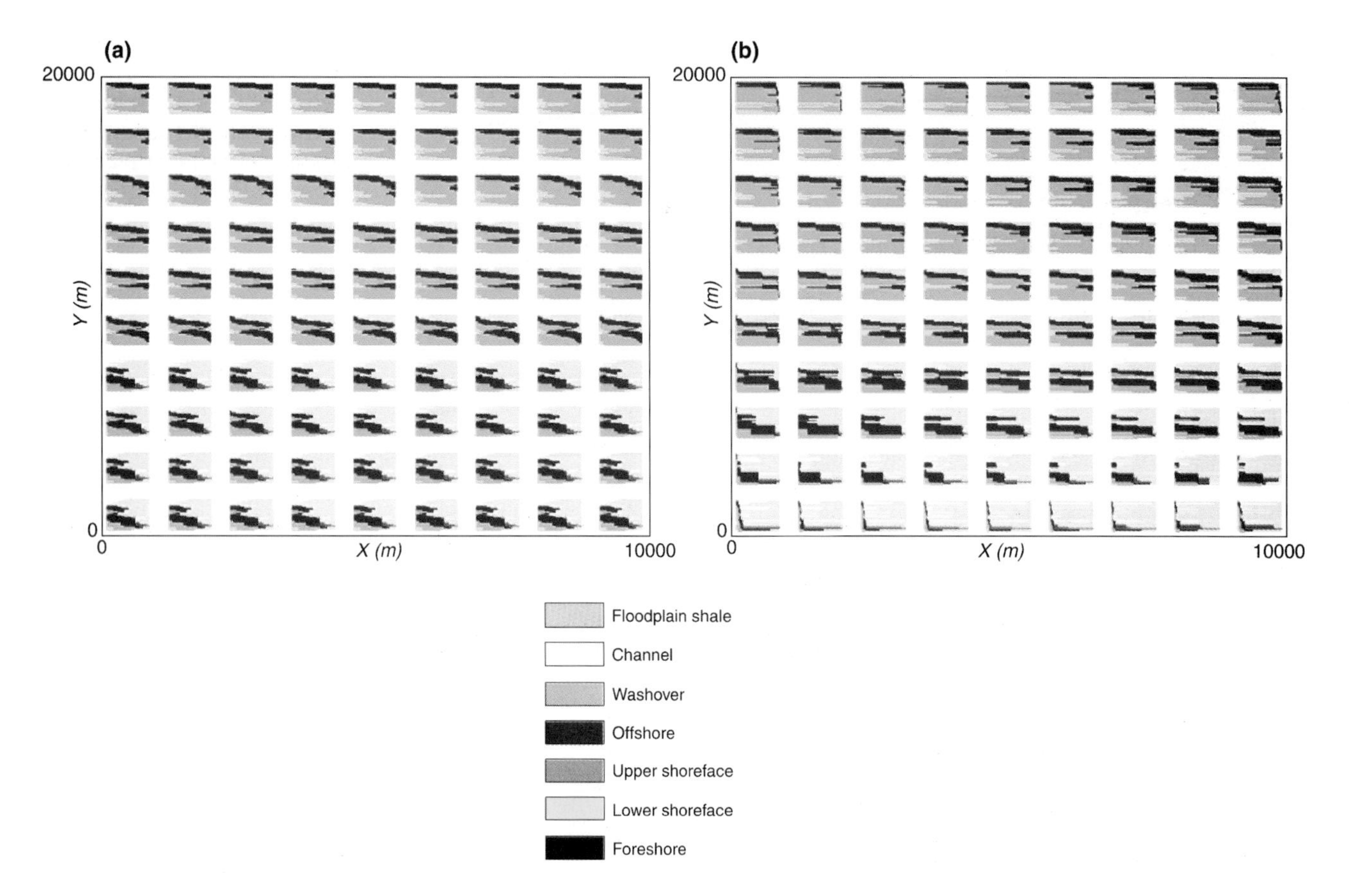

Figure 6. Three-dimensional blocks of proportions for the Mesa Verde example computed (a) from a partition of the area and (b) by kriging of well data with aggregation constraint.

geological information into petrophysical values and then to homogenize these properties into coarser cells that will be the framework for the fluid flow simulations.

Porosities and permeabilities distributions and regression laws obtained from experimental values are assigned to the fine-scale grid depending on each lithofacies. The last step is to define a reservoir grid and to scale-up the petrophysical properties in this coarser grid using algebraic and numeric methods in order to get an adequate grid for fluid flow computations.

In this process the petrophysical properties heterogeneities are directly linked to the lithofacies heterogeneities. The integration of seismically derived information to constrain the lithofacies simulations also has an influence on the petrophysical heterogeneities.

APPLICATIONS TO THE MESA VERDE EXAMPLE

Stationary Simulations

This case study is extracted from an outcrop database collected in the Mesa Verde area in the framework of a joint Elf, GDF, IFP, and Total project. The geological interpretation of the 3D outcrop (Campanian series of Colorado) in terms of chronostratigraphy and lithostratigraphy allowed us to define the main genetic units, their limits and reference levels, and the lithofacies corresponding to the three main depositional environments (Eschard et al., 1993; Homewood et al., 1992). For purpose of demonstration we focused on a seaward-stepping wedge of littoral sediment that corresponds to an overall progradation punctuated by transgressive events. In such a sequence the lateral transition between coastal plain, foreshore, shoreface, and offshore presents a typical nonstationary configuration. The conceptual seaward stepping stacking pattern model is shown in Figure 1a.

Ten representative wells have been selected from field sections in the database. They have been described using seven lithofacies corresponding to two main depositional environments, a marine sequence (offshore, upper and lower shoreface, and foreshore) and a continental sequence (washover, channels, and floodplain shales). Using these ten discretized wells as a numerical database, we computed the main geostatistical parameters for modeling purposes, the horizontal and vertical variograms, and the vertical proportion curve shown in Figure 4. The vertical and lateral transitions in the sedimentary system from a marine pole to a continental pole may be analyzed in Figure 4. Using this set of conditioning wells and geostatistical parameters, we performed the lithofacies simulation of the selected area (10 km × 20 km) first using the stationary algorithm. The display of a north-south vertical cross section in Figure 5a allows us to compare the results to the interpreted geological model (Figure 5c). In detail, the quality of

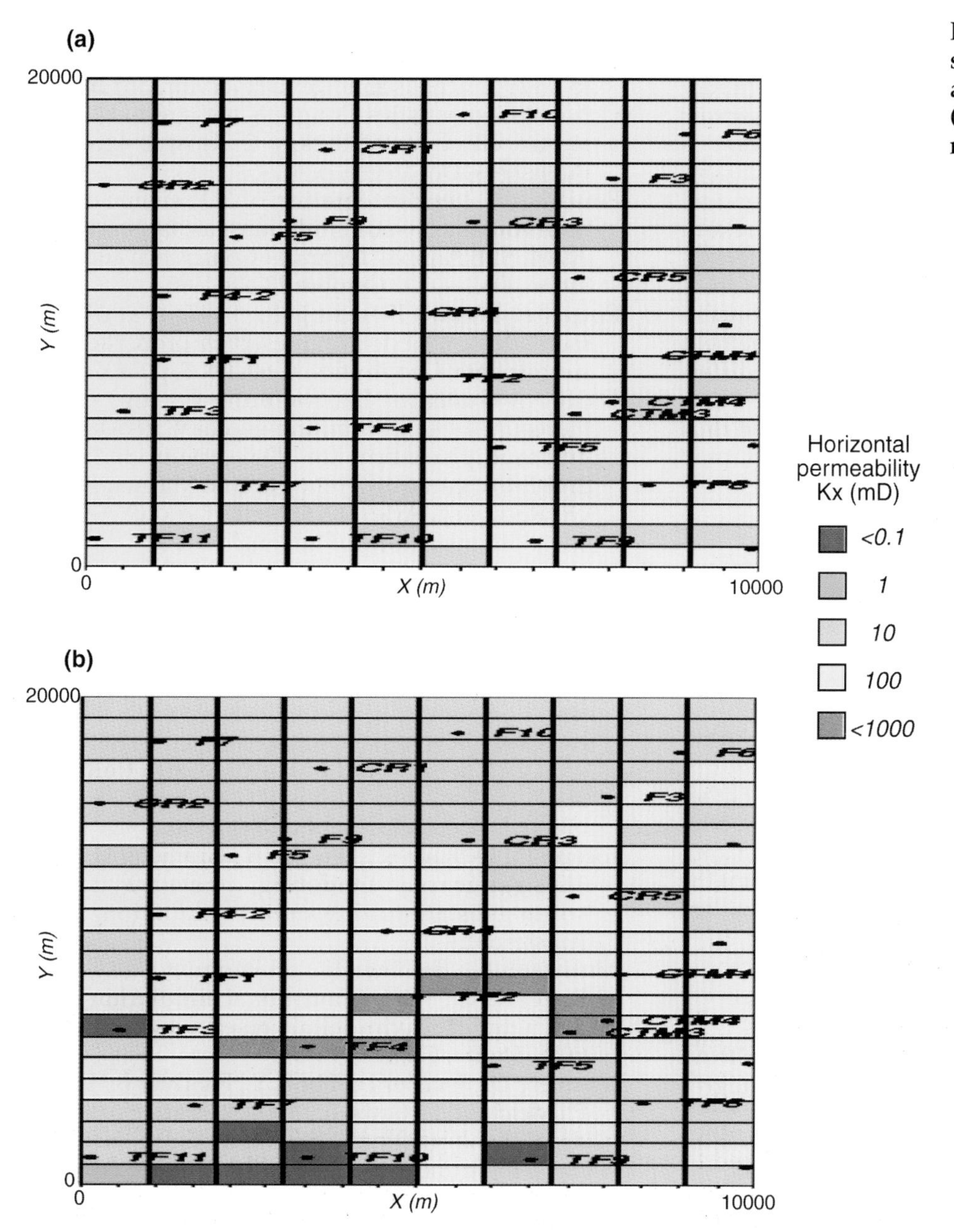

Figure 7. Comparison of scaled-up horizontal permeability maps resulting from (a) the stationary and (b) the nonstationary approaches.

the results is not satisfactory. The geometry of the clinoforms and of the shales is not well restored, and most of the main permeability barriers are not simulated, especially in the north-south direction.

Petrophysical values have been attributed to each cell of this 3D block using a model that has been defined for each lithofacies using core analysis. Then a scaling-up has been computed on a coarser grid.

Comparison Between Stationary and Nonstationary Simulations

To illustrate the impact of the nonstationary approach on the lithofacies simulations, we performed two other sets of simulations using the nonstationary algorithm from the same data set. First we used the geological interpretation from outcrop to define a partition of the area linked to the floodplain shales cumulated thickness, and we assigned a representative vertical proportion curve to each zone. The resulting 3D block of proportions is displayed in Figure 6a. We also computed another 3D block of proportions using the set of ten initial wells. The facies proportions computed from the wells have been kriged level by level with an aggregation constraint on the floodplain shales thickness. The result is displayed in Figure 6b. Figure 6 contains horizontal maps of the area that have been divided into regular blocks, each of which is filled with a local vertical proportion curve. Although the two blocks are different in details, the main vertical and horizontal evolution of the facies proportions are similar in both cases. If we compare on Figure 5 the same north-south cross section obtained using the stationary algorithm and one single vertical proportion curve (Figure 5a), and the nonstationary algorithm and one of the previous 3D blocks of proportions (Figure 5b), it appears that this last result is now much closer to the geological conceptual model and respects the main facies transitions.

After attribution of the same petrophysical model on both images and scaling-up on a simple reservoir grid, the comparison of the horizontal permeabilities maps on a chosen layer of the reservoir from a stationary model (Figure 7a) and from a nonstationary model (Figure 7b) reveals the differences in the final result. With a stationary approach the permeability heterogeneities are scattered throughout the field, whereas with the nonstationary simulation, a trend is shown in the low permeabilities in the southern part of the field to the higher permeabilities in the north. This difference points out the impact of the initial fine-scale model heterogeneities on the reservoir model and thus on the fluid flow simulations.

CONCLUSIONS

New algorithms in geostatistical simulations allow better integration of more and more external constraints and data to obtain more precise images of a reservoir under investigation. In particular, the use of 3D functions of proportions of lithofacies instead of a single proportion curve allows us to take into account and to integrate both qualitative and quantitative information derived from seismic or geological models.

The use of the nonstationary algorithm in the lithofacies simulation of a reservoir becomes very powerful because the heterogeneities are more accurately simulated. The scaling-up process will take into account the petrophysical heterogeneities at the fine scale that are directly linked to the lithofacies heterogeneities.

REFERENCES CITED

Doligez B., Ravenne C., Lemouzy P., and Galli A., 1992: Une méthodologie pour une étude intégrée des réservoirs: des données de puits aux simulations d'écoulement en utilisant des outils géostatistiques. Pétroles et Techniques, no. 372, p. 43-47.

Eschard R., Doligez B., Rahon D., Ravenne C., Leloch G., Heresim Group, 1991: A new approach for reservoir description and simulation using geostatistical methods. In Proceedings: Advances in reservoir technology, characterization, modeling and management, Edinburg, 3 p.

Eschard R., Tveiten B., Desaubliaux G., Lecomte J. C., Van Buchem F., 1993: High resolution sequence stratigraphy and reservoir prediction of the Brent group (Tampen Spur area) using an outcrop analogue (Mesaverde group, Colorado). In: Subsurface reservoir characterization from outcrop observations. R. Eschard, B. Doligez, eds., Editions Technip, Paris, p. 35-52.

de Fouquet C., Beucher H., Galli A., Ravenne C., 1988: Conditional simulation of random sets. Application to an argillaceous sandstone reservoir. 3rd International Geostatistics Congress, Avignon, M. Armstrong, ed.: Geostatistics, v. 2, Kluwer Academic Publishers, Dordrecht, p. 517-530.

Fournier F., 1995: Integration of 3D seismic data in reservoir stochastic simulations: a case study. SPE 30764. SPE Annual Technical Conference and Exhibition, Dallas, TX, October 22–25, 1995.

Fournier F., and Derain J. F., 1997: A statistical methodology for deriving reservoir properties from seismic data. Geophysics, v. 60, no. 5, p. 1437-1450.

Galli A., Guerillot D., Ravenne C., and Group Heresim, 1990: Combining geology, geostatistics and multiphase flow for 3D reservoir studies. In: 2nd European Conference on the Mathematics of Oil Recovery, Arles. D. Guerillot and O. Guillon, eds, Editions Technip, Paris, p. 11-19.

Homewood P., Guillocheau F., Eschard R., Cross T., 1992: Corrélations haute résolution et stratigraphie génétique: une démarche intégrée. Bull. Elf Aquitaine Production, F-31360, Boussens. 6, No. 2, p. 357–381.

Johann P., Fournier F., Souza O., Eschard R., Beucher H., 1996: 3D stochastic modeling constrained by well and seismic data on a turbidite field. SPE 36501. SPE Annual Technical Conference and Exhibition, Denver, CO, October 6–9, 1996.

Joseph P., Eschard R., Doligez B., Granjeon D., 1996: 3D stratigraphic modeling, a new way to constrain geostatistical reservoir simulation. EAEG Amsterdam.

Matheron G., Beucher H., de Fouquet C., Galli A., Ravenne C., 1987: Conditional simulation of the geometry of fluvio-deltaic reservoirs. SPE 16753. Ω/Σ, Dallas, TX, p. 591–599.

Mouliere D., Beucher H., Hu L. Y., Fournier F., Terdich P., Melchiori F., Grifi G., 1996: Integration of seismic derived information in reservoir stochastic modeling using truncated Gaussian approach. Geostatistical Congress of Wollongong, Australia. E.Y Boufi and N.A. Schofield (Eds.), Kluwer, p. 211–222.

Ravenne C., Galli A., Beucher H., Eschard R., Guérillot D., and Heresim Group, 1990: Outcrop studies and geostatistical modeling of a Middle Jurassic Brent analogue. In: Proceedings of The European Oil and Gas Conference, Altavilla Milicia, Palermo, Sicily, Italia. Imarisio et al. (Eds.), London: Chapman & Trotman, p. 497–520.

Ravenne C., and Beucher H., 1988: Recent development of sedimentary bodies in a fluvio-deltaic reservoir and their 3D conditional simulations. SPE 18310. In Proc. 63rd Annual Technical Conference and Exhibition of SPE, p. 463–476.

Al-Qahtani, M. Y., and I. Ershaghi, Characterization and estimation of permeability correlation structure from performance data, 1999, *in* R. Schatzinger and J. Jordan, eds., Reservoir Characterization-Recent Advances, AAPG Memoir 71, p. 343–358.

Chapter 24

Characterization and Estimation of Permeability Correlation Structure from Performance Data

Mohammed Y. Al-Qahtani
Saudi ARAMCO
Dhahran, Saudi Arabia

Iraj Ershaghi
University of Southern California
Los Angeles, California, U.S.A.

ABSTRACT

In this study, the influence of permeability structure and correlation length on the system effective permeability and recovery factors of 2-D (two-dimensional) cross-sectional reservoir models under waterflood is investigated. Reservoirs with identical statistical representation of permeability attributes are shown to exhibit different system effective permeability and production characteristics, which can be expressed by a mean and a variance. The mean and variance are shown to be significantly influenced by the correlation length. Detailed quantification of the influence of horizontal and vertical correlation lengths for different permeability distributions is presented.

The effect of capillary pressure, P_c, on the production characteristics and saturation profiles at different correlation lengths is also investigated. It is observed that neglecting P_c causes considerable error at large horizontal and short vertical correlation lengths. The effect of using constant as opposed to variable relative permeability attributes is investigated at different correlation lengths, and is shown to cause minimal effect.

Finally, a procedure for estimating the spatial correlation length from performance data is presented. Both the production performance data and the system's effective permeability are required in estimating correlation length.

INTRODUCTION

The ability to accurately model subsurface reservoirs, and those features that are likely to control and influence the displacement processes associated with hydrocarbon production, is essential if we are to improve our ability to predict reservoir performance. To achieve this goal, steps have to be undertaken to minimize the uncertainty on several fronts. With the help of geostatistics, multiples of equiprobable realizations of reservoirs can be generated for simulation purposes. A set of equiprobable stochastic reservoir realizations for a certain geologic model produces different performance data that reflect the influence of the spatial model assumed and the uncertainty associated with such a model. The purpose of this study is to examine the influence of spatial correlation length on the production characteristics and to examine the benefit of including performance data to estimate the system's spatial correlation properties.

In this study, the effect of reservoir geology on performance behavior is scrutinized. More specifically, we focus on the role of reservoir heterogeneity. One of the characteristics of geologic data sets is the presence of spatial correlation. The nature of the spatial continuity and the structure of rock properties affect the oil displacement process. Over the last decade, many publications have proposed the use of geostatistics to account for spatial structure of reservoirs (Haldorsen and Damsleth, 1990; Haldorsen and Lake, 1984; Al-Qahtani and Ershaghi, 1995). Different geostatistical techniques and algorithms have been developed to estimate unknown reservoir parameters at various locations while honoring the available data points at their corresponding locations (Isaaks and Srivastava, 1989; Deutsch and Journel, 1992).

Stochastic representations of reservoir models offer the opportunity to examine reservoir performance under many equiprobable geological images or realizations; moreover, they provide the necessary tools for conducting controlled experiments for uncertainty assessment and sensitivity studies associated with specific geological models and assumed scenarios. Influence of the geological parameters, such as correlation length, on sweep efficiency, for example, needs to be studied and quantified in order to have a better understanding of reservoir behavior. Quantification of error and assessment of uncertainty associated with assuming spatial models should be investigated.

The recovery process considered in this study is be based on immiscible displacement; i.e., water displacing oil. The fluid flow simulator used in this study has the capability of considering specific rock properties, such as k, ϕ, $k_{ro}(S_w)$, $k_{rw}(S_w)$, and $P_c(S_w)$, for each grid block.

EFFECTIVE PERMEABILITY OF CORRELATED RESERVOIRS

There is a need to quantify the relationship between the spatial continuity and the magnitude of system permeability. Spatial correlations that extend areally to distances much greater than the well-to-well spacing require validity checks; however, spatial correlations that represent interwell spacing are difficult to evaluate. This is especially true in the absence of an outcrop. The lack of or incomplete knowledge of the nature of spatial continuity can introduce serious errors. In this section, we present a critical evaluation of the influence of correlation lengths on the effective permeability of 2-D (two-dimensional) cross-sectional stochastic representation of reservoir models.

To conduct this study, a fast and reliable analytical technique for calculating system effective permeability must be developed. The technique is in part based on the work of Aasum (1992) and on the concept of a real space renormalization scheme suggested by King (1989a, b). Combining the two models constitutes the essence of the analytical technique. An overview of this analytical technique and its accuracy is presented by Al-Qahtani (1996).

The first part of this study examines the influence of correlation lengths on the calculated system effective permeability for a given permeability distribution. The second part of the study examines the influence of permeability distributions on the calculated system effective permeability at various correlation lengths.

Sets of realizations of porosity fields at different horizontal correlation lengths, corresponding to different vertical correlation lengths, are generated. The realizations are generated using the simulated annealing method (Deutsch, 1992; Leverett, 1941) for a specific normal porosity distribution with $\mu_\phi = 0.2$ and $\sigma_\phi = 0.05$. The spherical variogram model is assumed here, as the spatial correlation model, for all realizations. Each set of realizations at a particular correlation length contains 20 random realizations. This is arbitrary, but nevertheless is believed to be a good set of representative samples. Each realization is divided into 32×32 grid blocks. The grid block aspect ratio, $\varepsilon = \Delta x/\Delta z$, is equal to 10 for all realizations.

Influence of Correlation Length

In this section, we focus on the influence of correlation lengths, both horizontal and vertical, on the calculated system effective permeability for a given permeability distribution. Ten sets of realizations were generated for five horizontal correlation lengths at two vertical correlation lengths. The two vertical correlation lengths are 2 and 6 grid block units, respectively. This is equivalent to the normalized vertical correlation lengths $\zeta = Z/H = 2/32 = 0.0625$ and $\zeta = 6/32 = 0.1875$, respectively. H is the thickness of the reservoir in grid block units; and Z is the vertical correlation length in grid block units as well. For each vertical correlation length ζ, five horizontal correlation lengths are considered: 4, 8, 16, 32, and 64 grid block units, respectively. The corresponding normalized correlation lengths $\zeta = R/L = 4/32 = 0.125$, 0.25, 0.5, 1.0, and 2.0, respectively. L is the length of the reservoir in grid block units; R is the horizontal correlation length in grid block units as well.

Figure 1 shows typical realizations with $\zeta = 0.0625$ corresponding to the five λ values considered. Also, Figure 2 shows typical realizations with $\zeta = 0.1875$ corresponding to the five λ values considered. The ten sets of porosity field realizations are used to generate the corresponding permeability field realizations. This is accomplished by using a specified relationship model between porosity and permeability. This relationship is modeled using the simple function given by

$$\log k = a\phi + b \tag{1}$$

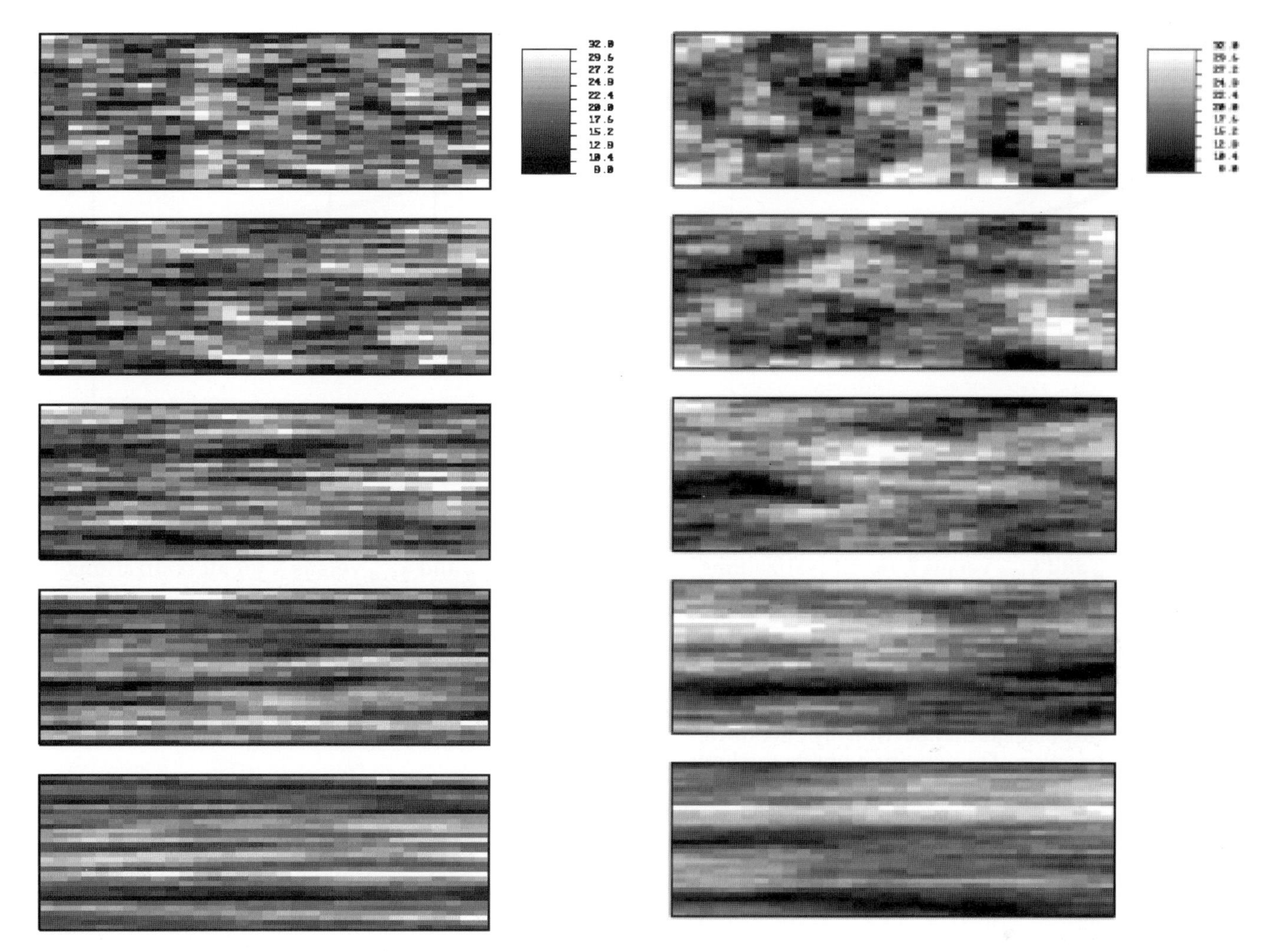

Figure 1. Five realizations with ζ = 0.0625. (a) λ = 0.125, (b) λ = 0.25, (c) λ = 0.5, (d) λ = 1.0, (e) λ = 2.0.

Figure 2. Five realizations with ζ = 0.1875. (a) λ = 0.125, (b) λ = 0.25, (c) λ = 0.5, (d) λ = 1.0, (e) λ = 2.0.

where a and b are constant. In this section, a and b are set to be 15.0 and –1.5, respectively. This results in a lognormal permeability distribution with a distinct mean, $\mu_{\log k}$ = 1.5 (or 31.6 md.) and standard deviation $\sigma_{\log k}$ = 0.75. Each grid block is assigned permeability values k_x and k_z with the ratio k_z/k_x = 0.1. Horizontal and vertical system effective permeability values, k_{Heff} and k_{Veff}, respectively, are calculated for each realization. Mean and standard deviation values are then calculated for each set of realizations.

Influence of Horizontal Correlation Length

The influence of λ on k_{Heff} and k_{Veff} is discussed in this section. Figures 3 and 4 show the system effective permeabilities, k_{Heff} and k_{Veff}, versus λ at ζ = 0.0625 and 0.1875, respectively. These values are plotted for all realizations to illustrate the nature of spread at each correlation length. The continuous line connects the mean system effective permeability for all realizations at a particular λ. It can be seen that λ has a direct relationship with k_{Heff} and an inverse relationship with k_{Veff}. We also observe that λ has a direct relationship with σ_{kH} and an inverse relationship with σ_{kV}; furthermore, it can be seen that the rate of change in the system effective permeability is high at small λ and low at large λ.

To explain the direct relationship between k_{Heff} and λ, consider the case at λ = 2.0. In this case, similar to infinite correlation, very long correlated paths are present. They resemble a layered system with low, moderate, and high permeability values. The effective permeability becomes, in effect, arithmetic average with minimal significance of low permeability values in the direction of pressure gradient (the x-direction). As λ is reduced, the occurrence of low permeability regions in the direction of pressure gradient increases, which influences the resistance and hence reduces k_{Heff}. Moreover, as λ decreases, more tortuous flow paths are created that have to pass through low permeability regions in the vertical direction created by the presence of vertical correlation. This is coupled with the fact that vertical permeability is usually low (in this case, k_z/k_x = 0.1).

In the case of k_{Veff}, in which λ has an inverse relationship with k_{Veff}, a different explanation is given. The pressure gradient is now in the vertical direction (the z-direction). Consider the case of large λ = 2.0. It can be stated that the flow directions follow extremely tortuous

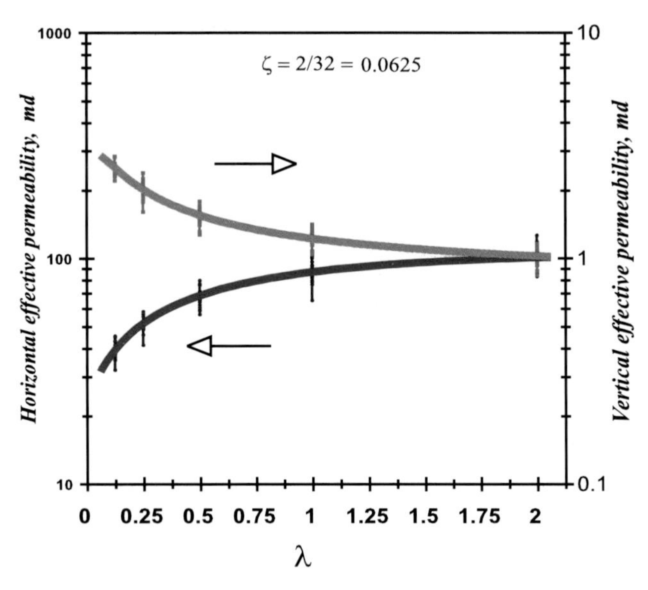

Figure 3. k_{Heff} **and** k_{Veff} **versus** λ **for all realizations with** $\zeta = 0.0625$**.**

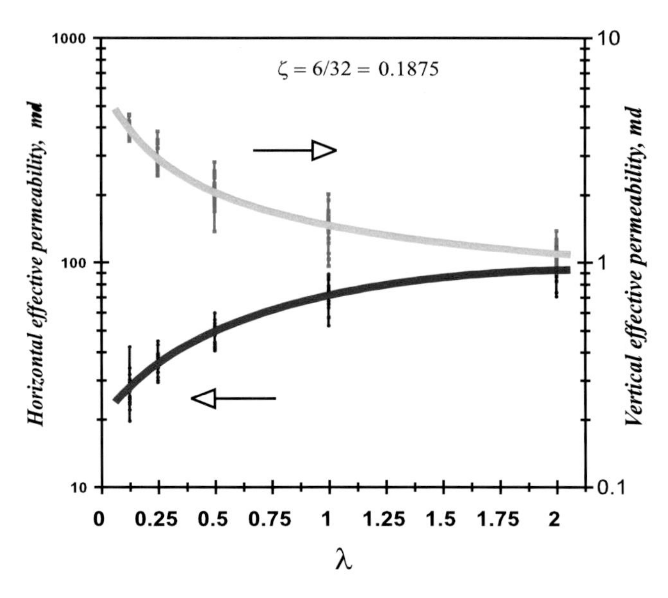

Figure 4. k_{Heff} **and** k_{Veff} **versus** λ **for all realizations with** $\zeta = 0.1875$**.**

paths coupled by the fact that vertical correlations are usually short ($\zeta = 0.0625$ and 0.1875). A sequence of low- and high-permeability sections is encountered along the *z*-direction. This situation yields the lowest k_{Veff}. As λ decreases gradually, relatively less tortuous paths are created, and consequently higher values of k_{Veff} are realized.

Influence of Vertical Correlation Length

The influence of ζ on k_{Heff} and k_{Veff} is discussed in this section. It is helpful to deal with mean system effective permeabilities when comparing the results obtained at various correlation length models. Figure 5 shows the normalized mean effective permeabilities, μ_{kHeff}/μ_{kH} and μ_{kVeff}/μ_{kV} versus λ for the two vertical correlation lengths $\zeta = 0.0625$ and 0.1875. It is observed that at a fixed λ, increasing ζ decreases k_{Heff} and hence μ_{kHeff}. In contrast, at a fixed λ, increasing ζ increases k_{Veff} and for that reason μ_{kVeff}.

These observations can be explained in the same manner. For the case of k_{Heff}, as ζ increases, the flow paths in both the *x*- and *z*-directions encounter thicker regions of low and high permeability values. This causes the paths to be more tortuous, which causes more resistance and consequently lower k_{Heff}. For the case of k_{Veff}, as ζ increases, less frequent encounters with low-permeability regions in the *z*-direction is more likely, which causes k_{Veff} to increase. Another observation that can be made is that, generally, the spread or standard deviation, σ_{kHeff}, of the k_{Heff} values of all realizations at a particular λ increases with increasing ζ. This is despite the fact that μ_{kHeff} decreases with increasing ζ, which, in turn, increases the degree of heterogeneity and consequently σ_{kHeff}.

Figure 5 is accurate for all permeability distributions with $\sigma_{\log k} = 1.5$ and any $\mu_{\log k}$. It can be observed that at $\lambda = 2.0$ and $\zeta = 0.0625$, k_{Heff} is approximately three times the magnitude of the mean μ_{kH}. Similarly, at $\lambda = 0.125$ and $\zeta = 0.1875$, k_{Heff} is only $0.9\ \mu_{kH}$. Similar observations can be drawn for k_{Veff}.

The Permeability Ratio k_{Veff}/k_{Heff}

The ratio k_{Veff}/k_{Heff} is a very important measure of reservoir conductivity. Correlation lengths influence the nature of this relationship significantly. In this section, the impact of correlation lengths on this important measure is discussed. Figures 6–8 show the ratio of permeabilities k_{Veff}/k_{Heff} versus λ for $\zeta = 0.0625$ and 0.1875, respectively. For both cases of ζ, as λ increases, the ratio of permeabilities decreases significantly. At large λ, the ratio $k_{Veff}/k_{Heff} \to 0.01$. In contrast, at small λ, the ratio $k_{Veff}/k_{Heff} \to 0.1$. It should be noted that $\mu_{kV}/\mu_{kH} = 0.1$. The influence is more pronounced when ζ increases. For the case of $\lambda = 0.125$ and $\zeta = 0.1875$, the permeability ratio is 0.15, which is greater than the 0.1 for a homogenous system. At large λ, the system is highly correlated in the horizontal direction and stratified in the vertical direction. This creates very favorable flow paths in the direction of longer correlations and least favorable ones in the direction of short correlations. This causes the apparent disparities between k_{Veff} and k_{Heff}. In contrast, as λ decreases, at short ζ, the system approaches randomness to a certain degree, which causes less favorable paths in the *x*-direction and more favorable paths in the *z*-direction. That explains why the ratio approaches 0.1. As ζ increases, at short λ, yet more variable paths are created in the *z*-direction, which causes the ratio of permeabilities to be greater than 0.1.

Influence of Permeability Distribution Parameters

In this section, we investigate the influence of permeability distribution, more specifically $\sigma_{\log k}$, on the

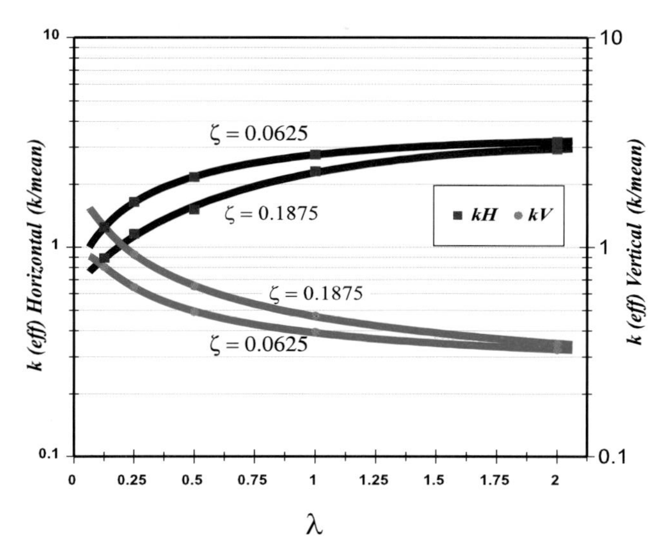

Figure 5. Normalized mean effective permeabilities μ_{kHeff}/μ_{kH} and μ_{kVeff}/μ_{kV} versus λ.

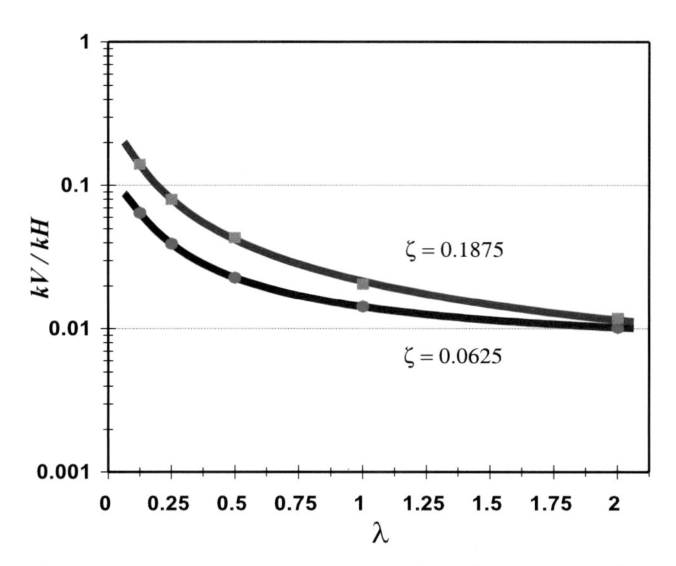

Figure 6. k_{Veff}/k_{Heff} versus λ for all realizations with $\zeta = 0.0625$ and 0.1875.

calculated effective permeabilities at different correlation lengths. Four permeability distributions are considered here, namely $\sigma_{\log k} = 0.25$, 0.5, 0.75, and 1.0, respectively. It should be noted that in the previous section $\sigma_{\log k} = 0.75$ was considered. The mean value in all cases is kept constant at $\mu_{\log k} = 1.5$ (or 31.6 md.). We consider three sets of realizations representing three correlation lengths $\lambda = 0.125$, 0.5, and 2.0, respectively, all at one vertical correlation length $\zeta = 0.0625$.

Horizontal and vertical system effective permeability values, k_{Heff} and k_{Veff}, respectively, are calculated for each realization at the four permeability distribution models. Mean and standard deviation values are then calculated for each set of realizations for each permeability distribution model. It is observed that $\sigma_{\log k}$ has a direct relationship with k_{Heff}

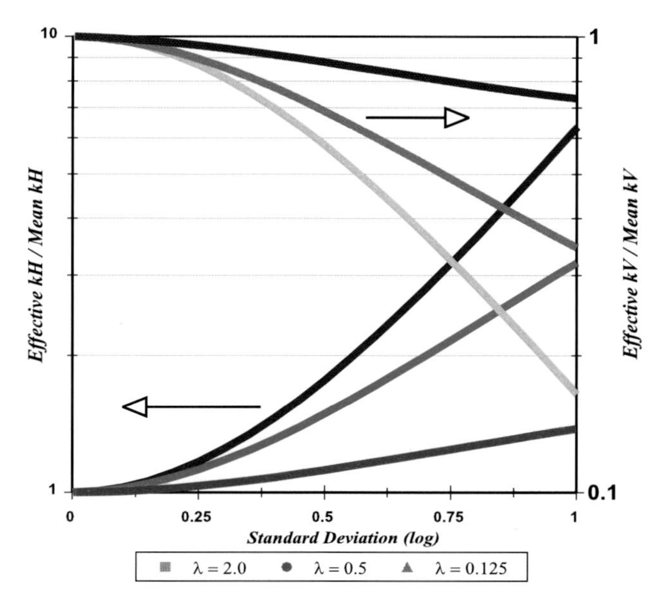

Figure 7. Normalized effective permeabilities μ_{kHeff}/μ_{kH} and μ_{kVeff}/μ_{kV} versus $\sigma_{\log k}$ with $\zeta = 0.0625$.

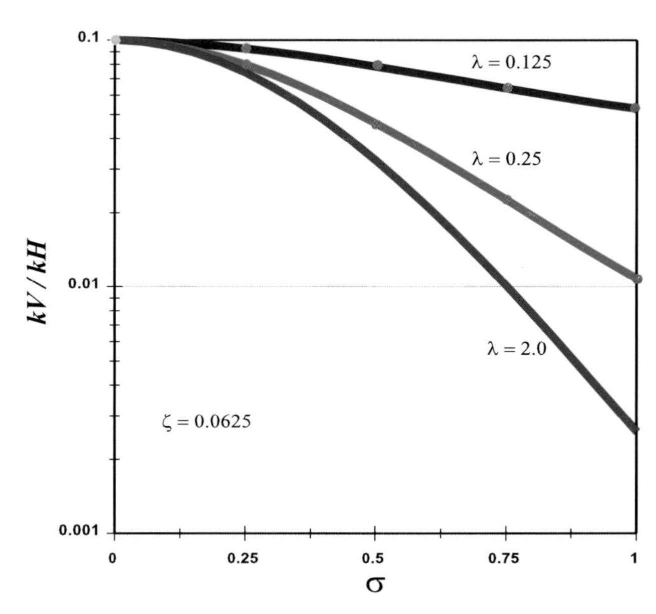

Figure 8. k_{Veff}/k_{Heff} versus $\sigma_{\log k}$ for all realizations with $\zeta = 0.0625$ and $\lambda = 0.125$, 0.5, and 2.0.

and an inverse relationship with k_{Veff}. It is also observed that the most substantial change arises at large λ. The rate of change in system effective permeabilities increases considerably as $\sigma_{\log k}$ increases.

Figure 7 shows the normalized system effective permeabilities k_{Heff}/μ_{kH} and k_{Veff}/μ_{kV} versus $\sigma_{\log k}$ for the correlation lengths under study. This figure is true for any mean value μ_k. In the case of $\sigma_{\log k} = 1.0$ at $\lambda = 2.0$, k_{Heff} is approximately six times the mean value μ_{kH}; k_{Veff} is a low 0.15 of the mean value μ_{kV}.

The above results indicate that increasing $\sigma_{\log k}$ generates extremely high and low permeability values. When they are distributed spatially to form long correlations (e.g., $\lambda = 2.0$), long correlated flow paths are created. As a result, the flow paths are less affected by low permeability values, hence higher k_{Heff} and lower k_{Veff}. By contrast, as λ decreases, the flow paths are more tortuous and resistive. In this situation, the flow paths are greatly affected by the low permeability values, hence lower k_{Heff} and higher k_{Veff}.

Figure 8 shows the ratio of permeabilities k_{Veff}/k_{Heff} versus $\sigma_{\log k}$ for the three λ under study. As $\sigma_{\log k}$ increases, the ratio of permeabilities decreases considerably. In the case of extreme variations (i.e., $\sigma_{\log k} = 1.0$, at large λ), the ratio $k_{Veff}/k_{Heff} \rightarrow 0.002$. This indicates a change of almost two orders of magnitude. By contrast, at small λ, the ratio $k_{Veff}/k_{Heff} \rightarrow 0.06$. It should be noted that $\mu_{kV}/\mu_{kH} = 0.1$. As $\sigma_{\log k}$ decreases to a low value (i.e., $\sigma_{\log k} = 0.25$), the ratio of permeabilities increases. Increasing λ at this low $\sigma_{\log k}$ has a negligible effect on the ratio of permeabilities. Ultimately, the ratio becomes 0.1 at $\sigma_{\log k} = 0$.

This study was conducted for the case of $\zeta = 0.0625$; however, the basic observations are true for other ζ values. The observations in this section and section 2.1 are sufficient enough to draw similar conclusions at other ζ. More detailed information is given by Al-Qahtani (1996).

RECOVERY FACTORS OF CORRELATED RESERVOIRS

Spatial continuity influences the sweep efficiency during immiscible displacement processes. Determination of spatial correlations is a very complicated process. Errors in representation of spatial correlations can translate into poor reservoir behavior forecasting. Efforts should be focused on reassessing the correctness and uncertainty of using these spatial models. Several investigators have attempted in a simplified approach to quantify such uncertainty (Hoiberg et al., 1990; Omre et al., 1991). We now extend the investigation in the previous section to include the influence of correlation lengths on the sweep efficiency and overall performance of 2-D cross-sectional stochastic reservoir models.

Methodology

The 10 sets of realizations generated in the preceding section are used to further extend the investigation to include recovery and performance in addition to system effective permeability. One additional set of realizations corresponding to a randomly correlated field or $\lambda = 0$ is generated for reference purposes. An overview of the displacement process for conducting this study is presented here. Mean and variance calculations of production profiles for each set of realizations are presented. Relationships between system effective permeability and breakthrough recovery are discussed.

For the purpose of this study, similar wettability characteristics and zero capillary pressure are assumed. The initial water saturation $S_{wi} = 0.184$ and the residual oil saturation $S_{or} = 0.138$ are assumed for all grid blocks. The intrinsic relative permeability functions k_{ro} and k_{rw} are given by Corey's functions (Honarpour et al., 1986)

$$k_{ro} = k'_{ro}\left[\frac{S_o - S_{or}}{1 - S_{or} - S_{wi}}\right]^{no} \quad (2)$$

$$k_{rw} = k'_{rw}\left[\frac{S_w - S_{wi}}{1 - S_{or} - S_{wi}}\right]^{nw} \quad (3)$$

where k'_{ro} and k'_{rw} are the end point relative permeability and are assumed to be 1.0 and 0.4, respectively. The exponents no and nw are assumed to be 3.0 and 2.0, respectively.

The displacement process in this section is an immiscible displacement in which water is displacing oil. The water is injected from one side (the left) and the oil is produced from the other side (the right). Figure 9 shows the reservoir simulation grid blocks and the location of the two wells. Water is injected at a constant rate and oil is produced at a constant pressure. All realizations are subjected to this waterflood scheme. The simulation process is terminated when the water cut exceeds 96%.

The simulated production results for each set of realizations as a function of f_w, $(r(f_w))_i$; $i = 1, 2, \ldots, N$, are then compiled together to generate an average response. In this case r denotes the recovery; N is the number of realizations in each set. For each set of realizations, the best predictor for the production characteristics can be obtained. The mean or

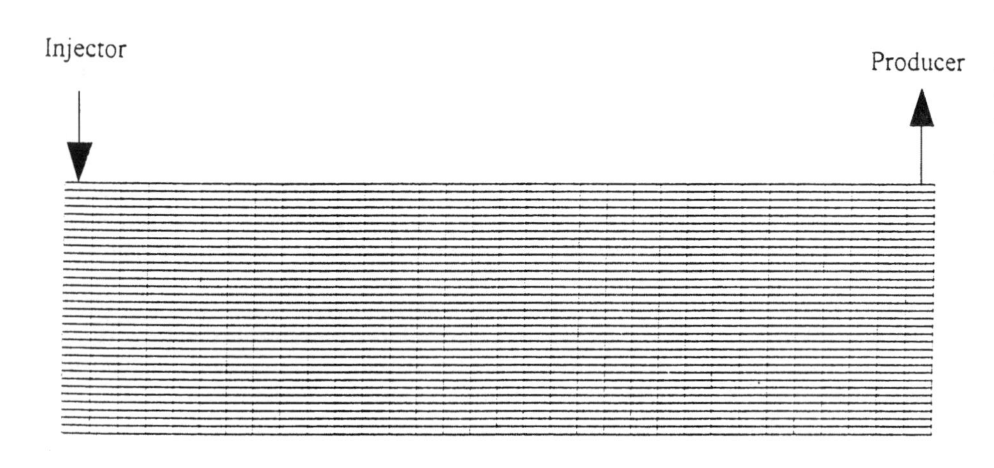

Figure 9. The reservoir cross-sectional model and the location of the injector and producer (32 × 32 grid blocks).

expected value predictor of recovery at an arbitrary f'_w is

$$\mu_R(f'_w) = \hat{E}\{R(f'_w)\} = \frac{1}{N}\sum_{i=1}^{N}\left[r(f'_w)\right]_i \tag{4}$$

with the corresponding variance

$$\hat{V}ar\{R(f'_w)\} = \frac{1}{N}\sum_{i=1}^{N}\left[\left(r(f'_w)\right)_i - \hat{E}\{R(f'_w)\}\right]^2 \tag{5}$$

The standard deviation is

$$\sigma_R(f'_w) = \sqrt{\hat{V}ar\{R(f'_w)\}}$$

Illustration of these results is presented in the next section. Breakthrough recovery factor, R_{bt}, is shown as a function of λ and ζ. System effective permeability values from the previous section are compared with their corresponding breakthrough recovery factors.

Results and Discussions

The influence of correlation lengths on the sweep efficiency and recovery factors of 2-D cross-sectional stochastic reservoir models is presented here. Ten sets of realizations corresponding to two vertical correlation lengths and five horizontal correlation lengths were considered for this study. These correlation lengths are λ = 0.125, 0.25, 0.5, 1.0, and 2.0, all with ζ = 0.0625 and 0.1875, respectively. An additional set of realizations corresponding to a random correlation was used as a reference model.

Influence of Correlation Length

Figure 10 shows the production profiles and their mean response for all realizations with ζ = 0.0625, all with λ = 0.125, 0.5, and 2.0, respectively. Also, Figure 11 shows the production profiles and their mean response for all realizations with ζ = 0.1875 at λ = 0.125, 0.5, and 2.0, respectively. It is clear that at any given λ, the realizations exhibit different behaviors. It is also clear that for systems with higher λ values, the breakthrough recovery decreases and variability increases. Probability calculations can be evaluated to within a certain level of confidence. All realizations are equiprobable; however, their responses are characterized by likelihood and confidence level parameters. This illustrates that if we were certain about the assumed correlation length, which is usually not the case, the behavior could be characterized at best by an expected value and variance. On the contrary, if λ is unknown the task is more difficult and the uncertainty is enormous.

Figure 12 shows f_w and the standard deviation, σ_R, versus the mean recovery or expected value, μ_R, for all λ with ζ = 0.0625 and 0.1875, respectively. It is clear that as λ increases, the recovery at any given f_w decreases.

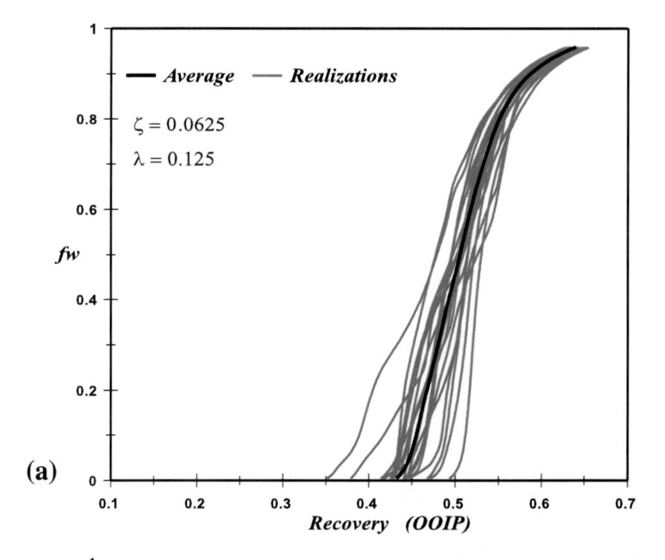

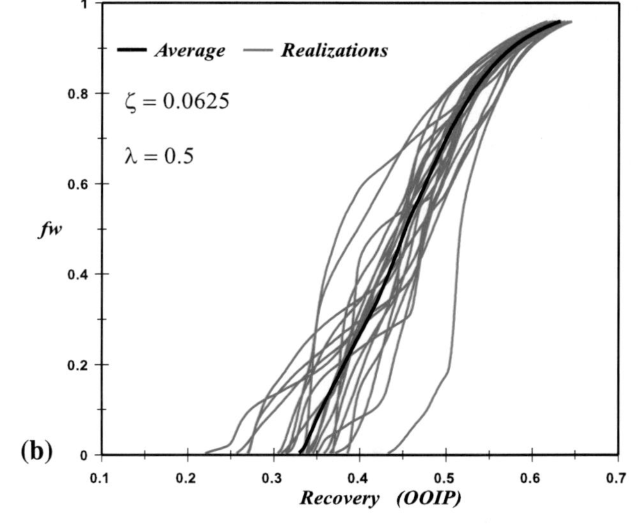

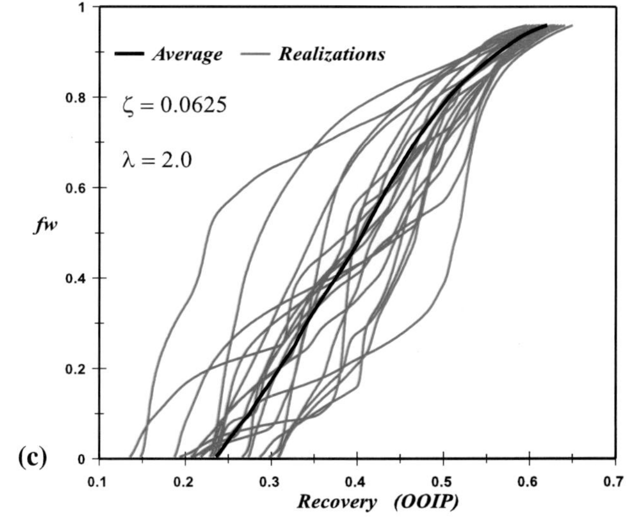

Figure 10. f_w versus recovery for all realizations with ζ = 0.0625. (a) λ = 0.125, (b) λ = 0.5, (c) λ = 2.0.

Longer correlation lengths in the direction of flow create very favorable paths, which causes water to travel faster and break through at an early time with less oil being recovered. As λ decreases, the flow paths become more

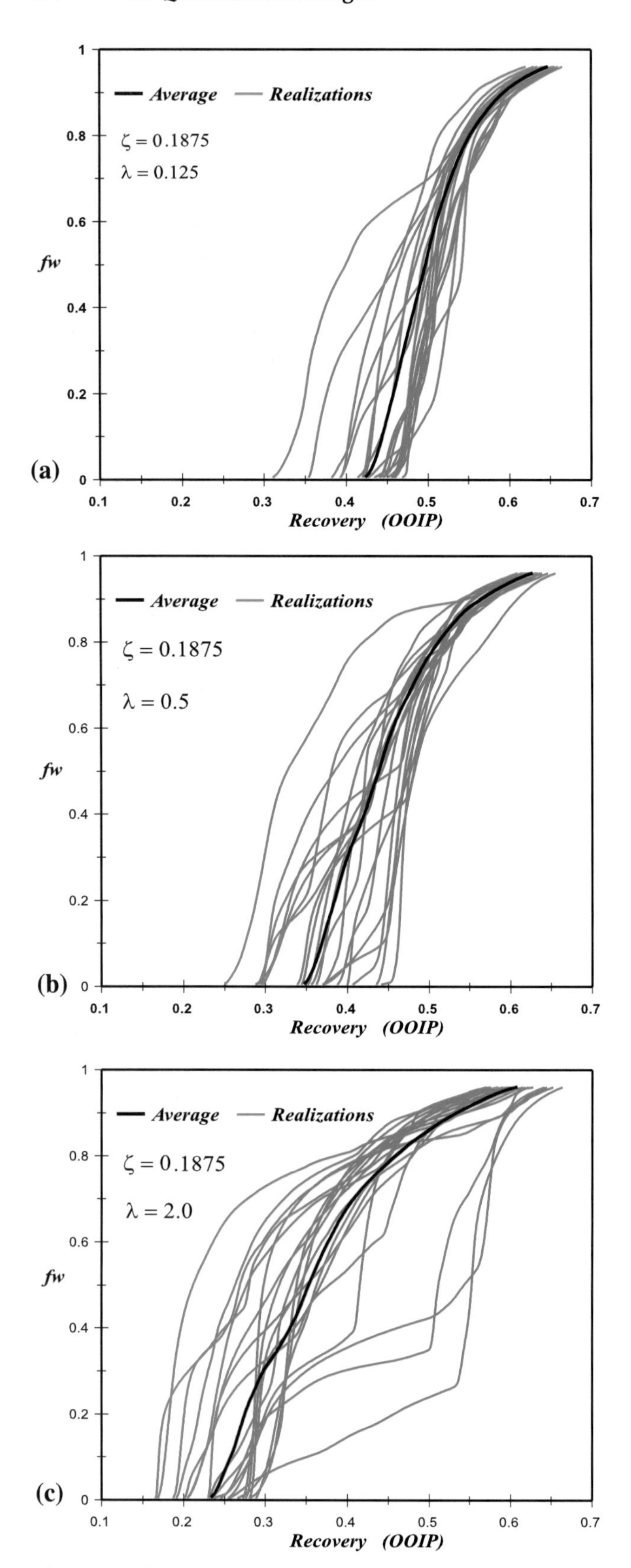

Figure 11. f_w versus recovery for all realizations with $\zeta = 0.1875$. (a) $\lambda = 0.125$, (b) $\lambda = 0.5$, (c) $\lambda = 2.0$.

tortuous and more resistant to flow, which delays T_{bt} and increases R_{bt}. With time, the water pushes the oil behind, and eventually similar ultimate recoveries are

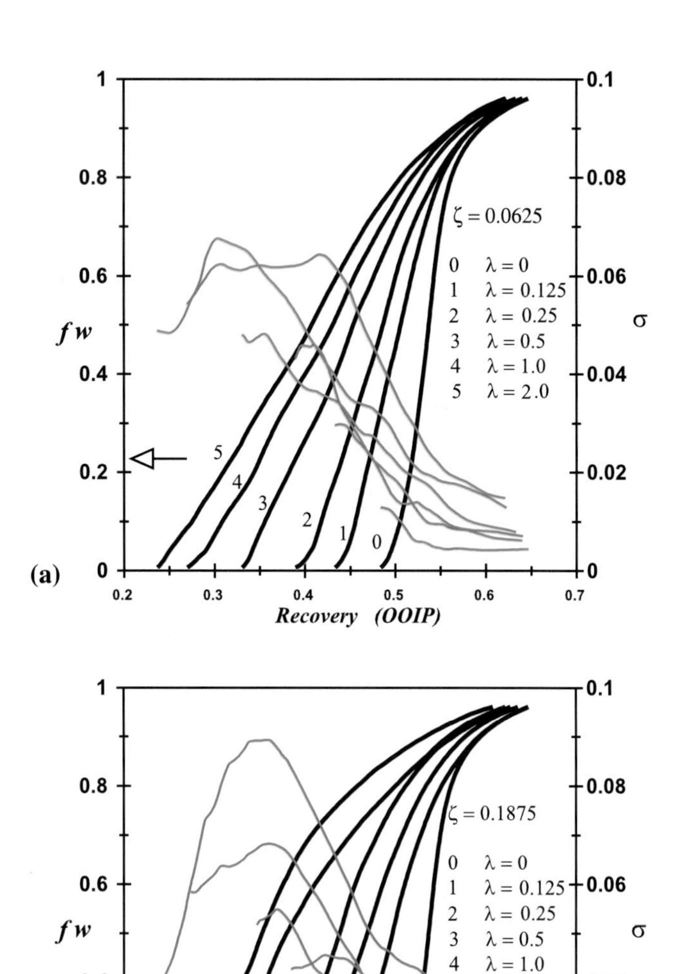

Figure 12. f_w **and the standard deviation,** σ_R**, versus the mean recovery or expected value** μ_R **for all** λ **(a) for** $\zeta = 0.0625$**, (b) for** $\zeta = 0.1875$**.**

reachable. The variability measure σ_R, however, generally increases as λ increases.

The differences between $\zeta = 0.0625$ and 0.1875 with regard to T_{bt} and R_{bt} at a particular λ are insignificant; however, the increases in f_w after breakthrough are sharper at $\zeta = 0.1875$. This is because the established paths within the system at a given λ have the same length. This makes it possible for the water to break through at approximately the same time and essentially with similar recovery; however, thicker paths in the case of $\zeta = 0.1875$ deliver higher volumes of water and consequently cause sharper increase in f_w. This also results in less recovery because less water is being used to flood low-permeability regions. Eventually, it would take longer time to reach similar recovery factors.

Figure 13 shows the breakthrough recovery factor R_{bt} versus λ. It also shows the variability measure

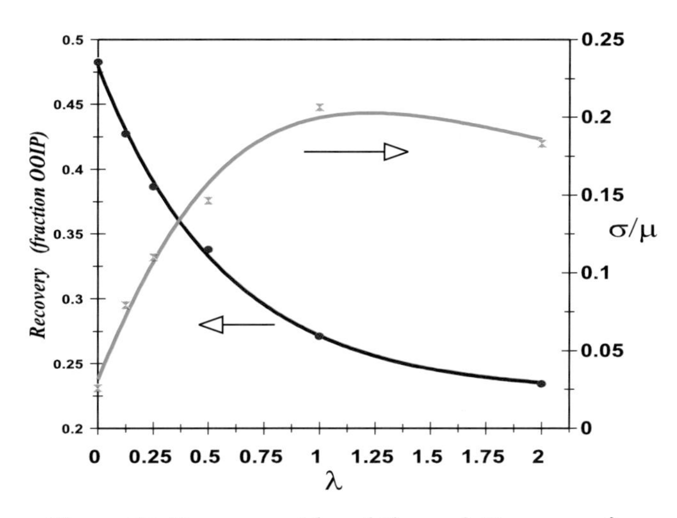

Figure 13. Recovery at breakthrough R_{bt} versus λ.

σ/μ versus λ. The figure represents the two ζ cases. As mentioned earlier, R_{bt} decreases as λ increases. The rate of decrease is high at small λ and low at large λ. The variability reaches maximum at $\lambda = 1.0$ and decreases thereafter. At small λ ($\lambda < 0.5$), the communication in the reservoir is good. Water contact with the reservoir rock is high and the sweep is more uniform. As a result, the variability is relatively small. At large λ ($\lambda \geq 2.0$), the flow paths are very long and their frequent appearance among different realizations is high, which causes the variability to start decreasing. On the contrary, at intermediate λ ($2.0 > \lambda > 0.5$), the flow paths are within the well spacing, and the variability of their frequent appearance at both wells is higher. This causes the variability at breakthrough to be the largest.

Effective Permeability versus R_{bt}

In this section, we use our knowledge of the system effective permeability values, both horizontal and vertical, to investigate the existence of relationship between them and recovery factors at breakthrough. Figure 14 shows k_{Heff} and k_{Veff} versus R_{bt} for all realizations with $\zeta = 0.0625$ and 0.1875, respectively. A clear trend is observed. As k_{Heff} increases, R_{bt} decreases. By contrast, as k_{Veff} increases, R_{bt} increases. The slope of best fit line is higher at $\zeta = 0.1875$. The scattered pattern of the data points is also higher at $\zeta = 0.1875$. Higher k_{Heff} values, which usually occur with increasing λ, cause channeling and nonuniform sweep, hence lower recovery factors. On the contrary, higher k_{Veff} values, which usually occur with decreasing λ, cause good communication and uniform sweep within the reservoir, and consequently higher recovery factors.

Effect of Assuming Different Variogram Models

Thus far, we have assumed that the spherical variogram models the spatial continuity. In this section, we examine the effect of assuming different variogram models for representing spatial heterogeneity at a given λ and ζ. These variogram models include, in addition to the spherical model, the exponential and the Gaussian. One case of correlation lengths λ and ζ is considered. The horizontal correlation length $\lambda = 0.5$ and the vertical correlation length $\zeta = 0.0625$ are considered for all realizations. For each case of variogram model, 20 realizations are generated using the simulated annealing method.

Figure 15 shows the mean and variance of the responses of the 20 realizations for each variogram model case. Note that the mean and variance are close. The behaviors of the spherical and exponential models show insignificant difference. The Gaussian model, however, shows a relatively small difference because the Gaussian model, unlike the other models, is a transition model that causes a higher degree of continuity. Consequently, a relatively favorable sweep is observed.

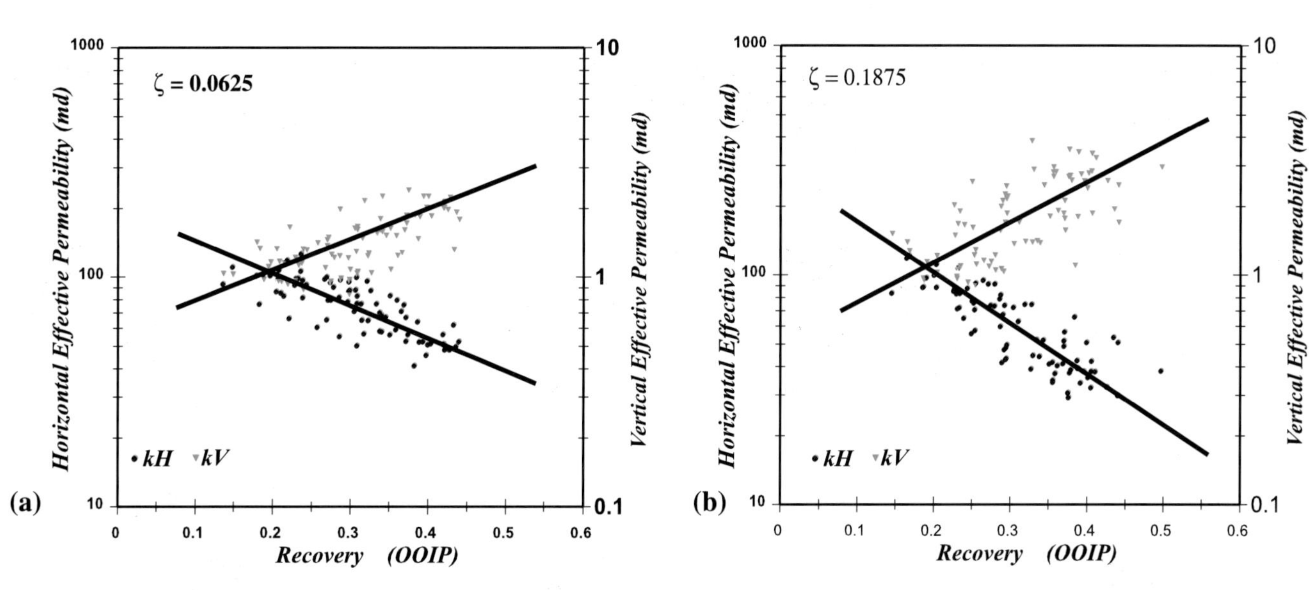

Figure 14. k_{Heff} and k_{Veff} versus recovery at breakthrough R_{bt} (a) for $\zeta = 0.0625$, (b) for $\zeta = 0.1875$.

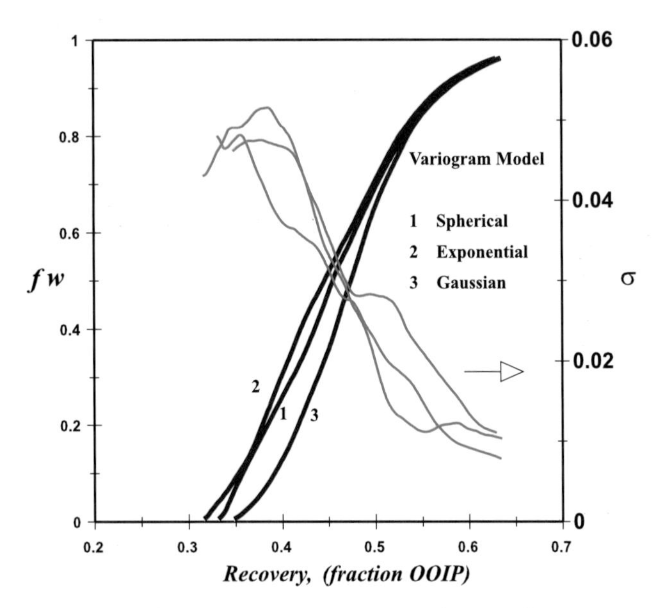

Figure 15. f_w **and** σ **versus recovery for reach variogram model with similar** λ **and** ζ**.**

RELATIVE PERMEABILITY AND CAPILLARY PRESSURE ISSUES

We need to examine several important issues related to modeling the performance of 2-D representation of stochastic reservoir images. The first issue is the effect of neglecting variable relative permeability and capillary pressure curves. The influence of correlation lengths on the recovery factors at these conditions is also another important issue. Also, the validity of considering constant instead of variable properties should be investigated. In this section, we discuss these issues in detail. The objective of this study is to examine and quantify the effect of permeability-dependent properties on the performance at different correlation lengths.

Multiphase Flow in Correlated Systems

The multiphase flow in heterogeneous porous media is, in part, governed by relative permeability and capillary pressure characteristics of the system. These characteristics have been found to vary with permeability heterogeneity. The importance of considering these properties has been recognized in the literature (Tjolsen et al., 1993; Chang and Yortsos, 1992; Chang and Mohanty, 1994); however, statistical quantification of the magnitude of error when they are neglected at different correlation lengths has not been studied in a systematic approach. In this section, we study the influence of the variability in relative permeability and capillary pressure curves. We also study the influence of correlation lengths on the performance and recovery factors when variable capillary pressure is assumed; moreover, the effect of using constant in contrast to variable properties at different correlation lengths is examined and an overview of the methodology followed in this study is presented.

Methodology

In this section, we present a systematic approach for studying the effect of relative permeability and capillary pressure, P_c, functions on the recovery factors at different correlation lengths. The first part examines the effect of variable relative permeability at zero P_c. The second part studies the effect of both variable relative permeability and P_c. Finally, we investigate the errors involved in assuming constant initial water saturation, S_{wi}, residual oil saturation, S_{or}, and similar relative permeability curves.

Heterogeneity is modeled here by spatially correlated equiprobable porosity fields parameterized in terms of its variance and a variogram model. The properties ϕ, k, S_{wi}, S_{or}, end point water relative permeability, k'_{rw}, and P_c are assumed to be correlated. For the purpose of this study, six sets of realizations are considered. They correspond to two vertical correlation lengths ζ and three horizontal correlation lengths λ. The three λ values are 0.125, 0.5, and 2.0, respectively; and the two ζ values are 0.0625 and 0.1875, respectively. These realizations are generated using the simulated annealing method. Each realization is divided into 32 × 32 grid blocks. The grid block aspect ratio, $\varepsilon = \Delta x/\Delta z$, is equal to 10 for all realizations.

The porosity field realizations are used to generate the corresponding permeability fields. Equation1 is used to generate the permeability data. This results in a lognormal permeability distribution with a distinct mean, $\mu_{\log k} = 1.5$ (or 31.6 md.) and standard deviation $\sigma_{\log k} = 0.75$. Each grid block is assigned permeability values k_x and k_z. The ratio k_z/k_x is equal to 0.1 for all grid blocks. All realizations are subjected to waterflood similar to that discussed in the previous section.

Effect of Variables S_{wi}, S_{or}, k'_{rw}, and $P_c = 0$

In this section, S_{wi}, S_{or}, and k'_{rw} are assumed to be correlated with permeability. The correlation modeling S_{wi} is given by the function

$$\log k = aS_{wi} + b \tag{6}$$

where a and b are constants. The mean value $\mu_{Swi} = 0.25$. The correlation modeling S_{or} is given by the function

$$\log k = cS_{or} + d \tag{7}$$

where c and d are constants. The mean value $\mu_{Sor} = 0.25$. Figure 16a shows S_{wi} and S_{or} as a function of permeability. The intrinsic relative permeability functions k_{ro} and k_{rw} are given by Corey's functions (King, 1989a) in equations 2 and 3, where k'_{ro} and k'_{rw} are the end-point relative permeability. The exponents no and nw are assumed to be 3.0 and 2.0, respectively. k'_{ro} is set at 1.0; however, k'_{rw} is assumed to be variable. The correlation modeling k'_{rw} is given by the function

$$\log k = ek'_{rw} + f \tag{8}$$

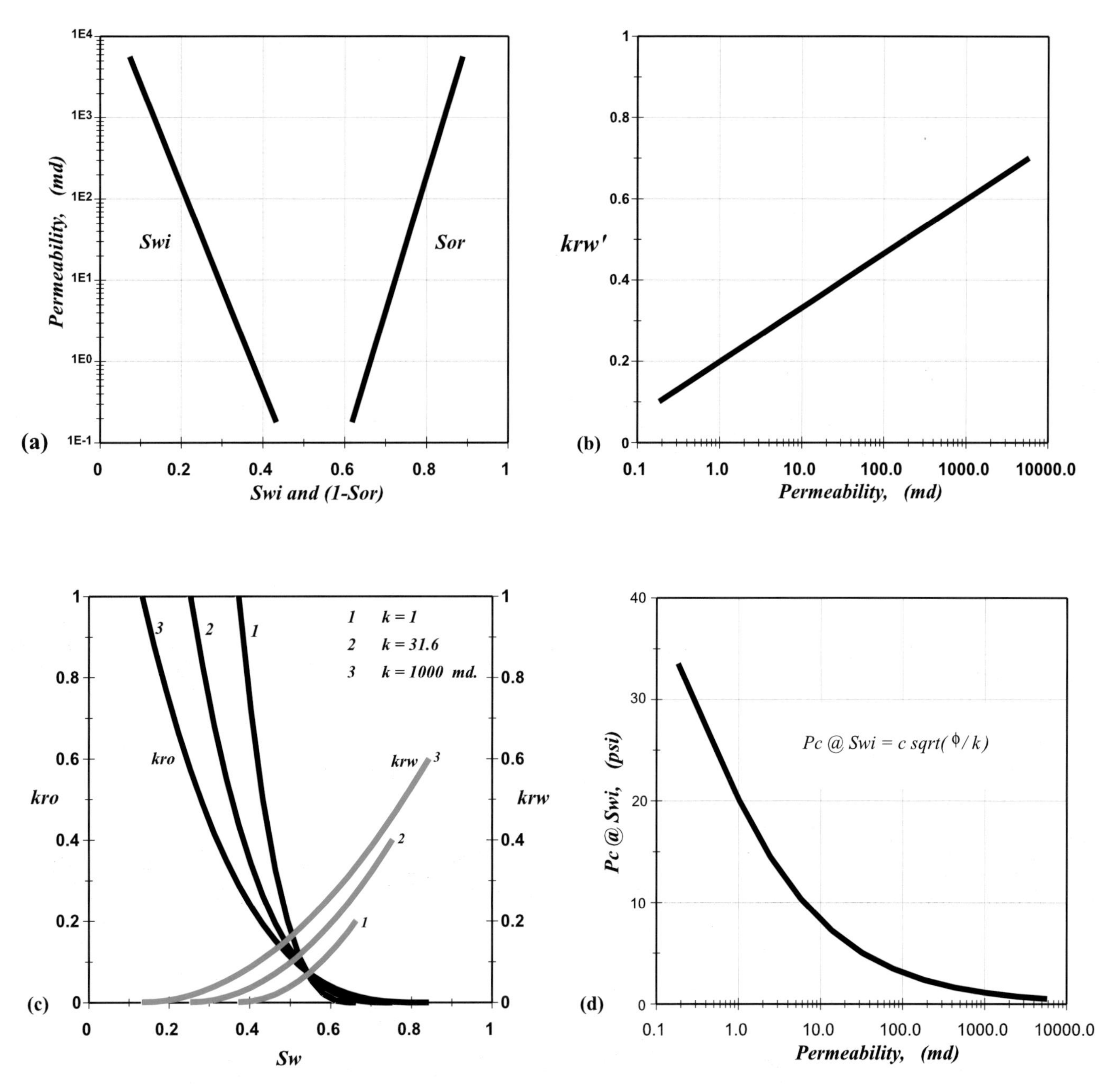

Figure 16. Basic properties. (a) S_{wi} and S_{or} as a function of permeability. (b) k'_{rw} versus permeability. (c) Three relative permeability curves for different permeability values. (d) $P_c(S_{wi})$ versus k using the *J*-functions.

where e and f are constants. The mean value $k'_{rw} = 0.4$. Figure 16b shows k'_{rw} versus permeability. Figure 16c shows three relative permeability curves for different permeability values. At this stage, P_c is assumed to be negligible and set to be zero. Figure 16d shows $P_c(S_{wi})$ versus permeability.

Figure 17a and b shows the average response and variance for all λ and ζ cases. The influence of correlation lengths is clear. It is consistent with the results of section 2. As λ increases, the recovery at breakthrough, R_{bt}, decreases and the variability increases. The effect of ζ is insignificant. Considering variable properties showed that the influence of correlation lengths on R_{bt} is present and in agreement with previous conclusions. This observation is compared to the results of the next two sections.

Effect of Variable S_{wi}, S_{or}, k'_{rw}, and P_c

In this section, the effect of further incorporating the variable P_c is investigated. The correlation among porosity, permeability, and capillary pressure known as the Leverett relationship or *J*-functions (Leverett, 1941) is assumed here. The *J*-function is given by

$$P_c = \sigma\sqrt{\frac{\phi}{k}}\,J(S_w) \qquad (9)$$

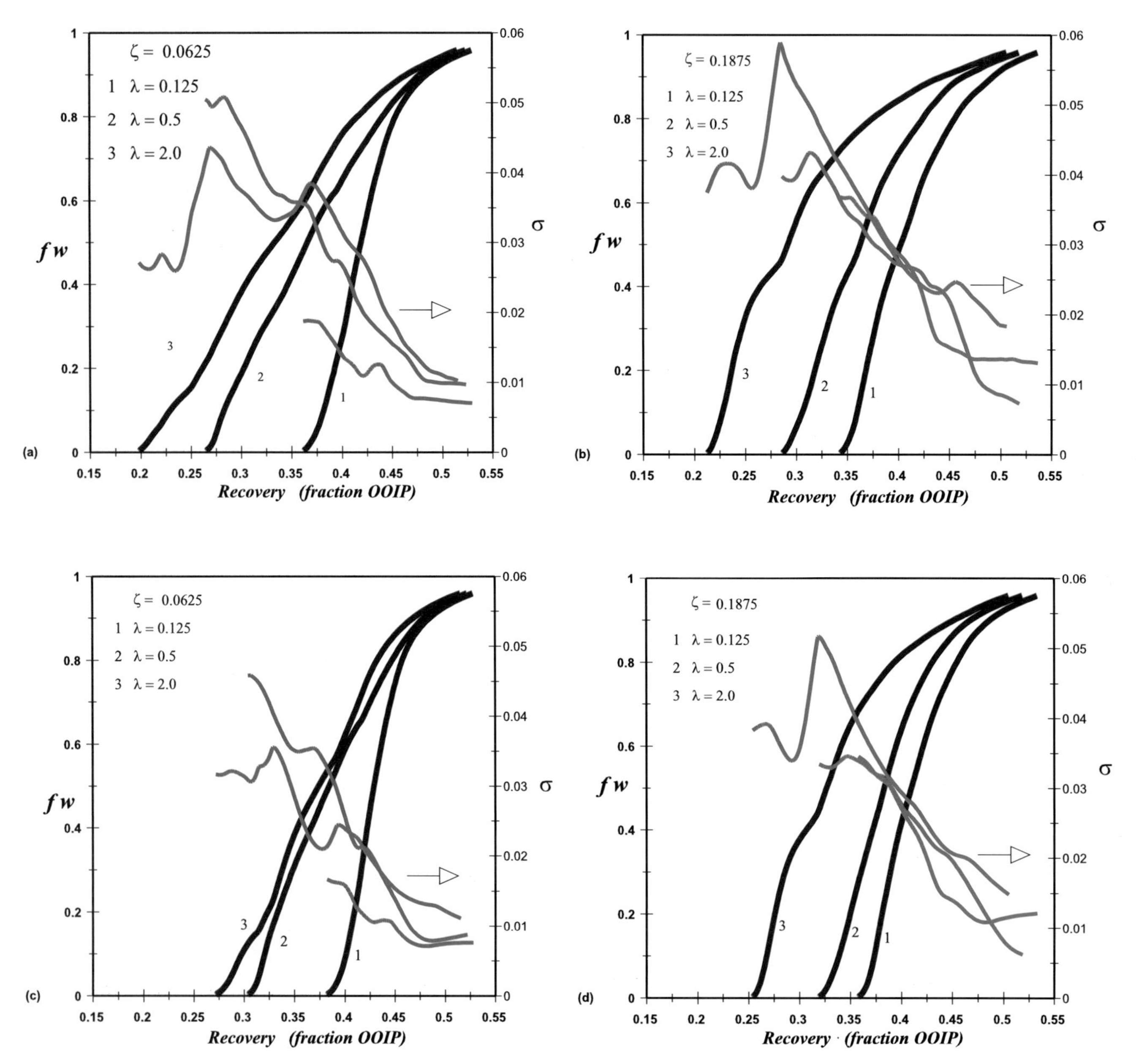

Figure 17. f_w **and** σ **versus recovery for three values of** $\lambda = 0.125$ **and 2.0. (a) The case of variables** S_{wi}, S_{or}, k'_{rw}, **and** $P_c = 0$ **at** $\zeta = 0.0625$. **(b) The case of variables** S_{wi}, S_{or}, k'_{rw}, **and** $P_c = 0$ **at** $\zeta = 0.1875$. **(c) The case of variables** S_{wi}, S_{or}, k'_{rw}, **and** P_c **at** $\zeta = 0.0625$. **(d) The case of variables** S_{wi}, S_{or}, k'_{rw}, **and** P_c **at** $\zeta = 0.1875$.

where σ is the interfacial tension. The $J(S_w)$ is a function of saturation only. The capillary pressure can be expressed as a function of porosity, permeability, and saturation

$$P_c(S_w) = \Gamma \sqrt{\frac{\phi}{k}} \left[\frac{S_w - S_{wi}}{1 - S_{or} - S_{wi}} \right]^{nd} \tag{10}$$

where Γ and nd are constants. The S_{wi} and S_{or} are given by the two models in the previous section. For example, for a block with $\sigma = 0.2$, these properties are estimated: $k = 31.6$ md., $S_{wi} = 0.25$, $S_{or} = 0.25$, and $P_c(S_{wi}) = 5.0$ psi.

Figure 17c and d shows the influence of correlation lengths, when variable P_c is considered, on the performance for all values of λ considered with $\zeta = 0.0625$ and 0.1875, respectively. It is noted that the trend is still evident. As λ increases, R_{bt} decreases and the variability increases; however, for the case of large λ ($\lambda > 0.5$) and $\zeta = 0.0625$, the influence is small. It is also noted that, when P_c is considered to vary with heterogeneity, the influence of correlation lengths became smaller. The disparity between R_{bt} values at $\lambda = 0.125$ and 2.0 in the case of variable P_c is much smaller than that when P_c is neglected. This is true regardless of ζ.

The most important observation is that significant changes in R_{bt} resulted when P_c is not neglected. This is observed to a certain degree for all λ and ζ cases considered. The most substantial change occurs for the case of large $\lambda = 2.0$ and small $\zeta = 0.0625$. In this case, the change in R_{bt} is around 37%. This is a considerable change. For the case of large $\lambda = 2.0$ and large $\zeta = 0.1875$, the change in R_{bt} is almost 20%. The change in recovery factors is significant at large λ; however, this change decreases rather substantially as ζ increases. On the contrary, the least significant change occurs for the case of small $\lambda = 0.125$ and large $\zeta = 0.1875$. In this case, the change in R_{bt} is approximately 5%. This is a small change. The changes in recovery factors are relatively insignificant at small λ.

We will assume that the "true" performance is that of the case when all properties as well as P_c are variable. Neglecting P_c would introduce an error. Comparing the results from this section and the previous section would quantify the magnitude of this error for all λ and ζ cases considered. Figure 18 shows the magnitude of error in R_{bt}, if P_c is neglected, as a function of correlation length. The smallest error is for the case of small λ and large ζ. On the contrary, the largest error is for the case of large λ and small ζ. For all cases of λ considered, the error is smaller at large ζ. The largest error is 27% for the case of large $\lambda = 2.0$ and small $\zeta = 0.0625$. This is because the correlated sections of high and low permeability are thin. This introduces a large surface area of low-permeability sections to be in contact with water. This process slows the front of the displacement and stabilizes the macroscopic saturation front movement.

Effect of Constant S_{wi}, S_{or}, k'_{rw}, and $P_c = 0$

In a preceding section, we considered variable S_{wi}, S_{or}, and k'_{rw} with zero P_c; however, in this section, we examine the effect of neglecting these variabilities. Constant S_{wi}, S_{or}, and k'_{rw} are considered instead. This will be investigated for two horizontal correlation lengths with $\zeta = 0.0625$, namely, $\lambda = 0.5$ and 2.0, respectively. The magnitude of change in recovery will reflect the importance of considering variable properties.

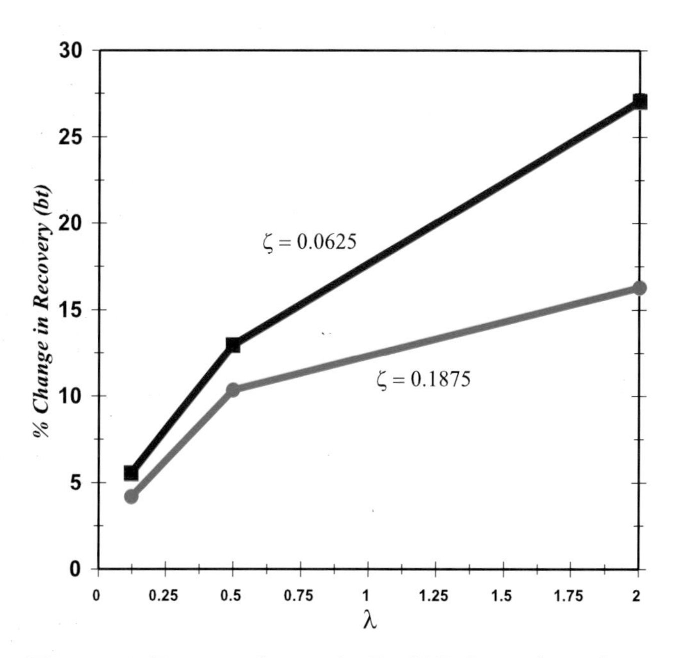

Figure 18. Error or change in R_{bt} if P_c is neglected versus λ for $\zeta = 0.0625$ and 0.1875.

Because porosity is not constant, initial water and residual oil saturations should be calculated. The average initial water saturation $\overline{S}_{wi}$ is given by

$$\overline{S}_{wi} = \frac{\sum_{n=1}^{N} \left(\phi S_{wi}\right)_n}{\sum_{n=1}^{N} \phi_n} \tag{11}$$

Similarly, the average residual oil saturation $\overline{S}_{or}$ is given by

$$\overline{S}_{or} = \frac{\sum_{n=1}^{N} \left(\phi S_{or}\right)_n}{\sum_{n=1}^{N} \phi_n} \tag{12}$$

where N is the number of grid blocks in each realization. The average end point water relative permeability $\overline{k'_{rw}} = \mu_{k'rw} = 0.4$.

In this section, $\overline{S}_{wi}$ and $\overline{S}_{or}$ were calculated for each realization with $\lambda = 2.0$ and $\zeta = 0.0625$. These values were assumed for all grid blocks. Furthermore, k'_{rw} was assumed to be constant, $\overline{k'_{rw}} = \mu_{k'rw} = 0.4$. The performance of these realizations under these constant properties is compared to the performance under variable properties. For the purpose of this particular study, capillary pressure is assumed to be zero.

Figure 19 shows the average response of the two cases of constant and variable properties (curves 1 and 2, respectively). It is shown that there is insignificant effect when these properties are assumed to be constant. This

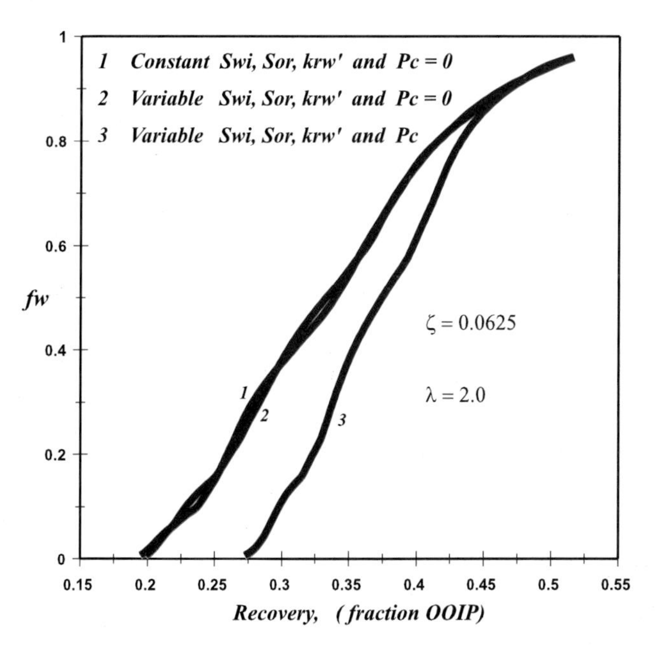

Figure 19. f_w versus R for the two cases of variable properties, S_{wi}, S_{or}, k'_{rw}, with and without P_c and for the case of constant properties, $\overline{S}_{wi}$, $\overline{S}_{or}$, and $\overline{k'_{rw}} = \mu_{k'rw}$, all with $\zeta = 0.0625$. (a) for $\lambda = 0.5$, (b) for $\lambda = 2.0$.

validates the study in the preceding section in which these properties were assumed to be constant. Moreover, this is in agreement with the study by Tjolsen et al. (1993). The choice of the constant properties is crucial. Average values of saturations are important and should be determined accurately. Also, the choice of k'_{rw} is also crucial. In this case, the choice of k'_{rw} to be the mean value $\bar{k}'_{rw} = \mu_{k'_{rw}}$ was accurate.

ESTIMATION OF CORRELATION LENGTH FROM PERFORMANCE DATA

In preceding sections we discussed the influence of correlation length on the system's effective permeability and production performance of correlated permeability fields. It was shown that there exist correlations between spatial correlation lengths and the effective permeability and performance of the reservoir systems. The implication of this is that for performance forecasting to be accurate, determination of correlation lengths is essential; however, the appraisal of horizontal correlation lengths is difficult, especially in the absence of an outcrop. In this section, a systematic procedure for identifying correlation lengths from performance data is suggested.

Methodology

For a given reservoir structure with a specific PDF (probability density function) of permeability and fluid and rock characteristics, type-curve diagnostic plots for the system's effective permeability and performance could be generated. First, several correlation lengths are assumed to represent the heterogeneity. For each case of correlation length, several equiprobable realizations are generated. The effective permeabilities of the systems in each correlation length case are calculated. In addition, waterflooding response is generated for each realization. Expected values are then calculated for each case of correlation length. The results of the system effective permeability calculations and waterflood responses are plotted versus correlation length. The resulting plots are used as type curve for the underlying reservoir structure and properties. If a waterflood response or system's effective permeability are available for any well in the reservoir, these plots help in estimating the correlation length with reasonable accuracy.

A Case Study

Consider the cases in sections 2 and 3 in which for a particular vertical correlation length, $\zeta = 0.0625$, five horizontal correlation lengths, $\lambda = 0.125, 0.25, 0.5, 1.0$, and 2.0, were assumed. The system's effective permeabilities were calculated and the waterflood responses were generated. The expected values were plotted versus λ. Figure 20 shows the type-curves of waterflood performance for this particular reservoir model. Figure 21 shows the type-curves for the system effective permeability for the same system. The plots are then used to delineate uncertainties associated with estimating the correlation lengths.

Now consider that we have three reservoir models randomly generated for different correlation lengths and that these correlation lengths are unknown; furthermore, consider that the system's effective permeabilities are known and that the waterflood responses are available. These models are model 1, 2, and 3, with the corresponding $k_{Heff} = 108$, 76, and 40 md, respectively. The task is to determine λ for each model as close as possible to the "true" correlation length.

Figure 20 shows the behavior of the three models on the performance type curve. Model 1 follows the curve corresponding to $\lambda = 2.0$ very closely. Due to the variability of recovery breakthrough at this large λ, it is helpful to confirm the observation with a similar one on the system effective permeability type curve. Figure 21 shows model 1 within the range of

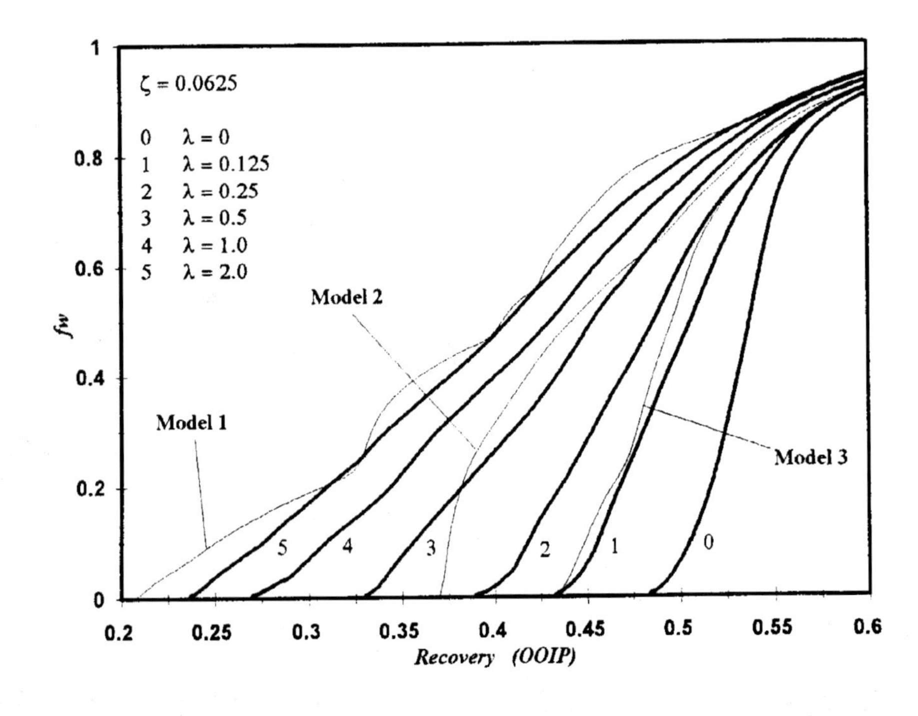

Figure 20. Type-curve for the performance under waterflooding for different correlation lengths λ. The performance of models 1, 2, and 3 also is shown.

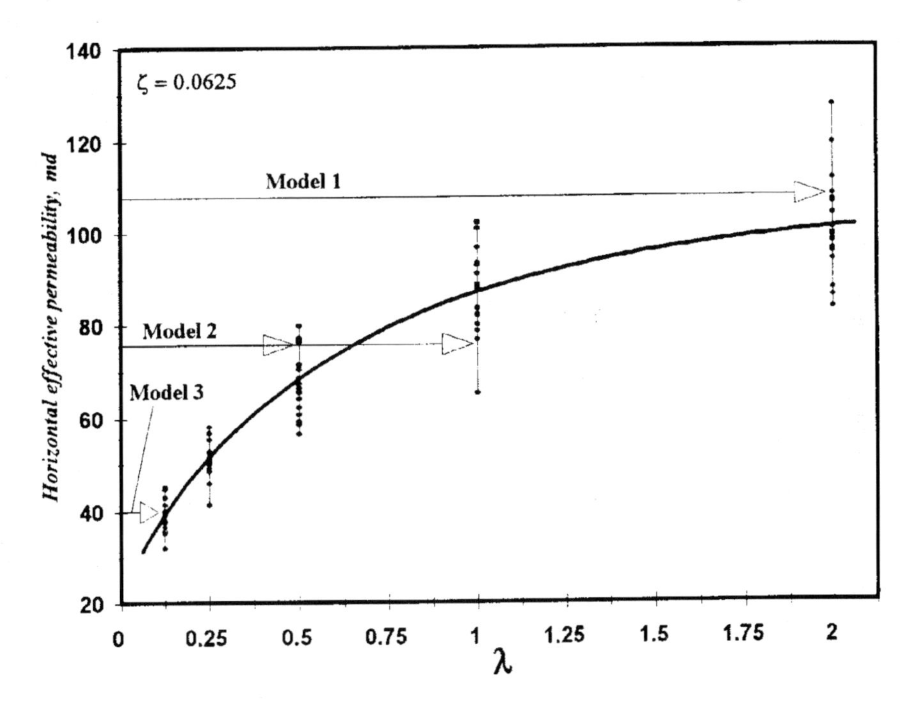

Figure 21. Type-curve for the system's effective permeability for different correlation lengths λ. The effective permeability values for models 1, 2, and 3 also are shown. k_{Heff} for models 1, 2, and 3 is 108, 76, and 40, respectively.

possibilities at λ = 2.0 and outside the range of possibilities at λ = 1.0. This additional information enhanced the confidence level and confirmed that the correlation length is λ = 2.0. The actual horizontal correlation length is 2.0.

For model 2, the waterflood performance is shown in Figure 20. From the recovery at breakthrough, the model is suspected to be between λ = 0.125 and 0.5; however, most of the performance follows the curve of λ = 0.5. The system's effective permeability in Figure 21 shows two possibilities of λ = 0.5 and 1.0. The inclusion of all of the performance data suggests a higher level of confidence for λ = 0.5, which is the exact value used in the construction of the model.

The third model is shown in the performance type curve, in Figure 20. The early part of the performance follows the curve with λ = 0.125; however, after f_w = 0.4, the performance deviates from the curve and follows the curve with λ = 0.25. The estimation of λ, relying on the performance alone, would be between 0.125 and 0.25; however, the system's effective permeability disputes the uncertainty and confirms λ to be 0.125. The actual correlation length for model 3 is λ = 0.125.

CONCLUSIONS

We have examined the nature of correlated permeability fields and their responses under immiscible recovery processes. In particular, the case of water injection was examined for both 2-D cross-sectional and areal representations of such fields. Based on the observations and analysis of results, we have reached the following conclusions.

(1) For correlated permeability fields with the same PDF of permeability, there exist relationships between the system effective permeabilities and correlation lengths in both the horizontal and vertical directions.

(2) The system effective horizontal permeability, k_{Heff}, is significantly influenced by correlation length. For large horizontal correlation length λ, the system effective permeability is high. For systems with small λ, low values of k_{Heff} are observed. The influence of λ is most significant at short correlation lengths (λ ≤ 0.5). The system effective vertical permeabilities, k_{Veff}, are also influenced by correlation length. High system effective vertical permeabilities are exhibited when λ is small.

(3) Vertical correlation length, *z*, influences the system effective permeability. For two systems with the same λ, the one with the shorter vertical correlation length has the higher system effective permeability value. The ratio of vertical to horizontal system effective permeability, k_{Veff}/k_{Heff}, is influenced by λ and ζ. At large λ and short ζ, this ratio becomes very small. On the contrary, at short λ and large ζ, this ratio becomes high.

(4) The permeability distribution parameter, σ_{logk}, influences the system effective permeability considerably for systems with large correlation lengths. This influence is small at short correlation lengths.

(5) For systems with the same PDF of permeability, under waterflood, there exists a relationship between the recovery factor at breakthrough and the correlation length. The highest breakthrough recovery factor is observed for uncorrelated systems. For two systems at different correlation lengths, the one with the higher correlation length results in a lower breakthrough recovery. The influence of λ on recovery factors at breakthrough is most significant at short correlation lengths (λ ≤ 0.5).

(6) The breakthrough time and recovery of reservoir models at a particular correlation length can be characterized by an expected value and variance. The variance values of the predicted breakthrough time and recovery are minimum for uncorrelated permeability fields. The variance increases as correlation length is increased and reaches maximum at intermediate correlation lengths (λ = 1.0). At larger correlation lengths the variance decreases.

(7) Similar reservoirs, but with different correlation lengths in the permeability field, respond differently to waterfloods. The implication of this observation is primarily for reservoirs with layering or compartmentalization. In such systems, each segment of the reservoir may have its own specific correlation lengths and response characteristics. It will assist history matching in reservoir simulation where the individual segments can point to the type of the correlation structure existing in the reservoir.

(8) Selection of spatial correlation models has insignificant influence on predicting the system performance.

(9) The effect of neglecting capillary pressure on the performance is substantial for reservoir systems with long horizontal and short vertical correlation lengths. Capillary pressure at these correlation lengths affect the water saturation profile within the system significantly as well. The effect is small for short horizontal and large vertical correlation lengths. The assumption of constant as opposed to variable irreducible water and oil saturations, S_{wi} and S_{or}, and end-point water relative permeability, k'_{rw}, was shown to have insignificant effect on the production response.

(10) A procedure for estimating spatial correlation length of correlated permeability fields is presented. For reservoir systems with the same PDF of permeability, and fluid and rock characteristics, type-curve plots for both performance data and system effective permeability can be generated at different correlation lengths. If waterflood performance data are available or if the system effective permeability is known from any source, determination of spatial correlation length, with a reasonable confidence level, is possible using these type-curves. The possibility of increasing the confidence level is enhanced if both types of information are available.

NOMENCLATURE

f_w = fractional flow, fraction
k = permeability, md
k_{Heff} = horizontal effective permeability, md
k_{Veff} = vertical effective permeability, md
OOIP = original oil in place
R_{bt} = breakthrough recovery, fraction
S_{wi} = initial water saturation
S_{or} = irreducible oil saturation
P_c = capillary pressure, psi
k'_{rw} = end point water relative permeability
λ = normalized horizontal correlation length
ζ = normalized vertical correlation length
μ = mean value
σ = standard deviation

ACKNOWLEDGMENTS

The first author wishes to thank Saudi Aramco for the sponsorship of his graduate studies at the University of Southern California and for support in publiing this paper. This study was supported by the DOE (Department of Energy)/USC/Tidelands Class III Project.

REFERENCES CITED

Aasum, Y.: Effective Properties of Reservoir Simulator Grid Blocks, Ph.D. Dissertation, The University of Tulsa, OK (1992), 241 p.

Al-Qahtani, M.Y., and I. Ershaghi: Improvements in the Stochastic Generation of Reservoir Images Using Performance Data, Paper SPE 29671 presented at the 1995 SPE Western Regional Meeting, Bakersfield, CA, March 8-10, p. 21–89.

Al-Qahtani, M.Y.: Characterization of Spatially Correlated Permeability Fields Using Performance Data, Ph.D. Dissertation, University of Southern California, Los Angeles, CA (1996), p. 21–89.

Chang, J., and Y.C. Yortsos: Effect of Capillary Heterogeneity on Buckley-Leverett Displacement, SPE Reservoir Engineering, May 1992, p. 285–293.

Chang, Y.C., and K.K. Mohanty: Stochastic Description of Multiphase Phase Flow in Heterogeneous Porous Media, Paper SPE 28443 presented at the 1994 SPE Annual Technical Conference and Exhibition, New Orleans, LA, September 25-28, p. 829–838.

Deutsch, C.: Annealing Techniques Applied to Reservoir Modeling and the Integration of Geological and Engineering (Well Test) Data, Ph.D. Dissertation, Stanford University, Stanford, CA (1992), 325 p.

Deutsch, C., and A. Journel: GSLIB: Geostatistical Software Library and User's Guide, Oxford University Press, New York (1992), 336 p.

Haldorsen, H.H., and E. Damsleth: Stochastic Modeling, JPT (April 1990), p. 404-412.

Haldorsen, H.H., and L.W. Lake: PA New Approach to Shale Management in Field-Scale Models, SPEJ (August 1984) p. 447-457.

Hoiberg, J., H. Omre, and H. Tjelmeland: A Stochastic Model for Shale Distribution in Petroleum Reservoirs, Proceedings from the Second CODATA Conference on Geomathematics and Geostatistics, 1990, Leeds, September 10-14.

Honarpour, M.M., L.F. Koederitz, and A.H. Harvey: Relative Permeability of Petroleum Reservoirs, CRC Press, Boca Raton (1986), p. 15–43.

Isaaks, E.H., and R.M. Srivastava: An Introduction to Applied Geostatistics, Oxford University Press, New York (1989), 561 p.

King, P.R.: Effective Values in Averaging, Mathematics in Oil Production, Oxford, 1989a, p. 217-234.

King, P.R.: The Use of Renormalization for Calculating Effective Permeability, Transport in Porous Media (February 1989b) p. 37-58.

Leverett, M.C.: Capillary Behavior in Porous Solids, Trans. AIME (1941) 142, p. 152-169.

Omre, H., H. Tjelmeland, Y. Qi, and L. Hinderaker: Assessment of Uncertainty in the Production Characteristics of a Sandstone Reservoir, 1991 Third International Reservoir Characterization Technical Conference, Tulsa, OK, November 3-5, p. 556–603.

Tjolsen, N., E. Damsleth, and T. Bu: Stochastic Relative Permeabilities Usually Have Neglectable Effect on Reservoir Performance, Paper SPE 26473 presented at the 1993 SPE Annual Technical Conference and Exhibition, Houston, TX, October 3-6, p. 591–600.

Campozana, F.P., L.W. Lake, and K. Sepehrnoori, How incorporating more data reduces uncertainty in recovery predictions, 1999, *in* R. Schatzinger and J. Jordan, eds., Reservoir Characterization-Recent Advances, AAPG Memoir 71, p. 359–368.

Chapter 25

How Incorporating More Data Reduces Uncertainty in Recovery Predictions

Fernando P. Campozana
Petrobras
Rio de Janeiro, Brazil

Larry W. Lake
Kamy Sepehrnoori
Center for Petroleum and Geosystems Engineering
The University of Texas at Austin
Austin, Texas, U.S.A.

ABSTRACT

From the discovery to the abandonment of a petroleum reservoir, there are many decisions that involve economic risks because of uncertainty in the production forecast. This uncertainty may be quantified by performing stochastic reservoir modeling (SRM); however, it is not practical to apply SRM to account for new data every time the model is updated.

This paper suggests a novel procedure to estimate reservoir uncertainty (and its reduction) as a function of the amount and type of data used in the reservoir modeling. Two types of data are analyzed: conditioning data and well-test data; however, the same procedure can be applied to other data types.

SRM is typically performed for the following stages: discovery, primary production, secondary production, and infill drilling. From those results, a set of curves is generated that can be used to estimate (1) the uncertainty for any other situation and (2) the uncertainty reduction caused by the introduction of new wells (with and without well-test data) into the description.

INTRODUCTION

Reservoir uncertainty results from a lack of information. Sample data taken at well locations (cores and well logs) are too scarce to allow a detailed description. Other types of information, such as (1) seismic data, (2) geological interpretation, (3) outcrop analogs, and (4) tracer, well-test, and production data, can reduce, but not eliminate, uncertainty in reservoir description. As a result, multiple, equiprobable models of the same reservoir can be generated, all of them honoring the available information. This process, called stochastic reservoir modeling (SRM), allows one to (1) quantify reservoir uncertainty, (2) verify fluctuations in the cash flow of projects, (3) find where unswept areas of the reservoir are probably located, and (4) better manage improved oil recovery projects or infill drilling.

To stochastically simulate reservoir performance, one generally needs to build a probability density function (pdf) for each of the reservoir simulation parameters. Assuming independence among all parameters, this can be achieved by performing the following steps (Haldorsen and Damsleth, 1990):

(1) Define a reference or most probable reservoir model
(2) Generate multiple realizations of each reservoir parameter of the reservoir model
(3 Perform repeated flow simulations varying only one reservoir parameter, keeping unchanged the others from the reference model parameters
(4) Define a performance parameter, such as oil recovery at a given time, water cut, or breakthrough time, to build the pdf of the parameters

This procedure provides an estimate of how reservoir performance varies with each parameter. This information is useful in determining the degree of accuracy necessary for each reservoir parameter. After estimating the pdfs of the reservoir parameters, multiple reservoir models can be generated by applying the Monte Carlo method, as shown in Figure 1. Each

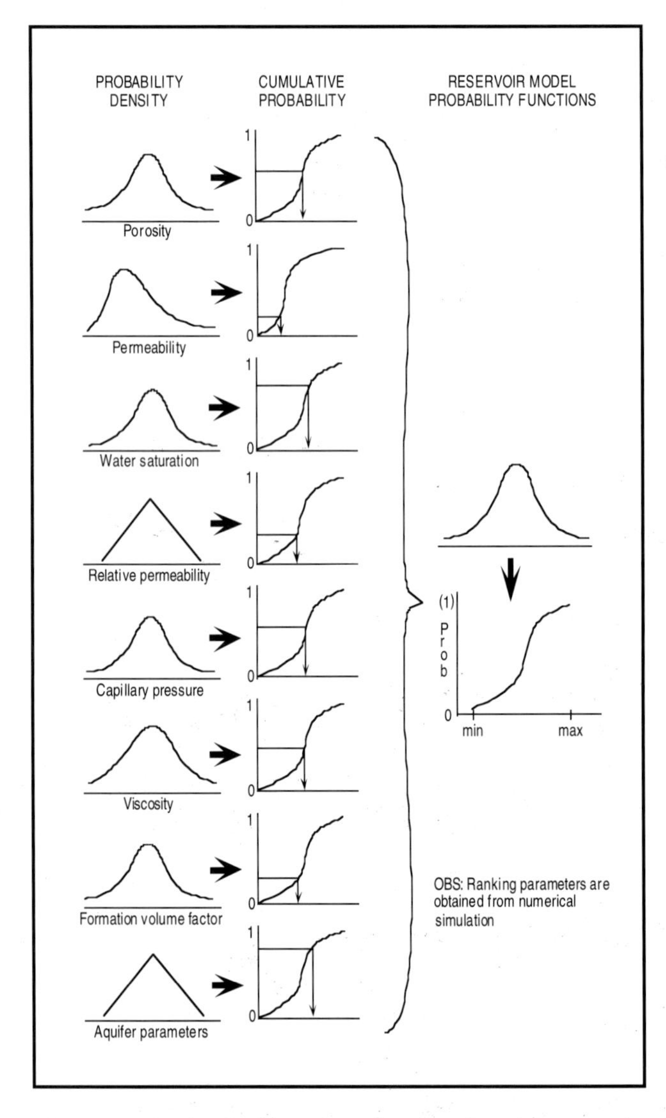

Figure 1. Monte Carlo approach to stochastic reservoir modeling.

reservoir model is defined by randomly selecting the reservoir parameters from their pdfs. Reservoir uncertainty is then quantified by using each reservoir model as input to a numerical flow simulator. Based on performance parameters obtained from the simulation results, a pdf can be built in a manner similar to that done for each individual reservoir parameter.

Stochastic reservoir modeling can be time consuming and sometimes computationally prohibitive. Because no satisfactory numerical simulator based on stochastic differential equations has yet been developed, each reservoir model must be input one at a time. To overcome this limitation, a number of techniques have been suggested. First, the sampling process can be minimized without losing information by using either the stratified sampling method or Latin HyperCube sampling (Ding et al., 1989); second, subjective pdfs (similar to triangular distributions) can be considered for those parameters that practical experience or previous sensitivity analysis have shown not to be critical (Øvreberg et al., 1990); and, finally, one can perform simplified numerical simulations to rank each reservoir description and then select those that correspond to the most important quantiles to run full numerical simulations (Ballin, 1992). The fast, simplified numerical simulations are usually done by using a coarse grid simulation, a tracer simulator, or a simplified flow model such as a streamline simulator.

It must be pointed out that usually reservoir parameters exhibit multivariate dependence, and the application of multivariate statistics is necessary. The calculated variance of production profiles, assuming independence among the variables, can be very different from that calculated considering multidependence; however, a complete study on the dependence of all variables is rarely feasible because it requires information that is not available. Bivariate statistics considering, for example, dependence between porosity and permeability, are strongly suggested.

Applying SRM every time new information is added to the reservoir description is not a viable option, even with the sampling techniques described. In this work, a procedure is recommended that allows one to infer uncertainty in production performance for any data configuration, as well as the uncertainty reduction caused by the incorporation of new data into the model, with a reasonable computational effort.

THE APPROACH

The more data that are included in a description, the better a resulting model should represent reality and reservoir uncertainty should decrease. To quantify reservoir uncertainty as a function of the amount and type of data used in the model without applying SRM after every model update, the following procedure is suggested: (1) generate a reference model that accounts for as much data as possible and use it as input of a numerical simulator to obtain a production forecast; (2) perform some samplings from the reference model (say, four) to mimic a typical sequence of data acquisition obtained from equally spaced wells (as in the example shown in Figure 2); (3) perform

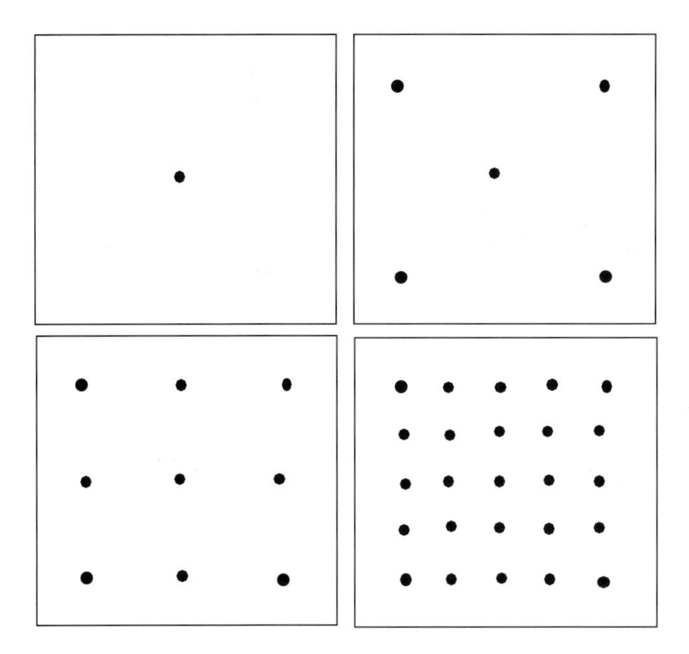

Figure 2. Four configurations of the conditioning data. From top to bottom and left to right: field discovery, primary recovery, secondary recovery, and infill drilling.

SRM for each data configuration assuming the only information known about the reference model is that provided by the samples; (4) choose one or more performance parameters to rank the realizations and quantify the uncertainty in the recovery for each data configuration; and (5) plot the performance parameter versus the number and type of data.

The uncertainty related to the reference model is bounded by two extremes: It is a maximum when no data are available, and it is zero when all block values of the reference model are known. With this information, plus the points obtained using the preceding procedure, a curve that describes reservoir uncertainty versus the number of wells can be generated. Interpolation may be used to estimate uncertainty for any number of wells.

Many performance parameters have been suggested in the literature (Ballin et al., 1993); most of them require the use of numerical flow simulation. An interesting parameter that does not rely on numerical simulation is the visual likeness factor (Ouenes and Saad, 1993), defined as

$$\Gamma = \frac{\sum_i (Z_i - \bar{Z})(Y_i - \bar{Y})}{\sqrt{\sum_i (Z_i - \bar{Z})^2 \sum_i (Y_i - \bar{Y})^2}} \tag{1}$$

where Z_i and Y_i are, respectively, the values of the variable at the ith block of the simulated image and the reference or base case. $\bar{Z}$ and $\bar{Y}$ are the average values. A perfect match yields Γ value of 1, whereas $\Gamma = 0$ means that there is no correlation between the base case and the generated image. The visual likeness factor is very similar to the well-known correlation coefficient, the difference being that Γ considers the spatial location of the samples. This is why Γ is sometimes called the spatial correlation coefficient.

Given a set of simulations, the uncertainty in the recovery prediction (URP) can be expressed as

$$URP_i^j = \frac{\left(RF_{opt} - RF_{pess}\right)_i^j}{RF_{base}} \tag{2}$$

where the subscript i refers to the number of conditioning wells, the superscript j refers to the algorithm used to generate the set of images, and RF_{opt}, RF_{pess}, and RF_{base} are, respectively, the most optimistic, the most pessimistic, and the base-case recovery factors after a given time of production. If the reference model is updated, this parameter can be adjusted accordingly.

Analyzing the boundaries, $URP = 0$ if all i realizations perfectly match the reference case, because both RF_{opt} and RF_{pess} coincide with RF_{base}. The upper bound of URP is obtained from equation 2 for the set of realizations that yields the largest or maximum recovery range

$$URP_{max} = \frac{\left(RF_{opt} - RF_{pess}\right)_{i_{max}}^j}{RF_{base}}$$

Generally, this set is the one that accounts for the least amount of data. A set of unconditional simulations could be used for this purpose. An alternative expression, URP^*, is independent of the base-case recovery and ranges between 0 and 1:

$$URP_i^{*j} = URP_i^j \frac{RF_{base}}{\left(RF_{opt} = RF_{pess}\right)_{max}} = \frac{\left(RF_{opt} = RF_{pess}\right)}{\left(RF_{opt} = RF_{pess}\right)_{max}} \tag{3}$$

A plot of URP versus the number of wells gives an estimate of the recovery range one would obtain for a generic configuration of wells, provided that they are approximately equally spaced. Obviously, if the wells are clustered, they will not reduce uncertainty as much as if they were equally spaced.

A third parameter measures the reduction of the uncertainty in the recovery prediction ($RURP$) as more data are added to the description:

$$RURP_i^j = \frac{\left(URP_{max} - URP\right)}{URP_{max}} = 1 - URP_i^{*j} \tag{4}$$

A plot of $RURP$ versus the number of wells quantifies the impact of new data on reservoir uncertainty.

AN EXAMPLE APPLICATION

A hypothetical reservoir submitted to waterflooding is stochastically modeled to observe how reservoir uncertainty decreases as more conditioning data and well-test information are added to the description. For the sake of simplicity, only absolute permeability is allowed to vary. All other reservoir parameters are assumed to be perfectly known. A synthetic base case is generated and used as input to a commercial ECLIPSE 100® numerical simulator to obtain dynamic data (well-test permeability and oil recovery). The results are compared to those obtained using several sets of geostatistical realizations that honor different amounts of conditioning and well-test data.

THE BASE CASE

A two-dimensional permeability field was generated using the matrix decomposition method (MDM) (Fogg and Lucia, 1989; Yang, 1990). The hypothetical reservoir is a square of 4500 ft × 4500 ft (1372.5 m × 1372.5 m) divided into 45 × 45 square gridblocks of 100 ft × 100 ft (30.5 m × 30.5 m) each. For purposes of numerical simulation, nine equally spaced wells are active in the field: four are water injectors and five are oil producers. Their location and rates are shown in Figure 3. The gray-scale map of the base-case permeability field is shown in Figure 4. The semivariogram model is isotropic, spherical, with no nugget, and has a range of 2250 ft (686 m) (half of the side length of the reservoir). The reservoir properties are given in Table 1.

The base case was then used as input to ECLIPSE 100. Two types of flow simulation were performed: a production forecast and well tests. The radius of investigation of the well-test simulations is equal to 1500 ft (457.5 m). These synthetic data are considered as reference or the truth case.

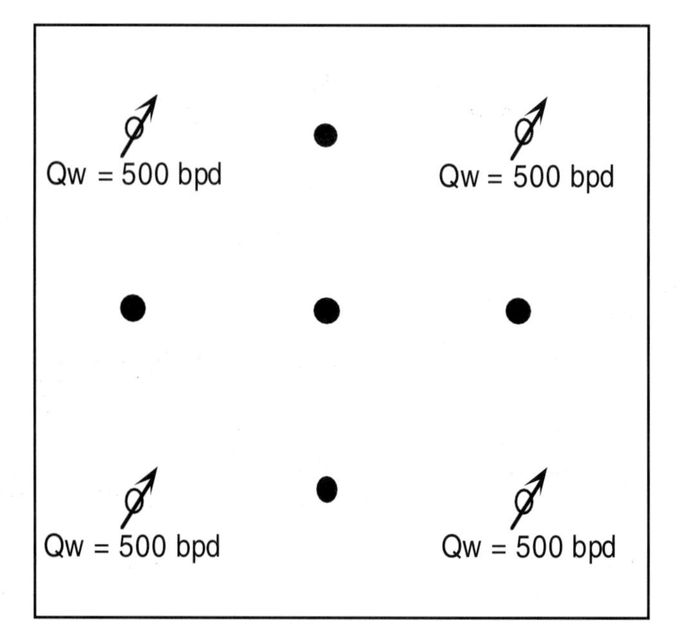

Figure 3. Well locations and water injection rates for all numerical simulations.

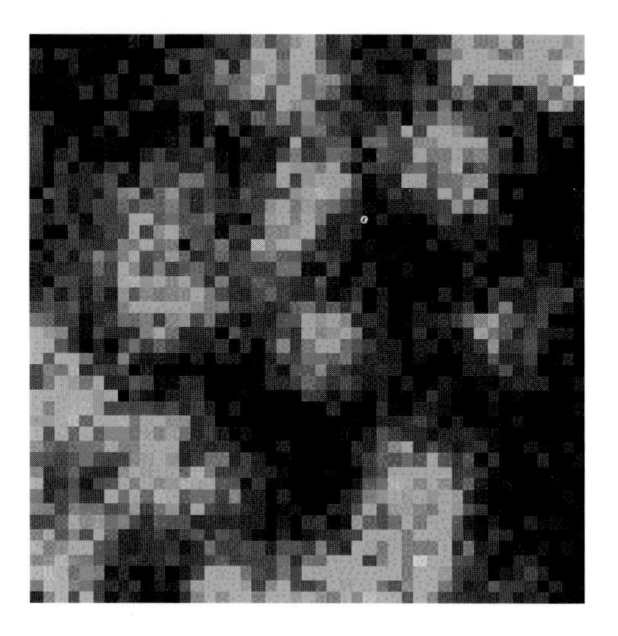

Figure 4. Base-case permeability distribution generated using sequential Gaussian simulation (SGS). The image is conditioned to the permeabilities at the nine wells shown in Figure 3 and to a spherical semivariogram. There is no nugget effect, and the range of correlation is one-half of the reservoir horizontal length.

Table 1. Reservoir Properties

Reservoir dimensions, ft	4500 × 4500
Number of gridblocks (n_x, n_y, n_z)	45 × 45 × 1
Block dimensions (Δx, Δy, Δz)	100 × 100 × 20
Porosity, fraction	0.2
Wellbore radius, ft	0.25
Total compressibility, 1/psi	4.7×10^{-7}
Initial water saturation	0.22
Oil formation volume factor, bbl/STB (at initial pressure, 2200 psia)	1.15
Oil viscosity at 2200 psia, cp	0.8
Initial pressure, psia	2200

STOCHASTIC MODELING

Three geostatistical algorithms were used to generate realizations conditioned to different types of data: turning bands method (TBM) (Journel and Huijbregts, 1978), simulated annealing (SA) (Kirpatrick et al., 1983; Farmer, 1989; Otten and Van Ginneken, 1989), and MTWELL (Campozana et al., 1996). The TBM algorithm was used to generate a set of unconditional simulations that was constrained only by the semivariogram. The SA algorithm was used to generate four sets of realizations conditioned to different numbers of wells (1, 5, 9, and 25) and to the semivariogram. Finally, the MTWELL algorithm was used to obtain four more sets of realizations conditioned to the semivariogram, to the

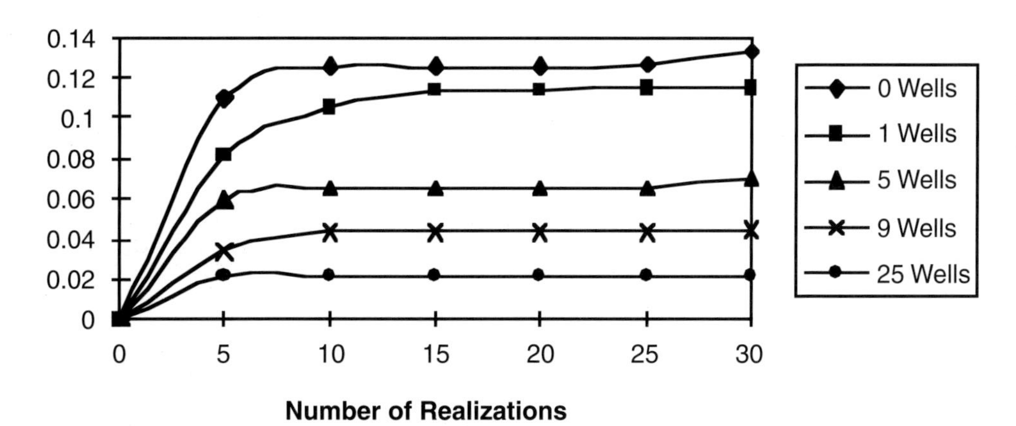

Figure 5. Variation of the recovery range with the number of simulated annealing (SA) realizations. The samplings are made in increments of 5.

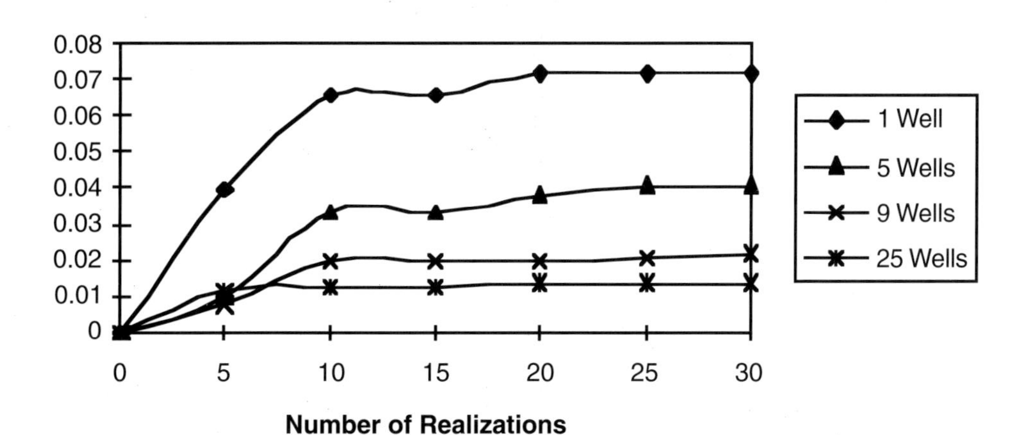

Figure 6. Variation of the recovery range with the number of MTWELL realizations. The samplings are made in increments of 5.

same wells used in the SA realizations, and to their respective well-test permeabilities derived from pressure-transient analysis. Each set is composed of 30 two-dimensional permeability realizations.

The configurations of the conditioning data used in the SA and MTWELL realizations are shown in Figure 2. These configurations can be associated with the discovery, primary production, secondary production, and infill drilling phases of the reservoir. Each well is representative of the block in which it is located. The semivariogram used in all realizations is identical to that of the base case; therefore, uncertainty in the semivariogram model is not being considered in this application.

A statistical procedure was followed to verify if 30 realizations were sufficient to yield meaningful results (Campozana et al., 1996). All sets of realizations were submitted to several samplings without replacement, and the recovery range ($OR_{opt} - OR_{pess}$) was calculated for each sample. Figures 5 and 6 show how the recovery range varies with the number of SA and MTWELL realizations, respectively. Stable values are reached after 25 realizations for all sets, indicating that 30 realizations are sufficient for the present example.

Figures 7–11 show, respectively, one TBM, two SA, and two MTWELL realizations randomly taken from their sets. Comparing these images with the base case (Figure 4), there is improvement as (1) the number of conditioning data increases and (2) well-test data are incorporated into the description. This improvement will be quantified in the following sections.

The Visual Likeness Factor

The visual likeness factor (equation 1) was calculated for all realizations, as well as the mean value $\overline{\Gamma}$ of each set. In the limiting case where all block values (2025) are

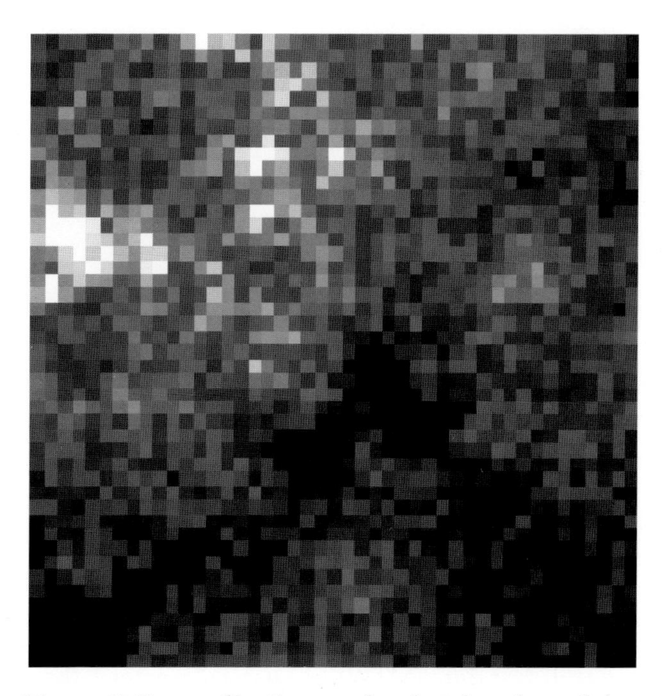

Figure 7. One realization randomly taken from the turning bands method (TBM) set (unconditional).

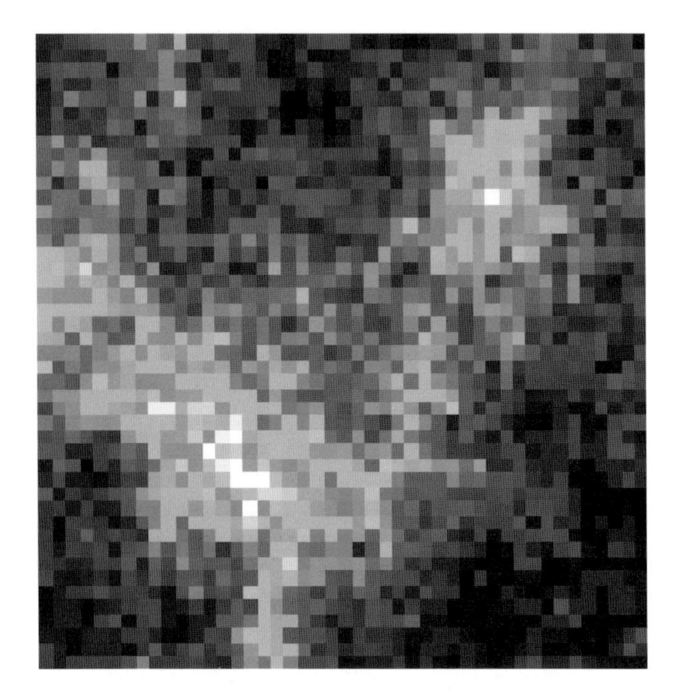

Figure 8. One realization randomly taken from the simulated annealing (SA) set conditioned to one well.

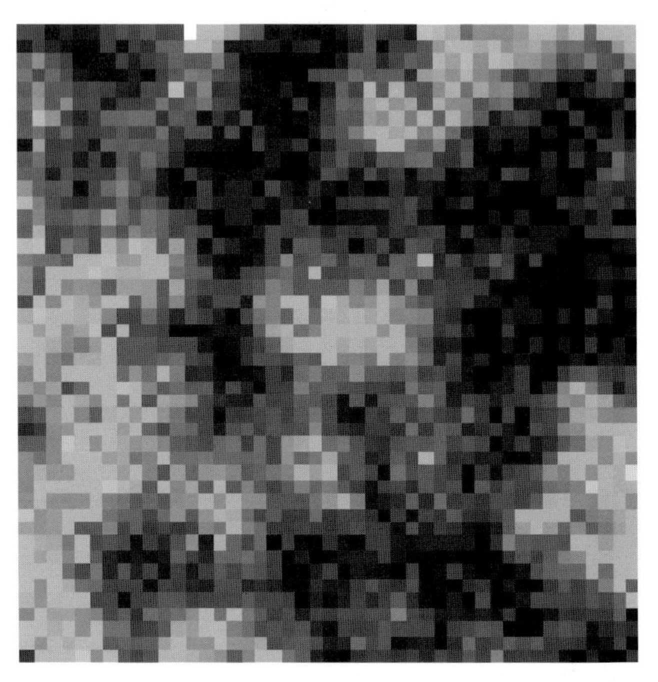

Figure 10. One realization randomly taken from the MTWELL set conditioned to one well plus its well-test permeability.

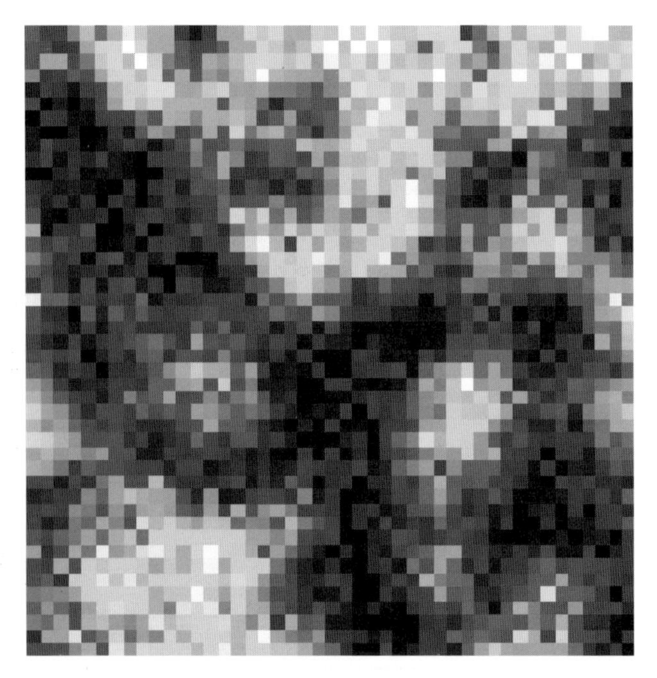

Figure 9. One realization randomly taken from the simulated annealing (SA) set conditioned to 25 wells.

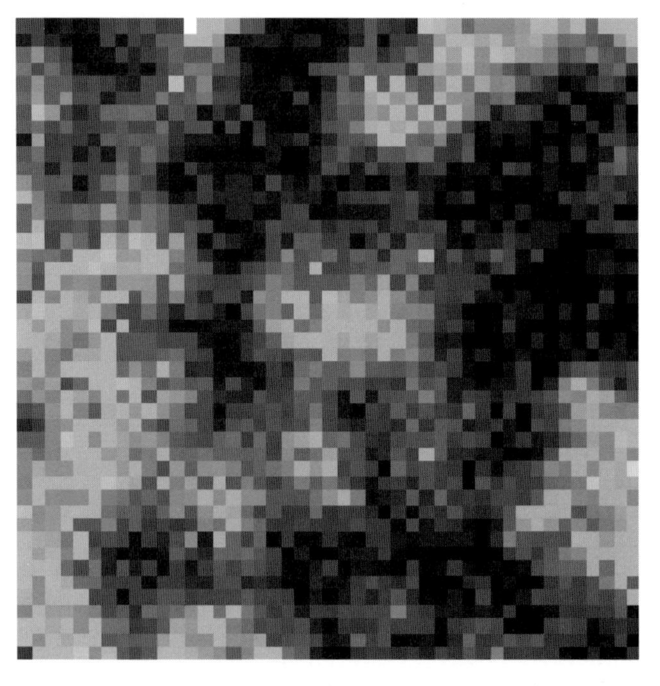

Figure 11. One realization randomly taken from the MTWELL set conditioned to 25 wells plus their well-test permeabilities.

known, $\bar{\Gamma} = 1$ for both the SA and MTWELL cases because all realizations are identical to the base case. A plot of $\bar{\Gamma}$ versus the number of wells (N_w) is shown in Figure 12. The visual likeness increases as N_w increases, but the impact of additional wells on $\bar{\Gamma}$ becomes smaller as N_w increases. The sets of realizations conditioned to the well-test data (MTWELL) have a greater $\bar{\Gamma}$ than those conditioned only to the data points (SA). The TBM set had the lowest value of $\bar{\Gamma}$.

For the cases run, the data points are close to a straight line on a semilog plot, and $\bar{\Gamma}$ can be reasonably well predicted for any number of equally spaced wells by using the equation $\bar{\Gamma} = 0.258 \log N_w + 0.145$ for the SA realizations and the equation $\bar{\Gamma} = 0.219 \log N_w + 0.281$ for the MTWELL realizations. Analyzing the difference between the two curves, one can see the improvement in $\bar{\Gamma}$ caused by the incorporation of

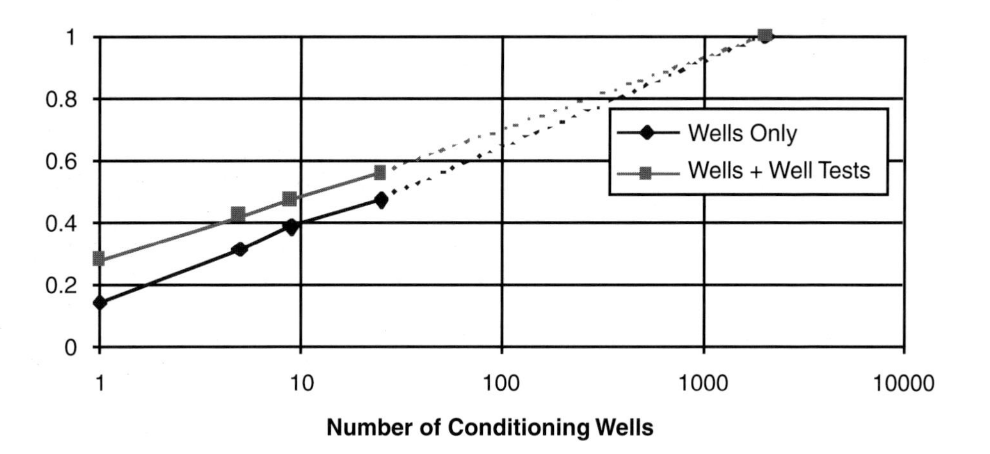

Figure 12. Variation of the visual likeness factor with the number of conditioning wells and type of data. The dotted lines are extrapolations.

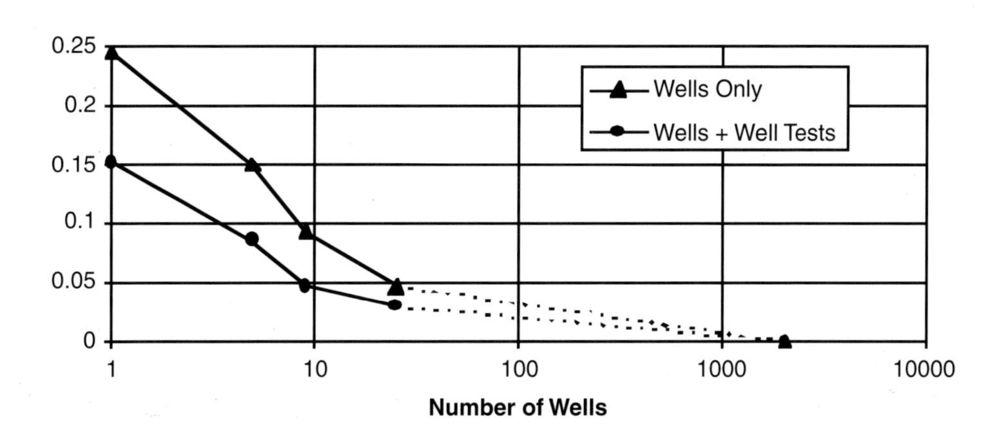

Figure 13. Variation of the uncertainty in the recovery prediction *(URP)* with the number of conditioning wells and type of data. The dotted lines are extrapolations.

well-test information into the description. Although accounting for well-test data always improves the description, its contribution diminishes as the number of conditioning wells increases.

Uncertainty in the Recovery Prediction

Numerical simulation was used to transfer uncertainty in the reservoir models to a production forecast. The nine sets of 30 permeability realizations described in the previous section were used as input to the commercial simulator ECLIPSE 100. All other reservoir properties were kept identical to those of the base case. Although the number of conditioning data ranged from zero to 25, the number of active wells (injectors and producers) was kept unchanged during the numerical simulation (see Figure 3).

The uncertainty in the recovery prediction (*URF*) was calculated for all sets of realizations (see equation 2). Figure 13 shows that, like the behavior of $\overline{\Gamma}$, *URP* decreases as N_w increases, but the impact of additional wells on *URF* decreases as N_w increases. Notice also the reduction in *URF* caused by the inclusion of well-test data into the description, especially for small values of N_w. If all 2025 permeability block values are known, *URF* must be zero, because all realizations would be identical to the base case. Unlike $\overline{\Gamma}$, *URF* is not linearly related to the logarithm of N_w.

The reduction of the uncertainty in the recovery prediction (*RURP*) can now be analyzed in detail. Since the TBM realizations had the largest recovery range of all sets, equation 4 can be rewritten as

$$RURP = \frac{(URP_{TBM} - URP)}{URP_{TBM}} \tag{5}$$

The results obtained using equation 5 are shown in Table 2 and plotted in Figure 14. For $N_w = 1$, the *RURP*

Table 2. Numerical Simulation Results

Algorithm*	N_w	*RF* Range**	*URP*	*RURP*
TBM	0	0.133	0.284	0
SA	1	0.115	0.245	0.135
SA	5	0.07	0.149	0.474
SA	9	0.044	0.094	0.669
SA	25	0.022	0.047	0.835
MTWELL	1	0.071	0.151	0.466
MTWELL	5	0.04	0.085	0.699
MTWELL	9	0.022	0.047	0.835
MTWELL	25	0.014	0.030	0.895
–	2025[†]	0	0	1

*TBM = turning bands method, SA = simulated annealing.
**Range of the recovery factor ($RF_{opt} - Rf_{pess}$).
[†]Extrapolation.

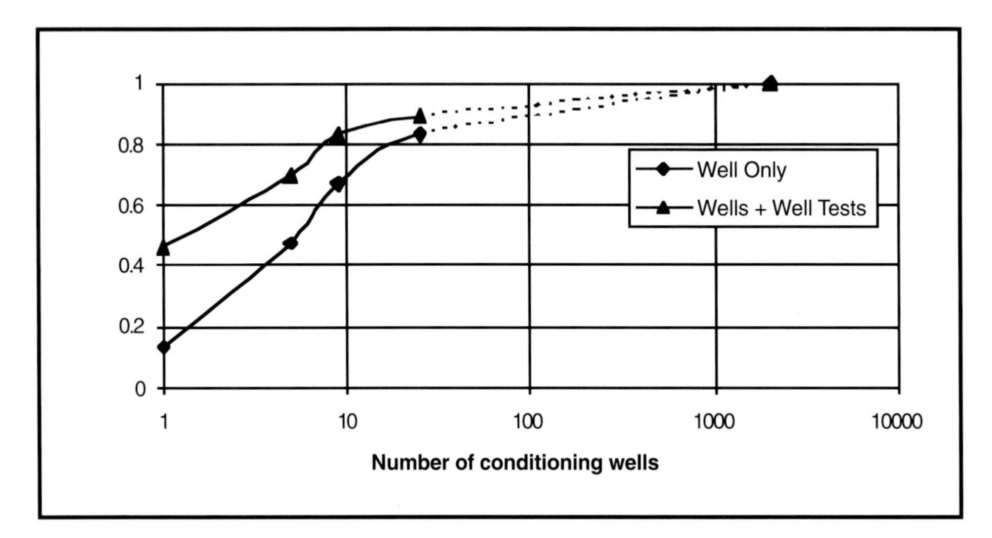

Figure 14. Reduction of the uncertainty in the recovery prediction *(RURP)* with the number of conditioning wells and type of data. The dotted lines are extrapolations.

of the MTWELL set is 47%, whereas that of the SA set is only 14%; however, to get the same *RURP* of 47% without considering well-test data, one needs five conditioning wells. Thus, in this situation, testing the single well is as valuable as adding four wells to the description. For N_w = 25, however, the *RURPs* of the MTWELL and SA realizations are not too different (89 and 83%, respectively).

CONCLUSIONS

A method based on stochastic reservoir modeling has been proposed to estimate reservoir uncertainty and its reduction as a function of the amount and type of data used in the description. This method allows one to assess (1) reservoir uncertainty for any data configuration and (2) the impact of new data on reservoir uncertainty. Three performance parameters are suggested to quantify geostatistical realizations. The application example shows how uncertainty decreases as more conditioning data and well-test data are incorporated into the model. The impact on reservoir uncertainty of adding well-test information to the description is greatest in the early stages of a field development, when only a few wells are available.

NOMENCLATURE

N_w	= number of wells
pdf	= probability density function
RF	= recovery factor (fraction)
RURP	= reduction of the uncertainty in the recovery prediction
SRM	= stochastic reservoir modeling
URP	= uncertainty in the recovery prediction
Y_i	= value of the stochastic variable at the *i*th block for a generalized realization
$\bar{Y}$	= mean value of the variable
Z_i	= value of the stochastic variable at the *i*th block for the base case realization
$\bar{Z}$	= mean value of the variable Z_i
Γ	=visual likeness factor
$\bar{\Gamma}$	= mean value of the visual likeness factor

Superscripts

j	= refers to the algorithm used to generate the realizations
*	= normalized

Subscripts

base	= base case
i	= refers to the number of conditioning data used in the realizations
max	= maximum
opt	= optimistic
pess	= pessimistic
TBM	= turning bands method

ACKNOWLEDGMENTS

The authors acknowledge the Enhanced Oil Recovery Research Program of the Center for Petroleum and Geosystems Engineering at The University of Texas at Austin for partial support of this work. We thank Geoquest for providing the ECLIPSE-100® simulator and the Deltas Industrial Affiliates Program of the Bureau of Economic Geology of The University of Texas at Austin for the computer time. Fernando Campozana thanks Petrobras for financial support during his stay at The University of Texas. Larry W. Lake holds the W.A. (Monty) Moncrief Centennial Endowed Chair.

REFERENCES CITED

Ballin, P.R., 1992, Approximation of flow simulation for uncertainty assessment, Ph.D. dissertation, Stanford University, 366 p.

Ballin, P.R., K. Azis, A.G. Journel, and L. Zuccolo, February 1993, Quantifying the impact of geological uncertainty on reservoir performing forecasts, Proceedings of the 12th Symposium on Reservoir Simulation of the Society of Petroleum Engineers, New Orleans, Louisiana, p. 47–57.

Campozana, F.P., L.W. Lake, and K. Sepehrnoori,

October 1996, Reservoir modeling constrained to multiple well-test permeabilities, Proceedings of the 71st Annual Technical Conference and Exhibition of the Society of Petroleum Engineers, Denver, Colorado, p. 851–860.

Ding, L.Y., R.K. Mehra, and J.K. Donnelly, February 1989, Stochastic modeling in reservoir simulation, Proceedings of the 10th Symposium on Reservoir Simulation of the Society of Petroleum Engineers, Houston, Texas, p. 303–320.

Farmer, C.L., July 1989, The mathematical generation of reservoir geology, Proceedings of the Joint IMA/SPE European Conference on Mathematical Oil Recovery, Robinson College, Cambridge University, p. 817.

Fogg, G.E., and F.J. Lucia, June 1989, Stochastic simulation of interwell-scale heterogeneity for improved prediction of sweep efficiency in a carbonate reservoir, Proceedings of the NIPER/DOE Second International Reservoir Characterization Technical Conference, Dallas, Texas, p. 355-381.

Haldorsen, H.H., and E. Damsleth, April 1990, Stochastic modeling, Journal of Petroleum Technology, p. 404–412.

Journel, A.G., and C.J. Huijbregts, 1978, Mining Geostatistics, Academic Press, London, 600 p.

Kirpatrick, S., C.D. Gelatt, Jr., and M.P. Vecchi, 1983, Optimization by Simulated Annealing, Science, v. 220, p. 671–680.

Otten, R., and L. Van Ginneken, 1989, The Annealing Algorithm, Kluwer Academic Publishers, Dordrecht, 224 p.

Ouenes, A., and N. Saad, 1993, A new, fast parallel simulated annealing algorithm for reservoir characterization, Proceedings of the 68th Annual Technical Conference and Exhibition of the Society of Petroleum Engineers, Houston, Texas p. 19–29.

Øvreberg, O., E. Damsleth, and H.H. Haldorsen, September 1990, Putting error-bars on reservoir engineering forecasts, Proceedings of the 65th SPE Annual Technical Conference and Exhibition of the Society of Petroleum Engineers, New Orleans, Louisiana, p. 399–410.

Yang, A.P., 1990, Stochastic heterogeneity and dispersion, Ph.D. dissertation, The University of Texas at Austin, 242 p.

de Sant' Anna Pizzaro, J. O., L. W. Lake, A simple method to estimate interwell autocorrelation, 1999, *in* R. Schatzinger and J. Jordan, eds., Reservoir Characterization-Recent Advances, AAPG Memoir 71, p. 369–380.

Chapter 26

A Simple Method to Estimate Interwell Autocorrelation

Jorge Oscar de Sant' Anna Pizarro
Petrobras
Rio de Janeiro, Brazil

Larry W. Lake
Center for Petroleum and Geosystems Engineering
The University of Texas at Austin
Austin, Texas, U.S.A.

ABSTRACT

The estimation of autocorrelation in the lateral or interwell direction is important when performing reservoir characterization studies using stochastic modeling. This paper presents a new method to estimate the interwell autocorrelation based on parameters, such as the vertical range and the variance, that can be estimated with commonly available data.

We used synthetic fields that were generated from stochastic simulations to provide data to construct the estimation charts. These charts relate the ratio of areal to vertical variance and the autocorrelation range (expressed variously) in two directions. Three different semivariogram models were considered: spherical, exponential, and truncated fractal.

The overall procedure is demonstrated using field data. We find that the approach gives the most self-consistent results when it is applied to previously identified facies; moreover, the autocorrelation trends follow the depositional pattern of the reservoir, which gives confidence in the validity of the approach.

INTRODUCTION

The importance of reservoir characterization methods has been established in the last decade by reservoir studies that are based on stochastic models that account for the heterogeneity of the porous media.

A reliable study depends on prior quantification of the heterogeneity of the reservoir (Srivastava, 1994). The geostatistical approach describes the heterogeneity through averages, variances, and autocorrelation. Although it has several advantages, when compared with deterministic approaches the confidence in geostatistical modeling will be a strong function of how well the input data represent reality. The horizontal autocorrelation, especially, is among the most critical of the parameters to be estimated; we propose a new procedure to estimate it in this paper. The method uses serial data from several vertical wells, as would exist from wells in mature projects.

MOTIVATION

There are several works in the literature stressing that the estimation of horizontal autocorrelation is important in achieving a good reservoir description. Lucia and Fogg (1989) stated that the principal difficulty in reservoir characterization is estimating the spatial distribution of petrophysical properties between vertical wellbores. These points also have been highlighted (Hewett, 1986; Lemouzy et al., 1995; Jensen et al., 1996).

Lambert (1981) calculated and tabulated Dykstra-Parsons coefficients (measures of heterogeneity) in both the horizontal and vertical directions from 689 wells in 22 fields. In 90% of the cases, the ratio of the coefficients in the horizontal and vertical directions was less than 1. This result, governed by the depositional trends observed in petroleum reservoirs, indicates a spatial dependence among these variables.

Figures 1 and 2 illustrate this idea in more detail. The figures show an idealized cross section (x,z) that is being penetrated by several vertical cored wells. We can calculate two types of variances for the cross section. The vertical variance is the arithmetic average of the variances for each individual well; the areal variance is the variance of the well averages. A cross section that is strongly autocorrelated in the x-direction will have a small areal variance; strong autocorrelation in the z-direction will lead to a small vertical variance. The ratio of these two is a measure of the extent of autocorrelation in the respective directions.

If we quantify the relationship between the autocorrelation and the variances in both directions, we will be able to estimate one parameter from the others; therefore, the main idea behind our method is that the horizontal autocorrelation must depend on the ratio of the areal-to-vertical variances and the autocorrelation in the vertical direction. With a chart that expresses this relationship, one can estimate autocorrelation in the horizontal direction.

METHOD OF STUDY

This section states the approach used to estimate autocorrelation in the interwell region of a reservoir. In the following derivation, we focus on the estimation of horizontal autocorrelation because, in most cases, vertical wells can provide information on autocorrelation in the vertical direction. We also restrict attention to the estimation of horizontal autocorrelation in permeability.

The autocorrelation will be a function of the type of model chosen to represent it. Three different theoretical semivariogram models were tested: spherical, exponential, and truncated fractal (power-law). The fractal model usually does not have a finite autocorrelation, but we will truncate at some upper cutoff; hence, the use of a truncated fractal.

The way we express autocorrelation depends on the semivariogram model being used. For the spherical semivariogram model,

$$\gamma(h) = \mathrm{cov}(0)\begin{cases} \dfrac{3}{2}\left(\dfrac{h}{\lambda}\right) - \dfrac{1}{2}\left(\dfrac{h}{\lambda}\right)^3 & 0 \le h \le \lambda \\ 1 & h \ge \lambda \end{cases} \tag{1}$$

autocorrelation is expressed by the range λ. For the vertical direction, $\lambda = \lambda_z$; for the lateral or horizontal, $\lambda = \lambda_x$. For the exponential model,

$$\gamma(h) = \mathrm{cov}(0)e^{-\frac{h}{\lambda}} \tag{2}$$

we use the autocorrelation length. As before, for the vertical direction, $\lambda = \lambda_z$; for the horizontal, $\lambda = \lambda_x$. For the truncated fractal fBm model,

$$\gamma(h) = \mathrm{cov}(0)\begin{cases} h^{2H} & 0 \le h \le \lambda \\ 1 & h \ge \lambda \end{cases} \tag{3}$$

we use the cut-off length, l. For the vertical direction, $l = l_z$; for the horizontal, $l = l_x$. We use the word "range" to generically express all three autocorrelation

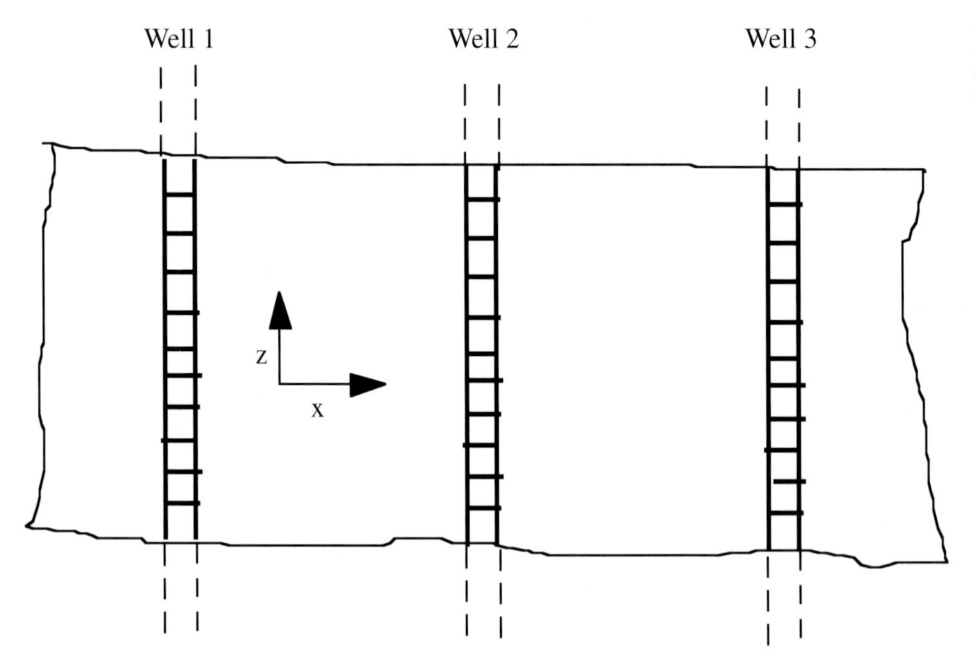

Figure 1. An idealized cross section (x,z) penetrated by vertical wells.

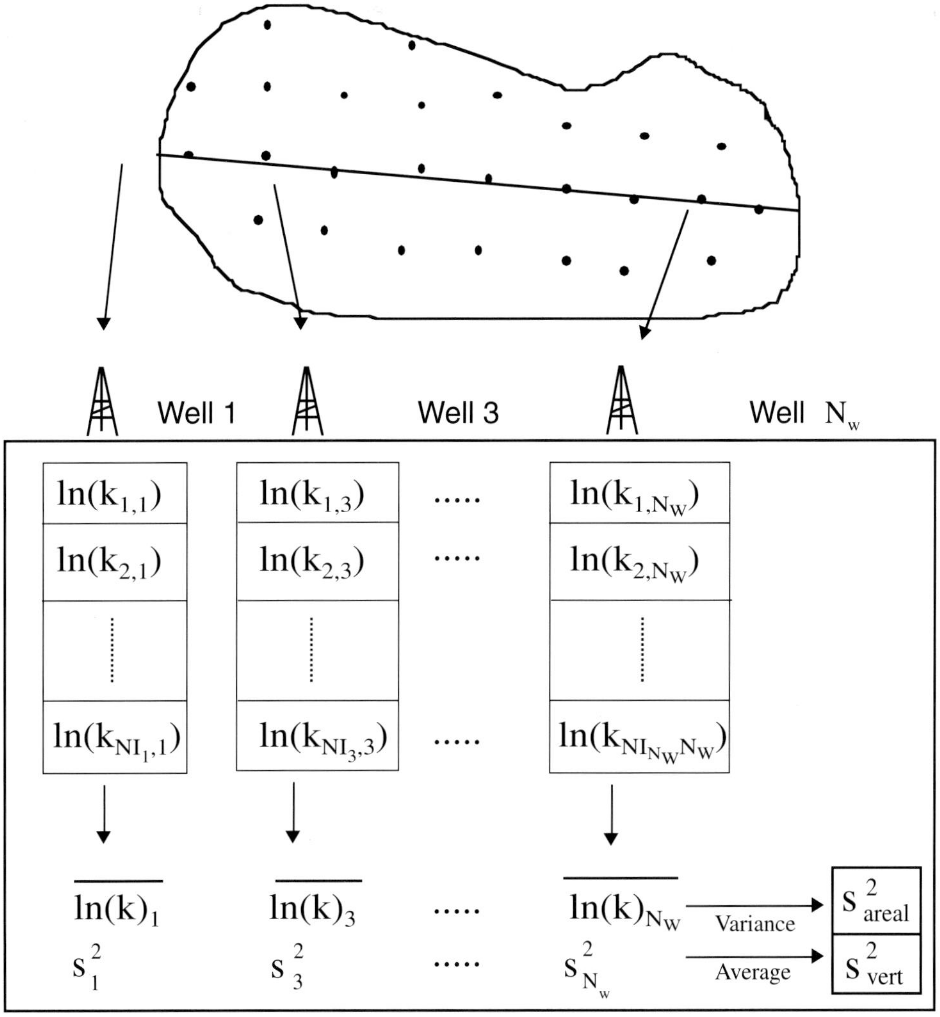

Figure 2. Schematic of the procedure to calculate the areal and vertical variances.

measures. Because a log-normal distribution is assumed for the permeability, all variances and averages are calculated on the logarithm of the permeability (ln$k$). In equations 1–3, cov(0) stands for the sill, h the lag distance, and H the Hurst coefficient. Note that the parameter λ expresses the autocorrelation of the permeability in all three models; however, because the models are different, λ is also different for each; for the spherical model λ is the range, for the exponential model λ is the autocorrelation length, and for the fractal model λ is the cutoff length. We seek, in this paper, to estimate λ_{zD} given certain variances and λ_{zD}.

Formally, the vertical variance is the expectation of the conditional variances of $Y(x,z) = \ln k(x,z)$:

$$\sigma^2_{vert} = E\left(\text{var}\left[Y(x_0, z)\right]\right) \tag{4}$$

where var $[Y(x_0,z)]$ indicates a z-direction variance conditioned to a fixed location x_0. As suggested in Figure 2, σ^2_{vert} is estimated as

$$s^2_{vert} = \frac{1}{N_w}\sum_{i=1}^{N_w}\left(\text{var}\left[Y(i,j)\right]\right) \qquad j = 1,\ldots,N_l \tag{5}$$

where i and j are the indices on the x- and z-directions, respectively. The areal variance is the variance of the conditional expectation of Y:

$$\sigma^2_{areal} = \text{var}\left(E\left[Y(x_0, z)\right]\right) \tag{6}$$

It is estimated as

$$s^2_{areal} = \text{var}\left[\frac{1}{N_l}\sum_{j=1}^{N_l} Y(i,j)\right] \qquad I = 1,\ldots,N_w \tag{7}$$

See Jensen et al. (1996) for details on conditional variances and expectations.

Our approach consists of the generation of synthetic, equiprobable permeability fields using stochastic simulation and a statistical treatment of the obtained data. There are several methods that can generate stochastic reservoir images (Srivastava, 1994). One is the matrix decomposition method (MDM). This method is an averaging technique based on the decomposition of the autocovariance matrix. The present study uses a program developed by Yang (1990) that performs simulations based on MDM.

The MDM can generate permeability fields for reservoir modeling, but several statistics must be known as input data. MDM requires that the geologic knowledge be quantified in terms of a few parameters that can characterize the static properties of the reservoir. In doing this, some assumptions are necessary. The permeability distribution must be by a single log-normal population that obeys second-order stationarity. The log-normal assumption was corroborated by the work of several authors, the results of which are summarized in a paper by Jensen et al. (1987).

The generation of synthetic permeability fields provided the necessary data to describe the relation among all investigated parameters. Averaging results from several realizations will provide a good estimate for the actual value of λ_x.

TYPE-CURVE PROCEDURE

To perform the numerical experiments, we established the following procedure. First, we chose a rectangular cross section with 4000 blocks, 200 blocks in the x-direction and 20 in the z-direction. The sampling procedure represents a reservoir with 11 vertical wells aligned in one direction having a constant spacing between them. We later concluded that the difference in results caused by staggered wells is small (Pizarro, 1998); therefore, the analysis can be used for irregularly spaced wells, provided the average spacing is used as the reference distance.

The ranges input to MDM were converted to dimensionless form to provide more generality to the results. The dimensionless horizontal autocorrelation was normalized by the interwell spacing; the total thickness is the reference distance in the vertical direction. We sample the generated fields at a constant interval to generate the estimated variances.

In the stochastic simulations, the dimensionless ranges ranged between 0.1 and 100 in the x-direction and between 0.1 and 2.0 in the z-direction. This range is believed to describe the autocorrelation existing in most petroleum reservoirs.

To make the results represent the expectations in equations 5 and 7 accurately, several realizations were performed. For each realization, Y values in each well were collected in a table in which there were 11 columns in the horizontal direction (each corresponding to one well) and 20 lines in the vertical direction representing layer values. For each well the arithmetic average of Y was calculated in the vertical direction as well as the variance in the vertical direction.

Once both areal and vertical variances are known, the calculation of the ratio between them is straightforward. Because all of the variances used in equations 5 and 7 are directly proportional to the variance, the ratio will be a function only of λ_{xD}, λ_{zD}, and the semivariogram model; therefore, if s^2_{areal} and s^2_{vert} are calculated, and λ_{zD} is estimated from fitting a specific semivariogram model to the vertical data, λ_{xD} can be estimated.

RESULTS

Here, we present the simulation results used to build the type-curve charts. The procedure previously stated was implemented, and several simulations were performed.

All fields were generated for $\sigma^2 = 1$. This population variance is equivalent to a Dykstra-Parsons coefficient of 0.63, a value that is within the range of the vertical V_{DP} values tabulated by Lambert (1981) from core. Because both areal and vertical variances are derived from the autocovariance, their ratio is independent of σ^2 (Pizarro, 1998). We used the minimal number of realizations (N_R) that would provide stable results, $N_R = 50$ (Pizarro, 1998).

The results obtained are expressed in Figures 3–5 for the spherical, exponential, and truncated fractal semivariogram models, respectively. Each figure is a plot of s^2_{areal}/s^2_{vert} versus λ_{z_D} with λ_{x_D} as a parameter. The truncated fractal fBm plot uses a Hurst coefficient of 0.25 because this seems to fit various types of field data (Neuman, 1994). Each figure contains results from 3500 MDM simulations because we calculate the average of 50 realizations.

Validation

The overall procedure cannot be validated because we have only analytical expressions for some limiting cases. We would need an extraordinary amount of field data to cover all the possible situations aside from the limiting cases; however, some insight into the results can be given by analyzing the available solutions.

The first result that can be compared is for the case when we have no autocorrelation in either the horizontal or vertical directions. This means that the permeability values are completely random, with no spatial correlation. The central limit theorem states that no matter what distribution a group of independent random variables are from, the sample mean of these variables is approximately normally distributed. So, if $Y_1, Y_2, Y_3, \ldots, Y_N$ denote independent random variables each having the same mean, μ, and variance, σ^2, and $\overline{Y}$ equals the mean of N of these random variables, we will obtain

$$\overline{Y} = \frac{Y_1}{N} + \frac{Y_2}{N} + \ldots \frac{Y_N}{N} = \frac{1}{N}\sum_{i=1}^{N} Y_i \qquad (8)$$

Applying the property that the variance of a sum of independent variables will be equal to the sum of the variance of each one, we obtain

$$\mathrm{Var}\left(\overline{Y}\right) = \frac{\sigma^2}{N} \qquad (9)$$

According to this procedure, the variance of the mean can be approximated by the areal variance (s^2_{areal}).

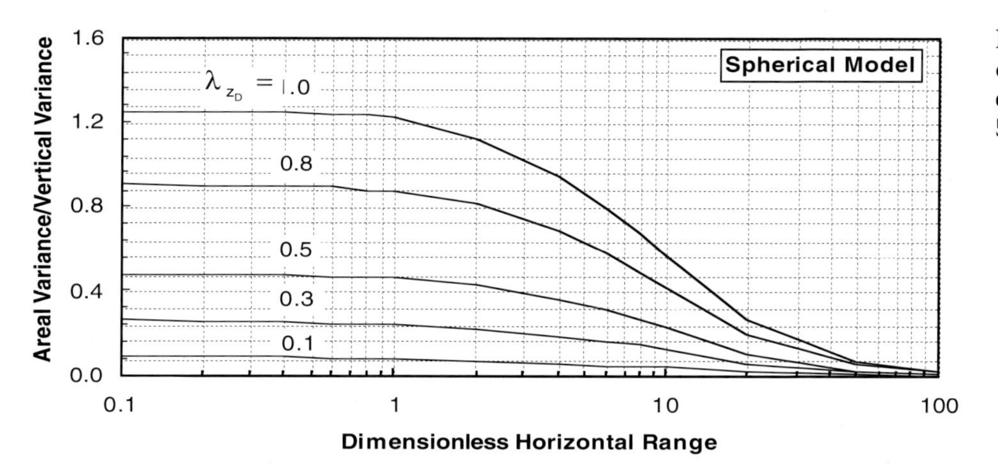

Figure 3. Autocorrelation chart for spherical semivariogram model obtained from 50 realizations.

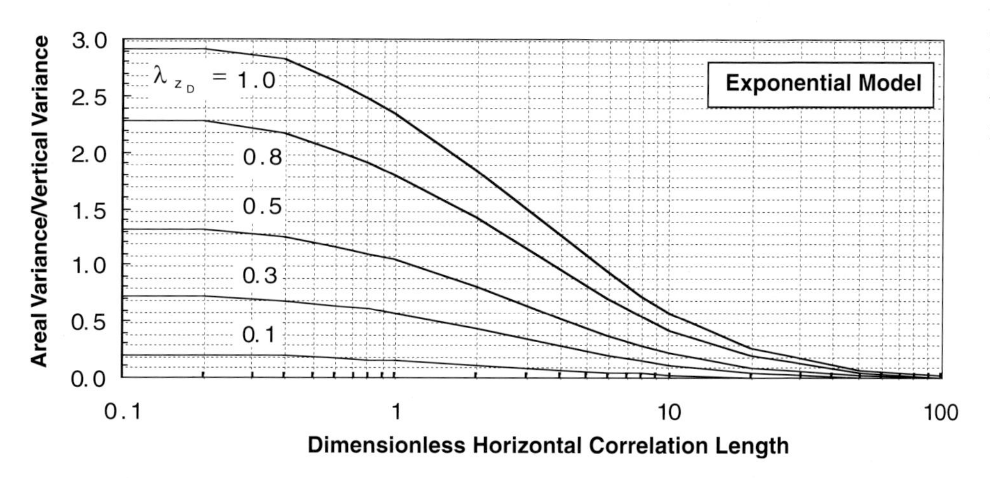

Figure 4. Autocorrelation chart for exponential semivariogram model obtained from 50 realizations.

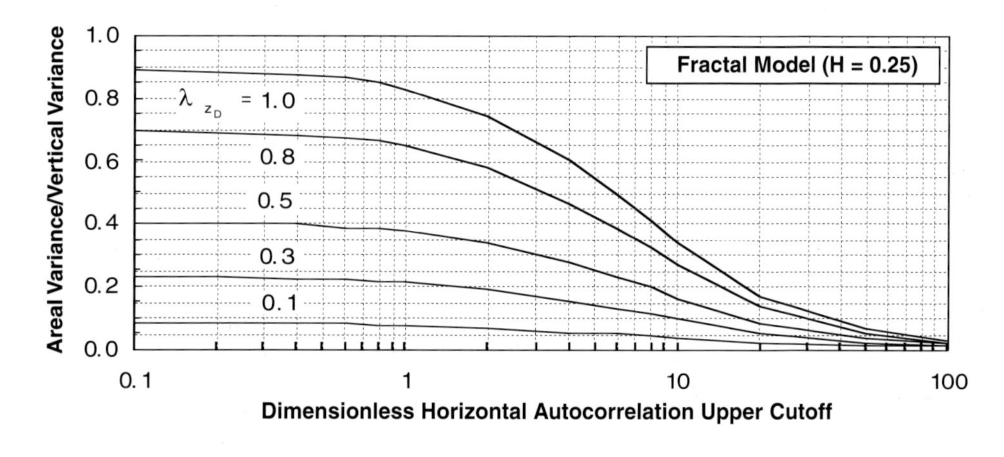

Figure 5. Autocorrelation chart for fractal semivariogram model obtained from 50 realizations.

Hence, the above relation states that, for the case of an uncorrelated field ($\lambda_{x_D} = \lambda_{z_D} = 0$), the areal variance should be equal to

$$\sigma^2_{areal} = \frac{\sigma^2}{N_l} \quad (10)$$

When the number of layers (N_l) equals 20, the above expression gives $\sigma^2_{areal} = 0.050$. In the numerical experiments with MDM, we obtained $s^2_{areal} = 0.047$, a good approximation.

Another way to validate the results is to make use of an analytical expression that describes the relationship between variances of properties measured at different scales. Several authors, including Neuman (1994), Lasseter et al. (1986), and Haldorsen (1986), have reported the effect of the scale on the measurement of heterogeneity. One of the strong points of geostatistics is its ability to represent this behavior. For instance, the permeability of a well can be measured from a series of core measurements representing blocks of a certain size, shape, and orientation. As the size of these blocks increases, the variance of the mean value within the blocks gets smaller, although they may have the same mean.

Krige's relationship (Journel and Huijbregts, 1978) can be adapted to the present application. If a mining deposit (D) is split up into blocks (V), the variance of the properties of these blocks will be called the block variance and be denoted by $\sigma^2_{(V/D)}$. The variance of point grades (O) within a block is expressed by $\sigma^2_{(O/V)}$. This relationship states that these variances can be related to the variance of point grades within a deposit ($\sigma^2_{(O/D)}$)

$$\sigma^2_{(V/D)} = \sigma^2_{(O/D)} - \sigma^2_{(O/V)} \tag{11}$$

In our application, the block represents a well, the point grades will be core samples within each well, and the deposit will correspond to the reservoir. With this analogy, Krige's relationship will be

$$s^2_{areal} + s^2_{vert} = \sigma^2 \tag{12}$$

Figure 6 illustrates how each variance behaves with λ_{x_D} increasing from 0.1 to 100. When σ^2 equals 1, an exponential model is used and $\lambda_{z_D} = 0.8$. The sum of s^2_{areal} and s^2_{vert} is constant and equals the value expected from Krige's relationship for all values of λ_{x_D}. The agreement between equation 12 and the numerical response is very good. The good agreement between analytical (equation 11) and numerical (equation 12) results in Figure 6 indicates that the dependence of our results on sample size is small. Equation 11 applies for essentially infinite sampling, whereas the terms in equation 12 must be estimated from finite samples.

The chart in Figure 6 reveals, also, how both s^2_{areal} and s^2_{vert} behave with increasing λ_{x_D}. As described, for small λ_{x_D}, s^2_{areal} reaches a maximum and s^2_{vert} a minimum. As λ_{x_D} becomes greater than 1, s^2_{areal} declines, while s^2_{vert} rises. For very large values of λ_{x_D}, the horizontal autocorrelation is so large that s^2_{areal} tends to zero, while s^2_{vert} approaches the population variance (σ^2).

Krige's equation also provides an alternative way to develop the relationships expressed in the charts of Figures 3–5. Knudsen and Kim (1978) showed that, using the definition of the autocovariogram, cov(h), the areal variance could be calculated by

$$\sigma^2_{(V/D)} = \frac{1}{V^2} \int_v \int_v \text{cov}(h)\, dv\, dv \tag{13}$$

where h stands for the distance between any two points in the volume V, and the integrals represent an integration over a volume; however, the amount of effort to be spent in these integrations is excessive, which motivated us to adopt the numerical approach using MDM.

Discussion

Here, we discuss the results and interpret some features of the type charts. The differences in results among the three semivariogram models are related with the degree of autocorrelation that each model incorporates. As s^2_{areal}/s^2_{vert} represents a ratio between two variances, the effect of the autocorrelation model will depend on which of the variances controls the final result. One way to illustrate this effect is by plotting the dependence of $\sigma^2_{areal}/\sigma^2_{vert}$ on the Hurst coefficient for the truncated fractal model. This parameter represents the degree of autocorrelation among data.

Figure 7 shows how s^2_{areal}/s^2_{vert} can vary with H and λ_{x_D} for $\lambda_{z_D} = 1$. For instance, considering a reservoir strongly autocorrelated in the horizontal direction ($\lambda_{x_D} = 100$), the larger the H, the smaller will be s^2_{areal}/s^2_{vert}. When dealing with smaller values of λ_{x_D}, however, the behavior of the vertical variance will control the ratio s^2_{areal}/s^2_{vert}; therefore, as H increases, s^2_{areal}/s^2_{vert} also increases.

The charts in Figures 3–5 also show that s^2_{areal}/s^2_{vert} increases when λ_{z_D} is greater than 0.1; however, in the horizontal direction, s^2_{areal}/s^2_{vert} decreases very slowly with the vertical autocorrelation until λ_{x_D} reaches 1.0. This behavior is the same for all three semivariogram models and shows that a field with $\lambda_{x_D} = 1$ will behave similarly to an uncorrelated field in this direction. This result is a consequence of the fact that s^2_{areal} depends strongly on horizontal autocorrelation only for ranges greater than the interwell spacing. Ranges smaller than 1 have only a slight effect on s^2_{areal}. This observation also

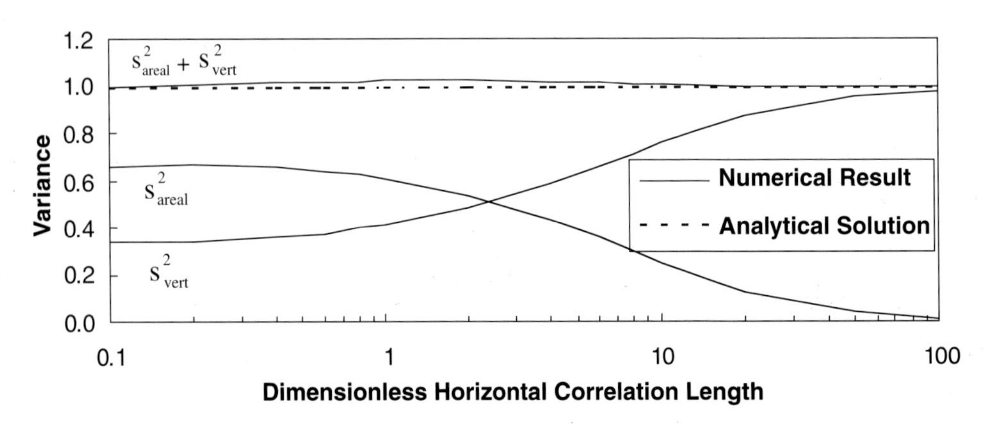

Figure 6. Analytical validation (exponential model, λ_{z_D} = 0.8, and 50 realizations).

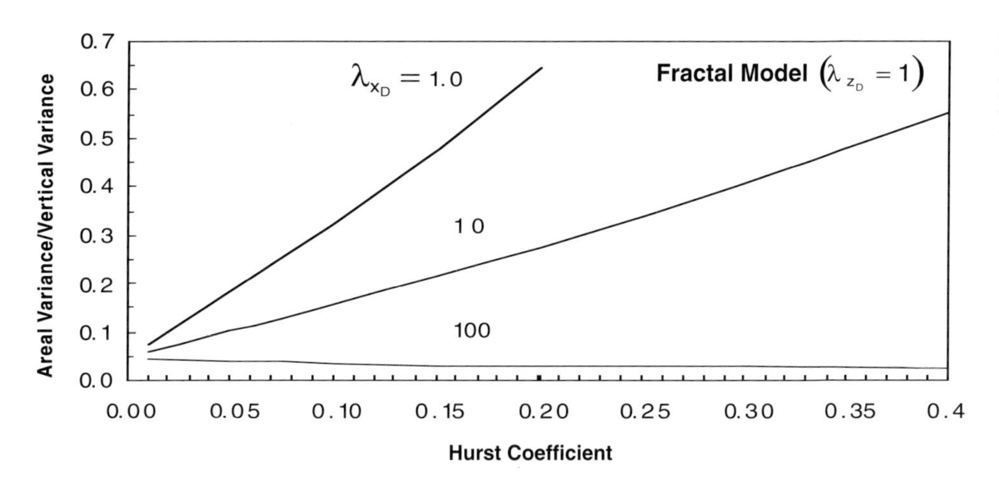

Figure 7. Estimation of s^2_{areal}/s^2_{vert} sensitivity to the Hurst coefficient for different truncated ranges.

illustrates that, as expected, it is impossible to estimate ranges smaller than the well spacing ($\lambda_{xD} < 1$).

One limitation of this method is the variability of s^2_{areal}/s^2_{vert} among the 50 realizations. Figure 8 expresses how the results for $\lambda_{zD} = 0.1$ and 0.5, obtained from the exponential model, vary with λ_{xD}. The figure shows the average of 50 realizations and also curves representing plus or minus one standard deviation about the mean. The standard deviation band shrinks slightly when lateral autocorrelation (as expressed by λ_{zD}) is large. The band is considerably smaller at $\lambda_{zD} = 0.1$ than at $\lambda_{zD} = 0.5$. A fact that is equally important because the curves become more horizontal for smaller λ_{zD} is that the error at a fixed variance ratio becomes quite large when lateral autocorrelation is small.

APPLICATION

We chose data from a particular field to demonstrate how the procedure works. The results represent an illustration of the method rather than a comprehensive analysis of the reservoir. A complete analysis requires additional effort and integration among geologists and reservoir engineers to interpret the results in the light of all the knowledge about the field.

We applied the procedure to the El Mar field, located in the Delaware basin of west Texas and New Mexico (Figure 9). This unit is currently operated by Burlington Resources Company.

The El Mar

This field is located 30 km north of Mentone, Texas. The entire field covers approximately 40 km^2, with two-thirds of the field located in the western portion of Loving County, Texas, and the remainder in southwest Lea County, New Mexico. In this study, we investigated only the data from the El Mar (Delaware) unit, which covers an area of 5×8 km and contains 175 wells.

The primary producing layer is the Ramsey sand, which is composed of an upper "A" sand and a lower "B" sand separated by shale laminae. These sandstones were deposited in deep water, probably by submarine-fan complexes formed by turbidity-current deposition during lowstands of sea level. Dutton et al. (1996) discussed this and other alternative models of Delaware sandstone deposition. For the purposes of this paper, the most important detail in Figure 9 is that the source of the turbiditic sands lies approximately north-northeast of the El Mar unit. The Ramsey formation lies at an average producing depth of 1500 m, and the formation's weighted average porosity and permeability are, respectively, 23% and 22 md. A description of other petrophysical properties of the Delaware formation can be found in Jenkins (1961). The difficulty in estimating

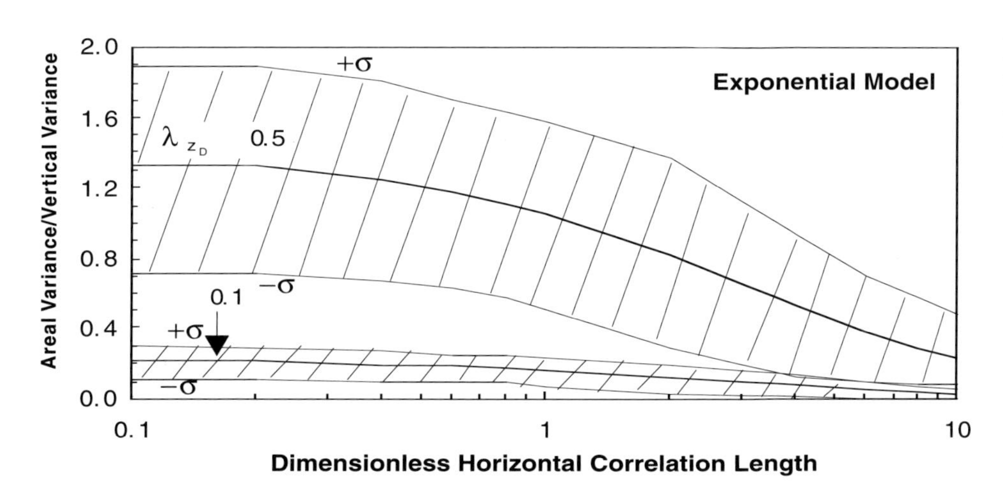

Figure 8. Variability (expressed as ±1 standard deviation) of s^2_{areal}/s^2_{vert} among the 50 realizations.

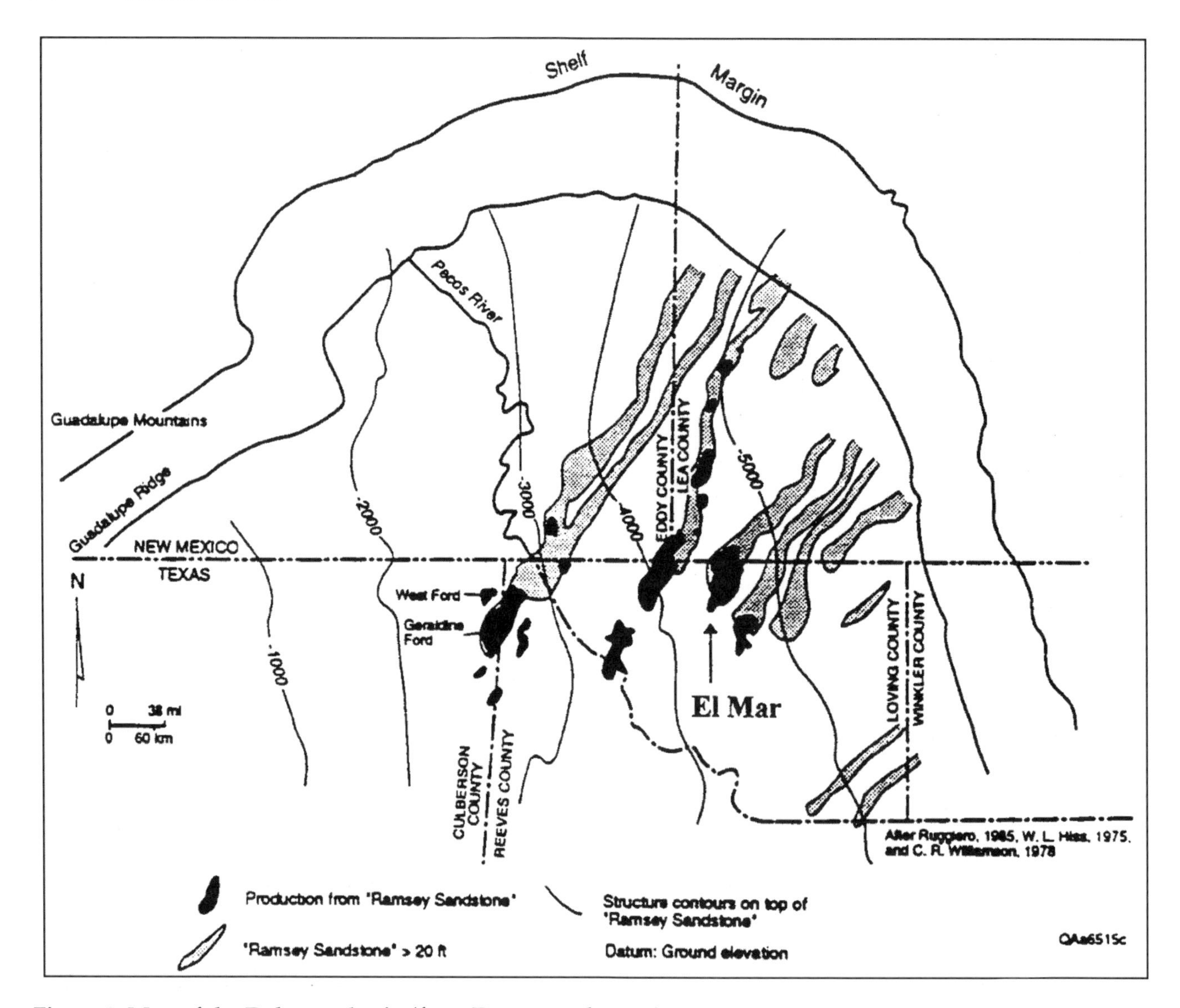

Figure 9. Map of the Delaware basin (from Dutton et al., 1996).

the water saturation caused a large number of cored wells to be taken. This made it possible to find several cored wells aligned in the orthogonal directions of the field, leaving it in a favorable condition to perform the present analysis. We selected six wells aligned in the north-south direction and two more in the east-west direction, for a total of three east-west wells.

We selected the data from the eight wells and built a vertical semivariogram of Y for each. Because the core sampled unit A indiscriminately, it is highly likely that the data are from multiple turbidite facies. Most of the data analyzed were from unit A because of the predominance of this unit in this part of the field. Our approach was to separate the data from both units because they seem to have different petrophysical properties. We performed the analysis only for Ramsey A because the amount of data for Ramsey B was less; only two wells, 1814 and 1824, have data that include core samples from both units A and B.

The goal here was simply a practical demonstration of the method; therefore, we assumed that there were no problems concerning data acquisition, and the further analysis was based on the semivariograms constructed with primary data. We also did not consider the possibility of nugget effects in the semivariogram fitting.

The procedure outlined was performed to calculate the average values and the s^2_{areal} and s^2_{vert} for each set of data. To calculate λ_{z_D}, the experimental semivariograms must be fitted by a theoretical model. The difficulties and the importance of this step are well described in the literature by several authors, including Journel and Huijbregts (1978), Isaaks and Srivastava (1989), and Olea (1994).

Estimating λ_{x_D}

We tested three different semivariogram models, trying to find the one that best described the experimental data.

The spherical semivariogram fit was considered poor. The second model tested was the truncated fBm fractal or power-law model. For the fitting, a value of

$H = 0.15$ gave a better adjustment than the value of $H = 0.25$; therefore, a new set of curves for the fractal model, analogous to those in Figure 5, was generated for $H = 0.15$. The final result was better than the spherical fitting for most wells. The exponential model gave a fit close to that obtained with the fractal for most wells, and reproduced a similar response for the ones in which the spherical model worked better. The results are summarized in Table 1 for the east-west section and in Table 2 for the north-south section.

Estimating λ_{z_D}

Using the λ_{z_D} values given in Table 1 for the appropriate model and the ratio s^2_{areal}/s^2_{vert} for each direction, we enter the charts and interpolate s^2_{areal}/s^2_{vert} as shown in Figure 10. This represents the autocorrelation in the x or lateral direction, the autocorrelation being expressed through an exponential semivariogram model. The results shown in Table 3 for the spherical, exponential, and fractal are, respectively, $\lambda_{z_{D|EW}} = 1.2$, 1, and 3 and $\lambda_{x_{D|NS}} = 4$, 2, and 6.

The results show anisotropy in the autocorrelation pattern of the El Mar field, at least along the directions indicated. The permeability range in the north-south direction is at least twice that in the east-west direction. This behavior roughly coincides with the depositional characteristic of the field because the major axis of the turbidite channel is aligned with the northeast-southwest direction (Figure 9).

A less rigorous approach would be to group the data from both units A and B together, instead of performing the analysis only for unit A. Additional data from two wells (1814 and 1824) must be considered. Because the wells in the east-west section do not contain the Ramsey B sand, only results from the north-south section will change. The main difference is that the value of s^2_{areal} changes from 0.23 to 0.36, making

Table 1. Summary of Results for the East-West Cross Section of the El Mar Field

	Wells			
	1514	1513	1512	Average
Average of ln(k) in vertical direction	1.15	2.25	2.17	1.86
Variance of ln(k) in vertical direction	1.96	2.44	2.81	2.40
Vertical correlation range with spherical model	0.21	0.18	0.17	0.19
Vertical correlation length with exponential model	0.09	0.07	0.08	0.08
Vertical truncated upper cutoff with fractal model ($H = 0.15$)	0.35	0.26	0.34	0.32

Table 2. Summary of Results for the North-South Cross Section of the El Mar Field

	Wells						
	1514	1524	1534	1532	1814	1824	Average
Average of ln(k) in vertical direction	1.15	1.89	2.37	2.40	1.85	1.58	1.87
Variance of ln(k) in vertical direction	1.96	1.47	0.46	2.09	4.00	1.71	1.95
Vertical correlation range with spherical model	0.21	0.20	0.23	0.20	0.22	0.24	0.22
Vertical correlation length with exponential model	0.09	0.08	0.12	0.09	0.11	0.11	0.10
Vertical truncated upper cutoff with fractal model ($H = 0.15$)	0.35	0.28	0.30	0.34	0.42	0.30	0.33

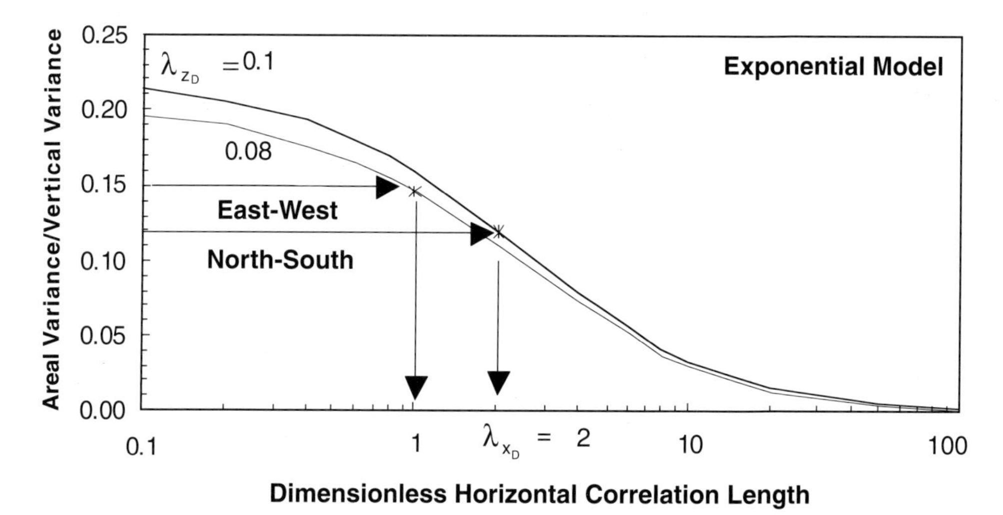

Figure 10. Determining λ_{x_D} for the north-south and the east-west sections of the El Mar field.

Table 3. Summary of the Results from Autocorrelation Analysis of the El Mar Field

Summary	North-South	East-West
Areal variance	0.23	0.37
V_{DP} areal	0.46	0.46
Vertical variance	1.95	2.40
V_{DP} vertical	0.75	0.79
Ratio areal/vertical variances	0.12	0.15
λ_{x_D} (spherical model)	4	1.2
λ_{x_D} (exponential model)	2	1
λ_{x_D} (fractal model)	6	3

$s^2_{areal}/s^2_{vert} = 0.18$ instead of 0.12. This change leads to the conclusion that the ranges are similar in both directions, in conflict with the depositional pattern. This illustrates the importance of prior geological analysis of the data to be used in making the estimation.

Final Remarks

The autocorrelation analysis of the El Mar field shows that results obtained by using different semivariogram models can be different. These results were expected because each model expresses the autocorrelation in a slightly different manner. For example, the spherical model (Figure 3) cannot be used to estimate λ_{z_D} values of less than 1, meaning that the spherical model contains no information about autocorrelation if the range is less than the well spacing. The exponential model, however, is discriminating to λ_{z_D} values less than the well spacing (Figure 4). Although this seems a little strange at first, it may not have a great effect on a simulated flow response. Indeed, Fogg et al. (1991) referred to a study in which two ranges of 120 and 390 m caused significantly different permeability patterns, but practically identical cumulative oil recoveries and water-oil rates, when these values were used in simulation.

When calculating the data to construct the charts in Figures 3–5, we also derived a chart to correct the vertical variance measured at the wells to the value that should be used in conditional simulation studies. Recall that the vertical variance measured at the wells would be equal to the population variance only when there was no autocorrelation. Figure 11 shows how s^2_{areal}/σ^2 varies with λ_{z_D} and λ_{x_D}. The more vertically autocorrelated the permeability field is, the more the s^2_{vert} is different from σ^2.

Figure 12 shows an analogous result, but it is expressed as the Dykstra-Parsons coefficient instead of the variance. To express the variance in terms of the V_{DP}, the chart is no longer general and depends on the population V_{DP}. Figure 12 considered a V_{DPpop} of 0.9.

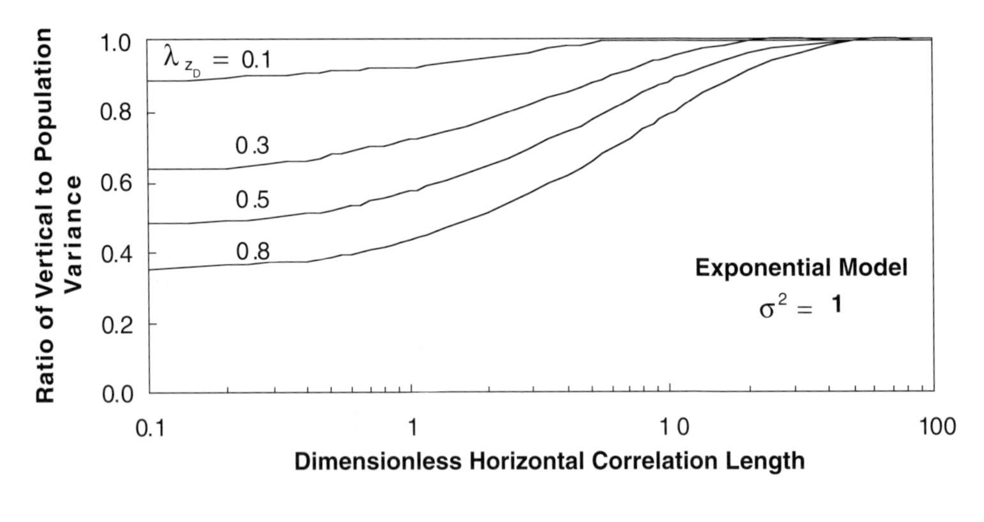

Figure 11. Differences between the vertical (sample) and population variance caused by autocorrelation.

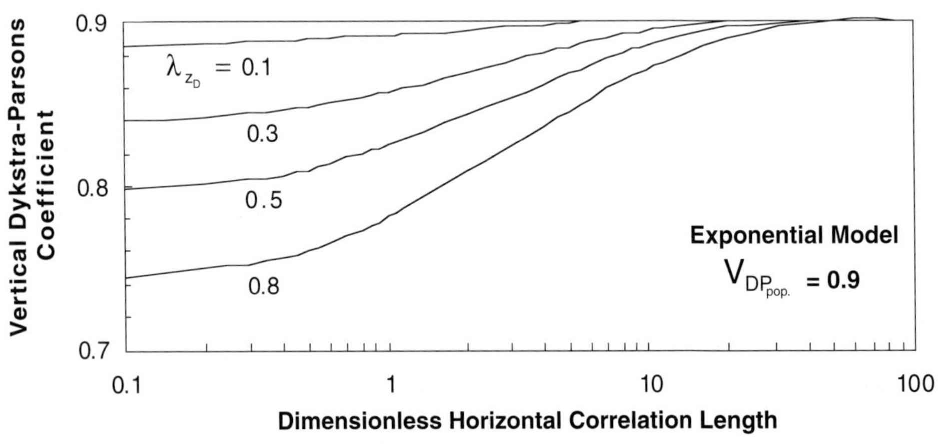

Figure 12. Differences between the vertical (sample) and population V_{DP} caused by autocorrelation.

These charts can be used to correct data for input into stochastic simulation. For instance, from Figure 12 we can conclude that a sample V_{DP} of 0.85 (s^2_{vert} = 3.6) calculated from a reservoir with λ_{z_D} = 0.3 and λ_{x_D} = 1.0, given by the exponential model, will have a population V_{DPpop} = 0.9 (σ^2 = 5.3).

SUMMARY AND CONCLUSIONS

We developed a set of charts for estimating autocorrelation in the horizontal direction based on the ratio of areal-to-vertical variances, the vertical autocorrelation, and the type of semivariogram model used.

The procedure was partially validated through a comparison of results; analytical solutions were available for some limiting situations. A field case was analyzed to demonstrate how the method could be applied. Data from eight wells in the El Mar field, Ramsey A sand, were used to estimate horizontal autocorrelation for two directions of the field.

Results from the El Mar analysis point out the anisotropy in the permeability distribution of this field. The autocorrelation ranges in the vertical direction were found to be between 0.1 and 0.3 of the sampled interval, depending on the type of model used. The horizontal range in the north-south direction ranged between two and six times the mean well spacing, being at least twice the autocorrelation in the east-west direction. These results can be used in conditional simulations to generate the expected permeability pattern of this field.

The present approach seems to be an effective tool to be used, along with the geological knowledge of the facies continuity, for estimating the autocorrelation for the interwell region between vertical wells.

NOMENCLATURE

a = semivariogram range
cov = covariance
D = mining deposit
E = expectation value
h = distance lag
H = Hurst coefficient
k = permeability
N_l = number of layers
N_r = number of realizations
N_w = number of wells
O = point grades
s^2 = estimation of the variance
V = volume of a block
V_{DP} = Dykstra-Parsons coefficient
var = variance
Y = natural logarithmic of the permeability
λ = autocorrelation range (general)
λ = range (spherical model)
λ = correlation length (exponential model)
λ = truncated upper cutoff (fractal model)
μ = population mean
σ = population standard deviation
σ^2 = population variance
σ^2_{areal} = areal variance
σ^2_{vert} = vertical variance

Subscripts

D = dimensionless variable
O/D = point grades within a mining deposit
O/V = point grades within a block
pop = population values
V/D = blocks within a mining deposit
x,y,z = Cartesian coordinate directions

ACKNOWLEDGMENTS

The authors acknowledge the Enhanced Oil Recovery Research Program of the Center for Petroleum and Geosystems Engineering at The University of Texas at Austin for partial support of this work. Jorge Pizarro thanks Petrobras S.A. for financial support during his stay at The University of Texas. Larry W. Lake holds the W.A. (Monty) Moncrief Centennial Endowed Chair. We thank I.H. Silberberg for his editorial comments and John Barnes with Burlington Resources for supplying the core data on the El Mar (Delaware) unit.

REFERENCES CITED

Dutton, S.P., S.D. Hovorka, and A.G. Cole, 1996, Application of advanced reservoir characterization, simulation, and production optimization strategies to maximize recovery in slope and basin clastic reservoirs, West Texas (Delaware basin), DOE Report No. 96-001244, 81 p.

Fogg, G.E., F.J. Lucia, and R.K. Senger, 1991, Stochastic simulation of interwell-scale heterogeneity for improved prediction of sweep efficiency in a carbonate reservoir, in L.W. Lake, H.E. Carroll, and T.C. Wesson, eds., Reservoir Characterization II: New York, Academic Press, p. 355-381.

Haldorsen, H.H., 1986, Simulation parameter assignment and the problem of scale in reservoir engineering, in L.W. Lake and H.E. Carroll, eds., Reservoir Characterization: New York, Academic Press, p. 293-340.

Hewett, T.A., 1986, Fractal distributions of reservoir heterogeneity and their influence on fluid transport, Paper SPE 15386 presented at the 61st Annual Technical Conference and Exhibition of the Society of Petroleum Engineers, New Orleans, Louisiana, 16 p.

Hiss, W.L., 1975, Stratigraphy and groundwater hydrology of the Capitan aquifer, southeastern New Mexico and western Texas, University of Colorado, Ph.D. dissertation, 396 p.

Isaaks, E.H. and R.M. Srivastava, 1989, Applied Geostatistics: New York, Oxford University Press, 561 p.

Jenkins, R.E., December 1961, Characteristics of the Delaware formation, Journal of Petroleum Technology, p. 1230-1236.

Jensen, J.L., D.V. Hinkley, and L.W. Lake, 1987, A statistical study of reservoir permeability: distributions, correlations, and averages, SPE Formation Evaluation, v. 2, p. 461-468.

Jensen, J.L., L.W. Lake, P.M.W. Corbett, and D.J. Goggin, 1996, Statistics for Petroleum Engineers and Geoscientists: Englewood Cliffs, New Jersey, Prentice Hall, 390 p.

Journel, A.G., and CH.J. Huijbregts, 1978, Mining Geostatistics: New York, Academic Press, 600 p.

Knudsen, H.P., and Y.C. Kim, 1978, A short course on geostatistical ore reserve estimation, Department of Mining and Geological Engineering, College of Mines, The University of Arizona, Tucson, Arizona, 224 p.

Lambert, M.E., 1981, A statistical study of reservoir heterogeneity, M.S. thesis, The University of Texas at Austin, 181 p.

Lasseter, T.J., J.R. Waggoner, and L.W. Lake, 1986, Reservoir heterogeneities and their influence on ultimate recovery, in L.W. Lake and H.E. Carroll, eds., Reservoir Characterization: New York, Academic Press, p. 545-560.

Lemouzy, P.M., J. Parpant, R. Eschard, C. Bachiana, I. Morelon, and B. Smart, 1995, Successful history matching of Chaunoy field reservoir behavior using geostatistical modeling, Proceedings of the Annual Technical Conference and Exhibition of the Society of Petroleum Engineers, Dallas, Texas, p. 23–38.

Lucia, F.J., and G.E. Fogg, 1989, Geologic/stochastic mapping of heterogeneity in a carbonate reservoir, 1989, Proceedings of the Annual Technical Conference and Exhibition of the Society of Petroleum Engineers, San Antonio, Texas, p. 275–283.

Neuman, S.P., March 1994, Generalized scaling of permeabilities: validation and effect of support scale, Geophysical Research Letters, v. 21, no. 5, p. 349-352.

Olea, R.A., 1994, Fundamentals of semivariogram estimation, modeling, and usage, in J.M. Yarus and R.L. Chambers, eds., Stochastic Modeling and Geostatistics: AAPG Computer Applications in Geology, No. 3, p. 27-36.

Pizarro, J.O.S., 1998, Estimating injectivity and lateral autocorrelation in heterogeneous media, Ph.D. dissertation, The University of Texas at Austin, 264 p.

Ruggiero, R.W., 1985, Depositional history and performance of a Bell Canyon sandstone reservoir, Ford-Geraldine field, west Texas, The University of Texas at Austin, M.S. thesis, 242 p.

Srivastava, R.M., 1994, An overview of stochastic methods for reservoir characterization, in J.M. Yarus and R.L. Chambers, eds., Stochastic Modeling and Geostatistics: AAPG Computer Applications in Geology, no. 3, p. 3-16.

Williamson, C.R., 1978, Depositional processes, diagenesis and reservoir properties of Permian deep-sea sandstones, Bell Canyon Formation, Texas-New Mexico, The University of Texas at Austin, Ph.D. dissertation, 262 p.

Yang, A.P., 1990, Stochastic heterogeneity and dispersion, Ph.D. dissertation, The University of Texas at Austin, 242 p.

Datta-Gupta, A., G. Xue, S. H. Lee, Nonparametric transformations for data correlation and integration: from theory to practice, 1999, *in* R. Schatzinger and J. Jordan, eds., Reservoir Characterization-Recent Advances, AAPG Memoir 71, p. 381–396.

Chapter 27

Nonparametric Transformations for Data Correlation and Integration: From Theory to Practice

Akhil Datta-Gupta
Guoping Xue
Sang Heon Lee
Department of Petroleum Engineering
Texas A&M University
College Station, Texas, U.S.A.

ABSTRACT

The purpose of this paper is two-fold. First, we introduce the use of nonparametric transformations for correlating petrophysical data during reservoir characterization. Such transformations are completely data driven and do not require an a priori functional relationship between response and predictor variables, which is the case with traditional multiple regression. The transformations are very general, computationally efficient, and can easily handle mixed data types; for example, continuous variables such as porosity, and permeability, and categorical variables such as rock type and lithofacies. The power of the nonparametric transformation techniques for data correlation has been illustrated through synthetic and field examples. Second, we use these transformations to propose a two-stage approach for data integration during heterogeneity characterization. The principal advantages of our approach over traditional cokriging or cosimulation methods are: (1) it does not require a linear relationship between primary and secondary data, (2) it exploits the secondary information to its full potential by maximizing the correlation between the primary and secondary data, (3) it can be easily applied to cases where several types of secondary or soft data are involved, and (4) it significantly reduces variance function calculations and thus greatly facilitates non-Gaussian cosimulation. We demonstrate the data integration procedure using synthetic and field examples. The field example involves estimation of pore-footage distribution using well data and multiple seismic attributes.

INTRODUCTION

During initial stages of data correlation, we often are interested in pursuing exploratory analysis. Rather than imposing our preconceived notions or models, we want to gain insight into the nature of the data set and, if possible, the underlying phenomena that might have produced the data set. Unfortunately, traditional multiple regression techniques for data correlation are limited in this respect because they require a priori assumptions of functional forms that relate the response (dependent) and predictor (independent) variables. This is a significant drawback for correlating rock or petrophysical properties because of the inexact nature of the underlying relationship. When used as a predictive tool for petrophysical data, conventional multiple regression suffers from several other limitations, as discussed by Jensen and Lake (1985), Wendt et al. (1986), and Xue et al. (1996).

Parametric transformations have been suggested for optimization of regression-based permeability-porosity predictions. Notable amongst these are the power transformations proposed by Jensen and Lake (1985). The underlying theory is that if the joint probability distribution function of two variables is binormal, their relationship will be linear (Hald, 1952). Several methods exist to estimate the exponents for power transformation. One method, described by Emerson and Stoto (1982) and adopted by Jensen and Lake (1985), is based on symmetrizing the probability distribution function (p.d.f.). Another method is a trial-and-error approach based on a normal probability plot of the data. By power transforming permeability and porosity separately, the authors are able to improve permeability-porosity correlation; however, using a trial-and-error method for selecting exponents for power transformation is time consuming, and symmetrizing the p.d.f. does not necessarily guarantee a binormal distribution of transformed variables. Most importantly, there are no indications as to whether power transformations will work for multivariate cases. This is a severe limitation because we are often interested in correlating permeability with multiple well log responses.

We propose here a more adaptive approach offered by nonparametric transformation and regression methods for correlating petrophysical data (Hastie and Tibshirani, 1990; Xue et al., 1996; Barman et al., 1998). The nonparametric transformation techniques generate regression relations in a flexible data-defined manner through the use of conditional expectations or scatterplot smoothers and in doing so, let the data itself suggest functional forms or detect inherent nonlinearities. Optimal nonparametric transformations can be shown to produce maximum correlation in the transformed space (Breiman and Friedman, 1985). The power of these methods lies in their ability to directly incorporate multiple and mixed variables, both continuous and categorical, into correlation. Moreover, the transformations are computationally efficient, easy to use and can provide significant insight during exploratory data analysis.

Next, we use these nonparametric transformations to propose a two-stage approach to integrate seismic or other secondary data into reservoir characterization. First, we calibrate seismic and well data using optimal nonparametric transformations. Stochastic cosimulation is then carried out in the transformed space to generate conditional realizations of reservoir properties. The principal advantages of this approach over traditional cosimulation methods are (1) it does not require a linear relationship between seismic and well data, (2) it exploits the secondary information to its full potential by maximizing the correlation between the primary and secondary data, and (3) it can be easily extended to cases where several types of soft data are involved. Moreover, the use of nonparametric transformations results in a significant reduction in variance function calculations and thus, greatly facilitates non-Gaussian cosimulation through the use of indicator approaches.

The organization of this paper is as follows. First, we briefly review the theory and motivation behind nonparametric transformations and regression. Next, we discuss application of such transformations for data correlation and integration. Finally, we present synthetic and field examples that demonstrate the power and utility of such transformations for correlating petrophysical properties using multiple regression. A field example involving integration of 3-D (three-dimensional) seismic data is also discussed to illustrate the data integration procedure.

NONPARAMETRIC TRANSFORMATION AND REGRESSION: THEORY

In general, the regression problem involves a set of predictors, for example, a p-dimensional random vector X and a random variable Y, which is called the response variable. The aim of regression analysis is to estimate the conditional expectation, $E(Y \mid X_1, X_2, \ldots, X_p)$. Conventional multiple regression requires a functional form to be presumed a priori for the regression surface, thus reducing the problem to that of estimating a set of parameters. Such a parametric approach can be successful provided the model assumed is appropriate. When the relationship between the response and predictor variables is unknown or inexact, as is frequently the case for reservoir rock or petrophysical properties, parametric regression can yield erroneous and even misleading results. This is the primary motivation behind nonparametric regression techniques, which make only few general assumptions about the regression surface (Friedman and Stuetzle, 1981).

The nonparametric transformations techniques generate regression relations in a flexible data-defined manner through the use of scatterplot smoothers, and in doing so let the data suggest the functionalities. The most extensively studied nonparametric regression techniques (kernel, nearest neighbor, or spline smoothing) are based on some sort of local averaging which take the form:

$$E(Y|X) = \sum_{i=1}^{N} H(x, x_i) y_i \qquad (1)$$

where $H(x, x')$ (the kernel function) usually has its maximum at $x' = x$, with its absolute value decreasing as $|x' - x|$ increases. A critical parameter in local averaging is the span $s(x)$, which is the interval, centered at x', over which most of the averaging takes place and thus, controls the bias-variance trade-off. Optimal span selection based on local cross-validation has been discussed by Friedman and Silverman (1989).

More recently, however, nonparametric regression techniques that are based on successive refinements have gained wide popularity in a variety of disciplines ranging from medical sciences to air pollution control (Hastie and Tibshirani, 1990). We will focus here on these nonparametric regression techniques that attempt to define the regression surface in an iterative fashion while remaining data-driven as opposed to model-driven. We can broadly classify them into those that do not transform the response variable (generalized additive models) and those that do (alternating conditional expectations and its variations). A brief discussion of these techniques follows. For further details, see Hastie and Tibshirani (1990), Buja et al. (1989), and Xue et al. (1996).

Generalized Additive Models

An additive regression model has the general form:

$$E\left(Y|X_1, X_2, \ldots, X_p\right) = \alpha + \sum_{l=1}^{p} \phi_l(X_l) + \varepsilon \qquad (2)$$

where X_l are the predictors and ϕ_l are functions of predictors. Thus additive models replace the problem of estimating a function of a p-dimensional variable X by one of estimating p separate one-dimensional functions, ϕ_l. Such models are attractive if they fit the data because they are far easier to interpret than a p-dimensional multivariate surface.

The technique for estimating ϕ_l is called the local scoring algorithm and uses scatterplot smoothers for example, a running mean, running median, running least squares line, kernel estimates, or a spline (see Buja et al. (1989) for a discussion of smoothing techniques). In order to motivate the algorithm, let us consider the following simple model:

$$E\left(Y|X_1, X_2\right) = \phi_1(X_1) + \phi_2(X_2) \qquad (3)$$

Given an initial estimate $\phi_1(X_1)$, one way to estimate $\phi_2(X_2)$ is to smooth the residual $R_2 = Y - \phi_1(X_1)$ on X_2. With this estimate of $\phi_2(X_2)$, we can get an improved estimate $\phi_1(X_1)$ by smoothing $R_1 = Y - \phi_2(X_2)$ on X_1. The resulting iterative smoothing procedure is called backfitting (Hastie and Tibshirani, 1990) and forms the core of additive models.

In general, an algorithm for fitting a generalized additive model consists of a hierarchy of three modules: (1) scatterplot smoothers which can be thought of as a general regression tool for fitting functional relationship between response and predictor variables, (2) a backfitting algorithm that cycles through the individual terms in the additive model and iteratively updates each using the Gauss-Siedel method by smoothing suitably defined partial residuals, and (3) a local scoring algorithm that uses an iteratively reweighted least squares procedure (Hastie and Tibshirani, 1990) to generate a new additive predictor. A step-by-step procedure for the generalized additive model can be found in Hastie and Tibshirani (1990).

Response Transformation Models: ACE Algorithm and Its Variations

The response transformation models generalize the additive model by allowing for a transformation of the response variable Y. The models have the following general form:

$$\theta(Y) = \alpha + \sum_{l=1}^{p} \phi_l(X_l) + \varepsilon \qquad (4)$$

The main motivation behind response transformation is that often a simple additive model may not be appropriate for $E(Y | X_1, X_2, \ldots, X_p)$, but may be quite appropriate for $E\theta(Y) | X_1, X_2, \ldots, X_p)$. An example of such a model is the alternating conditional expectation (ACE) algorithm and its modifications.

The ACE algorithm, originally proposed by Breiman and Friedman (1985), provides a method for estimating optimal transformations for multiple regression that result in a maximum correlation between a dependent (response) random variable and multiple independent (predictor) random variables. Such optimal transformations can be derived by minimizing the variance of a linear relationship between the transformed response variable and the sum of transformed predictor variables. For a given set of response variable Y and predictor variables $X_1, \ldots, X_p$, the ACE algorithm starts out by defining arbitrary measurable mean-zero transformations $\theta(Y), \phi_1(X_1), \ldots, \phi_p(X_p)$. The error ($e^2$) that is not explained by a regression of the transformed dependent variable on the sum of transformed independent variables is (under the constraint, $E[\theta^2(Y)] = 1$)

$$e^2\left(\theta, \phi_1, \ldots, \phi_p\right) = E\left\{\left[\theta(Y) - \sum_{l=1}^{p} \phi_l(X_l)\right]\right\}^2 \qquad (5)$$

The minimization of e^2 with respect to $\phi_1(X_1), \ldots, \phi_p(X_p)$ and $\theta(Y)$ is carried out through a series of single-function minimizations, resulting in the following equations

$$\phi_l(X_l) = E\left[\theta(Y) - \sum_{j \neq l} \phi_j\left(X_j\right)\middle| X_l\right]$$
$$\theta(Y) = E\left[\sum_{l=1}^{p} \phi_l(X_l)\middle|Y\right] \Big/ \left\| E\left[\sum_{l=1}^{p} \phi_l(X_l)\middle|Y\right]\right\| \qquad (6)$$

Two basic mathematical operations involved here are conditional expectations and iterative minimization, and hence the name alternating conditional expectations. The final $\phi_l(X_l)$, $l = 1,...,p$ and $\theta(Y)$ after the minimization are estimates of optimal transformation $\phi^*_l(X_l)$, $l = 1,...,p$ and $\theta^*(Y)$. In transformed space, the response and predictor variables will be related as follows

$$\theta^*(Y) = \sum_{l=1}^{p} \phi_l^*(X_l) + \xi \quad (7)$$

where ξ is the misfit.

The optimal transformations are derived solely based on the data sets and can be shown to result in a maximum correlation in the transformed space (Breiman and Friedman, 1985). The transformations do not require *a priori* assumptions of any functional form for the response or predictor variables and thus provide a powerful tool for exploratory data analysis and correlation.

Although ACE is a potent and versatile approach for building correlations, it suffers from some anomalies when one views it as a regression tool. Such anomalies become particularly prominent in low-correlation settings. A modification of ACE designed primarily for regression problems was proposed by Tibshirani (1988) and differs from ACE in that it chooses $\theta(Y)$ to achieve a special asymptotic variance stabilizing feature. The goal here is to estimate transformations θ and ϕ_l that have the following properties:

$$\begin{aligned} E\left\{\theta(Y)\middle|X_1, X_2, \ldots, X_p\right\} &= \sum_{l=1}^{p} \phi_l(X_l) \\ \operatorname{var}\left\{\theta(Y)\middle|\sum_{l=1}^{p} \phi_l(X_l)\right\} &= \text{constant} \end{aligned} \quad (8)$$

The transformation θ is assumed to be strictly monotone (and thus, invertible) and the conditional expectations are approximated using the scatterplot smoothing algorithm supersmoother (Friedman and Stuetzle, 1982). In the examples that follow, we use this modification of the ACE algorithm. A step by step procedure for the ACE model and its modification can be found in Hastie and Tibshirani (1990).

NONPARAMETRIC TRANSFORMATION: APPLICATION

In this section we discuss how the nonparametric transformation techniques discussed can be applied for data correlation and integrating diverse data types. We focus on the response transformation models because they are more general in nature and thus, encompass other nonparametric techniques.

Data Correlations and Estimation

Non-parametric transformations techniques offer a flexible and data-driven approach to building correlation without *a priori* assumptions regarding functional relationship between response and predictor variables. The following equation is used to estimate or predict dependent variable, y_i^{pre} for any given data point $\{x_{1i},\ldots,x_{pi}\}$ involving p-independent variables

$$y_i^{pre} = \theta^{*-1}\left[\sum_{l=1}^{p} \phi_l^*(x_{li})\right] \quad (9)$$

The calculation involves p forward transformations of $\{x_{1i},\ldots,x_{pi}\}$ to $\{\phi_1^*(x_{1i}),\ldots,\phi_p^*(x_{pi})\}$, and a backward transformation, equation 9. By restricting the transformation of the response variable to be monotone, we can ensure that θ^* is invertible.

The power of nonparametric transformations as a tool for correlation lies in their ability to handle variables of mixed type. For example, we can easily incorporate categorical variables such as rock types and lithofacies into the correlation and also handle missing data values without additional complications (Breiman and Friedman, 1985).

Our experience has shown that for most of the applications considered by us (petrophysical and pressure-volume-temperature [PVT]), the nonparametric transformations $\phi_l(X)$ and $\theta(Y)$ can be fitted by simple functions such as polynomials, power functions, or cubic splines (Xue et al., 1996; McCain et al., 1998). This allows for a rapid and powerful alternative to traditional multiple regression for building correlation for a variety of applications, particularly in the presence of several predictor variables.

Data Integration

One critical aspect of integrating different data types during reservoir characterization is the calibration between primary and secondary data, for example, correlating well and seismic data. Cokriging or cosimulation has traditionally been used for data integration in which such calibration is accomplished by modeling cross covariance functions. Given a primary variable (hard data) $y(u_\alpha)$, sampled at n locations and secondary variables (soft data) $x_1(u_\beta),\ldots,x_p(u_\beta)$, all assumed sampled at the same m locations the full cokriging estimator of $y(u)$ is

$$\begin{aligned} y(u) = \sum_{\alpha=1}^{n} \lambda_\alpha y(u_\alpha) + \sum_{\beta=1}^{m} \mu_{1\beta} x_1(u_\beta) + \ldots \\ + \sum_{\beta=1}^{m} \mu_{p\beta} x_p(u_\beta) \end{aligned} \quad (10)$$

where λ_α is the weights for the primary variable y, and $\mu_{l\beta}$ is the weights associated with the secondary variables, x_l $(l = 1,...,p)$; however, in practice, several limitations restrict the application of full cokriging for data integration. First, the implementation of equation 10 requires modeling of $(p + 1)^2$ variance functions, which include $(p + 1)$ covariance functions and $p(p + 1)$ cross covariance functions. Modeling of variance functions becomes extremely tedious when several secondary variables are involved, for example multiple seismic attributes. Second, cokriging matrix may become unstable (close to singular) because of the sparse primary and dense secondary data samples (Almeida, 1993). Third, because the cross covariance functions can only capture the linearity between the primary and secondary data samples, the influence of secondary data samples on the cokriging estimator can be reduced significantly in the presence of nonlinearity. This is particularly critical for integration of seismic data, since in general the link between reservoir and seismic properties can be expected to be nonunique, multivariate, and nonlinear (Xue and Datta-Gupta, 1996).

We propose here a two-stage approach to integrating seismic or other secondary data into reservoir characterization. First, we use the nonparametric transformational approach to calibrate the seismic and well data to maximize correlation between the two data sets. This leads to a set of transformations θ^*, ϕ_l^*, $l = 1, p$. Cokriging or stochastic cosimulation is then carried out in the transformed space to generate conditional realizations of reservoir properties. The cokriging equations now take the following form

$$\theta^*(y(u)) = \sum_{\alpha=1}^{n} \lambda'_\alpha \theta^*(y(u_\alpha)) + \sum_{\beta=1}^{m} \mu'_\beta \phi_s^*(x(u_\beta)) \quad (11)$$

where ϕ_s^* is the sum of transformed secondary data samples (for example, seismic attributes) as follows

$$\phi_s^*(x(u_\beta)) = \phi_1^*(x_1(u_\beta)) + \ldots + \phi_p^*(x_p(u_\beta)) \quad (12)$$

Notice that the new formulation for cokriging estimator with optimal transformations (equation 11) now contains only two terms compared to $(p + 1)$ terms in equation 10. As a result, it reduces the number of variance functions required by cokriging from $(p + 1)^2$ to 4 regardless of the number of secondary variables involved. A further simplification can be made by introducing collocated cokriging algorithm with Markov hypothesis (Almeida, 1993) into equation 11, resulting in the following collocated cokriging estimator

$$\theta^*(y(u)) = \sum_{\alpha=1}^{n} \lambda'_\alpha \theta^*(y(u_\alpha)) + \mu' \phi_s^*(x(u)) \quad (13)$$

The underlying assumption in equation 13, in addition to Markov screening hypothesis, is that the secondary data samples must be available at every location where the primary variable is to be estimated. This is always satisfied for seismic data.

Stochastic cosimulation algorithm provides another important tool for data integration. Such cosimulation generates multiple realizations of random field conditioned to prior information, allowing assessment of model uncertainty. Indicator cosimulation has the capability of integrating soft and hard data without the assumption of a multi-Gaussian distribution. The limitations of multi-Gaussian assumptions are well established in the literature (Journel and Alabert, 1990). The use of Markov-Bayes (Zhu and Journel, 1992) algorithm in conjunction with optimal transformation can significantly simplify the modeling of indicator variance function, especially when several types of soft data are involved.

In transformed space, we define hard indicator as

$$I(u_\alpha, \theta_j^*) = \begin{cases} 1, & \text{if } \theta^*(y(u_\alpha)) \le \theta_j^* \\ 0, & \text{otherwise} \end{cases} \quad (14)$$

where θj^* j= 1,...,k are cutoffs for transformed hard data. Similarly, we define local soft indicator data, originating from the calibration between transformed hard data and the sum of transformed soft data

$$z(u_\beta, \theta_j^*) = \Pr ob\{\theta^*[y(u_\beta)] \le \theta_j^* | \phi_s^*(u_\beta)\} \in [0,1] \quad (15)$$

Having indicator coded the transformed data, we follow the procedure outlined by Zhu and Journel (1992). After simulation, we back transform to the original data space.

RESULTS: SYNTHETIC AND FIELD EXAMPLES

In this section we describe application of the concepts discussed to synthetic and field examples. The synthetic examples are designed to test the validity of our approach and to compare with the methods currently in practice. The field examples serve to illustrate the versatility to handle field scale applications.

Data Correlation: A Synthetic Example

This synthetic example is designed to demonstrate the ability of nonparametric transformations to identify functional relationship during multiple regression and correlation. Our example involves 300 observations generated using the following model

$$y_i = x_{1i} + x_{2i}^2 + x_{3i}^3 + 0.1\varepsilon_i \quad (16)$$

where x_{1i}, x_{2i}, and x_{3i} are independently drawn from a uniform distribution $U(-0.5,0.5)$, and ε_i is drawn from a standard normal distribution $N(0,1)$. Figure 1a–c shows plots of y_i versus x_{1i}, x_{2i}, and x_{3i}, respectively. Except for y_i versus x_{1i}, the functional relationships between the dependent variable y_i and independent variables x_{2i}, and x_{3i} cannot be identified from the scatterplots.

The optimal transformations for y_i and x_{1i}, x_{2i}, and x_{3i} derived using ACE are plotted in Figure 1d–g. The transformations for both y_i and x_{1i} yield essentially straight lines. The transformation for x_{2i} reveals a quadratic function, and the transformation for x_{3i} reveals a cubic function. Thus, ACE is able to identify

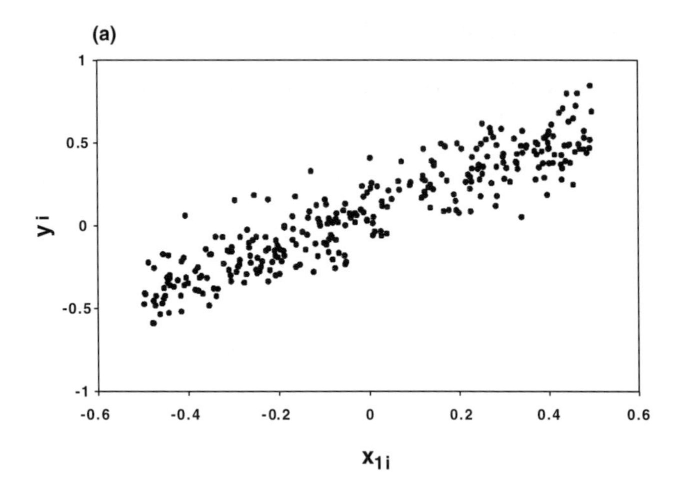

Figure 1a. Scatterplot of y_i versus x_{1i} simulated from multivariate model $y_i = x_{1i} + x_{2i}^2 + x_{3i}^3 + 0.1\varepsilon_i$, where x_{1i}, x_{2i}, and x_{3i} are independently drawn from uniform distribution $U(-0.5,0.5)$, and ε_i is independently drawn from standard normal distribution $N(0,1)$.

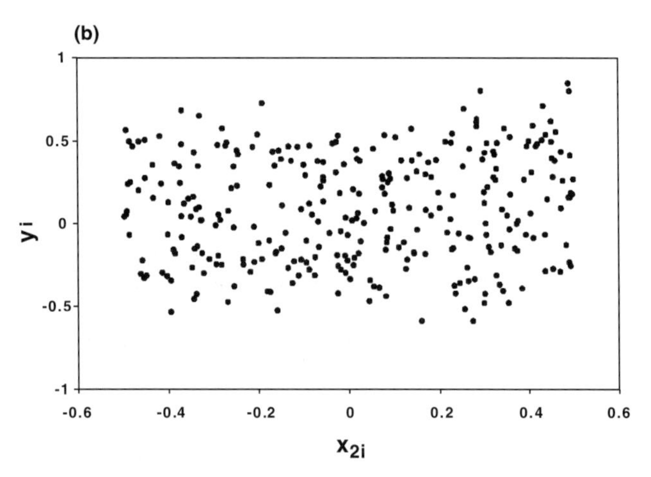

Figure 1b. Scatterplot of y_i versus x_{2i} simulated from multivariate model $y_i = x_{1i} + x_{2i}^2 + x_{3i}^3 + 0.1\varepsilon_i$.

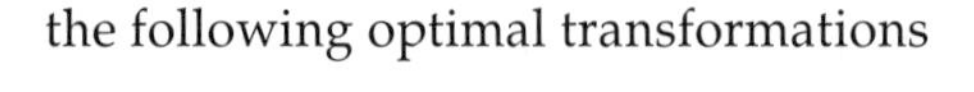

the following optimal transformations

$$\begin{aligned} &\theta^*(y_i) \cong y_i \\ &\phi_1^*(x_{1i}) \cong x_{1i}, \qquad \phi_2^*(x_{2i}) \cong x_{2i}^2, \qquad \phi_3^*(x_{3i}) \cong x_{3i}^3 \end{aligned} \tag{17}$$

This is, indeed, remarkable considering that the individual scatterplots hardly reveal any such relationships. A plot of transformed y_i versus the sum of transformed x_{1i}, x_{2i}, and x_{3i} is shown in Figure 1h. The relationship can be fitted approximately by

$$\theta^*(y_i) = \phi_1^*(x_{1i}) + \phi_2^*(x_{2i}) + \phi_3^*(x_{3i}) \tag{18}$$

which is exactly optimal.

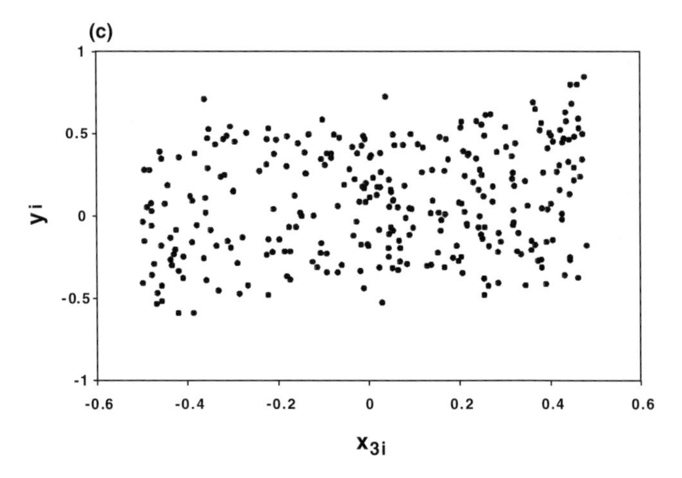

Figure 1c. Scatterplot of y_i versus x_{3i} simulated from multivariate model $y_i = x_{1i} + x_{2i}^2 + x_{3i}^3 + 0.1\varepsilon_i$.

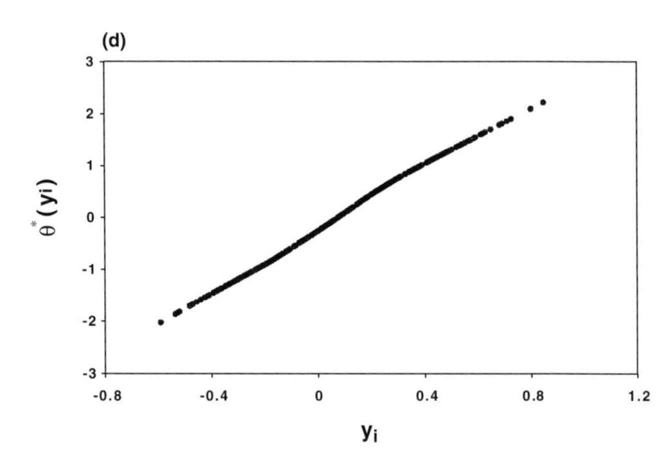

Figure 1d. Optimal transformation of y_i by ACE (alternating conditional expectation) algorithm.

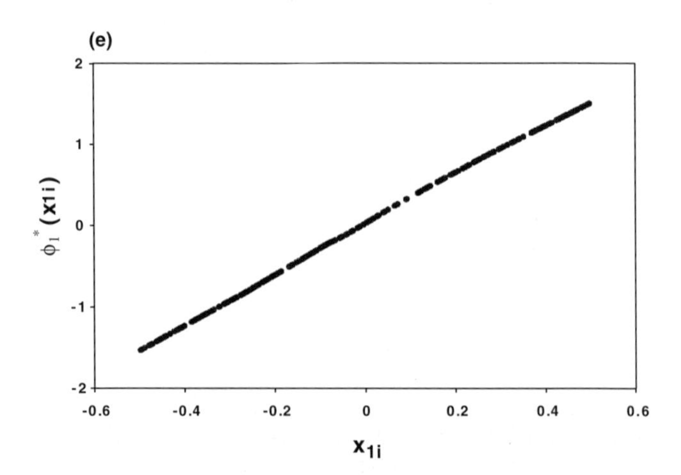

Figure 1e. Optimal transformation of x_{1i} by ACE.

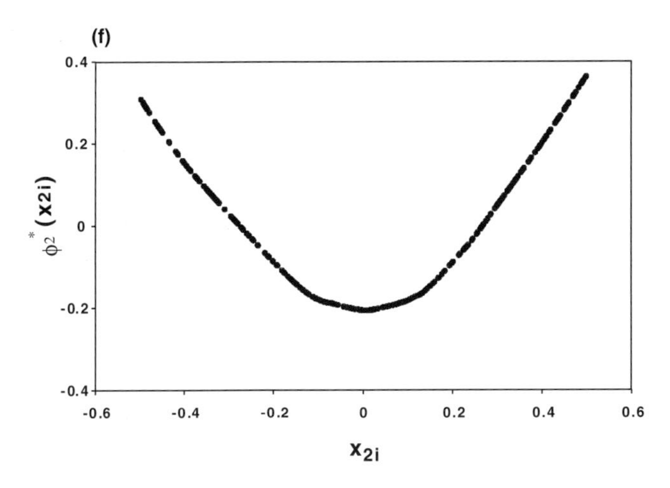

Figure 1f. Optimal transformation of x_{2i} by ACE.

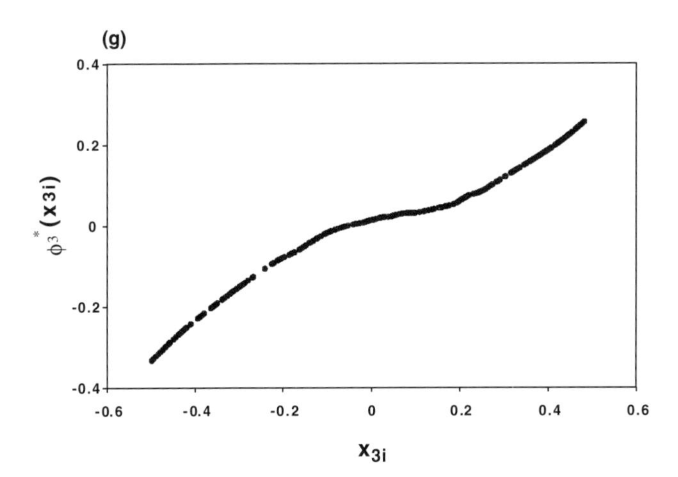

Figure 1g. Optimal transformation of x_{3i} by ACE.

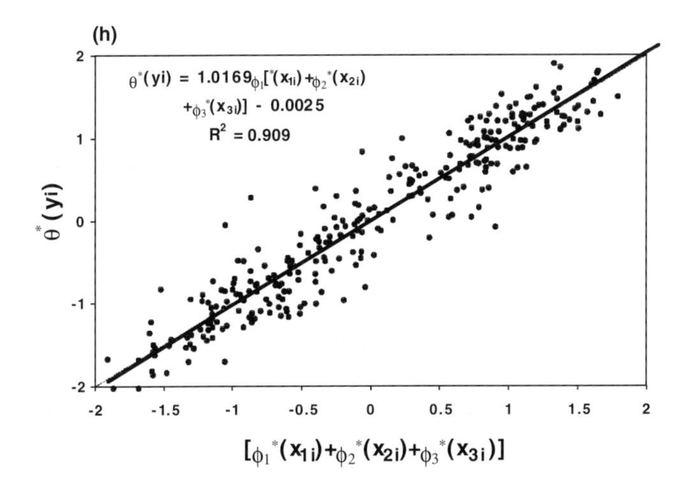

Figure 1h. Optimal transformation of y_i versus the sum of optimal transformations of x_{1i}, x_{2i}, and x_{3i}. The solid straight line represents a linear regression of the data.

We also applied generalized additive model (GAM) to the same synthetic example. The results were almost identical to ACE and, hence, are not shown here. The predictive ability of these models is indicated by the bootstrap prediction errors as summarized in Table 1. The bootstrap approach to estimating prediction error involves generating B bootstrap samples for each sample drawing independently and with replacement from the original data. The correlation model is rebuilt using each sample and then applied to the original, as well as to the bootstrap sample, to obtain estimates of apparent error and bias (optimism). The details can be found in Efron and Tibshirani (1993). Because of the similar performance characteristics of ACE and GAM, we have restricted to application of the ACE model to only the examples discussed in the following paragraphs.

Correlating Petrophysical Data: North Robertson Unit, West Texas

This field example serves to illustrate the versatility of nonparametric transformations to incorporate mixed data types, categorical and continuous, into correlation. The data belong to the North Robertson unit (NRU) located in Gaines County, west Texas. This is a mature, highly heterogeneous, shallow-shelf carbonate reservoir. The reservoir interval is about 1400 ft (427 m) in gross thickness, with 90% of the interval being dolostone having a complex pore structure. Because of the diagenetic modification of the pore structure, no obvious relationship between porosity and permeability can be established at NRU even when the data are separated based on the depositional environment (Davies and Vessel, 1996); however, definition of rock types based on pore geometry analysis shows good relationship between permeability and porosity within each rock type (Figure 2a–c). Figure 2d shows the data for all three rock types combined and, as expected, the correlation R^2 is reduced significantly.

Next, we apply nonparametric transformations to the data. Instead of having three separate correlations for the rock types, we can now correlate permeability directly to porosity and rock type. The transformations are shown in Figure 3a–c. The optimal correlation is shown in Figure 3d, with an R^2 of 0.74 as compared to 0.45 as shown in Figure 2d. On fitting the individual transformations with simple functions, we can derive an equation describing permeability as a

Table 1. Bootstrap Estimates of Prediction Error*

Prediction Error	ACE	GAM
Apparent error	0.009720	0.009308
Optimism	0.000379	0.000607
Total error	0.010099	0.009915

*50 bootstrap replications; synthetic example: $Y_i = x_{1i} + x_{2i}^2 + x_{3i}^3 + 0.1\varepsilon_i$. ACE = alternating conditional expectation algorithm, GAM = generalized additive model.

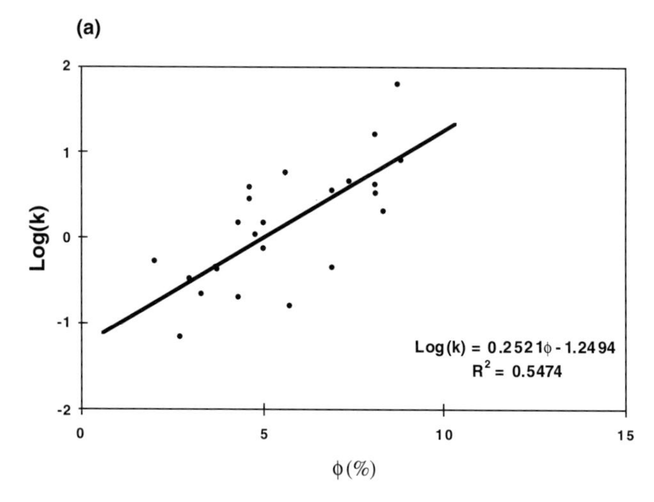

Figure 2a. Logarithmically transformed permeability versus core porosity of rock type 1. The solid straight line represents a linear regression of the data.

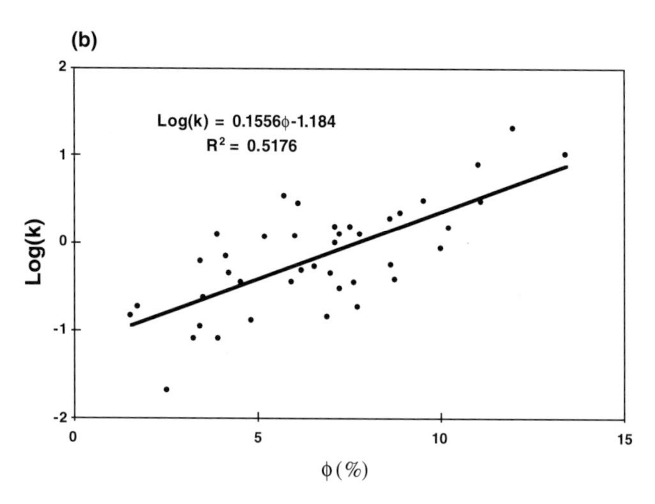

Figure 2b. Logarithmically transformed permeability versus core porosity of rock type 2. The solid straight line represents a linear regression of the data.

function of porosity and rock type:

$$\theta^*(k) = 1.0258\left[\phi_1^*(RT) + \phi_2^*(\phi)\right] \tag{19a}$$

where the functional forms describing $\theta^*(k)$, $\phi_1^*(RT)$ and $\phi_2^*(\phi)$ are shown in Figure 3a–c. Given a porosity and rock type, we use equation 19(a) to compute $\theta^*(k)$. The corresponding permeability can then be obtained from Figure 3c or using the fitted equation

$$k^{0.5} = 0.077 + 0.559\theta^*(k) + 0.752\left[\theta^*(k)\right]^2 \tag{19b}$$

Table 2 compares bootstrap prediction error associated with permeability estimates using different methods. Three choices have been compared: a single

Table 2. Bootstrap Estimates of Prediction Error*

Prediction Error	Logarithmic Model with All Rock Types	Logarithmic Models for Individual Rock Types	ACE** with All Rock Types
Apparent error	28.9099	22.2025	23.5064
Optimism	0.5808	5.0077	2.3199
Total error	29.4907	27.2102	25.8263

*100 bootstrap replications, North Robertson unit, west Texas.
**ACE = alternating conditional expectation algorithm.

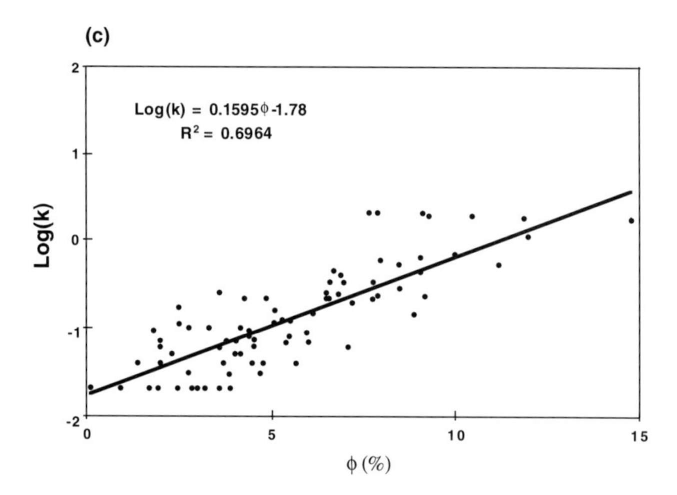

Figure 2c. Logarithmically transformed permeability versus core porosity of rock type 3. The solid straight line represents a linear regression of the data.

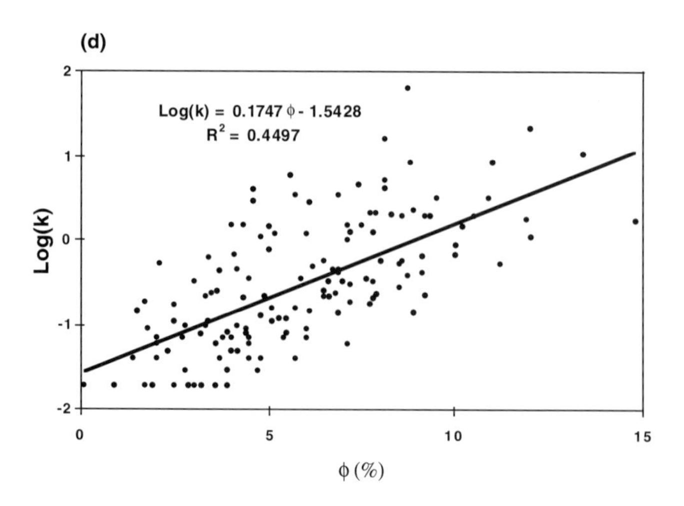

Figure 2d. Logarithmically transformed permeability versus core porosity of all rock types (rock types 1, 2, and 3). The solid straight line represents a linear regression of the data.

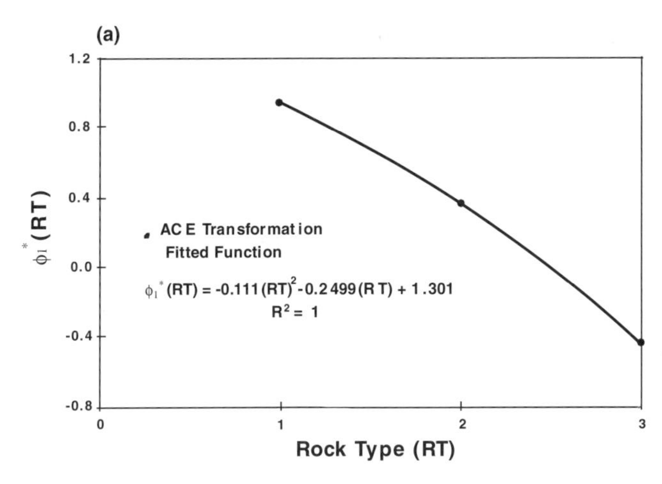

Figure 3a. Optimal transformation of rock type by ACE (alternating conditional expectation) algorithm. The solid line represents a fitted function.

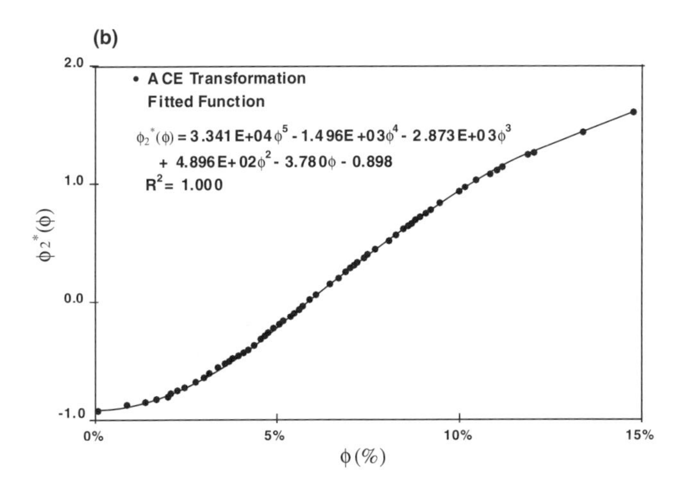

Figure 3b. Optimal transformation of porosity by ACE. The solid line represents a fitted function.

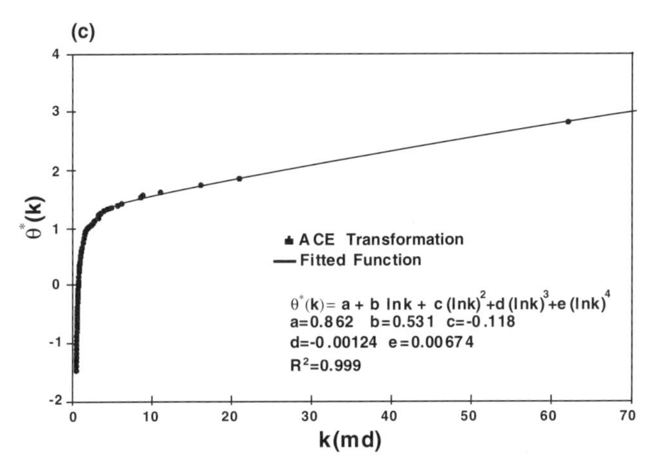

Figure 3c. Optimal transformation of permeability by ACE. The solid line represents a fitted function.

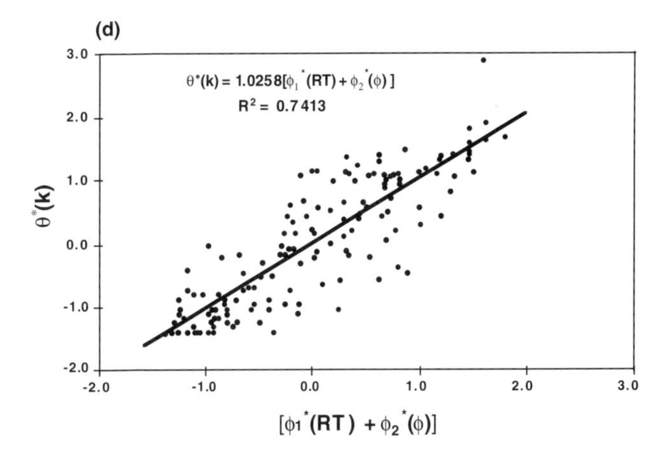

Figure 3d. Optimal transformation of permeability versus the sum of optimal transformations of rock type and porosity. The solid line represents a linear regression of the data.

correlation using all rock types, separate correlation for each rock type, and the correlation equation developed using the nonparametric approach. As expected, the correlation combining all rock types (Figure 2d) performs the worst. The correlation based on optimal transformation (Figure 3d) not only outperforms individual rock type correlation, but also collapses them into a single convenient equation.

Data Integration: A Synthetic Example

We simulated a 2-D (two-dimensional) synthetic case that includes one primary and two secondary variables having the following nonlinear relationship

$$y(u) = x_1^{3.5}(u) + x_2(u) + \varepsilon \tag{20}$$

where u is the location in 2-D space and ε is random Gaussian noise. The simulation grid size is 60×40. The secondary variables x_1 and x_2 are generated using the following models

$$\begin{aligned} x_1(u) &= 0.4[0.75t_1(u) + 0.25t_2(u)] + 1.2 \\ x_2(u) &= 2.0[0.25t_1(u) + 0.75t_2(u)] + 6.0 \end{aligned} \tag{21}$$

In equation 21, t_1 and t_2 are two mutually orthogonal realizations both with zero mean and unit variance. They are generated by unconditional sequential Gaussian simulations using a spherical semivariogram model. The constants in equation 21 were selected such that the two secondary variables will have a balanced effect on the primary variable.

Figure 4 shows grayscale maps of the simulated x_1, x_2, and y. We use this simulated y as known exhaustive reference. Next, 120 data points (5% of total) are obtained by sampling the exhaustive y data at a regular spacing. These sampled y data, together with the exhaustive secondary data (x_1 and x_2 at 2400

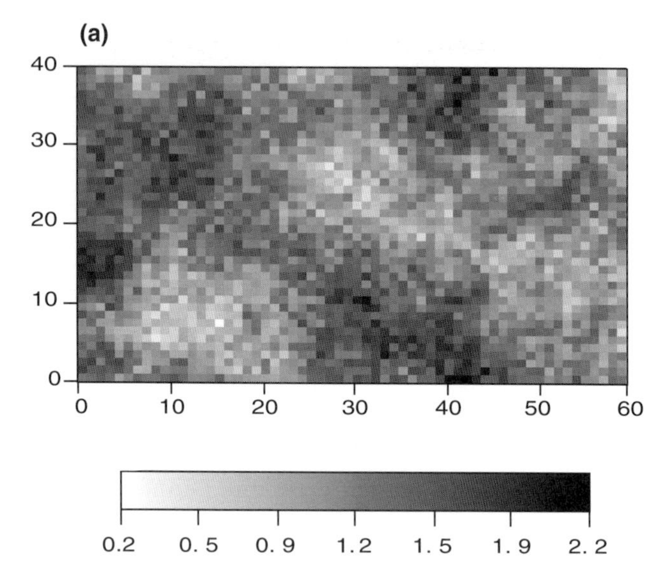

Figure 4a. Grayscale map of simulated x_1. The simulation grid is 60×40.

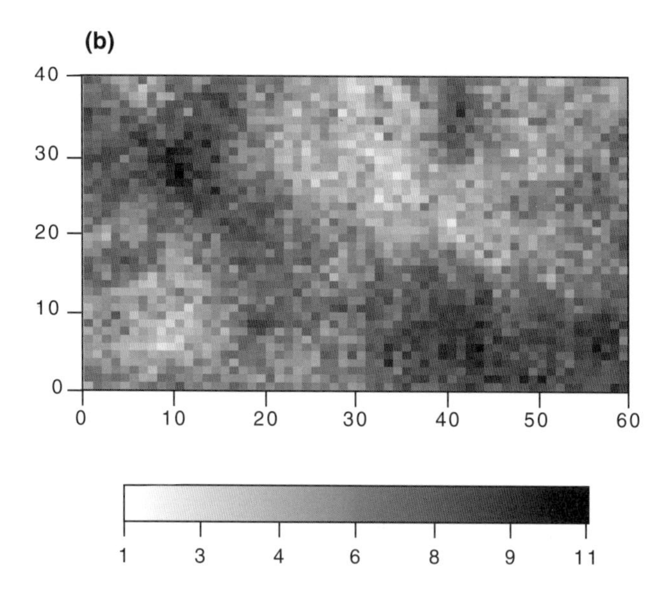

Figure 4b. Grayscale map of simulated x_2.

locations) are then used to estimate y values at unsampled locations.

The first step in using our proposed approach for data integration is to derive optimal transformations using the ACE algorithm. In the transformed space, the nonlinearity between primary and secondary variables is virtually eliminated, resulting in a maximal correlation $R^2 = 0.912$, as shown in Figure 5.

The second and third steps are application of optimal transformations to the sampled y data and to all secondary data. Figure 6 shows a grayscale map of the sum of transformed x_1 and x_2. Notice that because of the strong correlation established between the primary and secondary variables through the optimal transformations, many of the features of the exhaustive y-data are already apparent here.

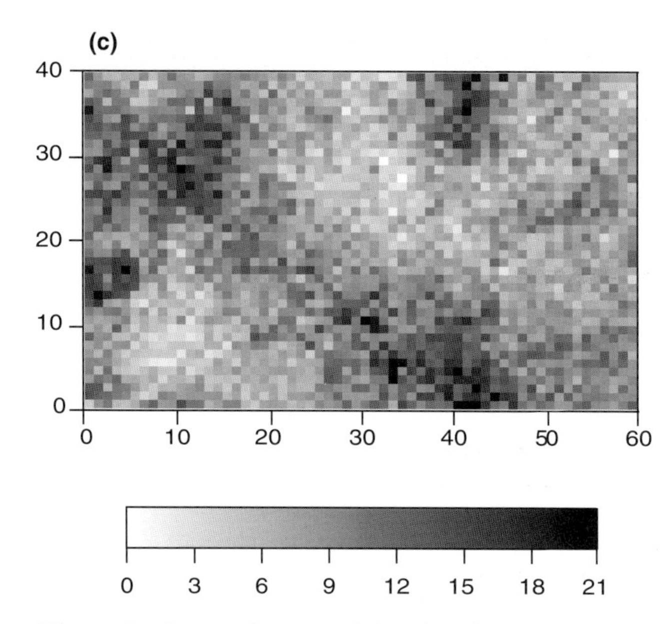

Figure 4c. Grayscale map of simulated y.

Next, we perform cokriging using equation 13 to estimate the transformed primary values at all locations. The final step is back transformation to the original space. Figure 7 shows the estimated primary values by cokriging using the optimal transformation approach. The correspondence with the original exhaustive reference (Figure 4c), indeed, is very good. For comparison purposes, we also conducted traditional ordinary kriging and cokriging estimation without using any transformations.

Figure 8a and b shows the scatterplots of true y (exhaustive reference) versus estimated y by traditional ordinary kriging and cokriging. Figure 8c shows the scatterplot of true y versus estimated y by cokriging using the optimal transformation approach. The power of the optimal transformation is quite evident from

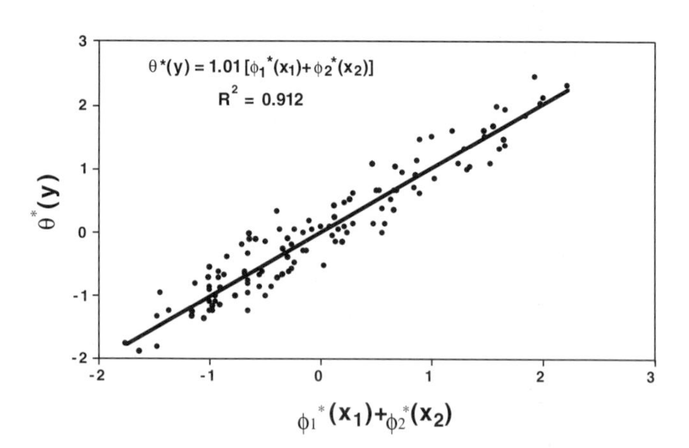

Figure 5. Optimal correlation between primary (y) and secondary variables (x_1 and x_2) as derived by ACE (alternating conditional expectation) algorithm. The solid line represents a linear regression of the transformed data.

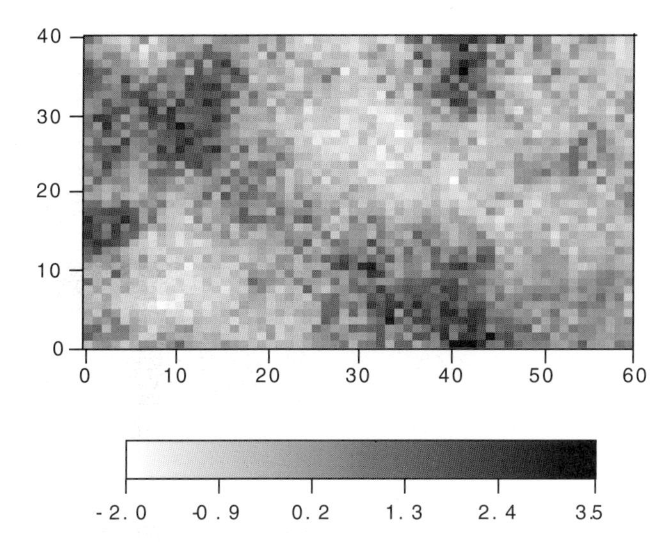

Figure 6. Grayscale map of the sum of transformed x_1 and x_2.

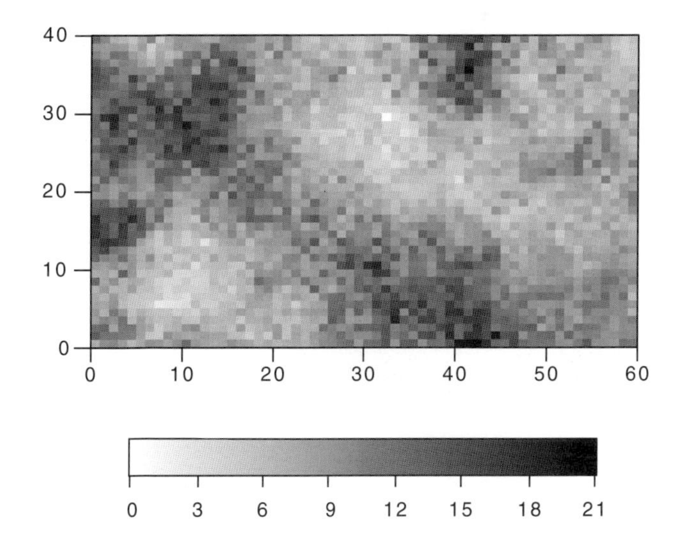

Figure 7. Grayscale map of the estimated y by collocated cokriging using optimal transformations.

these results. By incorporating optimal transformations, we have improved the correlation R^2 to 0.86, compared to 0.72 obtained by using traditional cokriging. As expected, ordinary kriging performs the worst since the secondary data sets are not used.

The cokriging estimation without transformations is affected by the nonlinearity between y and x_1. This is reflected in the concave-upward shape in Figure 8b; furthermore, to use the cokriging estimator (equation 10), six variance functions are required for this case as opposed to three when optimal transformations are used, both assuming symmetry in cross covariance functions. Table 3 presents a quantitative comparison of the statistics of error distribution. The results clearly indicate the superiority of our proposed approach as evidenced by the small standard deviation of error distribution obtained using the optimal transformations.

We also performed indicator cosimulation using Markov-Bayes algorithm in conjunction with ACE transformations. Five cutoffs, corresponding to 10% 30%, 50%, 70%, and 90% of the cumulative distribution function (c.d.f.) of the transformed dependent variables, were used in these simulations. Figure 9 shows two realizations of y. Both reproduce the features and statistics of the reference y very well. The use of ACE transformations greatly facilitates such an indicator cosimulation process by reducing variance function calculations.

Table 3. Summary Statistics for Error Distributions of y Estimated by Various Estimators

$\lvert y_{TRUE} - y_{EST} \rvert$	Ordinary Kriging	Cokriging	Cokriging Using ACE*
Minimum	0.00	0.00	0.00
Maximum	11.07	7.95	5.91
Mean	1.86	1.44	1.00
Standard Dev.	1.58	1.20	0.80

*ACE = alternating conditional expectation algorithm.

Integration of Seismic Data: Stratton Field, South Texas

The Stratton field (Levey et al., 1993) is located on the onshore south Texas Gulf Coast Basin. The Oligocene Frio Formation is one of the largest gas-productive intervals. The middle Frio Formation is characterized

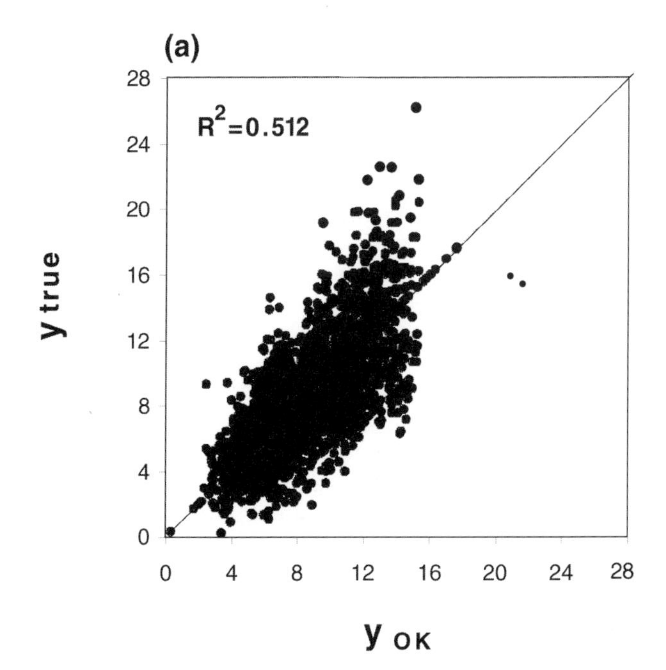

Figure 8a. True y (reference) versus the estimated y by ordinary kriging without using transformations. Secondary variables are not used.

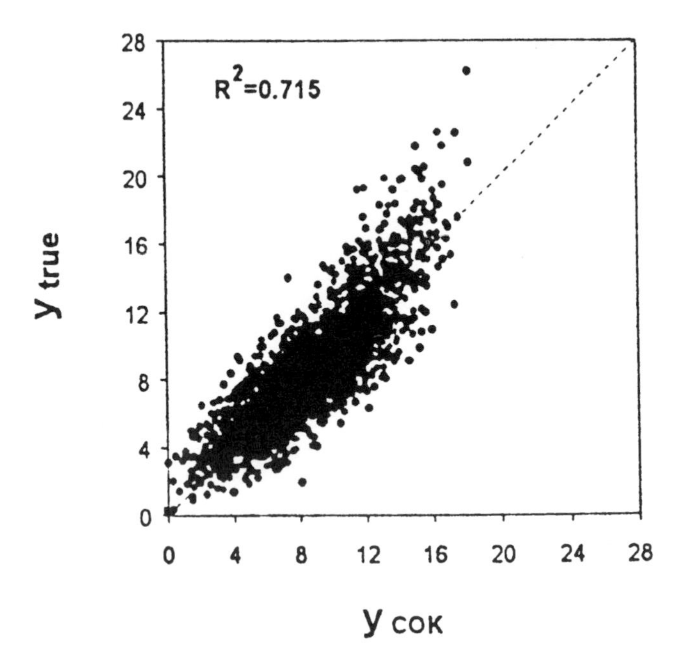

Figure 8b. True y (reference) versus the estimated y by cokriging without using transformations.

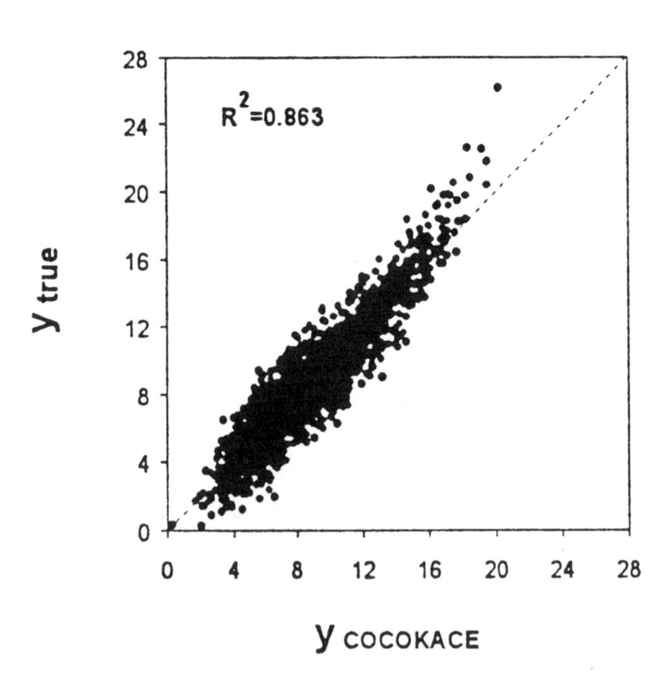

Figure 8c. True y (reference) versus the estimated y by collocated cokriging using optimal transformations by ACE (alternating conditional expectation) algorithm.

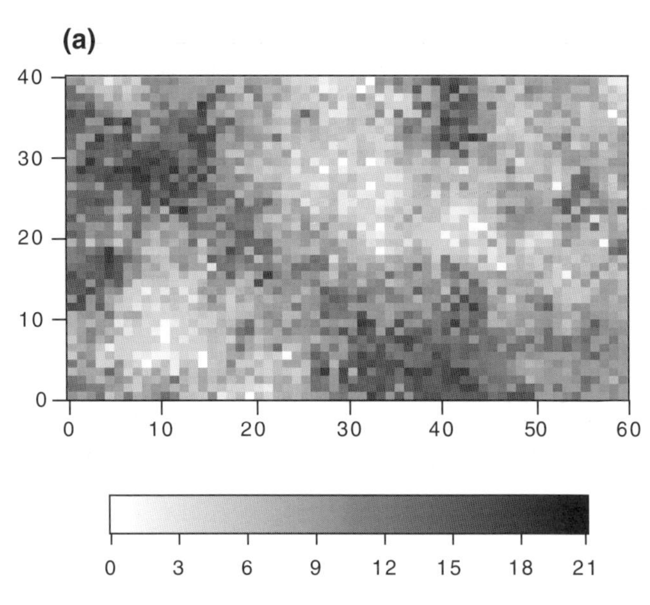

Figure 9a. Grayscale map of y by indicator cosimulation using Markov-Bayes algorithm in conjunction with ACE (alternating conditional expectation) algorithm. Realization 1.

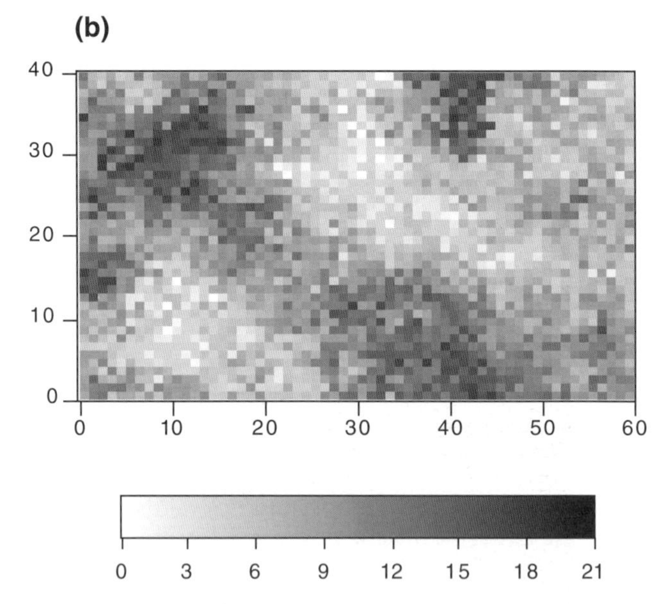

Figure 9b. Grayscale map of y by indicator cosimulation using Markov-Bayes algorithm in conjunction with ACE. Realization 2.

by a relatively gentle subsurface domal closure. It contains multiple stacked pay sandstones within a series of vertically stacked reservoir sequences respectively referred to as the B, C, D, E, and F series.

The available data for this study include 3-D seismic reflections and well log data from a 2 mi^2 (5 km^2) area of the Stratton field (Figure 10). The seismic data consist of 100 inlines and 200 crosslines with a trace spacing of 55 ft (16 m) in each direction. The well log data are from 21 wells. A zero offset vertical seismic profile (VSP) is available was and used for establishing a correlation between the stratigraphic depth and seismic traveltime (Figure 11). The seismic and well log data are used to derive an integrated description of

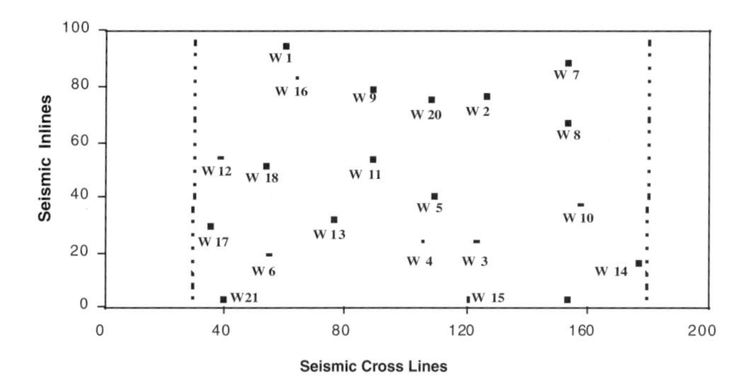

Figure 10. Three-dimensional seismic survey and well location for the study area, Stratton field, south Texas. The dotted framed area is the mapping area.

reservoir properties in the subject area using the proposed approach.

We selected F11 reservoir (Figure 11) for our detailed study because it is thick and easy to trace from seismic reflections in the study area. Our objective is to estimate pore footage ($h^*\phi$) in the study area using the data from well logs as the primary data set and multiple seismic attributes from 3-D seismic reflections as the secondary data set.

The reservoir properties, mainly reservoir thickness and porosity, for various reservoir facies in the middle Frio Formation in 21 wells are estimated using SP, neutron porosity, and density logs following the approach reported in Levey et al. (1993). Time horizons corresponding to reservoir zones are picked from the 3-D seismic data. Three type of seismic attributes, average seismic amplitude A_{AVG}, maximum amplitude A_{MAX}, and root mean square amplitude A_{RMS}, are extracted. All together, 72 pairs of pore-footage data from middle Frio reservoirs (B–F series) in 21 wells with corresponding nearby seismic attributes are selected for the data calibration. Figure 12a–c shows the scatterplot of pore footage from wells versus seismic A_{AVG}, A_{MAX}, and A_{RMS}. The highest linear correlation $R^2 = 0.302$ is between pore footage and average amplitude. Such a low correlation between pore footage and seismic attributes is not unusual considering the thickness of the sandstone zone, variations in lithology and fluid content, and data noise.

Optimal transformations are derived based on this data set using the ACE algorithm. Figure 12d is a scatterplot of the transformed pore footage versus the sum of transformed seismic attributes. The correlation R^2 is improved to 0.465 ($\rho = 0.68$) after transformations. This is quite significant in view of the scatter in the original data.

Figure 13 shows the grayscale map of one of the seismic attributes, A_{AVG}, corresponding to reservoir F11. Three such seismic attributes, A_{AVG}, A_{MAX}, and A_{RMS}, comprise our secondary data set, whereas the pore footage at the wells are the primary data. We transformed all these seismic attributes and 21 pore-footage data using optimal transformations. Finally, we conducted a collocated cokriging estimation for transformed $h^*\phi$ at all grid locations. A grayscale map of the estimated $h^*\phi$ by collocated cokriging after back transformation is shown in Figure 14.

For comparison, we also conducted an ordinary kriging on $h^*\phi$ data from the 21 wells, and the results are shown in Figure 15. Notice the severe smoothing

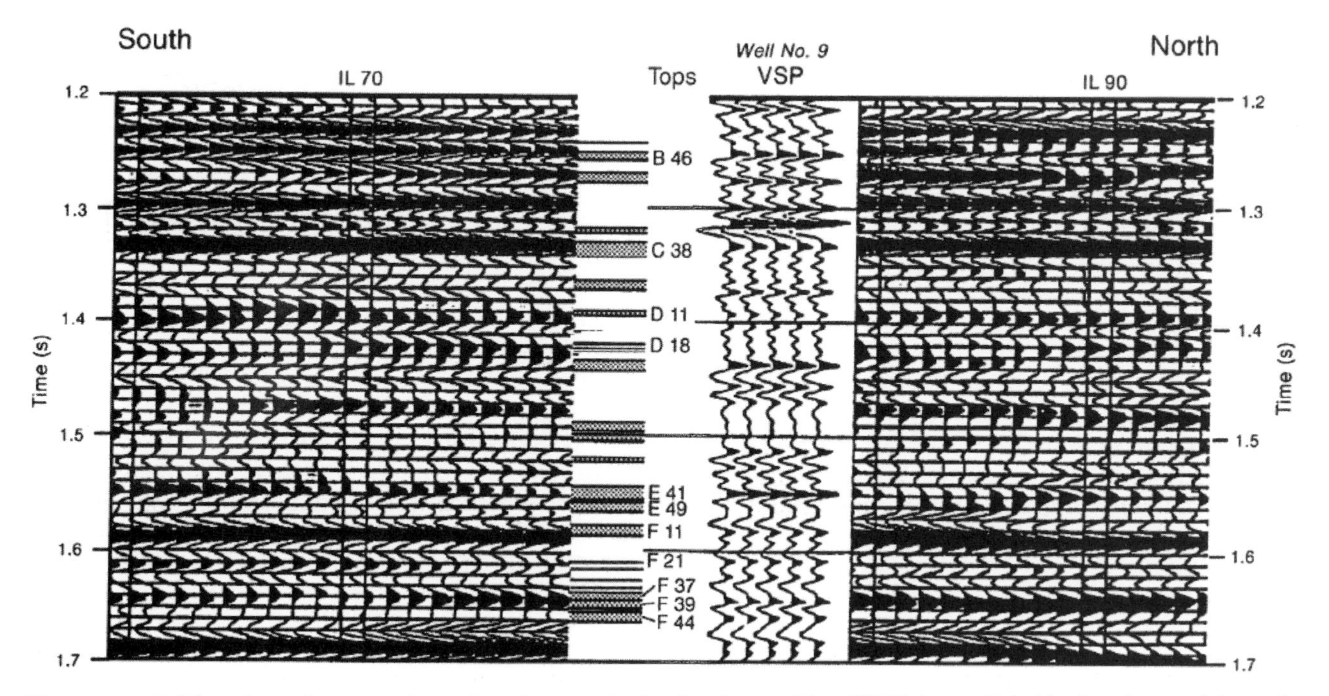

Figure 11. Calibration of reservoirs using the vertical seismic profiles (VSP) in well 9. Notice the position of F11 reservoir used for the detailed study (Levey et al., 1993).

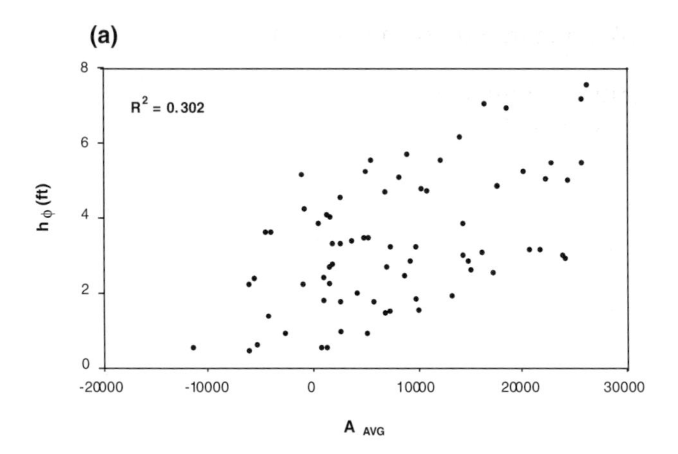

Figure 12a. Pore footage versus average seismic amplitude A_{AVG}.

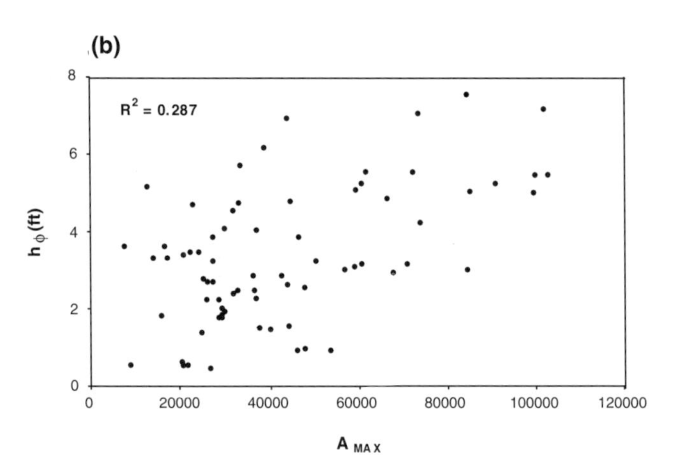

Figure 12 b. Pore footage versus maximum seismic amplitude A_{MAX}.

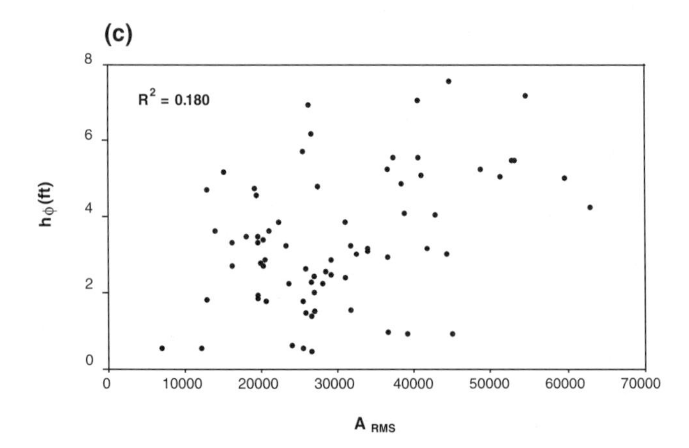

Figure 12c. Pore footage versus RMS seismic amplitude A_{RMS}.

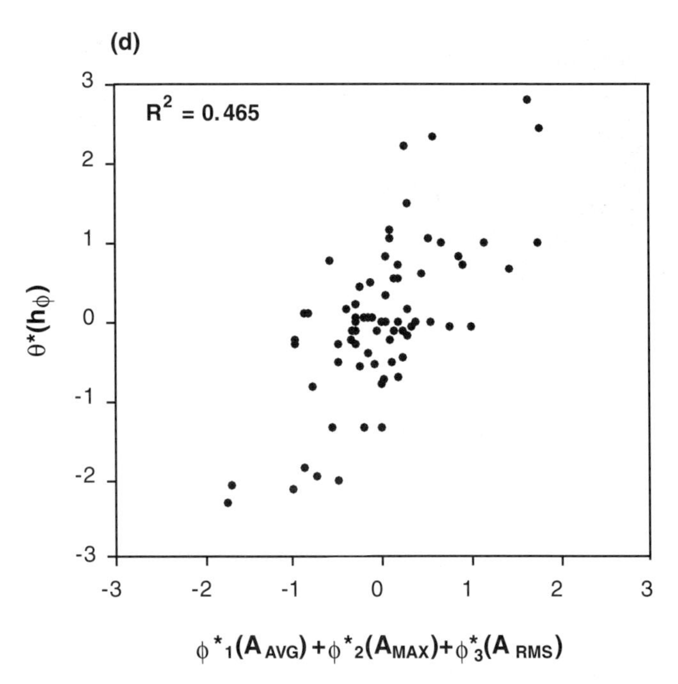

Figure 12d. Optimal correlation of pore footage versus seismic attributes as derived by ACE (alternating conditional expectation) algorithm.

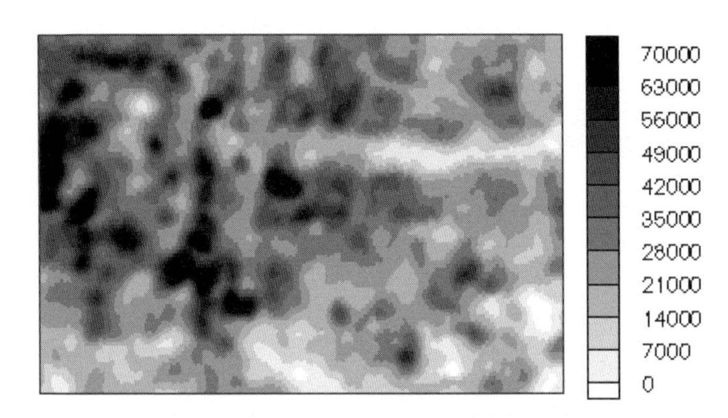

Figure 13. Grayscale map of the average seismic amplitude A_{AVG} distribution corresponding to F11 reservoir.

effects of the ordinary kriging and the lack of detail between the wells because of the absence of secondary information.

SUMMARY AND CONCLUSIONS

(1) Nonparametric transformation techniques offer a powerful, versatile, and fully automated tool for building correlations for petrological variables. Such transformations, being totally data driven, provide a direct approach to identifying functional relationships between dependent and independent variables during multiple regression.

(2) The power of nonparametric technique lies in its ability to directly incorporate multiple and mixed

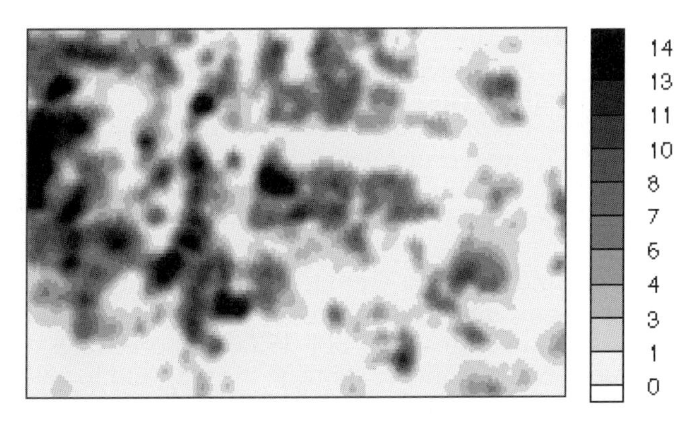

Figure 14. Grayscale map of the estimated pore footage by collocated cokriging using optimal transformation by ACE (alternating conditional expectation) algorithm.

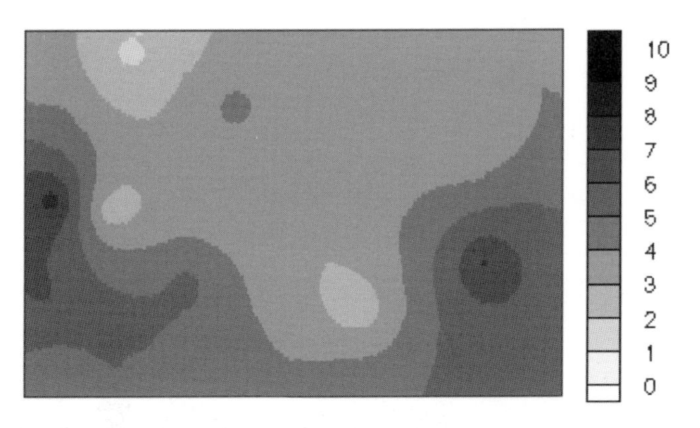

Figure 15. Grayscale map of the estimated pore footage by ordinary kriging from well data alone.

variables, both continuous and categorical, into correlation. Moreover, the transformations are computationally efficient, easy to use, and can provide significant insight during exploratory data analysis.

(3) We have presented synthetic and field examples to demonstrate the application of nonparametric transformation techniques for data correlation. A comparison of bootstrap prediction error clearly reveals the superiority of such techniques compared to conventional methods.

(4) Cokriging or cosimulation of multiple attributes is considerably simplified when carried out in conjunction with nonparametric transformations. The use of optimal transformations exploits the secondary data to its fullest potential and also allows for nonlinearity between reservoir properties and seismic attributes.

(5) The proposed data integration method greatly facilitates non-Gaussian cosimulation through the use of indicator approaches, particularly when multiple secondary variables are involved because of a significant reduction in variance function calculations.

(6) Synthetic case study clearly shows that cokriging and collocated cokriging using optimal transformation is far superior to ordinary kriging and cokriging in reproducing exhaustive reference data. The field case study demonstrates its capability of integrating multiple seismic attributes with well data for reservoir characterization.

ACKNOWLEDGMENTS

We would like to thank Mohan Kelkar and Xuri Huang at the University of Tulsa for their help in acquiring some of the field data. This work has been partially funded by a grant from the Mathematical Sciences Division of the National Science Foundation.

REFERENCES CITED

Almeida, A., 1993, Joint simulation of multiple variable with a Markov-type coregionalization model: Ph.D Dissertation, Stanford University, Stanford, CA. 199 p.

Barman, I., et al., 1998, Permeability Predictions in Carbonate Reservoirs Using Optimal Nonparametric Transforms: An Application at the Salt Creek Field, Kent County, TX, SPE 39667 in Proceedings of the 1998 SPE/DOE Symposium on Improved Oil Recovery, p. 113–125.

Breiman, L., and J. H. Friedman, 1985, Estimating optimal transformations for multiple regression and correlation: Journal of the American Statistical Association, v. 80, No. 391, p. 580.

Buja, A., T. Hastie, and R. Tibshirani, 1989, Linear smoothers and additive models: The Annals of Statistics, v. 17, p. 453-510.

Davies, D. K., and R. K. Vessel, 1996, Flow unit characterization of a shallow shelf carbonate reservoir: North Robertson Unit, West Texas, SPE/DOE 35433 in Proceedings of the SPE/DOE 10th Symposium on Improved Oil Recovery, p. 295–304.

Efron, B., and R. Tibshirani, 1993, An introduction to the Bootstrap: New York, Chapman and Hall, 436 p.

Emerson, J., and M. Stoto, 1982, Exploratory methods for choosing power transformations: Journal of the American Statistical Association, v. 77, No. 377, p. 103-108.

Friedman, J. H., and W. Stuetzle, 1981, Projection pursuit regression, Journal of the American Statistical Association, v. 76, No. 376, p. 817.

Friedman, J. H., and W. Stuetzle, 1982, Smoothing of scatterplots," Technical Report ORION006, Dept. of Statistics, Stanford University, California, 47 p.

Friedman, J. H., and B. W. Silverman, 1989, Flexible parsimonious smoothing and additive modeling, Technometrics, v. 31, No. 1, p. 3-20.

Hald, A., 1952, Statistical theory with engineering applications, John Wiley and Sons, New York, 97 p.

Hastie, T., and R. Tibshirani, 1990, Generalized Additive Models: London, Chapman and Hall, 335 p.

Jensen, J. L., and L. W. Lake, 1985, Optimization of regression-based porosity-permeability predictions: Transactions of the 10th Formation Evaluation Symposium, Calgary, September 29–October 2, Paper R, 22 p.

Journel, A. G., and F. G. Alabert, 1990, New method for reservoir mapping, Journal of Petroleum Technology (February), p. 212-218.

Levey, R. A., et al., 1993, Secondary natural gas recovery: targeted technology applications for infield reserve growth in fluvial reservoirs, Stratton Field, South Texas: Topical Report, GRI Contract No. 5088-212-1718, Gas Research Institute, Chicago, Illinois, 244 p.

McCain, W. D., R. Soto, P. Valko, and T. A. Blasingam, 1998, SPE51086 presented at the SPE Eastern Regional Conference and Exhibition, Pittsburgh, PA.

Tibshirani, R., 1988, Estimating optimal transformations for regression via additivity and variance stabilization, Journal of American Statistical Association, v. 82, p. 559-568.

Wendt, W. A., S. Sakurai, and P. H. Nelson, 1986, Permeability prediction from well logs using multiple regression: in Reservoir Characterization, Edited by Lake, L. W., and Carroll, H. B., Jr., Academic Press, Inc. Orlando, Florida, 659 p.

Xue, G., A. Datta-Gupta, P. Valko, and T. Blasingame, 1997, Optimal transformations for multiple regression: application for permeability estimation from well logs, SPE Formation Evaluation, v. 12, No. 2, p. 85–93.

Xue, G., and A. Datta-Gupta, 1996, A new approach for seismic data integration using optimal nonparametric transformations, SPE 36500 in Proceedings of the 1996 SPE Annual Technical Meeting and Exhibition, p. 37–50.

Zhu, H., and A. Journel, 1992, Formating and integrating soft data: stochastic imaging via the Markov-Bayes algorithm: in Geostatistics Troia '92, edited by A. Soares, Kluwer Academic Publishers, Dordrecht, The Netherlands, p. 1-12.

Index